Markus Wetzel

Influence of fully miscible lubrication oil on flow boiling of CO_2 inside horizontal evaporator tubes

Influence of fully miscible lubrication oil on flow boiling of CO_2 inside horizontal evaporator tubes

by
Markus Wetzel

KIT Scientific Publishing

Dissertation, Karlsruher Institut für Technologie
KIT-Fakultät für Chemieingenieurwesen und Verfahrenstechnik

Tag der mündlichen Prüfung: 22. September 2017
Referenten: Prof. Dr.-Ing. Thomas Wetzel
Prof. Dr.-Ing. Steffen Grohmann

Impressum

KIT Scientific Publishing

Karlsruher Institut für Technologie (KIT)
KIT Scientific Publishing
Straße am Forum 2
D-76131 Karlsruhe

www.ksp.kit.edu

Print on Demand 2017 – Gedruckt auf FSC-zertifiziertem Papier

ISBN 978-3-7315-0727-7
DOI 10.5445/KSP/1000074487

Influence of fully miscible lubrication oil on flow boiling of CO_2 inside horizontal evaporator tubes

zur Erlangung des akademischen Grades eines
DOKTORS DER INGENIEURWISSENSCHAFTEN (Dr.-Ing.)

der Fakultät für Chemieingenieurwesen und Verfahrenstechnik des
Karlsruher Institut für Technologie (KIT)

genehmigte
DISSERTATION

von
Dipl.-Ing. Markus Wetzel
aus Gernsbach

Referent: Prof. Dr.-Ing. Thomas Wetzel
Korreferent: Prof. Dr.-Ing. Steffen Grohmann
Tag der mündlichen Prüfung: 22. September 2017

Vorwort

Die vorliegende Arbeit entstand während meiner Tätigkeit als wissenschaftlicher Mitarbeiter am Institut für Thermische Verfahrenstechnik des Karlsruher Instituts für Technologie (KIT) in der Zeit von August 2009 bis September 2014. Zum Gelingen dieser Arbeit haben zahlreiche Personen beigetragen, denen mein ganz besonderer Dank gilt.
Meinem Doktorvater, Prof. Dr.-Ing. Thomas Wetzel, weder verwandt noch verschwägert, danke ich ganz herzlich für die Ermöglichung dieser Arbeit, einhergehend mit dem entgegengebrachten Vertrauen und der Unterstützung von Beginn an, sowohl in einfachen als auch in schwierigen Zeiten. Er ermöglichte mir einen tiefen Einblick in ein interessantes Forschungsgebiebt sowie meine persönliche Weiterentwicklung durch das Übertragen verschiedener Aufgabenbereiche und Verantwortlichkeiten, was zu einem erfolgreichen Start ins Berufsleben, nach meiner Institutszeit, maßgeblich beigetragen hat.
Herrn Prof. Dr.-Ing. Steffen Grohmann danke ich für die Übernahme des Korreferats, das Interesse an meiner Arbeit sowie die wertvolle fachliche Unterstützung.
Ein spezieller Dank gilt Herrn Prof. Dr.-Ing. Matthias Kind, Institutsleiter des TVT, sowie Herrn Prof. Dr.-Ing. Yasushi Saito, Universität Kyoto, für die Unterstützung vor und während meiner Promotionszeit sowie die zahlreichen Gespräche.
Frau Sonja Weise danke ich für die Unterstützung bei den letzten experimentellen Messungen, den wertvollen fachlichen Diskussionen, auch über meine Zeit am Institut hinaus, und die gemeinsamen Veröffentlichungen.
Meinen guten und langjährigen Freunden Markus Schlegel, Begleiter durch sämtliche Phasen des Studiums ab der ersten Minute und Berater in allen Lebenslagen, sowie Martin Wallenstein und Yabai Wang, ehemalige Bürokollegen und stets ehrliche Diskussionspartner, danke ich in besonderem Maße.
Ich danke auch meinem langjähringen Wegbegleiter am TVT, Benjamin Dietrich, für die schöne Zeit und Unterstützung vor und während meiner Promotion.
Ich bedanke mich bei allen Kollegen und Mitarbeitern des Instituts, die in ihrer freundlichen und hilfsbereiten Weise zu einem besonders angenehmen Arbeitsklima beigetragen haben. Ich blicke mit Freude auf eine schöne Zeit am TVT zurück. Für die tatkräftige Unterstützung in der Planung und Umsetzung von Anlagenänderungen danke ich an dieser Stelle auch den Mitarbeitern der Werkstatt, insbesondere Steffen Haury, Michael Wachter, Stefan Fink, Stefan Knecht, Andreas Roth, Markus Keller, sowie den Mitarbeitern des technischen

Büros, Eugen Mengesdorf und Sebastian Schwarz. Frau Annette Schucker danke ich für die Unterstützung bei Labortätigkeiten.
Ein besonderer Dank gilt dem Institutssekretariat um Frau Gisela Schimana, Frau Margit Morvay, Frau Rebecca Golling, Frau Tamara Cataldo und Frau Nicole Feger für die hervorragende Unterstützung bei Dienstreisen und allen weiteren Verwaltungsarbeiten.
Ganz herzlich danke ich auch meinen ehemaligen Studenten Anna-Lena Riegel, Sebastian Meinicke, Daniel Weber, Ariane Bisten, Igor Costa, Ariane Pigosso, Christopher Franke, Sebastian Siewert, Paul Knipper, Tobias Denz, Luciana Barros-Groß, Janine Witt, Ulrich Kraut, Tim Bornmann, Natalie Birk-Braun, Patrick-Simon Klug und Tobias Nachtrab für ihre Unterstützung und großes Engagement.
Mein Dank gilt dem Forschungsrat Kältetechnik e.V. für die unterstützenden Diskussionen und der Arbeitsgemeinschaft industrieller Forschungsvereinigungen (AiF) für die finanzielle Unterstützung dieser Arbeit.
Mein ganz besonderer Dank gilt meiner gesamten Familie und meiner lieben Beatrice für die Liebe und Unterstützung, ohne welche ich diesen langen Weg nicht hätte gehen können. Schmerzlich blicke ich auf den Tod meines Opas zurück. Er ist seitdem allgegenwärtig in meinen Gedanken, doch liebend gerne hätte ich ihm diese Arbeit als Wertschätzung und in Dank noch persönlich überreichen dürfen.

Markus Wetzel

Neustadt an der Weinstraße,
im September 2017

Kurzbeschreibung

Im Rahmen dieser Arbeit wurde der Einfluss von Kältemaschinenöl auf das Strömungssiedeverhalten von Kohlendioxid (CO_2, R744) untersucht.
Experimentelle Untersuchungen zum Wärmeübergang, Druckverlust und zur Strömungsform wurden in einer Glattrohr- (14,0 mm Innendurchmesser, Nickel) und einer Rippenrohrmessstrecke (8,62 mm Kernrohrdurchmesser, 60 Rippen, 0,25 mm Rippenhöhe, 18 ° Drallwinkel, Kupfer) in einem weiten Parameterbereich durchgeführt. Dieser umfasste Messungen bei Drücken von 14 – 40 bar, Massenstromdichten bis 300 $\mathrm{kg\,m^{-2}\,s^{-1}}$ im Glatt- bzw. 500 $\mathrm{kg\,m^{-2}\,s^{-1}}$ im Rippenrohr, Dampfgehalten von 0.1 – 0.9 sowie Wärmestromdichten bis 150 $\mathrm{kW\,m^{-2}}$. Beide Messstrecken wurden elektrisch beheizt, wobei durch Segmentierung des Rohres eine isotherme Rohrinnenwand als Randbedinung ermöglicht und eingestellt wurde. Als Kältemaschinenöl wurde ein im Versuchsbereich vollständig lösliches Polyolesteröl (Reniso™ C 85 E, Fuchs Petrolub SE) verwendet. Der Ölgehalt im unterkühlten CO_2 wurde zwischen 0 – 3% (massenbezogen) variiert. Dabei konnte der lokale Ölgehalt in der Messtrecke in Abhängigkeit von den Betriebsbedingungen Werte $> 50\%$ annehmen. Die Strömungsform wurde in einem der Messstrecke nachgeschalteten, nicht-beheizten Schauglas optisch mittels Hochgeschwindigkeitsvideoaufnahmen ermittelt.
Zur Feststellung des Öleinflusses wurden Messdaten aus den Gemisch-Versuchen mit denen aus Reinstoff-Versuchen verglichen. Aus der Literatur bekannte Zusammenhänge bzw. Erklärungsansätze wurden dabei zur Beurteilung der Beobachtungen herangezogen. Die experimentellen Messdaten wurden auf Basis von Gemischstoffwerten ausgewertet und dienten der Bewertung und Modifikation etablierter sowie neuer Vorausberechnungsmethoden. Neben dem Einfluss auf Gemischstoffwerte wurden Auswirkungen von Öl-induzierten Veränderungen der Strömungsform auf den Wärmeübergang und Druckverlust untersucht. Hierzu wurde auch ein neues Verfahren zur Unterstützung der Bestimmung der Strömungsform entwickelt. Diese Erkenntnisse sind wichtig, um den Öleinfluss verstehen und in Vorhersagemodellen zutreffend berücksichtigen zu können.

Abstract

In the scope of this work, the influence of lubricating oil on the flow boiling characteristics of Carbon dioxide (CO_2, R744) was investigated.

Heat transfer, pressure drop and flow pattern were experimentally investigated inside smooth (14.0 mm inside diameter, nickel) and micro-fin (8.62 mm core diameter, 60 fins, 0.25 mm fin height, 18 ° helix angle, copper) test tubes. The range of operation included pressures of 14 – 40 bar, mass flow velocities up to $300\,\mathrm{kg\,m^{-2}\,s^{-1}}$ in the smooth and $500\,\mathrm{kg\,m^{-2}\,s^{-1}}$ in the micro-fin tube, respectively, vapor qualities of 0.1 – 0.9 and heat fluxes up to $150\,\mathrm{kW\,m^{-2}}$. Both test sections were electrically heated. Segmentation of the tube wall into individually heatable sections allowed for realizing isothermal conditions at the inside tube surface. Fully miscible polyol ester oil (Reniso™ C 85 E, Fuchs Petrolub SE) was used as lubricant. Oil mass fractions in the sub-cooled refrigerant were varied between 0 – 3 wt.-%, whereby local oil mass fractions of $>$ 50 wt.-% could result, depending on the actual conditions of operation. Throughout all flow boiling experiments, flow patterns were determined inside a smooth, non-heated sight glass tube preceding the measurement section by means of high speed video recordings.

For determining of the influence of oil, results from measurements with CO_2-lubricant mixtures were compared to pure CO_2 data. Proposed explanations found in literature were consulted to interpret the observations made. The experimental data was evaluated based on mixture properties and used to evaluate and modify established predictive methods. Besides regarding the influence on mixture properties, the impact of oil-induced changes in flow pattern on the flow boiling heat transfer and two-phase pressure drop were investigated. For this purpose, a new measurement technique for supporting the identification of flow patterns was developed. The knowledge obtained from these studies contributes to understanding the influence of oil, which is essential for the development of precise predictive methods.

Contents

Vorwort . . . . **I**

Kurzbeschreibung . . . . **III**

Abstract . . . . **V**

List of abbreviations and symbols . . . . **XI**

1 Introduction **1**

2 State-of-the-art **5**

2.1 Flow pattern 5

2.2 Pressure drop 16

2.3 Heat Transfer 25

2.4 CO_2 and CO_2-lubricant mixtures 32

2.5 Lubricant influence on material properties 45

2.5.1 Bubble point temperature & vapor pressure 46

2.5.2 Viscosity 48

2.5.3 Density 49

2.5.4 Heat capacity 51

2.5.5 Thermal conductivity 51

2.5.6 Surface tension 54

3 Experimental procedure **57**

3.1 Test facility 57

3.1.1 Chiller 58

3.1.2 Brine circuits 58

3.1.3 Test loop 60

3.1.4 Oil loop 61

3.1.5 Measurement sections 61

3.1.6 Pre-heaters 63

3.2 Measurement technique and data acquisition 65

3.2.1 Data acquisition 65

3.2.2 Temperature measurement 66
3.2.3 Pressure and pressure drop measurement 67
3.2.4 Power measurement 67
3.2.5 Mass flow measurement 67
3.2.6 Flow pattern observation 68
3.3 Start-up and operation of the test facility 69
3.4 Experimental procedure 70
3.4.1 Heat transfer measurements 70
3.4.2 Pressure drop measurements 71
3.4.3 Flow pattern investigation 71
3.4.4 Hydrodynamic and thermal flow development 72
3.4.5 Oil mass fraction measurements 72
3.5 Evaluation of Measurements 73
3.5.1 Heat transfer 73
3.5.2 Vapor quality 74
3.5.3 Oil mass fraction 76
3.5.4 Mass flow velocity 76

4 Smooth tube results 77
4.1 Flow patterns 77
4.1.1 Flow pattern classification of CO_2-POE oil mixtures 77
4.1.2 Comparison of experimental flow patterns to flow pattern maps . . 79
4.1.3 Conclusion 84
4.2 Pressure drop 85
4.2.1 Frictional pressure drop 85
4.2.2 Total pressure drop 88
4.2.3 Conclusion 90
4.3 Heat transfer 91
4.3.1 Influence of the heat flux 91
4.3.2 Influence of the mass flow velocity 93
4.3.3 Influence of the vapor quality 94
4.3.4 Influence of the flow pattern 94
4.3.5 Comparison to predictive methods 96
4.3.6 Conclusion 98

5 Micro-fin tube results 101
5.1 Flow patterns 101
5.1.1 Flow pattern classification and distinction from smooth tube observations 101

5.1.2 Comparison of experimental flow patterns to flow pattern maps . . 103
5.1.3 Conclusion 105
5.2 Pressure drop 106
5.2.1 Frictional pressure drop 107
5.2.2 Total pressure drop 111
5.2.3 Conclusion 113
5.3 Heat transfer 114
5.3.1 Influence of the heat flux 114
5.3.2 Influence of the reduced pressure 117
5.3.3 Influence of the mass flow velocity 118
5.3.4 Influence of the vapor quality 120
5.3.5 Influence of the flow pattern 121
5.3.6 Comparison to predictive methods 123
5.3.7 Conclusion 126

6 Conclusion and outlook 129
6.1 Contents and results of this work 129
6.1.1 Flow pattern 130
6.1.2 Pressure drop 131
6.1.3 Heat transfer 132
6.2 Outlook 134

Bibliography 137

Appendix A 155
A.1 Measurement devices 155
A.2 Definitions for representing the error in measurement 155

Appendix B 159
B.1 Pure CO_2 investigations 159
B.2 CO_2-oil mixture investigations 163

Appendix C 167
C.1 Measurement principle 167
C.2 Preliminary setup and smooth tube test runs 168
C.3 Development of a flexible setup and static validation experiments 170
C.4 Test section integrated measurement setup and sight glass tube modification 172
C.5 Conclusion 172

Appendix D **175**
D.1 Fluid properties of CO_2 175
D.1.1 Critical and triple point data of CO_2 175
D.1.2 Saturated properties 175
D.1.3 Vapor/liquid substance properties 176
D.2 Fluid properties of CO_2-lubricant mixtures 179
D.2.1 Bubble point temperature and vapor pressure 179
D.2.2 Viscosity 180
D.2.3 Density 180
D.2.4 Heat capacity 180
D.2.5 Thermal conductivity 181
D.2.6 Surface Tension 182
D.3 Solid material properties 183

Appendix E **185**
E.1 Prediction methods for flow pattern 185
E.2 Two-phase pressure drop models 193
E.2.1 Pure/mixed refrigerant flow: Frictional pressure drop 193
E.2.2 Acceleration pressure drop 201
E.2.3 Refrigerant-lubricant mixtures 201
E.3 Heat transfer models 202
E.3.1 General established correlations 202
E.3.2 Flow boiling pure/mixed refrigerants 204
E.3.3 Flow boiling refrigerant-lubricant mixtures 212

Appendix F **215**
F.1 Smooth tube measurement data 215
F.1.1 Flow patterns 215
F.1.2 Pressure drop 218
F.1.3 Heat transfer 222
F.2 Micro-fin tube measurement data 223
F.2.1 Flow patterns 223
F.2.2 Pressure drop 231
F.2.3 Heat transfer 241

Appendix G **245**
G.1 Thermal conductivity measurement data 245
G.2 Oil mass fraction measurement data 247
G.3 Flow boiling experimental data 250

List of abbreviations and symbols

Latin symbols

A	area (cross section, surface)	m^2
C_F	fluid-dependent parameter	
$C_\text{micro-fin}$	fluid-dependent parameter	
$\dot{M}$	mass flow rate	$\mathrm{kg\,s^{-1}}$
$\dot{m}$	mass flow velocity	$\mathrm{kg\,m^{-2}\,s^{-1}}$
$\dot{Q}$	heat flow	W
$\dot{q}$	heat flux	$\mathrm{W\,m^{-2}}$
$\dot{x}$	vapor quality	
$\dot{x}_\mathrm{dry}$	dryout vapor quality	
$\tilde{M}$	molar mass	$\mathrm{g\,mol^{-1}}$
U	Perimeter	m
R_a	roughness	m
n	refractive index	
$\tilde{x}$	liquid mole fraction	$\mathrm{mol\,mol^{-1}}$
$B_\mathrm{L}(w_\mathrm{oil})$	individual coefficient	
c_p	specific isobaric heat capacity	$\mathrm{J\,kg^{-1}\,K^{-1}}$
d, D	diameter	m
d_fr	fin-root or core diameter	m
d_ft	fin-top diameter	m
EF_A	area enhancement factor	
EF_α	heat transfer enhancement factor	
$F(\tau^*)$	universal crossover function	
h_f	fin height	m
h_L	height of liquid	m
k	overall heat transfer coefficient	$\mathrm{W\,m^{-2}\,K^{-1}}$
L	length	m
l_f	axial pitch from fin to fin	m
n_f	number of fins	
OCR	oil circulation rate $(= w_\mathrm{oil,0} \times 100\,\%)$	$\%$
P	power	W

p	pressure	Pa
p_c ($\bar{p}_c$)	(pseudo-)critical pressure (of ROS)	Pa
$PF_{\Delta p}$	penalty factor for pressure drop	
T	temperature	K
T_c ($\bar{T}_c$)	(pseudo-)critical temperature (of ROS)	K
w_{oil}	local oil mass fraction	$kg\,kg^{-1}$
$w_{oil,0}$	nominal oil mass fraction	$kg\,kg^{-1}$

Greek symbols

$\alpha(\tau)$	universal function	
α	heat transfer coefficient	$W\,m^{-2}\,K^{-1}$
α_c	Friedel's parameter	
α_{Ch}	heat transfer coefficient according to correlation by Chen (1966)	$W\,m^{-2}\,K^{-1}$
α_{Co}	heat transfer coefficient according to correlation by Cooper (1984)	$W\,m^{-2}\,K^{-1}$
α_{DB}	heat transfer coefficient according to correlation by Dittus-Boelter (McAdams, 1942)	$W\,m^{-2}\,K^{-1}$
α_{FZ}	heat transfer coefficient according to correlation by Forster and Zuber (1955)	$W\,m^{-2}\,K^{-1}$
α_{Gn}	heat transfer coefficient according to correlation by Gnielinski (1975)	$W\,m^{-2}\,K^{-1}$
α_j	local heat transfer coefficient at segment j	$W\,m^{-2}\,K^{-1}$
α_{SA}	heat transfer coefficient according to correlation by Stephan and Abdelsalam (1980)	$W\,m^{-2}\,K^{-1}$
β_0	contact angle	$^\circ$
β	individual coefficient	
β	spiral/helix/lead angle	$^\circ$
$\Delta(w_{oil})$	coefficient	
Δh_v	heat of vaporization	$J\,kg^{-1}$
δ_L	liquid film thickness	m
ε	void fraction	
γ	apex/fin tip angle	$^\circ$
λ	thermal conductivity	$W\,m^{-1}\,K^{-1}$
λ_{mix}	pure refrigerant thermal conductivit of ROS	$W\,m^{-1}\,K^{-1}$
λ_{oil}	pure lubricant thermal conductivity	$W\,m^{-1}\,K^{-1}$
λ_{ref}	pure refrigerant thermal conductivity	$W\,m^{-1}\,K^{-1}$
$\mu_0(w_{oil})$	coefficient	

μ_{mix}	dynamic liquid mixture viscosity of ROS	Pa s
μ_{oil}	dynamic liquid viscosity of pure lubricant	Pa s
μ_{ref}	dynamic liquid viscosity of pure refrigerant	Pa s
Φ_P	pressure drop two-phase multiplier corresponding to phase $P = L, V, LO, VO$	
ρ_{mix}	liquid mixture density of ROS	$kg\,m^{-3}$
ρ_{oil}	liquid density of pure lubricant	$kg\,m^{-3}$
ρ_{ref}	liquid density of pure refrigerant	$kg\,m^{-3}$
σ	surface tension	$N\,m^{-1}$
σ_{mix}	surface tension of ROS	$N\,m^{-1}$
σ_{oil}	surface tension of pure lubricant	$N\,m^{-1}$
σ_{ref}	surface tension of pure refrigerant	$N\,m^{-1}$
τ	reduced temperature	
θ	dry/wetted angle	rad
ϑ	temperature	°C
ξ	friction factor	
τ^*	logarithmic reduced temperature	

Indices

α	(related to) heat transfer
Δp	(related to) pressure drop
A	(related to) surface area, annular
A-D	transition annular flow-partial dryout
A-M	transition annular-mist flow
B	bubbly flow
bub	bubble point
c	critical
cb	convective boiling
Ch	by Chen (1966)
char	characteristic
Co	by Cooper (1984)
cor	correction
D	dryout region
D-M	transition partial dryout-mist flow
DB	by Dittus-Boelter (McAdams, 1942)
De	dryout completion
Di	dryout inception
dry	dry, dryout

eff	effective
el	electrical
exp	experimental, determined from measurements
f	fin
fb	fin root, core
flow	flow
fric	friction
ft	fin tip
FZ	by Forster and Zuber (1955)
Gn	by Gnielinski (1975)
h	homogeneous or hydraulic
heat	heated
I	intermittent flow
i	in, inner; interface
I-A	transition intermittent-annular flow
I-B	transition intermittent-bubbly flow
in	inlet
ins	insulation
j	segment *j*
L	liquid phase
LO	all-liquid, assuming that all fluid is liquid
loss	loss
lt	laminar-turbulent (L-V)
M	mist flow
micro-fin	micro-fin tube
mix	mixture
mod	modified
mom	momentum, acceleration
nb	nucleate boiling or normal boiling point
o	out, outer
oil	oil, lubricant
onb	onset of nucleate boiling
out	outlet
p	isobaric
PP	by Petukhov and Popov (1963)
pred	predicted
r	reduced
ref	refrigerant
S	stratified flow

s	saturated
S-W	transition stratified-wavy flow
SA	by Stephan and Abdelsalam (1980)
SLUG	slug flow
SLUG+I	Slug + annular flow
SLUG+W	Slug + wavy flow
Slug-I	transition slug-intermittent flow
smooth	smooth tube
strat	stratified
t	triple point
tot	total
tp	two-phase
tt	turbulent-turbulent (L-V)
V	vapor phase
v	vaporization or isochoric
Vm	mean vapor phase
VO	all-vapor, assuming that all fluid is vapor
W	wavy flow
w	tube wall
W-A	transition wavy-annular flow
W-I	transition wavy-intermittent flow
W-Slug	transition wavy-slug flow
wet	wet, wetted

Numbers

Bd	Bond number $\left(=\left(\frac{\mathrm{Fr}}{\mathrm{We}}\right)_{\mathrm{L}}=\left(\frac{g\,d_{\mathrm{i}}^2\,\rho_{\mathrm{L}}}{\sigma}\right)_{\mathrm{L}}\right)$
Bo	Boiling number $=\left(\frac{\dot{q}}{\dot{m}\,\Delta h_{\mathrm{v}}}\right)$
Co	Convection number $\left(=\left(\frac{1-\dot{x}}{\dot{x}}\right)^{0.8}\left(\frac{\rho_{\mathrm{V}}}{\rho_{\mathrm{L}}}\right)^{0.5}\right)$
Eu	Euler number
Fa	parameter, specified by Fang (2013)
Fr	Froude number
$\mathrm{Fr}_{\mathrm{L\|V}}$	liquid/vapor Froude number $\left(=\frac{\dot{m}^2}{\rho_{\mathrm{L\|V}}^2\,g\,d_{\mathrm{i}}}\right)$
$\mathrm{Fr}_{\mathrm{Vm}}$	mean vapor Froude number $\left(=\frac{\dot{m}^2\dot{x}^2}{\rho_{\mathrm{L}}\,\rho_{\mathrm{L}}\,g\,d_{\mathrm{i}}}\right)$
Ja	Jakob number $=\left(\frac{c_{\mathrm{p,L}}\,\Delta T}{\Delta h_{\mathrm{v}}}\right)$
La	Laplace constant

Nu	Nusselt number
$Pr_{L\|V}$	liquid/vapor Prandtl number $\left(= \frac{\mu_{L\|V}\, c_{p,L\|V}}{\lambda_{L\|V}}\right)$
Re	Reynolds number
Re_L	liquid Reynolds number $\left(= \frac{\dot{m}\,(1-\dot{x})\,d_i}{\mu_L}\right)$
$Re_{L\|V}$	all-liquid/vapor Reynolds number $\left(= \frac{m\,d_i}{\mu_{L\|V}}\right)$
Re_{tp}	two-phase Reynolds number $(= Re_L\, F)$
Re_V	vapor Reynolds number $\left(= \frac{\dot{m}\,\dot{x}\,d_i}{\mu_V}\right)$
We	Weber number
X	Martinelli parameter $\left(= \frac{(\Delta p/\Delta L)_L}{(\Delta p/\Delta L)_V}\right)$

Abbreviations

AB	alkylbenzene (lubricant)
AE	absolute error
APE	absolute percentage error
CO_2	carbon dioxide
DAQ	data acquisition (system)
fps	frames per second
L	liquid
MAE	mean absolute error
MAPE	mean absolute percentage error
ME	mean error
MO	mineral oil
MPE	mean percentage error
MS	measurement section
PAG	polyalkylene glycol (lubricant)
PE	percentage error
POE	polyolester (lubricant)
RMSE	root mean squared error
ROS	refrigerant oil solutions
TC	thermocouple
TR	throttle (valve)
V	vapor

1 Introduction

The usage of carbon dioxide (CO_2, R744) as refrigerant dates back to the late 19th century, the first CO_2 compression refrigeration system in Europe built by Carl von Linde in 1881 (Pachai, 2008). CO_2 belonged to the most frequently used refrigerants (besides NH_3 and SO_2) in the early 20th century. With the introduction of synthetic chlorofluorocarbons (CFC) and hydrochlorofluorocarbons (HCFC) in the 1930s and 1940s, only NH_3 remained in practical use (Lorentzen, 1994). CO_2 was rediscovered as refrigerant by Gustav Lorentzen in 1988 and, triggered by restrictions relating the ozone depletion potential (Montreal protocol, 1989) and global warming potential (Kyoto protocol, 1997) of technical gases, the number of research studies and applications concerning CO_2 as refrigerant greatly increased (Javerschek, 2008). Applications include water heaters, heat pump systems, automotive A/C systems, small commercial applications and the low temperature stage of cascade systems (Hrnjak, 2006), e.g., NH_3-CO_2.

The rising attractiveness of CO_2 as refrigerant lead to numerous studies on flow boiling in recent years, summarized in the upcoming chapter. With increasing research done, the range of operation investigated was greatly extended and the precision in predicting flow boiling heat transfer coefficients or two-phase pressure drops of CO_2 significantly improved. Schael (2009) presented predictive methods for precisely predicting heat transfer, pressure drop and flow pattern of CO_2, covering a wide range of operation inside smooth and micro-fin tubes.

Pure refrigerant investigations that were mainly focused on, however, do not give information on practical issues such as the influence of lubricating oil used in vapor compression systems. Oil used for lubricating the moving parts of a compressor unavoidably becomes part of the circulating fluid, ranging from 0.1 % to 8.0 % in concentration, depending on the type of installation (Bandarra Filho et al., 2009). Affecting thermodynamic and transport properties of the fluid, such as bubble point temperature, viscosity, surface tension and thermal conductivity, the presence of oil can lead to a significant influence on flow boiling characteristics (Thome, 1996). While mixture properties can be taken into account, effects such as lubricant-induced foaming are not captured by correlations based on pure refrigerant or refrigerant-mixture data.

In a first experimental study published on the influence of lubricant on flow boiling characteristics of CO_2, Hassan (2004) showed that heat transfer coefficients can be deteriorated by up to 50 % with respect to pure CO_2. Such strong influence was confirmed in preceding studies,

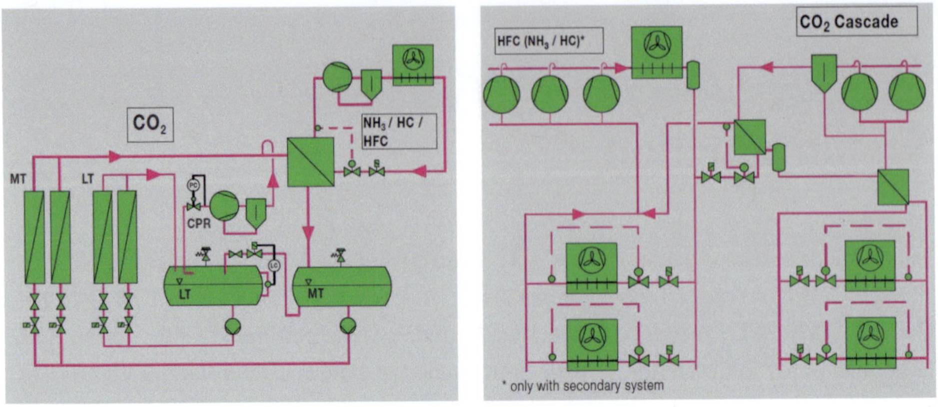

Figure 1.1: CO_2 applications for refrigeration (Bitzer Kühlmaschinenbau GmbH, 2015) –
Cascade system for industrial applications (left), Conventional system combined with CO_2 cascade (right)

however, the influence is closely linked to the operating conditions and especially the type of lubricant used. Observations reported for fully miscible CO_2-POE oil mixtures (Kim and Hrnjak, 2012) differ distinctly from partial miscible CO_2-PAG oil mixture characteristics (Dang et al., 2013). Full miscibility generally leads to a steady change in heat transfer and pressure drop. Almost immiscible lubricants result in an early formation of a separate oil-rich phase, especially affecting heat transfer by covering the heated tube surface and, hence, inducing a significant mass and heat transfer resistance and change in surface-near bubble point temperature even at very low oil concentrations.

Several publications on the influence of lubricant on flow boiling CO_2 followed since, pointing out the relevance in research on the one hand. On the other hand, dissimilarities in observations reported by different authors that could result from the huge variety of operating parameters as well as missing consolidation in the precision accuracy of established predictive methods reported, reveal the limited knowledge on the effect of lubricant on flow boiling characteristics of CO_2 and that further research is essential.

Concerning the geometry of the evaporator tube, focus in recent research was mainly laid on micro-fin tubes. These tubes generally feature a significant enhancement in tube wetting and turbulence in the liquid. The enhancement in heat transfer realized is typically greater than the penalty due to increasing pressure drop (Cavallini et al., 1999). Dang et al. (2010) reported that the dryout quality for micro-fin tubes is larger than for smooth tubes, thus the micro-fin evaporator tube is used almost completely yielding high efficiency. Yet, already small amounts of lubricating oil can influence heat transfer characteristics in a micro-fin tube (Ha and Bergles, 1993). The number of publications on the effect of lubricating oil on flow boiling behavior of CO_2 inside enhanced tubes is limited, while the tube geometry implies further parameter variation.

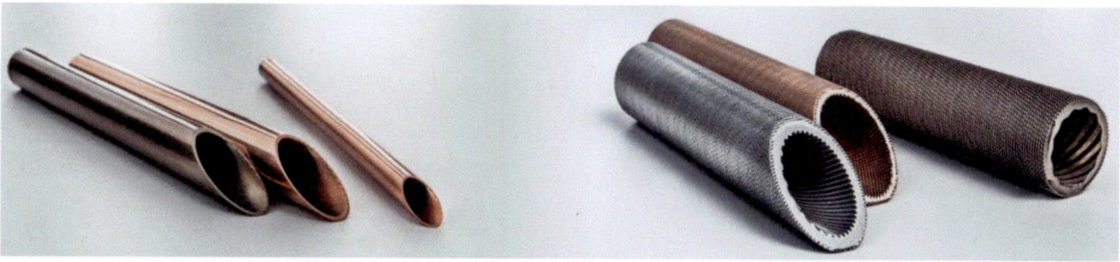

Figure 1.2: Enhanced tubes for boiling and condensing heat transfer (Wieland Werke AG, 2016)

In the scope of this work, heat transfer coefficients, pressure drops and flow patterns of flow boiling CO_2 were investigated in both, a smooth tube and a micro-fin tube of 14.0 mm and 8.6 mm in inner diameter, respectively. Owing to its significance in technical application, investigations in the micro-fin tube were pursued with main interest. As lubricant mass fractions typically yield values between 0.5 wt.-% to 3 wt.-% in the sub-cooled liquid of refrigeration systems (Thome, 1996), investigations focused on CO_2-lubricant mixtures with nominal oil mass fractions of 1 wt.-%, 2 wt.-% and 3 wt.-%. Providing excellent miscibility with CO_2 and being widely used in CO_2-based applications of refrigeration, POE-85 oil (Reniso™ C 85 E, Fuchs Petrolub SE) was selected as lubricant for the experimental investigation. The aim of the investigations was to identify oil-induced influencing factors on flow boiling characteristics and to evaluate the applicability of existing predictive methods for the boiling heat transfer coefficient, two-phase pressure drop and flow pattern. Accuracy in prediction of these design parameters affects the efficiency of designing evaporators.

2 State-of-the-art

Flow boiling in horizontal tubes is a phenomenon that attracted the attention of researchers for many decades by now. Extensive investigations - both experimental and theoretical - have contributed to the present state-of-the-art. Due to the highly complex nature of flow boiling (including the rate of tube inclination), however, there is still no commonly applicable approach for its description and understanding. The complexity, lack of knowledge and uncertainty of prediction greatly increases when regarding refrigerant-lubricant mixtures in real vapor compression refrigeration systems, where lubricating oil enters the refrigerant loop by lubrication of the compressor. Consequently, the literature survey on the state-of-the-art on flow boiling in tubes presented in this chapter emphasizes on publications about flow boiling of CO_2-oil mixtures, refrigerant-oil mixtures in general and prediction methods for thermodynamic properties of refrigerant-oil solutions (ROS). Besides, latest reports on boiling in micro-fin tubes are summarized.
An extensive study on the flow boiling characteristics of CO_2 was made by Schael (2009). Heat transfer, pressure drop and flow pattern were investigated at saturation conditions ranging from 10 bar (-40 °C) to 40 bar ($+5$ °C). Operating conditions included mass flow velocities up to 500(600) $\mathrm{kg\,m^{-2}\,s^{-1}}$ and heat fluxes up to 80(120) $\mathrm{kW\,m^{-2}}$ in the smooth tube (micro-fin tube). In his work, Schael provides a comprehensive review on recent investigations on CO_2 flow boiling in smooth and micro-fin tubes. The most accurate prediction methods for CO_2 as concluded by Schael are reviewed in the following. This knowledge base is complemented by literature sources depicting the latest published work. The following Tab. 2.1-2.2 list the analyzed papers on flow boiling research of refrigerant-lubricant mixtures and CO_2, respectively.

2.1 Flow pattern

Pure/mixed refrigerants

In flow boiling, there are (at least) two phases encountered: a liquid and a vapor phase, flowing at different velocities. Thereby, phase distribution varies according to the operating conditions. These flow distributions are classified by *flow patterns*, such as pictured in Fig. 2.1 for a horizontal evaporator (Collier and Thome, 1994). In the following a description of flow patterns is given.

Table 2.1: Publications on experimental flow boiling investigations of Refrigerant-lubricant mixtures

Authors	Refrigerant/ Lubricant	Measurement section	d_i/m	ϑ_s/°C	$\dot{m}$/kg m^{-2} s^{-1}	$\dot{q}$/kW m^{-2}	w_{oil}/wt.-%
Schlager et al. (1989)	R22/MO (150 SUS; 300 SUS) [a]	water-heated smooth and micro-fin tubes	8.0; 8.9	+3	125 … 400	NA	0 … 5
Ha and Bergles (1993)	R12/3GS (MO20) [a]	indirect electrically heated copper micro-fin tube	8.0	+1	25 … 100	5, 10	0 … 5
Eckels et al. (1994)	R134a/POE (169 SUS; 369 SUS) [a]	water-heated smooth and micro-fin tubes	8.0; 8.9	+1	85 … 375	NA	0 … 5
Hambraeus (1995)	R134a/ POE29 [a], POE124 [b], POE32 [a]	copper smooth tube with wrapped film heater	12	−10 … +10	40 … 170	2 … 8	0 … 4.5
Zürcher et al. (1997)	R134A/ POE68	fluid-heated copper smooth tubes	10.9	+4.4	100 … 300	5 … 10	0 … 5
Zürcher et al. (1998a)	R407C/ POE68 [a]	fluid-heated copper micro-fin tubes	11.9	+4 … +11	100 … 300	1 … 15	0 … 5
Zürcher et al. (1998b)	R407C/ POE68 [a]	fluid-heated copper smooth tubes	10.9	+4 … +11	100 … 300	1 … 15	0 … 5
Eckels et al. (1994)	R134a/POE (150 SUS) [a]	water-heated smooth and micro-fin tubes	8.0; 8.9	+1	125 … 375	NA	0 … 5
Mermond et al. (1999a)	R134A/ POE68	fluid-heated stainless steel smooth tubes	12	+11.2	200	5 … 20	0 … 6
Wongwises et al. (2002)	R134A/ PAG180	smooth tube	7.8	−6 … +30	150 … 590	0	5
Barbosa et al. (2004)	R12/ 1GS MO [a]	metal and glass smooth tubes	2.86	+30	395 … 1451	-	53 … 81
Castro et al. (2004)	R134A/ POE10 [a]	metal and glass smooth tubes	3.22	+28 … +39	N/A	-	47 … 78
Wei et al. (2007)	R22/ NM56 [a]	smooth tube with wrapped film heater	6.34	+5	200 … 400	3 … 14	0 … 5
Targanski and Cieslinski (2007)	R407C/ N/A	fluid-heated copper and stainless steel smooth (1), corrugated (2) and micro-fin (3) tubes	8.0 (1), 8.8 (2), 8.92 (3)	0	250 … 500	N/A	0 … 5
Hu et al. (2008c)	R410A/ POE68 [a]	smooth tube with wrapped film heater	6.34	+5	200 … 400	7 … 15	0 … 5
Hu et al. (2008d)	R410A/ POE68 [a]	smooth tube with wrapped film heater	6.34	+5	200 … 400	7 … 15	0 … 5

continued on next page

Authors	Refrigerant/ Lubricant	Measurement section	d_i/m	ϑ_s/°C	$\dot{m}$/kg m^{-2} s^{-1}	$\dot{q}$/kW m^{-2}	w_{oil}/wt.-%
Hu et al. (2008b)	R410A/ POE68 [a]	micro-fin tube with wrapped film heater	6.4	+5	200 … 400	7 … 15	0 … 5
Castro et al. (2009)	R134A/ POE10 [a]	metal and glass smooth tubes	3.22	+30	2000 … 3000	N/A	60 … 80
Ding et al. (2009)	R410A/ POE68 [a]	micro-fin tube with wrapped film heater	4.6	+5	200 … 400	7 … 15	0 … 5
Hu et al. (2009a)	R410A/ POE68 [a]	micro-fin tube with wrapped film heater	4.18	+5	200 … 400	7 … 15	0 … 5
Hu et al. (2009b)	R410A/ POE68 [a]	smooth tubes with wrapped film heater	2.00, 4.18	+5	200 … 400	7 … 14	0
Han et al. (2013)	HFO-1234yf/ POE	micro-fin tube with wrapped film heater	6.44	+5 … +15	100 … 400	4 … 12	0 … 5
Hu et al. (2014)	R410A/ POE68 [a]	micro-fin tubes with wrapped film heater	3.56, 4.60	+5	200 … 400	7 … 17	0 … 5

[a] fully miscible
[b] partially miscible

Table 2.2: Publications on flow boiling of CO_2

Authors	Lubricant	Measurement section	d_i/m	$\vartheta_s/°C$	$\dot{m}/kg\,m^{-2}\,s^{-1}$	$\dot{q}/kW\,m^{-2}$	$w_{oil}/wt.\text{-}\%$
Hassan (2004)	N/A [b]	stainless steel smooth tube, heated by R22 shell-side condensation	10.06	$-30\ldots-10$	$90\ldots125$	$5\ldots17$	$0\ldots7$
Park and Hrnjak (2006)	-	fluid-heated copper smooth tube;	6.1	$-30, -15$	$100\ldots400$	$5\ldots15$	-
Hashimoto and Kiyotani (2006)	-	directly heated copper smooth tube	5.2	$0\ldots+10$	$200\ldots700$	$10\ldots40$	-
Dang et al. (2006)	PAG100 [b]	directly heated stainless steel smooth tube;	$2\ldots6$	$+15$	360, 720, 1440	$4\ldots36$	$0\ldots5$
Koyama et al. (2006)	PAG	fluid-heated copper smooth (1) and micro-fin (2) tubes	4.42 (1); 4.90 (2)	$-5\ldots+15$	$360\ldots650$	< 80	$0.04\ldots07$
Gao and Honda (2006)	PAG100 [b]	directly heated smooth tube; micro-fin tube with wrapped film heater	3.04	$+10$	$190\ldots1300$	$5\ldots30$	$0\ldots0.7$
Cho and Kim (2007)	-	directly heated smooth and micro-fin tubes	4.00, 7.72; 4.06, 8.92	$0\ldots+20$	$212\ldots565$	$6\ldots20$	-
Park and Hrnjak (2007a)	-	fluid-heated copper smooth tube; multi-port aluminum tube with 10 round channels	3.5, 6.1; 0.89	$-30, -15$	$100\ldots400$	$2\ldots15$	-
Zhao and Bansal (2007)	-	directly heated stainless steel smooth tube	4.57	-30	$140\ldots230$	$12\ldots19$	-
Gao et al. (2007)	PAG100 [b]	directly heated smooth tube; micro-fin tube with wrapped film heater	3.04	$-10\ldots+10$	$190\ldots1300$	$5\ldots30$	$0\ldots0.7$
Kim et al. (2008b)	-	smooth and micro-fin tubes	3.48	$-25, -15$	$200\ldots800$	0	-
Higashiiue et al. (2008)	-	fluid-temperated copper micro-fin tube	5.38	$-5, +5, +15$	$300\ldots540$	N/A (0)	-
Gao et al. (2008b)	PAG100 [b]	directly heated smooth tube; micro-fin tube with wrapped film heater	3.75	$+10$	$100\ldots500$	$10\ldots30$	$0\ldots1$
Katsuta et al. (2008)	PAG100	directly heated stainless steel smooth tube	3.0	$-10\ldots+10$	200, 400	$5\ldots15$	$0\ldots4$
Müller and Eggers (2008)	POE85 [a], PAO68[b]	fluid-heated smooth tube	9.4	$-20; -10; 0$	96-292	N/A	$0\ldots4$

continued on next page

Authors	Lubricant	Measurement section	d_i/m	$\vartheta_s/°\mathrm{C}$	$\dot{m}/\mathrm{kg\,m^{-2}\,s^{-1}}$	$\dot{q}/\mathrm{kW\,m^{-2}}$	$w_{oil}/\mathrm{wt.\text{-}\%}$
Mastrullo et al. (2009)	-	directly heated stainless steel smooth tube	6.0	−7.8 … +5.8	200 … 349	10 … 21	-
Ozawa et al. (2009)	-	directly heated smooth tube	1, 2, 3	+5 … +27	100 … 700	5 … 50	-
Mastrullo et al. (2010)	-	directly heated stainless steel smooth tube	6.0	−7.8 … +5.8	200 … 349	10 … 21	-
Dang et al. (2010)	-	directly heated copper micro-fin tube;	2.0	+15	360 … 720	4 … 18	-
Ono et al. (2010)	PAG100 [b]	directly heated smooth tube; micro-fin tube with wrapped film heater	3.75	+10	100 … 500	0 … 30	0 … 1
Pehlivanoglu et al. (2010)	POE85 [a]	fluid-heated copper smooth tube	6.1	−30, −15	100 … 400	2 … 15	0.5 … 2
Kim et al. (2010)	POE85 [a]	fluid-heated copper smooth tube	11.2	−30, −10	40 … 200	2 … 15	0.5 … 4
Oh et al. (2011)	-	directly heated stainless steel smooth tube	0.5, 1.5, 3.0	0 … +15	50 … 600	4 … 40	-
Oh and Son (2011)	-	directly heated stainless steel smooth tube	4.5	0 … +40	200 … 1000	10 … 40	-
Bansal (2011)	POE60 [a]	directly heated stainless steel smooth tube	4.35	−40 … −30	259	16	0 … 4-
Mastrullo et al. (2012b)	-	directly heated stainless steel smooth tube	6.0	+5	200 … 350	5, 20	-
Mastrullo et al. (2012a)	-	directly heated stainless steel smooth tube	6.0	+7, +12	150 … 500	5, 20	-
Zhao and Bansal (2012)	-	directly heated copper micro-fin tube	7.33	−30	100 … 250	10 … 25	-
Kim and Hrnjak (2012)	POE85 [a]	fluid-heated smooth and micro-fin tubes	11.2	−15	40 … 200	0.5 … 10	0.5 … 2
Grauso et al. (2013a)	-	directly heated stainless steel smooth tube	6.0	+7, +12	150 … 500	5, 20	-
Dang et al. (2013)	PAG100 [b]	electrically heated stainless steel smooth tube	2, 4, 6	+15	360, 720, 1440	4 … 36	0.5 … 5
Grauso et al. (2014)	-	stainless steel smooth tube	6.0	+7, +12	150 … 500	0	-
Hrnjak and Kim (2014)	POE85 [a]	fluid-heated smooth and micro-fin tubes	11.2	−15	40 … 200	0.5 … 10	0.5 … 2

[a] fully miscible
[b] partially miscible

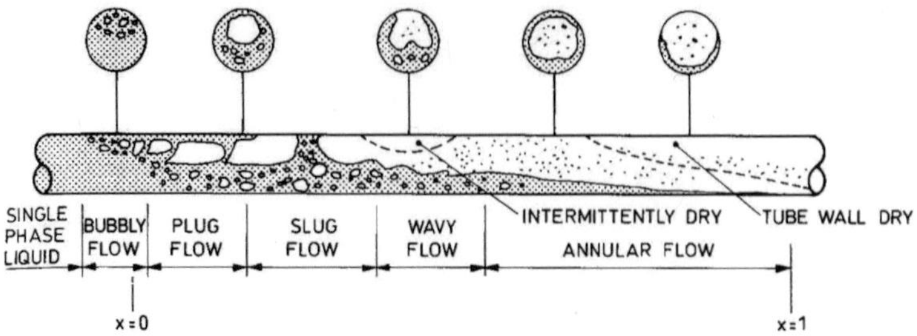

Figure 2.1: Flow patterns in a horizontal evaporator tube, published in Collier and Thome (1994)

- *Bubbly flow* contains vapor bubbles dispersed in the liquid phase. Due to gravity, bubbles tend to flow at the upper part of the tube.
- *Plug flow* is characterized by bubbles traveling at the upper part of the tube coalescing to large bubbles of approximately tube diameter in size.
- *Stratified flow* occurs at very low flow velocities. Both phases are separated by a planar interface. The upper part of the tube wall is dry.
- At increasing vapor velocity, the interface gets disturbed and *wavy flow* develops with waves traveling in flow direction.
- A further increase of vapor velocity may lead to *slug flow*. Liquid slugs travel in flow direction, periodically wetting the whole tube wall.
- In *annular flow*, the vapor phase flows with high velocity in the center of the tube while the liquid flows as thin film at the tube wall.

It is of importance to know that the exact classification of flow pattern may vary from author to author. Slug and plug flow are often summarized as *intermittent flow*. When liquid film during annular flow begins to break, partial dryout of the tube occurs, generally named *dryout region*. Further, combined flow patterns were distinguished by some authors, such as wavy-annular flow.
Flow patterns are classically predicted by means of flow pattern maps. Herein, flow regimes are separated by transition curves. The location of flow regimes in those maps is greatly affected by fluid properties, flow conditions and tube geometry. According to investigations by Schael (2009), the flow pattern maps by Steiner (2002) and Thome and El Hajal (2003) are in fairly good agreement to observed flow patterns for CO_2 in smooth tubes, while deviations in the micro-fin tube are larger. The Steiner flow pattern map actually represents a modification of the Taitel and Dukler (1976) map and is of dimensionless form. The transition from stratified to wavy flow is given by the product of liquid Reynolds and vapor Froude number, $(\mathrm{Re_L\,Fr_V})^{0.5}$. An average vapor Froude number, $\mathrm{Fr_{Vm}}$, was introduced to

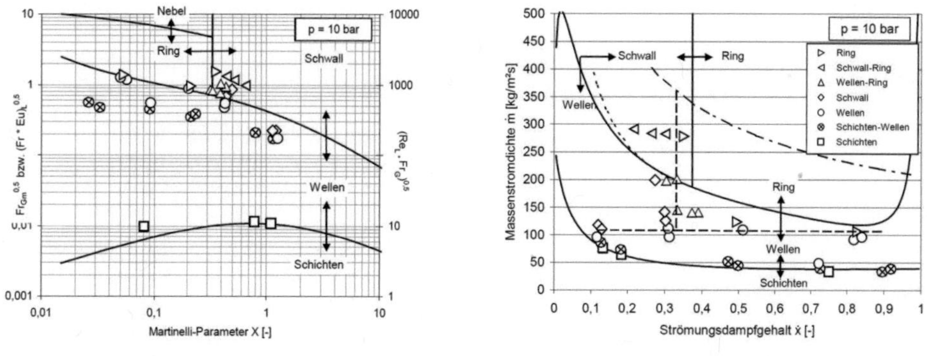

Figure 2.2: Flow pattern maps of Steiner (left) and Thome and El Hajal (right) with experimental data of CO_2 (symbols) by Schael (2009)

express the wavy to intermittent/annular flow transition. Transition from intermittent to bubbly flow occurs above a limiting value for the product of liquid Euler and Froude number, $(\mathrm{Fr\,Eu})_{\mathrm{L}}^{0.5}$. Transition criteria are given in dependency of Martinelli parameter, X.

$$X = \left(\frac{1-\dot{x}}{\dot{x}}\right)^{(2-n)/2} \left(\frac{\rho_{\mathrm{V}}}{\rho_{\mathrm{L}}}\right)^{0.5} \left(\frac{\mu_{\mathrm{L}}}{\mu_{\mathrm{V}}}\right)^{n/2} \tag{2.1}$$

where $\dot{x}$ is the vapor quality and $\rho_{\mathrm{L|V}}$ and $\mu_{\mathrm{L|V}}$ are the liquid/vapor density and viscosity, respectively.

Being not handy from a practical point of view, Kattan et al. (1998a) transferred the original Steiner map to the dimensionful graphic rendition of mass flow velocity versus vapor quality. Besides, slight modifications of wavy-intermittent/annular and annular-dryout transitions were performed. Starting from this map, several authors proposed updated versions, among those Thome and El Hajal. The Steiner and Thome and El Hajal maps are pictured in Fig. 2.2 as presented in Schael (2009). Calculated transitions are shown as solid lines, symbols represent experimentally determined flow patterns of CO_2 at 10 bar.

Flow pattern maps depend on the correct prediction of void fraction, for which a variety of different models exist. Taking the two-fluid model as basis assuming stratified flow and using a friction factor correlation for vapor flow pressure drop, Taitel and Dukler (1976) and following Steiner (1983) derived an iterative method solving the momentum equation.

$$X^2 = \left\{ \left(\frac{\pi}{\tilde{U}_{\mathrm{V}}+\tilde{U}_{\mathrm{i}}}\right)^{-n} \left(\frac{\pi^2}{64\tilde{A}_{\mathrm{V}}^2}\right) \left(\frac{\tilde{U}_{\mathrm{V}}+\tilde{U}_{\mathrm{i}}}{\tilde{A}_{\mathrm{V}}}+\frac{\tilde{U}_{\mathrm{i}}}{\tilde{A}_{\mathrm{L}}}\right) - \frac{1}{(\mathrm{Fr\,Eu_V})} \right\} \times \\ \times \left\{ \left(\frac{\pi}{\tilde{U}_{\mathrm{L}}}\right)^{-n} \left(\frac{\pi^2 \tilde{U}_{\mathrm{L}}}{64\tilde{A}_{\mathrm{L}}^3}\right) \right\} \tag{2.2}$$

where $\tilde{A}_{L|V} = A_{L|V}/d_i^2$ is the dimensionless liquid/vapor flow area, $\tilde{U}_{L|V} = U_{L|V}/d_i$ is the dimensionless perimeter of wetted and dry tube wall, $\tilde{U}_i = U_i/d_i$ is the dimensionless perimeter of liquid-vapor interface and $n = 0.25$ is the Reynolds exponent of Blasius law. For non-inclined tubes, the product of vapor Froude and Euler number yields $(\mathrm{Fr}\,\mathrm{Eu})_V = 0$. Steiner recommended using the Rouhani and Axelsson (1970) void fraction correlation derived from applying a drift flux model. Its good accuracy has been further confirmed by Wojtan et al. (2004) and recently by Saito et al. (2011) for CO_2 and CO_2-lubricant mixture flow boiling inside a horizontal smooth tube.

$$\varepsilon = \frac{\dot{x}}{\rho_V} \left\{ (1 + 0.12\,(1-\dot{x})) \left(\frac{\dot{x}}{\rho_V} + \frac{1-\dot{x}}{1-\rho_L} \right) + \frac{1.18\,(1-\dot{x})}{\dot{m}\,\rho_L^{0.5}} \left(g\,\sigma\,(\rho_L - \rho_V) \right)^{0.25} \right\}^{-1} \tag{2.3}$$

where $\dot{m}$ is the mass flux, σ is the surface tension and g is the gravitational constant.
As the flow pattern significantly impacts on pressure drop and mostly on heat transfer, several state-of-the-art prediction models for these design parameters have yet been developed taking the effects of flow patterns into account. These models are usually of high accuracy, but they demand for a precise flow pattern prediction. That is why flow pattern map development is an ongoing process, especially for new refrigerants applied. Updated flow pattern maps for CO_2 and various other refrigerants have been proposed since and were investigated within the scope of this work.
The flow pattern maps of Kattan et al. (1998a) and Thome (2004) were subject to several flow pattern maps developed by Thome and co-workers henceforth. Wojtan et al. (2005a) developed a diabatic flow pattern map (Fig. 2.3) based on R22 and R410A measurements that has widely been used for flow pattern prediction in literature. Supported by void fraction measurements, the stratified-Wavy and Wavy-Annular transition curves were given

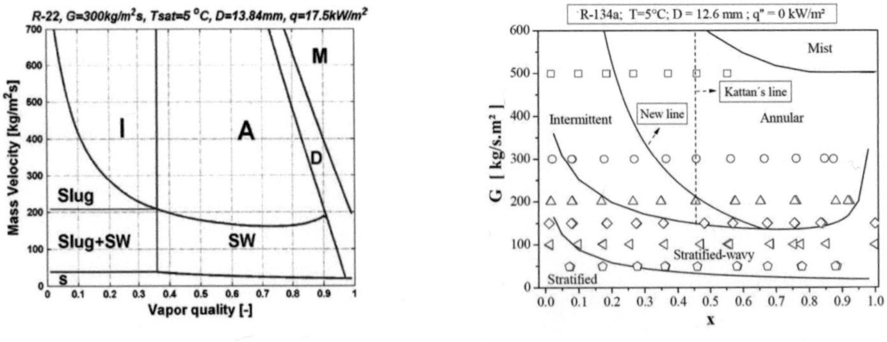

Figure 2.3: Wojtan et al. (2005a) (left) and Barbieri et al. (2008) (right) flow pattern maps, published ibidem

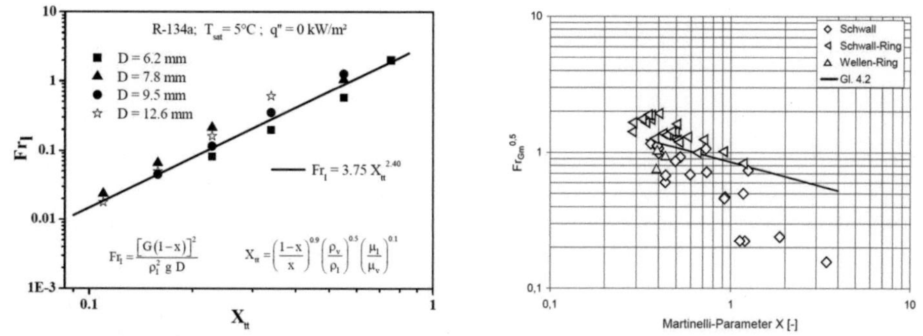

Figure 2.4: Intermittent to annular flow transition criteria for R134a (left) and CO_2 (right) two-phase flow in smooth tubes, published in Barbieri et al. (2008) and Schael (2009), respectively

as proposed by Kattan et al. for adiabatic flow. The effect of heat input leading to partial dryout of the tube was separated from these expressions and captured by a new criterion for the onset of dryout. In doing so, the mist flow curve was modified as well.

Barbieri et al. (2008) presented a new transition curve based on R134a flow boiling experiments in smooth tubes, supported by observations of Schael (2009) encountered for the intermittent to annular flow transition for CO_2. Tests were performed at a fixed saturation temperature of $+5$ °C varying the tube diameter from 6.2 to 12.6 mm, hence corresponding to the range of diameters investigated in the scope of this work. The transition curve was proposed in terms of liquid Froude number ($\mathrm{Fr_L}$) and Martinelli parameter (X_tt), while Schael gave an expression for the mean vapor Froude number ($\mathrm{Fr_{Vm}}$), see Fig. 2.4.

$$\mathrm{Fr_{L,I\text{-}A}} = 3.75 X_\mathrm{tt}^{2.40} \tag{2.4}$$

$$\mathrm{Fr_L} = \frac{((1-\dot{x})\,\dot{m})^2}{\rho_\mathrm{L}^2\, g\, d_\mathrm{i}} \tag{2.5}$$

$$X_\mathrm{tt} = \left(\frac{1-\dot{x}}{\dot{x}}\right)^{0.9} \left(\frac{\rho_\mathrm{V}}{\rho_\mathrm{L}}\right)^{0.5} \left(\frac{\mu_\mathrm{L}}{\mu_\mathrm{V}}\right)^{0.1} \tag{2.6}$$

where Eq. 2.4 represents the intermittent-annular transition criterion. The other transition boundaries were assumed to correspond to the Kattan et al. map.

Rollmann and Spindler (2011) determined flow patterns of R407C flow boiling in a 8.92 mm micro-fin tube. A comparison of the experimental data to the flow pattern map of Wojtan et al. (2005a) was presented, showing poor agreement. Due to the helix micro-fin structure, annular flow regime was observed at significant lower mass flow rates than predicted. The authors suggested, that the Barbieri et al. intermittent-annular transition curve by reflecting the influence of diameter and mass flow rate sounds promising. However, the special structure of the micro-fin tube has to be taken into account as well.

Refrigerant-lubricant mixtures

The presence of lubricant can change flow patterns in various ways (Wang et al., 2014) (Bandarra Filho et al., 2009). As a matter of fact, flow pattern transitions are dependent on fluid properties, which in turn are affected by the liquid oil mass fraction. Increasing surface tension and viscosity, for instance, may lead to an enlarged wetted perimeter and thicker liquid film. Another very important fact is the possibility of foam formation. Due to foaming, the wetting of the tube might as well increase, especially for low mass flow rates and height of the stratified liquid flow. On the other hand, the presence of foam can exert a significant influence on pressure drop. Hence, accounting for flow pattern when discussing heat transfer or pressure drop measurements is essential for a comprehensive understanding of the flow boiling behavior of refrigerant-lubricant mixtures.

Comparing flow patterns of R12-lubricant mixtures flow boiling in a micro-fin tube to those observed in a smooth tube, Ha and Bergles (1993) demonstrated the oil effect of enhancing the tube wetting at low mass flow rates. At elevated vapor qualities, the liquid film observed at the tube perimeter became highly viscous. Foaming suppression due to the micro-fin structure was reported.

Flow patterns of refrigerant-lubricant mixtures can vary from pure/mixed refrigerants, especially when foaming occurs. Such effects are not considered in classical flow pattern definitions nor in flow pattern maps. For that reason, Wongwises et al. (2002) conducted visual studies for two-phase flow of mixtures of R134a and fully miscible lubricating oil. It was found out, that the presence of lubricant induced an early annular flow formation, attributed to mixture properties. Besides the beforehand mentioned flow patterns, mixed patterns including froth-wavy and froth-wavy-annular flow as well complete froth flow were distinguished. Examples of these are shown in Fig. 2.5.

Wei et al. (2007) conducted flow boiling experiments for R22-NM56 mixtures inside 6.34 and 2.50 mm smooth tubes. The mineral oil used was fully miscible in the range of operation. Inside the 6.34 mm tube it was observed that the presence of oil (> 3 wt.-%) could shift the wavy-annular transition to lower (mass flow) velocities and delay annular-mist flow due increasing viscosity and decreasing surface tension, hence promoting annular flow. Contrarily, annular/slug flow patterns were continuously been observed throughout all experiments conducted in the small diameter tube. It was explained by confinement of bubble release and capillary effects due to the small tube dimensions.

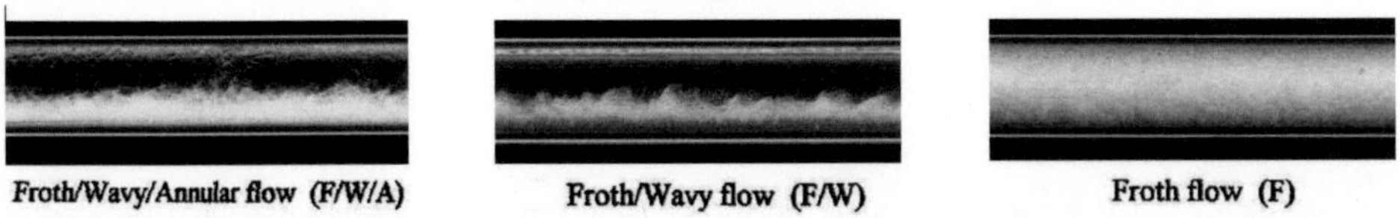

Figure 2.5: High-speed video recordings of R134a-lubricant mixture flow patterns by Wongwises et al. (2002)

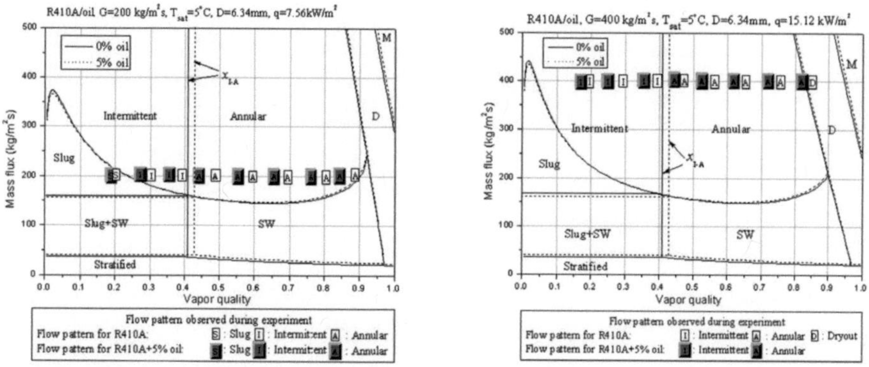

Figure 2.6: Flow patterns observed for R410a-POE oil mixtures inside 6.34 mm smooth tube and comparison to Wojtan et al. flow pattern map calculated using mixture properties, published in Hu et al. (2008c)

Using the same test facility, Hu et al. (2008c) reported experimental results from flow boiling experiments of R410a-oil mixtures inside a 6.34 mm smooth tube. The POE oil used was fully miscible. Observed flow patterns at $T_s = 5\,°C$ and mass flow velocities of 200, 300 and 400 $kg\,m^{-2}\,s^{-1}$ were compared to the flow pattern map of Wojtan et al. (2005a), which was calculated using mixture properties. The map perfectly reflected the experimental data points, see Fig. 2.6. However, data points as shown for the pure refrigerant and refrigerant-lubricant mixture with $w_{oil} = 5$ wt.-% basically coincide and the transition of slug-annular flow shifts to higher vapor qualities. Both facts do not match the described observations given in Hu et al. (2008d), where the same author denotes the increase in pressure drop in low and high vapor quality regions to a promotion of annular flow due to the presence of oil. Similar results were presented by Hu et al. (2009a) for measurements of R410a-oil mixtures inside a 4.18 mm smooth tube. The flow pattern map of Wojtan et al. (2005a) was found to predict mixture flow patterns with high accuracy.

Visual flow pattern observations of oil rich R134a-POE oil mixture flow inside a small diameter tube ($d_i = 3.22$ mm) were reported by Castro et al. (2004), see flow patterns in Fig. 2.7. Experiments were carried out for oil mass fractions ranging from 47 to 78 wt.-%. Flow patterns were observed at different axial locations of the unheated tube, changing from single phase liquid flow at the inlet to bubbly flow and further to foam flow near the outlet of the tube. Upon reaching foam flow conditions, pressure drop gradient increases significantly.

Dias et al. (2011) developed a two-phase flow model for calculation of axial temperature and vapor pressure profiles. Basic momentum, mass and energy conservation equations are solved numerically in dependency of the flow regime as observed by Castro et al. (2004).

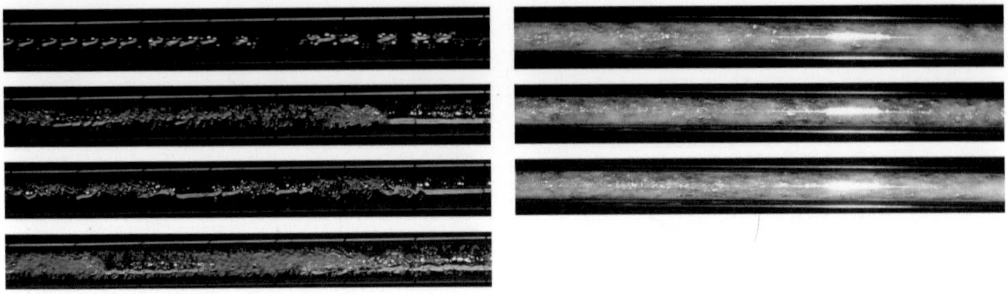

Figure 2.7: Flow patterns observed for oil-rich R134a-POE oil pipe flow near the inlet (left) and outlet (right) of a 3.22 mm non-heated smooth tube, published in Castro et al. (2004)

Transition from single phase to bubbly flow occurs, when the refrigerant concentration dissolved in the oil exceeds its solubility. Foam flow develops after reaching a limiting void fraction. Experimental data by Castro et al. (2009) was used for validation, yielding accurate predictions in the first two regions and larger deviations in foam flow. However, the model was developed for oil-rich mixtures ($w_{\text{oil}} > 60$ wt.-%) with the refrigerant dissolved in the oil out-gassing along the tube. Its range of validity does not match the experimental conditions within the scope of this work.
R410a flow boiling measurements by Oh et al. (2011) inside a 3 mm smooth tube showed that the transition vapor quality intermittent-annular flow ($\dot{x}_{\text{I-A}}$) was a strong function of mass flow rate, congruent to most recent investigations in literature and contrary to the prediction by Wojtan et al. (2005a).

2.2 Pressure drop

Pure/mixed refrigerants

An early and widely investigated frictional pressure drop model was developed by Storek and Brauer (1980). The model is based on homogeneous fluid properties, i.e., homogeneous fluid density (ρ_h) and homogeneous dynamic viscosity (μ_{h}).

$$\rho_{\text{h}} = \frac{\rho_{\text{V}}\,\rho_{\text{L}}}{\dot{x}\,(\rho_{\text{L}} - \rho_{\text{V}}) + \rho_{\text{V}}} \tag{2.7}$$

$$\mu_{\text{h}} = \frac{\mu_{\text{V}}\,\mu_{\text{L}}}{\dot{x}\,(\mu_{\text{L}} - \mu_{\text{V}}) + \mu_{\text{V}}} \tag{2.8}$$

where $\dot{x}$ is the vapor quality and $\rho_{\text{L}|\text{V}}$ and $\mu_{\text{L}|\text{V}}$ are the liquid/vapor density and dynamic viscosity, respectively.
Pressure drop is calculated in analogy to the Darcy-Weisbach equation, substituting the Darcy friction factor by a modified homogeneous friction factor (ξ_{h}). This friction factor

accounts for the friction pressure drop due to fictive homogeneous flow and influences of the two-phase flow by means of a correction factor.

$$\left(\frac{dp}{dL}\right)_{\mathrm{tp}} = \xi_{\mathrm{tp}} \frac{\dot{m}^2}{2\,\rho_{\mathrm{L}}\,d_{\mathrm{i}}} \tag{2.9}$$

$$\xi_{\mathrm{tp}} = \xi_{\mathrm{h}}\,\xi_{\mathrm{cor}} \tag{2.10}$$

where $\dot{m}$ is the mass flow velocity, d_{i} is the inner pipe diameter, ξ_{h} is the fictive homogeneous flow friction factor and ξ_{cor} is the correction factor due to the influence of two-phase flow, in particular of phase interface deformation turbulence.

The aim of Müller-Steinhagen and Heck (1986) was to develop an easy-to-use frictional pressure drop correlation since models often required a great effort for implementing, e.g., the flow pattern based method by Bandel (1973). The to date well established correlation corresponds to an empirical approach of interpolating between all-liquid ($\dot{x} = 0$) and all-vapor ($\dot{x} = 1$) pressure drop.

$$\left(\frac{dp}{dL}\right)_{\mathrm{LO|VO}} = \xi_{\mathrm{LO|VO}} \frac{\dot{m}^2}{2\,\rho_{\mathrm{L|V}}\,d} \tag{2.11}$$

where $\xi_{\mathrm{L|V}}$ is the liquid/vapor friction factor, respectively, $\dot{m}$ the total mass flow velocity, d the pipe inner diameter and $\rho_{\mathrm{L|V}}$ the respective liquid/vapor density.

The authors identified three general characteristics: (1) frictional pressure drop increases linearly within $0.0 < \dot{x} < 0.7$, (2) yielding all-vapor pressure drop at $\dot{x} \approx 0.5$ and (3) reaching its maximum at $\dot{x} \approx 0.85$. These trends are well portrayed by pressure drop measurements, as can be seen in Fig. 2.8 from results by Mancin et al. (2014b) and Katsuta et al. (2008), e.g.. From superposition of both vapor quality regions ($\dot{x} < 0.7$ and $\dot{x} > 0.7$) Eq. 2.12 was obtained.

$$\frac{dp}{dL} = \underbrace{\left[\left(\frac{dp}{dL}\right)_{\mathrm{L}} + 2\left(\left(\frac{dp}{dL}\right)_{\mathrm{V}} - \left(\frac{dp}{dL}\right)_{\mathrm{L}}\right)\dot{x}\right](1-\dot{x})^{1/3}}_{A} + \underbrace{\left(\frac{dp}{dL}\right)_{\mathrm{V}}}_{B}\dot{x}^3 \tag{2.12}$$

where term A represents the near linear increase of pressure drop in the lower vapor quality region and term B the closing condition to all-vapor pressure drop, superimposed by factors $(1-\dot{x})^{1/3}$ and $\dot{x}^3$, respectively. Despite its simplicity, it is among the most established and accurate two-phase pressure drop correlations at present. This has been confirmed by latest investigations, e.g., by Mastrullo et al. (2010), Xu et al. (2012), Grauso et al. (2013a) and Grauso et al. (2014) and by Rollmann et al. (2011) for a micro-fin tube using a correction multiplier.

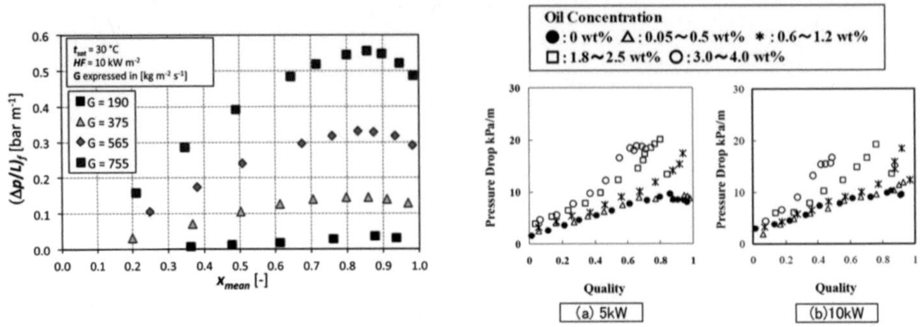

Figure 2.8: Pressure drops of R134a inside 3.6 mm micro-fin tube (left) and CO_2-lubricant mixtures inside 3.0 mm smooth tube (right), published in Mancin et al. (2014b) and Katsuta et al. (2008), respectively

A widely used frictional pressure correlation for micro-fin tubes was developed by Cavallini et al. (1997). In fact, this two-phase multiplier-type correlation embodies an adaptation of Friedel correlation to the special geometry of micro-fin tubes. Herein, the micro-fin tube is treated as tube with roughness of fin height.

The direct measurement of frictional pressure drop in horizontal tubes actually requires adiabatic flow conditions, since the evaporation of liquid causes an additional pressure loss due to acceleration of liquid molecules to the vapor velocity. Hence, this pressure drop contribution must be taken into account when conducting pressure drop measurements under diabatic conditions.

$$(\Delta p)_{\mathrm{tp}} = (\Delta p)_{\mathrm{fric}} + (\Delta p)_{\mathrm{mom}} \tag{2.13}$$

where $(\Delta p)_{\mathrm{tp}}$ is the total two-phase pressure drop and $(\Delta p)_{\mathrm{fric}}$ and $(\Delta p)_{\mathrm{mom}}$ are the frictional and acceleration pressure drop contributions.

To account for the acceleration pressure drop, usually a heterogeneous model is applied (Moreno Quibén and Thome, 2007a) (Cheng et al., 2008a) (Katsuta et al., 2008) (Da Silva Lima et al., 2009) (Hu et al., 2009b).

$$(\Delta p)_{\mathrm{mom}} = \dot{m} \left\{ \left(\frac{\dot{x}}{\rho_{\mathrm{V}}\, \varepsilon} + \frac{1-\dot{x}}{\rho_{\mathrm{L}}\,(1-\varepsilon)} \right)_{\mathrm{i}} - \left(\frac{\dot{x}}{\rho_{\mathrm{V}}\, \varepsilon} + \frac{1-\dot{x}}{\rho_{\mathrm{L}}\,(1-\varepsilon)} \right)_{\mathrm{o}} \right\} \tag{2.14}$$

where $\dot{m}$ is the total mass flow velocity, $\dot{x}$ is the vapor quality, ε is the void fraction and $\rho_{\mathrm{L|V}}$ are the liquid/vapor densities.

From diabatic and adiabatic pressure drop measurements with R22, R134a and R410a inside 8 and 13.8 mm smooth tubes, Moreno Quibén and Thome (2007b) developed a frictional pressure drop model widely used in literature. The proposed method segregates pressure drop calculation into different models valid for a certain flow pattern and, hence, relies on a

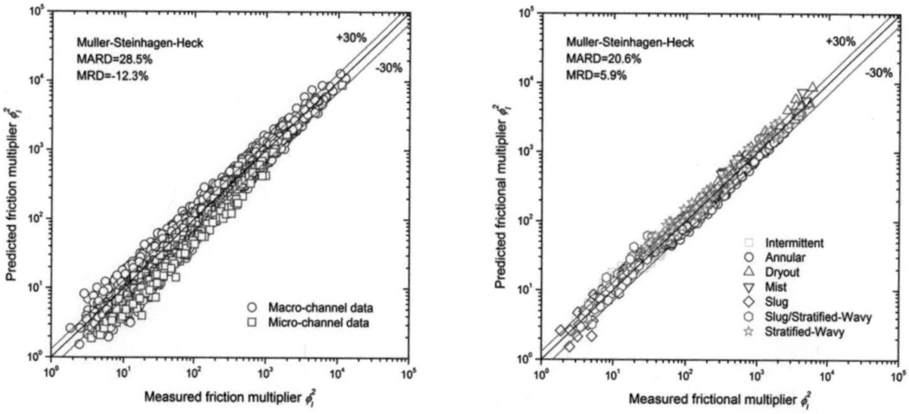

Figure 2.9: Performance of Müller-Steinhagen and Heck (1986) correlation compared to experimental data, published by Xu and Fang (2012)

method for predicting flow pattern at operating conditions. Flow pattern characterization and prediction follows Wojtan et al. (2005a). Further modifications have been proposed to extend validity to flat tubes (Moreno Quibén et al., 2009) and to fit CO_2 in particular (Cheng et al., 2008a). The latter is described in Sec. 2.4.

Recently, Fang and co-workers performed several investigations on predictive methods for pressure drop. A large number of experimental flow boiling (Xu and Fang, 2012) and flow condensation (Xu and Fang, 2013) pressure drops collected from literature were compared to a variety of predictive methods. In both cases, the Müller-Steinhagen and Heck (1986) model performed best and yielded an overall outstanding accuracy, see Fig. 2.9. Its prediction accuracy was also compared in dependency of observed flow pattern (only for data sets were flow patterns were presented), as shown in the right diagram. Yet, new correlations for flow evaporation and condensation pressure drop were proposed, using an improved correlation for single-phase frictional factor (Fang et al., 2011).

Refrigerant-lubricant mixtures

From experiments with R22-lubricant mixtures inside smooth and micro-fin tubes (Schlager et al., 1989), Schlager et al. (1990) proposed a general method for predicting refrigerant-lubricant mixture pressure drops based on a penalty factor ($PF_{\Delta p}$) expression.

$$\left(\frac{\Delta p}{\Delta L}\right)_{\text{mix}} = \left(\frac{\Delta p}{\Delta L}\right)_{\text{ref}} PF_{\Delta p} \tag{2.15}$$

This factor should embody the lubricant effect, i.e., the relative pressure drop increase with respect to the pure refrigerant pressure drop, so that no mixture properties are required

for using this prediction method. The penalty factor was given in form of two different empirical correlations.
Starting from the idea of Schlager et al., Eckels et al. (1994) and Eckels et al. (1998b) proposed a modified easy-to-use prediction method for pressure drops in smooth and micro-fin tubes. Two different kinds of empirical correlations were proposed for the penalty factor, yielding a polynomial

$$PF_{\Delta\mathrm{p}} = a_0 + a_1\left(w_{\mathrm{oil,0}}\right) + a_2\left(\dot{m}'\right) + a_3\left(w_{\mathrm{oil,0}}\dot{m}'\right) + a_4\left(w^2_{\mathrm{oil,0}}\dot{m}'\right) + \\ + a_5\left(w_{\mathrm{oil,0}}\dot{m}'^2\right) + a_6\left(w^2_{\mathrm{oil,0}}\dot{m}'^2\right) + a_7\left(w^2_{\mathrm{oil,0}}\right) + a_8\left(\dot{m}'^2\right) \quad (2.16)$$

and a logarithmic

$$\ln(PF_{\Delta\mathrm{p}}) = b_0 + b_1\left(w_{\mathrm{oil,0}}\right) + b_2\left(\dot{m}'\right) + b_3\left(w_{\mathrm{oil,0}}\dot{m}'\right) + b_4\left(w^2_{\mathrm{oil,0}}\dot{m}'\right) + \\ + b_5\left(w_{\mathrm{oil,0}}\dot{m}'^2\right) + b_6\left(w^2_{\mathrm{oil,0}}\dot{m}'^2\right) + b_7\left(w^2_{\mathrm{oil,0}}\right) + b_8\left(\dot{m}'^2\right) \quad (2.17)$$

expression, where $\dot{m}' = (\dot{m}/250\,\mathrm{kg\,m^{-2}\,s^{-1}})$, $w_{\mathrm{oil,0}}$ is the nominal oil mass fraction and a_{1-8} and b_{1-8} are constants that need to be fitted. Again, mixture properties were neither required nor considered.
Nidegger et al. (1997), Zürcher et al. (1997), Zürcher et al. (1998a) and Zürcher et al. (1998b) investigated pressure drops of R134a and R407C-oil mixtures inside smooth and micro-fin tubes of 10.92 – 11.90 mm in inner diameter. In both cases, the oil was completely miscible. In all tubes, increasing the oil mass fraction and, thus, liquid viscosity caused a continuous raise in pressure drop, especially at high vapor qualities. A modification of the Friedel (1979) frictional pressure drop model to account for the change in viscosity was developed by Zürcher et al. (1998b) by introducing a deterioration factor (f_{oil}), see Eq. 2.18. Here, the exponent (n) of the viscosity has to be fitted individually to the experimentally investigated refrigerant, such as by using Eq. 2.19.

$$\left(\frac{\Delta p}{\Delta L}\right)_{\mathrm{fric}} = \left(\Phi^2_{\mathrm{LO}}\right)_{\mathrm{ref}} 2 f_{\mathrm{LO}} \frac{\dot{m}}{\rho_{\mathrm{L}}\, d_{\mathrm{i}}} f^n_{\mathrm{oil}} \quad (2.18)$$

$$n = \frac{\ln\left(\left(\Phi^2_{\mathrm{LO}}\right)_{\mathrm{mix}}\right) - \ln\left(\left(\Phi^2_{\mathrm{LO}}\right)_{\mathrm{ref}}\right)}{w_{\mathrm{oil}} \ln\left(f_{\mathrm{oil}}\right)} \quad (2.19)$$

where $\left(\Phi^2_{\mathrm{LO}}\right)_{\mathrm{ref}}$ and $\left(\Phi^2_{\mathrm{LO}}\right)_{\mathrm{mix}}$ are the two-phase multipliers for all-liquid pure refrigerant and mixture flow and $f_{\mathrm{oil}} = (\mu_{\mathrm{oil}}/\mu_{\mathrm{ref}})$ is the lubricant-refrigerant viscosity ratio.
Hu et al. (2008d) conducted two-phase pressure drop measurements of R410a-lubricant mixtures inside a 6.34 mm smooth tube. Pressure drop was found to increase due to the presence of oil at all conditions. Three main characteristics of the oil influence were

addressed. (1) Oil promotes early annular flow in the low vapor quality region, thus the increase in pressure drop with respect to the pure refrigerant was significant. (2) At medium vapor qualities, annular flow was observed for both pure refrigerant and refrigerant-lubricant mixture. Hence, the influence of oil on pressure drop was observed being comparatively small. (3) In the high vapor quality region, the presence of lubricant delays dryout and maintains the wetting of tube, anew leading to a distinct increase in pressure drop with oil mass fraction. The authors presented a frictional pressure drop correlation based on vapor two-phase multiplier (ϕ_V) and mixture properties.

$$\left(\frac{\Delta p}{\Delta L}\right)_{tp} = \phi_V^2 \left(\frac{\Delta p}{\Delta L}\right)_V \tag{2.20}$$

The pressure drop of pure vapor flow (Δp_V) was calculated using the Fanning friction factor (f_V) given by Jung et al. (1989), see Eq. 2.22.

$$\left(\frac{\Delta p}{\Delta L}\right)_V = 2 f_V \frac{(\dot{m}\dot{x})^2}{\rho_V d_i} \tag{2.21}$$

$$f_V = \frac{0.046}{\mathrm{Re}_V^{0.2}} \tag{2.22}$$

where Re_V is the vapor Reynolds number.

$$\mathrm{Re}_V = \frac{\dot{m}\dot{x} d_i}{\mu_V} \tag{2.23}$$

ϕ_V was determined from experimental data (see Fig. 2.10, left) and predicted by an empirical exponential function of Martinelli parameter (X_{tt}).

$$\Phi_V = 1 + 3.52 X_{tt}^{0.46} \tag{2.24}$$

Hu et al. (2009b) continued pressure drop investigations of R410a-lubricant mixtures inside 2.00 and 4.18 mm smooth tubes, resulting in coinciding observations. The distinct pressure drop increase in the low vapor quality region was further attributed to foaming. A modified expression for ϕ_V was proposed to capture the dependency on tube diameter as shown in Fig. 2.10 (right).

$$\phi_V = 1 + f_1(d_i^*) X_{tt}^{f_2(d_i^*)} \tag{2.25}$$

where $f_1(d_i^*) = 0.095 + 1.38 d_i^* - 0.132 d_i^{*2}$ and $f_2(d_i^*) = 0.795 - 0.051 d_i^*$ are empirical functions of inner diameter ($d_i^* = d_i/$ mm).

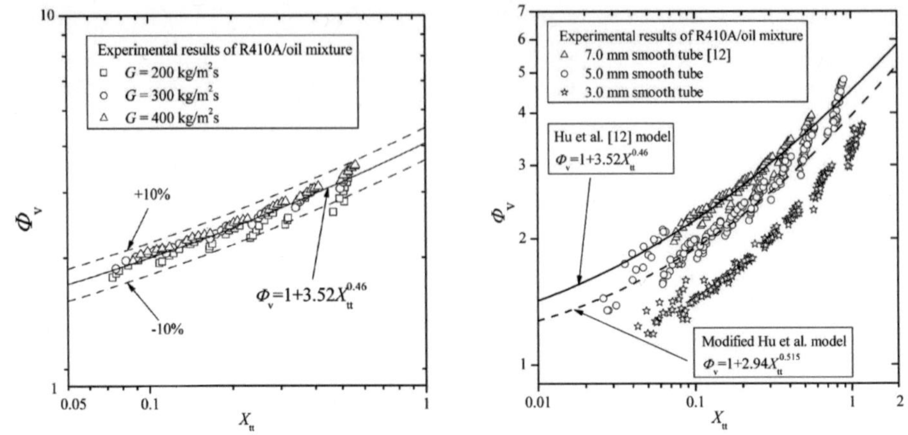

Figure 2.10: Vapor two-phase multiplier for R410a-lubricant flow in 6.34mm smooth tube (left) and dependency of φ_V on tube diameter (right), published in Hu et al. (2008d) and Hu et al. (2009b), respectively

Under same operating conditions as Hu et al. (2008d), Hu et al. (2008b) performed two-phase pressure drop measurements for R410a-lubricant inside a 6.50 mm micro-fin tube. Observations basically match those from smooth tube experiments. However, penalty factors ($PF_{\Delta p} = \Delta p_{\text{mix}} / \Delta p_{\text{ref}}$) for the micro-fin tube are slightly smaller, especially in the low vapor quality region. This could be explained by annular flow promotion in the micro-fin tube. Two Lockhart-Martinelli-type correlations with respect to sole vapor (Δp_{V}) and sole liquid (Δp_{L}) were proposed.

$$\left(\frac{\Delta p}{\Delta L}\right)_{\text{tp}} = \phi^2_{V/L} \left(\frac{\Delta p}{\Delta L}\right)_{\text{L|V}} \tag{2.26}$$

The pressure drop of liquid/vapor assumed to flow alone in the pipe ($\Delta p_{\text{L|V}}$) was calculated using the Fanning friction factor ($f_{\text{L|V}}$) determined in previous tests, in Eq. 2.29. Liquid mixture properties were considered. Further, the hydraulic diameter (d_{h}) was used instead of inner diameter.

$$\left(\frac{\Delta p}{\Delta L}\right)_{\text{V}} = 2 f_{\text{V}} \frac{(\dot{m}\dot{x})^2}{\rho_{\text{V}} d_{\text{h}}} \tag{2.27}$$

$$\left(\frac{\Delta p}{\Delta L}\right)_{\text{L}} = 2 f_{\text{L}} \frac{(\dot{m}\,(1-\dot{x}))^2}{\rho_{\text{mix}} d_{\text{h}}} \tag{2.28}$$

$$f_{\text{L|V}} = \frac{0.02}{\text{Re}_{\text{L|V}}^{0.104}} \tag{2.29}$$

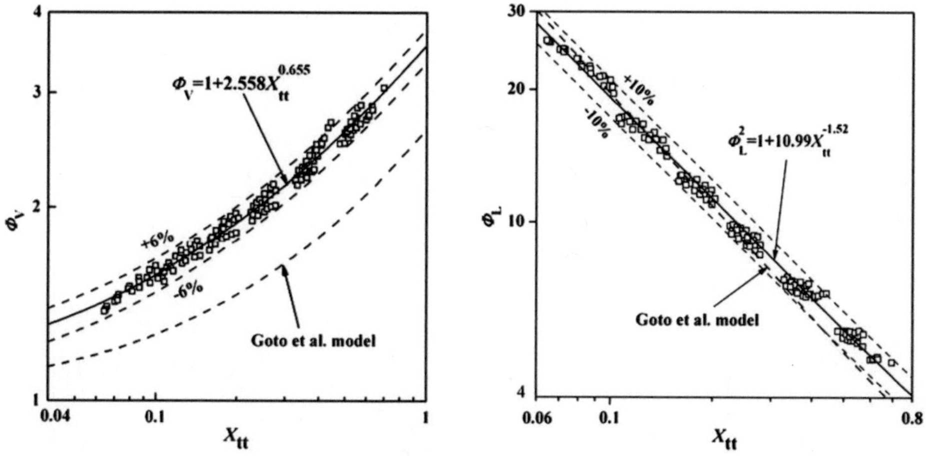

Figure 2.11: Determined vapor (left) and liquid (right) two-phase multipliers ($\varphi_{V|L}$) for R410a-oil mixture flow inside 6.50 mm micro-fin tube as function of Martinelli parameter, published in Hu et al. (2008b)

where $\mathrm{Re}_{L|V}$ is calculated using the hydraulic diameter (d_h) and mixture properties for the liquid phase.

$$\mathrm{Re}_V = \frac{\dot{m}\dot{x}d_h}{\mu_V} \tag{2.30}$$

$$\mathrm{Re}_L = \frac{\dot{m}\,(1-\dot{x})\,d_h}{\mu_{mix}} \tag{2.31}$$

Two-phase multipliers were fitted to the observed trends pictured in Fig. 2.11 by exponential functions of Martinelli parameter.

$$\Phi_V = 1 + 2.558\,X_{tt}^{0.655} \tag{2.32}$$

$$\Phi_L = \left(1 + 10.99\,X_{tt}^{-1.52}\right)^{0.5} \tag{2.33}$$

From pressure drop measurements inside a 4.60 mm micro-fin tube, Ding et al. (2009) proposed modified expressions for the friction factor and vapor two-phase multiplier to be used in conjunction with the vapor phase correlation by Hu et al. (2008b).
Huang et al. (2010) adopted Eq. 2.34 for flow condensation pressure drop. A graphical comparison of friction factors following Eq. 2.29 and 2.34 is shown in Fig. 2.12.

$$f_V = \frac{0.128}{\mathrm{Re}_V^{0.267}} \tag{2.34}$$

$$\Phi_V = 1 + 1.892\,X_{tt}^{0.587} \tag{2.35}$$

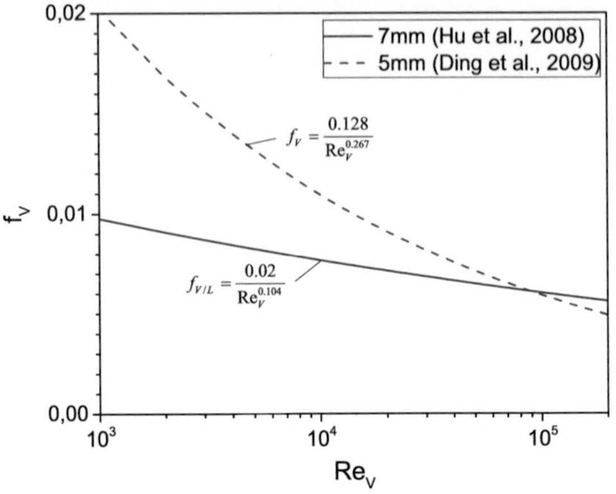

Figure 2.12: Fanning friction factors determined experimentally inside 6.50 mm and 4.60 mm micro-fin tubes by Hu et al. (2008b) and Ding et al. (2009)

The range of applicability of above mentioned correlations based on a modified two-phase multiplier model, however, is limited by the investigated range of operation and mathematically. As Awad (2010) insisted, there is a discontinuity of the vapor phase expression at high values of Martinelli parameter when using Eq. 2.32. The opposite would be the case for the liquid phase expression at low values of Martinelli parameter using Eq. 2.33. This problem generally arises from definition if polynomial functions for Φ_V^2 and Φ_L^2 are not of orders of 2 or -2, respectively.

Barbosa et al. (2004) and Castro et al. (2009) carried out experiments with refrigerant-oil mixtures of high oil mass fractions ($w_{oil} > 50$ wt.-%) and the refrigerant out-gassing along the unheated small diameter smooth tube. Barbosa et al. developed a pressure drop correlation for mineral oil-R12 mixture flow in a 2.88 mm tube based on vapor two-phase multiplier given in Eq. 2.36.

$$\Phi_V^2 = 1.24 \frac{\dot{x}}{1-\dot{x}} w_{oil} + \left(1.95 \exp\left\{6.94 \times 10^{-3} \frac{Re_V}{Re_L}\right\}\right) X + X^2 \qquad (2.36)$$

where Re_V and Re_L are the vapor and liquid Reynolds numbers, respectively, and X is the Martinelli parameter.

Castro et al. modified the expression for Φ_V^2 to fit their pressure drop data of POE oil-R134a mixture flow inside a 3.22 mm tube.

$$\Phi_V^2 = -5,38 \frac{\dot{x}}{1-\dot{x}} w_{oil} + \left(1.36 \exp\left\{7.96 \times 10^{-5} \left(\frac{Re_V}{Re_L}\right)^{2.19}\right\}\right) X + X^2 \qquad (2.37)$$

Flow boiling measurements by Han et al. (2013) for homogeneous R1234yf-POE oil mixtures inside a 6.44 mm micro-fin tube revealed typical trends of two-phase pressure drop with respect to mass flow rate, saturation temperature, vapor quality and heat flux. The pressure drop increased by up to 30 % when increasing the oil mass fraction to 5 wt.-%, related to changes in viscosity and surface tension.

2.3 Heat Transfer

Pure/mixed refrigerants

Schael (2009) concluded that the Steiner (2002) correlation predicts flow boiling heat transfer coefficients in smooth and micro-fin tubes accurately by specifying both, fluid-dependent parameter and heat flux exponent. The two-phase heat transfer coefficient is defined by cubic superposition of nucleate (α_{nb}) and convective (α_{cb}) boiling contributions.

$$\alpha_{tp} = \left(\alpha_{nb}^3 + \alpha_{cb}^3\right)^{1/3} \tag{2.38}$$

Due to its explicit development for micro-fin tubes, the correlation of Cavallini et al. (1999) predicted micro-fin tube heat transfer coefficients equally well, later confirmed by Rollmann et al. (2011) for R407C flow boiling in a similar tube as used in this work. Cavallini et al. used a linear superposition of nucleate (α_{nb}) and convective (α_{cb}) boiling contributions.

$$\alpha_{tp} = \alpha_{nb} + \alpha_{cb} \tag{2.39}$$

where α_{nb} is given as slightly modified Cooper correlation (Eq. E.150, App. E.3), but the general heat transfer behavior is enclosed in α_{cb}.

$$\alpha_{cb} = \left\{0.023\,\mathrm{Re}_L^{0.8}\,\mathrm{Pr}_L^{1/3}\,F\,\frac{\lambda_L}{d}\right\}\left\{R_a^s\,(\mathrm{Bo}\,\mathrm{Fr}_V)^t\left(\frac{d_0}{d}\right)^v\left(\frac{\dot{m}_0}{\dot{m}}\right)^z\right\} \tag{2.40}$$

where the first combined term corresponds to the heat transfer performance in a smooth tube and the second term accounts for the influence of the micro-fin structure.
From flow boiling measurements with five refrigerants including CO_2, Oh et al. (2011) developed a new Chen-type heat transfer correlation valid for small diameter tubes (0.5…3.0 mm), see App. E.3. Convective boiling multiplier (F, Eq. 2.41) and nucleate boiling suppression (S, Eq. 2.42) factors were both given as functions of liquid pressure drop two-phase multiplier (Φ_L) for considering laminar flow in small tubes.

$$F = \max\left\{1;\ 0.76 + 0.023\,\Phi_L^{2.2}\right\} \tag{2.41}$$

$$S = 0.279\,\Phi_L^{-0.058}\,\mathrm{Bo}^{-0.098} \tag{2.42}$$

Refrigerant-lubricant mixtures

Ha and Bergles (1993) investigated R12-3GS lubricant mixtures flow boiling heat transfer inside a micro-fin tube. Heat transfer coefficients of R12 strongly depended on mass flow velocity, hence indicating convective boiling dominance. From that point, it was obvious that heat transfer was significantly deteriorated by increasing the oil mass fraction in the mixture, since a rising in liquid viscosity leads to lower Reynolds numbers and tends towards laminar flow. This effect can clearly be seen in the left diagram of Fig. 2.13 in dependency of vapor quality. Pure R12 convective heat transfer coefficients are greatly augmented with the vapor quality, whereas the most oil-rich mixture even revealed a steady decrease along the evaporation progress.
Using the same type of correlation procedure as for the pressure drop, Eckels et al. (1994) and Eckels et al. (1998b) proposed a method for predicting heat transfer coefficients in smooth and micro-fin tubes. Analogous to Eq. 2.15, mixture heat transfer coefficients were related to pure refrigerant heat transfer coefficients by means of an enhancement factor (E_α). Empirical correlations were suggested, refer to Eq. 2.16 and 2.17 in Sec. 2.2.

$$\alpha_{\text{mix}} = \alpha_{\text{ref}} E_\alpha \tag{2.43}$$

Especially for modeling heat transfer of refrigerant-oil solutions and comparing experimental data to predicted values by correlations, using the "right" temperature is important. Affecting mixture properties, the bubble point temperature and hence wall superheat is dependent on local oil mass fraction. Consequently Thome (1995) proposed a thermodynamically sound approach, including the substitution of pure refrigerant saturation temperature by mixture bubble point temperature for correct specification of the heat transfer coefficient.

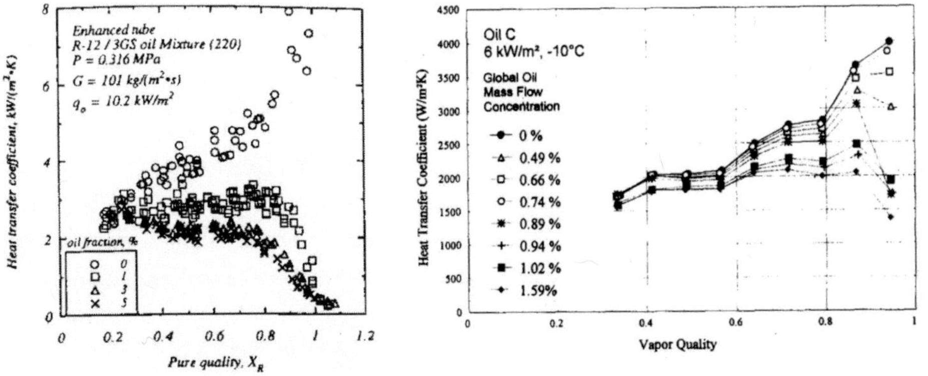

Figure 2.13: Influence of lubricant mass fraction on R12-mineral oil (left) and R134a-POE oil (right) mixture heat transfer coefficients, presented by Ha and Bergles (1993) and Hambraeus (1995), respectively

Hambraeus (1995) studied the effect of different ester lubricants on flow boiling heat transfer of R134a. While two oils used had comparable properties (miscible with R134a, similar viscosities), the third oil was not miscible and significantly more viscous. It was found out that the viscosity had a major impact on heat transfer performance, while partial miscibility was of no importance. Heat transfer coefficients decreased under nearly all conditions, see Fig. 2.13 (right), for instance. The correlation by Shah (1982) along with using mixture properties was identified as best predictive method. The observed trend of decreasing heat transfer with increase in oil mass fraction was correctly captured.

Nidegger et al. (1997), Zürcher et al. (1997), Zürcher et al. (1998a) and Zürcher et al. (1998b) conducted flow boiling heat transfer experiments using R134a and R407C oil mixtures inside smooth and micro-fin tubes of 10.92 – 11.90 mm in inner diameter. The oil was completely miscible in both fluids. Smooth tube results of both refrigerants were generally similar, revealing a heat transfer enhancement at low mass flow velocities and intermediate vapor qualities and a steady decrease with increasing oil mass fraction at higher mass flow rates. At high vapor qualities, a sharp drop in heat transfer coefficient occurred, attributed to the increase of liquid viscosity with local oil mass fraction and associated to the strong convective boiling behavior observed. Inside the micro-fin tube, a controversial influence of oil at intermediate vapor qualities was observed in some extend. Heat transfer coefficients of R134a significantly diminished at low mass flow velocities, which was suggested to be related to oil hold-up in the micro-fins. At higher mass flow velocities, heat transfer coefficients were partially enhanced at intermediate vapor qualities. For R407C-oil mixtures, heat transfer coefficients did not show that sharp decrease in heat at low mass flow rates, instead heat transfer steadily decreased with increasing oil mass fraction over the full range of operation. In addition, foaming was observed for R407C-oil mixtures with increasing intensity as local oil mass fraction increased.

Zürcher et al. (1998b) presented a general approach of modeling heat transfer coefficients of refrigerant-lubricant mixtures based on the Kattan et al. (1998b) model. Strong convective boiling behavior and the liquid viscosity as main influencing parameter were identified. Therefore, an adaptation of the model was proposed by either using a multiplication pre-factor inside the convective boiling contribution term (Eq. 2.44) or, even more comprehensive, incorporating the liquid refrigerant-lubricant mixture viscosity in the correlation (Eq. 2.45). This would yield the additional benefit of accounting for changes in transition boundary curves of the underlying flow pattern map. Predictions agreed well with the experimental data, ascribed to the strong convective boiling contribution.

$$\alpha_{\mathrm{cb}} = C \mathrm{Re}_{\mathrm{L}}^{n} \mathrm{Pr}_{\mathrm{L}}^{0.4} \left(\frac{\mu_{\mathrm{oil}}}{\mu_{\mathrm{ref}}} \right)^{0.4-n} \tag{2.44}$$

$$\alpha_{\mathrm{cb}} = C \mathrm{Re}_{\mathrm{mix}}^{n} \mathrm{Pr}_{\mathrm{mix}}^{0.4} \tag{2.45}$$

where Re_L is the Reynolds number of the liquid film according to the model and calculated using pure refrigerant properties, Pr_L is the liquid Prandtl number and μ_{ref} and μ_{oil} are the viscosities of the pure refrigerant and pure lubricant.
In flow boiling experiments of R22-NM56 mixtures inside a 6.34 mm smooth tube, Wei et al. (2007) observed an enhancement in heat transfer with increasing oil mass fraction in the low vapor quality region. This could be related to the promotion of annular flow due to the presence of oil. The converse effect was observed in the high vapor quality region. Further, these effects are reduced with increasing mass flow velocity. In the small diameter tube (2.50 mm), the enhancement effect was markedly reduced. At high mass flow rates (300 $\mathrm{kg\,m^{-2}\,s^{-1}}$) heat transfer was deteriorated in the whole range of operation. This was explained by a change from turbulent to laminar flow, lower liquid flow velocity and annular flow throughout all experiments conducted. Reviewing available heat transfer correlations, the authors suggested a new correlation for small diameter tubes based on two-phase multiplier (E) as first proposed by Shah (1982) and modified by Kandlikar (1990). The proposed correlation, given by Eq.2.46-2.48, requires using mixture properties for the liquid phase.

$$\alpha_{\mathrm{tp}} = E\,\alpha_{\mathrm{L}} \tag{2.46}$$

$$E = a_0 + \left(a_1\,\mathrm{Co}^{n_1} + a_2\,\mathrm{Bo}^{n_2}\right)\mathrm{Fr}_{\mathrm{L}}^{n_3} + w_{oil}\,a_3\,\frac{\left(\dot{x}\,(1-\dot{x})\right)^{n_4}}{\dot{m}^{n_5}} \tag{2.47}$$

$$\alpha_{\mathrm{L}} = \max\left\{4.364\,\frac{\lambda_{\mathrm{mix}}}{d_{\mathrm{i}}},\ 0.023\,\mathrm{Re}_{\mathrm{L}}^{0.8}\,\mathrm{Pr}_{\mathrm{L}}^{0.4}\,\frac{\lambda_{\mathrm{mix}}}{d_{\mathrm{i}}}\right\} \tag{2.48}$$

At the same test facility, Hu et al. (2008c) conducted heat transfer experiments inside a 6.34 mm smooth tube using mixtures of R410a and fully miscible POE68. In the low vapor quality region it was found out that the heat transfer coefficient was enhanced continuously by increasing the oil mass fraction. With increase in vapor quality, the maximum in heat transfer coefficient was shifted towards lower oil mass fractions. Increasing significantly with mass flow velocity, heat transfer coefficients further showed a strong convective boiling behavior. No further discussion of their results was given.
To accurately correlate the experimentally obtained heat transfer data, the heat transfer model by Kattan et al. (1998c) was applied using mixture properties for the liquid phase. Herein, both liquid and vapor phase are regarded separately in calculating the overall two-phase flow boiling heat transfer coefficient (α_{tp}).

$$\alpha_{\mathrm{tp}} = \frac{\theta_{\mathrm{dry}}}{2\pi}\,\alpha_{\mathrm{V}} + \frac{1-\theta_{\mathrm{dry}}}{2\pi}\left(\alpha_{\mathrm{nb}}^3 + \alpha_{\mathrm{cb}}^3\right)^{1/3} \tag{2.49}$$

where θ_{dry} is the dry angle of unwetted tube perimeter, α_{V} (calculated using Eq. E.144) is the vapor heat transfer coefficient at the dry tube perimeter and α_{nb} (calculated using

Eq. E.150) and α_{cb} are the nucleate and convective boiling heat transfer contributions at the liquid-wetted tube wall, respectively. α_{cb} was correlated by the Nusselt correlation

$$\mathrm{Nu} = \alpha_{\mathrm{cb}} \frac{\lambda_{\mathrm{L}}}{\delta} = C\,\mathrm{Re}_{\mathrm{L},\delta}^{m}\,\mathrm{Pr}_{\mathrm{L}}^{0.4} \tag{2.50}$$

where $\lambda_{\mathrm{L}}\,(=\lambda_{\mathrm{mix}})$, is the heat conductivity of the liquid (mixture), δ is the liquid film thickness, $\mathrm{Re}_{\mathrm{L},\delta}$ and Pr_{L} are the liquid film Reynolds and Prandtl numbers.

$$\mathrm{Re}_{\mathrm{L},\delta} = \frac{4\,\dot{m}\,\delta\,(1-\dot{x})}{\mu_{\mathrm{L}}\,(1-\varepsilon)} \tag{2.51}$$

$$\delta = \frac{\pi\,d_{\mathrm{i}}\,(1-\varepsilon)}{2\,(2\,\pi-\theta_{\mathrm{dry}})} \tag{2.52}$$

$$\theta_{\mathrm{dry}} = \begin{cases} \theta_{\mathrm{strat}} & \text{stratified flow} \\ \theta_{\mathrm{strat}} \left(\dfrac{\dot{m}_{\text{W-I|A}} - \dot{m}}{\dot{m}_{\text{W-I|A}} - \dot{m}_{\text{S-W}}} \right) & \text{wavy flow} \\ 0 & \text{annular flow} \end{cases} \tag{2.53}$$

Prediction of dry angle, film thickness and thus liquid Reynolds number requires the knowledge/prediction of flow pattern. As reviewed in Sec. 2.1, Hu et al. recommended using the flow pattern map by Wojtan et al. (2005a) in conjunction with refrigerant-lubricant mixture properties. By rearranging Eq. 2.49 and 2.50 to Eq. 2.54 and fitting C and m to the experimental R410-POE mixture data, $C = 0.027$ and $m = 0.635$ were obtained.

$$\left(\left(\frac{2\,\pi\,\alpha_{\mathrm{tp}} - \theta_{\mathrm{dry}}\,\alpha_{\mathrm{V}}}{2\,\pi - \theta_{\mathrm{dry}}} \right)^{3} - \alpha_{\mathrm{nb}}^{3} \right)^{1/3} \mathrm{Pr}_{\mathrm{L}}^{-0.4}\,\frac{\delta}{\lambda_{\mathrm{mix}}} = C\,\mathrm{Re}_{\mathrm{L},\delta}^{m} \mathrel{\hat{=}} Y \tag{2.54}$$

Hu et al. (2009a) continued heat transfer measurements inside a smaller sized 4.18 mm smooth tube. Similarly to the 6.34 mm tube experiments, heat transfer coefficients were greatly enhanced in the low and medium vapor quality region ($\dot{x} < 0.7$). The prediction accuracy of the Kattan et al. (1998c) heat transfer model in junction with the Wojtan et al. (2005a) flow pattern map was investigated. However, simple linear fit on logarithmic scale of Eq. 2.54 using the same constants C and m over the full range of investigated oil mass fractions (see Fig. 2.14) did not yield a satisfying accuracy. Instead, 2nd order polynomial functions of the nominal oil mass fraction ($w_{\mathrm{oil},0}$) were proposed.

$$C = \mathrm{e}^{-4.18-21.04\,w_{\mathrm{oil},0}+377.5\,w_{\mathrm{oil},0}^{2}} \tag{2.55}$$

$$m = 0.686 + 5.35\,w_{\mathrm{oil},0} - 69.6\,w_{\mathrm{oil},0}^{2} \tag{2.56}$$

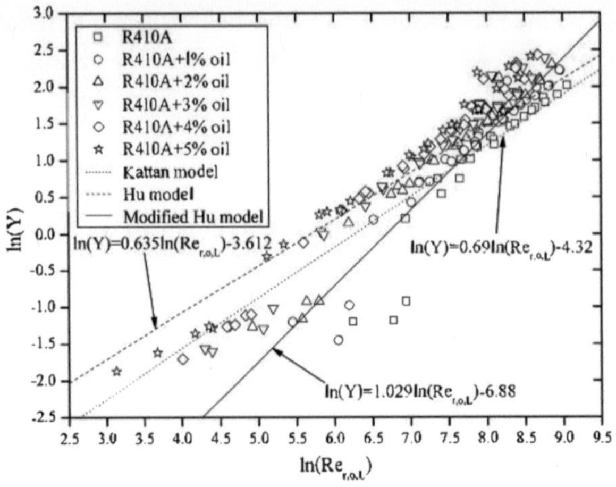

Figure 2.14: Plot of $\ln(Y)$ as function of $\ln(\mathrm{Re_L})$, published in Hu et al. (2009a)

Hu et al. (2008a) proposed a correlation based on the Gungor and Winterton (1986) model for predicting R410a-lubricant mixture flow boiling heat transfer in a 6.50 mm micro-fin tube. Herein the fin-bottom diameter of the micro-fin tube was chosen as characteristic length. Further, the authors introduced mixture properties to be used for the liquid phase.

$$E = 1 + 33686.87\,\mathrm{Bo}^{1.16} + 1.169\,X_{\mathrm{tt}}^{-0.86} \tag{2.57}$$

$$S = 1 + 2.53 \times 10^{-6}\,E^{1.489}\,\mathrm{Re}_{\mathrm{L}}^{1.17} \tag{2.58}$$

$$\alpha_{tp} = E\,(E_{\mathrm{RB}}\,\alpha_{\mathrm{DB}}) + S\,\alpha_{\mathrm{Co}} \tag{2.59}$$

where E and S are the convective boiling enhancement and nucleate boiling suppression factors, E_{RB} is the ribbed tube enhancement factor, α_{DB} and α_{Co} the heat transfer coefficients calculated by Dittus-Boelter and Cooper correlations.

$$E_{\mathrm{RB}} = \left\{ 1 + \left(2.64\,\mathrm{Re}_{\mathrm{L}}^{0.036}\,\mathrm{Pr}_{\mathrm{L}}^{-0.024} \left(\frac{h_{\mathrm{f}}}{l_{\mathrm{f}}}\right)^{0.21} \left(\frac{\beta}{90\,^{\circ}}\right)^{0.29} \right)^{7} \right\}^{1/7} \tag{2.60}$$

where Bo is the Boiling number, X_{tt} is the Martinelli parameter, h_f is the fin height, l_f is the axial fin pitch and β is the helix angle. Reynolds ($\mathrm{Re_L}$) and Prandtl ($\mathrm{Pr_L}$) numbers are calculated using mixture properties.

$$\mathrm{Re_L} = \dot{m}\,(1-\dot{x})\,d_{\mathrm{fr}}\,\mu_{\mathrm{mix}}^{-1} \tag{2.61}$$

$$\mathrm{Pr_L} = c_{\mathrm{p,mix}}\,\mu_{\mathrm{mix}}\,\lambda_{\mathrm{mix}}^{-1} \tag{2.62}$$

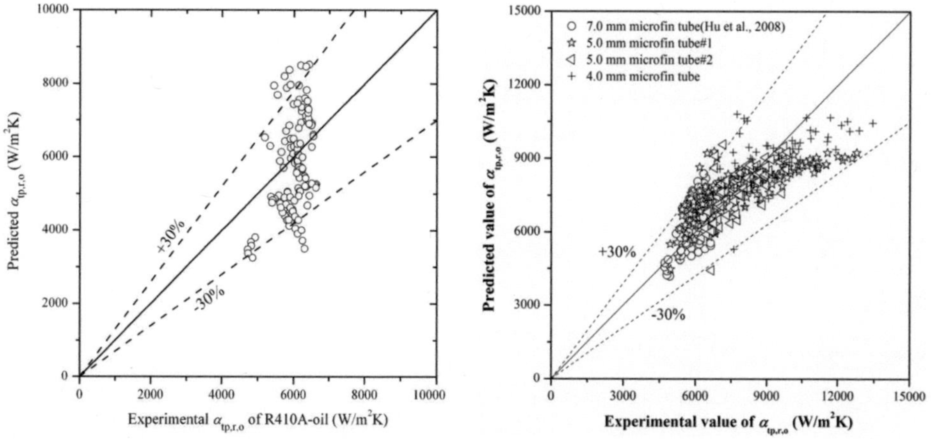

Figure 2.15: Comparison of measured to predicted heat transfer coefficients of R410a-lubricant mixtures by Hu et al. (2008a) correlation (left) and Hu et al. (2014) correlation (right), published ibidem

Even though predicting most data points within a deviation of ±30 %, the correlation does not capture the experimental trend precisely, as shown in Fig. 2.15.
Hu et al. (2014) extended the heat transfer database to smaller sized micro-fin tubes of 4.60 and 3.56 mm in inner diameter and different fin-geometries. In general, observations matched those inside the conventional sized 6.50 mm micro-fin tube. Increasing fin height and number lead to an increase in heat transfer coefficient of 5 – 25% under most conditions. The previously developed heat transfer correlation by Hu et al. (2008a) was adopted to the enlarged database by (1) substituting the Cooper correlation by Rohsenow correlation for nucleate boiling and (2) re-fitting the parameters of Eq. 2.57 and 2.58.

$$\alpha_{\mathrm{nb}} = \frac{\dot{q}}{\Delta T} \tag{2.63}$$

$$\Delta T = \frac{C_{\mathrm{sf}}\,\Delta h_{\mathrm{v}}}{c_{\mathrm{p,mix}}} \left(\frac{\dot{q}}{\mu_{\mathrm{mix}}\,\Delta h_{\mathrm{v}}} \left(\frac{\sigma_{\mathrm{mix}}}{g\,(\rho_{\mathrm{mix}} - \rho_{\mathrm{V}})} \right)^{0.5} \right)^{0.33} \left(\frac{c_{\mathrm{p,mix}}\,\mu_{\mathrm{mix}}}{\lambda_{\mathrm{mix}}} \right)^{1.73} \tag{2.64}$$

$$C_{\mathrm{sf}} = 0.0517 + 4.24\,w_{\mathrm{oil}} + 0.066\,(SP_{\mathrm{wet}}/SP_{\mathrm{bf}}) \tag{2.65}$$

$$E = 1 + 16921\,\mathrm{Bo}^{1.16} + 1.231\,X_{\mathrm{tt}}^{-0.86} \tag{2.66}$$

$$S = 1 + 1.203 \times 10^{-4}\,E^{1.25}\,\mathrm{Re}_{\mathrm{L}}^{1.17} \tag{2.67}$$

where $\dot{q}$ is the heat flux, ΔT is wall superheat, C_{sf} is a boiling fluid-surface dependent parameter and SP_{wet} and SP_{bf} are not further discussed tube perimeters.
Experimentally determined flow boiling heat transfer coefficients of R1234yf-POE oil mixtures inside a 6.44 mm micro-fin tube were presented by Han et al. (2013), agreeing well to predicted values by Schlager et al. (1990) and Eckels et al. (1998a) correlations.

Heat transfer coefficients were mainly found to be deteriorated due to the presence of oil. Heat transfer enhancement was reported for a limited range of operation, such as at high mass flow rate along with low vapor qualities. Further, the presence of oil influenced the occurrence of dryout yielding a sharp drop in heat transfer at a certain quality. Dryout quality decreased with increasing oil mass fraction.

2.4 CO_2 and CO_2-lubricant mixtures

Pure CO_2

Schael (2009) investigated flow boiling characteristics of pure CO_2 in a wide range of operation, using the same 14 mm smooth tube and 8.62 mm micro-fin measurement sections as in the scope of this work. The typical strong nucleate boiling behavior of CO_2 was clearly observed. From visual flow observations as well as from peripheral local heat transfer coefficients, the effect of flow pattern on heat transfer performance and the significance of its precise predictability was illustrated. Hence, flow patterns were compared to available predictive methods in literature. A fairly good agreement for smooth tube data was achieved with the flow pattern maps of Steiner (2002) and Thome and El Hajal (2003). Flow patterns in the micro-fin tube especially revealed a wider annular flow regime as predicted, but were still predictable by Steiner to some extend. Significant differences, for both tubes, were encountered at the intermittent-annular transition. This transition has been regarded as independent of mass flow velocity (i.e., at a constant Martinelli parameter and thus vapor quality) throughout all modifications of the original Taitel and Dukler (1976) map made up to this point. Schael as well as several other authors in literature, among others Park and Hrnjak (2007a), Barbieri et al. (2008), Higashiiue et al. (2008), Ozawa et al. (2009), Mastrullo et al. (2012a) and Grauso et al. (2013a), found this transition to be inaccurate. Schael derived the following transition condition from his data, which was found to apply for the micro-fin tube as well.

$$\mathrm{Fr_{Vm,I-A}}^{0.5} = \left(\frac{\dot{m}^2 \dot{x}^2}{g\, d_i\, \rho_\mathrm{L}\, \rho_\mathrm{V}} \right)^{0.5} = a X_\mathrm{tt}^b \qquad (2.68)$$

where $\mathrm{Fr_{Vm,I-A}}$ represents the I-A transition averaged vapor Froude number. The empirical parameters yield $a = 0.85$ and $b = -0.35$ for the smooth tube and $a = 0.55$ and $b = -0.50$ for the mirco-fin tube.

Heat transfer coefficients were well predicted in trend and value by the Steiner (2002) correlation, that was introduced in Sec. 2.3. Performing modifications for the fluid dependent parameter C_F and heat flux exponent $n(p_\mathrm{r})$, 93.1 % (smooth tube) or 82.4 % (micro-fin tube) of the measured data was predicted within ± 30 % error with deviations of MAPE = 13.5 % and MAPE = 17.3 %, respectively. Moreover, experimental heat transfer coefficients in the micro-fin tube were also in good agreement to predicted values by Cavallini et al. (1999) correlation, which was specifically developed for micro-fin tubes by considering the special tube geometry.

$$C_F = 1.57 \tag{2.69}$$

$$n(p_r) = \begin{cases} 0.75 - 0.027 \times 10^{1.26 p_r} & \text{smooth tube} \\ 0.44 & \text{micro-fin tube} \end{cases} \tag{2.70}$$

Smooth tube frictional pressure drops were predicted with high accuracy by the empirical correlation of Müller-Steinhagen and Heck (1986) and the homogeneous model of Storek and Brauer (1980). Besides, both correlations were modified to fit micro-fin pressure drop data by introducing correction factors ($C_{\text{micro-fin}}$, see Eq. 2.71-2.72), yielding outstanding prediction accuracy.

$$C_{\text{micro-fin}} = 1.63 \qquad \text{(Müller-Steinhagen and Heck)} \tag{2.71}$$

$$C_{\text{micro-fin}} = 2.6 - 1.52\dot{x} \qquad \text{(Storek and Brauer)} \tag{2.72}$$

Based on a large CO_2 heat transfer data base, Cheng et al. (2008a) presented an updated flow pattern map of Cheng et al. (2006), see App. E.1. As in preceding flow pattern maps, e.g., by Steiner (2002), Kattan et al. (1998a), Zürcher et al. (2002), El Hajal et al. (2003) and Wojtan et al. (2005a), transition from intermittent to annular flow was assumed at a constant value of Martinelli parameter. Based on the updated flow pattern map, a colligated frictional pressure drop model was proposed (given in App. E.2). It was developed by modifying the Moreno Quibén and Thome (2007b) pressure drop model, which was validated for R22, R410a and R134a, using CO_2 pressure drop data from five different sources in literature as basis for validation. Frictional pressure drops are calculated from correlations derived for each of the eight distinguished flow patterns. That implies an inevitably connection to the Cheng et al. (2008a) flow pattern map and relies on its accurate prediction. The same counts for the updated Cheng et al. (2006) heat transfer model by Cheng et al. (2008b), compiled in App. E.3.

Cho and Kim (2007) determined heat transfer coefficients and pressure drop of CO_2 by experiments inside 4.0 – 8.9 mm micro-fin and smooth tubes. The flow boiling behavior described mainly followed the trend as typically observed in literature. Pressure drops and heat transfer coefficients increased with decreasing tube diameter. Heat transfer was dominated by nucleate boiling. The trend in saturation temperature, however, was different than expected as it reversed upon passing +5 °C, see Fig. 2.16 (left). The right diagram in Fig. 2.16 shows data by Oh and Son (2011) under similar operating conditions.

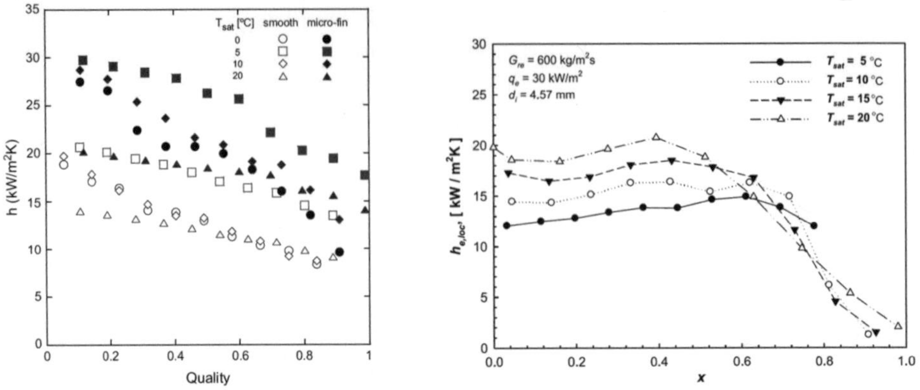

Figure 2.16: Effect of saturation temperature on CO_2 flow boiling heat transfer coefficient in 4.0 – 4.5 mm smooth tubes under similar conditions, published in Cho and Kim (2007) and Oh and Son (2011)

Recently, Mastrullo and Grauso extensively studied CO_2 flow boiling characteristics inside a 6 mm smooth tube. Mastrullo et al. (2009) reported of first heat transfer experiments carried out at $-8\,°C < T_s < +6\,°C$. The well known strong nucleate boiling contribution of CO_2 was confirmed. The influence of flow pattern on peripheral local heat transfer coefficients was shown by measurements varying the mass flow rates and thus the flow regime, especially in the medium and high vapor quality region. For the heat transfer database obtained, Mastrullo et al. (2010) conducted a comparison to available correlations in literature. The correlation by Cheng et al. (2008b) was the only one that could follow the trend of heat transfer coefficients and yielded good accuracy. Mastrullo et al. (2012b) and Mastrullo et al. (2012a) investigated two-phase flow pattern and its influence on local heat transfer. Visual observations and information from peripheral local heat transfer coefficients were compared to the Cheng et al. (2008a) flow pattern map, ascertaining poor agreement of measured and predicted flow regimes. Particularly, transitions from slug/intermittent to annular flow and further to dryout did not match the authors' observations, see Fig. 2.17. New transition lines were proposed, given in App. E.1. Associated to the flow visualization experiments, Grauso et al. (2013b) presented local heat transfer measurements and elucidated the effect by flow pattern. Comparing experimental frictional pressure drops to predictive models in literature, Mastrullo et al. (2010) and Grauso et al. (2014) determined great deviations for the Cheng et al. (2008a) pressure drop model, while models by Müller-Steinhagen and Heck and Friedel (1979) were found to deliver a precise prediction.

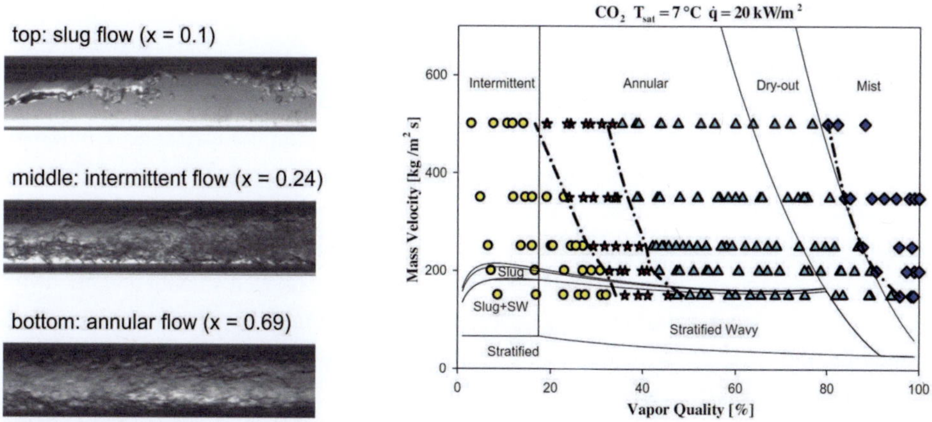

Figure 2.17: High-speed video recording of flow patterns of pure CO_2inside a 6 mm smooth tube (left) and comparison of observed flow patterns (○slug | ★intermittent | △annular | ◆dryout) to Cheng et al. (2008a) flow pattern map (right), published in Mastrullo et al. (2012a)

Dang et al. (2010) investigated flow boiling heat transfer of CO_2 in a small-sized micro-fin tube ($\bar{d_i} = 2.0$ mm) at an elevated saturation temperature of +15 °C. Heat transfer coefficients strongly increased by increasing the heat flux, while being deteriorated by an increase in mass flow velocity. Thus, nucleate boiling was dominant. At higher heat fluxes, deterioration with mass flow velocity is weakened, indicating a stronger nucleate boiling suppression by convection in the micro-fin tube. Based on a cylindrical surface area with mean inner diameter of the micro-fin tube, heat transfer coefficients increased by a factor of 1.8 – 2.2 with respect to smooth tube experiments, attributed to an area enlargement of factor 2. The onset of dryout was shifted to much higher vapor qualities in the micro-fin ($\dot{x}_{\mathrm{dry}} > 0.90$) tube than compared to the smooth tube ($0.60 < \dot{x}_{\mathrm{dry}} > 0.85$).

A new Nusselt correlation for flow boiling heat transfer was proposed by Fang (2013), introducing a new dimensionless parameter specified as Fa. According to the author, it accounts for the relative influence of surface tension and buoyancy forces on bubble formation and bubble departure, respectively.

$$Fa = \frac{(\rho_{\mathrm{L}} - \rho_{\mathrm{V}})\,\sigma}{\dot{m}^2\, d_{\mathrm{h}}} \tag{2.73}$$

The correlation was specifically developed for CO_2 and comprises a large database of nearly 3000 experimental data points from 13 studies found in literature. A wide range of operating conditions is covered, however, its validity is limited to a maximum tube diameter of 7.75 mm and most data points (64.6 %) were obtained from experiments in tubes having hydraulic diameters < 3 mm.

CO_2-PAG/PAO mixtures

As can be seen from Tab. 2.2, most investigations for CO_2-lubricant mixtures were carried out using either PAG or POE lubricating oils. POE lubricants and CO_2 are generally well miscible at common oil mass fractions and a certain lower temperature range. Contrarily, mixtures of CO_2 and PAG lubricants usually exhibit a strongly pronounced miscibility gap. This leads to the formation of three phases: a pure CO_2 vapor phase (1), a CO_2-rich liquid phase (2) and an oil-rich layer with dissolved CO_2 (3).
Gao and Ono conducted flow boiling experiments inside small-sized smooth and micro-fin tubes ($3.00 \geq d_i/\text{ mm} \geq 3.75$) using CO_2-PAG100 lubricant mixtures at constant saturation temperature of $+10\,°C$. Measurement sections were heated directly by Joule-heat or by wrapped film-heater, respectively.
Heat transfer measurements of Gao and Honda (2006) and Gao et al. (2007) showed that pure CO_2 heat transfer was dominated by nucleate boiling contribution in both tubes. In contrast to the smooth tube, the micro-fin tube remained wetted upon increasing the vapor quality, thus yielding high and constant heat transfer coefficients. The addition of oil resulted in a sharp drop in heat transfer due to the formation of an oil rich phase at the tube wall. Measurements inside the micro-fin tube revealed, that convective boiling gained greater importance at the high mass flow velocity ($770\,\mathrm{kg\,m^{-2}\,s^{-1}}$).
Narrowing the range of mass flow rates to $100 - 500\,\mathrm{kg\,m^{-2}\,s^{-1}}$, Gao et al. (2008b) and Ono et al. (2010) continued flow boiling measurements investigating heat transfer, pressure drop and flow pattern. Heat transfer coefficients showed the same trends as observed before,

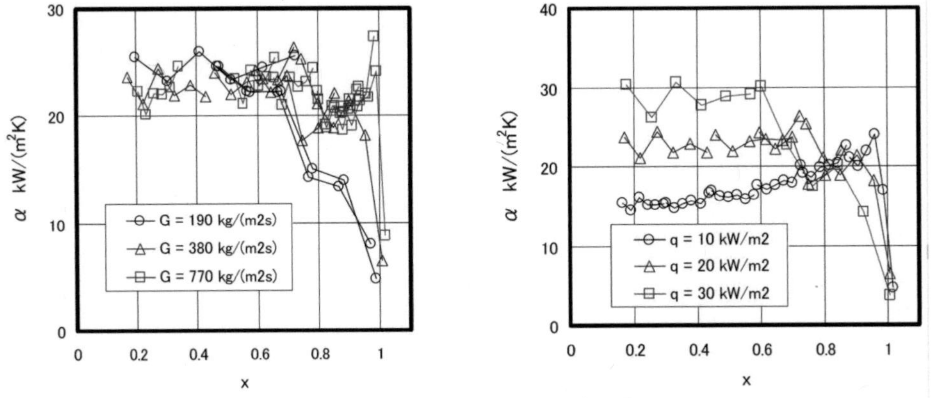

(a) Effects of mass velocity ($q = 20$ kW/ m^2) (b) Effects of heat flux ($G = 380$ kg/(m^2 s))

Figure 2.18: Effect of mass flow velocity on flow boiling heat transfer of pure CO_2 (left) and CO_2-PAG oil mixture (right, $w_{oil} = 0.7$ wt.-%) in a 3.04 mm micro-fin tube at $+10\,°C$, published in Gao and Honda (2006)

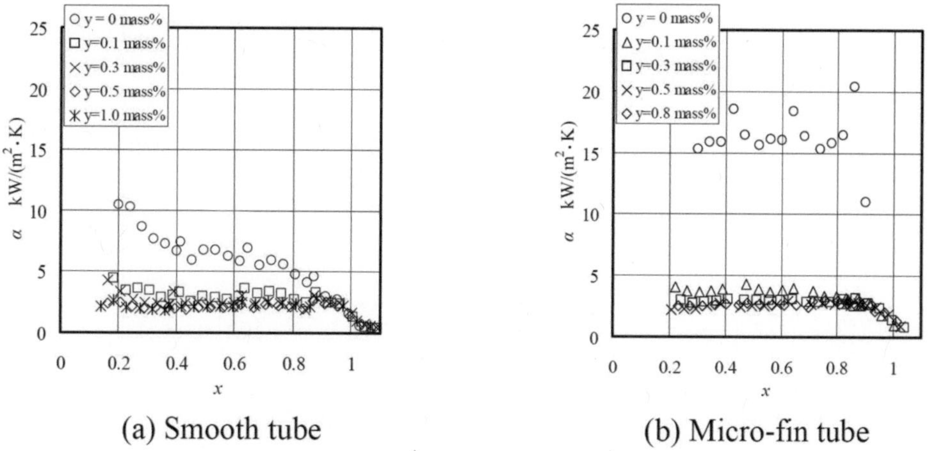

Figure 2.19: Effect of PAG oil mass fraction on CO_2 flow boiling heat transfer inside 3.75 mm smooth (left) and micro-fin (right) tubes (+10 °C, 190 kg m^{-2} s^{-1}, 10 kW m^{-2}), published in Gao et al. (2008b)

i.e., nucleate boiling dominance and the effect of partial dryout for the smooth tube while the micro-fin tube remained wetted for all vapor qualities. The addition of oil caused a sharp drop at an oil mass fraction of 0.1 wt.-%, while further raising the oil mass fraction did not show any effect. While smooth tube heat transfer coefficients dropped by up to 50 % with respect to pure CO_2, deterioration in the micro-fin tube turned out to be > 75 %. The authors concluded that heat transfer was dominated by convective boiling in the case of CO_2-PAG oil mixtures, to the contrary of pure CO_2. Heat transfer characteristics of CO_2-PAG oil mixtures were reasoned by an early formation of oil-rich film stagnating at the bottom of the tube wall. Experimental results published are shown in Fig. 2.19

Pressure drops were found to increase by up to 70 % with respect to the smooth tube. At vapor qualities of $\dot{x} < 0.7$ no significant influence of PAG oil on pressure drop was observed. However, further rising the vapor quality yielded an increase in pressure by up to 100 % in comparison to pure CO_2, attributed to an increase in thickness of the liquid oil film and the reduction of effective cross-sectional flow area. In case of pure CO_2, the homogeneous model provided an accurate prediction for both tubes.

In the smooth tube, slug and wavy flow patterns dominated for both CO_2 and CO_2-PAG oil mixtures. The spiral grooves of the micro-fin tube caused an early transition to annular flow, i.e., slug-to-annular and wavy-to-annular transitions shifted to lower mass flow rates and vapor qualities. Predicted flow patterns by Cheng et al. (2008a) showed good accordance to observations made in the micro-fin tube but did not agree well with smooth tube flow patterns, as can be seen in Fig. 2.20.

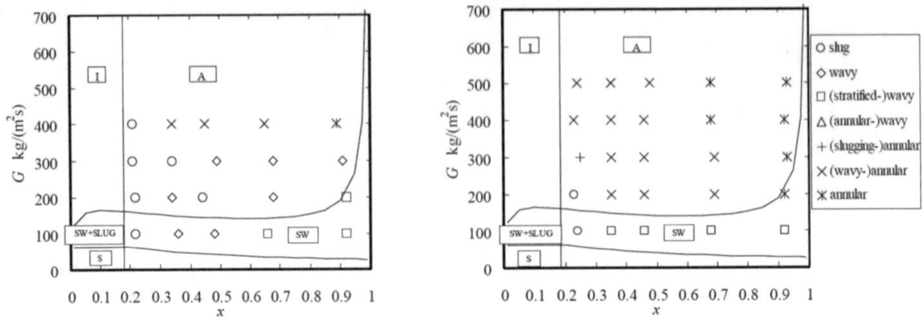

(a) Smooth tube, Pure CO_2 (b) Micro-fin tube, Pure CO_2

Figure 2.20: Adiabatic flow patterns observed for CO_2 flowing inside 3.75 mm smooth (left) and micro-fin (right) tubes at a saturation temperature of $+10\,°C$, published in Gao et al. (2008b)

Under similar conditions using the same type of oil, Katsuta et al. (2008) investigated flow boiling heat transfer, pressure drop and dryout quality of CO_2-PAG mixtures inside a smooth tube. From measurements it was concluded that pressure drop increases with oil mass fraction and vapor quality while being independent of heat flux. The vapor quality at which pressure drop exhibits a steep rise tends towards lower values with increasing oil mass fractions. Heat transfer coefficients are severely reduced with presence of very small oil amounts, attributed to the formation of an oil-rich layer at the tube wall. However, further increasing the oil mass fraction does not affect heat transfer significantly. The vapor quality of dryout is continuously reduced with rise in oil mass fraction.

Based on their results, the authors modified pure refrigerant correlations for use with CO_2-lubricant mixtures by means of the oil contamination approach. Friction pressure drop was predicted by a Lockhart-Martinelli-type correlation, adjusting the liquid two-phase multiplier (ϕ_{LO}) to flow conditions (Eq. 2.74-2.76).

$$\left(\frac{\Delta p}{\Delta L}\right)_{\mathrm{fric}} = \left(\frac{\Delta p}{\Delta L}\right)_{\mathrm{LO}} \phi_{\mathrm{LO}}^2 \tag{2.74}$$

$$\phi_{\mathrm{LO}} = \begin{cases} \left(0.374 \ln\left(X_{\mathrm{tt}}^{-1}\right) + 0.908\right) \dfrac{\rho_{\mathrm{L}}}{\rho_{\mathrm{V}}}\, \omega & \text{annular} \\ \left(0.309 \ln\left(X_{\mathrm{tt}}^{-1}\right) + 0.773\right) \dfrac{\rho_{\mathrm{L}}}{\rho_{\mathrm{V}}}\, \omega & \text{slug/froth} \end{cases} \tag{2.75}$$

$$\omega = \begin{cases} 1.0 & \mathrm{Re}_{\mathrm{LO}} < 2500 \\ 1.3 X_{\mathrm{tt}}^{0.197} & \mathrm{Re}_{\mathrm{LO}} \geq 2500 \end{cases} \tag{2.76}$$

where $(\Delta p/\Delta L)_{\mathrm{LO}}$ is the all-liquid pressure drop, $\rho_{\mathrm{L|V}}$ is the liquid/vapor density, X_{tt} is the Martinelli parameter and $\mathrm{Re}_{\mathrm{LO}}$ is the all-liquid Reynolds number.

For heat transfer, a modified Chen-type correlation was proposed. Hereafter, two-phase flow boiling heat transfer coefficient (α_{tp}) is calculated from linear superposition of forced convective (α_{cb}) and nucleate boiling (α_{nb}) heat transfer contributions.

$$\alpha_{tp} = \alpha_{cb} + \alpha_{nb} \tag{2.77}$$

Contrary to Chen (1966), the Stephan-Abdelsalam Equation and a newly introduced oil-factor (φ_{oil}) were introduced for predicting nucleate boiling heat transfer.

$$\alpha_{cb} = \alpha_{DB}\, F \quad = 0.023\, \mathrm{Re}_L^{0.8}\, \mathrm{Pr}_L^{0.4} \left(\frac{\lambda_L}{d_i}\right) F \tag{2.78}$$

$$\alpha_{nb} = \alpha_{SA}\, S\, \varphi_{oil} = 207 \left(\frac{\dot{q}\, d_i}{\lambda_L\, T_s}\right)^{0.745} \left(\frac{\rho_V}{\rho_L}\right)^{0.581} \mathrm{Pr}_L^{0.533}\, S\, \varphi_{oil} \tag{2.79}$$

where $\dot{q}$ is the heat flux, T_s is the saturation temperature and Re_L and Pr_L are the liquid Reynolds and Prandtl numbers. Fluid properties (λ_L, $\rho_{L|V}$) correspond to those of pure CO_2. Forced convection and nucleate boiling term multipliers (F, S and φ_{oil}) are given by Eq. 2.80-2.82.

$$F = 1 + 0.258 \left(\frac{1}{X_{tt}}\right)^{0.886} + 92.32 \left(\frac{\rho_V}{\rho_L}\right)^{3} \left(\frac{1}{X_{tt}}\right)^{0.9} \tag{2.80}$$

$$S = \ln\left(2.332\, \mathrm{Bo}^{0.518}\, \mathrm{Bd}^{1.27}\, \mathrm{Fr}^{0.964}\, \mathrm{Re}_{tp}^{-0.834}\right) \tag{2.81}$$

$$\varphi_{oil} = \left(1 + OCR^{0.698}\, \mathrm{Bo}^{0.207}\, \mathrm{Bd}^{0.912}\right)^{-1} \tag{2.82}$$

At last, the authors gave an adapted expression for the prediction of dryout quality ($\dot{x}_{dry}$), regarding the influence of oil by introducing a dryout factor ($\eta_{dry,oil}$), see Eq. 2.83-2.84.

$$\dot{x}_{dry} = 0.269\, \mathrm{Re}^{0.0571}\, \mathrm{Fr}^{0.0697}\, \mathrm{Bo}^{-0.0519}\, \eta_{dry,oil} \tag{2.83}$$

$$\eta_{dry,oil} = \begin{cases} 1.0 & OCR < 1 \\ 1.169 \exp\{-0.17\, OCR\} & OCR \geq 1 \end{cases} \tag{2.84}$$

Dang et al. (2006) and Dang et al. (2013) conducted flow boiling experiments in horizontal smooth tubes, investigating heat transfer and pressure drop of CO_2-lubricant mixtures containing 0 . . . 5 wt.-% partial miscible PAG oil at a saturation temperature of +15 °C. Concordantly to observations for CO_2-PAG oil mixtures discussed before, heat transfer coefficients sharply decreased at a low oil mass fraction, see Fig. 2.21. This critical oil mass fraction was found to be depended on tube diameter ($w_{oil,c} = 1$ wt.-% for $d_i = 6$ mm).

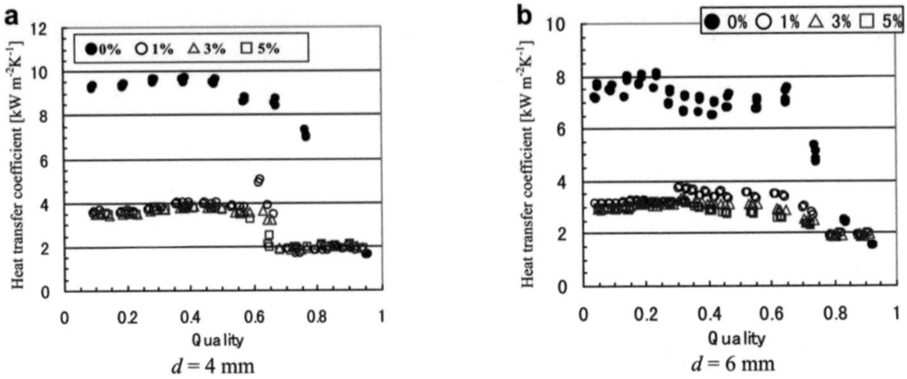

Figure 2.21: Effect of PAG oil on CO_2 flow boiling heat transfer inside 4 mm and 6 mm smooth tubes (conditions: $+15\,°C$, $720\,kg\,m^{-2}\,s^{-1}$, $18\,kW\,m^{-2}$), published in Dang et al. (2013)

Further increasing the oil mass fraction did not influence heat transfer. Heat flux was found to positively influence heat transfer only for low mass flow velocities and vapor qualities, while in the other case no influence was observed. The authors concluded that lubricant suppresses nucleate boiling in pre-dryout region, thus leading to a greater influence of convective boiling at high mass flow velocities and vapor qualities. Diabatic pressure drop increased with oil mass fraction, attributed to a higher liquid viscosity and the formation of an oil-rich layer along the inner tube wall. At high oil mass fraction of 5 wt.-%, pressure drop increase was up to $+80\,\%$ in comparison to pure CO_2. Observations made by the authors apply well to those by Ono et al. (2010) for the same type of CO_2-lubricant mixtures.
Li et al. (2014) developed a correlation for the two-phase heat transfer coefficient (α_{tp}) based on cubic superposition of convective (α_{cb}) and nucleate boiling (α_{nb}) contributions as suggested by Steiner and Taborek (1992).

$$\alpha_{tp} = \left[(\alpha_{cb})^3 + (\alpha_{nb})^3 \right] \tag{2.85}$$

α_{cb} and α_{nb} are given as proposed by Chen (1966), i.e., by Dittus-Boelter and Forster-Zuber Equations extended by two-phase forced convection multiplier and nucleate boiling suppression factor, respectively. In contrast to other approaches in literature (e.g., by Aiyoshizawa et al. (2006), Katsuta et al. (2008) and Gao et al. (2008a)), these terms are expressed by using CO_2-lubricant mixture properties instead of those for pure refrigerant. This step was concluded to be essential in order to correctly cover the lubricant influence.

$$\alpha_{cb} = \alpha_{DB}\, F = 0.023\, \mathrm{Re}_{mix}^{0.8}\, \mathrm{Pr}_{mix}^{0.4} \left(\frac{\lambda_{mix}}{d_i} \right) F \tag{2.86}$$

$$\alpha_{nb} = \alpha_{FZ}\, S = 0.00122 \left(\frac{\lambda_{mix}^{0.79}\, c_{p,mix}^{0.45}\, \rho_{mix}^{0.49}}{\sigma_{mix}^{0.5}\, \mu_{mix}^{0.29}\, \Delta h_V^{0.24}\, \rho_V^{0.24}} \right) \Delta T^{0.24}\, \Delta p^{0.75}\, S \tag{2.87}$$

where $\Delta T = T_\text{w} - T_\text{s}(p)$ is the wall superheat, $\Delta p = p_\text{s}(T_\text{w}) - p$ is the vapor pressure difference at the wall superheat, p and T_s $(= f(p))$ are the actual system pressure and corresponding saturation temperature, respectively, T_w is the wall temperature, $p_\text{s}(T_\text{w})$ is the saturation pressure of the fluid in near tube wall surface region and Re_mix is the two-phase Reynolds-number (see Eq. 2.88)

$$\text{Re}_\text{mix} = \dot{m}\,(1 - \dot{x})\,d_\text{i}\,\mu_\text{mix}^{-1} \tag{2.88}$$

Two-phase multiplier (F) for convective and suppression factor (S) for nucleate boiling heat transfer contributions were modified to fit the experimental data of Dang et al. (2013).

$$S = S_\text{oil}\left(0.9 + 0.4\left(\text{Re}_\text{tp,mix} \times 10^{-3}\right)^{0.3} / \left(\text{Bo} \times 10^{3}\right)^{0.23}\right)^{-1} \tag{2.89}$$

$$F = \begin{cases} 1.0, & X_\text{tt} \geq 10 \\ 2.55\,(0.413 + 1/X_\text{tt})^{1.05}, & X_\text{tt} < 10 \end{cases} \tag{2.90}$$

$$\text{Re}_\text{tp,mix} = \text{Re}_\text{mix}\,F^{1.25} \tag{2.91}$$

where S_oil was introduced as specific suppression factor to capture the influence induced by the presence of lubricating oil and $\text{Re}_\text{tp,mix}$ as modified two-phase Reynolds number. Even for mixtures of very low oil mass fraction ($w_\text{oil} = 0.005$), nucleate boiling heat transfer diminished by $> 50\ \%$ compared to pure CO_2. This observation was explained by formation of an oil-rich layer ($w_\text{oil} > 0.5$) at all operating conditions when using immiscible CO_2-PAG oil mixtures. Hence, for flow boiling heat transfer measurements by Dang et al. S_oil follows Eq. 2.92, ranging from $0 < S_\text{oil} < 0.5$. As stated by Dang (2014) for miscible CO_2-POE oil mixtures, no significant influence should be observed at low oil mass fractions.

$$S_\text{oil} = -\frac{w_\text{oil}}{2\ln(1 - w_\text{oil})} \qquad (CO_2\text{-PAG oil mixtures}) \tag{2.92}$$

$$S_\text{oil} = 1 \qquad (CO_2\text{-POE oil mixtures}) \tag{2.93}$$

CO_2-POE mixtures

Often cited in literature, Hassan (2004) investigated flow boiling of CO_2-lubricant mixtures at low temperatures ($-30\ldots-10\,°\text{C}$) and mass flow velocities ($90\ldots125\,\text{kg}\,\text{m}^{-2}\,\text{s}^{-1}$) inside a 10 mm smooth tube. Heat transfer coefficients were reported for oil mass fractions ranging from $0.2 - 7.0$ wt.-%. The oil was completely miscible with CO_2. At a constant oil mass fraction, the overall trends with mass flow velocity, saturation temperature and heat flux were analogous to those observed for pure CO_2. When increasing the oil mass fraction, however, heat transfer coefficients were continuously diminished over the whole range of discussed

data. Experimental data was compared to correlations by Shah (1982) and Kandlikar (1990). Additionally, the influence of viscosity on the prediction was investigated. Calculations were performed using the pure refrigerant property in one case and mixture viscosity in the other case. As a result, the overprediction of heat transfer coefficients was reduced. The Kandlikar correlation yielded an acceptable accuracy with RMSE = 15 % for all data.
Müller and Eggers (2008) investigated heat transfer characteristics of CO_2-lubricant mixtures in a vertical 9.4 mm smooth tube. Two lubricants were used: fully miscible POE and partially miscible PAO oil. However, the authors stated that no difference in flow heat transfer behavior was noticed. Only one diagram was presented showing flow boiling heat transfer coefficients of CO_2-POE oil mixtures as function of vapor quality. Heat transfer coefficients decreased monotonously as the oil mass fraction was increased from 1 to 4 wt.-%.
Between 2007 and 2012, Zhao and Bansal reported on experimental flow boiling heat transfer investigations of CO_2 and CO_2-oil mixtures at low temperatures (< -25 °C) inside smooth and micro-fin tubes. Zhao and Bansal (2007) carried out experiments with pure CO_2 inside a 4.57 mm smooth tube, presenting only very few data points at -30 °C. The database was extended by Zhao and Bansal (2009) to saturation temperatures between -15 and -40 °C. Analogous to Schael (2009), the authors determined a strong nucleate boiling dominance on heat transfer, while a contribution by convective boiling was detectable only at very low temperatures due to an increase in liquid-vapor density ratio. Comparable results were obtained by Zhao and Bansal (2012) from measurements in a 7.31 mm micro-fin tube as shown in Fig. 2.22 (left). However, convective boiling heat transfer was more pronounced, see Fig. 2.22 (right). As previously found out by Schael, heat transfer coefficients were well predicted by the correlation of Liu and Winterton (1991) in the case of the smooth tube, and predicted values by Cavallini et al. (1999) were most accurate for the micro-fin tube.

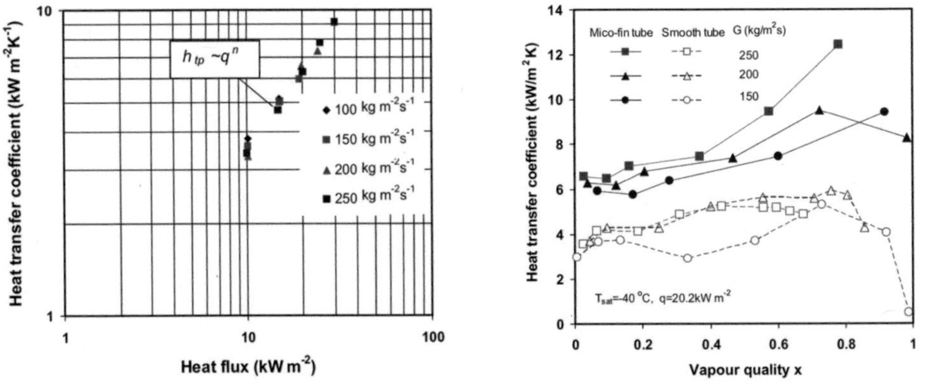

Figure 2.22: Heat transfer coefficients of pure CO_2 in a 7.31 mm micro-fin tube at $\dot{x} < 0.05$ (left) and comparison of smooth and micro-fin tube data (right), published in Zhao and Bansal (2012) and Bansal (2012)

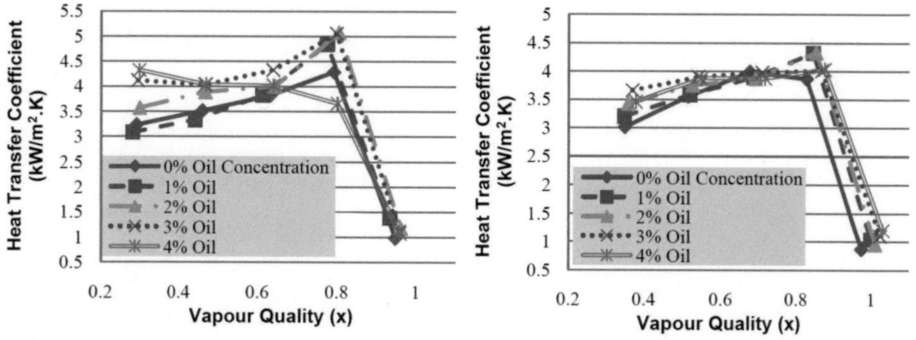

Figure 2.23: Heat transfer coefficients of CO_2-POE oil mixtures in a 4.55 mm smooth tube at $\dot{m} = 259\,\mathrm{kg\,m^{-2}\,s^{-1}}$, $\dot{q} \approx 16\,\mathrm{kW\,m^{-2}}$ and $T_s = -30\,°C$ (left) and $T_s = -40\,°C$ (right), published in Bansal (2011)

Bansal (2011) presented preliminary results for CO_2-POE60 oil mixtures inside a 4.55 mm smooth tube, shown in Fig. 2.23. An increase in heat transfer coefficient with increasing oil mass fraction was depicted, especially at low and medium vapor qualities. At high oil mass fractions ($w_{oil} = 4$ wt.-%), heat transfer coefficients were diminished again. Further, heat transfer enhancement was more pronounced at the higher temperature.

Since the transition from stratified to non-stratified (intermittent, annular) flow is of great importance, Saito et al. (2011) conducted a literature survey identifying several transition criteria. Emphasizing on elevated reduced pressures and liquid film flow, two important facts were identified: (1) the liquid velocity often neglected due to $\rho_L >> \rho_V \rightarrow u_V >> u_L$ and (2) the Kelvin-Helmholtz instability criterion is applied. At high pressures, neglecting the liquid velocity leads to a growing uncertainty as the ratio of liquid to vapor density becomes smaller. Further, the underlain transition criterion does not apply for thin liquid films, i.e., high vapor qualities, resulting in a faulty solution. For this condition, a theory was derived for the onset of *disturbance wave* (Hanratty, 1983), see Fig. 2.24 for a schematic representation.

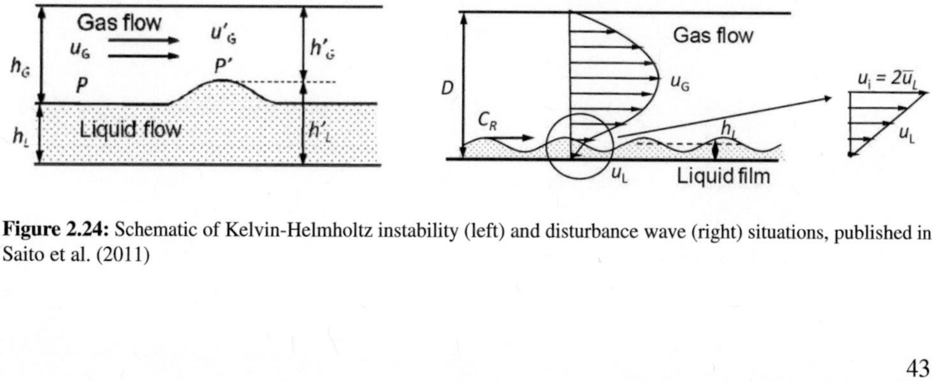

Figure 2.24: Schematic of Kelvin-Helmholtz instability (left) and disturbance wave (right) situations, published in Saito et al. (2011)

Based on these ideas, a modified wavy to intermittent/annular transition criterion for the Steiner flow pattern map was proposed, supported by adiabatic two-phase flow observations of CO_2-POE oil mixtures in a smooth tube.

$$\bar{u}_\mathrm{V} - C_2\,\bar{u}_\mathrm{L} \geq C_1 \left\{ \left(\frac{\varepsilon\,\pi\,g\,d\,(\rho_\mathrm{L}-\rho_\mathrm{V})}{4\,\rho_\mathrm{V}\,\sqrt{1-(2\,h_\mathrm{L}/d-1)^2}} \right) \times \left(1 + \frac{\pi^2\,d^2}{25\,h_\mathrm{L}^2} \left(\frac{\mathrm{We}}{\mathrm{Fr}} \right)_\mathrm{L} \right) \right\}^{0.5} \tag{2.94}$$

where $\bar{u}_\mathrm{V}$ is the mean vapor velocity, h_L is the height of the flowing liquid. Further, $C_1 = 0.5\,(\cos(\pi\,h_\mathrm{L}/d)+1)$ and $C_2 = 0.5\,(\cos(\pi\,h_\mathrm{L}/d)+1)+1$.

To date, a group of researchers headed by Hrnjak actively investigated flow boiling of CO_2. Park and Hrnjak (2005), Park and Hrnjak (2006), Park and Hrnjak (2007b), Kim et al. (2008b) and Oh et al. (2011) conducted flow boiling experiments with pure CO_2. Continuing CO_2 flow boiling experiments, Kim et al. (2010), Pehlivanoglu et al. (2010), Kim and Hrnjak (2012) and Hrnjak and Kim (2014) recently investigated the influence of fully soluble lubricating oil (POE Reniso C 85 E) on flow boiling characteristics.

Adiabatic two-phase pressure drops of pure CO_2 inside 3.5 mm smooth and micro-fin tubes were measured by Kim et al. (2008b). Further, a comparison to measurements by Zilly et al. (2003) that used 6.2 mm tubes is given. For both tubes, nearly identical trends were observed ($\Delta p \uparrow$ for $\dot{m} \uparrow$, $T_\mathrm{s} \downarrow$, $\dot{d} \downarrow$). Pressure drop penalty factors yielded $1.3 < PF_{\Delta\mathrm{p}} < 1.6$. Best results in pressure drop prediction were achieved by Friedel (1979) correlation (smooth tube) and Cavallini et al. (1997) correlation (micro-fin tube).

Kim et al. (2010) and Pehlivanoglu et al. (2010) carried out experiments inside smooth tubes of 6.1 mm and 11.2 mm in inner diameter, respectively. Results for oil mass fractions ranging from 0.5 to 2.0 wt.-% were discussed. In heat transfer measurements by Kim et al. at mass flow velocities $\leq 200\,\mathrm{kg\,m^{-2}\,s^{-1}}$, heat transfer coefficients in the range of stratified flows were deteriorated with increasing vapor quality. This was explained by the dominance of nucleate boiling and a reduction in tube wetting (partial dryout), demonstrated by peripheral local heat transfer measurements. At higher mass flow velocities, convective boiling became more dominant and annular flow appeared. As nucleate boiling is suppressed by an increase in surface tension, oil was found to deteriorate flow boiling heat transfer in general. However, in the case of stratified flow regimes, oil could also enhance heat transfer since increasing the surface tension leads to an increase of wetted perimeter of the tube. Pehlivanoglu et al. confirmed the trend of increasing convective boiling dominance for mass flow velocities up to $400\,\mathrm{kg\,m^{-2}\,s^{-1}}$. Further, friction pressure drop was found to be almost independent of oil mass fraction.

Kim and Hrnjak (2012) and Hrnjak and Kim (2014) conducted flow boiling measurements with CO_2-POE mixtures inside a smooth and an enhanced tube of both 11.2 mm in inner diameter. The test sections were designed as cross-flow heat exchangers. The oil mass fraction was varied from 0.5 to 2.0 wt.-%.. Flow patterns, frictional pressure drops and

heat transfer coefficients were determined experimentally. In the smooth tube CO_2-oil mixtures tended to an early formation of annular-like flow pattern, explained by an increase in viscosity and surface tension due to the oil. Further, foam formation was observed and found to be a more critical influencing factor on flow pattern. Heat transfer coefficients were generally deteriorated with increasing oil mass fraction due to suppression of nucleate boiling. However, at high flow velocities ($\dot{m} \uparrow$ and $\dot{x} \uparrow$) increasing the oil mass fraction could lead to an enhancement in heat transfer attributed to an increase of convective heat transfer contribution. In micro-fin tube experiments on the other hand, no distinct change in flow pattern was observed when increasing the oil mass fraction. Further, heat transfer coefficients were similar to those of pure CO_2, related to increased turbulence induced by the micro-fin structure and thus the dominance of convective heat transfer.
Pressure drops were found to be unaffected by the oil mass fraction in both measurement sections. This result is non-congruent with most observations in literature and remarkable in so far as the mixture viscosity increases, especially with increasing vapor quality. Moreover, foaming was observed due to the presence of oil. Measurements revealed that the foam velocity was much lower than those of vapor and liquid phase, see Fig. 2.25. Hence, pressure drop would be expected to increase. No predictive methods were presented.

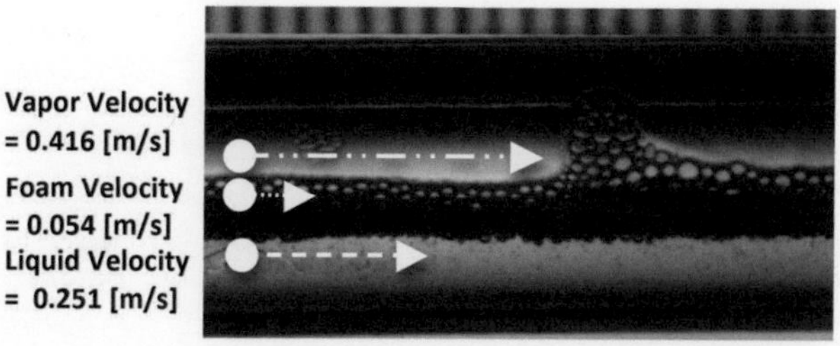

Figure 2.25: Flow velocities of CO_2-POE mixtures in a smooth tube, published in Kim and Hrnjak (2012)

2.5 Lubricant influence on material properties

In literature two basic approaches for dealing with refrigerant-lubricant mixtures are known. Affiliated to usually small amounts of oil in the refrigerant, the **oil contamination approach** treats refrigerant-lubricant mixtures as oil-contaminated refrigerants. Pure refrigerant material properties are used when evaluating experimental data from heat transfer and pressure drop experiments. To account for the discrepancy to pure refrigerant measurement data, purely empirical contamination parameters are introduced and fitted to the experimental results. This approach reflects an easy-to-use method. However, it does not contribute to a better understanding of the thermodynamics behind the boiling process.

Thome (2006) suggested using a **thermodynamic approach** for modelling lubricant-refrigerant mixtures by substituting the pure refrigerant by mixture properties. This approach was chosen for describing the CO_2-lubricant mixtures investigated within the scope of this work. It is suitable for capturing the real thermodynamics, but it requires for the knowledge of - in many cases unknown - refrigerant-lubricant mixture properties. Among others, Jensen and Jackman (1984), Conde (1996), Mermond et al. (1999b), Shen and Groll (2003), Wei et al. (2008), Bandarra Filho et al. (2009) and Zhao and Bansal (2009) reviewed methods for predicting mixture properties. The following sections provide a summary of established methods. Suitability for the examined CO_2-lubricant mixtures is discussed.

2.5.1 Bubble point temperature & vapor pressure

For mixtures of R22, R113, R114 and R134a with various oil types Thome (1995) suggested a generalized empirical correlation for the vapor pressure curve, developed by Takaishi and Oguchi for R22-AB oil mixtures. The correlation is given in Eq. 2.95, rearranged for the bubble point temperature (T_{bub}) as function of vapor pressure (p_s).

$$\frac{T_{bub}}{\mathrm{K}} = \frac{A(w_{oil})}{\ln\left(\frac{p_s}{\mathrm{MPa}}\right) - B(w_{oil})} \tag{2.95}$$

where $A(w_{oil})$ and $B(w_{oil})$ are empirical functions of the oil mass fraction.

$$A(w_{oil}) = a_0 + a_1\, w_{oil} + a_3\, w_{oil}^3 + a_5\, w_{oil}^5 + a_7\, w_{oil}^7 \tag{2.96}$$

$$B(w_{oil}) = b_0 + b_1\, w_{oil} + b_3\, w_{oil}^3 + b_5\, w_{oil}^5 + b_7\, w_{oil}^7 \tag{2.97}$$

The empirical parameters are listed in Tab. 2.3. However, a_0 and b_0 have to be adapted to the respective refrigerant, i.e., by first using a precise vapor pressure equation of the pure refrigerant and then evaluating Eq. 2.95 for $w_{oil} = 0$. To increase accuracy, a_0 and b_0 should be evaluated at each design vapor pressure regarded.

Table 2.3: Empirical constants for the vapor pressure curve by Takaishi and Oguchi (Thome, 1995)

$a_0 = -2394.5$	$a_1 = 182.52$	$a_3 = -724.21$	$a_5 = 3868.0$	$a_7 = -5268.9$
$b_0 = 8.0736$	$b_1 = 0.72212$	$b_3 = 2.3914$	$b_5 = -13.779$	$b_7 = 17.066$

Zhelezny et al. (2001) investigated the vapor pressure of fully soluble NH_3-oil solutions containing oil mass fractions of 5-95 wt.-%. Zhelezny (2007) adapted the correlation to CO_2-oil mixtures using manufacturer's data for Reniso C 85 E.

$$\ln\left(\frac{\bar{p}_c}{p_s}\right) = \alpha_c \ln\left(\frac{\bar{T}_c}{T_{bub}}\right) + 5.957 \ln\left(\frac{\bar{T}_c}{T_{bub}}\right)^{2.64} \tag{2.98}$$

Table 2.4: Constants for vapor pressure curve for mixtures of CO_2 and Reniso C 85 E (Zhelezny, 2007)

$a_1 = 73.7537$	$a_2 = -36.4423$	$a_3 = -37.0377$	$a_4 = -1.816879$	$a_5 = 0.973128$
$b_1 = 303.897$	$b_2 = 303.666$	$b_3 = 1.22524$	$b_4 = 0.225542$	
$c_1 = 6.79411$	$c_2 = -5.6879$	$c_3 = -0.624642$		

where $\bar{p}_c$ and $\bar{T}_c$ are the pseudo-critical pressure and temperature of the mixture and α_c is the Riedel's parameter. Empirical parameters are listed in Tab. 2.4.

$$\frac{\bar{p}_c}{\text{bar}} = \frac{a_1 + a_2\, w_{oil} + a_3\, {w_{oil}}^2}{1 + a_4\, w_{oil} + a_5\, {w_{oil}}^2} \tag{2.99}$$

$$\frac{\bar{T}_c}{\text{K}} = \frac{b_1 + b_2\, w_{oil}}{1 + b_3\, w_{oil} + b_4\, {w_{oil}}^2} \tag{2.100}$$

$$\bar{\alpha}_R = (c_1 + c_2\, w_{oil})\,(1 + c_3\, w_{oil})^{-1} \tag{2.101}$$

A graphical representation of the correlations with respect to CO_2-POE data is shown in Fig. 2.26. For comparison, the correlations by Thome (1995) and Zhelezny (2007) are shown as bright and dark lines, respectively. Symbols represent measured data. On a rough scale, Zhelezny's correlation generally provides the most accurate prediction of vapor pressure over the whole range of pictured saturation conditions ($-20\,°C \leq T \leq +30\,°C$, $0 \leq w_{oil} \leq 1$). In practice, however, local oil mass fractions below 0.5 are considered (Thome, 1995). Thus, higher accuracy of prediction in the lower range of oil mass fraction is of outmost importance. Basic validation experiments along with a recommendation of the author of this work, examining both methods, are described in Sec. D.2.1.

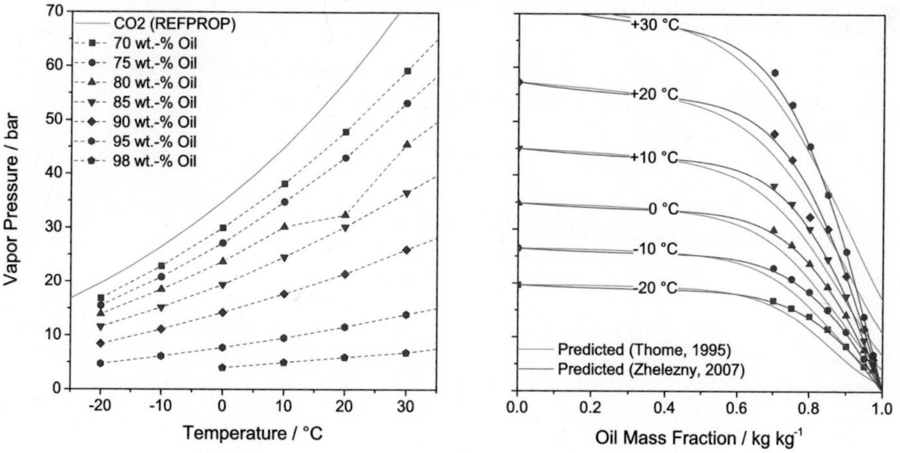

Figure 2.26: Vapor Pressure as function of temperature and oil mass fraction

2.5.2 Viscosity

Based on measurements with R113-oil mixtures, Jensen and Jackman (1984) developed a well-established correlation for the dynamic mixture viscosity (μ_{mix}) as function of oil mass fraction (w_{oil}) and pure substance viscosities of refrigerant (μ_{ref}) and lubricant (μ_{oil}).

$$\ln\left(\frac{\mu_{mix}}{\mathrm{Pa\,s}}\right) = \ln\left(\frac{\mu_{ref}}{\mathrm{Pa\,s}}\right) + w_{oil}\left(\frac{\mu_{oil}}{\mu_{ref}}\right)^{0.3} \tag{2.102}$$

Yokozeki (1994) introduced an Arrhenius-type correlation (i.e., applying a log-linear law) for calculating the dynamic mixture viscosity (μ_{mix}) as function of so called effective oil weight fraction (ξ_{oil}), see Eq. 2.103-2.104. This correlation was also adapted by Choi et al. (1999) and Yokozeki et al. (2000) for various mixtures of alternative refrigerants and lubricants.

$$\ln\left(\frac{\mu_{mix}}{\mathrm{Pa\,s}}\right) = (1-\xi_{oil})\ln\left(\frac{\mu_{ref}}{\mathrm{Pa\,s}}\right) + \xi_{oil}\ln\left(\frac{\mu_{oil}}{\mathrm{Pa\,s}}\right) \tag{2.103}$$

$$\xi_{oil} = \left(\tilde{M}_{oil}\tilde{x}_{oil}\right)\left(\tilde{M}_{ref}\,(1-\tilde{x}_{oil}) + \tilde{M}_{oil}\tilde{x}_{oil}\right)^{-1} \tag{2.104}$$

where $\tilde{M}_{ref}$ and $\tilde{M}_{oil}$ are the molar masses of pure refrigerant and oil, respectively, and $\tilde{x}_{oil}$ is the molar oil fraction of the mixture.
Zürcher et al. (1998b) carried out experiments with R407C-oil mixtures. Analogous to Yokozeki (1994), they proposed a similar Arrhenius-type correlation (Eq. 2.105) for the dynamic mixture viscosity (μ_{mix}) as function of the oil mass fraction (w_{oil}).

$$\ln\left(\frac{\mu_{mix}}{\mathrm{Pa\,s}}\right) = (1-w_{oil})\ln\left(\frac{\mu_{ref}}{\mathrm{Pa\,s}}\right) + w_{oil}\ln\left(\frac{\mu_{oil}}{\mathrm{Pa\,s}}\right) \tag{2.105}$$

Adapted to manufacturer's data for mixtures of CO_2 and Reniso C 85 E, Zhelezny (2007) proposed a method for CO_2-POE mixtures as function of oil mass fraction (w_{oil}) and reduced temperature ($\tau = 1 - T/\bar{T}_c$). Empirical parameters are listed in Tab. 2.5.

$$\ln\left(\frac{\mu_{mix}}{\mathrm{Pa\,s}}\right) = \ln\left(\mu_0(w_{oil})\right) + \left(\alpha(\tau) + \Delta(w_{oil})\right)\ln(\tau) \tag{2.106}$$

$$\mu_0(w_{oil}) = \exp\left(\frac{a_1 + a_2\,(1-w_{oil})^{0.5}}{1 + a_3\,(1-w_{oil})^{0.5}}\right) \tag{2.107}$$

$$\Delta(w_{oil}) = \left(\frac{b_1 + b_2\,(1-w_{oil}) + b_3\,(1-w_{oil})^2}{1 + b_4\,(1-w_{oil}) + b_5\,(1-w_{oil})^2}\right)^2 \tag{2.108}$$

$$\alpha(\tau) = \left(c_1 + c_2\,\tau - c_3\,\tau^2\right)\left(1 + c_4\,\tau +_5\,\tau^2 + c_6\,\tau^3\right)^{-1} \tag{2.109}$$

$$\frac{\bar{T}_c}{\mathrm{K}} = \frac{d_1 + d_2\,(1-w_{oil})}{1 + d_3\,(1-w_{oil})} \tag{2.110}$$

Table 2.5: Parameters for dynamic viscosity correlation by Zhelezny (2007) for mixtures of CO_2 and Reniso C 85 E

$a_1 = 12.5438$	$a_2 = -8.58805$	$a_3 = -0.46019$			
$b_1 = 2.27770$	$b_2 = 0.1486$	$b_3 = -18.6209$	$b_4 = 12.3145$	$b_5 = -1.25934$	
$c_1 = 0.61538$	$c_2 = 16.155$	$c_3 = 33.098$	$c_4 = 10.591$	$c_5 = -33.506$	$c_6 = 24.077$
$d_1 = 770.252$	$d_2 = -212.747$	$d_3 = 0.833120$			

A comparison of predicted values by Zhelezny's correlation to measured data in the range of applicability ($-20\ °C \leq T_{bub} \leq +30\ °C$) is shown in the right graph of Fig. 2.27 in dependency of the oil mass fraction. Data points are represented by dark symbols and predicted values by bright gray solid lines. Symbols at $w_{oil} = 0$ representing pure CO_2 were determined using REFPROP, i.e., the correlation of Span and Wagner (1996). Both, pure refrigerant and mixture data are predicted with high accuracy. Hence, the correlation is recommended and was used for CO_2-POE85 mixtures as investigated in this work.

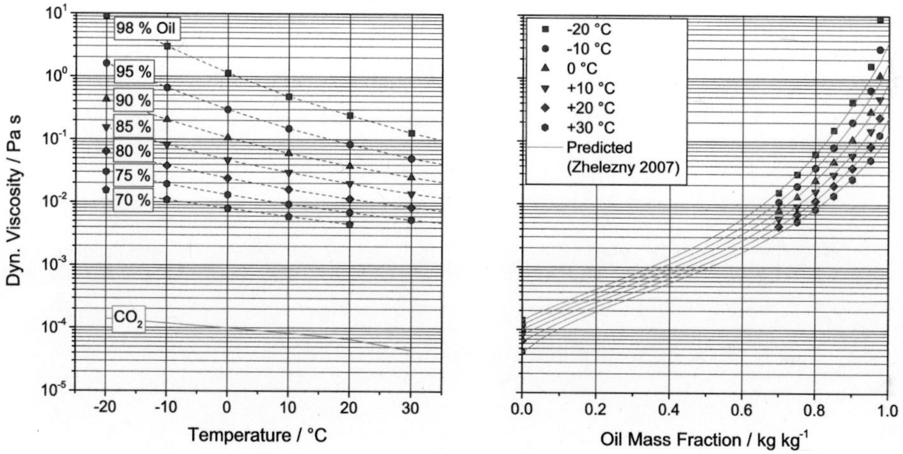

Figure 2.27: Dynamic Viscosity as function of temperature and oil mass fraction

2.5.3 Density

Due to its simplicity and fairly good agreement to compared measured data, the ideal mixture correlation (Eq. 2.111) by Jensen and Jackman (1984) is among the most adapted methods in literature for calculating the mixture density (ρ_{mix}) of refrigerant-oil mixtures.

$$\frac{1}{\rho_{mix}} = \frac{1 - w_{oil}}{\rho_{ref}} + \frac{w_{oil}}{\rho_{oil}} \tag{2.111}$$

where $\rho_{ref/oil}$ is pure refrigerant/oil liquid densitiy and w_{oil} is the oil mass fraction.

An actual oil mass fraction and temperature-dependent correlation for the density of CO_2-oil mixtures was derived by Seeton and Hrnjak (2006), see Eq. 2.112. Due to its form, however, applicability to mixtures other than the investigated CO_2-POE32 mixtures is doubtful.

$$\frac{\rho_{mix}}{\mathrm{kg\,m^{-3}}} = \sum_{i=0}^{2} (1-w_{oil})^i \left(a_{1,i} + a_{2,i} \left(\frac{T}{\mathrm{K}}\right) + a_{3,i} \left(\frac{T}{\mathrm{K}}\right)^2 \right) \tag{2.112}$$

Medvedev et al. (2004) developed a universal method for predicting the liquid density of refrigerant-oil solutions based on thermodynamic similarity within the technically relevant range of $0 \leq w_{oil} \leq 0.7$, validated by experiments with various types of refrigerant-oil solutions (e.g., R134a and NH_3 with different lubricants). Zhelezny (2007) adapted the correlation to CO_2-oil mixtures using manufacturer's data for Reniso C 85 E. Empirical parameters are listed in Tab. 2.6.

$$\ln\left(\frac{\rho_{mix}}{\mathrm{kg\,m^{-3}}}\right) = \ln\left(\frac{\bar{\rho}_c(w_{oil})}{\mathrm{kg\,m^{-3}}}\right) + B_L(w_{oil})\,\tau^{*d_1 F(\tau)} \tag{2.113}$$

$$B_L(w_{oil}) = a_1 + a_2\,(1-w_{oil}) + a_3\,(1-w_{oil})^2 + a_4\,(1-w_{oil})^3 \tag{2.114}$$

$$F(\tau^*) = 1 - 1.113\,\tau^{*0.4}\,\ln(\tau^*)^{-1} \tag{2.115}$$

$$\frac{\bar{T}_c(w_{oil})}{\mathrm{K}} = \frac{b_1 + b_2\,(1-w_{oil})}{1 + b_3\,(1-w_{oil})} \tag{2.116}$$

$$\frac{\bar{\rho}_c(w_{oil})}{\mathrm{kg\,m^{-3}}} = (c_1 + c_2\,(1-w_{oil}))^{-1} \tag{2.117}$$

where $\tau^* = \ln(T_c/T)$ is the logarithmic reduced temperature.

Table 2.6: Parameters for density correlation for mixtures of CO_2 and Reniso C 85 E (Zhelezny, 2007)

$a_1 = 1.64685$	$a_2 = -0.021096$	$a_3 = -0.187221$	$a_4 = 0.211344$
$b_1 = 770.252$	$b_2 = -212.747$	$b_3 = 0.833120$	
$c_1 = 0.00311927$	$c_2 = -0.0009806887$		
$d_1 = 0.325$			

The graph on the right side of Fig. 2.28 shows a comparison of predicted densities by Zhelezny (2007), represented by solid gray lines, to manufacturer's data (dark symbols). Densities are plotted as function of oil mass fraction at seven design temperatures in total. Apparently, the correlation yields a satisfying precision. At $w_{oil} = 0$ (pure CO_2), predicted values match well those for pure refrigerant densities calculated by the method of Span and Wagner (1996) (REFPROP). Thus, the correlation can be recommended for use with CO_2-POE mixtures and was applied within the scope of this work.

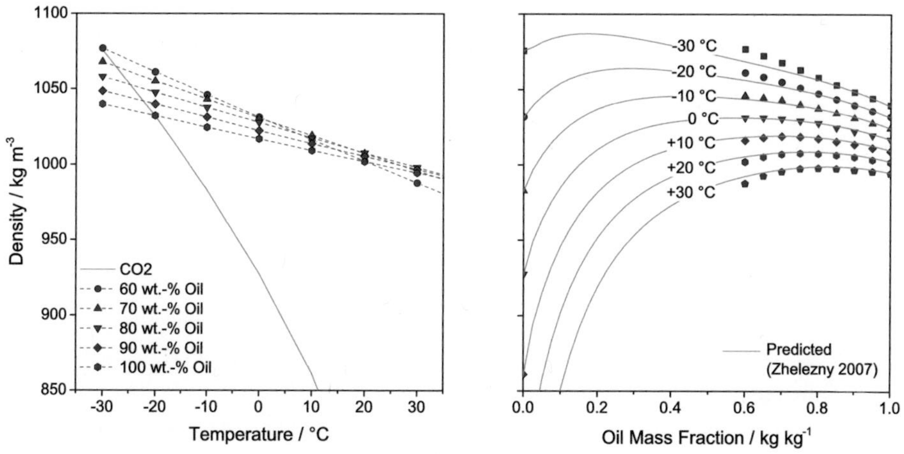

Figure 2.28: Liquid Density as function of temperature and oil mass fraction

2.5.4 Heat capacity

For the prediction of heat capacity of refrigerant-oil mixtures, Jensen and Jackman (1984) and Thome (1995) concordantly suggested a simple linear interpolation of pure substance properties, see Eq. 2.118. This method was also recommended by Baustian et al. (1986) and Conde (1996), and used without modification in this work.

$$c_{p,mix} = (1 - w_{oil})\, c_{p,ref} + w_{oil}\, c_{p,oil} \tag{2.118}$$

2.5.5 Thermal conductivity

Jensen and Jackman (1984) and Shen and Groll (2003) recommended using the correlation by Filippov and Novoselova for predicting the mixture thermal conductivity (λ_{mix}).

$$\lambda_{mix} = (1 - w_{oil})\, \lambda_{ref} + w_{oil}\, \lambda_{oil} - 0.72\, w_{oil}\, (1 - w_{oil})\, \left(\lambda_{oil} - \lambda_{ref}\right) \tag{2.119}$$

where $\lambda_{ref/oil}$ is the pure refrigerant/oil heat conductivity.
Another mixture thermal conductivity correlation is given by Saksena and Harminder (1975), see Eq. 2.120.

$$\lambda_{mix} = (1 - \tilde{x}_{oil})\, \lambda_{ref} + \tilde{x}_{oil}\, \lambda_{oil} - 0.72 \tilde{x}_{oil}\, (1 - \tilde{x}_{oil})\, \left(\lambda_{oil} - \lambda_{ref}\right) \tag{2.120}$$

where $\rho_{ref/oil}$ is the pure refrigerant/oil liquid density and $\tilde{x}_{oil}$ is the molar oil fraction.

A more complex correlation for the mixture thermal conductivity as function of molar oil fraction was derived by Baroncini et al. (1981), see. Eq. 2.121.

$$\frac{\lambda_{mix}}{\mathrm{W\,m^{-1}\,K^{-1}}} = \left[\tilde{x}_{oil}^2 A_2 + (1-\tilde{x}_{oil})^2 A_1 + n\tilde{x}_{oil}\,(1-\tilde{x}_{oil})\,\sqrt{A_1 A_2}\right] F(t) \tag{2.121}$$

where A_i are characteristic parameters of the mixture, n is a pure characteristic mixture number from 1.0 to 2.0, $F(t) = (1-t)^{0.38} t^{-1/6}$ is a a function of the reduced temperature $t = T_{bub}/T_{c,mix}$ and T_{bub} and $T_{c,mix}$ are the bubble point and critical mixture temperatures, respectively.

For this work, contract measurements for the thermal conductivity of POE Reniso C 85 E, CO_2 and mixtures of these components were conducted by Feja and Römer (2008), Ihmels (2008a) and Ihmels (2008b). The experimental data is given in the appendix, see Sec. G.1. Feja and Römer (2008) used a steady-state cylinder gap method. While the inner cylinder embodying the inner wall of the gap is constantly heated, temperatures in the surrounding liquid filling a cylindrical gap of 1 mm are measured by PT100 resistance thermometers. The measurement cell was validated by measurements of the thermal conductivities of R134a, water and glycerol. Following, measurements with CO_2 and Reniso C 85 E were carried out. Experimentally determined thermal conductivities were correlated by simple linear regression method, see Eq. 2.122.

$$\frac{\lambda_{mix}}{\mathrm{W\,m^{-1}\,K^{-1}}} = 0.43171 - 0.00111\,\frac{T}{\mathrm{K}} - w_{oil}\left(0.2706 - 0.00101\,\frac{T}{\mathrm{K}}\right) \tag{2.122}$$

Thermal conductivity measurements of pure Reniso C 85 E POE (Ihmels, 2008a) and mixtures of CO_2 and Reniso C 85 E (Ihmels, 2008b) were carried out by transient heated wire method. System pressure (60 bar $\leq p \leq$ 160 bar) and liquid temperature (-40 °C $\leq T \leq +10$ °C) were varied. Fitted to the experimental data, Ihmels derived Eq. 2.123 for the mixture thermal conductivity. Fitted parameters are listed in Tab. 2.7. Further details are given in Sec. D.2.5.

$$\frac{\lambda_{mix}}{\mathrm{mW\,m^{-1}\,K^{-1}}} = A_0 + A_1\,w_{oil} + (B_0 + B_1\,w_{oil})\,\frac{T}{\mathrm{K}} + (C_0 + C_1\,w_{oil})\,\frac{p}{\mathrm{bar}} \tag{2.123}$$

Table 2.7: Parameters for the thermal conductivity correlation for mixtures of CO_2 and Reniso C 85 E reported by Ihmels (2008b)

$A_0 = 450.374$	$B_0 = -1.205751$	$C_0 = 0.101799$
$A_1 = -2.949013$	$B_1 = 0.0114649$	$C_1 = -0.00082012$

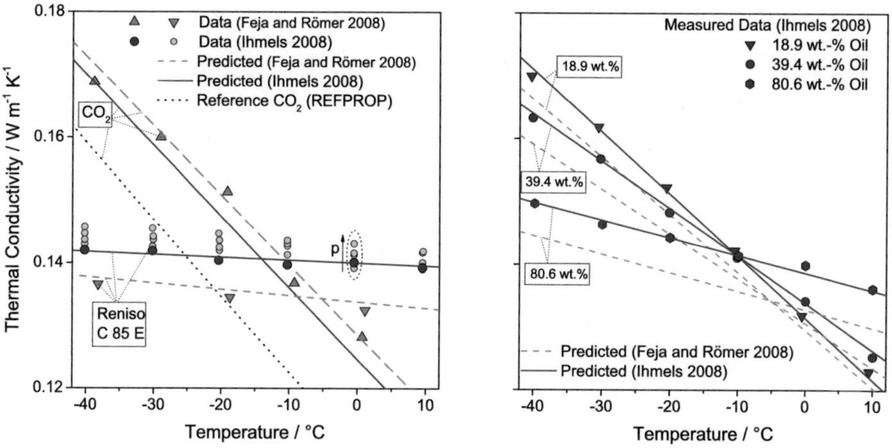

Figure 2.29: Comparison of different predictive methods for the thermal conductivity of CO_2-POE oil mixtures

A comparison of the data and predicting methods by Feja and Römer (2008) and Ihmels (2008b) is given in Fig. 2.29. Pure substance data is shown in the left diagram. Triangle symbols represent experimental data for CO_2 (△) and Reniso C 85 E (▽) by Feja and Römer (2008). Measurements by Ihmels (2008b) with Reniso C 85 E (◯) were conducted at different pressures. Of those, dark symbols characterize measurements closest to saturation pressure condition. Predicted values are represented by bright gray and dark gray solid lines, respectively. Pure CO_2 values calculated by Span and Wagner (1996) are shown as black solid line. Experimental (symbols) and predicted (lines) mixture data at saturation condition is shown in the right diagram for three oil mass fractions.

Both methods yielded similar experimental results and comparable slopes of fitted prediction curves. One possible reason for the slightly lower measurement results of Feja and Römer could be the influence of pressure on thermal conductivity. Measurements with Reniso C 85 E were done at a constant pressure of 2 bar. Ihmels (2008b) investigated the thermal conductivity at different pressures ranging from 5 – 160 bar, see small symbols for each temperature. It can be noticed that the experimental values decrease with decreasing pressure. This fact can also explain the slight difference in the slope of the predicted pure oil lines. The increase of the vapor pressure with the temperature is taken into account by Eq. 2.123 while it is not by Eq. 2.122. Since mixture properties are to be determined at the respective vapor pressure of the fluid, knowledge of the pressure dependency is of importance.

Specifically fitted to experimental measurement data of the respective mixtures investigated in this work, the regression curve given by Ihmels was used.

2.5.6 Surface tension

Brock and Bird (1955) proposed a prediction method for the surface tension of pure fluids based on the principle of corresponding states, see Eq. 2.124. According to the authors, Pandey et al. (1999) successfully applied this method to mixtures of R133 with benzene, toluene, p-xylene, acetone and cyclohexane. The authors substituted the implied critical parameters by pseudo-critical values for the regarded binary mixtures.

$$\frac{\sigma}{\mathrm{mN\,m^{-1}}} = (-0.281 + 0.133\,\alpha_c)\left(\frac{p_c}{\mathrm{atm}}\right)^{2/3}\left(\frac{T_c}{\mathrm{K}}\right)^{1/3}(\tau)^{11/9} \tag{2.124}$$

where p_c and T_c are the (pseudo-)critical pressure and temperature (of the mixture), respectively, and $\tau = 1 - T/T_c$ is the reduced temperature.
For calculating Riedel's parameter at the (pseudo-)critical point, $\alpha_c = (d\ln(p_s)/d\ln(T_s))_c$, Miller and Thodos (1963) proposed Eq. 2.125.

$$\alpha_c = 0.9076\left(1 + \frac{t_{r,nb}}{1 - t_{r,nb}}\ln\left(\frac{p_c}{\mathrm{atm}}\right)\right) \tag{2.125}$$

where $t_{r,nb} = T_{nb}/T_c$ is the reduced temperature at normal boiling point (T_{nb}).
The probably most widely used correlation for the surface tension (σ) of refrigerant-lubricant mixtures is given by Eq. 2.126, proposed and validated by Jensen and Jackman (1984) for various R113-oil mixtures. For instance, it was used by Spindler and Hahne (2009), Zhao and Bansal (2009), Zhu et al. (2013) and Hu et al. (2013). Hu et al. (2008c) used the same type of expression slightly modifying the oil mass fraction's exponent from 0.5 to 0.51. For the required surface tension of the lubricant in Eq. 2.126, Barbosa et al. (2004) and Zhao and Bansal (2009) recommended using Eq. 2.124.

$$\sigma_{mix} = \sigma_{ref} + \left(\sigma_{oil} - \sigma_{ref}\right) w_{oil}^{0.5} \tag{2.126}$$

Jensen and Jackman's established correlation represents a simple interpolation method, lacking a broad data base for reliable validation. As Wang et al. (2014) recalled observations by Wallner and Dick, the surface tension of refrigerant-lubricant mixtures can show three general types of behavior, see Fig. 2.30. These different behaviors are also reflected by different prediction methods, see Sec. D.2.6. Thus, despite its general approval, applicability of Eq. 2.126 to CO_2-POE85 mixtures must be doubted.
There was no clear description of the surface tension of CO_2-POE mixtures nor trustworthy experimental data found. Blindly guessing the mixture surface tension, however, can lead to remarkable deviations in both directions as was previously illustrated. Therefore, the pure refrigerant surface tension was assumed unchanged.

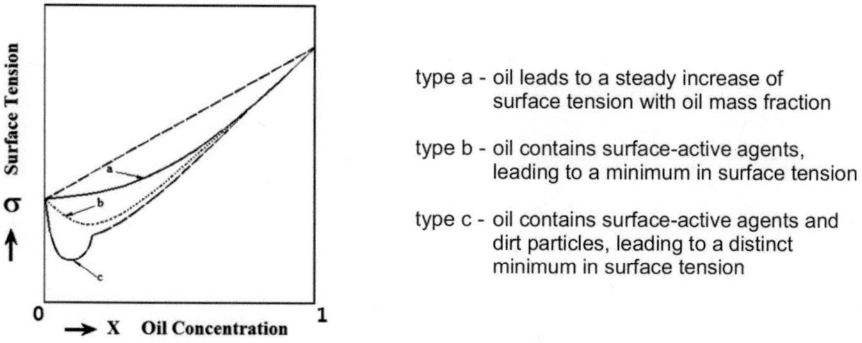

Figure 2.30: Different types of surface tension behavior with oil mass fraction, published in Wang et al. (2014)

3 Experimental procedure

Experiments on heat transfer, pressure drop and flow pattern observation of flow boiling CO_2 and CO_2-oil mixtures were carried out at the KIT/TVT flow boiling test facility. As lubrication oil POE85 (Reniso C 85 E by Fuchs Europe) was used. Under the specified test conditions, investigated CO_2-lubricant mixtures do not exhibit a miscibility gap. Two measurement sections were used for measurements: a smooth nickel tube of 14.0 mm inner diameter and an micro-fin copper tube of 8.62 mm core diameter. The measurement sections, constructed by Niederkrüger (1991) and Schael (2009) respectively, are further described in Sec. 3.1.5. Table 3.1 summarizes the range of operation of the flow boiling test facility.

Table 3.1: Operation range of flow boiling test facility

System pressure (max.)	65 bar
Mass flow rate (max.)	200 $kg\,h^{-1}$
Heat flux (max.)	200 $kW\,m^{-2}$
Pre-heater power (max.)	20 kW
Temperature range	$-50\ldots+50$ °C
Refrigeration capacity (min.)	35 kW (at -50 °C)

3.1 Test facility

The facility was originally installed by Schmidt (1986). Modifications have been made since by Niederkrüger (1991), Wettermann (1999) and Schael (2009). Last, extensive pure CO_2 flow boiling measurements inside a smooth and an micro-fin tube were conducted by Schael. As both measurements sections are used for flow boiling experiments within the scope of this work, a detailed description is given in Sec. 3.1.5. Further, Saito performed first experiments with CO_2-oil mixtures inside the smooth tube measurement section.
At the outset, the test facility was extensively reconstructed including the modification of the implemented oil loop, installation of new measurement technique and migration of the whole test loop to an Aluminum frame. Further on, the original R410/R23 cascade chiller had to be substituted at an early stage due to problems by leakage, controlling and lack in capacity. In coherence, existing difficulties in controlling the temperatures of the condenser and sub-cooler of the CO_2 test loop as well as the highly inefficient operation had to be solved. These problems mainly arose from the fact of a missing buffer tank for the brine and the realization of the thermal coupling by a single brine circuit. The condenser and sub-cooler

sub-circuits were connected to a main brine supply circuit, both affecting each mode of operation. Further, the brine had to be heated electrically to maintain continuous operation at low demand for refrigeration power (i.e. for low heat input in the test loop), resulting in high energy consumption. As solution, two separate brine circuits each connected to a main brine storage were constructed. Accordingly, the modified flow boiling test facility now consists of four main fluid circuits described in the following. Schematic diagrams are presented in Fig. 3.1-3.3.

3.1.1 Chiller

Refrigeration power is supplied by a chiller manufactured by L+R Kältetechnik. It consists of a single stage R507A refrigerant circuit with tandem compressor and achieves a refrigeration capacity of > 30 kW at a temperature of -50 °C. Heat is removed by a water-cooled tube condenser. The evaporator is realized as brazed plate heat exchanger, in which heat is transferred from the brine to the evaporating refrigerant. The brine is pumped from an internal insulated storage tank to the heat exchanger and back to tank, thus ensuring a constant brine temperature in the tank. The brine storage tank has a capacity of 300 l and serves as thermal buffer. It is equipped with six flanges for flow and return allowing the connection of external brine circuits, such as the condenser and sub-cooler brine circuits of the test facility. A schematic of the chiller is shown in Fig. 3.1.

3.1.2 Brine circuits

Two brine circuits link the test loop to the temperature-regulated brine storage tank of the chiller for thermal coupling, see Fig. 3.2. The heat exchanger of the test loop for shell-side condensation of the CO_2 is connected to the first circuit. Brine circulates by means of a brine pump at constant rotation speed. A pneumatic regulation valve connects the brine circuit to the storage tank on the suction side of the pump. In combination with a by-pass valve, the condensing temperature of the CO_2 can be set by controlling the flow of cold brine from the storage tank into the brine circuit. This temperature controlling method offers the advantage of a highly constant brine flow rate through the condenser and hence low retention time, leading to quick and precise controller responses. This fact especially refers to the condenser circuit since heat input for evaporation in the test loop varies significantly. Hence, flow rates and retention time of the brine would vary similarly resulting in the need of finding new control parameters for each operation condition. The sub-cooler circuit was constructed similarly to the condenser circuit described prior. However, the demand on cooling power for sub-cooling of the CO_2/CO_2-oil mixture is comparatively low and constant for all measurement conditions. Therefore, a low power rotational speed controlled brine pump was chosen differing from controlling the inlet flow of cold brine.

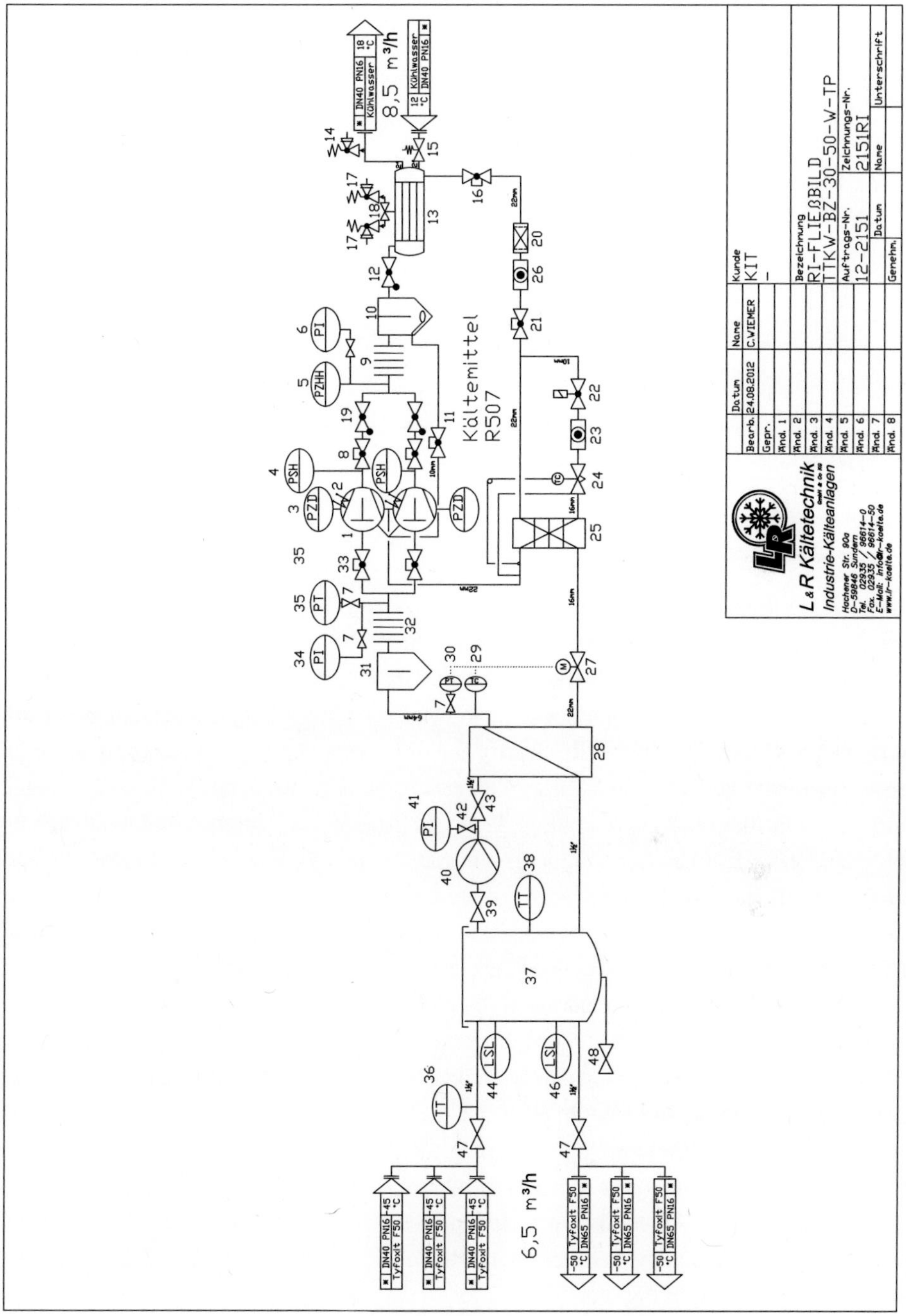

Figure 3.1: Schematic of water-cooled chiller (L+R Kältetechnik)

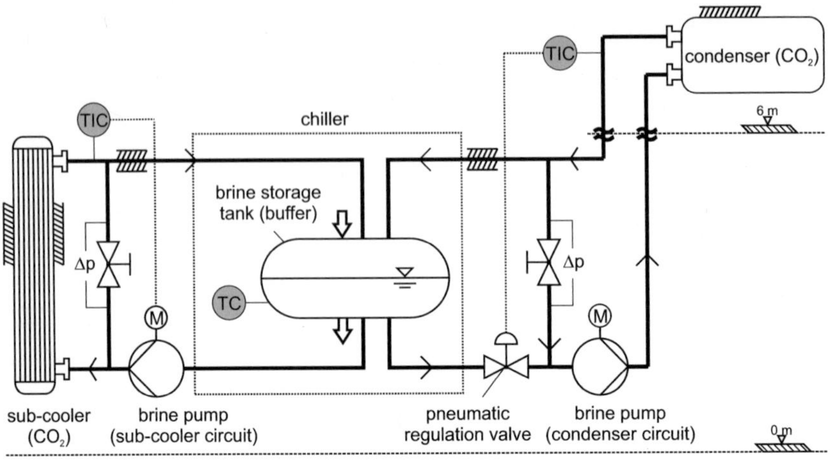

Figure 3.2: Schematic of brine circuits

3.1.3 Test loop

In operation mode, the fluid (CO_2 or CO_2-oil mixture) is pumped from the storage tank into the test loop. Following, it is sub-cooled inside a straight-tube heat exchanger to avoid cavitation at the refrigerant pump and expansion into the two-phase region at the inlet into the test section. The latter is of especial importance for a precise adjustment of vapor quality at small mass fluxes. The mass flow rate of the sub-cooled liquid is measured by a Coriolis mass flow meter and adjusted by use of an electronic regulation valve. Three pre-heaters allow the adjustment of vapor quality. These are realized as straight tubes with each two integrated electrical heating conductors in bottom-near axial grooves. Further, heating conductors are integrated into the tube wall of the measurement section and guard heater for conducting heat transfer measurements and to avoid an axial temperature profile, respectively. Thyristor controller are used to control the electrical power input of each heating conductor, whereby the control signal is determined by means of manually adjustable potentiometers. Flow patterns are determined at the sight glass by means of high-speed video recording at 500 fps (frames per second). In addition, a new measurement technique based on confocal laser displacement measurement to determine the film thickness of the flowing liquid at the crest of the tube was developed and integrated into the sight glass section (see Sec. 3.2.6 and App. C). In the shell and tube condenser of the test loop the system pressure is adjusted and evaporated CO_2 is re-liquefied. The system pressure is limited to 65 bar by a pair of safety relief valves installed at the condenser and the inlet of the test section. A P&I diagram of the test loop is shown in Fig. 3.3.

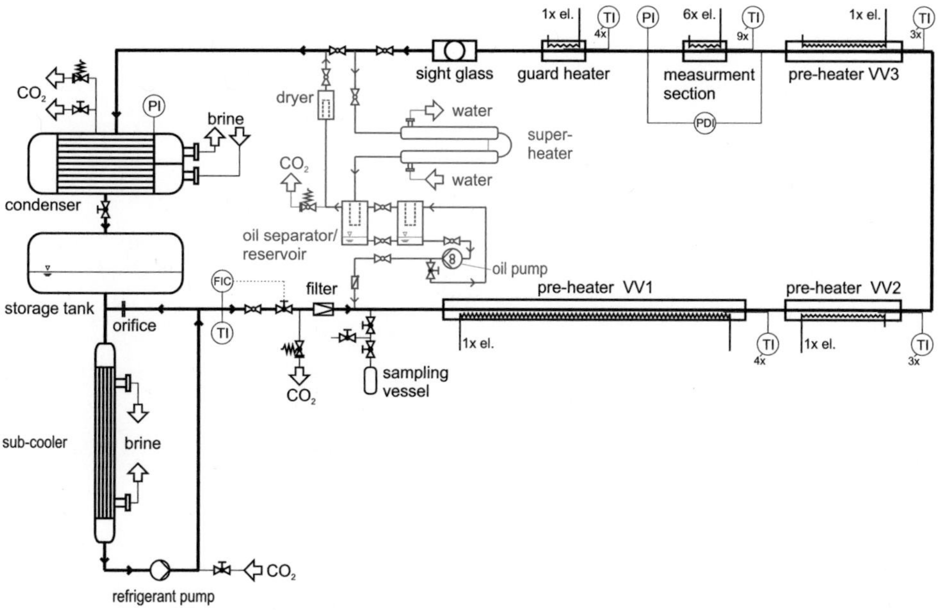

Figure 3.3: P&I diagram of test loop with oil circuit

3.1.4 Oil loop

An oil loop has been installed allowing the adjustment of oil mass fraction in the liquid refrigerant (see Fig. 3.3, gray lines and elements). Oil separator and attached vessel form a combined reservoir for the lubrication oil. A gear pump delivers the lubricant into the test loop. Back flow into the oil loop is prevented by a non-return valve. The amount of lubricant flow into the test loop can be estimated by observing the change in liquid level in a sight glass attached to the oil separator or the reading value of an integrated volume reader. Further, a super-heater was implemented into the oil loop for lubricant removal. Inside the water-heated tubular heat exchanger CO_2 can be evaporated. Water is supplied by an electrically heated peripheral water circuit. Liquid (oil-rich phase) and vapor (CO_2) are separated in the oil separator. While liquid lubricant accumulates in the reservoir, gaseous CO_2 flows back into the test loop. Any moisture in the on-flowing gas is removed in a filter dryer.

3.1.5 Measurement sections

For the experimental investigation of flow boiling heat transfer and pressure drop of CO_2-oil mixtures two different measurement sections were used, see Tab. 3.2 and Fig. 3.4 for specifications. Both test sections are constructed similarly for comparability and easy interchangeability.

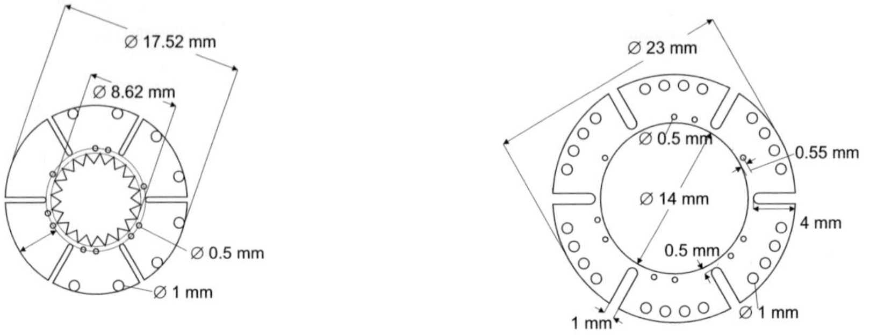

Figure 3.4: Cross sections of micro-fin (left) and smooth (right) measurement section

The nickel smooth tube measurement section was constructed by Niederkrüger (1991) and was used since for heat transfer and pressure drop experiments for CO_2, mixtures of R116, R134a and R846 and mixtures of R12 and SF6. It measures 23×4.5 mm in diameter and 256 mm in total length, while the heated length equals 200 mm. The ring chambers of the inlet and outlet flanges consist of six equally distributed bores for inlet pressure and pressure drop measurements. The tube wall is axially divided by six grooves of 1 mm in width and 4 mm in depth, leaving a residual wall thickness of 0.5 mm at the groove bottom. In each such tube segment a heating conductor is integrated in the tube wall close to the outside surface. All six heating conductors of the measurement section are controllable independently. In a distance of 0.5 mm from the inner wall surface, miniature thermocouples (0.5 mm) are integrated centered into the tube wall of each segment for measuring the wall temperature. Referred to inlet of the measurement section the axial measurement position equals (150 mm). The described set-up allows local heat transfer measurements under uniform heat flux ($\dot{q}(\varphi) = \text{const.}$) and uniform wall temperature ($T_w(\varphi) = \text{const.}$).

Table 3.2: Measurement sections

Measurement section:	Smooth tube	Micro-fin tube
Material:	Nickel	Copper
Total length:	256 mm	267 mm
Total heated length:	200 mm	200 mm
Heated length up to measurement position:	122 mm	100 mm
Distance from inlet to measurement position:	150 mm	133 mm
Outside diameter:	23.0 mm	17.52 mm
Inside (core) diameter:	14.0 mm	8.62 mm
Mean roughness height:	0.15 µm	-
No. of fins:	-	60
Fin height:	-	0.20 mm
Apex angle:	-	30 °
Lead angle:	-	18 °

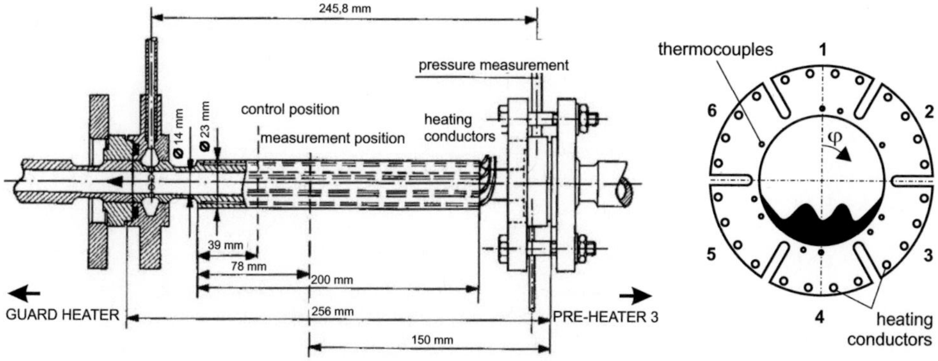

Figure 3.5: Smooth tube measurement section; axial view (left) and cross-sectional view (right)

The material joining of thermocouples and heating conductors into the tube wall was accomplished by galvanoplastic coating (Niederkrüger, 1991). Due to equal material properties (heat conductivity, e.g.) of the base tube and the galvanic material, the tube wall is treated as uniform in material. Meander-type laying of the heating conductors in the tube wall permits the negligence of inductive effects by the electrical current. Thermal boundary conditions could be verified by a total of four wall-integrated control thermocouples. Their positions are shifted axially by 39 mm and distorted by 15 °. Axial and cross-sectional views of the smooth tube measurement section are shown in Fig. 3.5.
For heat transfer and pressure drop investigation inside micro-fin tube Schael (2009) constructed a measurement section analogous to the smooth tube, see specifications given in Tab. 3.2. It consists of a micro-fin base tube (Cuprofin EDX® by Wieland) of 9.52×0.45 mm in diameter and 256 mm in length. Along the heated length (200 mm), the wall thickness of the copper tube was increased to a total of 4 mm by galvanoplastic coating confining miniature thermocouples (∅ 0.5 mm, type E) and heating conductors (∅ 1.0 mm). Alike, heating segments are separated by axial grooves. The remaining wall thickness at the groove bottom equals the thickness of the base tube (0.45 mm).

3.1.6 Pre-heaters

Pre-heaters are used for the adjustment of vapor quality in the measurement section. Similar to the measurement section, these are pipes with integrated heating conductors at the bottom. Since the bottom of the tube is always wetted with liquid, high heat fluxes and thus change in vapor quality are realizable this way. However, at least one pre-heater had to be replaced in an early stage due to irreversible damage. Failure of heating conductors was a frequent problem in the past. With the aim of prolonged lifetime a re-construction and substitution of all pre-heaters in the test loop was realized in this work.

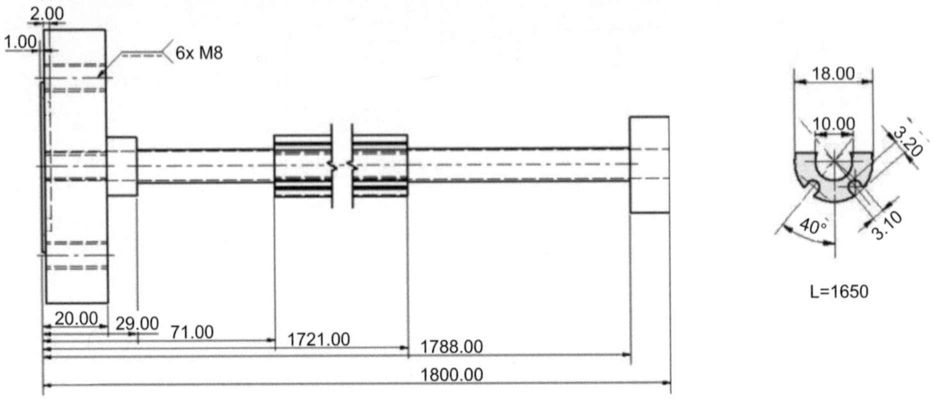

Figure 3.6: Micro-fin pre-heater, axial view (left) and copper shell (right)

The first two pre-heaters (VV1, VV2) of the test loop consist of 15 × 5 mm copper base tube with (3600 mm) and (1800 mm) in length, respectively. At the bottom of each tube two axial grooves of 4.1 mm in diameter are provided for laying of the heating conductors (∅ 4.0 mm). They are connected to the tube by hard-soldering. Both pre-heaters are positioned in the inlet section of the test loop ahead of the U-turn connecting pipe and are in operation for experiments with both measurement sections, see Fig. 3.3.
The third pre-heater (VV3) is positioned in-between the U-turn and inlet pipe and the measurement section. Therefore, its geometry is adapted to the respective measurement section in use. The pre-heater for use in combination with the smooth tube is identical to pre-heater VV2. For micro-fin tube measurements, a pre-heater consisting of a straight micro-fin base tube (Cuprofin EDX®) was constructed by Siewert (2013), see Fig. 3.6 for constructional drawings. As the wall of the tube is too thin (0.45 mm) for complete enclosure of the heating conductors, an additional half-cylinder shaped copper shell with 10 mm in inner diameter and 4 mm in wall thickness was hard-soldered on the micro-fin tube. The shell contains two axial grooves (∅ 3.1 mm) at the bottom of the tube that were milled in the outer tube wall, displaced by 40 ° with respect to the perpendicular, see right image of Fig. 3.6. A heating conductor (∅ 3 mm) was hard-soldered in each groove.
For the transition from smooth tube pre-heater to the micro-fin pre-heater preceding the measurement section, a new connecting tube was constructed. In its inlet flange the flow diameter is reduced to that of the micro-fin tube (8.6 mm), see Fig. 3.7.

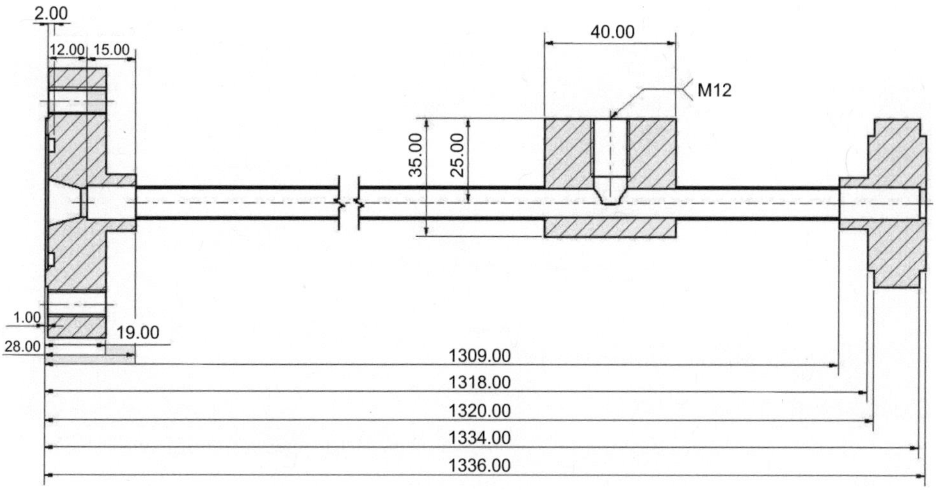

Figure 3.7: Inlet tube preceding pre-heater VV3

3.2 Measurement technique and data acquisition

For the experimental evaluation temperatures, pressures, pressure drop, electrical power input, liquid mass flow rate and sample weight for oil mass fraction determination were measured. The measurement devices used in this work are listed in Tab. A.2 and determined measurement uncertainties are given in Tab. A.1, see Appendix A. Measurement values were recorded using the commercial software LabVIEW™.

3.2.1 Data acquisition

For first measurements, a digital multimeter KE2750 DMM in combination with two 20-port 7700 modules by Keithley was used for data acquisition (DAQ). It was connected to a PC over GPIB interface allowing a reading rate of $\approx$ 10 fps. In the following, the DAQ was exchanged by a modular CompactDAQ® system by National Instruments. It consists of an 8-port USB-chassis equipped with modules for measurements of voltage (2x, for TC), resistance (1x, for RTD) and current (1x, for measuring transducers). Advantages of the new system include an easy implementation in LabVIEW, high reading rates (>1 samples per second for all channels in high resolution mode) and a compact construction. The system was installed in an electrical cabinet to ensure steady environmental conditions and protection towards the outside. It was connected to a PC by USB.

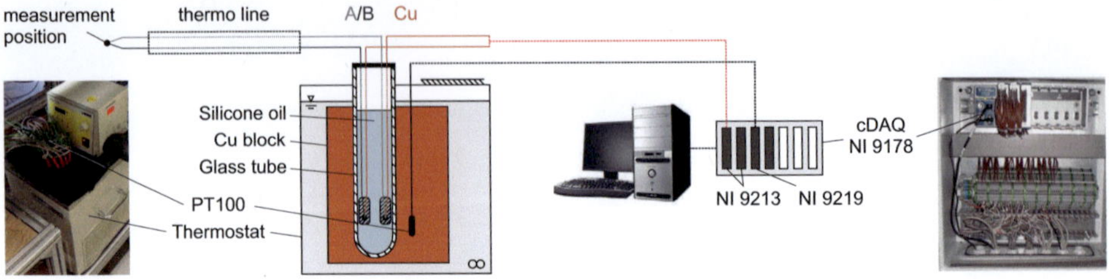

Figure 3.8: Thermocouple DAQ

3.2.2 Temperature measurement

Thermocouples of type E (NiCr/CuNi, $S_{\mathrm{NiCr/CuNi}} > 55\,\mu\mathrm{V\,K^{-1}}$) of 0.5 mm in outer diameter were used for measuring the tube wall temperatures of the measurement section and the guard heater. A total of ten thermocouples are integrated in the tube wall of the measurement section for determination of the inner tube wall temperature (see Sec. 3.1.5). In addition, two thermocouples were placed on the tube to measure the top and bottom outer wall temperatures. The outer wall temperature of the guard heater was measured by four thermocouples distributed on the tube. Type K thermocouples (NiCr/Ni, $S_{\mathrm{NiCr/Ni}} > 40\,\mu\mathrm{V\,K^{-1}}$) of 1 mm were used for measuring the outer tube wall temperature of the pre-heaters as well as the outside insulation temperature at all sections. Each wire end of a thermocouple was connected to a copper wire inside a glass tube, filled with liquid (silicone oil) for electrical insulation and improved temperature equalization. All glass tubes were placed in a copper body inside a thermostat bath that is held at a constant temperature of 40 °C (= reference junction temperature). See Fig. 3.8 for a schematic set-up of thermocouple DAQ.

PT100 resistance thermometers were installed in the test loop to measure the sub-cooled fluid temperature in front of the regulation valve (inlet to the test section), the fluid temperatures in front of pre-heater VV3 and behind the measurement section as well as the temperature of the reference junction. Diameters of the used thermometers range from 2 to 3 mm. Each PT100 was connected in 4-wire mode to an insulated channel of the NI 9219 module. The resistance was measured by applying an excitation current of 1 mA.

All temperature sensors were calibrated before and during measurements, whereby PT100 were calibrated first and thermocouples second in order. For calibration, PT100 thermometers were placed in a thermostat bath filled with glycol-water mixture. A mixing ratio of 50 : 50 allowed setting-up temperatures down to −40 °C. As reference a high precision PT25 was used for measuring the bath temperature. Second, all free thermocouples (thermocouples not integrated in the tube wall of the measurement section) were calibrated by means of a second thermostat. The temperature of the first thermostat was held at 40 °C while the temperature of second thermostat was varied in the range of operation. Bath temperatures were measured with the calibrated PT100 of the reference junction and the high precision PT25 reference. Last, the thermocouples in the measurement section were calibrated. Since any moving of the measurement section could have caused irreversible damage to the integrated and not replaceable thermocouples, calibration was accomplished during operation of the test loop with the saturation temperature as reference.

3.2.3 Pressure and pressure drop measurement

Absolute pressures at the inlet of the measurement section and in front of the regulation valve at the inlet to the test section were measured using pressure transducers by Burster (model 8106). Equipped with signal amplifier, the output signal (4...0 mA) was adapted to the range of pressure (10...40 bar). Further, an internal bridge circuit of resistances allowed setting-up the signal offset. Both, amplifier and offset could be adjusted by two screws on the outside of the transducer. A differential pressure transducer by Rosemount (model 3051C) was used for the determination of pressure drop along the measurement section. Via HART protocol the output signal (4...20 mA) was scalable to any pressure drop range within ±623 mbar. For flow boiling measurements, pressure drop ranges of 0...20, 0...100 and 0...500 mbar were chosen. All three transducers were connected to the NI 9219 cDAQ module. Connection to the measurement joints at the test facility was realized by 6 mm stainless steel tubing. At the beginning and during measurements, differential and absolute pressure transducers were calibrated within the denoted ranges using a differential pressure gauge by Desgranges et Hout (model DH 5501). For a precise reference, all masses were re-measured by DH Budenberg (2010).

3.2.4 Power measurement

Electrical heating power in pre-heaters, measurement section and guard heater were regulated by thyristor controllers at line voltage of 230 V. A change in phase angle control was accomplished manually by potentiometers. The electrical power was measured using four power meters by Yokogawa (WT 1030), each equipped with three signal inlets and read-out via GPIB interface. From measured values (P_{el}), power loss along the wiring and cold ends of heating conductors (P_{loss}) were subtracted to yield the effective heating power (P_{eff}).

$$P_{eff} = P_{el} - P_{loss} \tag{3.1}$$

3.2.5 Mass flow measurement

For measuring the liquid flow rate, a Coriolis flow meter by Endress & Hauser (Promass 60) was installed before the regulation valve in the sub-cooled inlet-tubing to the test section. The range of measurement equals 0...200 kg h^{-1}, producing an output signal of 4...20 mA measured by a NI 9219 module.

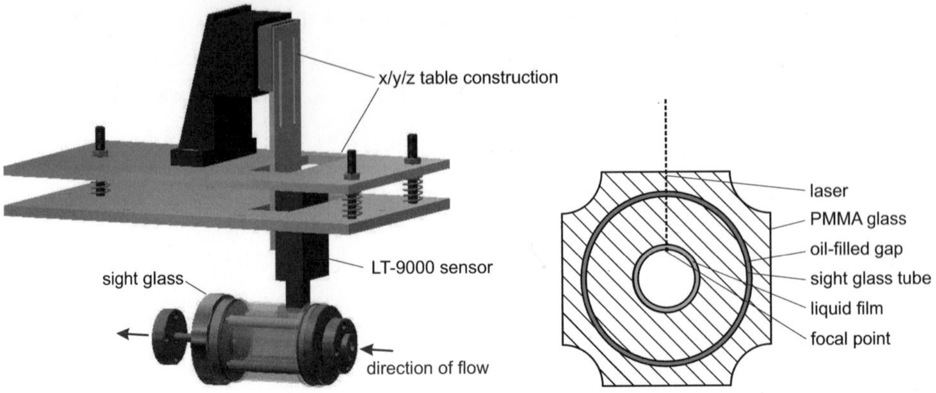

Figure 3.9: Sight glass construction for flow pattern observation

3.2.6 Flow pattern observation

Flow patterns were determined classically by high speed video recording of the two-phase flow. For this purpose, a Motion Corder Analyzer by Kodak was installed at the sight glass of the test section. It has a video recording capacity of 1092 frames with a resolution of 512×240 pixel, which yields a single video length of 2.1 s at 500 fps.
Besides evaluating video recordings, recently developed flow pattern maps partially (Mastrullo et al., 2012a) or fully (Cheng et al., 2006) relied on local heat transfer measurements for determining flow pattern transitions. In order to provide a more precise determination of these transitions, especially from stratified to intermittent or to fully developed annular flow and the onset of dryout, a new film thickness measurement technique was introduced (Wetzel et al., 2009) (Wetzel et al., 2011). It allows for measuring the liquid film thickness at the crest of the sight glass based on the principle of confocal laser displacement. A laser displacement meter by Keyence (model LT-9000) was installed perpendicular above the sight glass, see Fig. 3.9. This measurement technique allows distance measurements through transparent media (see right image). However, to avoid scattering of the laser signal at phase-interfaces the sight glass construction had to be modified. A PMMA glass body with planer polished outside surfaces enclosed the round tube. The gap (0.5 mm) in between was filled with lubrication oil as the refraction indices of the three materials are similar. Before starting a film thickness measurement, the focal point of the measuring device had to be adjusted to the inner glass tube surface. A change of signal with respect to this surface indicates a liquid film. The measurement technique including its applicability and limitations in use is described in detail in App. C.

3.3 Start-up and operation of the test facility

Prior to filling with CO_2, the test loop was evacuated to a remaining pressure of 0.1 mbar. The temperatures of the condenser and sub-cooler were set to −40 °C. After evacuation, CO_2 was filled from a dip-tube bottle at the inlet valve near the refrigerant pump (see Fig. 3.3). Shutoff valves to the test section were closed in order that the inflowing liquid is stored in condenser, storage tank and sub-cooler.

For taking the test loop into operation the test section was first evacuated to 0.1 mbar. Temperatures of the condenser and sub-cooler were set to the desired operating values. Upon opening the shutoff valves the test section was pressurized by the influx of CO_2. Zeroing of the differential pressure transducer was done at operating pressure. For the time of the calibration procedure, the transducer was disconnected from the test loop by shutoff valves to avoid any pressure fluctuations. The refrigerant pump was started and the mass flow rate set-up at the controller of the regulation valve. After flow stabilization powers at the pre-heaters were adjusted in order to obtain the desired vapor quality. Target temperatures at the condenser and sub-cooler were adjusted if needed.

The addition of oil to the CO_2 occurred discontinuously, i.e. the oil mass fraction in the liquid bulk desired for the upcoming measurement series was set to fixed value. For this purpose, lubrication oil was filled into the reservoir of the oil loop. The oil-containing reservoir and oil loop were evacuated to 0.1 mbar. Heating the reservoir by means of an electrical heating wire on the outside of the vessel ensured moisture-free conditions. The oil loop was pressurized by opening the shutoff valve between test section and refrigerant dryer. Following, dry lubrication oil was pumped out of the lubricant reservoir into the test loop to raise the oil mass fraction in liquid bulk. A gearwheel flow meter by Kobold (model DZR) measured the amount of oil pumped into the test loop. When the approximated oil volume was reached the oil pump was turned off. Homogeneity of the CO_2-oil mixture was achieved after circulation for several hours.

For oil removal out of the CO_2-oil mixture, first the oil loop was evacuated and then pressurized by CO_2 to the operation conditions as described before. The mass flow rate in the test loop was set to $50\,\mathrm{kg\,h^{-1}}$ and the water temperature of the super-heater 30 K higher than the condenser temperature. By opening the shutoff valve to the super-heater and closing the valve to the condenser the oil containing fluid flowed through the super-heater for evaporation of CO_2. The remaining oil-rich liquid phase was separated from the gaseous refrigerant inside the oil separator.

The oil loop was solely used for the purpose of introducing or removing oil. In measurement mode the oil loop was completely bypassed and not in operation. Thus, the target oil mass fraction in the bulk for the following measurement series was set to a constant value. It was verified by regular checks, see the description of gravimetric oil mass fraction measurements in Sec. 3.4.5.

Table 3.3: Experimental test conditions

Test fluid	mixtures of CO_2 (4.5, purity: 99.995 %) and POE lubrication oil (Reniso C 85 E, Fuchs Europe)
Nominal oil mass fraction (w_{oil})	0...3 wt.-%
Saturation condition (p_s, T_s)	10.0...39.7 bar (−40...+5 °C)
Vapor quality ($\dot{x}$)	0.10...0.95
Heat flux ($\dot{q}$)	0...120 kW m^{-2}
Test section	smooth tube (nickel), micro-fin tube (copper)
Inner diameter (d_i)	14.0 mm (smooth tube), 8.62 mm (micro-fin tube)
Thermal boundary condition	constant wall temperature

3.4 Experimental procedure

Flow boiling experiments were carried out for mixtures of CO_2 and POE lubrication oil. CO_2 was used in quality of 4.5. The lubricant (Reniso C 85 E) was supplied in sealed containers in pre-dry, near moisture-free condition by Fuchs Europe. For starting flow boiling measurements, the test loop was taken in operation as described in Sec. 3.3. Test conditions for the experimental investigations are summarized in Tab. 3.3.

3.4.1 Heat transfer measurements

Flow boiling heat transfer measurements were performed at constant wall temperature as boundary condition. Starting, system pressure (by condenser temperature) and mass flow velocity (calculated from mass flow rate and cross section of the respective test section geometry) were preset. Heating power of pre-heaters VV1 and VV2 were adjusted to yield the desired vapor quality at the inlet of the measurement section. Set values for these powers were calculated automatically by the monitoring LabVIEW routine, taking the preset values for system pressure, mass flow rate and fluid properties at saturation condition into account. Next, the heating power of the bottom segment (segment 4) was slowly adjusted to a preset value. This value was automatically calculated from predefined values for the heat flux at segment 4 (based on application relevance and previous measurements) and displayed on the screen. The remaining segments were adjusted to reach equal temperatures. To avoid the influence of hysteresis effects and to assure reproducibility to other measurements, experiments were carried out starting at the highest heat flux and proceeding to near-adiabatic condition.

Besides adjusting the heating power of the measurement section, set values for the pre-heater powers were automatically renewed based on the actual heat input and the desired vapor quality in the measurement section. To avoid the formation of an axial temperature profile along the measurement section including inlet and outlet tubing, the electrical powers of pre-heater VV3 and the guard heater were adjusted to approximately result in the same heat flux of the measurement section. Thus, these values were simultaneously calculated from

actual set-up values. Powers for the preceding pre-heaters VV1 and VV2 were readjusted with respect to the desired vapor quality in the measurement section.
Once wall temperatures in the measurement section were equalized, the respective powers to pre-heaters and guard heater applied and flow conditions stabilized, recording of measurement data was started. 20 consecutive measured values were recorded and averaged for each measurement point. During data acquisition, measurement values should not vary more than ± 25 mbar in system pressure, $\pm 5\,\mathrm{kg\,m^{-2}\,s^{-1}}$ in mass flow rate and all wall temperatures of the measurement section should be within the range of 0.1 K. Besides, flow pattern and pressure drop were recorded.

3.4.2 Pressure drop measurements

Pressure drops were measured under diabatic and adiabatic conditions. Each measurement point results from averaging of 20 consecutive measured values. Diabatic measurements comprise the determination of pressure drop during heat transfer measurements, see Sec. 3.4.1. Here, pressure drop results from friction and acceleration of vapor. Sole frictional pressure drops were determined from adiabatic measurements. For these measurements no power was applied to pre-heater VV3, the measurement section or the guard heater. Vapor qualities were set-up by solely heating with pre-heaters VV1 and VV2. Prior to starting pressure drop experiments, a zeroing and range configuration of the differential pressure transducer was accomplished, see Sec. 3.2.3 and 3.3.

3.4.3 Flow pattern investigation

Analogous to pressure drop measurements, flow patterns were investigated during flow boiling experiments at diabatic and adiabatic boundary conditions. Flow patterns were captured by high speed video recording immediately after starting data acquisition and recording. Classification into flow regimes follows subjective interpretation of the visualized flow pattern, supported by film thickness measurements at the crest of the tube as described in Sec. 3.2.6. Under dry conditions, the sensor was adjusted in position so that the inner surface of the sight glass was detected with highest signal intensity just within the lower range of measurement. Its distance to the sensor was recorded and served as baseline for film thickness determination. As investigated to determine its influence on pressure drop and heat transfer, observations should match the flow pattern inside the test section. Fins of the micro-fin tube have a severe effect on two-phase flow pattern. For this purpose, video recording of the flow pattern and film thickness measurements during micro-fin tube experiments were done as close to the outlet of the guard heater as possible.

3.4.4 Hydrodynamic and thermal flow development

Schmidt (1986) conducted a literature study on two-phase flow development in pipes. Two-phase flow was considered hydrodynamically and thermally developed once pressure drop and temperature profile of the tube wall remain unchanged, respectively. Full hydrodynamic development of two-phase flow after 90° elbow pipes was found to be completed typically after 100 diameters. In few cases, authors report of up to 200 diameters. Thermal development of two-phase was considered to take up to 25 diameters. For adiabatic measurements, length to diameter ratio of the straight unheated inlet before the measurement section yields $(L/d) > 220$. The ratio for the heated length during diabatic measurements was $(L/d) > 120$. Thus, two-phase flow was assumed to be hydrodynamically and thermally developed.

3.4.5 Oil mass fraction measurements

The mass fraction of lubricant in the liquid bulk (nominal oil mass fraction) is equal to the oil mass fraction at the inlet of an evaporator. It is determined gravimetrically from liquid samples extracted under operation conditions. Initially, the sampling vessel (see Fig. 3.10) is evacuated and its tare weight determined by weighing. To extract liquid samples, the sampling vessel is reconnected to the test loop and the connecting pipes are evacuated. The valve to the test loop is opened, followed by quickly closing the valve at the sampling vessel to entrap the entire fluid sample. The total weight of the fluid (CO_2 and lubricant) and vessel is measured gravimetrically. Following, CO_2 is released slowly by carefully opening the valve of the pressurized sampling vessel. It must be ensured that no liquid droplets are entrained by the out-flowing gas. When pressure equalization with the surroundings is completed, the vessel is evacuated once more for several minutes to remove any CO_2 dissolved in the lubricant. The total mass of vessel and lubricant is determined. Knowing the tare weight of the vessel, the fluid masses and hence the oil mass fraction can be determined. Measurements of the oil mass fraction are repeated periodically to assure a non-changing oil mass fraction.

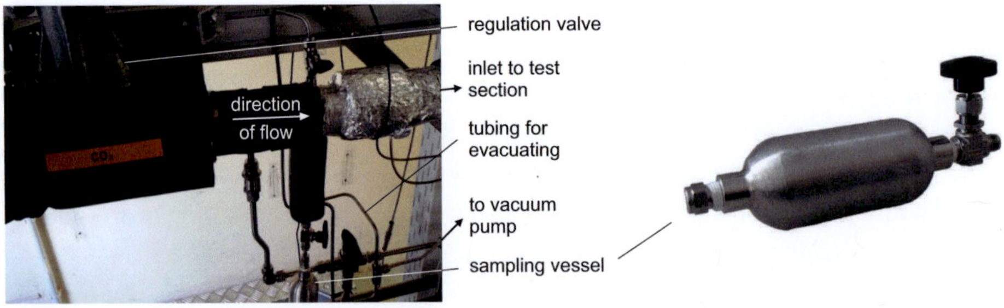

Figure 3.10: Setup for oil concetration measurements

3.5 Evaluation of Measurements

3.5.1 Heat transfer

Heat transfer coefficients (α_j) were determined locally for each segment j with respect to the center of the measurement section. With the local heat flux ($\dot{q}_j$) and inside wall surface temperature ($T_{w,j}$) at segment j and the saturation temperature of the pure refrigerant (T_s), calculation in general follows Eq. 3.2.

$$\alpha_j = \frac{\dot{q}_j}{T_{w,j} - T_s} \tag{3.2}$$

For refrigerant-oil mixtures, however, this equation does not apply. The addition of lubricant to the refrigerant leads to a change in boiling temperature, thus requiring the bubble point temperature (T_b) as function of local oil mass fraction (w_{oil}) as effective liquid temperature instead (Thome, 1995).

$$\alpha_j = \frac{\dot{q}_j}{T_{w,j} - T_b(w_{oil})} \tag{3.3}$$

The local heat flux ($\dot{q}_j$) results from electrical heating power ($P_{el,j}$) corrected by reactive power in the connection lines and cold ends of the heating conductors ($P_{loss,j}$) and heat loss to the surroundings ($\dot{Q}_{loss,j}$), with reference to the surface area of segment j.

$$\dot{q}_j = \frac{P_{el,j} - P_{loss,j} - \dot{Q}_{loss}}{\frac{1}{6}A_i} \tag{3.4}$$

Reference area for heat transfer (A_i) of the smooth tube equals the geometrical inner surface along the heated length (= 200 mm for both measurement sections). Concerning definitions for micro-fin tubes by authors in literature, reference surface areas are defined differently with respect to inner diameter. In this work, experimental heat transfer coefficients and thus heat flux are referred to the lateral surface of a cylindrical tube with diameter that equals the core diameter ($d_i = 8.62$ mm, maximum inner diameter) of the micro-fin tube.
Wall temperatures ($T_{w,j}$) in Eq. 3.3 refer to the inner tube surface, calculated by accounting for heat conduction through the tube wall.

$$T_{w,j} = T_{TC,j} - \frac{6\dot{Q}_j \ln\left(\frac{d_{TC}}{d_i}\right)}{2\pi L_{heat}\,\lambda_{MS}(T_{MS})} = T_{TC,j} - \frac{\dot{q}_j d_i \ln\left(\frac{d_{TC}}{d_i}\right)}{2\lambda_{MS}(\overline{T}_{MS})} \tag{3.5}$$

where $T_{TC,j}$ is the temperature measured by the respective thermocouple in segment j, d_{TC} the diameter with respect to the thermocouple position, d_i the inner (core) diameter of the

tube and λ_{MS} the heat conductivity of the tube material (see Eq. D.29 or D.30 in App. D.3 with $\overline{T}_{MS} = \overline{T}_{Ni/Cu} = \frac{1}{2}(T_{TC,j} + T_{w,j})$).
The effective heating power $(P_{el,j} - P_{loss,j})$ is determined from manufacturer specification on resistance and length of the hot part of the heating conductors and the overall measured circuit resistance, see Appendix. Heat loss results from conduction through the tube insulation. For a precise determination along the measurement section, heat loss is calculated from measured temperatures at the inside ($T_{ins,i}$) and outside ($T_{ins,o}$) surface of the insulation.

$$\dot{Q}_{loss} = \frac{1}{6}\left(\frac{2\pi L_{heat}\lambda_{ins}(\overline{T}_{ins})}{\ln\left(\frac{d_{ins,o}}{d_{ins,i}}\right)}(T_{ins,i} - T_{ins,o})\right) \tag{3.6}$$

where $L_{heat}(= 200\text{ mm})$ is the heated length, $d_{ins,i/o}$ the inner/outer diameter and λ_{ins} the heat conductivity of the insulation (see Eq. D.27 in the Appendix).
Local heat transfer coefficients are determined by Eq. 3.3, calculating heat flux first (Eq. 3.4 and 3.6) and then the respective inner wall temperature by iteration (Eq. 3.5 and D.29 or D.30). Circumference-averaged heat transfer coefficients at the measurement position result from averaging all six local heat transfer coefficients, according to Eq. 3.7.

$$\overline{\alpha} = \frac{1}{6}\sum_{j=1}^{6}\alpha_j \tag{3.7}$$

3.5.2 Vapor quality

The thermodynamic vapor quality ($\dot{x}$) is defined as the ratio of vapor ($\dot{M}_V$) to total ($\dot{M} = \dot{M}_G + \dot{M}_L$) mass flow rate. While the vapor phase was assumed as pure CO_2 ($\dot{M}_V = \dot{M}_{CO_2,V}$), the liquid phase consisted of liquid CO_2 and oil ($\dot{M}_L = \dot{M}_{CO_2,L} + \dot{M}_{oil}$). It can be determined from energy balance around the test section, i.e. from the point of sub-cooled liquid in front of the regulation (= throttling) valve to the measurement position in the measurement section, to be more precisely.

$$\dot{x} \equiv \frac{\dot{M}_V}{\dot{M}} = \frac{\dot{M}_V}{\dot{M}_V + \dot{M}_L} \tag{3.8}$$

Anticipating Sec. 3.5.3, an increase in vapor quality automatically leads to an accumulation of lubricant in the liquid phase as the non-volatile lubricant remains in the liquid phase. As result, the local oil mass fraction and the bubble point temperature of the mixture changes.

As stated by Thome (1995) for refrigerant-oil mixtures, sensible heat to vapor and liquid phases contributes to a change in enthalpy (dh) and, thus, a change in vapor quality ($d\dot{x}$).

$$dh = \Delta h_v \, d\dot{x} + (\dot{x} c_{p,V} + (1 - \dot{x}) \, c_{p,L}) \, dT_b = \frac{d\dot{Q}}{\dot{M}} \tag{3.9}$$

where Δh_v is the specific heat of vaporization, $c_{p,V/L}$ is the specific vapor/liquid isobaric heat capacity, dT_b is the change in bubble point temperature, $d\dot{Q}$ is the heat input and $\dot{M}$ is the mass flow rate.

Determining the total enthalpy change (Δh) from sub-cooled inlet up to the position of measurement, sensible heat to the sub-cooled liquid for reaching the bubble point had to be taken into account.

$$\Delta h = h'(T_{b,\dot{x}=0}) - h_L(T_{TR}, p_{TR}) + \sum_{d\dot{x}}^{\dot{x}} dh \tag{3.10}$$

where $h'(T_{b,\dot{x}=0})$ is the specific enthalpy on the saturation line at inlet bubble point temperature ($T_{b,\dot{x}=0}$) and $h_L(T_{TR}, p_{TR})$ is the specific enthalpy of sub-cooled liquid in front of the regulation (= throttling, TR) valve.

As bubble point temperature and isobaric heat capacity are dependent on the local oil mass fraction, vapor qualities were not solved explicitly from Eq. 3.10. Instead, heat release curves ($\Delta h(\dot{x})$) were prepared as proposed by Thome (1995). Calculation proceeded by increasing the vapor quality until the total change in enthalpy just exceeded the respective heat input, i.e. $\left(\Delta h \dot{M}\right) > \sum d\dot{Q} \, (= \dot{Q})$. Step size for change in vapor quality was chosen to $d\dot{x} = 10^{-4}$, hence yielding the accuracy of vapor quality calculation.

Total heat input was determined from measured powers applied to pre-heaters, measurement section and guard heater with respect to the reference position.

$$\dot{Q} = \sum (P_{el} - P_{loss} - \dot{Q}_{loss}) \tag{3.11}$$

where P_{el} is the electrical power measured by power meter, P_{loss} the reactive power in the connection lines and cold ends of the heating conductor and $\dot{Q}_{loss}$ the heat loss for each tube section passed downstream up to the reference position of vapor quality determination.

For practical reasons, heat loss along each tube section was determined from averaged heat transfer coefficients ($\bar{k}_i$) with respect to the inner surface (A_i). These combined $\bar{k}_i A_i$-values comprise heat loss at all existing thermal bridges along the respective tube section.

$$\dot{Q}_{loss} = \bar{k}_i A_i \, (T_{ins,i} - T_{ins,o}) \tag{3.12}$$

where $T_{ins,i/o}$ are mean temperatures at the inside and outside of the insulation.

3.5.3 Oil mass fraction

The inlet oil mass fraction is defined as mass fraction of oil in the liquid bulk (i.e. for $\dot{x} \leq 0$) and corresponds to the ratio of oil circulation rate ($\dot{M}_{oil}$) to total (liquid) flow rate ($\dot{M}$).

$$w_{oil,in} = \frac{\dot{M}_{oil}}{\dot{M}} \tag{3.13}$$

While CO_2 evaporates, non-volatile lubricant accumulates in the liquid phase. Hence, the local oil mass fraction is a function of vapor quality as well. With $\dot{M}_L = \dot{M}\,(1 - \dot{x})$ following Eq. 3.8 and $\dot{M}_{oil} = \dot{M}\, w_{oil,in}$ from Eq. 3.13, an expression for the local oil mass fraction (w_{oil}) in terms of local vapor quality ($\dot{x}$) and inlet oil mass fraction ($w_{oil,in}$) results.

$$w_{oil} = \frac{w_{oil,in}}{1 - \dot{x}} \tag{3.14}$$

3.5.4 Mass flow velocity

The total mass flow velocity ($\dot{m}$) is determined from the measured mass flow rate of subcooled flow (= total mass flow rate, $\dot{M}$) at the inlet with respect to the cross-sectional area of flow (A_{flow}) in the respective measurement section.

$$\dot{m} = \frac{\dot{M}}{A_{flow}} \tag{3.15}$$

The smooth tube cross-sectional area equals the geometrical cross-section cylinder, i.e. $A_{flow,smooth} = \frac{\pi}{4}\, d_i^2 = 153.9\ \mathrm{mm}^2$. The cross sectional area of the micro-fin tube was estimated by Schael (2009) by means of microscopic surface imagery and validated by experiment. The value given, equaling $A_{flow,micro-fin} = 55.3\ \mathrm{mm}^2$, was used for the mass velocity calculation by Eq. 3.15.

4 Smooth tube results

In this chapter results from flow boiling experiments inside the 14.0 mm smooth tube test section are presented. The experimental investigation included flow pattern observations, pressure drop and heat transfer measurements. Measurements were carried out under adiabatic and pseudo-diabatic boundary conditions with respect to the test section. In the case of pseudo-diabatic measurements, the power input to the test section for heating was adjusted to yield a near-isothermal wall condition. The sight glass for the visualization of flow patterns preceding the guard heater of the test section, however, could not be heated at any time. Thus, when referring to pseudo-diabatic measurements in the following, the diabatic condition implies to the tubing up to the entry of the sight glass only.
Pressure drop and heat transfer experiments inside the smooth tube were primarily carried out for the sake of reproducibility (refer to App. B) and gaining basic information about the effect of lubricant on flow boiling pressure drop of CO_2. The experimental data base compared to micro-fin tube experiments (see Sec. 5.2) is comparatively small and focused on experiments at 26.4 bar.

4.1 Flow patterns

The investigation of flow patterns during flow boiling experiments is related to the strong influence of phase distribution on heat transfer and pressure drop. In addition to system pressure, flow velocity, vapor fraction and heat flux to the evaporating liquid, the local oil mass fraction severely impacts on flow patterns of refrigerant-lubricant mixtures. The presence of oil effects fluid properties, facilitates the formation of foam and can promote the wetting of the tube. Heat transfer and pressure drop measurements must be connected to the present flow pattern in order to understand and construe the results. In the first place, a reasonable and systematic definition of the observed flow patterns is essential for comparisons to predictive methods and to facilitate unambiguous communication with other researchers.

4.1.1 Flow pattern classification of CO_2-POE oil mixtures

Classifying flow patterns of CO_2-POE oil mixtures in a horizontal evaporator tube, five main categories were identified and defined as listed below. Visual observations in form of high-speed video recordings are illustrated in Fig. 4.1.

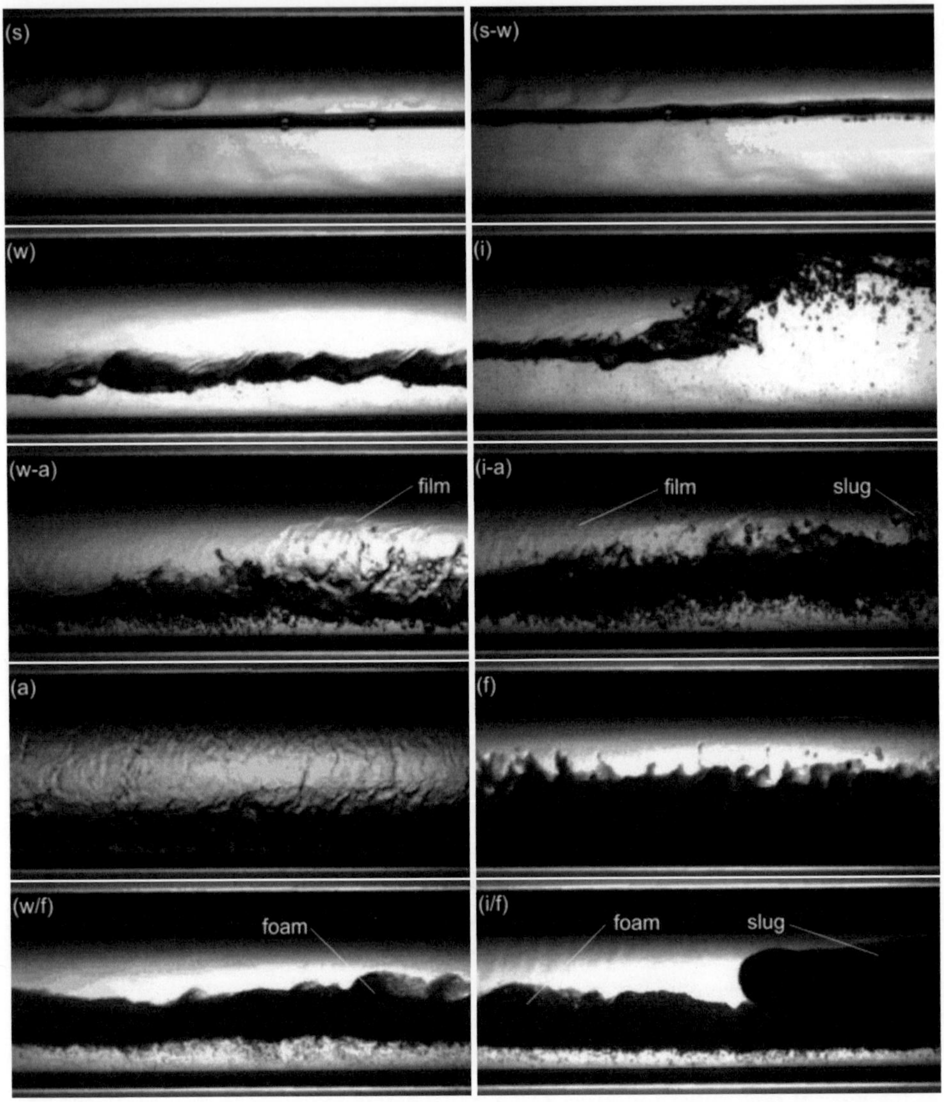

Figure 4.1: Observed flow patterns of CO_2-POE oil mixtures containing 3 wt.-% lubricant; (s) stratified, (s-w) stratified-wavy, (w) wavy, (i) intermittent (slug/plug), (w-a) wavy-annular, (i-a) intermittent-annular, (a) annular and (f) foam flow

- A clear horizontal, smooth separation of vapor (top) and liquid (below) phase is designated as ***stratified (s)*** or ***stratified-wavy flow (s-w)***. The latter denotes the transition pattern with first small waves forming at the surface. Both flow patterns feature a permanently liquid-wetted bottom of the tube while the the crest is dry.
- ***Wavy flow (w)*** also consists of a horizontal phase separation with the crest of the tube being unwetted. However, rolling waves at the gas-liquid interface traveling at higher velocity than the core liquid phase lead to a steady fluctuation of local liquid level and thus alternating grade of tube wetting.
- ***Intermittent flow (i)*** comprises slug and plug flow, as both flow pattern are characterized by alternation of partial to total tube wetting. With increasing vapor quality, slugs merge into annular flow.
- Flow regimes containing an annular liquid film are subdivided into three categories. ***Annular flow (a)*** implies a fully developed, stable liquid film around the whole tube perimeter. The film thickness diminishes towards the crest of the tube due to gravity. ***Wavy-annular (w-a)*** and ***intermittent-annular flow (i-a)*** designate the respective transition regimes from wavy or intermittent to annular flow, respectively.
- If foaming occurs intensively and the flow regime is hardly to be distinguished due to heavy foam formation, the pattern is classified as ***foam or froth flow (f)***. It is characterized by complete tube wetting and foam filling the tube, thus blocking the passage for gas flow and leading to an increase in pressure drop.

Besides the above mentioned main categories, flow patterns (except for foam flow) are additionally marked as ***foaming (/f)*** in case of significant foam formation being observed. The last two images of Fig. 4.1 give an example. Foaming occurs due to the evaporation of refrigerant in refrigerant-lubricant mixtures, and it is intensified by increasing the local heat input to the liquid (e.g., inside the measurement section during diabatic flow boiling experiments). The lubricant concentration enriches at the gas-liquid interface of bubbles formed, leading to a stabilization of the enclosing liquid film. The formation of a foam layer may have different consequences, such as increasing the grade of liquid tube wetting, effecting the shear stress and inducing a further flow resistance.

4.1.2 Comparison of experimental flow patterns to flow pattern maps

The first aspect that needs to be discussed here - prior to actually comparing experimentally determined flow patterns to flow patterns maps available in literature - are the transition regions from wavy to intermittent (slug/plug) or annular flow and intermittent to annular flow, i.e., from partial to full tube wetting (un)steady state of flow.

The flow pattern map of Steiner is the most cited one and up-to-date general flow pattern maps available. As numerous authors stated, however, it lacks accuracy in predicting the technical relevant wavy to slug/plug or annular flow transition. Thus, most effort as been spent in developing or modifying prediction methods for this transition region, e.g., by Kattan et al. (1998a) in simple manner. It was found out that common flow pattern maps and modifications generally are valid in limited ranges of operation (e.g., certain refrigerants, pressures, mass flow velocities, tube geometries). However, none of the available flow pattern maps could predict the flow patterns observed in the experiments carried out in this work satisfactorily, e.g., see graphs in App. F.

A more general modification of the Steiner transition curve was developed in the course of this work (Saito et al., 2011), see Eq. 2.94 in Sec. 2.4 and Eq. E.30-E.33 in App. E.1. Following Taitel and Dukler (1976), a multiplier C_1 accounts for the liquid film height that a wave has to grow in order to reach the transition state to intermittent/annular flow. Further, the relative velocity of vapor to liquid phase was regarded as transition criterion, i.e., the effective velocity of vapor over the surface of the liquid film. The respective velocity of the liquid film at the liquid-vapor interface is estimated by a second multiplier C_2.

At conditions of low liquid level or low vapor qualities, respectively, the precise calculation of the vapor fraction is crucial when regarding the relative phase velocity. Even slight deviations promptly lead to unsound negative values, i.e., lower velocity of vapor than the liquid. The author of this work therefore proposes to apply C_1 according to Saito et al., but to modify C_2 (see Eq. 4.1). The trends of C_1 and C_2 as functions of the height of liquid are illustrated in Fig. 4.2.

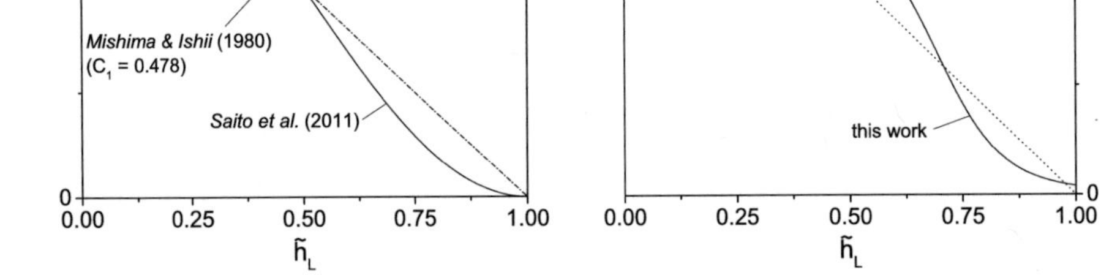

Figure 4.2: Wavy to intermittent or annular flow transition curve parameters C_1 and C_2 as functions of the dimensionless liquid height ($\tilde{h}_L = h_L/d$); literature citations from Saito et al. (2011)

$$C_2 = 2 - \sum_{i=1}^{2} \frac{1}{1 + \exp\left(m_i \cdot \left(\tilde{h}_{\mathrm{L}} - c_i\right)\right)} \tag{4.1}$$

where $\tilde{h}_{\mathrm{L}} = h_{\mathrm{L}}/d$ is dimensionless liquid height, d is the tube diameter and $m_{i=1,2}$ and $c_{i=1,2}$ are empirical constants (given in Tab. 4.1).

Table 4.1: Empirical parameters in Eq. 4.1 for calculating C_2

$c_1 = 0.3$	$m_1 = -5$	$c_2 = 0.7$	$m_2 = -15$

The transition from intermittent to annular flow is generally described by means of a transition mean vapor ($\mathrm{Fr}^{0.5}_{\mathrm{Vm,I\text{-}A}}$) or liquid ($\mathrm{Fr}_{\mathrm{L,I\text{-}A}}$) Froude number as function of Martinelli parameter (X_{tt}). The first approach provides the possibility of implementation into the dimensionless flow pattern map by Steiner. Schael presented an expression for $\mathrm{Fr}^{0.5}_{\mathrm{Vm,I\text{-}A}}$ based on CO_2 measurements (see Fig. 2.4 (left) in Sec. 2.1). Using the same correlation procedure, Eq. 4.2 was developed from adiabatic CO_2-oil mixture data gained in this work. Figure 4.3 (left) depicts that the observed transition regions for CO_2 (dashed line) and CO_2-POE oil mixtures (solid line) under adiabatic conditions hardly differ from each other.

$$\mathrm{Fr}^{0.5}_{\mathrm{Vm,I\text{-}A}} = 0.8 X_{\mathrm{tt}}^{-0.4} \tag{4.2}$$

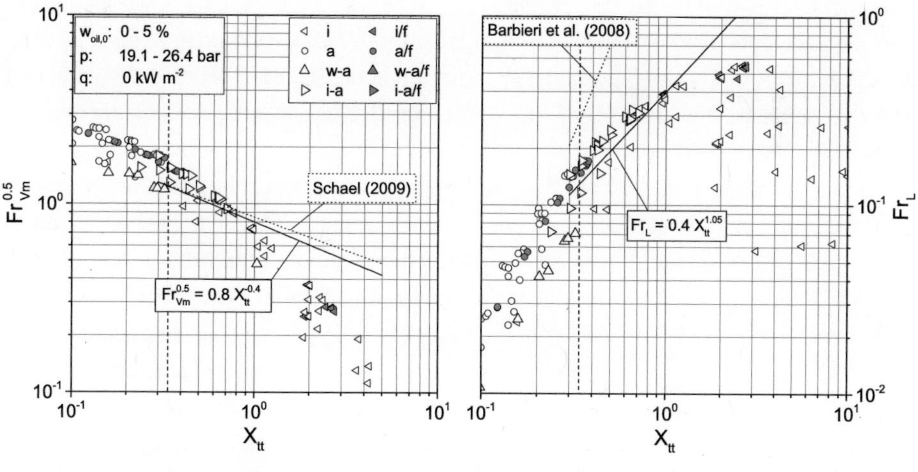

Figure 4.3: Transition from partial to complete tube wetting in the smooth tube in terms of $\mathrm{Fr}^{0.5}_{\mathrm{Vm}}$ (left) and Fr_{L} (right) over X_{tt}; Symbols represent data points from this work and solid lines predicted flow transitions

A comparison of the experimental data to the transition criterion in terms of $Fr_{L,I\text{-}A} = f(X_{tt})$ is shown in Fig. 4.3 (right). The dashed line represents the transition curve developed by Barbieri et al. (2008) based on data from flow boiling experiments with R134a. It differs notably from the new transition curve (solid line) derived from the CO_2-lubricant data mentioned above. A new transition equation for CO_2-lubricant flow is proposed.

$$Fr_{L,I\text{-}A} = 0.4 X_{tt}^{1.05} \tag{4.3}$$

Agreeing well experimental data, Eq. 4.2 is recommended for estimating the transition from wavy or intermittent to fully developed annular flow of adiabatic two-phase flow of CO_2-POE oil mixtures inside smooth tubes. The range of operation ensured by experimental data gained within this work includes $w_{oil,0} \leq 5$ wt.-%, 19.1 bar $\leq p \leq$ 26.4 bar and $X_{tt} < 0.8$.
Flow patterns for CO_2 under adiabatic conditions are presented in Fig. 4.4 (top) at pressures of 19.1 bar (left) and 26.4 bar (right). Observations from flow boiling experiments carried out by Saito et al. within the scope of this research project are represented by symbols. Solid lines indicate transition curves by the modified Steiner flow pattern map. The beforehand introduced wavy to intermittent/annular and intermittent to annular transition conditions developed in this work are shown as dot-dashed lines.
Within the investigated range shown, the modified Steiner flow pattern map describes the observed flow patterns precisely. The first modified wavy to intermittent/annular flow transition curve (solid line, Saito et al. (2011)) yields a slightly better agreement to the experimental data than the newly developed curve (dot-dashed line). However, the transition criterion developed here assures better applicability to higher pressures (e.g., at 39.7 bar, see Fig. F.1 in App. F.1), i.e., when the vapor to liquid density ratio is reduced significantly. Likewise, adiabatic flow patterns of CO_2-POE oil mixtures are well predicted by means of the modified flow pattern map, see Fig. 4.4 (middle). Foaming is observed progressively as the oil concentration increases, but flow patterns in general match the predicted regimes.
In the case of heat transfer measurements, i.e., with increasing heat flux, the predictability of flow patterns observed diminishes. Strong foam formation and the development of complete foam flow are the consequence of the local heat input. The more the local heat input is increased, the more significant the foam formation becomes. Figure 4.4 (bottom) indicates the advancing foam formation and deviation to flow pattern prediction at a low (5 $kW\,m^{-2}$) and moderate (15 $kW\,m^{-2}$) heat flux. For visual demonstration, recorded flow patterns for pure CO_2 (top images) and a CO_2-POE oil mixture containing 3 wt.-% of oil (bottom images) are shown in Fig. 4.5. In all cases, pressure (26.4 bar), mass flow rate (150 $kg\,m^{-2}\,s^{-1}$) and vapor quality (50 %) are equal. From left to right, the average heat flux inside the measurement section is increased from 0 $kW\,m^{-2}$ (adiabatic condition) to 15 $kW\,m^{-2}$. While there is no change in flow pattern noticeable for pure CO_2, strong foam formation is clearly visible in the case of the CO_2-oil mixture.

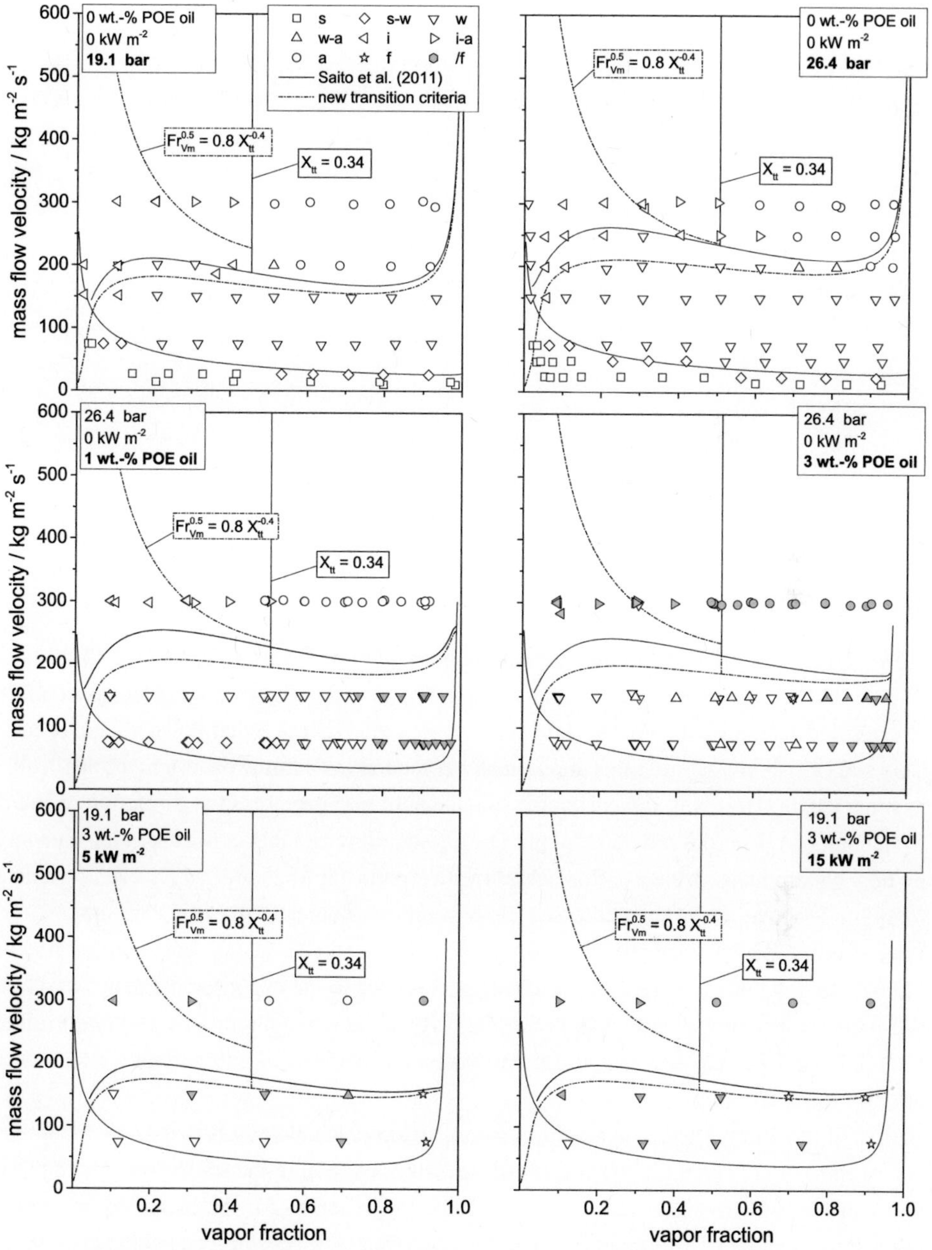

Figure 4.4: Adiabatic (top and middle) and diabatic (bottom) flow patterns, modified Steiner flow pattern map by Saito et al. (2011) and new transition curves for pure CO_2 (top) and CO_2-POE oil mixtures containing 1 – 3 wt.-% lubricant (middle and bottom)

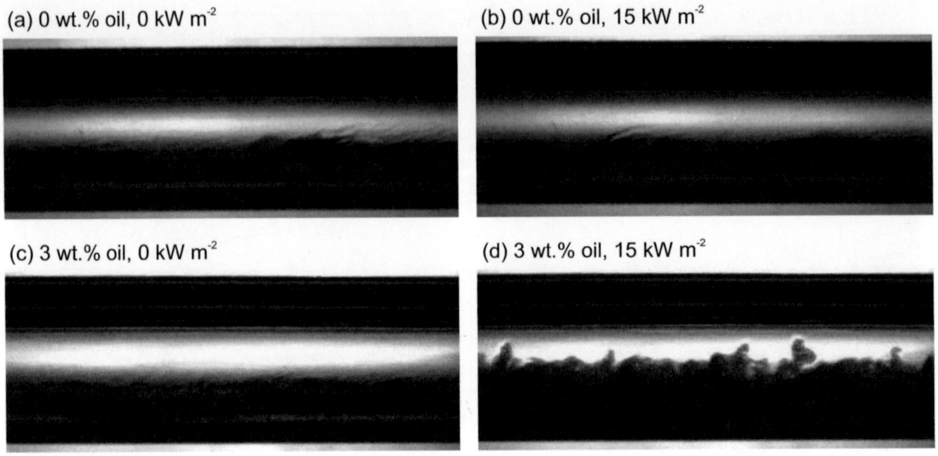

Figure 4.5: Recorded flow patterns at 26.4 bar, 150 $\mathrm{kg\,m^{-2}\,s^{-1}}$ and 50 % in vapor quality for pure CO_2 (top) and a CO_2-oil mixture of 3 wt.-% (bottom) under adiabatic (left) and diabatic (right) conditions

4.1.3 Conclusion

Flow patterns of CO_2-POE oil mixtures of nominal oil mass fractions up to 5 wt.-% inside a smooth tube were investigated, illustrating the effect of oil in a conventional and profoundly investigated evaporator geometry. Adiabatic experiments revealed that flow patterns of pure CO_2 and CO_2-lubricant mixtures are basically comparable and flow pattern maps applicable to some extent. However, increasing the oil concentration promotes the formation of foam, a phenomenon that does not occur in pure refrigerant flow and that is not taken into account by flow pattern maps. Hence, a flow pattern with significant foam present (or even complete foaming flow) cannot not be addressed in conventional flow pattern maps correctly.

The available flow pattern data was compared to the well established flow pattern map of Steiner, among others. A general promising accordance to the experimental data was assessed. Greater deviations were observed at the wavy to intermittent/annular and intermittent to annular transitions. In parts, these deviations and various modifications were already addressed in literature.

In case of the intermittent to annular transition, an approach already focused on by Barbieri et al. (2008) and Schael (2009) was used. Instead of expecting the transition to take place at a constant Martinelli parameter, a new transition criterion for the mean vapor Froude number as function of Martinelli parameter was derived. The resulting equation is similar to the one developed previously by Schael for pure CO_2.

Describing the transition from wavy to intermittent/annular flow, the transition criterion developed by Steiner was modified. This approach was introduced by Saito et al. (2011). It implies the occurrence of two separate effects, depending on the state of flow: the Kelvin-

Helmholtz instability due to the Bernoulli effect and the situation of disturbance wave for shallow liquid films (i.e., for low liquid level in the tube). The first approach mentioned is incorporated in commonly used flow pattern maps, such as the maps by Taitel and Dukler and Steiner. The second effect was introduced for special treatment of a small liquid film, for which a laminar velocity profile is assumed. Additionally, the effect of liquid level was taken into account, similar to Taitel and Dukler. As the liquid level increases and the vapor-liquid interface approaches the crest of the tube, the disturbance necessary to induce intermittent flow reduces. Regarding both aspects, the transition criterion of Steiner was modified to describe the flow patterns of CO_2-POE oil mixtures observed with good agreement.

4.2 Pressure drop

Pressure drop during flow boiling experiments was measured along the length of the measurement section. In general, pressure drop results from frictional forces acting on the two-phase flow and the acceleration of the evaporating liquid. In order to determine the frictional pressure drop directly from experiments, adiabatic experiments (no evaporation) were carried out. For identifying the pressure drop contribution by acceleration, the total pressure drop during diabatic heat transfer experiments was investigated.

4.2.1 Frictional pressure drop

For determining the frictional two-phase pressure drop, experiments were carried out under near adiabatic conditions. These conditions include that no heat input was applied to neither the last pre-heater (VV3), the measurement section nor the guard heater between measurement section and sight glass tube. Hence, measurements from adiabatic flow pattern investigation and from heat transfer experiments (last measurement point of each series) were used in determining the frictional pressure drop.
The effect of lubricating oil mass fraction on the frictional pressure drop of CO_2 is demonstrated in the left image of Fig. 4.6 for a pressure of 26.4 bar and a mass flow velocity of 300 $kg\,m^{-2}\,s^{-1}$. Frictional pressure drops of CO_2-POE oil mixtures containing 1 and 3 wt.-% lubricant are shown as light and dark gray symbols, respectively. As reference, open circular symbols represent measurement data of Schael (2009) for pure CO_2. At vapor qualities below 0.7 the effect of lubricant is almost negligible and within the range of uncertainty under the given conditions. Changes in local oil mass fraction and liquid properties are comparatively small, e.g., for a nominal oil mass fraction of 3 wt.-% the local oil mass fraction rises to 10 wt.-% and the liquid viscosity increases by 84 %. Despite the considerable increase in liquid viscosity, the pressure drop is almost unaffected by the presence of lubricant. This fact is related to the diminishing influence of the liquid phase on the two-phase pressure drop with vapor quality. Further increasing the vapor quality

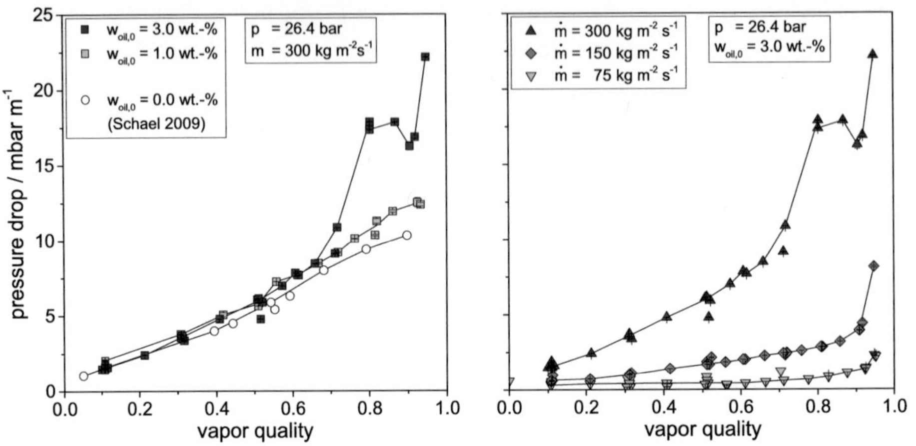

Figure 4.6: Frictional pressure drops in the 14.0 mm smooth tube at 26.4 bar for a mass flow velocity of 300 $\mathrm{kg\,m^{-2}\,s^{-1}}$ (left) and an oil mass fraction of 3 wt.-% (right) as function vapor quality

to 0.9 results in a local oil mass fraction of 30 wt.-% and an increase in liquid viscosity by approximately 600 %. Accompanied by an increased tendency to foam formation, the frictional pressure drop was found to increase at vapor qualities > 0.7 with respect to pure CO_2 measurement data.

The right image of Fig. 4.6 represents the effect of mass flow velocity for a constant oil mass fraction of 3 wt.-% and a pressure of 26.4 bar. As expected and previously found for pure CO_2 by Schael, frictional pressure drop increases with mass flow velocity. In addition, the steep increase in pressure drop at high vapor qualities (>0.7) as discussed before can be seen for all mass flow velocities. Such behavior was observed for CO_2-lubricant mixtures only. Frictional pressure drop of pure CO_2 yields a maximum in the range of $0.85 < \dot{x} < 0.97$, further decreasing to the pressure drop of pure vapor phase. Regarding CO_2-lubricant mixtures, there is always a (lubricant-rich) liquid film present as the lubricant is not evaporated.

Figure 4.7 further demonstrates the effect of mass flow velocity on frictional pressure drop at different vapor qualities. As for any refrigerant, frictional pressure drop increases approximately by $\dot{m}^2$.

Experimentally determined frictional pressure drops were compared to several contemporary and established frictional pressure drop correlations in literature. Table 4.2 summarizes the error statistics. Best agreement to measurements was found for the correlation of Xu and Fang (2012) with 73.7 % of all data points predicted with APE < 30 % and MAPE $= 24.5$ %, closely followed by the correlations of Müller-Steinhagen and Heck (1986) and Storek and Brauer (1980). Parity plots for the latter correlations are shown in Fig. 4.8. With increasing pressure drop, values are increasingly under-estimated. The correlation of Xu and Fang was

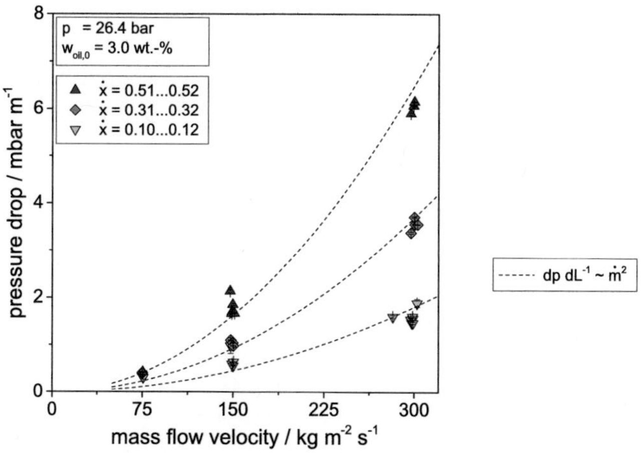

Figure 4.7: Frictional pressure drops in the smooth tube as function of mass flow velocity and vapor quality at a pressure of 26.4 bar and a constant nominal oil mass fraction of 3.0 wt.-%

the most recent prediction method investigated. Due to its close relation to the correlation of Müller-Steinhagen and Heck, the error statistics determined for both correlations are very similar, revealing no significant improvement. All other correlations investigated do not yield acceptable results, partially resulting in MAPE > 100 %. Most correlations over-predicted the frictional pressure drop.
Parity plots of all correlations are given in App. F, Sec. F.1.2.

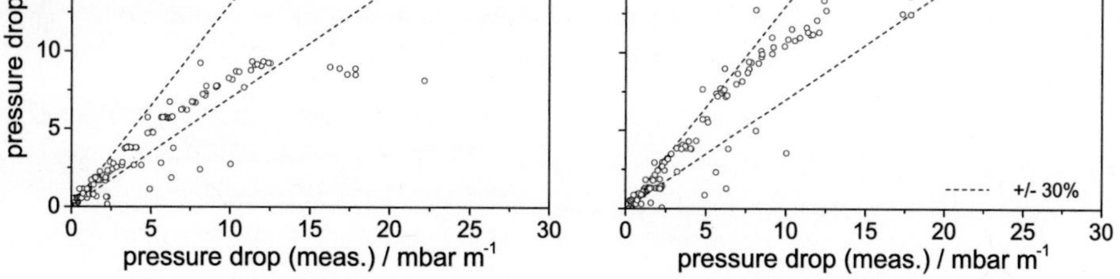

Figure 4.8: Comparison of predicted to measured frictional pressure drops at 26.4 bar for CO_2-POE oil mixtures containing nominal oil mass fractions of 0 – 3 wt.-%

Table 4.2: Error statistics of evaluated pressure drop correlations with respect to adiabatic frictional pressure drops measured in the smooth tube

Correlation by author	MPE	MAPE	Data that yields APE $\leq$ 30 %
Xu and Fang (2012)	−0.3 %	24.5 %	73.7 %
Hu et al. (2008c)	+176.1 %	179.6 %	10.5 %
Katsuta et al. (2008)	+144.7 %	150.6 %	12.8 %
Cheng et al. (2008a)	+47.0 %	60.9 %	48.1 %
Moreno Quibén and Thome (2007b)	−3.3 %	29.6 %	61.7 %
Yoon et al. (2004)	+29.9 %	47.9 %	31.6 %
Müller-Steinhagen and Heck (1986)	−4.3 %	25.4 %	73.7 %
Storek and Brauer (1980)	+3.5 %	25.1 %	71.4 %
Friedel (1979)	+34.1 %	46.7 %	49.6 %
Chisholm (1973)	+60.3 %	79.6 %	24.8 %
Grönnerud (1972)	+12.5 %	32.9 %	51.1 %
Lockhart and Martinelli (1949)	+136.3 %	150.8 %	14.3 %

4.2.2 Total pressure drop

As is known, the total pressure drop of a refrigerant two-phase flow is described by linear contributions due to friction and acceleration of the evaporating liquid to the vapor velocity. Depending on the underlain friction pressure drop model, a homogeneous or heterogeneous acceleration pressure drop model is used (see App. E, Sec. E.2.2). Pressure drops of CO_2-POE oil mixtures containing a nominal oil mass fraction of 3 wt.-% measured at 19.1 bar and 150 $\mathrm{kg\,m^{-2}\,s^{-1}}$ for different inlet vapor qualities and varying heat fluxes applied to the measurement section are presented in Fig. 4.9 by symbols. Solid and dashed lines represent the acceleration pressure drop contribution predicted by either homogeneous or

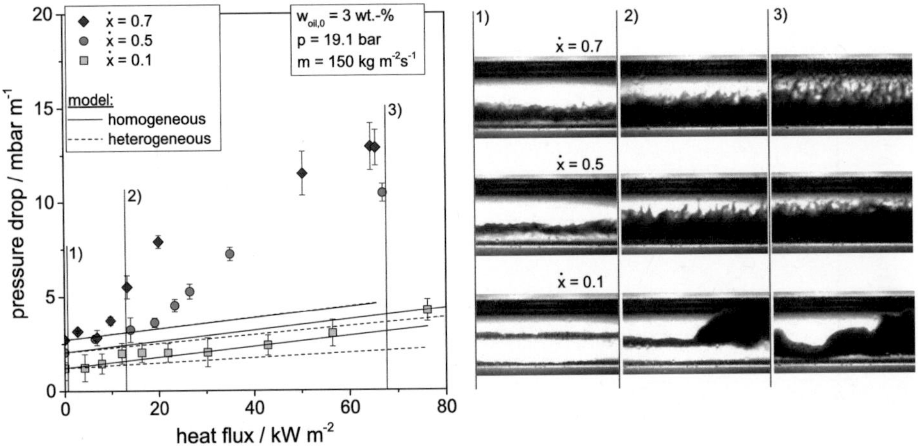

Figure 4.9: Total pressure drop as function of heat flux for $w_{oil,0} = 3.0$ wt.-%, $p = 19.1$ bar, $\dot{m} = 150\ \mathrm{kg\,m^{-2}\,s^{-1}}$ and $\dot{x} = 0.1 - 0.7$ (left) and the respective flow patterns observed (right)

heterogeneous model, respectively, normalized to the frictional pressure drop measured under adiabatic conditions. On the right, images depict the flow patterns observed.
With increasing heat flux the difference of measured to predicted pressure drop increases, especially with increasing vapor quality. Even for pure CO_2 Schael (2009) pointed out that both models under-predict the acceleration pressure drop contribution. He concluded that the increase in total pressure drop is more pronounced at low heat fluxes ($\leq 10\,\mathrm{kW\,m^{-2}}$), declining towards higher heat fluxes. This assessment is contrary to observations made for CO_2-POE oil mixtures. Here, a steep increase in pressure drop was observed at moderate heat fluxes ($10\,\mathrm{kW\,m^{-2}} < \dot{q} < 40\,\mathrm{kW\,m^{-2}}$). This behavior is probably related to a change in flow pattern, as flow pattern visualizations on the right side of Fig. E.2.2 suggest. The evaporation of liquid increases with heat flux, promoting the formation of foam since the non-volatile lubricant remains in the liquid film, stabilizing and retaining vapor bubbles. As the local oil mass fraction increases with vapor quality, further improving the tendency of foaming.
The error statistics of combined frictional and acceleration pressure drop prediction to measured total pressure drops is presented in Tab. 4.3. As expected from the findings discussed above, total pressure drops are significantly under-predicted by almost all correlations. Correlations that resulted in an overall over-prediction of frictional pressure drop profit from the under-prediction of the acceleration pressure drop contribution. However, no correlation was found suitable to adequately predict the total pressure drop of CO_2-POE oil mixtures.
Parity plots of the general well suitable correlations of Müller-Steinhagen and Heck (1986) and Storek and Brauer (1980) are displayed in Fig. 4.10. In contrast to the frictional pressure drop (see Fig. 4.8), a comparison to experimentally determined total pressure drops reveals a significant under-prediction, especially for higher pressure drop. Parity plots of all other correlations are given in App. F, Sec. F.1.2.

Table 4.3: Error statistics of evaluated pressure drop correlations with respect to total pressure drops measured in the smooth tube

Correlation by author	MPE	MAPE	Data that yields APE $\leq 30\,\%$
Xu and Fang (2012)	−39.7 %	45.7 %	23.3 %
Hu et al. (2008c)	+47.5 %	76.2 %	30.6 %
Katsuta et al. (2008)	+50.1 %	79.2 %	26.7 %
Cheng et al. (2008a)	−22.5 %	47.4 %	25.8 %
Moreno Quibén and Thome (2007b)	−41.5 %	48.8 %	20.2 %
Yoon et al. (2004)	−29.6 %	47.9 %	19.8 %
Müller-Steinhagen and Heck (1986)	−41.8 %	46.8 %	20.7 %
Storek and Brauer (1980)	−40.6 %	48.0 %	23.5 %
Friedel (1979)	−22.3 %	41.4 %	31.2 %
Chisholm (1973)	−2.8 %	48.8 %	29.9 %
Grönnerud (1972)	−36.3 %	48.6 %	20.8 %
Lockhart and Martinelli (1949)	+34.6 %	71.2 %	28.5 %

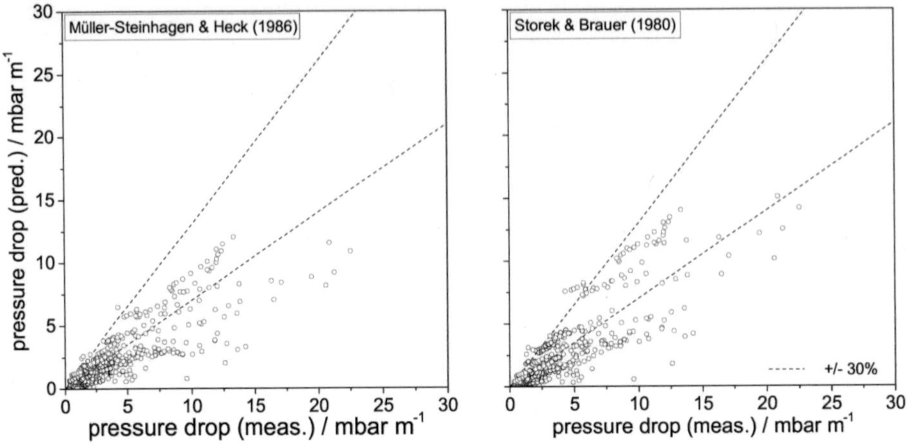

Figure 4.10: Comparison of predicted to measured total pressure drops at 26.4 bar for CO_2-POE oil mixtures containing nominal oil mass fractions of $0-3$ wt.-%

4.2.3 Conclusion

Frictional pressure drops measured during two-phase flow of CO_2-POE oil mixtures were similar in magnitude and behavior compared to those determined for pure CO_2 by Schael (2009). Due to a strong increase in local oil mass fraction and, hence, liquid viscosity accompanied by an increased tendency to foaming, frictional pressure drops measured at high vapor qualities ($\dot{x} > 0.7$) increased significantly for mixtures of higher nominal oil mass fraction. A comparison to frictional pressure drop models in literature revealed best agreement of predicted to experimental values using the correlations of Xu and Fang (2012), Müller-Steinhagen and Heck (1986) and Storek and Brauer (1980) to experimental data, yielding MAPE < 25.5 % and predicting > 71 % of all data with APE ≤ 30 %. Other correlations over-predicted the frictional pressure drop significantly.
In heat transfer experiments, total pressure drops were measured varying the heat flux applied to the measurement section. A strong influence of the applied heat flux was observed, especially at elevated vapor qualities. In the range of low to medium heat fluxes ($10\,\mathrm{kW\,m^{-2}} < \dot{q} < 40\,\mathrm{kW\,m^{-2}}$), the total pressure drop was found to leap up significantly. This behavior is contrary to observations previously made for pure CO_2 (Schael, 2009) and it was related to foaming, as demonstrated by flow pattern visualizations. Foam formation is favored by increasing the local oil mass fraction, i.e., by increasing the nominal oil mass fraction and vapor quality, as well as by increasing the rate of evaporation, i.e., the heat flux applied to the measurement section. Experimentally determined pressure drops were compared to predicted frictional pressure drops combined by either homogeneous or heterogeneous acceleration pressure drop contribution, depending on the underlain type of

frictional pressure drop model. The comparison illustrated that neither homogeneous nor heterogeneous model was suitable for predicting the acceleration pressure drop contribution due to the evaporating liquid, strongly under-predicting the change of pressure drop with heat flux. That fact lead to a general under-prediction of the combined total pressure drop for most correlations. Yielding MAPE $= 41.4$ % and predicting 31.2 % of the experimental data with APE ≤ 30 %, the correlation of Friedel (1979) was found most accurate of all correlations tested. It confirms that total pressure drops of CO_2-POE oil mixtures cannot be predicted with satisfying accuracy. As experimental investigations focused on identifying fundamental effects of lubricant on the flow boiling behavior of CO_2, reducing the experimental effort in favor of an enlarged data base for the micro-fin tube, no pressure drop model was derived.

4.3 Heat transfer

Circumference-averaged heat transfer coefficients were determined from averaging local heat transfer coefficients at six peripheral locations around the circumference of the tube under isothermal wall conditions. Following, the influence of operating conditions on the flow boiling heat transfer of the investigated CO_2-POE oil mixtures is discussed.

4.3.1 Influence of the heat flux

Flow boiling heat transfer is generally characterized by convective and nucleate boiling contributions, see Sec. 2.3. When convective boiling dominates, heat transfer is mainly affected by mass flow velocity and almost independent of heat flux. In the nucleate boiling dominated regime, the heat flux denotes the main influence on heat transfer as the bubble formation on the heated surface increases, improving the heat transfer. However, nucleate boiling requires a minimum heat flux (q_{onb}) for bubbles to be generated.

As shown experimentally by Schael (2009) for CO_2, nucleate boiling plays a dominant role over a large range of heat fluxes, especially at higher vapor pressures. Schael substantiated his observation estimating the onset of nucleate boiling by means of Eq. E.155, App. E.3.2. At 26.4 bar and a mass flow velocity of 300 $\text{kg}\,\text{m}^{-2}\,\text{s}^{-1}$, the onset of nucleate boiling was determined to $q_{\text{onb}} = 0.8\ \text{kW}\,\text{m}^{-2}$. With respect to the range of heat flux investigated, observing convective boiling dominated heat transfer conditions was not expected for the CO_2-POE oil mixture flow boiling experiments carried out at 26.4 bar in this work.

In Fig. 4.11, the influence of heat flux on the circumference-averaged heat transfer coefficient at a pressure of 26.4 bar at different mass flow rates and vapor qualities is illustrated. For pure CO_2 (open symbols), the heat transfer coefficient significantly increases with heat flux, indicating the dominance of nucleate boiling. These observations correspond to those reported by Schael. In general, this also applies to CO_2-POE oil mixtures (filled symbols) by quality. At low heat fluxes, heat transfer coefficients are normally in the range of those for pure CO_2 and show a comparable behavior as the heat flux gently rises. However, in the range of elevated heat fluxes, the boiling curve flattens, reaching a maximum in heat transfer at a certain limiting heat flux. Further increasing the heat flux leads to declination of the heat transfer coefficient while the wall temperature rapidly increases.

The location of the heat flux maximum is mainly dependent on nominal oil mass fraction and vapor quality, see bottom plots of Fig. 4.11 at a mass flow velocity of 150 $\mathrm{kg\,m^{-2}\,s^{-1}}$ and vapor qualities of 0.3 (left) and 0.7 (right). With progressing evaporation of the refrigerant the local oil concentration in close proximity to the heated wall surface of the tube rises. Subsequently, the mass flow resistance in the lubricant-rich liquid layer increases, hindering both bubbles in leaving and liquid refrigerant in reaching the heated surface. In addition, the bubble point temperature and the surface temperature rise.

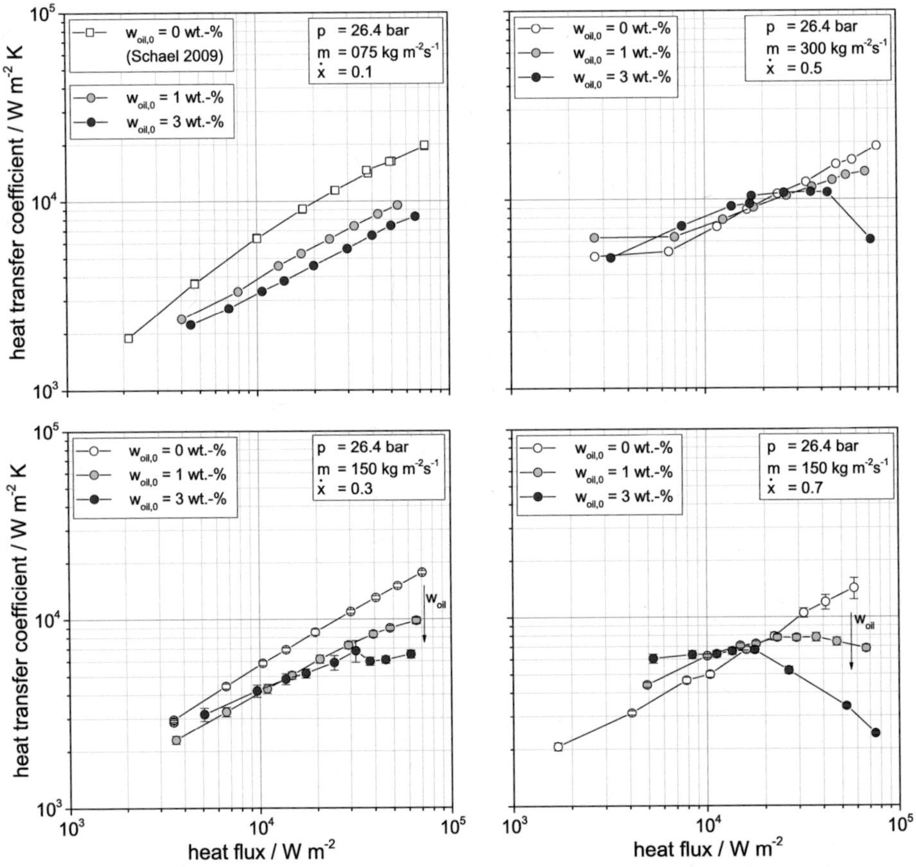

Figure 4.11: Influence of heat flux on the circumference-averaged heat transfer coefficient at a pressure of 26.4 bar for mass flow velocities of 75 – 300 $\mathrm{kg\,m^{-2}\,s^{-1}}$ and vapor qualities of 0.1 – 0.7

4.3.2 Influence of the mass flow velocity

Figure 4.12 portrays the influence of the mass flow velocity on the circumference-averaged heat transfer coefficient. In contrast to the obtained data for nominal oil mass fractions of 1 wt.-% and 3 wt.-% (light and dark gray symbols, respectively), experimental data from pure CO_2 measurements reported by Schael (2009) are presented (open symbols). From the measurement points for CO_2-POE oil experiments shown, an increase of the heat transfer coefficient with mass flow velocity becomes obvious, despite the scattering related to the influence of other operating parameters (e.g., vapor quality, flow pattern) that are not considered here.

Steiner (2002) described the dependency of heat transfer coefficient from mass flow velocity by power law (Eq. 4.4) that was confirmed by Schael for pure CO_2 measurements. In Fig. 4.12, this dependency is indicated by dashed and dotted lines for measurements yielding average heat fluxes of $6-8\ \mathrm{kW\,m^{-2}}$ and $48-52\ \mathrm{kW\,m^{-2}}$, respectively, roughly following the observed trend.

$$\alpha \propto \left(\frac{\dot{m}}{\dot{m}_0}\right)^{0.25} \tag{4.4}$$

where $\dot{m}_0 = 100\ \mathrm{kg\,m^{-2}\,s^{-1}}$ corresponds to the substance-specific standard mass flow velocity reported for CO_2.

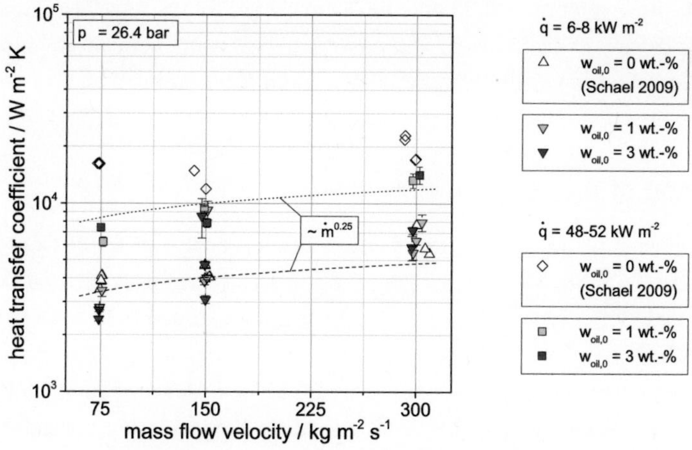

Figure 4.12: Influence of the mass flow velocity on the heat transfer at 26.4 bar

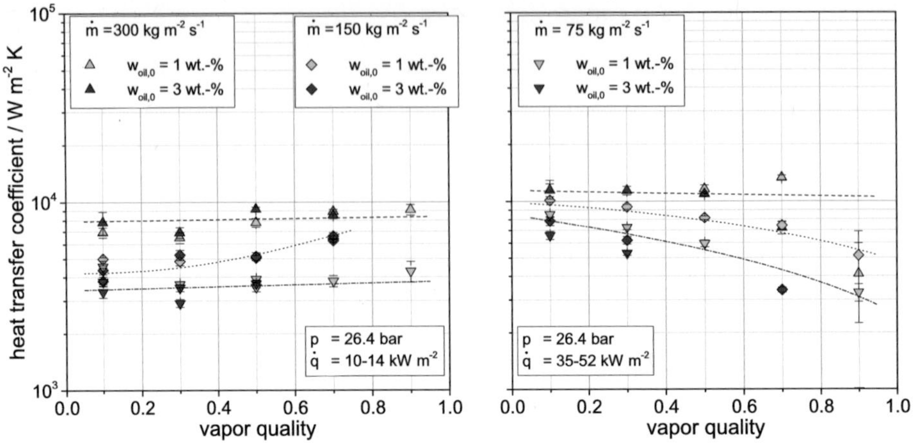

Figure 4.13: Influence of vapor quality on the circumference-averaged heat transfer coefficient at low (left) and high (right) heat fluxes

4.3.3 Influence of the vapor quality

The influence of the vapor quality on heat transfer is demonstrated in Fig. 4.13 in dependency of the mass flow velocity at low (left) and high (right) heat fluxes. Observed trends are represented by dashed lines (at $\dot{m} = 300\ \mathrm{kg\,m^{-2}\,s^{-1}}$), dotted lines (at $\dot{m} = 150\ \mathrm{kg\,m^{-2}\,s^{-1}}$) and dash-dotted lines (at $\dot{m} = 75\ \mathrm{kg\,m^{-2}\,s^{-1}}$).

At low heat fluxes, the circumference-averaged heat transfer coefficient almost remains constant at mass flow velocities of 75 and 300 $\mathrm{kg\,m^{-2}\,s^{-1}}$. At 150 $\mathrm{kg\,m^{-2}\,s^{-1}}$ a slight but noticeable enhancement in heat transfer was observed. This could indicate an improvement in tube wetting and, thus, a positive effect on flow pattern. Concerning high heat fluxes, heat transfer is generally deteriorated with vapor quality. Reasons were discussed in Sec. 4.3.1.

4.3.4 Influence of the flow pattern

The general influence of flow pattern on the flow boiling heat transfer of CO_2 was well discussed by Schael (2009). This section covers oil-induced effects.

The influence of flow pattern on the heat transfer is demonstrated in Fig. 4.14 for conditions corresponding to $w_{\mathrm{oil}} = 0 - 3$ wt.-%, $p = 26.4$ bar, $\dot{m} = 150\ \mathrm{kg\,m^{-2}\,s^{-1}}$ and $\dot{x} = 0.5$. Circumference-averaged heat transfer coefficients are plotted vs. heat flux in the left image. The boiling curves follow the behavior as discussed for Fig. 4.11. For measurements applying an average heat flux of 12 $\mathrm{kW\,m^{-2}}$, visualized flow patterns are shown and local heat transfer coefficients determined for each tube segment are illustrated in the right image.

From the flow patterns shown for pure CO_2 (open symbols) it can be concluded that liquid fills less than half the tube, covering the bottom segments (3, 4, 5) only. For these segments,

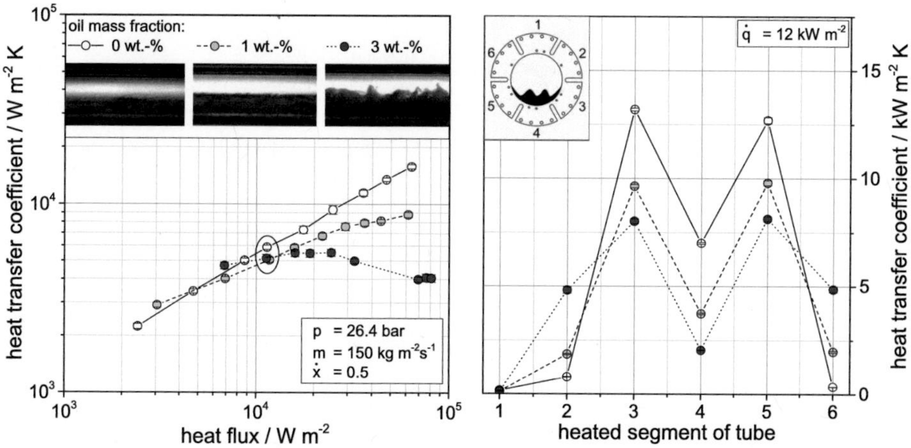

Figure 4.14: Influence of flow pattern on the heat transfer – circumference-averaged heat transfer coefficient vs. heat flux (left) and local heat transfer coefficients for $\dot{q} = 12\,\mathrm{kW\,m^{-2}}$ (right)

local heat transfer coefficients are at least one order in magnitude higher than for the mostly or even complete dry upper segments (1, 2, 6). As already observed in previous research (Wettermann (1999), Schael (2009)), the heat transfer at the bottom tube segment (4) is usually slightly lower than at the adjacent segments (3, 4). Mesler and Mailen (1977) related this phenomenon to the presence of a small liquid film. For such film with thickness in the range of bubble size, the liquid surface is very close to the heated wall surface, influencing the process of bubbles departure. These conditions could enable smaller bubbles to leave that otherwise would grow.

Alike observations were made for CO_2-POE oil mixtures with nominal oil mass fractions of 1 wt.-% (bright gray symbols) and 3 wt.-% (dark gray symbols), despite two main characteristics clearly recognizable. Due to reasons discussed before, heat transfer coefficients at the liquid-covered tube segments deteriorated as the oil mass fraction increases. Surface effects and foaming induced by the presence of lubricant, however, promote the wetting of the tube, yielding a noticeably increase in local heat transfer of the upper side segments (2, 6).

These observations clarify the complexity in predicting the circumference-averaged heat transfer coefficients. As seen from the encircled points in the left diagram, heat transfer intensifying and weakening effects for both mixtures are neutralizing each other. In comparison to the pure CO_2 measurements, the reduction in heat transfer at the bottom tube segments is not compensated by the heat transfer enhancement at both adjacent segments. Yet, under certain operating conditions, circumference-averaged heat transfer coefficient are equalized or even slightly enhanced, as seen before in Fig. 4.11 (bottom, right) for low heat fluxes. Figure 4.15 illustrates such apparently contrary influence of lubricant on the local heat transfer under different operating conditions.

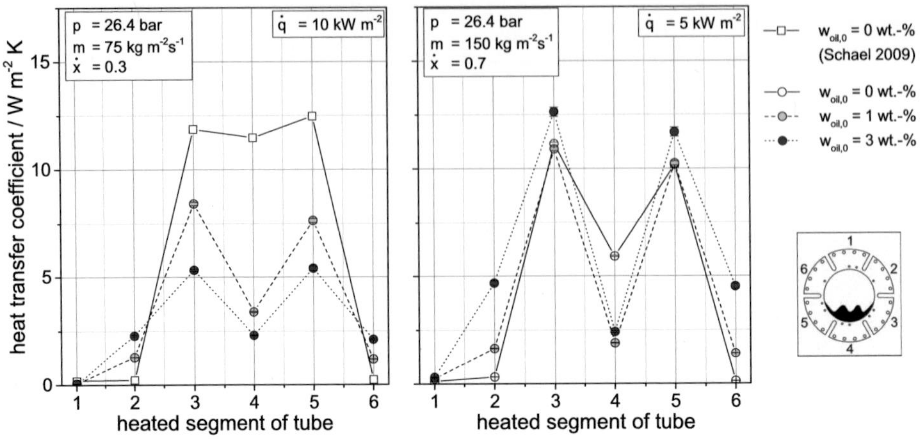

Figure 4.15: Influence of flow pattern on the local heat transfer coefficients at different operating conditions

4.3.5 Comparison to predictive methods

Experimentally determined circumference-averaged heat transfer coefficients were compared to predictions by established correlations found in literature. Evaluated error statistics for the predictive methods tested are listed in Tab. 4.4.

According to Schael (2009), the correlation by Steiner (2002) is best suitable for predicting heat transfer of pure CO_2. Schael proposed slight modifications of the original correlation to achieve an outstanding agreement with 93.1 % of all data points predicted yielding APE < 30 %, MAPE $= 13.5$ % and MPE $= +0.2$ %. Reproducing pure CO_2 measurements within the scope of this work, the quality of prediction was confirmed with 93.5 % of all new data points predicted yielding APE < 30 %, MAPE $= 10.3$ % and MPE $= -0.6$ %. With increasing oil concentration, however, the accuracy of prediction significantly drops. As

Table 4.4: Error statistics of evaluated heat transfer correlations with respect to heat transfer coefficients measured in the smooth tube

Correlation by author	MPE	MAPE	Data that yields APE ≤ 30 %
Li et al. (2014)	+117.7 %	152.0 %	28.4 %
Fang (2013)	−2.1 %	32.4 %	57.1 %
Schael (2009)	+67.2 %	71.8 %	45.9 %
Gao et al. (2008a)	−65.6 %	70.3 %	5.3 %
Cheng et al. (2008b)	−35.6 %	43.2 %	40.2 %
Cheng et al. (2008b) (S=1)	+6.7 %	39.6 %	66.3 %
Wojtan et al. (2005b)	+6.2 %	38.9 %	59.2 %
Thome and El Hajal (2004)	+10.4 %	35.3 %	65.7 %
Steiner (2002)	+35.2 %	47.4 %	69.5 %
Kattan et al. (1998c)	+28.2 %	42.3 %	68.3 %

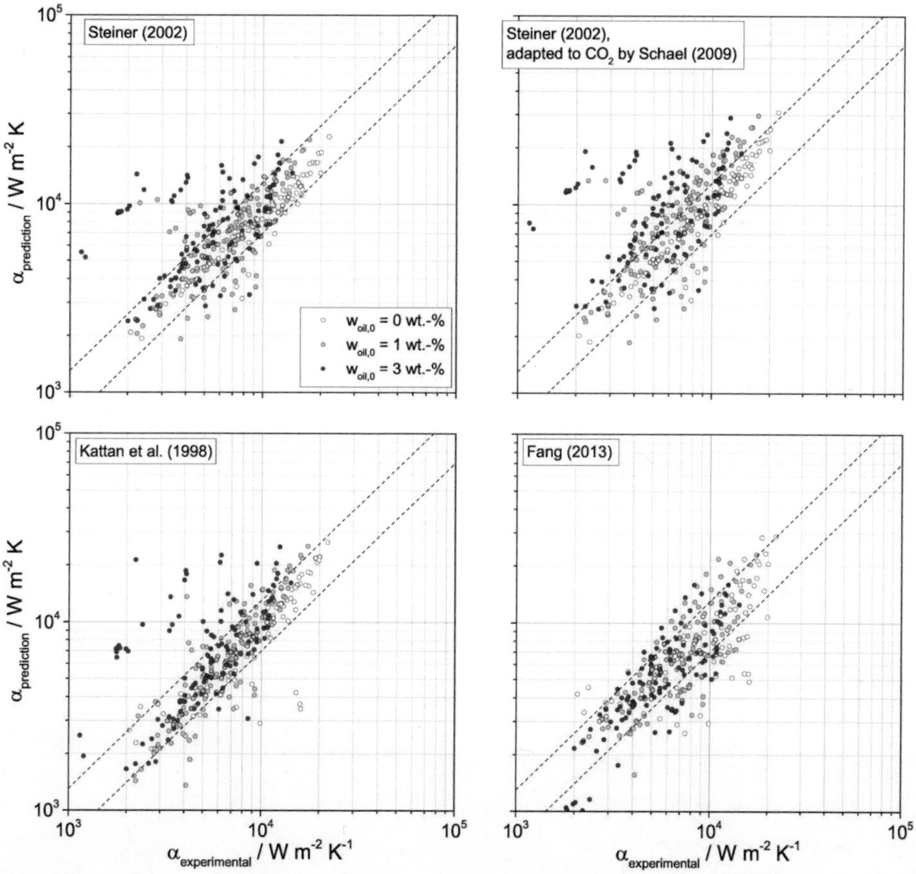

Figure 4.16: Comparison of predicted to measured heat transfer coefficients at 26.4 bar for CO_2-POE oil mixtures inside the smooth tube

can be seen in Tab. 4.4, less than 50 % of the experimental data containing nominal oil mass fraction of 0 – 3 wt.-% are predicted within a range of error of ±30 %. The parity plot shown in Fig. 4.16 (top, left) illustrates the accuracy in prediction according to the respective oil mass fraction. Heat transfer coefficients determined for pure CO_2 (open symbols) are well predicted by the modified Steiner correlation. As the oil mass fraction increases, heat transfer coefficients are increasingly over-predicted.

The established CO_2-based correlation of Cheng et al. (2008b) resulted in great scattering of predicted to experimental values and a significant under-prediction of heat transfer. As main reason for the disagreement of the experimental data obtained in the scope of this work the suppression factor for nucleate boiling introduced by the authors was identified. Related to the annular liquid film thickness, the suppression factor reduces the predicted influence of

nucleate boiling in the annular flow regime. Neglecting nucleate boiling suppression ($S = 1$) greatly improved the accuracy of prediction.
The correlations by Gao et al. (2008a) and Li et al. (2014) show poor agreement to the experimental data. Particularly, both correlations were developed based on experimental data of CO_2-PAG oil mixtures. In contrast to POE lubricants, PAG-type lubricants typically show distinct miscibility-gaps with CO_2, leading to a formation of an additional liquid phase highly rich on lubricant, covering the wall surface. This fact could be responsible for the sharp drop in heat transfer even at a very low oil mass fraction (refer to Fig. 2.21 in Sec. 2.4) due to suppression of nucleate boiling, as reported by Dang et al. (2013) and Li et al. (2014). Consequently, applied to CO_2-POE oil measurements both correlations performed worst of all correlations tested. A third correlation based on data of CO_2-PAG oil mixtures proposed by Katsuta et al. (2008) could not be applied accurately due to missing information in the publication.
None of the evaluated correlations could predict heat transfer coefficients of CO_2-POE oil mixtures satisfactorily. Parity plots shown in Fig. 4.16 reveal a pronounced scattering of predicted to measured values. Best results were achieved for the correlation of Fang (2013), yielding MPE $= -2.1$ % and MAPE $= 32.4$ %. Still, less than 60 % of all data points are predicted within a range of error of ± 30 %. Additional parity plots are shown in App. F.1.3.

4.3.6 Conclusion

The influence of oil on flow boiling heat transfer of CO_2 inside a smooth tube was discussed. The accuracy of up-to-date predictive methods was presented.
At low vapor qualities and heat fluxes, CO_2-POE oil mixtures show similar behavior in flow boiling heat transfer as pure CO_2. Under most conditions, however, heat transfer coefficients are reduced quantitatively when lubricant is added. Slight improvements of heat transfer with increasing oil mass fractions were observed at operating conditions yielding low liquid hold-up (i.e., vapor qualities of $0.5 \leq \dot{x} \leq 0.7$) and low heat fluxes. Here, foaming due to the presence of oil could be the reason for an enhanced wetting of the tube and, thus, an enhancement in heat transfer. Local heat transfer coefficients determined at different locations on the perimeter of the tube support the assumption of a positive lubricant-induced effect on the flow pattern. Increasing the heat flux generally lead to a decrease in slope of the boiling curve and, eventually, in a deterioration in heat flux. The limiting heat flux was found to be dependent on vapor quality and nominal oil mass fraction, moving to lower values as both parameters increase. This fact could be attributed to an enrichment of lubricant in the proximity of the wall surface, affecting mass transport of the refrigerant and bubble point temperature.

The experimental flow boiling data base at 26.4 bar and oil mass fractions of $0-3$ wt.-% was compared to predicted values by established heat transfer correlations found in literature. Among those, the correlations of Li et al. (2014) and Gao et al. (2008a) were developed for CO_2-PAG oil mixtures. Due to different thermo-physical properties (in particular miscibility), both correlations performed worst of all methods tested. A recent correlation developed by Fang (2013) based on pure CO_2 data showed least scattering of predicted to experimental values. The modified Steiner correlation for CO_2 proposed by Schael (2009) revealed outstanding agreement to reproduced pure CO_2 data. However, it was found to greatly over-estimate heat transfer coefficients of CO_2-POE mixtures. The established CO_2-based correlation of Cheng et al. (2008b) showed poor agreement to the experimental data, whereby accuracy in prediction greatly improved by neglecting the nucleate boiling suppression.

Summing up, however, no correlation could predict heat transfer coefficients of flow boiling CO_2-POE oil mixtures satisfactorily. Experimental investigations focused on identifying fundamental effects of lubricant on the flow boiling heat transfer of CO_2. In favor of an enlarged heat transfer data base for the micro-fin tube, the experimental effort for the smooth tube was strongly reduced and no modeling of the heat transfer coefficient was performed.

5 Micro-fin tube results

In the following, results from flow boiling experiments in the 8.62 mm micro-fin tube are presented, including flow patterns, two-phase pressure drop data and heat transfer measurements conducted under adiabatic and pseudo-diabatic boundary conditions.

5.1 Flow patterns

Flow patterns of flow boiling CO_2-POE oil mixtures inside a micro-fin tube were investigated. Determination of flow patterns and prediction by flow pattern maps followed the procedure of the smooth tube, evaluating high speed video records. As no flow pattern maps are presently available for micro-fin tubes, flow patterns identified were compared to established smooth tube flow patterns maps. Note that the non-heated sight glass tube succeeding the measurement section and guard heater was designed as smooth tube of inner diameter equal to the fin-root diameter of the micro-fin tube. Flow patterns thus had to be determined close to the outlet of the guard heater in order to observe the swirling effect of the micro-fins.

5.1.1 Flow pattern classification and distinction from smooth tube observations

In general, flow patterns observed for CO_2-POE oil mixtures inside the micro-fin tube correspond well to those observed in the smooth tube. Flow pattern classification hence follows Sec. 4.1.1.

Owing to the micro-fins, stratified flow was practically not observed in the micro-fin tube. The induced swirl interrupts flow stratification, leading to wavy flow formation even at very low mass flow velocities. Further, annular flow was frequently overlaid by slug or wavy flow characteristics. Smooth and uniform annuli were hence seldom observed. Schael (2009) previously made similar observations during pure CO_2 experiments as well.

It was observed that significant foaming already occurred at low mass flow velocities and vapor qualities under adiabatic conditions. Foam produced in the opposing pre-heaters hence endured longer in the micro-fin test section. This fact is demonstrated by the images shown in Fig. 5.1, representing observed flow patterns in the smooth (left) and micro-fin (right) tube under equal operating conditions. The multi-bubble containing foam layer present on the liquid interface occupies a significant part of the flow cross section, particularly affecting the transition from wavy to slug flow.

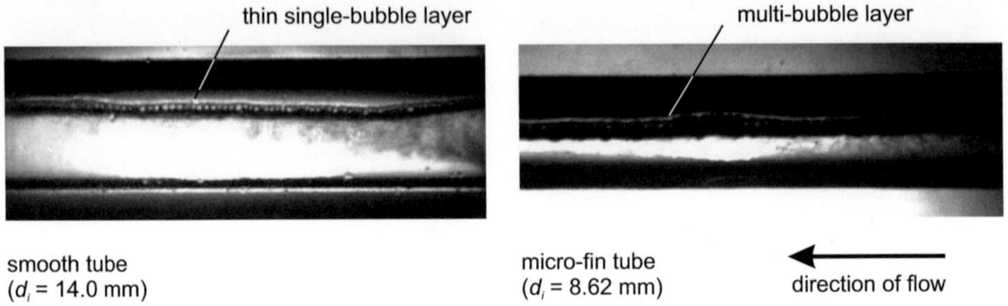

Figure 5.1: Wavy flow observed for CO_2-POE oil mixtures with a nominal oil mass fraction of 3 wt.-% in the smooth (left) and micro-fin (right) tube at a pressure of 26.4 bar, a mass flow velocity of 75 $\mathrm{kg\,m^{-2}\,s^{-1}}$ and a vapor quality of 0.1 under adiabatic condition

Another effect induced by the presence of oil observed in the micro-fin tube was the formation of eruptive-like slugs during diabatic heat transfer experiments, see Fig. 5.2. Slightly similar to the phenomenon known in micro-channels, frequency and intensity of these slugs were incomparably weaker pronounced than in micro-tube two-phase flow. This flow pattern is characterized by a drying-out liquid flow (1), often leading to a complete dryout of the tube, followed by an eruptive and heavyly foaming slug (2-3). After the flow had settled (4), foaming wavy flow (5) reappeared. These slugs were observed under weak convective boiling conditions, i.e., at low mass flow velocities and vapor qualities. They seem to occur due to partial blocking of the micro-fin tube by the foam, which is formed in the heated tube segments. With increased pressure drop over the assumed foam front, the liquid accelerates, causing an explosively slug-like re-wetting of the tube. The duration between two such slugs was in the range of a few seconds. Consequently, distinctive fluctuations of pressure drop and wall temperatures (the latter due to the wall dryout) were observed.

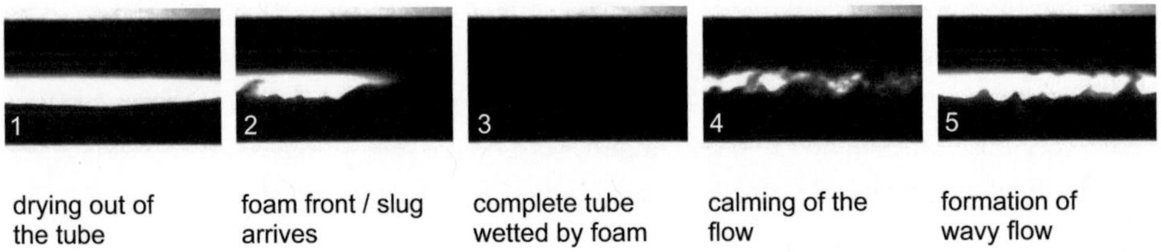

Figure 5.2: Eruptive slug formation in the micro-fin tube

5.1.2 Comparison of experimental flow patterns to flow pattern maps

As mentioned initially, flow pattern maps based on micro-fin tube experiments are not available. Thus, established smooth tube flow pattern maps are used as reference for the precision in prediction. The flow pattern map found to be most suitable for predicting two-phase flow patterns of CO_2-POE oil mixtures is the map of Steiner modified by Saito et al.. Transition curves were calculated regarding the fin-root diameter of the tube.
From experiments with pure CO_2, Schael (2009) ascertained that the transition from intermittent to annular flow was similar to that in the smooth tube, however generally taking place at low mass flow velocities. The transition could be described analogous to the smooth tube. Adapted to the flow boiling data obtained in this work, the transition criterion yields Eq. 5.1. Alike, the approach by Barbieri et al. (2008) to determine the intermittent to annular transition was used and fitted to the observations made, see Eq. 5.2.

$$\mathrm{Fr}_{\mathrm{Vm,I\text{-}A}}^{0.5} = 0.7 X_{\mathrm{tt}}^{-0.35} \tag{5.1}$$

$$\mathrm{Fr}_{\mathrm{L,I\text{-}A}} = 0.3 X_{\mathrm{tt}}^{1.5} \tag{5.2}$$

Figure 5.3 illustrates the predicted transition (solid lines) and experimentally determined flow patterns (symbols) at adiabatic conditions. The transition for CO_2-POE oil mixtures differ but slightly from that observed for pure CO_2. Equation 5.1 is recommended for estimating the intermittent to annular transition for CO_2-POE oil mixtures in the range of $12\,\mathrm{bar} \leq p \leq 40\,\mathrm{bar}$, $50\,\mathrm{kg\,m^{-2}\,s^{-1}} \leq \dot{m} \leq 500\,\mathrm{kg\,m^{-2}\,s^{-1}}$ and $w_{oil,0} \leq 3$ wt.-%.

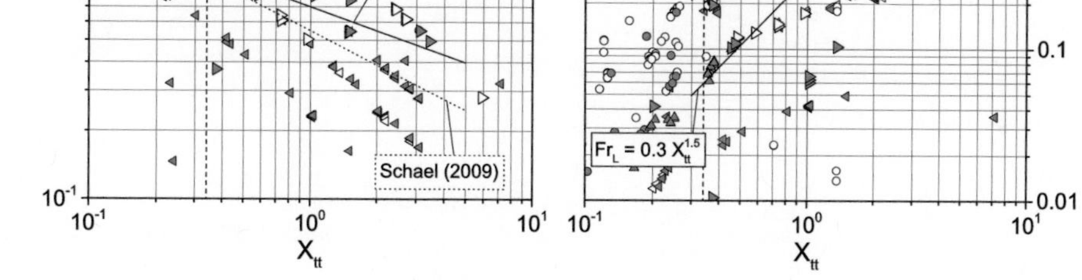

Figure 5.3: Transition from partial to complete tube wetting in the micro-fin tube in terms of $\mathrm{Fr}_{\mathrm{Vm}}^{0.5}$ (left) and Fr_{L} (right) over X_{tt}; Symbols represent experimental data and solid lines predicted transitions by Eq. 5.1 and Eq. 5.2

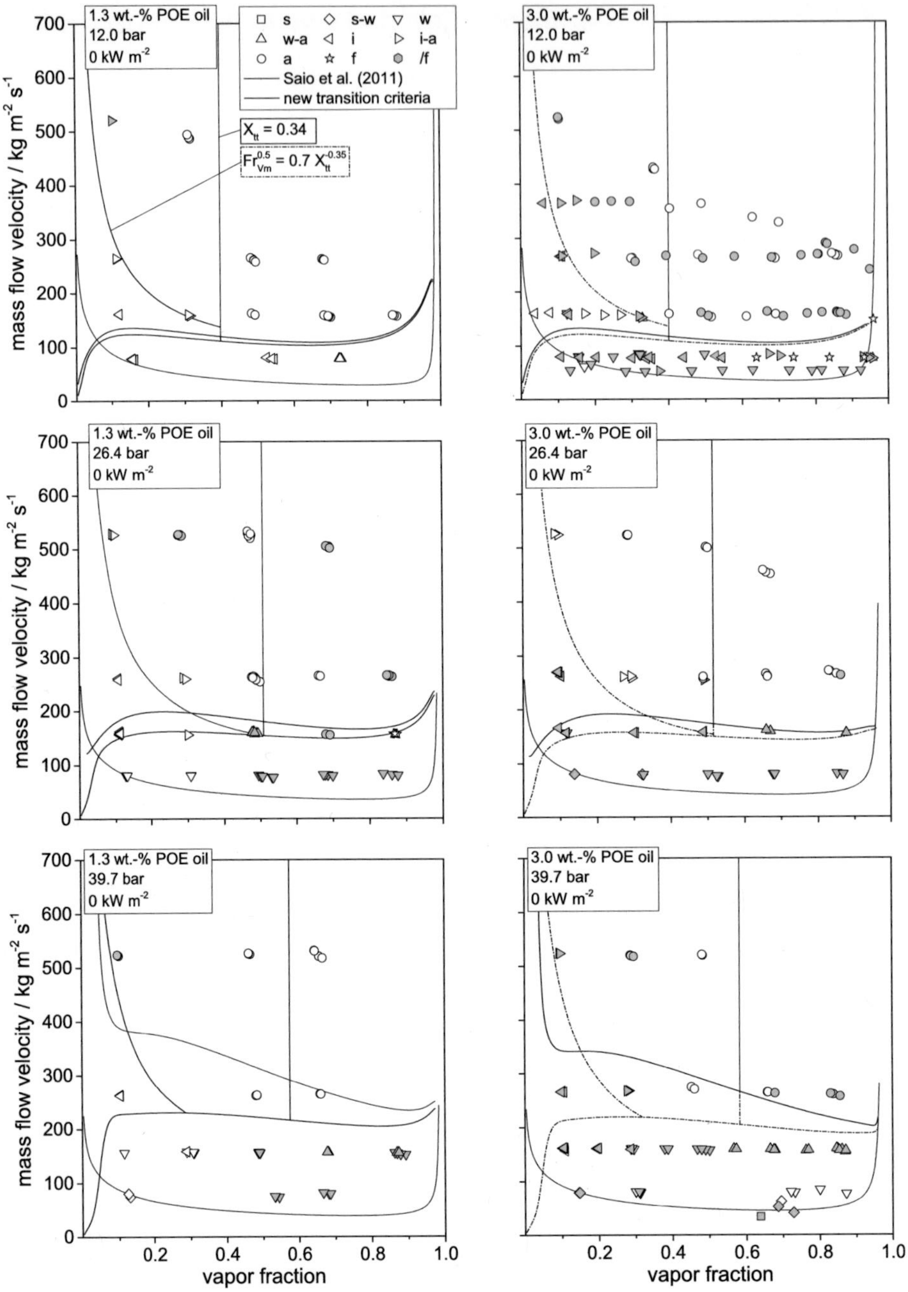

Figure 5.4: Adiabatic flow patterns of CO_2-POE oil mixtures with nominal oil mass fractions of 1.3 and 3.0 wt.-% inside the micro-fin tube at pressures of 12.0 – 39.7 bar, mass flow velocities of 50 – 500 $kg\,m^{-2}\,s^{-1}$ and vapor qualities of 0.10 – 0.95

In Fig. 5.4, flow patterns experimentally determined under adiabatic conditions are plotted as symbols in the modified Steiner flow pattern map. Predicted flow pattern transitions are shown as lines in dimensionless form ($\dot{m} = f(\dot{x})$). The wavy to intermittent/annular flow transition represented as solid lines corresponds to the criterion derived by Saito et al. (2011), whereas all other solid lines are equivalent to the original Steiner map. Dash-dotted lines represent the intermittent-annular (Eq. 5.1) and wavy-intermittent/annular (see Sec. 4.1.2) transitions, respectively, developed based on the experimental data of this work.
Following the trend of observations, the intermittent to annular flow transition is shifted clearly towards lower mass flow velocities, compared to the smooth tube. The same accounts for the wavy to intermittent/annular flow transition. One reason is reduction of the tube diameter (14.0 mm to 8.52 mm), consistently noticeable when comparing the flow pattern maps calculated for the smooth and the micro-fin tube diameter. Still, the wavy to intermittent/annular transition is predicted at slightly higher mass flow velocities than observed. The latter is attributed to the swirling effect of the micro-fins favoring annular flow formation. Summarizing, flow patterns in the micro-fin tube are actually predicted in good accordance to the observations made, when using smooth tube flow pattern maps with some modifications in the transition lines.

5.1.3 Conclusion

Flow patterns of flow boiling CO_2-POE oil mixtures inside a micro-fin tube were investigated. For this purpose, high speed-speed video recordings at the sight glass succeeding the measurement section and guard heater were evaluated. Determination of the flow pattern followed the classification introduced for the smooth tube. In general, flow patterns observed in both tubes were alike.
Differences in behavior with respect to the smooth tube were outlined as a frequent overlay of wavy flow induced by the micro-fins of the tube, the formation of a more stable and longer enduring foam layer and the appearance of eruption-like slugs when increasing the heat flux in diabatic measurements at low mass flow velocities. The promotion of foam formation may be related to the micro-fins, as the swirling effect enlarges the liquid interface and enhances the vapor-liquid mixing. Due to the early presence of foam and the smaller tube diameter, partial blocking of the tube may occur for flows of low convective character (i.e., low mass flow velocity and low vapor quality). When the pressure drop increases along the obstruction, it is blown out in the form of an eruptive slug. During this state of flow, the tube wall frequently dries out, followed by complete re-wetting by the foam-filled slug. Significant fluctuations in wall temperatures and pressure drop are the measurable consequence.

Flow patterns observed in experiments at adiabatic conditions were compared to the Steiner flow pattern map, modified by Saito et al. (2011) in wavy to intermittent/annular flow transition. A further update of the criterion parameters (C_1, C_2) presented before in Sec. 4.1.2 along with a modified intermittent to annular transition criterion based on the observations made in the micro-fin tube yielded a significant improvement in flow pattern prediction. Largest deviations are found in the transition region from wavy to intermittent/annular flow, as the swirling effect of the micro-fins promoting annular flow is not regarded by the smooth tube criterion. The intermittent to annular flow transition is successfully predicted by Eq. 5.1, developed in the scope of this work based on the micro-fin tube data. Besides being developed based on smooth tube experimental data and regarding a smooth wall surface, the modified Steiner flow pattern map was found to predict the observed flow patterns in relatively good accordance.

5.2 Pressure drop

Pressure drop experiments in the micro-fin tube were carried out analogously to the smooth tube, see Sec. 4.2. Frictional pressure drop and total pressure drop were measured during adiabatic and diabatic experiments, respectively. With regard to measurements of pure CO_2 in the smooth tube, the pressure drop of CO_2-POE oil mixtures investigated in the micro-fin tube is influenced by the roughness of the tube and the presence of oil, affecting the physical properties of the fluid.

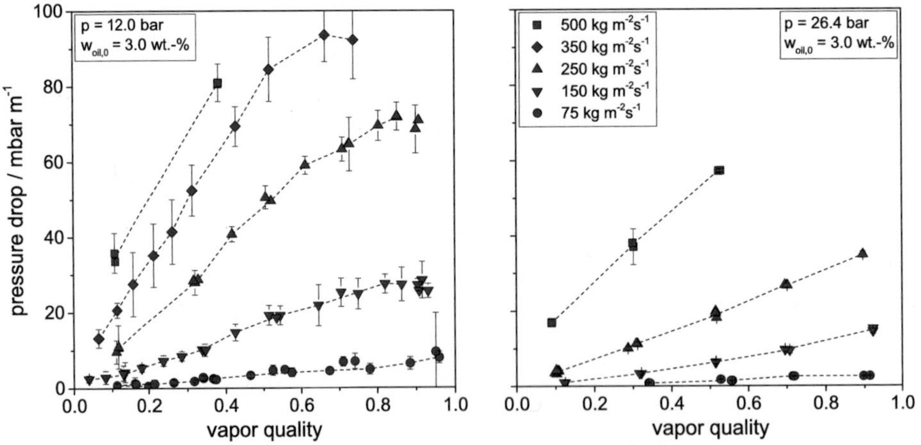

Figure 5.5: Frictional pressure drops in the micro-fin tube as function of mass flow velocity and vapor quality at a constant nominal oil mass fraction of 3.0 wt.-% and pressures of 12.0 bar (left) and 26.4 bar (right)

5.2.1 Frictional pressure drop

Frictional pressure drop measurements yield from adiabatic flow pattern experiments as well as from heat transfer measurements, i.e., from measurements taken at the end of each heat transfer measurement series under adiabatic condition. In both cases, no heating power was supplied to either the measurement section or the succeeding guard heater. The desired vapor quality at the inlet of the measurement section was set-up primarily by the heat input to pre-heaters 1-2. In the case that two-phase flow was very unstable, pre-heater 3 was used in parts as well.

In Fig. 5.5, the frictional pressure drop inside the micro-fin tube for a CO_2-POE oil mixture with an oil mass fraction of 3.0 wt.-% at pressures of 12.0 bar (left) and 26.4 bar (right) are plotted versus vapor quality. The mass flow velocity for the measurements series shown varies from 75 to 500 $\mathrm{kg\,m^{-2}\,s^{-1}}$. As expected and previously observed for pure CO_2 by Schael (2009), frictional pressure drop increases almost linearly in the range of low to medium vapor qualities, reaching a maximum at a vapor quality above 0.6. This trend can be seen best at low pressures (left plot) due to the larger density ratio of liquid to vapor phase. Increasing the system pressure impedes depicting that trend as foam formation induces further frictional resistance, hence leading to an increase in pressure drop. Furthermore, pressure drop increases with mass flow velocity. As shown in Fig. 5.6 for measurements at a constant nominal oil mass fraction of 3.0 wt.-%, a system pressure of 26.4 bar and average vapor qualities of 0.1, 0.3 and 0.5, the increase in frictional pressure drop was found to be nearly proportional to $\dot{m}^2$.

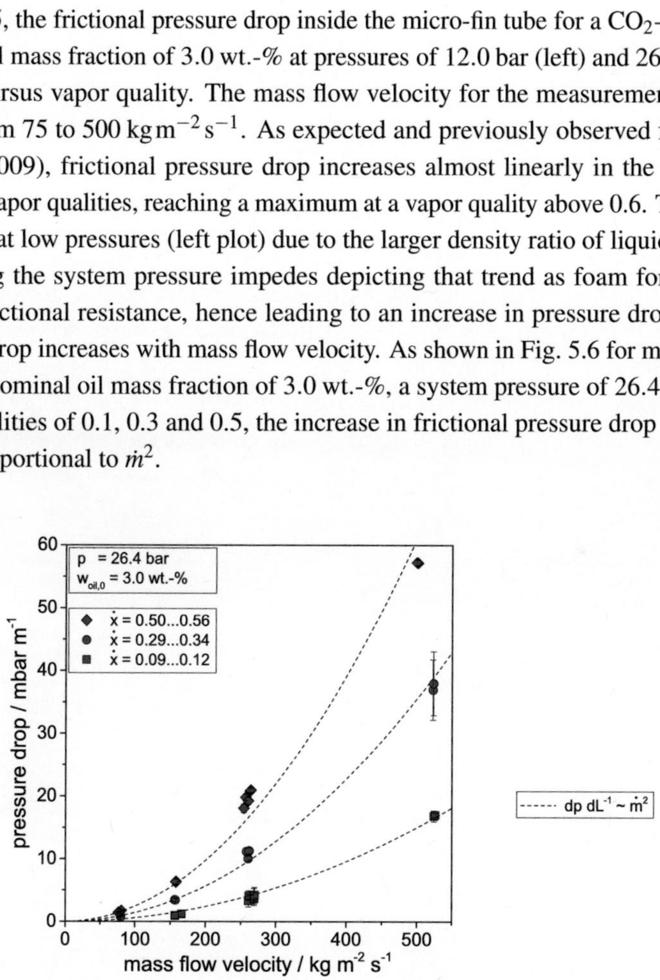

Figure 5.6: Frictional pressure drops in the micro-fin tube as function of mass flow velocity and vapor quality at a pressure of 26.4 bar and a constant nominal oil mass fraction of 3.0 wt.-%

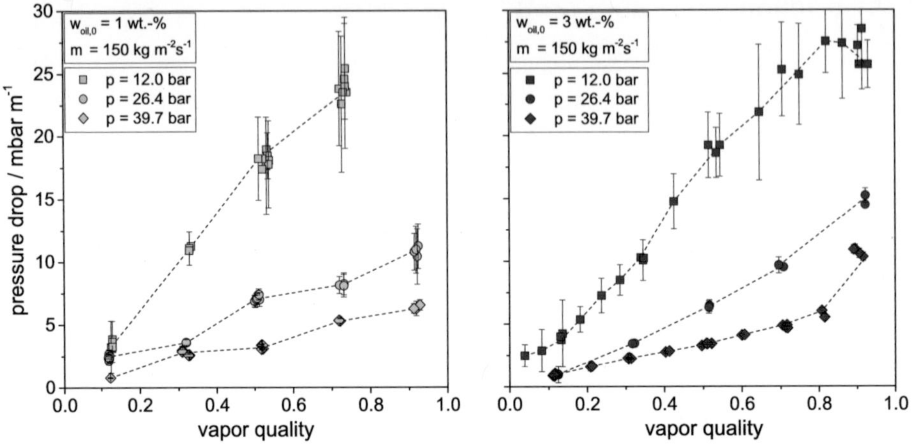

Figure 5.7: Frictional pressure drops in the micro-fin tube for a mass flow velocity of 150 $kg\,m^{-2}\,s^{-1}$ and oil mass fractions of 1 (left) and 3 wt.-% (right) as function of system pressure and vapor quality

The influence of the system pressure on the frictional pressure drop is demonstrated in Fig. 5.7 for CO_2-POE oil mixtures containing 1 (left) and 3 wt.-% (right) lubricant. The mass flow velocity was hold constant at 150 $kg\,m^{-2}\,s^{-1}$. With increasing system pressure, the frictional pressure drop clearly decreases and the dependency on vapor quality diminishes. This fact is related to the lower liquid to vapor density ratio at higher pressures and, hence, lower vapor velocities.

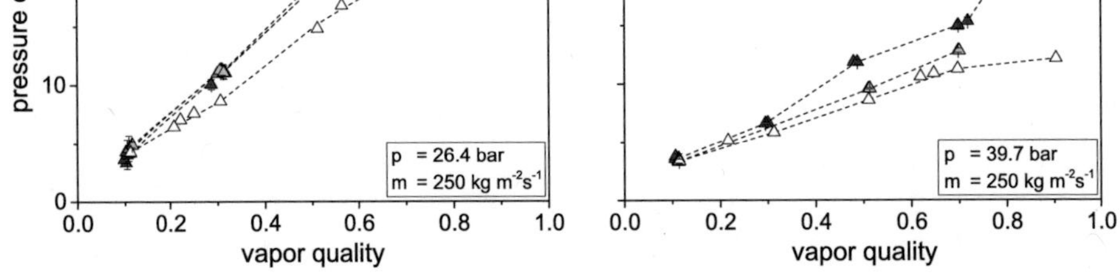

Figure 5.8: Frictional pressure drops in the micro-fin tube as function of oil mass fraction and vapor quality at a constant mass flow velocity of 250 $kg\,m^{-2}\,s^{-1}$ and pressures of 26.4 bar (left) and 39.7 bar (right)

The effect of lubricant on the frictional pressure drop is demonstrated in Fig. 5.8 for two pressures (26.4 bar, left, and 39.7 bar, right) and a constant mass flow velocity of 250 $kg\,m^{-2}\,s^{-1}$. The nominal oil mass fraction of the CO_2-POE oil mixture was varied from 1 to 3 wt.-%. At low vapor qualities the influence of lubricating oil on the pressure drop was almost negligible within the range of oil mass fractions investigated. The local oil mass fraction scarcely differs from the nominal oil mass fraction and the liquid mixture properties are close to those of pure CO_2. Increasing the vapor fraction results in a higher local oil mass fraction. Hence, the viscosity of the liquid phase and the frictional pressure drop increase. At high vapor quantities, foam was commonly observed, significantly contributing to flow resistance and pressure drop.

Frictional pressure drops experimentally determined in the micro-fin tube were compared to predictive methods found within a literature survey. Correlations especially developed from micro-fin data are those of Hu et al. (2008a), Kedzierski and Goncalves (1999), Choi et al. (1999) and Cavallini et al. (1997). Choi et al. assumed that their correlation is also valid for refrigerant-lubricant mixtures when using the mixture liquid viscosity and the correct vapor quality. The recently published correlation of Xu and Fang (2012) is based on hydraulic diameter and generally applicable to in-tube flow, comprising experimental results in circular and rectangular pipes as measurement basis. Correlations of Hu et al. (2008a), Hu et al. (2008c) and Katsuta et al. (2008) were developed especially for refrigerant-lubricant mixtures. Yoon et al. (2004) developed a correlation for pure CO_2 boiling in smooth tubes. However, predicted values did not agree satisfactorily to experimental data by Schael (2009). Besides the few micro-fin correlations available, Schael evaluated selected smooth tube correlations as well based on fin-root diameter. Best agreement to experimental micro-fin tube data for pure CO_2 was found for the correlations of Müller-Steinhagen and Heck (1986) and Storek and Brauer (1980), if multiplying penalty factors of constant $PF_{\Delta p} = 1.63$ or $PF_{\Delta p} = 2.6 - 1.52\dot{x}$ were applied, respectively. The introduction of penalty factors of that magnitude implies that prediction yielded large deviations when using the fin-root diameter as characteristic dimension.

Kedzierski and Goncalves concluded that the hydraulic diameter should be chosen as characteristic dimension for predicting pressure drops in enhanced tubes. The authors recommended an expression for micro-fin tubes, including structural characteristics of the micro-fins.

$$d_{\mathrm{h}} = \frac{4A\cos(\beta)}{n_{\mathrm{f}} U_{\mathrm{f}}} \tag{5.3}$$

where A is the free cross section for flow, β is the helix angle of the micro-fins, n_{f} is the number of fins and U_{f} is the perimeter of one fin.

Table 5.1: Error statistics of evaluated pressure drop correlations with respect to adiabatic frictional pressure drops measured in the micro-fin tube (hydraulic diameter $d_h = 5.18$ mm)

Correlation by author	$PF_{\Delta p}$	MPE	MAPE	Data that yields APE ≤ 30 %
Xu and Fang (2012)	-	+9.5 %	22.7 %	82.8 %
Hu et al. (2008c)	-	+185.2 %	185.3 %	1.4 %
Katsuta et al. (2008)	-	+199.4 %	199.9 %	5.4 %
Cheng et al. (2008a)	-	+107.0 %	112.3 %	28.1 %
Moreno Quibén and Thome (2007b)	-	+86.6 %	90.6 %	18.3 %
Yoon et al. (2004)	-	+34.6 %	56.7 %	36.6 %
Müller-Steinhagen and Heck (1986)	-	+8.8 %	24.0 %	80.8 %
	0.96	+4.2 %	23.1 %	83.8 %
Storek and Brauer (1980)	-	+19.5 %	25.7 %	67.0 %
	$1-0.36\dot{x}$	−3.5 %	14.7 %	91.5 %
Friedel (1979)	-	+66.9 %	69.3 %	45.3 %
Chisholm (1973)	-	+92.3 %	100.1 %	22.9 %
Grönnerud (1972)	-	+51.4 %	56.1 %	40.7 %
Lockhart and Martinelli (1949)	-	+147.5 %	155.5 %	24.5 %
Hu et al. (2008a) (V)	-	+85.7 %	86.8 %	24.3 %
Hu et al. (2008a) (L)	-	+71.7 %	76.9 %	28.1 %
Choi et al. (1999)	-	−46.5 %	47.7 %	11.0 %
Cavallini et al. (1997)	-	+32.5 %	37.7 %	62.2 %

Table 5.1 summarizes the results of comparing the experimental data base to the predictive methods investigated within the scope of this work. With the exception of the beforehand mentioned micro-fin correlations, the remaining pressure drop correlations originated from measurements in conventional pipes. It occurs that the correlations of Xu and Fang, Müller-Steinhagen and Heck and Storek and Brauer could describe experimental results best among all correlations evaluated, on average just slightly over-predicting the frictional pressure drop by $8.8-19.5$ % and yielding an absolute deviation of ≤ 30 % for $67.0-82.8$ % of data. Introducing a multiplying factor of $PF_{\Delta p} = 1-0.36\dot{x}$ dependent on the vapor fraction, prediction accuracy of the correlation Storek and Brauer could be increased significantly. With 91.5 % of frictional pressure drop data predicted with APE ≤ 30 % and MAPE $= 14.7$ %, the modified Storek and Brauer correlation performed best of all correlations tested and is well suitable to predict adiabatic frictional pressure drops of CO_2-POE oil mixtures inside the micro-fin tube.

The quality of prediction obtained for the correlations of best performance is shown graphically in Fig. 5.9. Here, predicted pressure drops from correlations of Xu and Fang (2012), Cavallini et al. (1999), Müller-Steinhagen and Heck (1986) and Storek and Brauer (1980) in modified form are plotted versus experimental values. The pressure drop correlation by Cavallini et al. developed from micro-fin data over-predicts experimental data on average and yields significant scattering compared to the other correlations displayed. Predicted values using the the modified Storek and Brauer correlation feature least scattering correspond best to the trend in frictional pressure drop observed.

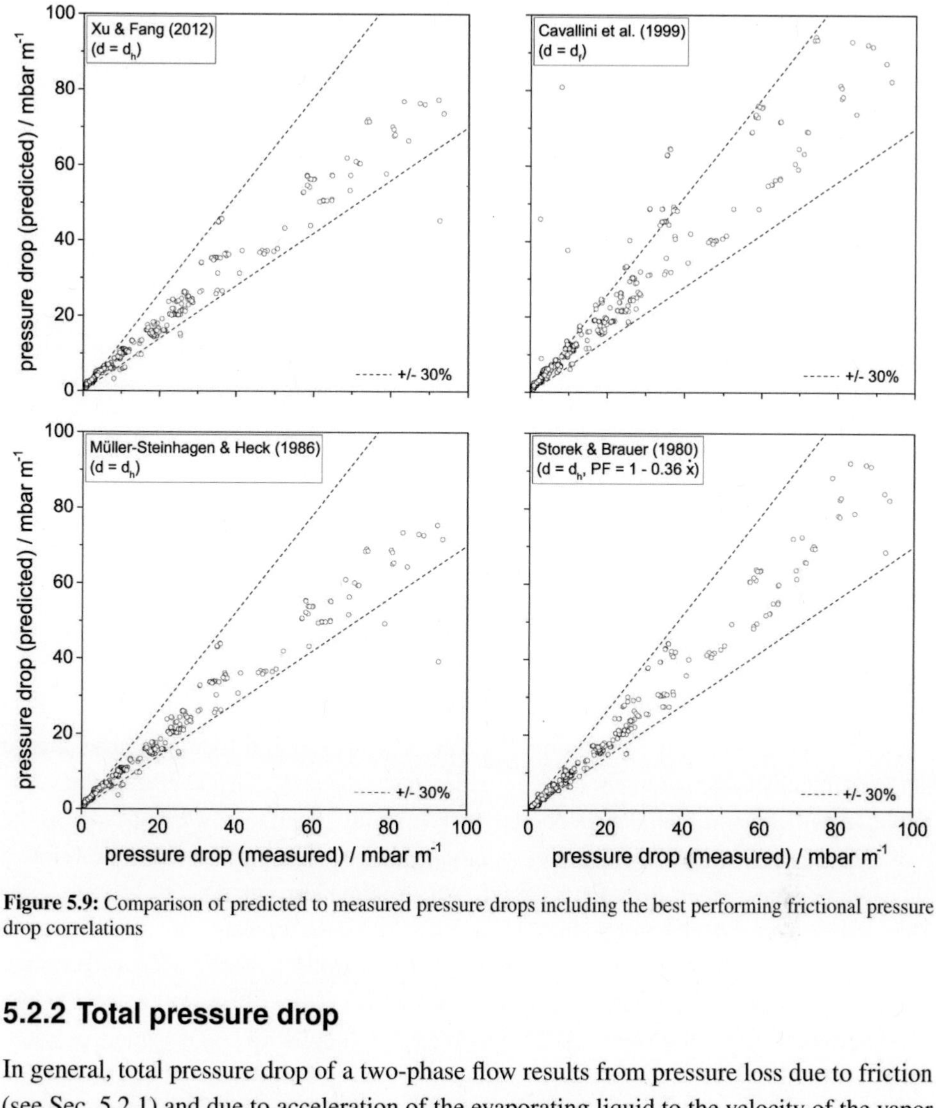

Figure 5.9: Comparison of predicted to measured pressure drops including the best performing frictional pressure drop correlations

5.2.2 Total pressure drop

In general, total pressure drop of a two-phase flow results from pressure loss due to friction (see Sec. 5.2.1) and due to acceleration of the evaporating liquid to the velocity of the vapor phase. Two types of models are available for calculating the acceleration pressure drop, either assuming homogeneous or heterogeneous flow. In Fig. 5.10, experimentally determined total pressure drops (symbols) are shown for a CO_2-POE oil mixture with a nominal oil mass fraction of 3 wt.-%. Further, a comparison to predicted values using the homogeneous (solid lines) and heterogeneous (dotted lines) model is given. Model calculation was normalized to the frictional pressure drop determined under adiabatic condition ($\dot{q} = 0\,\mathrm{kW\,m^{-2}}$). Flow pattern visualizations shown on the right illustrate the influence.

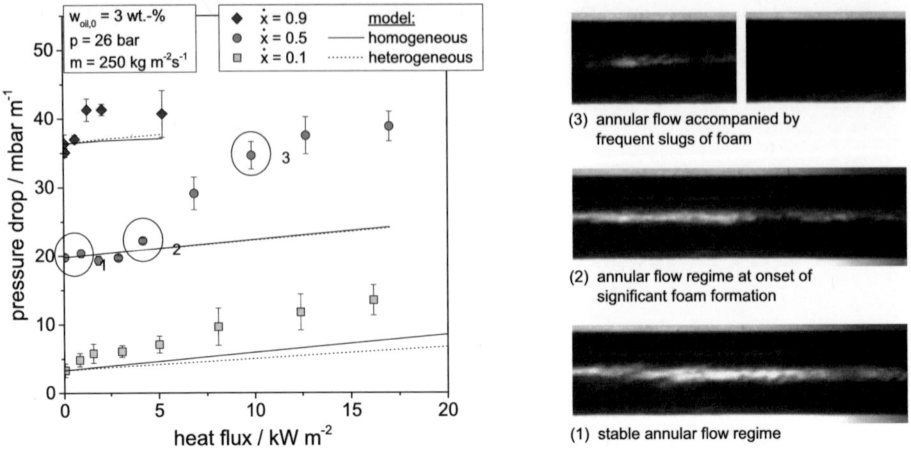

Figure 5.10: Total pressure drop as function of heat flux for $w_{oil,0} = 3.0$ wt.-%, $p = 26.4$ bar, $\dot{m} = 250\,\mathrm{kg\,m^{-2}\,s^{-1}}$ and $\dot{x} = 0.1 - 0.9$

In contrary to observations made for pure CO_2 (Schael, 2009), none of the acceleration pressure drop models could capture the influence of heat flux, especially at increased vapor qualities. For the most part, total pressure drop is underestimated significantly. The steeper progress with respect to the model and, hence, in accordance with pure CO_2 data is most likely related to the formation of foam. The presence of foam produced due to the evaporation of liquid connotes an additional flow resistance. That applies to the vapor phase, in particular, as foam layers observed were mainly found to travel at speed near the estimated liquid velocity. Under certain conditions, the accompanied foam flow turns into a slug-like superposition of flow. On its onset, pressure drop leaps up ($2 \rightarrow 3$).

The combined error statistics of predicted frictional and acceleration pressure drop is given in Tab. 5.2. Depending on the type of frictional pressure drop model, the respective acceleration pressure drop model was chosen. The correlation of Choi et al. (1999) already includes the acceleration pressure drop contribution due to evaporating liquid. As expected, pressure drops predicted by Xu and Fang (2012), Müller-Steinhagen and Heck (1986) and Storek and Brauer (1980) correlations yield lower accuracy while other correlations overestimating frictional pressure drop benefit by the underestimation of acceleration pressure drop. Among those, the correlations of Friedel (1979), Choi et al. (1999)and Cavallini et al. (1999) predict most data with APE ≤ 30 %. The non-modified Storek and Brauer correlation yielded the smallest mean absolute error with MAPE $= 27.3$ %, predicting 68.3 % of all data with a percentage error of ≤ 30 %. Error diagrams for all correlations are found in App. F.2.2.

Table 5.2: Error statistics of evaluated pressure drop correlations with respect to total pressure drops measured in the micro-fin tube at diabatic conditions (hydraulic diameter $d_h = 5.18$ mm)

Correlation by author	$PF_{\Delta p}$	MPE	MAPE	Data that yields APE ≤ 30 %
Xu and Fang (2012)	-	-15.2 %	29.2 %	64.1 %
Hu et al. (2008c)	-	$+112.3$ %	124.9 %	16.6 %
Katsuta et al. (2008)	-	$+114.7$ %	124.8 %	22.2 %
Cheng et al. (2008a)	-	$+57.0$ %	82.2 %	38.0 %
Moreno Quibén and Thome (2007b)	-	$+46.6$ %	65.2 %	32.4 %
Yoon et al. (2004)	-	$+4.2$ %	44.2 %	47.2 %
Müller-Steinhagen and Heck (1986)	-	-15.7 %	29.9 %	63.8 %
	0.96	-18.9 %	30.9 %	61.1 %
Storek and Brauer (1980)	-	-5.1 %	27.3 %	68.3 %
	$1-0.36\dot{x}$	-22.2 %	28.8 %	60.7 %
Friedel (1979)	-	$+18.3$ %	41.3 %	67.0 %
Chisholm (1973)	-	$+42.1$ %	61.8 %	42.6 %
Grönnerud (1972)	-	$+11.5$ %	41.6 %	49.7 %
Lockhart and Martinelli (1949)	-	$+87.9$ %	110.6 %	26.2 %
Hu et al. (2008a) (V)	-	$+42.0$ %	59.1 %	43.1 %
Hu et al. (2008a) (L)	-	$+33.4$ %	55.8 %	43.1 %
Choi et al. (1999)	-	-22.3 %	29.1 %	63.1 %
Cavallini et al. (1997)	-	-0.8 %	31.8 %	67.5 %

5.2.3 Conclusion

Two-phase frictional pressure drops of CO_2-POE oil mixtures with nominal oil mass fractions of up 3 wt.-% were determined. Differences to observations previously made for pure CO_2 by Schael (2009) were discussed. The lubricant-induced promotion of foam formation at high vapor qualities was found leading to significant fluctuations in pressure drop. This effect was more pronounced at low system pressure. The presence of lubricating oil especially affected the total pressure drop measured at elevated heat fluxes. With increasing heat flux, the intensity of foam formation increased, leading to additional flow resistance and change in flow behavior. As consequence, the total pressure drop was found to leap up and fluctuations in pressure drop increased.
Experimental data of frictional pressure drop were compared to predicted vales by several general predictive methods as well as to frictional pressure drop correlations especially developed for two-phase flow in micro-fin tubes. As geometrical dimension in general methods, the hydraulic diameter as proposed by Kedzierski and Goncalves (1999) was used. The correlations of Müller-Steinhagen and Heck (1986) and Xu and Fang (2012) yielded good agreement to experimental data, predicting > 81 % of all adiabatic pressure drop data with APE ≤ 30 %. The Xu and Fang correlation actually represents the correlation of Müller-Steinhagen and Heck in slightly modified form, explaining the near identical results. Best agreement to experimental data was found using the Storek and Brauer (1980) correlation in slight modification. Amended by a multiplication factor of $PF_{\Delta p} = 1 - 0.36\dot{x}$,

91.5 % of all data was predicted with APE $\leq$ 30 %. Further, the error diagram in Fig. 5.9 and a MAPE = 14.7 % underline the good accordance to the trend of measurement.
Total pressure drops were calculated by superposition of frictional pressure drop and pressure drop due to acceleration of the evaporating liquid. The acceleration pressure drop model was chosen with respect to the homogeneous or heterogeneous type of frictional pressure drop model used. Due to the above specified influence of the lubricant, total pressure drops could not be predicted with adequate precision. Pressure drops generally increased stronger than predicted by either acceleration pressure drop model.

5.3 Heat transfer

The experimental determination of circumference-averaged heat transfer coefficients in the micro-fin tube follows the procedure for the smooth tube (refer to Sec. 4.3) in analogous manner. However, defining of the reference surface is not comparably trivial. Several definitions are found in literature, e.g., the total effective area, the area of a smooth tube with maximum/minimum inside diameter or the smooth tube area of respective mean inside diameter (Cavallini et al., 1999). Experimental heat transfer coefficients reported in the following were determined corresponding to the area of a smooth tube of maximum inside (= fin-bottom) diameter, providing convenient comparability to a non-finned tube of respective size and representing most common practice (Nidegger et al., 1997).

5.3.1 Influence of the heat flux

Heat transfer regimes observed in the micro-fin tube equal those in smooth tubes, i.e., convective and nucleate boiling. Flow boiling of CO_2 is strongly characterized by nucleate boiling. However, related to the surface structure of the micro-fin tube, convection is enhanced (Schael, 2009). Figure 5.11 depicts the influence of heat flux on the circumference-averaged heat transfer coefficient of CO_2-POE oil mixtures ($w_oil = 1$ wt.-%).
At low pressure (12.0 bar), convective boiling ($\dot{q} < 20\,\mathrm{kW\,m^{-2}}$, dotted line) and nucleate boiling ($\dot{q} > 20\,\mathrm{kW\,m^{-2}}$, dashed line) regimes are clearly visible. With increasing pressure the number of nucleation sites increases, leading to nucleate boiling dominated heat transfer. According to Gorenflo (2002), the required excess vapor pressure Δp of a bubble nucleus for bubble growth induced by super-heating ΔT of the wall surface is proportionally dependent on surface tension σ, hence, decreasing with temperature and saturation pressure.

$$\Delta p_{\mathrm{onb}}(\Delta T) = \sigma r^{-1} \tag{5.4}$$

where r is the radius of curvature of the interface to the liquid.

These observations and orders of magnitude correspond well to those reported for pure CO_2 (Schael, 2009). However, due to the presence of lubricant, the overall heat transfer is suppressed at conditions of high heat flux, progressing with increasing heat flux. The non-volatile lubricant enriches at the location of evaporation, i.e., at the vapor-liquid interface, affecting the surface tension and inducing mass flow resistance for the onflowing liquid refrigerant as well as departing bubbles in the near-wall liquid layer. Hence, maximum bubble density at the surface of the tube wall is reached more quickly. This effect was already observed in smooth tube experiments, see Sec. 4.3.
Figure 5.12 exemplifies the influence of heat flux on the circumference-averaged heat flux for two mass flow velocities at a constant pressure of 26.4 bar and vapor quality of 0.3, varying the nominal oil mass fraction from 0 wt.-% to 3 wt.-%. It is striking to observe how heat transfer progressively deteriorates as the nominal oil mass fraction is increased, counting for the low (75 $\mathrm{kg\,m^{-2}\,s^{-1}}$) as well as for the high (250 $\mathrm{kg\,m^{-2}\,s^{-1}}$) mass flow velocity. Differing to the pure substance, the boiling curves of the CO_2-lubricant mixtures at the low mass flow velocity show less inclination than at the high mass flow velocity. This effect could be related to an increased suppression of nucleate boiling by the augmented oil content in the liquid, which would also be in line with the nearly constant trend for heat fluxes below 12 $\mathrm{kW\,m^{-2}}$.
The pressure dependent influence of heat flux (power law with exponent $n(p_r)$) on nucleate boiling heat transfer (α_{nb}) was determined from boiling curves in double logarithmic representation, equaling the slope of the curve in the nucleate boiling dominated regime.

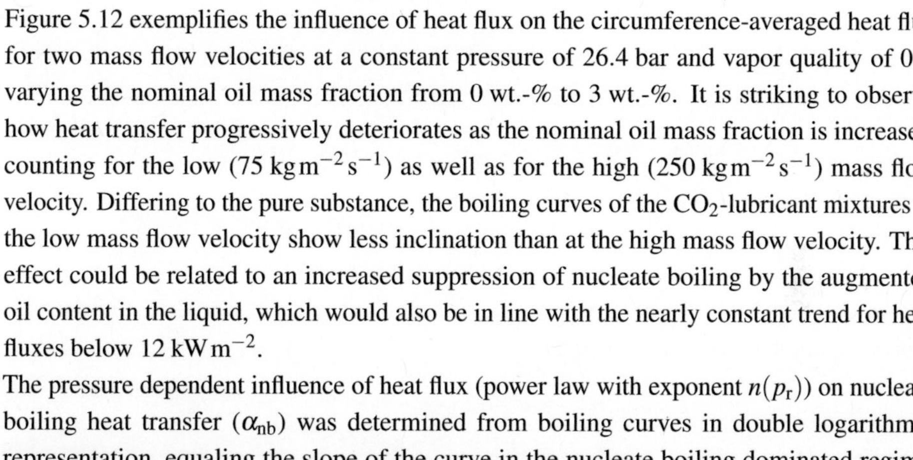

Figure 5.11: Influence of heat flux on the circumference-averaged heat transfer coefficient in the micro-fin tube at conditions of strong convection

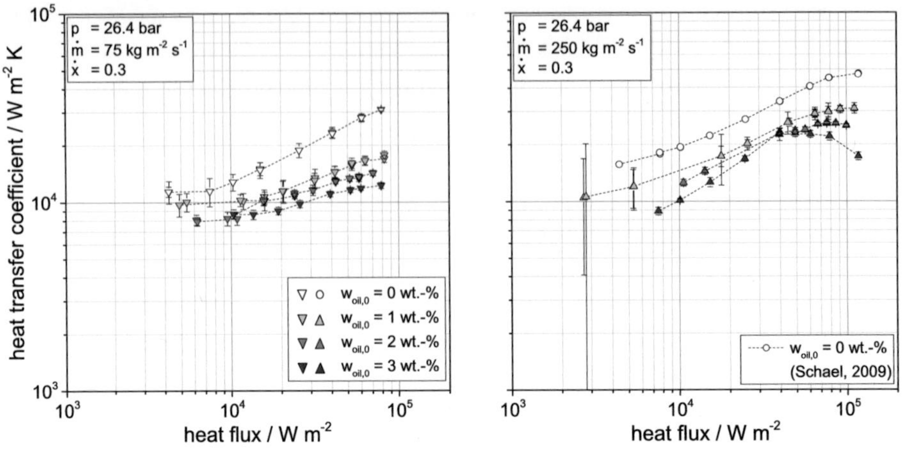

Figure 5.12: Influence of oil on the circumference-averaged heat transfer coefficient at a pressure of 26.4 bar, a vapor quality of 0.3 and mass flow velocities of 75 $\mathrm{kg\,m^{-2}\,s^{-1}}$ (left) and 250 $\mathrm{kg\,m^{-2}\,s^{-1}}$ (right)

However, heat transfer coefficients had to be modified (α_{mod}) in order to subtract the combined influence of vapor quality.

$$\alpha_{\mathrm{nb}} \propto \dot{q}^{n(p_{\mathrm{r}})}\,\mathrm{F}(\dot{x}) \tag{5.5}$$

$$\alpha_{\mathrm{mod}} = \alpha_{\mathrm{exp}}\,\mathrm{F}(\dot{x})^{-1} \propto \dot{q}^{n(p_{\mathrm{r}})} \tag{5.6}$$

$$\mathrm{F}(\dot{x}) = 1 - \dot{x}\,p_{\mathrm{r}}^{0.1}\left(\frac{\dot{q}}{\dot{q}_{\mathrm{c}}}\right)^{0.3} \tag{5.7}$$

where $\mathrm{F}(\dot{x})$ is the vapor fraction dependency according to Steiner (2002), confirmed for the micro-fin tube by Schael, $p_{\mathrm{r}} = p\,p_{\mathrm{c}}^{-1}$ is the reduced pressure and $\dot{q}_{\mathrm{c}}$ corresponds to a reference heat flux (see Eq. E.169 in App. E.3).

Determined values of $n(p_{\mathrm{r}})$ for CO_2-POE oil mixtures containing a nominal oil mass fraction of 1 wt.-% and mass flow velocities raging from 150 to 500 $\mathrm{kg\,m^{-2}\,s^{-1}}$ are presented in Fig. 5.13 as function of reduced pressure. The dotted line represents the averaged exponent ($n(p_{\mathrm{r}}) = 0.51$) experimentally determined by Schael from measurements with pure CO_2 in the micro-fin tube.

At low reduced pressures, the relation derived by Schael was found to be in very good agreement to values measured in this work. With increasing pressure, a progressing decline in $n(p_{\mathrm{r}})$ and a steady decrease with mass flow velocity can be seen. This observation is linked to the progressing degradation of heat transfer in the high heat flux region, as discussed previously.

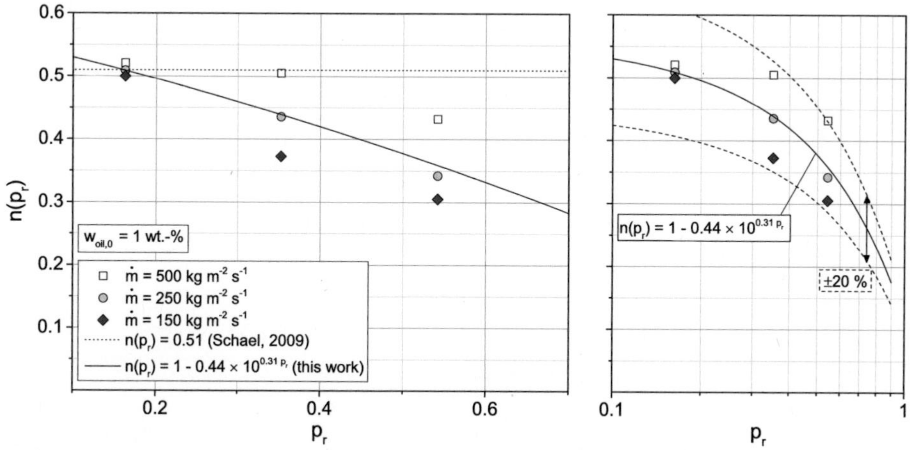

Figure 5.13: Exponent $n(p_\mathrm{r})$ of heat flux as function of reduced pressure determined for $w_{oil} = 1$ wt.-%

Based on the Steiner expression for the exponent $n(p_\mathrm{r})$, a new correlation to describe the trend of measurement data was derived, see Eq. 5.8.

$$n(p_\mathrm{r}) = 1 - 0.44 \times 10^{0.31\,p_\mathrm{r}} \tag{5.8}$$

Plotted as solid line, a comparison of the newly derived relationship to measured data is given in Fig. 5.13. The right diagram illustrates the correlation's quality on a single-logarithmic scale, indicating a deviation of ± 20 % by dashed lines. The correlation roughly describes the observed trend of $n = f(p_\mathrm{r})$.

5.3.2 Influence of the reduced pressure

The boiling curves shown in Fig. 5.11 indicate the influence of pressure on heat transfer. In the convective boiling region, heat transfer is reduced as the pressure increases. In contrast, nucleate boiling heat transfer is enhanced with pressure, however, solely for heat fluxes below a limiting heat flux (see Sec. 4.3.1).

For CO_2, the dependency of heat transfer on reduced pressure was discussed by Schael (2009), showing similar behavior as observed in the smooth tube. The observed trend was reproduced successfully by applying the pool boiling correction function by Gorenflo (2002) for finned surfaces.

$$\mathrm{F}(p_\mathrm{r})_{\text{micro-fin tube}} = EF_\mathrm{A}{}^{0.5}\,\mathrm{F}(p_\mathrm{r})_{\text{smooth tube}} \tag{5.9}$$

$$\mathrm{F}(p_\mathrm{r})_{\text{smooth tube}} = 2.692\,p_\mathrm{r}^{0.43} + \frac{1.6\,p_\mathrm{r}^{6.5}}{1 - p_\mathrm{r}^{4.4}} \tag{5.10}$$

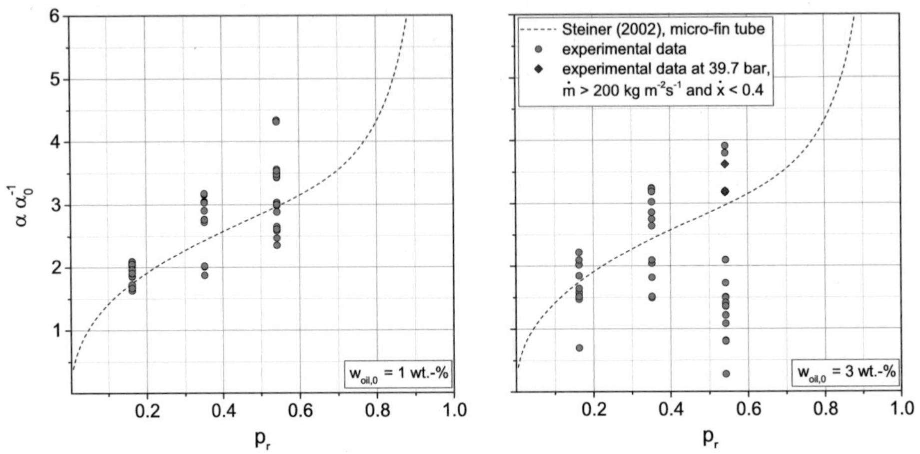

Figure 5.14: Influence of reduced pressure on the circumference-averaged heat transfer coefficient in the micro-fin tube for nominal oil concentrations of 1 wt.-% (left) and 3 wt.-% (right)

where $EF_A = 2$ is the area enhancement factor for the micro-fin tube and $F(p_r)_{\text{smooth tube}}$ is the Steiner pressure function for flow boiling in smooth tubes.
Figure 5.14 illustrates the influence of reduced pressure on heat transfer at different nominal oil mass fractions, normalized to $\alpha_0 = 18900\,\mathrm{W\,m^{-2}\,K^{-1}}$ (CO_2 specific value according to Steiner). Dashed lines represent Eq. 5.9. Since the pressure function of Steiner refers to the standard condition of $\dot{q}_0 = 150000\,\mathrm{W\,m^{-2}\,K^{-1}}$ for CO_2, determined ratios $\alpha\,\alpha_0^1$ were extrapolated according to the heat flux applied (see dependency given in Eq. 5.5). Equation 5.8 was applied to determine the exponent of the heat flux.
Equation 5.9 applies satisfactorily to experimental values. As for pure CO_2, relative large scattering of data points was noticed, in case of the CO_2-lubricant mixtures investigated even stronger pronounced due to reasons quoted before. Scattering, and deviations in general, increase as the reduced pressure and local oil concentration increases. Further, as described in Sec. 5.3.1, the slope of boiling curve progressively deteriorates with heat flux under these conditions. Values determined for high mass flow velocities and low vapor qualities agree better to Eq. 5.9, see dark filled diamond symbols in the right plot at $p_r = 0.54$.

5.3.3 Influence of the mass flow velocity

The influence of the mass flow velocity on the circumference-averaged heat transfer coefficient is illustrated in Fig. 5.15, taking the influence of vapor quality, heat flux and pressure into account. For low ($\leq 200\,\mathrm{kg\,m^{-2}\,s^{-1}}$) and most medium mass flow velocities ($\leq 300\,\mathrm{kg\,m^{-2}\,s^{-1}}$) the heat transfer coefficient generally increases with mass flow velocity, on average showing a stronger increase compared to smooth tube experiments ($\alpha \propto \dot{m}^{0.25}$).

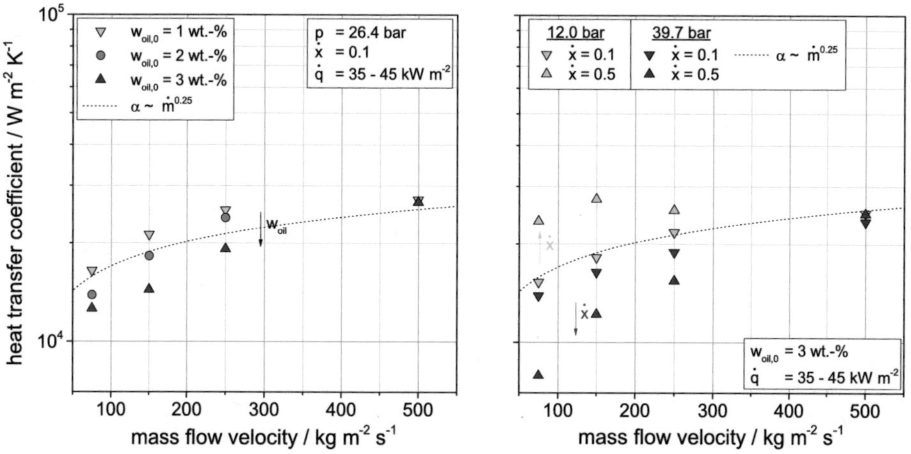

Figure 5.15: Influence of the mass flow velocity on the circumference-averaged heat transfer coefficient in the micro-fin tube for heat fluxes from 35 to 45 $\mathrm{kW\,m^{-2}}$

For mass flow velocities $> 300\,\mathrm{kg\,m^{-2}\,s^{-1}}$ and typically lower pressures, a reverse trend was recurrently observed. This phenomenon was already discovered in pure CO_2 experiments (Schael, 2009). As stated there, reasons could be a turnover from laminar to turbulent flow in the micro-fins, suppressing nucleate boiling by the increased convection, or the observed *over-flowing* of the micro-fins intensified with increasing mass flow velocity. The latter can be interpreted as a reduction of surface efficiency. The observations concord to the principle of operation of micro-fins, enhancing heat transfer by promoting tube wetting at low mass flow velocities (turnover from stratified to annular flow), whereas no further improvement by change in flow pattern is to be expected beyond the point of fully developed annular flow.
Comparable behavior of heat transfer with respect to the mass flow velocity was confirmed for CO_2-POE oil mixtures containing different nominal oil mass fractions. This circumstance can be inferred from the right plot in Fig. 5.15, showing determined heat transfer coefficients as function of mass flow velocity and nominal oil concentrations at a constant pressure of 26.4 bar and a vapor quality of 0.1. However, and as already discussed previously, the influence of heat flux on the nucleate boiling heat transfer is related to the local oil mass fraction and varies dependent on pressure and mass flux. This implies that, within a certain range of elevated heat fluxes (dependent on the condition of operation), heat transfer deteriorates as the vapor quality or mass flow velocity is increased. For instance, an increase in the heat transfer with vapor quality was observed under low pressure (light gray symbols), while the opposite behavior was encountered under high pressure (dark gray symbols). Thus, the dependency of the heat transfer coefficient on the mass flow velocity is related in a very complex manner to the combination of operating parameters. Further plots affirming this circumstance are shown in App. F.2.3.

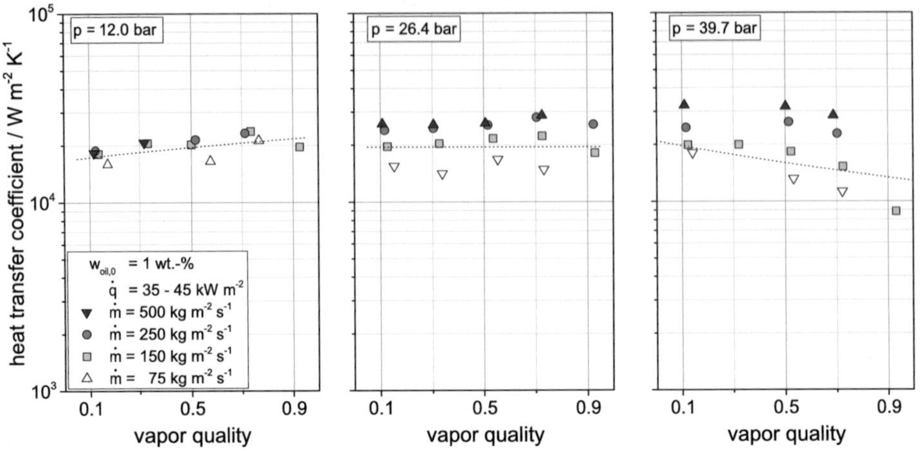

Figure 5.16: Influence of vapor quality on the circumference-averaged heat transfer coefficient in the micro-fin tube for CO_2-POE oil mixtures with $w_{oil} = 1$ wt.-% at heat fluxes of 35 to 45 $\mathrm{kW\,m^{-2}}$

5.3.4 Influence of the vapor quality

Figure 5.16 depicts the influence of vapor quality on the circumference-averaged heat transfer coefficient for CO_2-POE oil mixtures with a nominal oil mass fraction of 1 wt.-% at an average heat flux of 40 $\mathrm{kW\,m^{-2}}$, i.e., in the nucleate boiling regime. With increasing vapor quality, convection is primarily enhanced. For conditions of advanced convective influence on the overall heat transfer (low pressures, left plot), increasing the vapor quality also yields a slight increase of the heat transfer coefficient. As the pressure is increased and the influence of nucleate boiling gains on overall weight, the tendency observed changed to decreasing heat transfer coefficients with vapor quality (see right diagram).
As the nominal oil concentration was increased, the lubricant-induced deterioration effect was found to be generally augmented at elevated heat fluxes. This observation is illustrated in Fig. 5.17 (left). Only at low pressures and low mass flow velocities the heat transfer coefficient slightly increases. This might be attributed to the stronger convective influence at these conditions and flow pattern effects, i.e., promoting the transition to slug and further to annular flow.
At low heat fluxes, therefore, a stronger influence by convective boiling, an enhancement in heat transfer with vapor quality was observed within the whole range of nominal oil mass fractions. From the right diagram in Fig. 5.17 a steady and comparable increase of the heat transfer coefficient with vapor quality can be seen. Likewise, this circumstance complies to observations made in the micro-fin tube using pure CO_2 (Schael, 2009).

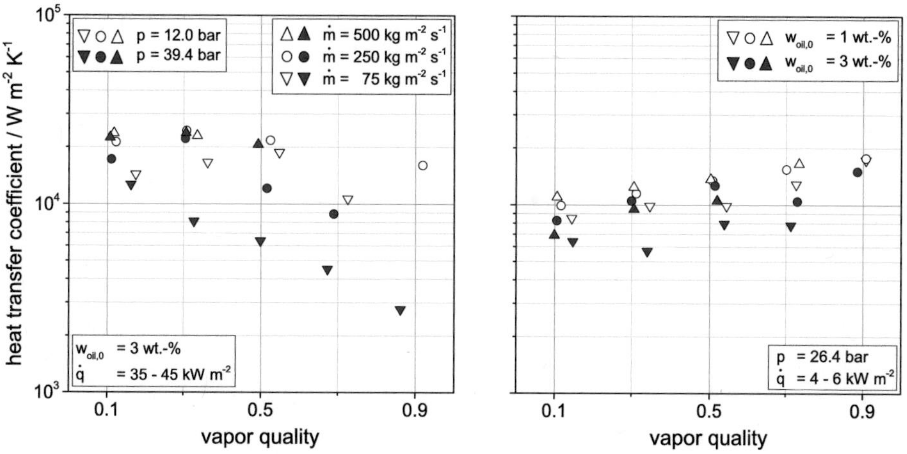

Figure 5.17: Influence of vapor quality on the circumference-averaged heat transfer coefficient in the micro-fin tube for CO_2-POE oil mixtures with $w_{oil} = 3$ wt.-% at heat fluxes of 35 to 45 $kW m^{-2}$ (left) and for $p = 26.4$ bar at heat fluxes of 4 to 6 $kW m^{-2}$ (right)

5.3.5 Influence of the flow pattern

Comparing both, smooth and micro-fin tube, the main differences regarding the geometry-induced influence on heat transfer are related to the area enhancement and the effect on flow pattern. As discussed in Sec. 5.1, micro-fins improve tube wetting by inducing a swirl, guiding the liquid along the helix direction towards the top of the tube, leading to an enhancement in tube wetting and a transition to annular flow at lower mass flow velocities. Therefore, the most pronounced positive effect of lubricant on heat transfer observed in the smooth tube, namely increasing the wetted perimeter of the tube for a stratified/wavy flow, drops in the micro-fin tube. This conclusion is affirmed by the boiling curves shown in Fig. 5.12 for the low mass flow velocity (left plot), revealing a progressive deterioration of heat transfer with increasing oil mass fraction. Local heat transfer coefficients for these measurements are shown in Fig. 5.18 (top). The asymmetric distribution of heat transfer coefficients around the circumference results from the induced swirl by the micro-fins. It also indicates the helix direction, that is in counter counting direction of segments. Looking at the order of magnitude of local heat transfer coefficients it is evident that the complete tube surface is wetted. The decrease of heat transfer coefficients for increasing oil mass fractions can be clearly seen.

The bottom plots show local heat transfer coefficients of CO_2-POE oil mixtures with $w_{oil} = 1$ wt.-% determined at a mass flow velocity of 250 $kg m^{-2} s^{-1}$, a vapor quality of 0.5 and pressures of 12.0 – 39.7 bar. Boiling curves determined under these conditions of operation were presented before, see Fig. 5.11 (left).

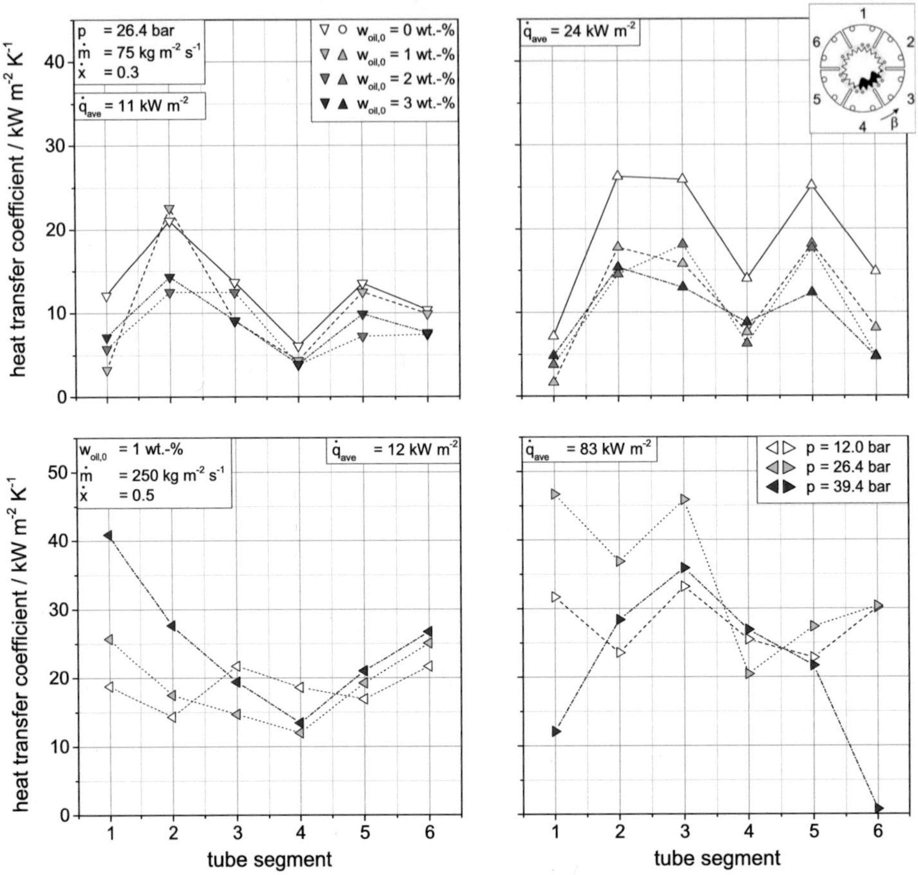

Figure 5.18: Influence of flow pattern on local heat transfer in the micro-fin tube

Local heat transfer coefficients at the moderate heat flux of $12\,\mathrm{kW\,m^{-2}}$ are highest on average at 39.4 bar, see bottom left diagram. This can be explained by the stronger influence of nucleate boiling, especially for segments 1, 2, 5 and 6. These segments are mainly wetted due to the guidance of liquid by the micro-fin helix. Due to gravity effects, the thickness of the liquid film in the upper part of the tube (segments 1, 2, 6) is smaller than at the bottom (segment 4, in the first instance). It was already observed in smooth tube experiments and described in literature that nucleate boiling heat transfer can be significantly enhanced as the liquid film becomes very thin (Mesler and Mailen, 1977). Additionally, the spatial position of these segments and the swirl flow may as well induce a kind of rinsing effect, promoting the removal of lubricant from the surface.

The strong decline of the overall heat transfer coefficient at 39.7 bar as the average heat flux equals $83\,\mathrm{kW\,m^{-2}}$ (bottom right diagram) can be related to a distinct change in flow pattern.

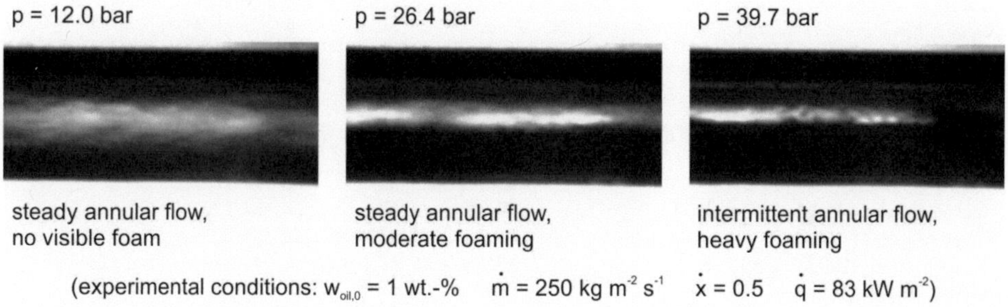

Figure 5.19: Flow patterns observed in the micro-fin tube for CO_2-POE oil mixtures with $w_{oil} = 1$ wt.-% at a mass flow velocity of 250 $kg\,m^{-2}\,s^{-1}$, vapor quality of 0.5 and an average heat fluxes of 83 $kW\,m^{-2}$

From the observation of flow patterns, a low amplitude pulsating flow was noticed, leading to perceivable periods of partial dryout. The latter can be connected to the an increasing foam formation at higher pressures. Flow patterns shown in Fig. 5.19 recorded under these conditions support the assumption, revealing significant foaming at 39.7 bar owing to the presence of lubricant.

5.3.6 Comparison to predictive methods

Experimentally determined circumference-averaged heat transfer coefficients were compared to predictions by established correlations using mixture properties. Besides available smooth tube correlations (refer to Sec. 4.3.5), predictive methods derived for enhanced tubes were tested. Calculation methods not referring heat transfer to the corresponding smooth tube surface of diameter d_i were converted respectively. Evaluated error statistics for all methods tested are listed in Tab. 5.3.
Smooth tube correlations were considered within the evaluation of micro-fin tube data since micro-fin tube correlations available in literature are rare. The statistical values in Tab. 5.3 and the parity plots in App. F.2.3 reveal that these methods are not applicable to the micro-fin tube. Solely the correlation of Thome and El Hajal (2004) shows relatively small scattering referring to its parity plot.
From the micro-fin tube correlations tested, the correlation of Cavallini et al. (1999) proved most accurate in comparison to the experimental data obtained in this work, followed by the correlations of Thome et al. (1997) and Koyama et al. (1995). The Steiner correlation adapted to CO_2 micro-fin tube data by Schael (2009) did not turn out to be applicable to the CO_2-POE oil mixture data, showing large deviations. Parity plots for these four correlations mentioned are shown in Fig. 5.20, visualizing considerable scattering of plotted data. That underlines again that flow boiling heat transfer of CO_2-POE oil mixtures can not be predicted precisely by present correlations.

Table 5.3: Error statistics of evaluated heat transfer correlations with respect micro-fin measurement data

Correlation by author	MPE	MAPE	Data that yields APE $\leq 30\,\%$
Micro-fin tube correlations			
Schael (2009) ($F(p_r)_{\text{micro-fin tube}}$, $n(p_r) = 0.51$)	+55.3 %	65.9 %	39.1 %
Hu et al. (2008a)	+280.8 %	290.9 %	20.6 %
Yun et al. (2002)	−34.6 %	55.2 %	22.1 %
Cavallini et al. (1999)	+39.0 %	47.2 %	61.1 %
Thome et al. (1997)	+63.3 %	72.8 %	52.4 %
Koyama et al. (1995)	+28.5 %	59.5 %	49.8 %
Smooth tube correlations			
Li et al. (2014)	−33.3 %	60.3 %	19.3 %
Fang (2013)	−25.8 %	38.7 %	38.5 %
Schael (2009)	−11.2 %	43.9 %	40.7 %
Gao et al. (2008)	−79.9 %	80.7 %	1.1 %
Cheng et al. (2008b)	−60.4 %	64.5 %	8.5 %
Cheng et al. (2008b) (S=1)	−41.4 %	51.9 %	16.5 %
Wojtan et al. (2005b)	−46.8 %	56.1 %	14.1 %
Thome and El Hajal (2004)	−46.1 %	52.3 %	11.9 %
Steiner (2002)	−31.0 %	44.8 %	29.0 %
Kattan et al. (1998c)	−38.0 %	52.1 %	18.9 %

To illustrate the influence of oil on the accuracy of prediction, error statistics for the Steiner correlation adapted to pure CO_2 micro-fin tube data by Schael and the correlation of Cavallini et al. are presented in Tab. 5.4, differentiated by nominal oil mass fractions investigated in this work. The comparison clearly shows that the deterioration in heat transfer with increasing oil content is not adequately captured by any correlation, hence, resulting in an over-estimation of the heat transfer coefficient. For both correlations, the mean percentage error (MPE) increases steadily, resulting in a significant reduction of data points within a range of error of $\pm 30\,\%$. The progressive over-estimation of heat transfer with oil coincides with the overall positive MPE values shown in Tab. 5.3.

Table 5.4: Influence of lubricant on the accuracy of prediction

Cavallini et al. (1999)				
Nominal oil mass fraction	0 wt.-%	1 wt.-%	2 wt.-%	3 wt.-%
MPE	+6.4 %	+21.1 %	+42.3 %	+71.3 %
MAPE	15.2 %	30.0 %	45.2 %	79.5 %
Data that yields APE $\leq 30\,\%$	91.9 %	71.0 %	42.7 %	44.8 %
Steiner (2002), adapted to CO_2 and micro-fin tube by Schael (2009)*				
Nominal oil mass fraction	0 wt.-%	1 wt.-%	2 wt.-%	3 wt.-%
MPE	+17.7 % (−4.7 %)	+33.5 % (+33.2 %)	+62.5 % (+62.5 %)	+93.8 % (+97.5 %)
MAPE	33.0 % (15.4 %)	47.3 % (47.4 %)	64.2 % (64.2 %)	100.7 % (104.7 %)
Data that yields APE $\leq 30\,\%$	69.4 % (91.0 %)	40.8 % (43.5 %)	27.7 % (27.7 %)	33.8 % (35.3 %)

* values in brackets only evaluated for $\dot{m} \leq 450\,\mathrm{kg\,m^{-2}\,s^{-1}}$

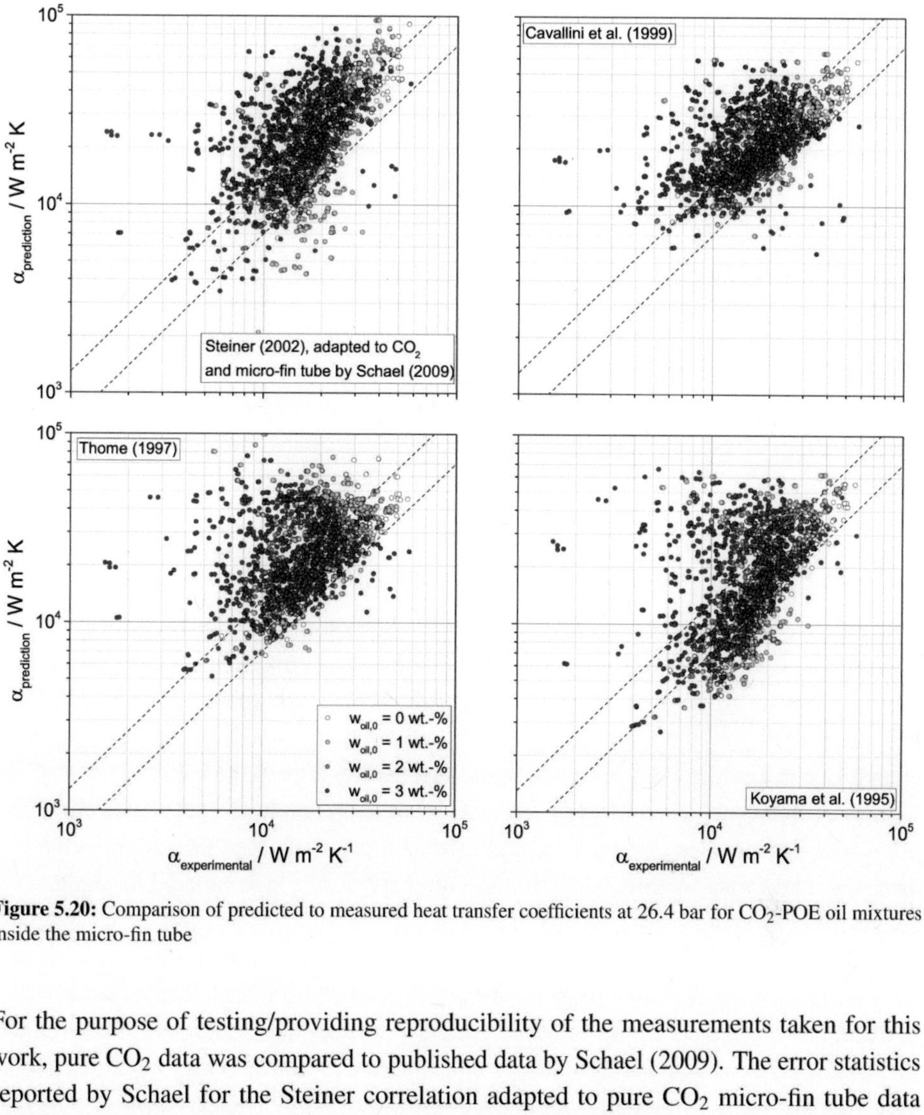

Figure 5.20: Comparison of predicted to measured heat transfer coefficients at 26.4 bar for CO_2-POE oil mixtures inside the micro-fin tube

For the purpose of testing/providing reproducibility of the measurements taken for this work, pure CO_2 data was compared to published data by Schael (2009). The error statistics reported by Schael for the Steiner correlation adapted to pure CO_2 micro-fin tube data (MPE $= -3.7$ %, MAPE $= 17.3$ %, 82.4 % of all data within range of error of ± 30 %, including only measurements at $\dot{m} < 260\ \mathrm{kg\,m^{-2}\,s^{-1}}$) and the correlation of Cavallini et al. (MPE $= +9.3$ %, MAPE $= 17.6$ %, 82.8 % of all data within range of error of ± 30 %) correspond well to those determined in this work, see values in column '0 wt.-%' in Tab. 5.4.

5.3.7 Conclusion

Results from flow boiling experiments of CO_2-POE oil mixtures containing nominal oil mass fractions of $1-3$ wt.-% conducted at pressures of $12.0-39.7$ bar, mass flow velocities of $75-500\,\mathrm{kg\,m^{-2}\,s^{-1}}$, vapor qualities of $0.1-0.9$ and heat fluxes of up to $120\,\mathrm{kW\,m^{-2}}$ were presented. The influences of lubricant and operating parameters on heat transfer were discussed.

The measurements presented reveal that the pressure-dependent influence of heat flux on the heat transfer coefficient did not match observations made by Schael (2009) for pure CO_2. While Schael reported of a nearly constant heat flux exponent n, which does not correspond to the trend described by Steiner (2002) or Gorenflo (2002), a different relationship for $n(p_r)$ was derived based on measurement data at a moderate nominal oil mass fraction of 1 wt.-%. Equation 5.8 is recommended to be used in junction with the Steiner heat transfer correlation, as it describes the experimental data of CO_2-POE oil mixtures in the range of operation investigated significantly better.

Convective and nucleate boiling regimes were identified in the micro-fin tube experiments, whereby convective boiling was primarily observed at the lowest pressure of 12.0 bar. General observations made for the heat transfer match well to those made for pure CO_2 (Schael, 2009), among these, the partial dryout of the tube surface under high heat flux conditions. Differing to the pure substance, CO_2-POE oil mixtures were affected by additional flow pattern changes, the presence of foam up to foam flow and a deteriorated heat transfer at high heat fluxes likely due to an advancing suppression of nucleate boiling. Although the lubricant exhibits no miscibility gap in the range of operation investigated, it accumulates at the evaporation sites of CO_2 (i.e., wall surface), primarily resulting in an increase in viscosity, mass transport resistance in the liquid and a rise in bubble point temperature (Bandarra Filho et al., 2009).

The positive effect of lubricant on tube wetting observed for stratified flow in the smooth tube, under certain conditions enhancing the overall heat transfer, did not apply to the results in the micro-fin tube. Local heat transfer coefficients determined for a mass flow velocity of $75\,\mathrm{kg\,m^{-2}\,s^{-1}}$ using pure CO_2 clearly indicated the presence of an evaporating liquid film, although the flow patterns determined inside the succeeding (smooth) sight glass tube suggested wavy flow (see Sec. 5.1). Hence, the enhanced tube geometry can already lead to a complete wetting of the tube at comparably low mass flow rates. With the negative effects of the presence of lubricant on heat transfer even stronger predominating compared to the smooth tube (especially the suppression of nucleate boiling) the presence of lubricant in micro-fin tubes leads to a more pronounced decrease of overall heat transfer coefficients.

Experimentally obtained heat transfer coefficients were used for testing the accuracy of established and new predictive methods, including correlations developed for smooth and micro-fin tubes. Of the recent correlations, the ones by Li et al. (2014), Hu et al. (2008a) and Gao et al. (2008a) were developed based on refrigerant-lubricant mixture data. In testing the applicability of available correlations to the experimental data gained in this work, the deteriorating effect of lubricant on heat transfer was confirmed. Correlations proven to be accurate when applied to pure CO_2 (Schael, 2009) yielded a general over-prediction. Pure CO_2 experiments conducted in this work affirmed the outstanding accuracies of the Cavallini et al. (1999) correlation and, for mass flow velocities $< 450\,\mathrm{kg\,m^{-2}\,s^{-1}}$, the Steiner correlation modified by Schael. With increasing nominal oil mass fraction, deviations and the mean percentage error increased, denoting an advancing over-prediction of heat transfer coefficients. Highest overall accuracy was determined for the Cavallini et al. (1999) correlation, predicting 61.1 % of the data within a range of error of ±30 %. Correlations developed based on data of other refrigerant-lubricant mixtures did not yield a good agreement to the experimental data base.

6 Conclusion and outlook

Known as low temperature refrigerant R744, carbon dioxide (CO_2) significantly re-gained importance as working fluid in compression-type refrigeration systems within the last two decades. Extensive work on flow boiling characteristics of CO_2, as for instance, the study of Schael, contributed to improving predictive methods for heat transfer and pressure drop. However, most investigations focused on pure refrigerant evaporation, neglecting the presence of lubricating oil required in actual refrigeration applications for lubricating the moving parts of the compressor.

Concomitant with the lubricant type and concentration representing further influencing variables, the number of operational parameters to be verified for ensuring an efficient designing of the evaporator increased considerably. Previous work on refrigerant-lubricant mixtures indicate that flow boiling characteristics differ from those of pure refrigerants or refrigerant-mixtures in many ways, including the tendency to foaming and vast differences in liquid viscosity and saturation temperature. The lack of knowledge concerning refrigerant-lubricant mixture properties was recently summarized by Bandarra Filho et al. (2009). Consequently, the applicability of established correlations to refrigerant-lubricant mixtures is to be questioned. Significant uncertainty in predicting the respective design parameters satisfactorily leads to an over-sized and uneconomical design of refrigeration units.

Assuming that correlations and predictive models approved for CO_2 cannot be applied to CO_2-lubricant mixtures satisfactorily, the aim of this work was to provide a broad experimental data basis related to the range of operation of typical applications in refrigeration. Main focus regarding the geometry of the evaporator was laid on highly efficient micro-fin evaporator tubes. Besides, experiments in a smooth evaporator tube were intended for validation and recognitional purposes. Established as well as lately published predictive methods were envisioned for examination in verifying the hypothesis made at the outset. Where appropriate, modifications should be developed.

6.1 Contents and results of this work

Experimental investigations of flow boiling CO_2-lubricant mixtures were conducted using the test facility of the Institute of Thermal Process Engineering at the Karlsruhe Institute of Technology (KIT). Two electrically heated measurement sections constructed and validated in previous work (Schael, 2009)(Niederkrüger, 1991) were used, namely a nickel smooth tube of 14.0 mm in inner diameter and a copper micro-fin tube having a core diameter of

8.62 mm. Both measurement sections feature a segmentation of the tube wall in circumferential direction allowing measurements under near-isothermal surface condition. That way, heat transfer coefficients were determined locally in dependence of the degree of surface wetting, closely related to the present flow pattern. Simultaneously, the pressure drop over the measurement section was measured and the flow pattern determined by high speed video recordings made at a non-heated sight glass tube succeeding the measurement section. POE oil with a standard viscosity of 85 $mm^2 s^{-1}$ at 40 °C and complete miscibility with CO_2 in the range of operation investigated (Reniso C 85 E) was used.
Flow boiling experiments were carried out at pressures of 10 – 40 bar, heat fluxes up to 120 $kW m^{-2}$, vapor qualities of 0.1 – 0.9 and mass flow velocities of 75 – 300 $kg m^{-2} s^{-1}$ in the smooth tube, respectively 75 – 500 $kg m^{-2} s^{-1}$ in the micro-fin tube. Total oil mass fractions set up amounted to 1 – 5 wt.-% with respect to the sub-cooled liquid mixture. Lubricant-induced influences on flow pattern, pressure drop and heat transfer observed were discussed.
Experimentally determined flow patterns, pressure drops and heat transfer coefficients were compared to established predictive methods. Based on the realizings modifications to improve the accuracy in prediction were developed. Mixture properties were determined using property methods adapted to available material data of CO_2 and Reniso C 85 E POE oil as well as contract measurements.

6.1.1 Flow pattern

Flow patterns were classified qualitatively into stratified, wavy, intermittent and annular flow as well as combinations of these. It was found that flow patterns of CO_2-POE oil mixtures are basically comparable to those determined for pure CO_2. The most significant difference to pure refrigerant observations relates to the tendency to foam formation. Foaming can induce a basic change in flow pattern and, for instance, enhance the degree of wetting of the tube surface. As well, foam denotes an additional resistance in mass transfer for the evaporating liquid and reduces the free cross-section of flow. The presence of foam was addressed when experimentally observed.
Schael (2009) identified the Steiner (2002) and Thome and El Hajal (2003) flow pattern maps as generally well applicable to CO_2 adiabatic two-phase flow in smooth and micro-fin tubes, respectively. Major discrepancies between experimentally determined and predicted flow patterns were reported for the transitions from wavy to intermittent or annular flow and non-annular to annular flow. Especially in the micro-fin tube annular flow occurs at significantly lower mass flow rates and vapor qualities than predicted by flow pattern maps developed from smooth tube data.
For both, the smooth and the micro-fin tube, the intermittent to annular flow transition for adiabatic two-phase flow of CO_2-POE oil mixtures containing oil mass fractions of up

to 5 wt.-% could be described successfully by mean vapor Froude number ($Fr_{V m,I\text{-}A}$) in dependency of the Martinelli parameter (X_{tt}). The transition criteria developed in this work are given by Eq. 4.2 and Eq. 5.1.
Based on flow patterns observed in the smooth tube, a modification of the wavy to intermittent transition of the Steiner flow pattern map was developed. Two boundary conditions were distinguished and considered in modeling the flow transition: high system pressure and high vapor qualities ($\dot{x} \rightarrow 1$) corresponding to low liquid levels ($\tilde{h}_L \rightarrow 0$) in the horizontal tube. Most researchers neglected the liquid phase velocity when referring to the relative velocity due to density ratio generally being $\rho_L \rho_V^{-1} >> 1$. This assumption is not true at elevated system pressure, leading to overestimating the relative velocity. For shallow liquid films under high vapor quality conditions, on the other hand, the superficial liquid velocity at the vapor-liquid interface does not correspond to the mean liquid phase velocity. In this case a linear velocity profile in the liquid film can be assumed. Following, the superficial liquid velocity corresponds to $u_L(z = \tilde{h}_L) = 2\,\bar{u}_L$. The transition criterion by Steiner was modified incorporating these considerations, showing good agreement to flow patterns experimentally determined in the smooth and the micro-fin tube.
The flow pattern transitions of special technical relevance, namely wavy to intermittent and non-annular to annular flow, could be predicted in good accordance to experimental observations under adiabatic conditions for both tube geometries by the modified transition criteria developed in the scope of this work. It must be noted, however, that an increasing disagreement is encountered as the intensity of foam formation increases. Foam flow or foam-dominated flow is not considered in available flow pattern maps. Such insufficiently captured conditions are encountered with increasing vapor quality, heat flux and oil concentration.

6.1.2 Pressure drop

The pressure drop was measured during adiabatic flow pattern investigations as well as during heat transfer experiments. Generally, the two-phase pressure drop consists of contributions due to friction and acceleration of the evaporating liquid to the vapor phase velocity. The frictional pressure (i.e., pressure drops determined under adiabatic conditions) of CO_2 was reported to be predictable with high accuracy using the established correlations of Müller-Steinhagen and Heck (1986) and Storek and Brauer (1980) (Schael, 2009). Regarding the micro-fin tube, Schael proposed penalty factors for the correlations mentioned. That way, pressure drops in the smooth and the micro-fin tube were predicted in very good accordance to experimentally determined values. Both pressure drop correlations were confirmed to perform well compared to other predictive methods tested. It was found that substituting the inside core diameter by an expression for the hydraulic diameter proposed by Kedzierski and Goncalves (1999) resulted in highly accurate predictions made for the pressure drop in the micro-fin tube using the correlations of Müller-Steinhagen and Heck

and of Xu and Fang (2012). In addition and analogous to Schael (2009), adaptations by introducing multiplying factors to the correlations of Müller-Steinhagen and Heck and Storek and Brauer were considered. The correlation of Storek and Brauer performed best of all tested correlations in predicting adiabatic pressure drops of CO_2-POE oil mixtures when applying a factor of $PF_{\Delta p} = 1 - 0.36\dot{x}$, yielding a MPE $= -3.5$ % and predicting over 91 % of all data within a range of error of ± 30 %.
Diabatic total pressure drops were found to be strongly influenced by the presence of lubricant. In contrast to flow boiling of pure CO_2 flow patterns of CO_2-POE oil mixtures often exhibit foaming, gaining in significance with increasing heat flux, vapor quality and nominal oil mass fraction. Foam as obstructing phase reduces the free cross-section of flow and leads to an increase in pressure drop. It does not affect the momentum pressure drop of the evaporating liquid. Moreover, foam adhering to the tube surface relates to further frictional forces acting against the flow of liquid. None of the tested correlations combined with the respective acceleration pressure drop model was able to predict the total pressure drop in acceptable agreement to the experimental data.
Applying the hydraulic diameter defined by Kedzierski and Goncalves to the three correlations mentioned in combination with momentum pressure drop models, diabatic pressure drops in the micro-fin tube were predicted in relatively good agreement. Best results were achieved using the micro-fin correlation proposed by Cavallini et al. (1997). With 67.5 % of all data predicted in a range of error of ± 30 % and a mean percentage error of MPE $= -0.8$ % it can be used to roughly estimate the total two-phase pressure drop of CO_2-POE oil mixtures.

6.1.3 Heat transfer

Heat transfer was investigated at isothermal wall condition including locally determined heat transfer coefficients for each tube segment and circumference-averaged values. From local measurements along the perimeter conclusions on the present flow pattern and degree of wetting can be drawn. The electrical power input was equivalent to heat fluxes of up to 120 $\mathrm{kW\,m^{-2}}$.
Experiments undertaken in the **smooth evaporator tube** aimed at identifying general lubricant-induced effects. From these measurements it was concluded that for low vapor qualities and heat fluxes CO_2-POE oil mixtures show similar behavior compared to pure CO_2, referred to the order of magnitude of determined heat transfer coefficients and the trend of boiling curves. However, heat transfer coefficients decreased by tendency with increasing oil mass fraction. In concordance to pertinent literature the degradation of heat transfer with oil mass fraction was assumed to be related in first place to an increase in mass flow resistance in the lubricant-rich liquid boundary layer. The most severe influence on heat transfer was encountered at high heat fluxes and local oil mass fractions, i.e., high vapor

quality and nominal oil mass fraction. Here, the mass flow resistance in the liquid due to the presence of non-evaporating lubricant remaining at the site of evaporating and foam further hindering bubbles in leaving the surrounding liquid become crucial. Evaporating refrigerant withheld in proximity of the tube surface results in a vapor/foam layer comparable to film boiling. The heat transfer coefficient decreases; at the onset gently and then at an increasing rate. Flow boiling in this region is characterized by notably elevated wall temperatures as well as highly fluctuating conditions resembling a type of boiling crisis as known for pure refrigerants. However, the increase in wall temperature is significantly less pronounced and semi-stable conditions at moderate wall temperatures could be reached for CO_2-POE oil mixtures. Contrarily, local heat transfer coefficients were also found to be enhanced by the presence of lubricant, for instance, under conditions of low liquid level. The positive influence on heat transfer could be related to lubricant-induced changes in flow pattern increasing the degree of wetting of the tube surface. The correlations of Steiner (2002) and Fang (2013) were identified to predict heat transfer coefficients with highest accuracy to experimental data. Accuracy in prediction progressively diminished with increasing oil mass fraction.

Many observations for the heat transfer made investigating the **micro-fin tube** matched well to those reported by Schael (2009) for pure CO_2. Belonging to these are enhanced heat transfer coefficients for low mass flow velocities and vapor qualities related to an increase in the degree of wetting and the partial dryout of the tube surface under high heat flux conditions. As described for the smooth tube, however, CO_2-POE oil mixtures were affected by changes in flow pattern. The overlaid presence of foam up to complete foam flow and the deterioration of heat transfer at high heat fluxes likely related to an advancing suppression of nucleate boiling.

A new expression for the exponent of heat flux ($n(p_r)$, see Eq. 5.8 in Sec. 5.3.1) describing the dependency of the heat transfer coefficient on heat flux was developed based on experimental data at a moderate nominal oil mass fraction of 1 wt.-%. The trend described by Eq. 5.8 is similar to the description by Steiner (2002). The influence of reduced pressure on nucleate boiling heat transfer is described fairly well by correcting the Steiner function according to Gorenflo (2002) for finned surfaces, see Eq. 5.10 and Eq. 5.9 in Sec. 5.3.2, respectively.

Numerous established and newly proposed prediction methods for the flow boiling heat transfer coefficient developed from smooth tube and micro-fin data were compared to the experimental data base acquired in this work. Following the conclusion drawn by Schael (2009) for experiments with pure CO_2, the correlation of Cavallini et al. (1999) was found to be best suitable for predicting heat transfer coefficients of CO_2-POE oil mixtures in the micro-fin tube. It captures over 61 % of the data within a range of error of ± 30 % and yielded an overall mean percentage error of MPE = 39 %. A summary addressing the influence of nominal oil mass fraction on the accuracy in prediction (Tab. 5.4) showed that heat transfer coefficients are increasingly over-predicted as the oil mass fraction increases. Taking only data containing a nominal oil mass fraction of 3 wt.-% less than 45 % of data was predicted within a range of error of ±30 %, while the mean the percentage error increased to over 70 %. Similar conclusion was drawn for the smooth correlation of Steiner adapted to experimental CO_2 micro-fin data by Schael (2009).

6.2 Outlook

The literature survey and the investigations conducted in this work showed that the influence of oil is versatile and substantial for a precise prediction of design parameters such as the heat transfer coefficient and pressure drop. Comparisons of experimental data to predictions made using established correlations developed for pure refrigerants as well as refrigerant-lubricant mixtures revealed that the influence of oil cannot be described satisfactorily as of the state of knowledge and, hence, is not yet understood. As the properties of refrigerant-lubricant mixtures vary considerably in parts, especially concerning the miscibility, different approaches are needed for modeling, increasing the effort in research. This includes extending the available data base as well as the development of new predictive methods.

Results of this work clarified the influence of flow pattern on heat transfer and the importance of accurate flow pattern prediction. In this work, a new but simple technique was developed to successfully determine the size of a liquid film inside an adiabatic sight glass tube, allowing a more precise estimation of the transition to full wetting flow. However, the technique based on confocal laser displacement contains several limitations in application. It allows for one-dimensional measurements (i.e., at fixed position) only, no conclusions about the actual flow pattern can be drawn and the technique (as any light-based measurement) is not applicable to foaming flow or flow containing a significant number of bubbles. An alternative measurement technique for a precise determination of flow pattern could be based on optical tomography, representing a low-cost and high-resolution but complex application.

At an increasing rate, enhanced geometries have been investigated in flow boiling experiments. The focus in research was mainly laid on internally micro-finned tubes. As presented by numerous authors, enhancement of the evaporating inner surface can lead to a significant enhancements in flow pattern and heat transfer. Simultaneously, a moderate increase in pressure drop is typically reported. For the evaluation of the overall efficiency of micro-finned tubes compared to the equivalent smooth tube, so-called enhancement ($EF_\alpha = \alpha_{\text{micro-fin}}/\alpha_{\text{smooth}}$) and penalty ($PF_{\Delta p} = \Delta p_{\text{micro-fin}}/\Delta p_{\text{smooth}}$) factors are generally defined and equally compared. Schael (2009) concluded from measurements with pure CO_2, revealing constant penalty factors of $PF_{\Delta p} = 1.8$, that the enhancement factor in the under the respective conditions of operations must yield $EF_\alpha > 1.8$ in order for the micro-fin tube to outperform the equivalent smooth tube. For a complete thermodynamic efficiency evaluation, an exergetic analysis has to be conducted.

Table 6.1: Publications on flow boiling in porous media (foams)

Authors	Test fluid	Measurement section
Topin et al. (2006)	n-Pentane	Copper foam filled rchannels
Kim et al. (2008a)	FC-72	Copper foam filled channel
Zhao et al. (2009), Lu and Zhao (2009), Du et al. (2011)	R134a	Copper foam filled tubes
Li and Leong (2011)	FV-72, Water	Aluminum foam filled channel
Dawidowicz and Cieśliński (2012)	R22, R134a, R407C	Tube coated with porous aluminum
Hu et al. (2013)	R410A with POE68 oil	Copper foam filled tube
Madani et al. (2013)	n-Pentane	Copper foam filled channel
Zhu et al. (2013)	R410A with POE68 oil	Copper foam filled tubes
Mancin et al. (2014a)	R134a, R1234ze(E)	Copper foam filled channel

In the late 2000s, first results from flow boiling research in tubes with metallic foam inserts were published, with increasing number of publications since (see Tab. 6.1). Through-flown highly porous metal foams as shown in Fig. 6.1 promise a comparatively low pressure drop increase and a significant increase in heat transfer. Reasons for the assumed enhancement of heat transfer could include the enlargement of surface area (= nucleation sites), an improvement in tube wetting and heat conduction through the high-conductive metal bridges of the foam. Reasoned by the great potential in application, a succeeding research project for investigating flow boiling of CO_2 in metal foams was recently started at the Institute of Thermal Process Engineering at KIT.

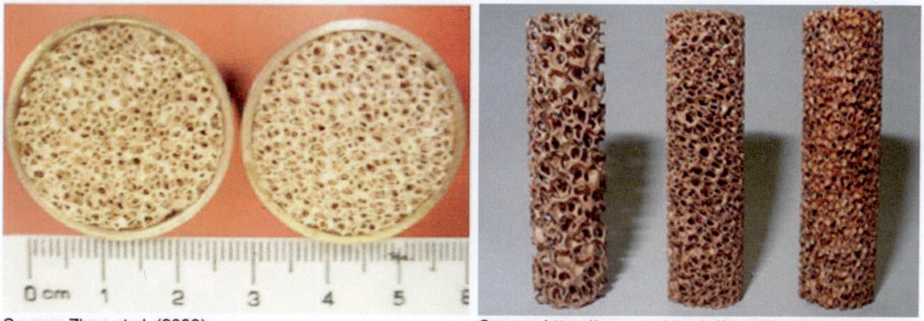

Source: Zhao et al. (2009) Source: https://www.tvt.kit.edu/21_2501.php

Figure 6.1: Metal foams used in flow boiling experiments

Bibliography

K. Aiyoshizawa, T. Yamada, N. Haraguchi, C. Dang, and E. Hihara. The influence of lubricating oil on boiling heat transfer characteristics of carbon dioxide in horizontal smooth tubes. In *Proc. 43rd National Heat Transfer Symposium of Japan*, volume 1. Nihon Dennetsu Gakkai, 2006.

M. M. Awad. Comments on "Measurement and correlation of frictional two-phase pressure drop of R410A/POE oil mixture flow boiling in a 7mm straight micro-fin tube" by H.-t. Hu, G.-l. Ding, and K.-j. Wang. *Appl Therm Eng*, 30(2-3):260–261, 2010.

H. D. Baehr and K. Stephan. *Wärme- und Stoffübertragung*. Springer, Berlin Heidelberg, 5th edition, 2006.

E. P. Bandarra Filho, L. Cheng, and J. R. Thome. Flow boiling characteristics and flow pattern visualization of refrigerant/lubricant oil mixtures. *Int J Refrig*, 32(2):185–202, 2009.

J. Bandel. *Druckverlust und Wärmeübergang bei der Verdampfung siedender Kältemittel im durchströmten waagerechten Rohr*. Ph.d. thesis, University of Karlsruhe, 1973.

P. K. Bansal. In-tube boiling heat transfer of CO2 lubricant mixture at low temperatures: Preliminary results. *ASHRAE Transactions*, 117(1):186–194, 2011.

P. K. Bansal. A review - status of CO2 as a low temperature refrigerant: fundamentals and R&D opportunities. *Appl Therm Eng*, 41:18–29, 2012.

P. E. L. Barbieri, J. M. S. Jabardo, and E. P. Bandarra Filho. Flow patterns in convective boiling of refrigerant R-134a in smooth tubes of several diameters. In *Proc. 5th European Thermal-Sciences Conference*, 2008.

J. R. Barbosa, V. T. Lacerda, and A. T. Prata. Prediction of pressure drop in refrigerant-lubricant oil flows with high contents of oil and refrigerant outgassing in small diameter tubes. *Int J Refrig*, 27(2):129–139, 2004.

C. Baroncini, P Di Filippo, G. Latini, and M. Pacetti. Organic liquid thermal conductivity: A prediction method in the reduced temperature range 0.3 to 0.8. *International Journal of Thermophysics*, 2(1):21–38, 1981.

J. J. Baustian, M. B. Pate, and A. E. Bergles. Poperties of oil-refrigerant liquid mixtures with applications to oil concentration measurement: Part 1 - Thermophysical and transport properties. *ASHRAE Transactions*, 92:55–73, 1986.

K. I. Bell, G. F. Hewitt, and S. D. Morris. Nucleate pool boiling of refrigerant/oil mixtures. *Experimental Heat Transfer*, 1(1):71–86, 1987.

Bitzer Kühlmaschinenbau GmbH. Refrigerant Report, 2015.

H. Blasius. Das Aehnlichkeitsgesetz bei Reibungsvorgängen. *Sonderabdruck aus der Zeitschrift des Vereines deutscher Ingenieure*, 1912.

J. R. Brock and R. B. Bird. Surface tension and the principle of corresponding states. *Aiche J*, 1(2):174–177, 1955.

H. O. S. Castro, J. L. Gasche, and W. P. Conti. Foam flow of oil-refrigerant R134a mixture in a small diameter tube. In *Proc. International Refrigeration and Air Conditioning Conference*, 2004.

H. O. S. Castro, J. L. Gasche, and A. T. Prata. Pressure drop correlation for oil–refrigerant R134a mixture flashing flow in a small diameter tube. *Int J Refrig*, 32(3):421–429, 2009.

A. Cavallini, D. Del Col, L. Doretti, G. A. Longo, and L. Rosetto. Pressure drop during condensation and vaporization of refrigerants inside enhanced tubes. *Heat and Technology*, 15(1):3–10, 1997.

A. Cavallini, D. Del Col, L. Doretti, G. A. Longo, and L. Rosetto. Enhanced intube heat transfer with refrigerants. In *Proc. 20th International Congress of Refrigeration*, 1999.

A. Cavallini, D. Del Col, L. Doretti, G. A. Longo, and L. Rossetto. Heat transfer and pressure drop during condensation of refrigerants inside horizontal enhanced tubes. *Int J Refrig*, 23(1):4–25, 2000.

J. C. Chen. Correlation for boiling heat transfer to saturated fluids in convective flow. *Industrial & Engineering Chemistry Process Design and Development*, 5(3):322–329, 1966.

L. Cheng, G. Ribatski, J. Moreno Quibén, and J. R. Thome. New prediction methods for CO2 evaporation inside tubes: Part I - A two-phase flow pattern map and a flow pattern based phenomenological model for two-phase flow frictional pressure drops. *Int J Heat Mass Tran*, 51(1-2):111–124, 2008a.

L. Cheng, G. Ribatski, and J. R. Thome. New prediction methods for CO2 evaporation inside tubes: Part II - An updated general flow boiling heat transfer model based on flow patterns. *Int J Heat Mass Tran*, 51(1-2):125–135, 2008b.

L. X. Cheng, G. Ribatski, L. Wojtan, and J. R. Thome. New flow boiling heat transfer model and flow pattern map for carbon dioxide evaporating inside horizontal tubes. *Int J Heat Mass Tran*, 49(21-22):4082–4094, 2006.

D. Chisholm and A. D. K. Laird. Two-phase flow in rough tubes. *Transactions of the American Society of Mechanical Engineers*, 80(2):276–286, 1958.

J. M. Cho and M. S. Kim. Experimental studies on the evaporative heat transfer and pressure drop of CO2 in smooth and micro-fin tubes of the diameters of 5 and 9.52mm. *Int J Refrig*, 30(6):986–994, 2007.

J. Y. Choi, M. A. Kedzierski, and P. A. Domanski. A generalized pressure drop correlation for evaporation and condensation of alternative refrigerants in smooth and micro-fin tubes. Technical Report NISTIR 6333, NIST, 1999.

J. G. Collier and J. R. Thome. *Convective boiling and condensation*. Oxford University Press, 3rd edition, 1994.

M. R. Conde. Estimation of thermophysical properties of lubricating oils and their solutions with refrigerants: An appraisal of existing methods. *Appl Therm Eng*, 16(1):51–61, 1996.

M. G. Cooper. *Saturation nucleate pool boiling - A simple correlation*, pages 785–793. Pergamon, 1984.

R. J. Da Silva Lima, J. Moreno Quibén, C. Kuhn, T. Boyman, and J. R. Thome. Ammonia two-phase flow in a horizontal smooth tube: Flow pattern observations, diabatic and adiabatic frictional pressure drops and assessment of prediction methods. *Int J Heat Mass Tran*, 52(9-10):2273–2288, 2009.

C. Dang. Addition to 'Flow boiling heat transfer of carbon dioxide with polyalkylene glycol-type lubricating oil in pre-dryout region inside horizontal tube', Jan 2014.

C. Dang, N. Haraguchi, T. Yamada, and E. Hihara. Effect of lubricating oil on flow boiling heat transfer of carbon dioxide . In *Proc. 7th Gustav Lorentzen Conference on Natural Working Fluids*, 2006.

C. Dang, N. Haraguchi, and E. Hihara. Flow boiling heat transfer of carbon dioxide inside a small-sized microfin tube. *Int J Refrig*, 33(4):655–663, 2010.

C. Dang, N. Haraguchi, T. Yamada, M. Li, and E. Hihara. Effect of lubricating oil on flow boiling heat transfer of carbon dioxide. *Int J Refrig*, 36(1):136–144, 2013.

P. S. Davis, F. Theeuwes, R. J. Bearman, and R. P. Gordon. Non-steady-state, hot wire, thermal conductivity apparatus. *The Journal of Chemical Physics*, 55(10):4776, 1971.

B. Dawidowicz and J. T. Cieśliński. Heat transfer and pressure drop during flow boiling of pure refrigerants and refrigerant/oil mixtures in tube with porous coating. *Int J Heat Mass Tran*, 55(9-10):2549–2558, 2012.

DH Budenberg. Kalibrierschein, Jun 2010.

J. P. Dias, J. L. Gasche, and A. L. Seixlack. Mathematical modeling of the ester oil-refrigerant R134a mixture two-phase flow with foam formation through a small diameter tube. *Journal of the Brazilian Society of Mechanical Sciences and Engineering*, 33(3), 2011.

G. Ding, H. Hu, X. Huang, B. Deng, and Y. Gao. Experimental investigation and correlation of two-phase frictional pressure drop of R410A-oil mixture flow boiling in a 5mm microfin tube. *Int J Refrig*, 32(1):150–161, 2009.

F. W. Dittus and L. M. Boelter. Heat transfer in automobile radiators of the turbulent type. *University of California Publications in Engineering*, 2:443–461, 1930.

Y. P. Du, C. Y. Zhao, Y. Tian, and Z. G. Qu. Analytical considerations of flow boiling heat transfer in metal-foam filled tubes. *Heat Mass Transfer*, 48(1):165–173, 2011.

S. J. Eckels, T. M. Doerr, and M. B. Pate. In-tube heat transfer and pressure drop of R-134a and ester lubricant mixtures in a smooth tube and a micro-fin tube: Part 1 - Evaporation. *ASHRAE Transactions*, 100(2):265–282, 1994.

S. J. Eckels, T. M. Doerr, and M. B. Pate. A comparison of the heat transfer and pressure drop performance of R-134a-lubricant mixtures in different diameter smooth tubes and micro-fin tubes. *ASHRAE Transactions*, 104(1):376–386, 1998a.

S. J. Eckels, T. M. Doerr, and M. B. Pate. Heat transfer coefficients and pressure drops for R-134a and an ester lubricant mixture in a smooth tube and a micro-fin tube. *ASHRAE Transactions*, 104(1):366–375, 1998b.

J. El Hajal, J. R. Thome, and A. Cavallini. Condensation in horizontal tubes, part 1: Two-phase flow pattern map. *Int J Heat Mass Tran*, 46(18):3349–3363, 2003.

X. Fang. A new correlation of flow boiling heat transfer coefficients for carbon dioxide. *Int J Heat Mass Tran*, 64:802–807, 2013.

X. Fang, Y. Xu, and Z. Zhou. New correlations of single-phase friction factor for turbulent pipe flow and evaluation of existing single-phase friction factor correlations. *Nuclear Engineering and Design*, 241(3):897–902, 2011.

S. Feja and S. Römer. Bestimmung der Wärmeleitfähigkeit von CO2-Öl-Gemischen. Technical report, ILK Dresden, Jan 2008.

A. Fenghour, W. A. Wakeham, and V. Vesovic. The viscosity of carbon dioxide. *Journal of Physical and Chemical Reference Data*, 27(1):31, 1998.

H. K. Forster and N. Zuber. Dynamics of vapor bubbles and boiling heat transfer. *Aiche J*, 1 (4):531–535, 1955.

L. Friedel. Improved friction pressure drop correlations for horizontal and vertical two phase pipe flow. *3R International*, 18(7):485–491, 1979.

L. Gao and T. Honda. Experiments on flow boiling heat transfer of pure CO2 and CO2-oil mixtures in horizontal smooth and micro-fin tubes. In *Proc. International Refrigeration and Air Conditioning Conference*, 2006.

L. Gao, T. Honda, and S. Koyama. Experiments on flow boiling heat transfer of almost pure CO2 and CO2-oil mixtures in horizontal smooth and microfin tubes. *HVAC&R Research*, 13(3):415–425, 2007.

L. Gao, Y. Matsusaka, M. Sato, and T. Honda. Correlations of evaporation heat transfer of CO2-PAG oil mixtures in a horizontal smooth tube. In *Proc. 2008 JSRAE Annual Conference*, pages 279–282. Japan Society of Refrigeration and Air Conditioning Engineers, 2008a.

L. Gao, T. Ono, and T. Honda. Flow patterns and heat transfer characteristics of flow boiling of CO2 and oil mixtures in horizontal smooth and micro-fin tubes. In *Proc. 8th Gustav Lorentzen Conference on Natural Working Fluids*, 2008b.

V. Gnielinski. Neue Gleichungen für den Wärme- und den Stoffübergang in turbulent durchströmten Rohren und Kanälen. *Forschung im Ingenieurwesen*, 41(1):8–16, 1975.

D. Gorenflo. *Behältersieden (Sieden bei freiher Konvektion)*, chapter Hab. Springer, Berlin, Heidelberg, 9th edition, 2002.

S. Grauso, R. Mastrullo, A. W. Mauro, J. R. Thome, and G. P. Vanoli. Flow pattern map, heat transfer and pressure drops during evaporation of R-1234ze(E) and R134a in a horizontal, circular smooth tube: Experiments and assessment of predictive methods. *Int J Refrig*, 36 (2):478–491, 2013a.

S. Grauso, R. Mastrullo, A. W. Mauro, and G. P. Vanoli. Flow boiling of R410A and CO2 from low to medium reduced pressures in macro channels: Experiments and assessment of prediction methods. *Int J Heat Mass Tran*, 56(1-2):107–118, 2013b.

S. Grauso, R. Mastrullo, A. W. Mauro, and G. P. Vanoli. Two-phase adiabatic frictional pressure gradients for R410A and CO2 in a macro channel: Experiments and a simplified predictive method for annular flow from low to medium reduced pressures. *Exp Therm Fluid Sci*, 52:79–87, 2014.

K. E. Gungor and R. H. S. Winterton. A general correlation for flow boiling in tubes and annuli. *Int J Heat Mass Tran*, 29(3):351–358, 1986.

S. Ha and A. E. Bergles. The influence of oil on local evaporation heat transfer inside a horizontal microfin tube. *ASHRAE Transactions*, 99(1):1244–1255, 1993.

K. Hambraeus. Heat transfer of oil-contaminated HFC134a in a horizontal evaporator. *Int J Refrig*, 18(2):87–99, 1995.

X. Han, P. Li, X. Yuan, Q. Wang, and G. Chen. The boiling heat transfer characteristics of the mixture HFO-1234yf/oil inside a micro-fin tube. *Int J Heat Mass Tran*, 67:1122–1130, 2013.

T. J. Hanratty. *Waves on Fluid Interface*. Interfacial instabilities caused by air flow over a thin liquid layer. Academic Press Inc., New York, 1983.

K. Hashimoto and A. Kiyotani. Experimental study of pure CO2 heat transfer during flow boiling inside horizontal tubes. In *Proc. 6th Gustav Lorentzen Conference on Natural Working Fluids*, 2006.

M. Hassan. *Flow boiling of pure and oil contaminated carbon dioxide as refrigerant*. Ph.d. thesis, Technical University of Denmark, 2004.

S. Higashiiue, S. Ikeda, K. Kuwahara, and S. Koyama. Visual observation of two-phase carbon dioxide flowing in a horizontal micro-fin tube. In *Proc. 8th Gustav Lorentzen Conference on Natural Working Fluids*, 2008.

P. S. Hrnjak. Carbon Dioxide Systems. *HVAC&R Research*, 12(1):1–2, 2006.

P. S. Hrnjak and S. Kim. Oil effects of in-tube boiling of CO2. In *Proc. 11th Gustav Lorentzen Conference on Natural Refrigerants*, 2014.

H. Hu, G. Ding, and K. Wang. Heat transfer characteristics of R410A-oil mixture flow boiling inside a 7mm straight microfin tube. *Int J Refrig*, 31(6):1081–1093, 2008a.

H. Hu, G. Ding, and K. Wang. Measurement and correlation of frictional two-phase pressure drop of R410A/POE oil mixture flow boiling in a 7mm straight micro-fin tube. *Appl Therm Eng*, 28(11-12):1272–1283, 2008b.

H. Hu, G. Ding, W. Wei, Z. Wang, and K. Wang. Heat transfer characteristics of R410A-oil mixture flow boiling inside a 7mm straight smooth tube. *Exp Therm Fluid Sci*, 32(3): 857–869, 2008c.

H. Hu, G. Ding, W. Wei, Z. Wang, and K. Wang. Measurement and correlation of frictional pressure drop of R-410A/oil mixture flow boiling in a 7 mm straight smooth tube. *HVAC&R Research*, 14(5):763–781, 2008d.

H. Hu, G. Ding, X. Huang, B. Deng, and Y. Gao. Measurement and correlation of flow-boiling heat transfer of a R-410A/oil mixture inside a 4.18 mm straight smooth tube. *HVAC&R Research*, 15(2):287–314, 2009a.

H. Hu, Y. Zhu, G. Ding, and S. Sun. Effect of oil on two-phase pressure drop of refrigerant flow boiling inside circular tubes filled with metal foam. *Int J Refrig*, 36(2):516–526, 2013.

H. Hu, H. Peng, G. Ding, Y. Zheng, Y. Gao, and J. Song. Influence of oil on heat transfer charactersitics of R410a flow boiling in conventional and small size microfin tubes. In *Proc. International Refrigeration and Air Conditioning Conference*, 2014.

H. T. Hu, G. L. Ding, X. C. Huang, B. Deng, and Y. F. Gao. Pressure drop during horizontal flow boiling of R410A/oil mixture in 5mm and 3mm smooth tubes. *Appl Therm Eng*, 29 (16):3353–3365, 2009b.

X. Huang, G. Ding, H. Hu, Y. Zhu, Y. Gao, and B. Deng. Flow condensation pressure drop characteristics of R410A-oil mixture inside small diameter horizontal microfin tubes. *Int J Refrig*, 33(7):1356–1369, 2010.

C. Ihmels. Experimentelle Bestimmung der Wärmeleitfähigkeiten von einem Öl (Reniso C85 E), 2008a.

C. Ihmels. Experimentelle Bestimmung der Wärmeleitfähigkeiten Kohlendioxid + Öl (Reniso C85 E) Mischungen, 2008b.

O. Javerschek. Commercial Refrigeration systems with CO2 as refrigerant. In *Proc. 8th Gustav Lorentzen Conference on Natural Working Fluids*, 2008.

M. K. Jensen and D. L. Jackman. Prediction of nucleate pool boiling heat transfer coefficients of refrigerant-oil mixtures. *J Heat Trans*, 106(1):184, 1984.

D. S. Jung, M. McLinden, R. Radermacher, and D. Didion. A study of flow boiling heat transfer with refrigerant mixtures. *Int J Heat Mass Tran*, 32(9):1751–1764, 1989.

S. G. Kandlikar. A General Correlation for Saturated Two-Phase Flow Boiling Heat Transfer Inside Horizontal and Vertical Tubes. *J Heat Trans*, 112(1):219, 1990.

M. Katsuta, K. Okuma, T. Hirade, and N. Miyachi. The effect of the lubricating oil fraction rate on the CO2 evaporating thermal and hydraulic characteristics. In *Proc. International Refrigeration and Air Conditioning Conference*, 2008.

N. Kattan, J. R. Thome, and D. Favrat. Flow boiling in horizontal tubes: Part 1 - Development of a diabatic two-phase flow pattern map. *J Heat Trans*, 120(1):140–147, 1998a.

N. Kattan, J. R. Thome, and D. Favrat. Flow boiling in horizontal tubes: Part 2 - New heat transfer data for five refrigerants. *J Heat Trans*, 120(1):148, 1998b.

N. Kattan, J. R. Thome, and D. Favrat. Flow boiling in horizontal tubes: Part 3 - Development of a new heat transfer model based on flow pattern. *J Heat Trans*, 120(1):156, 1998c.

M. A. Kedzierski and J. M. Goncalves. Horizontal convective condensation of alternative refrigerants within a micro-fin tube. *Enhanced Heat Transfer*, 6:161–178, 1999.

D. W. Kim, A. Bar-Cohen, and B. Han. Forced convection and flow boiling of a dielectric liquid in a foam-filled channel. In *Proc. ITHERM*, pages 86–94, 2008a.

S. Kim and P. S. Hrnjak. Effect of oil on flow boiling heat transfer and flow patterns of CO2 in 11.2 mm horizontal smooth and enhanced tube. In *Proc. International Refrigeration and Air Conditioning Conference*, 2012.

S. Kim, N. Pehlivanoglu, and P. S. Hrnjak. R744 flow boiling heat transfer with and without oil at low temperatures in 11.2 mm horizontal smooth tube. In *Proc. International Refrigeration and Air Conditioning Conference*, 2010.

Y. J. Kim, J. Jang, P. S. Hrnjak, and M. S. Kim. Adiabatic horizontal and vertical pressure drop of carbon dioxide inside smooth and microfin tubes at low temperatures. *J Heat Trans*, 130(11):111001, 2008b.

S. Koyama, S. M. Lee, D. Ito, K. Kuwahara, and H. Ogawa. Experimental study on flow boiling of pure CO2 and CO2-oil mixtures inside horizontal smooth and micro-fin copper tubes. In *Proc. 6th Gustav Lorentzen Conference on Natural Working Fluids*, 2006.

U. Kraut. *Zum Einfluss der Gemischstoffwerte auf die Vorausberechnung des Wärmeübergangs und Druckverlusts beim Strömungssieden von CO2-Öl-Gemischen*. Bachelor thesis, Karlsruhe Institute of Technology (KIT), 2013.

H. Y. Li and K. C. Leong. Experimental and numerical study of single and two-phase flow and heat transfer in aluminum foams. *Int J Heat Mass Tran*, 54(23-24):4904–4912, 2011.

M. Li, C. Dang, and E. Hihara. Flow boiling heat transfer of carbon dioxide with polyalkylene glycol-type lubricating oil in pre-dryout region inside horizontal tube. *Int J Refrig*, 2014.

Z. Liu and R. H. S. Winterton. A general correlation for saturated and subcooled flow boiling in tubes and annuli, based on a nucleate pool boiling equation. *Int J Heat Mass Tran*, 34(11):2759–2766, 1991.

R. W. Lockhart and R. C. Martinelli. Proposed correlation of data for isothermal two-phase, two-component flow in pipes. *Chemical Engineering Progress*, 45(1):39–48, 1949.

G. Lorentzen. Revival of carbon dioxide as a refrigerant. *Int J Refrig*, 17(5):292–301, 1994.

W. Lu and C. Y. Zhao. Numerical modelling of flow boiling heat transfer in horizontal metal-foam tubes. *Advanced Engineering Materials*, 11(10):832–836, 2009.

B. Madani, L. Tadrist, and F. Topin. Experimental analysis of upward flow boiling heat transfer in a channel provided with copper metallic foam. *Appl Therm Eng*, 52(2): 336–344, 2013.

S. Mancin, A. Diani, L. Doretti, and L. Rossetto. R134a and R1234ze(E) liquid and flow boiling heat transfer in a high porosity copper foam. *Int J Heat Mass Tran*, 74:77–87, 2014a.

S. Mancin, A. Diani, and L. Rossetto. R134a flow boiling heat transfer and pressure drop inside a 3.4mm ID microfin tube. *Energy Procedia*, 45:608–615, 2014b.

R. Mastrullo, A. W. Mauro, A. Rosato, and G. P. Vanoli. Carbon dioxide local heat transfer coefficients during flow boiling in a horizontal circular smooth tube. *Int J Heat Mass Tran*, 52(19-20):4184–4194, 2009.

R. Mastrullo, A. W. Mauro, A. Rosato, and G. P. Vanoli. Carbon dioxide heat transfer coefficients and pressure drops during flow boiling: Assessment of predictive methods. *Int J Refrig*, 33(6):1068–1085, 2010.

R. Mastrullo, A. W. Mauro, J. R. Thome, D. Toto, and G. P. Vanoli. Flow pattern maps for convective boiling of CO2 and R410A in a horizontal smooth tube: Experiments and new correlations analyzing the effect of the reduced pressure. *Int J Heat Mass Tran*, 55(5-6): 1519–1528, 2012a.

R. Mastrullo, A. W. Mauro, J. R. Thome, and G. P. Vanoli. CO2 and R410A: Two-phase flow visualizations and flow boiling measurements at medium (0.50) reduced pressure. *Appl Therm Eng*, 49:2–8, 2012b.

W. H. McAdams. *Heat Transmission*. McGraw-Hill, New York, 2nd edition, 1942.

O. O. Medvedev, P. V. Zhelezny, and V. P. Zhelezny. Prediction of phase equilibria and thermodynamic properties of refrigerant/oil solutions. *Fluid Phase Equilibria*, 215(1): 29–38, 2004.

Y. Mermond, M. Feidt, and C. Marvillet. Flow boiling of refrigerant-oil mixtures: Measurements of heat transfer coefficients. In *Proc. 20th International Congress of Refrigeration*, volume 2. International Institute of Refrigeration, 1999a.

Y. Mermond, M. Feidt, and C. Marvillet. Propriétés thermodynamiques et physiques des mélanges de fluides frigorigènes et d'huiles. *Int J Refrig*, 22(7):569–579, 1999b.

R. Mesler and G. Mailen. Nucleate boiling in thin liquid films. *Aiche J*, 23(6):954–957, 1977.

D. G. Miller and G. Thodos. On the reduced Frost-Kalkwarf vapor presure equation. *Industrial & Engineering Chemistry Fundamentals*, 2(1):78–80, 1963.

M. Müller and R. Eggers. Heat transfer characteristics of CO2-oil-mixtures. In *Proc. 8th Gustav Lorentzen Conference on Natural Working Fluids*, 2008.

H. Müller-Steinhagen and K. Heck. A simple friction pressure drop correlation for two-phase flow in pipes. *Chemical Engineering and Processing: Process Intensification*, 20 (6):297–308, 1986.

J. Moreno Quibén and J. R. Thome. Flow pattern based two-phase frictional pressure drop model for horizontal tubes. Part I: Diabatic and adiabatic experimental study. *International Journal of Heat and Fluid Flow*, 28(5):1049–1059, 2007a.

J. Moreno Quibén and J. R. Thome. Flow pattern based two-phase frictional pressure drop model for horizontal tubes, Part II: New phenomenological model. *International Journal of Heat and Fluid Flow*, 28(5):1060–1072, 2007b.

J. Moreno Quibén, L. Cheng, R. J. da Silva Lima, and J. R. Thome. Flow boiling in horizontal flattened tubes: Part I - Two-phase frictional pressure drop results and model. *Int J Heat Mass Tran*, 52(15-16):3634–3644, 2009.

E. Nidegger, J. R. Thome, and D. Favrat. Flow boiling and pressure drop measurements for R-134a/oil mixtures part 1: Evaporation in a microfin tube. *HVAC&R Research*, 3(1): 38–53, 1997.

M. Niederkrüger. *Strömungssieden von reinen Stoffen und binären zeotropen Gemischen im waagerechten Rohr bei mittleren und hohen Drücken*. Ph.d. thesis, University of Karlsruhe, 1991.

H. K. Oh and C. H. Son. Flow boiling heat transfer and pressure drop characteristics of CO2 in horizontal tube of 4.57-mm inner diameter. *Appl Therm Eng*, 31(2-3):163–172, 2011.

J. T. Oh, A. S. Pamitran, K. I. Choi, and P. S. Hrnjak. Experimental investigation on two-phase flow boiling heat transfer of five refrigerants in horizontal small tubes of 0.5, 1.5 and 3.0mm inner diameters. *Int J Heat Mass Tran*, 54(9-10):2080–2088, 2011.

T. Ono, L. Gao, and T. Honda. Heat transfer and flow characteristics of flow boiling of CO2-oil mixtures in horizontal smooth and micro-fin tubes. *Heat Transfer-Asian Research*, 39 (3):195–207, 2010.

M. Ozawa, T. Ami, I. Ishihara, H. Umekawa, R. Matsumoto, Y. Tanaka, T. Yamamoto, and Y. Ueda. Flow pattern and boiling heat transfer of CO2 in horizontal small-bore tubes. *Int J Multiphas Flow*, 35(8):699–709, 2009.

A. C. Pachai. CO2 cycles in a historic perspective. In *Proc. 8th Gustav Lorentzen Conference on Natural Working Fluids*, 2008.

J. D. Pandey, G. P. Dubey, and N. Tripathi. Surface tension and ultrasonic velocity of binary liquid mixtures at 298.15 K. *Proc. Indian Acad. Sci. (Chem. Sci.)*, 111(2):361–367, 1999.

C. Y. Park and P. S. Hrnjak. Flow boiling heat transfer of CO2 at low temperatures in a horizontal smooth tube. *J Heat Trans*, 127(12):1305, 2005.

C. Y. Park and P. S. Hrnjak. CO2 and R410A flow boiling heat transfer and pressure drop at low temperatures in a horizontal smooth tube. In *Proc. International Refrigeration and Air Conditioning Conference*, 2006.

C. Y. Park and P. S. Hrnjak. Carbon dioxide and R410a flow boiling heat transfer, pressure drop, and flow pattern in horizontal tubes at low temperatures. Technical Report ACRC TR-258, University of Illinois, 2007a.

C. Y. Park and P. S. Hrnjak. CO2 and R410A flow boiling heat transfer, pressure drop, and flow pattern at low temperatures in a horizontal smooth tube. *Int J Refrig*, 30(1):166–178, 2007b.

N. Pehlivanoglu, S. Kim, and P. S. Hrnjak. Effect of oil on heat transfer and pressure drop of R744 in 6.1 mm horizontal smooth tube. In *Proc. International Refrigeration and Air Conditioning Conference*, 2010.

B. S. Petukhov and V. N. Popov. Theoretical calculation of heat exchanger in turbulent flow in tubes of an incompressible fluid with variable physical properties. *High Temperature Heat Physics*, 1(1):69–83, 1963.

W. Rathjen and J. Straub. Die Temperaturabhängigkeit der Oberflächenspannung von reinen Kältemitteln vom Tripelpunkt bis zum kritischen Punkt. *Wärme- und Stoffübertragung*, 14(1):59–73, 1980.

A. L. Riegel. *Messunsicherheits-Analyse der Messkette der Versuchsanlage zum Strömungssieden von CO2 durch Anwendung des Verfahrens nach GUM*. Seminar work, Karlsruhe Institute of Technology (KIT), 2011.

H. M. Roder. A transient hot wire thermal conductivity apparatus for fluids. *Journal of Research of the National Bureau of Standards*, 88(5):457–493, 1981.

P. Rollmann and K. Spindler. Wärmeübergang und Druckverlust beim Strömungssieden von R407C. *KI Kälte - Luft - Klimatechnik*, 2011.

P. Rollmann, K. Spindler, and H. Müller-Steinhagen. Heat transfer, pressure drop and flow patterns during flow boiling of R407C in a horizontal microfin tube. *Heat Mass Transfer*, 47(8):951–961, 2011.

S. Z. Rouhani and E. Axelsson. Calculation of void volume fraction in the subcooled and quality boiling regions. *Int J Heat Mass Tran*, 13(2):383–393, 1970.

Y. Saito. Material properties of mixtures of CO2 and Reniso C85E, Nov 2014.

Y. Saito, M. Wetzel, M. Kind, and T. Wetzel. A new flow pattern map for CO2 two-phase flow in a horizontal smooth tube. In *Proc. 23rd IIR International Congress of Refrigeration*, volume 1, pages 501–508. International Institute of Refrigeration (IIF/IIR), 2011.

M. P. Saksena and Harminder. Thermal conductivity of binary liquid mixtures. *Chemical Physics Letters*, 25(3):445–448, 1975.

A. E. Schael. *Über das Strömungsverdampfen von CO2 im glatten und innen berippten Rohr - Hydrodynamik, Wärmeübergang, Druckverlust*. Ph.d. thesis, University of Karlsruhe, 2009.

L. M. Schlager, M. B. Pate, and A. E. Bergles. Heat transfer and pressure drop performance of smooth and internally finned tubes with oil and refrigerant 22 mixtures. *ASHRAE Transactions*, 95(2):160–169, 1989.

L. M. Schlager, M. B. Pate, and A. E. Bergles. Performance predictions of refrigerant-oil mixtures in smooth and internally finned tubes - Part II: Design Equations. *ASHRAE Transactions*, 96(1):170–182, 1990.

H. Schmidt. *Beitrag zum Verständnis des Wärmeübertrags im horizontalen Verdampferrohr*. Ph.d. thesis, University of Karlsruhe, 1986.

C. J. Seeton and P. S. Hrnjak. Thermophysical properties of CO2-lubricant mixtures and their affect on 2-phase flow in small channels (less than 1mm). In *Proc. International Refrigeration and Air Conditioning Conference*, 2006.

M. M. Shah. Chart correlation for saturated boiling heat transfer: Equations and further study. *ASHRAE Transactions*, 82(1):165–196, 1982.

B. Shen and E. A. Groll. Critical literature review of lubricant influence on refrigerant heat transfer and pressure drop. Technical Report ARTI-21CR/611-20080, AHRI, 2003.

S. Siewert. *Umbau der Versuchsanlage zum CO2-Strömungssieden*. Seminar work, Karlsruhe Institute of Technology (KIT), 2013.

R. Span and W. Wagner. A new equation of state for carbon dioxide covering the fluid region from the triple-point temperature to 1100 K at pressures up to 800 MPa. *Journal of Physical and Chemical Reference Data*, 25(6):1509, 1996.

K. Spindler and E. Hahne. The influence of oil on nucleate pool boiling heat transfer. *Heat Mass Transfer*, 45(7):979–990, 2009.

D. Steiner. Zweiphasenströmung in Apparateelementen. Technical report, Forschungs-Gesellschaft Verfahrens-Technik e.V., 1983.

D. Steiner. *Strömungssieden gesättigter Flüssigkeiten*, chapter Hbb. Springer, Berlin, Heidelberg, 9th edition, 2002.

D. Steiner and J. Taborek. Flow boiling heat transfer in vertical tubes correlated by an asymptotic model. *Heat Transfer Eng*, 13(2):43–69, 1992.

K. Stephan and M. Abdelsalam. Heat-transfer correlations for natural convection boiling. *Int J Heat Mass Tran*, 23(1):73–87, 1980.

H. Storek and H. Brauer. *Reibungsdruckverlust der adiabaten Gas/Flüssigkeit-Strömung in horizontalen und vertikalen Rohren*, volume 599 of *Verein Deutscher Ingenieure: VDI-Forschungsheft*. VDI-Verlag, Düsseldorf, 1980.

Y. Taitel and A. E. Dukler. A model for predicting flow regime transitions in horizontal and near horizontal gas-liquid flow. *Aiche J*, 22(1):47–55, 1976.

W. Targanski and J. T. Cieslinski. Evaporation of R407C/oil mixtures inside corrugated and micro-fin tubes. *Appl Therm Eng*, 27(13):2226–2232, 2007.

J. R. Thome. Comprehensive thermodynamic approach to modeling refrigerant-lubricating oil mixtures. *HVAC&R Research*, 1(2), 1995.

J. R. Thome. Boiling of new refrigerants: A state-of-the-art review. *Int J Refrig*, 19(7): 435–457, 1996.

J. R. Thome. Two-phase heat transfer using no-phase flow models? *Heat Transfer Eng*, 25 (6):1–2, 2004.

J. R. Thome. Update on advances in flow pattern based two-phase heat transfer models. *Exp Therm Fluid Sci*, 29(3):341–349, 2005.

J. R. Thome. State-of-the-art overview of boiling and two-phase flows in microchannels. *Heat Transfer Eng*, 27(9):4–19, 2006.

J. R. Thome and J. El Hajal. Two-phase flow pattern map for evaporation in horizontal tubes: Latest version. *Heat Transfer Eng*, 24(6):3–10, 2003.

J. R. Thome and J. El Hajal. Flow boiling heat transfer to carbon dioxide: general prediction method. *Int J Refrig*, 27(3):294–301, 2004.

F. Topin, J. P. Bonnet, B. Madani, and L. Tadrist. Experimental analysis of multiphase flow in metallic foam: flow laws, heat transfer and convective boiling. *Advanced Engineering Materials*, 8(9):890–899, 2006.

V. Vesovic, W. A. Wakeham, G. A. Olchowy, J. V. Sengers, J. T. R. Watson, and J. Millat. The Transport Properties of Carbon Dioxide. *Journal of Physical and Chemical Reference Data*, 19(3):763, 1990.

C. C. Wang, A. Hafner, C. S. Kuo, and W. D. Hsieh. Influence of lubricant on the nucleate boiling heat transfer performance of refrigerant - A review. *Heat Transfer Eng*, 35(6-8): 651–663, 2014.

W. Wei, G. Ding, H. Hu, and K. Wang. Influence of lubricant oil on heat transfer performance of refrigerant flow boiling inside small diameter tubes. Part I: Experimental study. *Exp Therm Fluid Sci*, 32(1):67–76, 2007.

W. Wei, G. Ding, H. Hu, and K. Wang. Models of thermodynamic and transport properties of POE VG68 and R410A/POE VG68 mixture. *Frontiers of Energy and Power Engineering in China*, 2(2):227–234, 2008.

M. Wettermann. *Wärmeübergang beim Sieden von Gemischen bei Zwangskonvektion im horizontalen Verdampferrohr*. Ph.d. thesis, University of Karlsruhe, 1999.

M. Wetzel, Y. Saito, M. Kind, and T. Wetzel. Einfluss von Kältemaschinenöl auf den Wärmeübergang beim Strömungssieden von CO2. In *Proc. Deutsche Kälte-Klima-Tagung*, volume 1, pages 555–562. Deutscher Kaelte- und Klimatechischer Verein (DKV), 2009.

M. Wetzel, Y. Saito, M. Kind, and T. Wetzel. Investigations on the flow boiling heat transfer of CO2 and CO2-oil-mixtures. In *Proc. 23rd IIR International Congress of Refrigeration*, pages 509–516. International Institute of Refrigeration (IIF/IIR), 2011.

Wieland Werke AG. www.wieland-industrierohre.de, 2016.

J. Witt. *Schauglas-Konstruktion zur optischen Untersuchung von Strömungsformen beim Strömungssieden von CO2 und CO2-Öl-Gemischen*. Bachelor thesis, Karlsruhe Institute of Technology, 2013.

L. Wojtan, T. Ursenbacher, and J. R. Thome. Interfacial measurements in stratified types of flow. Part II: Measurements for R-22 and R-410A. *Int J Multiphas Flow*, 30(2):125–137, 2004.

L. Wojtan, T. Ursenbacher, and J. R. Thome. Investigation of flow boiling in horizontal tubes: Part I - A new diabatic two-phase flow pattern map. *Int J Heat Mass Tran*, 48(14): 2955–2969, 2005a.

L. Wojtan, T. Ursenbacher, and J. R. Thome. Investigation of flow boiling in horizontal tubes: Part II - Development of a new heat transfer model for stratified-wavy, dryout and mist flow regimes. *Int J Heat Mass Tran*, 48(14):2970–2985, 2005b.

Somchai Wongwises, Tawatchai Wongchang, Jatuporn Kaewon, and Chi-Chuan Wang. A Visual Study of Two-Phase Flow Patterns of HFC-134a and Lubricant Oil Mixtures. *Heat Transfer Eng*, 23(4):13–22, 2002.

Y. Xu and X. Fang. A new correlation of two-phase frictional pressure drop for condensing flow in pipes. *Nuclear Engineering and Design*, 263:87–96, 2013.

Y. Xu and X. D. Fang. A new correlation of two-phase frictional pressure drop for evaporating flow in pipes. *Int J Refrig*, 35(7):2039–2050, 2012.

Y. Xu, X. Fang, X. Su, Z. Zhou, and W. Chen. Evaluation of frictional pressure drop correlations for two-phase flow in pipes. *Nuclear Engineering and Design*, 253:86–97, 2012.

A. Yokozeki. Solubility and viscosity of refrigerant-oil mixtures. In *Proc. International Compressor Engineering Conference*, 1994.

A. Yokozeki, K. Takigawa, and S. I. Sandler. Solubility and viscosity of hydrofluorocarbon/alkylbenzene oil mixtures. In *Proc. International Refrigeration and Air Conditioning Conference*, 2000.

S. H. Yoon, E. S. Cho, Y. W. Hwang, M. S. Kim, K. Min, and Y. Kim. Characteristics of evaporative heat transfer and pressure drop of carbon dioxide and correlation development. *Int J Refrig*, 27(2):111–119, 2004.

R. Yun, Y. Kim, K. Seo, and H. Y. Kim. A generalized correlation for evaporation heat transfer of refrigerants in micro-fin tubes. *Int J Heat Mass Tran*, 45(10):2003–2010, 2002.

C. Y. Zhao, W. Lu, and S. A. Tassou. Flow boiling heat transfer in horizontal metal-foam tubes. *J Heat Trans*, 131(12):121002, 2009.

X. Zhao and P. K. Bansal. Flow boiling heat transfer characteristics of CO2 at low temperatures. *Int J Refrig*, 30(6):937–945, 2007.

X. Zhao and P. K. Bansal. Critical review of flow boiling heat transfer of CO2-lubricant mixtures. *Int J Heat Mass Tran*, 52(3-4):870–879, 2009.

X. Zhao and P. K. Bansal. Flow boiling heat transfer analysis of new experimental data of CO2 in a micro-fin tube at -30 °C. *International Journal of Thermal Sciences*, 59:38–44, 2012.

V. P. Zhelezny. Material properties of CO2/POE oil solutions, 2007.

V. P. Zhelezny, R. F. Abed, and M. Rybnikov. Saturated vapour pressure of solutions of synthetic compressor oils (XMPA) in ammonia. *High Temperatures-High Pressures*, 33 (6):707–713, 2001.

Y. Zhu, H. Hu, G. Ding, S. Sun, and Y. Jing. Influence of metal foam on heat transfer characteristics of refrigerant-oil mixture flow boiling inside circular tubes. *Appl Therm Eng*, 50(1):1246–1256, 2013.

J. Zilly, J. Jang, and P. S. Hrnjak. Condensation of CO2 at low temperature inside horizontal microfinned tubes. Technical report, University of Illinois, Mar 2003.

O. Zürcher, J. R. Thome, and D. Favrat. Flow boiling and pressure drop measurements for R-134a/oil mixtures part 2: Evaporation in a plain tube. *HVAC&R Research*, 3(1):54–64, 1997.

O. Zürcher, J. R. Thome, and D. Favrat. In-tube flow boiling of R-407C and R-407C/oil mixtures part I: Microfin tube. *HVAC&R Research*, 4(4):347–372, 1998a.

O. Zürcher, J. R. Thome, and D. Favrat. In-tube flow boiling of R-407C and R-407C/oil mixtures part II: Plain tube results and predictions. *HVAC&R Research*, 4(4):373–399, 1998b.

O. Zürcher, D. Favrat, and J. R. Thome. Development of a diabatic two-phase flow pattern map for horizontal flow boiling. *Int J Heat Mass Tran*, 45(2):291–301, 2002.

Own publications and conference contributions

Quotable publications

S. Weise, **M. Wetzel**, B. Dietrich and T. Wetzel. Influence of fully miscible lubricating oil on the pressure drop during flow boiling of CO_2 inside an enhanced tube. *Exp. Therm Fluid Sci*, 81:223-233, 2016.

S. Weise, **M. Wetzel**, M. Hornberger, B. Dietrich and T. Wetzel. Experimental Investigation of of the influence of lubricating oil on the flow boiling heat transfer and pressure drop of CO_2 inside an enhanced tube. In *Proc. 24th International Congress of Refrigeration*, 2015.

M. Wetzel, B. Dietrich and T. Wetzel. Influence of oil on heat transfer and pressure drop during flow boiling of CO_2 at low temperatures. *Exp. Therm Fluid Sci*, 59:202-212, 2014.

B. Dietrich, **M. Wetzel**, S. Weise and T. Wetzel. Strömungssieden von CO_2-Öl-Gemischen im glatten und innenstrukturierten Rohr. In *Proc. Deutsche Kälte-Klima-Tagung*, 2014.

M. Wetzel, B. Dietrich and T. Wetzel. Influence of oil on heat transfer and pressure drop during flow boiling of CO_2 at low temperatures. In *Proc. 8th World Conference on Experimental Heat Transfer, Fluid Dynamics and Thermodynamics*, 2013.

M. Wetzel and T. Wetzel. Strömungsverdampfen von CO_2-Öl-Gemischen im horizontalen Rohr. *Chemie Ingenieur Technik*, 84(8):1424, 2012.

M. Wetzel, B. Dietrich and T. Wetzel. Wärmeübergang und Druckverlust beim Strömungssieden von CO_2-Öl-Gemischen. In *Proc. Deutsche Kälte-Klima-Tagung*, 2012.

M. Wetzel, Y. Saito, M. Kind and T. Wetzel. Investigations on the flow boiling heat transfer of CO_2 and CO_2-oil-mixtures. In *Proc. 23rd IIR International Congress of Refrigeration*, 1:509-516, 2011.

Y. Saito, **M. Wetzel**, M. Kind and T. Wetzel. A new flow pattern map for CO_2 two-phase flow in a horizontal smooth tube. In *Proc. 23rd IIR International Congress of Refrigeration*, 1:501-508, 2011.

M. Wetzel Y. Saito, M. Kind and T. Wetzel. Einfluss von Kältemaschinenöl auf den Wärmeübergang beim Strömungssieden von CO_2. In *Proc. Deutsche Kälte-Klima-Tagung*, 1:555-562, 2009.

Conference contributions

S. Weise, **M. Wetzel**, B. Dietrich and T. Wetzel. Druckverlust und Wärmeübergang beim Strömungssieden von CO_2 - Einfluss von POE-Öl im horizontalen innenstrukturierten Rohr. *Deutsche Kälte-Klima-Tagung*, Nov 16-18, Kassel, 2016 (**talk**).

S. Weise, **M. Wetzel**, M. Hornberg, B. Dietrich and T. Wetzel. Experimental investigation of the influence of lubricating oil on the flow boiling heat transfer and pressure drop of CO_2 inside an enhanced tube. *24th IIR International Congress of Refrigeration*, Aug 16-22, Yokohama, 2015 (**talk**).

M. Wetzel, B. Dietrich and T. Wetzel. Wärmeübergang und Druckverlust beim Strömungssieden von CO_2-Öl-Gemischen im innenstrukturierten Rohr. *ProcessNet Fachausschuss Mehrphasenströmungen*, Mar 24-25, Fulda, 2014 (**poster**).

M. Wetzel, B. Dietrich and T. Wetzel. Influence of oil on heat transfer and pressure drop during flow boiling of CO_2 at low temperatures. *8th World Conference on Experimental Heat Transfer, Fluid Mechanics and Thermodynamics*, Jun 16-20, Lisbon, 2013 (**talk**).

M. Wetzel, B. Dietrich, T. Wetzel and M. Kind. Wärmeübergang und Druckverlust beim Strömungssieden von CO_2-Öl-Gemischen. *ProcessNet Fachausschuss Wärme- und Stoffübertragung*, Mar 20-21, Baden-Baden, 2013 (**poster**).

M. Wetzel and T. Wetzel. Wärmeübergang und Druckverlust beim Strömungssieden von CO_2-Öl-Gemischen. *Deutsche Kälte-Klima-Tagung*, Nov 21-23, Würzburg, 2012 (**talk**).

M. Wetzel, B. Dietrich and T. Wetzel. Strömungsverdampfen von CO_2-Öl-Gemischen im horizontalen Rohr. *ProcessNet Jahrestagung*, Sep 10-13, Karlsruhe, 2012 (**poster**).

M. Wetzel, Y. Saito, M. Kind and T. Wetzel. Experimentelle Untersuchung des Wärmeübergangs beim Strömungsverdampfen von CO_2 und CO_2-Öl-Gemischen. *ProcessNet Fachausschuss Wärme- und Stoffübertragung*, Mar 21-22, Frankfurt a.M., 2011 (**talk**).

M. Wetzel, Y. Saito, M. Kind and T. Wetzel. Investigations on the flow boiling heat transfer of CO_2 and CO_2-oil-mixtures. *23rd IIR International Congress of Refrigeration*, Aug 21-26, Prague, 2011 (**talk**).

Y. Saito, **M. Wetzel**, M. Kind and T. Wetzel. New flow pattern map for CO_2 two-phase flow in horizontal smooth tube. *23rd IIR International Congress of Refrigeration*, Aug 21-26, Prague, 2011 (**talk**).

M. Wetzel, Y. Saito, M. Kind and T. Wetzel. Einfluss von Kältemaschinenöl auf den Wärmeübergang beim Strömungssieden von CO_2. *ProcessNet Fachausschuss Wärme- und Stoffübertragung*, Mar 08-10, Hamburg, 2010 (**poster**).

M. Wetzel, Y. Saito, M. Kind and T. Wetzel. Einfluss von Kältemaschinenöl auf den Wärmeübergang beim Strömungssieden von CO_2. *Deutsche Kälte-Klima-Tagung*, Nov 18-20, Berlin, 2009 (**talk**).

Appendix A

A.1 Measurement devices

In order to determine non-directly accessible measurement data (e.g., temperature, pressure, mass flow rate), the respective measurement devices were calibrated prior conducting flow boiling experiments. Thermocouples and pressure transducers were controlled repeatedly to ensure the accuracy of measurement. For thermocouples polynomial calibration curves of 4th order were determined, giving the temperature as function of thermal voltage recorded. The temperature dependent resistance of PT100 platinum thermometers was correlated by 2nd order polynomial functions of the temperature. For all devices with integrated measuring transducers, linear relationships of the output signal ($4-20$ mA) to the respective measurand were determined. To maximize the accuracy of pressure drop measurements, two different set of parameters were determined for the differential pressure transducer dependent on the range of measurement.
A complex analysis of the uncertainty in measurement based on the *Guide to the expression of Uncertainty in Measurement (GUM)* (International Organization for Standardization, 1993) was carried out (Riegel, 2011), investigating the interaction of all accessible measurement uncertainties on the distinct measurement parameter. Based on the analysis, extended measurement uncertainties ($k = 2$) for the most influential measurement parameters, i.e., temperature and pressure, were determined, see Tab. A.1. Table A.2 lists the measurement devices used for recording the experimental data values during experiments and devices used for calibration.

A.2 Definitions for representing the error in measurement

Experimental data values for each measuring device were recorded multiply (20 times) during experiments in order to determine the stability and standard deviation of measurements. Error bars in diagrams shown represent the standard deviation (σ) according to Eq. A.1.

$$\sigma^2 = \frac{1}{n-1}\sum_{i=1}^{n}(X_i - \bar{X})^2 \tag{A.1}$$

where $n = 20$ is the number of measurements, X_i is the measured value no. i of parameter X and $\bar{X}$ is the mean value of parameter X.

For evaluation of the accuracy of predictive methods with respect to the experimental data base obtained in this work, the errors defined as follows were determined.

$$\mathrm{AE} = |(X_i)_{\mathrm{pred}} - (X_i)_{\mathrm{exp}}| \tag{A.2}$$

$$\mathrm{ME} = \frac{1}{n}\sum_{i=1}^{n}\left((X_i)_{\mathrm{pred}} - (X_i)_{\mathrm{exp}}\right) \tag{A.3}$$

$$\mathrm{MAE} = \frac{1}{n}\sum_{i=1}^{n}\left|\left((X_i)_{\mathrm{pred}} - (X_i)_{\mathrm{exp}}\right)\right| \tag{A.4}$$

$$\mathrm{PE} = \frac{(X_i)_{\mathrm{pred}} - (X_i)_{\mathrm{exp}}}{(X_i)_{\mathrm{exp}}} \tag{A.5}$$

$$\mathrm{APE} = \left|\frac{(X_i)_{\mathrm{pred}} - (X_i)_{\mathrm{exp}}}{(X_i)_{\mathrm{exp}}}\right| \tag{A.6}$$

$$\mathrm{MPE} = \frac{1}{n}\sum_{i=1}^{n}\left(\frac{(X_i)_{\mathrm{pred}} - (X_i)_{\mathrm{exp}}}{(X_i)_{\mathrm{exp}}}\right) \tag{A.7}$$

$$\mathrm{MAPE} = |\mathrm{MPE}| \tag{A.8}$$

where AE is the absolute error, ME is the mean error, MAE is the mean absolute error PE is the percentage error, APE is the absolute percentage error, MPE is the mean percentage error and MAPE is the mean absolute percentage error.

Table A.1: Extended measurement uncertainties

Thermocouple (type K, E)	± 0.17 K
Platinum resistance thermometer (PT100)	± 0.023 K
Pressure transducer	± 0.025 bar

Table A.2: Measurement devices

Description	**Type**	**Manufacturer**	**Technical data**
DAQ Multimeter	KE 2750	Keithley	5-port, GPIB interface
DAQ module for general-purpose	KE 7700	Keithley	20-ch differential multiplexer
cDAQ USB-chassis	NI 9178	National Instruments	8-port, USB interface
cDAQ module for thermovoltage	NI 9213	National Instruments	16-ch voltage rng.: $U = \pm 78\ \mu$V res.: 24 bit rate (high res.): 16 $\mathrm{S\,s^{-1}}$ err.: $< \pm 0.15$ rdg.
cDAQ module for current	NI 9217	National Instruments	4-ch universal rng.: $I = \pm 25$ mA res.: 24 bit rate (high res.): 8 $\mathrm{S\,s^{-1}}$ err.: $< \pm 0.6$ rdg. ± 0.01 rng.
cDAQ module for resistance	NI 9219	National Instruments	4-ch resistance (4-wire) rng.: $R = 0 - 400\ \Omega$ res.: 24 bit rate (high res.): 5 $\mathrm{S\,s^{-1}}$ sensitivity (RTD): < 3 mK
Pressure transducer	8106	Burster	rng.: $I = 4 - 20$ mA ($0 < p/\mathrm{bar} < 100$) err.: $< \pm 0.5$ % rdg.
Differential pressure transducer	3051C	Rosemount	rng.: $I = 4 - 20$ mA err.: $< \pm 0.37$ mbar ($0 - 100$ mbar) $< \pm 0.27$ mbar ($0 - 10$ mbar)
Flow meter	Promass 60	Endress & Hauser	rng.: $I = 4 - 20$ mA ($\dot{m} \leq 2000\ \mathrm{kg\,h^{-1}}$) err.: $< \pm 0.15$ rdg. ± 0.1 rng.
Digital power meter	WT 1030	Yokogawa	res.: 16 bit err.: $< \pm 0.2$ rdg. $\pm 0.225\ \mathrm{kg\,h^{-1}}$
High speed video camera	Motion Corder Analyzer	Kodak	—
Laser confocal displacement meter	LT-9000	Keyence	—
DC measuring bridge	MKT 25	Anton Paar	—
High precision RTD	Pt 25	Rosemount	—
Differential pressure gauge	DH 5501	Desgranges et Hout	—

Appendix B

For validating the test facility described in Sec. 3.1, flow boiling experiments with pure CO_2 in both measurement sections were carried out initially. The determined heat transfer coefficients and total pressure drops were compared to data published by Schael (2009) for verifying the reproducibility of previously conducted pure refrigerant experiments. The experiments by Schael were carried out in the time of 2002 to 2005. Newly obtained experimental data is listed in App. G. In addition, repeatability and reproducibility of CO_2-oil mixture flow boiling measurements inside the micro-fin tube have been performed incessantly. The time span between two series as well as the operating person at the test facility were varied.

B.1 Pure CO_2 investigations

Counting for all figures in this section, bright gray symbols represent experimental data points obtained by Schael and dark gray symbols represent reproduced data within the scope of this work.

Smooth tube

Reproducibility tests inside the smooth tube were run at a pressure of 26.4 bar, mass flow velocities from 150 to 300 $\mathrm{kg\,m^{-2}\,s^{-1}}$, vapor qualities from 0.1 to 0.9 and for heat fluxes of up to 100 $\mathrm{kW\,m^{-2}}$. Exemplary measurement results are presented in Fig. B.1-B.3.
Peripheral averaged heat transfer coefficients at two different operating conditions are plotted versus heat flux in Fig. B.1. As the figures depict, previously measured heat transfer coefficients by Schael could be reproduced with good accuracy. For a more detailed analysis including the effect of flow pattern, corresponding peripheral local heat transfer coefficients are shown in Fig. B.2 for two average heat fluxes of approximately 30 and 65 $\mathrm{kW\,m^{-2}}$. For a mass flow velocity of 150 $\mathrm{kg\,m^{-2}\,s^{-1}}$ (left image), wavy flow was observed over the whole range of range vapor qualities. The presence of wavy flow during the the heat transfer measurements pictured is revealed by the sharp drop in heat transfer coefficient at the upper segments (60 ° and 300 °). During slug flow, occuring at a mass flow velocity of 300 $\mathrm{kg\,m^{-2}\,s^{-1}}$ and a vapor quality of 10 % (right image), all segments were periodically wetted by liquid slugs. Thus, heat transfer coefficients along the tube perimeter are almost equal. Here, too, previous and new data points agree well in trend and magnitude.

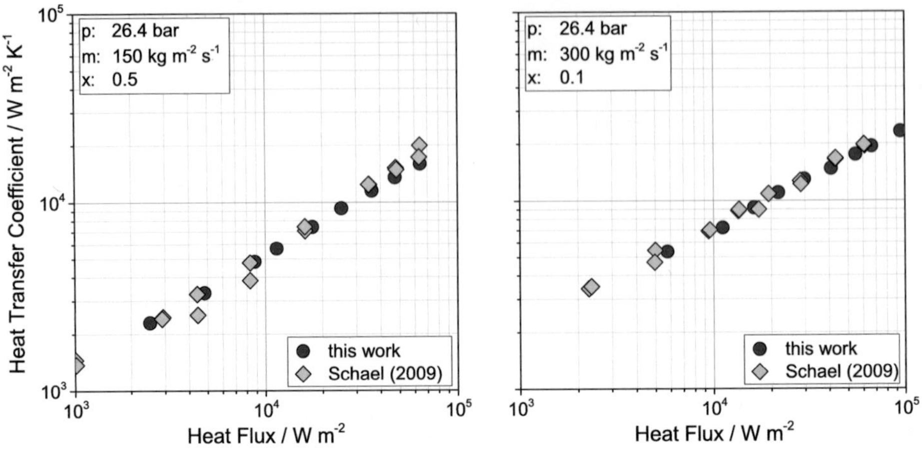

Figure B.1: Peripheral averaged heat transfer coefficients of two measurement series inside the smooth tube using pure CO_2

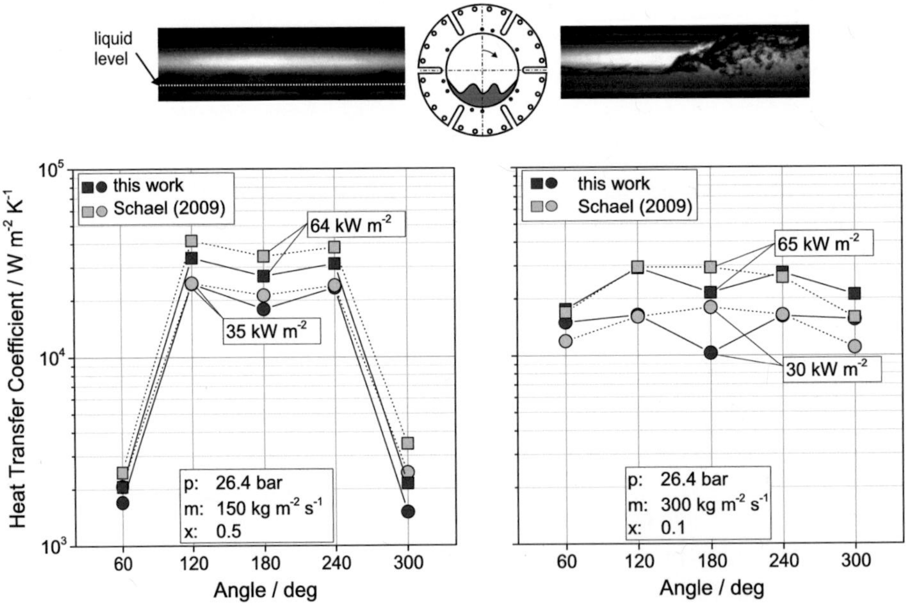

Figure B.2: Peripheral local distribution of heat transfer coefficients, corresponding to the test conditions in Fig. B.1

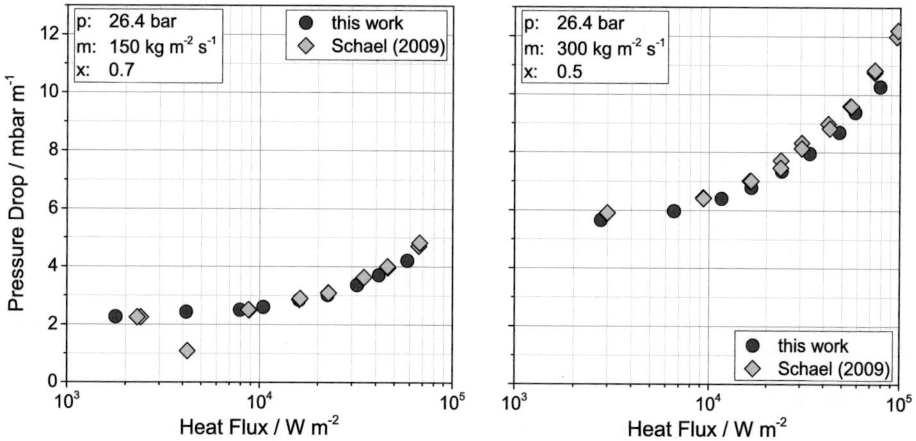

Figure B.3: Total pressure drops from flow boiling experiments of pure CO_2 under diabatic conditions inside the smooth tube

Simultaneously, the total pressure drop along the measurement section was measured. Figure B.3 shows experimentally determined pressure drops versus heat flux for two measurement series. In both cases, the experimental pressure drop data by Schael could be reproduced accurately.

Micro-fin tube

Reproducibility of the experimental determination of flow boiling heat transfer coefficients and diabatic total pressure drops inside the micro-fin tube is addressed in Fig. B.4-B.6. The range of operation of these test runs included pressures from 19.1 to 39.7 bar, mass flow velocities from 75 to 500 $\mathrm{kg\,m^{-2}\,s^{-1}}$, vapor qualities from 0.1 to 0.9 and heat fluxes up to 120 $\mathrm{kW\,m^{-2}}$.

Peripheral averaged heat transfer coefficients (see Fig. B.4) and corresponding local heat transfer coefficients (see Fig. B.5) for pure CO_2 are compared to data published by Schael. Both measurement series that exemplarily shown depict a general good agreement of new and old data points.

As for the smooth tube, diabatic total pressure drops were compared to the previously obtained data as well. A general good agreement in trend and value of the data was found out, as the measurement series plotted in Fig. B.6 confirm.

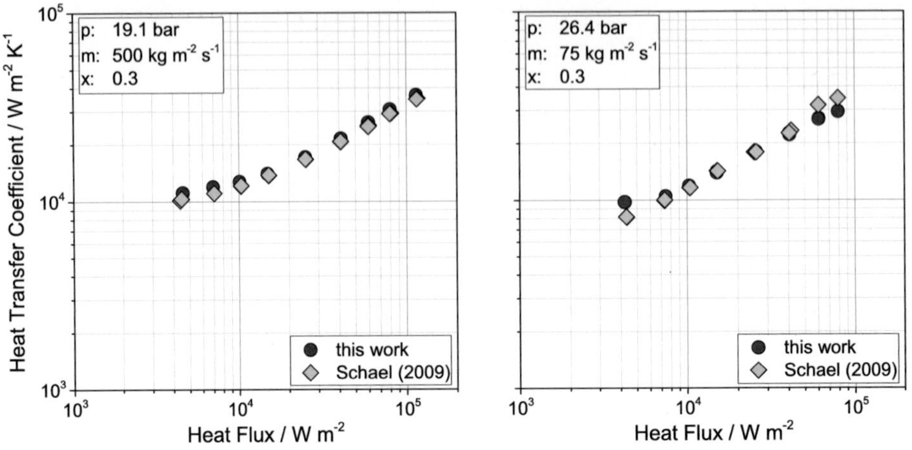

Figure B.4: Peripheral averaged heat transfer coefficients of two measurement series inside the micro-fin tube using pure CO_2

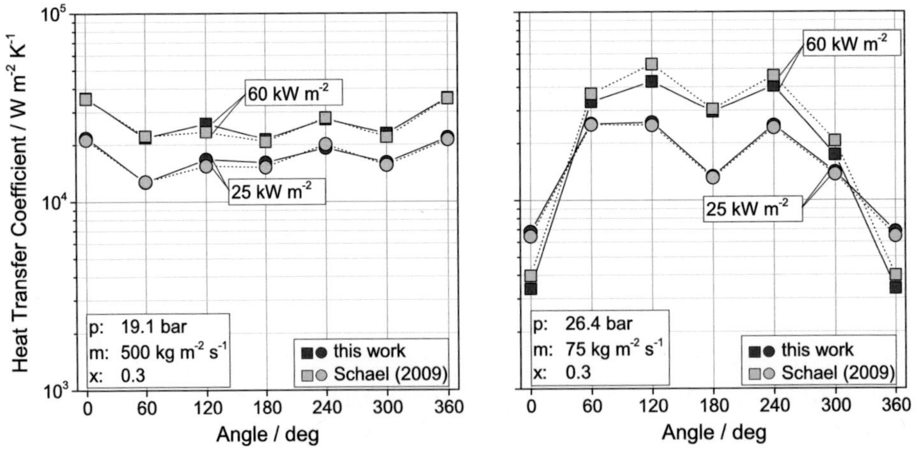

Figure B.5: Peripheral local distribution of heat transfer coefficients, corresponding to the test conditions in Fig. B.4

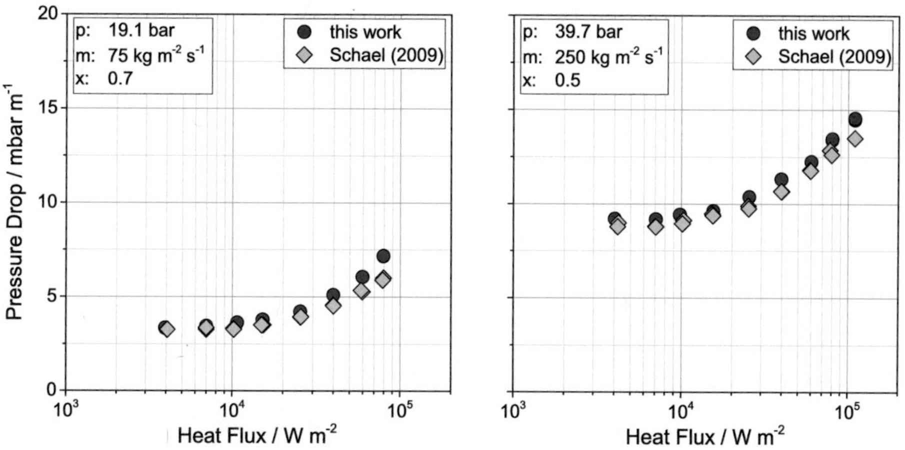

Figure B.6: Total pressure drops from flow boiling experiments of CO_2 under diabatic conditions inside the micro-fin tube

B.2 CO_2-oil mixture investigations

Analogous to pure CO_2 measurements, reproducibility of own experimental results using CO_2-POE oil mixtures was verified by repeating several measurement series at various test conditions. As exemplarily shown in Fig. B.7 by average heat transfer coefficients versus heat flux for two test series and mixtures containing 3 wt.-% lubricant, heat transfer experiments conducted by different operating persons (abbreviations TN, LB, MW) and different points in time were generally in good agreement. The same conclusion counts for local heat transfer measurements, as can be seen Fig. B.8. Displayed peripheral local heat transfer coefficients (left) and wall temperature distribution (right) correspond to the averaged heat transfer coefficients in Fig. B.7 (right).
Total pressure drops yielded similar reliable results. In general, measurements agreed well in trend and magnitude. This fact is illustrated in Fig. B.9 by two reproduced measurement series. However, measurements revealed that pressure drops are highly sensitive to the flow pattern, especially regarding foam or foam-containing flow. Large deviations are the consequence. Hence, indicating the stability of measurement conditions by supplying error bars in graphs as well as analyzing flow patterns during flow boiling experiments is of essential need.

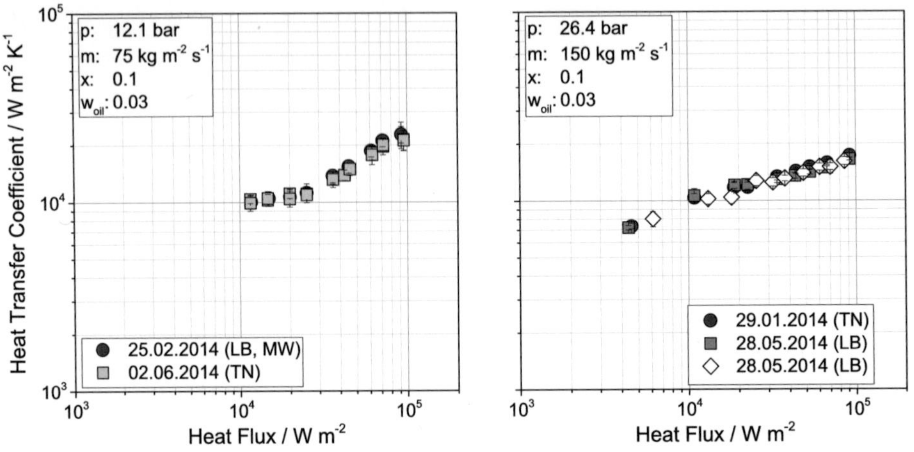

Figure B.7: Peripheral averaged heat transfer coefficients of two measurement series inside the micro-fin tube using a CO_2-oil mixture of 3 wt.-% in oil content

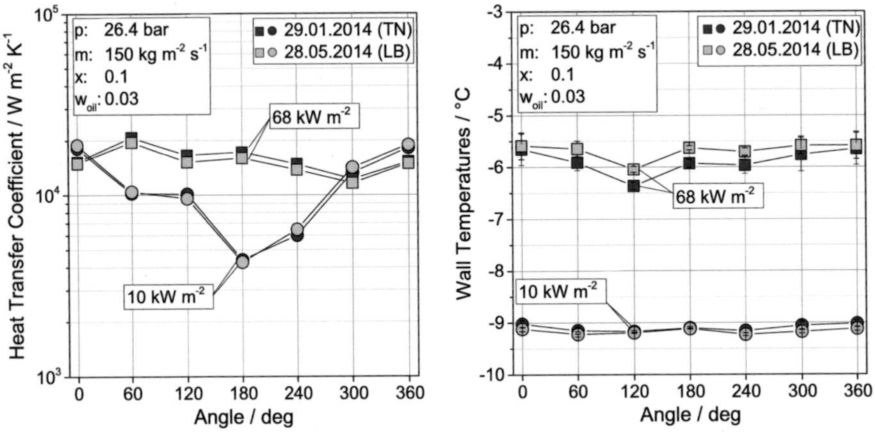

Figure B.8: Peripheral local distribution of heat transfer coefficients (left) and inner wall temperatures (right), corresponding to the test conditions in Fig. B.7 (right)

In Fig. B.10, total pressure drops of four measurements under nearly the same conditions are presented (right). For the first three series, a good agreement in trend and value of the determined pressure drops was noticed. Error bars representing standard deviations were comparatively large. This fact indicated a strong intermittent and foaming flow, affirmed by the flow patterns observed (see top right image in Fig. B.10). Out of line, pressure

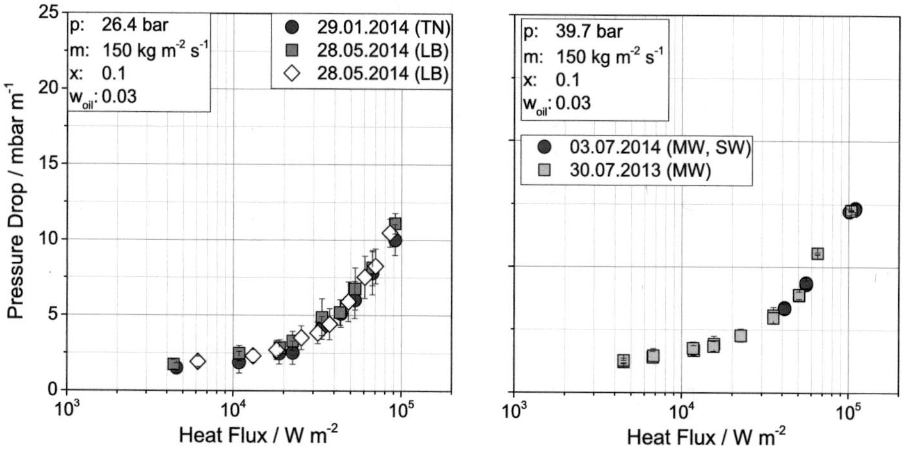

Figure B.9: Total pressure drops from flow boiling experiments of CO_2-oil mixtures with an oil mass fraction of 3 wt.-% under diabatic conditions inside the micro-fin tube

drops of the fourth series conducted were almost steady and heat flux independent for $\dot{q} > 30\,\mathrm{kW\,m^{-2}}$, see white symbols in the left diagram. Here, froth flow completely filling out the tube was observed (middle right image). No liquid slugs breaking through nor tube dryout was observed. Due to the steady-like flow, deviations of measured pressure drops decreased remarkably. By lowering the heat flux to $\dot{q} = 26\,\mathrm{kW\,m^{-2}}$, flow pattern changed to intermittent flow (bottom right image).

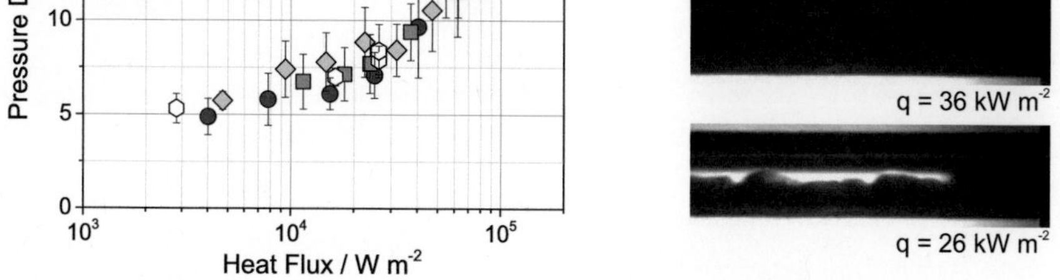

Figure B.10: Total pressure drop (left) and flow patterns observed (right) in the micro-fin tube

The displayed test series at 250 $kg\,m^{-2}\,s^{-1}$ and a vapor quality of 0.1 indicate the sensitiveness of CO_2-oil mixture flow boiling results. The reason for the occurrence of different pressure drop behavior for the fourth measurement series could not be clearly identified. Regarding absolute values and deviations of pressure, mass flow rate and vapor quality, no differences were found. As only difference in the procedure of operation, the rate of subcooling was varied. In case of the fourth measurement series, subcooling of the liquid entering the test section was increased by $-2.9\,°C$ with respect to the first three series. Due to the greater subcooling, the essential power input through the pre-heaters increased as well to assure the same inlet vapor quality. Hence, a possible explanation is that the transition of foaming-intermittent flow to fully foam flow was transcended by the increase in total heat input. The effect of subcooling was addressed in other test series as well. For instance, inlet temperatures of the test series at 150 $kg\,m^{-2}\,s^{-1}$ and a mean vapor quality of 0.1 presented in Fig. B.7 (right), Fig. B.8 and Fig. B.9 (left) were varied in a similar range. Here, no influence could be identified.

Appendix C

A new technique for measuring the film thickness at a fixed point (here: crest of the sight glass tube) with a spatio-temporal resolution of $0,1$ mm and 500 fps based on confocal laser displacement was introduced and first test runs performed. With its development, a more definite determination of the state of flow in future measurements to refine prediction methods for flow patterns was pursued. In the following, the principle of measurement, constructed measurement setups and conducted experiments are presented.

C.1 Measurement principle

The technique for measuring the thickness of a small (≤ 2 mm) liquid film or transparent solid phase relies on the principle of confocal laser displacement. As schematically shown in Fig. C.1 (left), the focal point of the laser ray emitted by the measurement device (Keyence LT-9000) is determined by a set of lenses oscillating in vertical direction. The amplitude of oscillation (± 1 mm) corresponds to the range of measurement. If the focal point impinges on an optical interface (i.e., the interface of two phases having different optical densities), the sensor registers a (local) maximum of back reflexion. The intensity of the signal along with the vertical position of the lens system is assigned to a measurement point being displayed as displacement from the reference distance ($L_0 = 30$ mm). The integrated A/D converter translates the displacement (Δs) into an analog output of ± 4 mA that is transmitted to a PC for recording.

For measurements through transparent media (see Fig. C.1, middle), multiple measurements points may be detected, i.e., in number as many interfaces as present. As illustrated for n > 1, the distance to the focal point is extended by light refraction. The reference distance from sensor to the focal point increases and the displacement of the lens system is no longer equal to the total displacement. Since refractive effects are not accounted for by the measurement system, manual correction is necessary. For a given distance L_1 to the top surface, Eq. C.1 accounts for the change in reference distance, yielding the real reference distance L_{0^*}.

$$L_{0^*} = L_1 + \frac{D}{2}\left(1 + \frac{L_1}{L_0}\right)\left(\left(\frac{4L_0^2}{D^2} + 1\right)n^2 - 1\right)^{0.5} \tag{C.1}$$

where D is the lens diameter, $L_0 = 30$ mm is the base reference distance and $n = \sin(\alpha)/\sin(\beta)$ is the refractive index of dense to surrounding (air) medium.

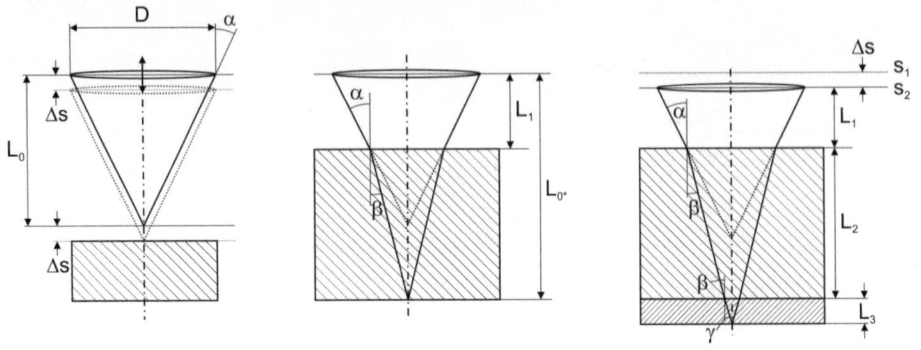

Figure C.1: Confocal laser displacement measurement in media consisting of one (left), two (middle) or three (right) optical differing phases

Figure C.1 (right) demonstrates the measuring of a liquid film on the bottom surface of a transparent phase (e.g., at the inside of a sight glass tube). The change in reference distance due to length and optical density of the run-through phase becomes irrelevant, if the thickness of liquid film (L_3) lies within the range of measurement. In that case, the bottom surface (2-3 phase interface) is be focused at the lower end of the range of measurement and set as reference. For calculating the film thickness from the measurement points s_1 (displacement at reference) and s_2 (displacement at film surface), Eq. C.2 was derived.

$$L_3 = \Delta s \cos(\alpha) \left(n^2 - \sin^2(\alpha)\right)^{-0.5} \tag{C.2}$$

where $\Delta s = s_2 - s_1$, $\alpha = \arctan(\frac{D}{2L_0})$, D is the lens diameter, $L_0 = 30$ mm is the base reference distance of the displacement sensor and n is refractive index.

C.2 Preliminary setup and smooth tube test runs

For checking the application of the confocal laser displacement system for measuring the liquid film of a flowing liquid inside a tube, a preliminary setup for implementation into the smooth tube test test section was developed (see Fig. C.2). It consisted of a PMMA plate with a rectangular notch, where the sensor was installed on a z-table for vertical adjustment. The composite structure was placed horizontally above the sight glass tube in such a way that the measurement point of the sensor was positioned at the crest of the tube. By means of screws functioning as a tripod, the tilting of the sensor with respect to the sight glass could be adjusted. The circular sight glass was encompassed by a rectangular prism made of PMMA. Its homogeneous body with glass-like optical density and flat surface was necessary in order to reduce scattering of the laser light on its way from to the liquid film at the inside

Figure C.2: Preliminary setup of film thickness measurement system

of the glass tube. The gap in between the glass tube and the surrounding PMMA block was filled with oil for further reduction of light scattering.

Preceding a film thickness measurement, the sensor was adjusted in all directions, i.e., positioned (1) orthogonally to the sight glass surface, (2) at the depth of the crest of the tube and (3) in height such that the laser focused on the inner wall surface of the sight glass tube at the outer range of measurement. The appropriate setup was achieved when a maximum in signal intensity (shortest distance, minimal light scattering) was reached. A measurement was conducted with the resulting distance displayed counting as reference state. The difference of reference state to the distance measured during operation corresponds to the (uncorrected) thickness of liquid film (δ_L) at the crest of the tube.

$$\delta_L = s_L - \bar{s}_W \tag{C.3}$$

where $\bar{s}_W$ is the averaged reference state of the inner wall surface and s_L is the displayed value representing the liquid-vapor interface.

In Fig. C.3, results from film thickness measurements of wavy (left) and annular (right) flow are presented. The liquid film thickness (δ_L) obtained from Eq. C.3 and corrected by Eq. C.2 is shown as solid line. The short dashed line represents the afore measured reference state (= inner wall surface, s_W), which averaged value was used for determining (δ_L). The average film thickness obtained from measurements of wavy flow yields 65 μm, which is in the range of uncertainty. It can be reasoned that there is no liquid film present at the crest of the tube. High speed video recordings of the state of flow substantiate the argumentation, see the image in the left diagram. For annular flow, a liquid film was visually observed and clearly detected, see Fig. C.3 (right). The film thickness averaged out to 520 μm. From the results it was concluded that the technique is suitable for measuring the film thickness.

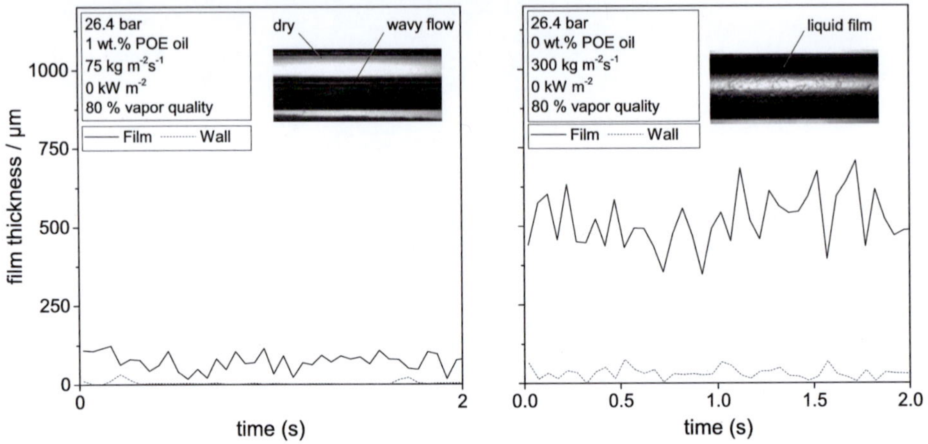

Figure C.3: Measurements of the liquid film thickness during wavy (left) and annular flow (right) inside the smooth tube

C.3 Development of a flexible setup and static validation experiments

A new setup was developed (Witt, 2013), applicable to basic validation experiments in a static test environment outside the test loop. In order to avoid constructive measures that could affect the measurement system afterwards, the complete measurement setup was designed to be transferable to the test section as is. Aluminum plates were chosen as rigid basis. A flexible tripod stand was realized by means of screw-spring composite legs with hemispheres for smooth movement. That way, a precise and smooth adjustment of the plane containing the measurement sensor was possible. The sensor itself was mounted on a y-z-table for vertical and horizontal (in the plane of cross section when referring to the installation in the test section) fine positioning.

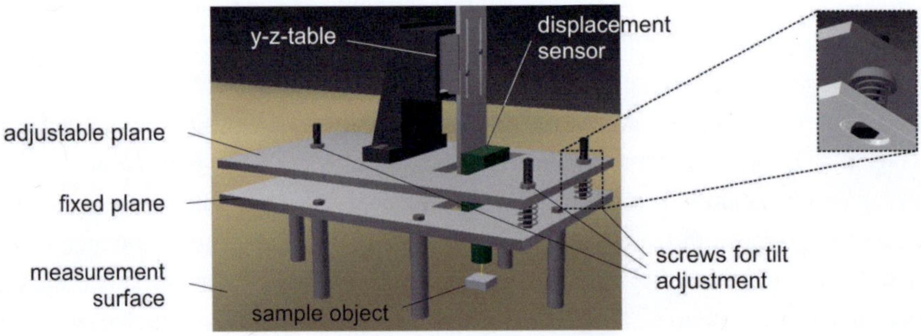

Figure C.4: CAD drawing of the measurement system

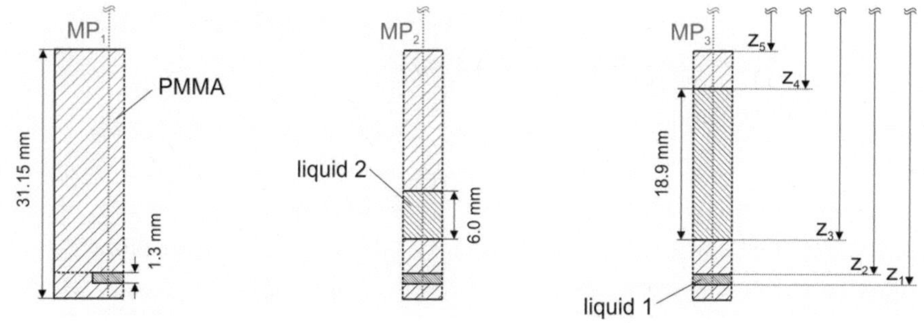

Figure C.5: Schematic of the constructed PMMA sample used for static laser displacement validation experiments

For the conduct of static experiments, the setup was installed on metal legs on a flat surface as shown in Fig. C.4. Different transparent samples were investigated using this setup. These tests were primarily done to determine the non-specified diameter of the sensor's lens and to validate the calculation specification to be used to determine the liquid film thickness.
The lens diameter was determined by measuring the cross sections of the light cone of the laser. The cross sections were projected on a flat plane in distinct spacings (10 mm), starting at the edge of the sensor and then receding along the axis of the laser. From the circular projections dimensioned, the lens diameter was identified as $D = 10$ mm. Further, the angle of the emitted rays to the perpendicular was found to be $\alpha = \arctan(\frac{D}{2L_0}) = 9.46\,°$.
Validating the calculation specification for measurements through optical dense media was done using the sample demonstrated in Fig. C.5. The distance from the sensor's lens to the bottom of the sample body was measured at three different measurement positions MP_1 – MP_3 (along the gray dotted lines). The gaps displayed were filled with ethanol (liquid 1) and oil (liquid 2). From own experiments and values published in literature, the refractive indices yielded $n_{ethanol} = 1.36$, $n_{oil} = 1.46$, $n_{PMMA} = 1.49$. The distances determined experimentally are listed in Tab. C.1. The results indicate that the relative error in displacement measurements due to ignoring light refraction in the optical dense media is significant (absolute > 25 %). Accounting for these effects, the resulting error becomes negligible (absolute ≤ 1.5 %).

Table C.1: Displacements measured in static experiments inside a PMMA sample block at three different locations

location (measurement point)	MP_1	MP_2	MP_3
real distance [mm]	39.70	39.70	39.70
displacement measured ignoring refraction [mm]	29.11	29.21	29.51
relative error [%]	−26.7	−26.4	−25.6
corrected displacement regarding refraction [mm]	39.09	39.37	40.07
relative error [%]	−1.5	−0.8	+0.9

C.4 Test section integrated measurement setup and sight glass tube modification

For integrating the measurement setup into the test section, a modification of the sight glass was made. A rectangular PMMA body with planar surfaces was designed to encase the glass tube for preventing scattering of the laser light and simplifying the device adjustment. A gap between the body's inner excavation surface and the outer surface of the glass tube was incorporated for easy mounting and to avoid temperature-induced tensions. By means of two bores, the gap was filled with oil of similar optical density as PMMA. The glass tube surrounded by the PMMA body was clamped between two brass flanges and insulated against fluid emission with PTFE gauntlets on either side. Placed within the two flanges, the sight glass was further surrounded by a circular PMMA tube of the size of the flanges. The enclosed space is flushed with dry gas (e.g., nitrogen) during flow boiling experiments in order to prevent condensation of moisture on the cold surfaces. The PMMA tube was issued a small opening for inserting the measurement sensor. A CAD drawing of the construction as integrated into the test section is shown in Fig. C.6. Additional drawings and information of the measurement setup and constructive measures are supplied in Appendix.

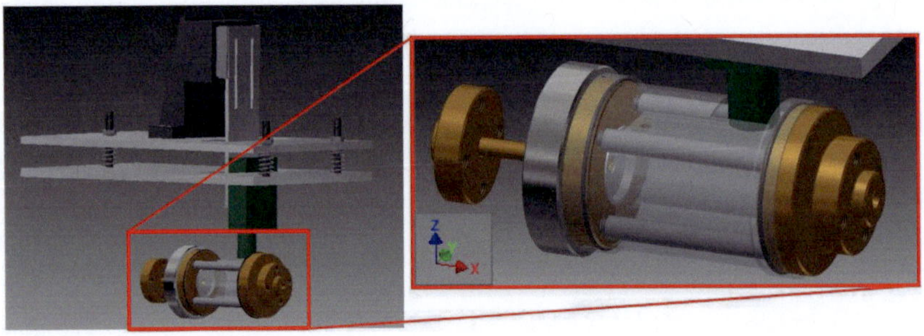

Figure C.6: CAD drawings of the the measurement setup and modified sight glass construction (Witt, 2013)

C.5 Conclusion

A new measurement technique based on the principle of confocal laser displacement was introduced to determine the thickness of a liquid film inside a round sight glass tube. It was successfully applied to measure the film thickness at the crest of the tube, supporting the distinction between non- and complete wetting flow patterns, i.e., wavy and annular flow, respectively. A specification for correcting the output displacement due to light refraction effects was proposed and validated by static experiments. This correction is necessary as light refraction is not regarded by the measurement device. In other words, homogeneous single phase medium is assumed for the environment of displacement measurement. This is

not the case, if, as done, the measurement of displacement takes place through a liquid film of optical density different than the surrounding. In case of CO_2, the estimated error in film thickness due to neglecting the refractive index would yield 20 %.
Despite its successful implementation, several limitations concerning applicability and constructive measures were encountered that have to be taken into account. First of all, only one-dimensional measurements (here: crest of the tube) of light transparent liquid films can be conducted. This fact mainly puts a limit to adiabatic (bubble-free) and non-foaming flows. Otherwise, determination of the film thickness is not possible as both, bubbles and foam, prevent the light from passing the liquid in straight way, either absorbing or reflecting it. To allocate constructive measures, the complete measurement system (sensor mounting as well as sight glass tube) inevitably must be fixed rigid and secured against any movement throughout the entire measurement series. As seen before in the left plot of Fig. C.3, even small movements - during the film thickness measurement as well as during capturing the reference displacement - of either sensor or test section lead to measurable deviations. For the series shown, uncertainty in measurement solely due to this effect yields to approximately 100 μm. As the size of the tube is reduced (14.0 mm smooth tube to 8.62 mm micro-fin tube), the addressed effect further gains weight. Due to this reason and the fact that the measurement device, from a constructive view, could not be placed exactly at the inlet of the sight glass, the technique could not be applied to film thickness measurements in the micro-fin tube. In order to achieve this in further experiments, constructive modifications must be realized based on the experience-gained instructions mentioned above.

Appendix D

D.1 Fluid properties of CO_2

A comprehensive survey on available correlations for predicting pure substance properties of CO_2 was carried out by Schael (2009). The following methods were suggested as most accurate and, hence, used in this work.

D.1.1 Critical and triple point data of CO_2

Table D.1 lists the critical and triple point data of CO_2 (Span and Wagner, 1996).

Table D.1: Critical and triple point data of CO_2 (Span and Wagner, 1996)

Triple point temperature	$T_t /$ K	$= 216.592 \pm 0.003$
Triple point vapor pressure	$p_t /$ MPa	$= 0.51795 \pm 0.0001$
Critical temperature	$T_c /$ K	$= 304.1282 \pm 0.0015$
Critical vapor pressure	$p_c /$ MPa	$= 7.3773 \pm 0.0030$
Critical density	$\rho_c / \mathrm{kg\,m^{-3}}$	$= 476.6 \pm 0.6$

D.1.2 Saturated properties

Equations for saturated properties were published by Span and Wagner (1996) in terms of reduced pressure ($p_r = p_s\, p_c^{-1}$), reduced temperature ($T_r = T\, T_c^{-1}$) and reduced density ($\rho_r = \rho_{V|L}\, \rho_c^{-1}$). Empirical constants are listed in Tab. D.2.
The **vapor pressure** curve of CO_2 reads

$$\ln(p_r) = \left(\sum_{i=1}^{4} a_i\, (1 - T_r)^{t_i} \right) T_r^{-1} \tag{D.1}$$

The saturated vapor and liquid **density** reads

$$\ln(\rho_r) = \left(\sum_{i=1}^{5} a_i\, (1 - T_r)^{t_i} \right) \tag{D.2}$$

Table D.2: Constants for saturated properties by Span and Wagner (1996)

	p_s		ρ_V		ρ_L	
i	a_i	t_i	a_i	t_i	a_i	t_i
1	−7.0602087	1	−1.7074879	0.34	1.9245108	0.34
2	1.9391218	1.5	−0.82274670	0.5	−0.62385555	0.5
3	−1.6463597	2	−4.6008549	1	−0.32731127	10/6
4	−3.2995634	4	−10.111178	7/3	0.39245142	11/6
5	–	–	−29.742252	14/3	0	0

D.1.3 Vapor/liquid substance properties

Vapor/liquid **dynamic viscosities** were calculated using the Fenghour et al. (1998) model.

$$\eta(\rho,T) = \eta^{o}(T) + \Delta\eta(\rho) + \Delta_{c}\eta(\rho,T) \tag{D.3}$$

$$\eta^{o} = 1.00697\, T^{1/2}\, \zeta_{\lambda}^{*-1} \times 10^{-6} \tag{D.4}$$

$$\Delta\eta(\rho,T) = e_1\,\rho' + e_2\,\rho'^2 + e_3\,\rho'^6\, T^{*-3} + e_4\,\rho'^8 + e_5\,\rho'^8\, T^{*-1} \tag{D.5}$$

$$\Delta_{c}\eta(\rho,T) \approx 0 \tag{D.6}$$

$$\zeta_{\lambda}^{*} = \sum_{i=0}^{7} a_i\,(T^{*})^{-i} \tag{D.7}$$

where $k\varepsilon^{-1} = 251.196$ K is an energy scaling parameters, ζ_{λ} is the effective cross section, $T^{*} = T\,k\varepsilon^{-1}$ is a reduced temperature and $\rho' = \rho\,\left(\mathrm{kg\,m^{-3}}\right)^{-1}$ is the density.

Differing to Schael (2009), the widely used method by Vesovic et al. (1990) was used for calculating both, liquid and vapor, **heat conductivities**.

$$\lambda(\rho,T) = \lambda^{o}(T) + \Delta\lambda(\rho) + \Delta_{c}\lambda(\rho,T) \tag{D.8}$$

$$\lambda^{o} = 0.475598\, T^{1/2}\,(1+r^2)\,\zeta_{\lambda}^{*-1} \times 10^{-3} \tag{D.9}$$

$$\Delta\lambda(\rho) = \sum_{i=1}^{4} d_i\,\rho^{-i} \times 10^{-3} \tag{D.10}$$

$$\Delta_{c}\lambda(\rho,T) \approx 0 \tag{D.11}$$

$$\zeta_{\lambda}^{*} = \zeta_{\lambda}\,\left(\pi\sigma^2\right)^{-1} = \sum_{i=0}^{7} b_i\,(T^{*})^{-i} \tag{D.12}$$

$$r = 2\,c_{\mathrm{int}}\,(5\,k)^{-1} \tag{D.13}$$

$$c_{\mathrm{int}}\,k^{-1} = 1.0 + \exp\left(\frac{-183.5}{T\,\mathrm{K}^{-1}}\right) \sum_{i=1}^{5} c_i \left(\frac{T\,\mathrm{K}^{-1}}{100}\right)^{2-i} \tag{D.14}$$

where $\sigma = 0.3751 \times 10^{-9}$ m is a length scaling parameter and c_{int} is the internal isochoric heat capacity of the molecule.

The empirical constants b_i, c_i, d_i and e_i are listed in Tab. D.3.

Table D.3: Model constants for viscosity (Fenghour et al., 1998) and heat conductivity (Vesovic et al., 1990)

i	a_i	b_i	c_i	d_i	e_i
0	0.235156	0.4226159		2.447164×10^{-2}	–
1	−0.491266	0.6280115	0.02387869	8.705605×10^{-5}	4.071119×10^{-3}
2	0.05211155	−0.5387661	4.3507904	-6.547950×10^{-8}	7.198037×10^{-5}
3	0.05347906	0.6735941	−10.33404	6.594919×10^{-11}	2.411697×10^{-17}
4	−0.01537102	0	7.981590	–	2.971072×10^{-23}
5	–	0	−1.940558	–	$-1.627888 \times 10^{-23}$
6	–	−0.4362677	–	–	–
7	–	0.2255388	–	–	–

The **surface tension** of CO_2 is calculated according to Rathjen and Straub (1980).

$$\sigma = 0.084497 \left(1 - T\, T_{\text{c}}^{-1}\right)^{1.28} \tag{D.15}$$

where T is the temperature and T_{c} is the critical temperature (in [K]).

Density, **specific enthalpy** and **specific isobaric heat capacity** were calculated by means of the equation of state of Span and Wagner (1996). Each property $\phi(\rho_{\text{r}}, \tau)$ is described by its ideal $\phi^{\text{o}}(\rho_{\text{r}}, \tau)$ and residual $\phi^{\text{r}}(\rho_{\text{r}}, \tau)$ part as function of reduced density $\rho_{\text{r}} = \rho\, \rho_{\text{c}}^{-1}$ and inverse reduced temperature $\tau_{\text{r}} = T_{\text{c}}^{-1}\, T^{-1}$.

$$\phi(\rho_{\text{r}}, \tau) = \phi^{\text{o}}(\rho_{\text{r}}, \tau) + \phi^{\text{r}}(\rho_{\text{r}}, \tau) \tag{D.16}$$

$$\phi^{\text{o}}(\rho_{\text{r}}, \tau) = \ln(\rho_{\text{r}}) + a_1^{\text{o}} + a_2^{\text{o}}\, \tau + a_3^{\text{o}}\, \ln(\tau) + \sum_{i=4}^{8} a_i \ln\left(1 - \mathrm{e}^{-\tau\, \theta_i^{\text{o}}}\right) \tag{D.17}$$

$$\begin{aligned}\phi^{\text{r}}(\rho_{\text{r}}, \tau) = {} & \sum_{i=1}^{7} n_i\, \rho_{\text{c}}^{d_i}\, \tau^{t_i} + \sum_{i=8}^{34} n_i\, \rho_{\text{c}}^{d_i}\, \tau^{t_i}\, \mathrm{e}^{-\rho_{\text{r}}^{c_i}} + \\ & \sum_{i=35}^{39} n_i\, \rho_{\text{c}}^{d_i}\, \tau^{t_i}\, \mathrm{e}^{-\alpha_i(\rho_{\text{r}}-\varepsilon_i)^2 - \beta_i(\tau-\gamma_i)^2} + \sum_{i=40}^{42} n_i \Delta^{b_i}\, \rho_{\text{r}}\, \psi\end{aligned} \tag{D.18}$$

$$\Delta = \theta^2 + B_i\left((\rho_{\text{r}} - 1)^2\right)^{a_i} \tag{D.19}$$

$$\theta = (1 - \tau) + A_i\left((\rho_{\text{r}} - 1)^2\right)^{(2\beta_i)^{-1}} \tag{D.20}$$

$$\psi = \mathrm{e}^{-C_i(\rho_{\text{r}}-1)^2 - D_i(\tau-1)^2} \tag{D.21}$$

The empirical constants (a_i^{o}, a_i, θ_i^{o}, n_i, d_i, t_i, c_i, α_i, β_i, γ_i, ε_i, b_i, A_i, B_i and C_i) are listed in Tab. D.4.

Table D.4: Constants for the EOS of Span and Wagner (1996)

i	n_i	d_i	t_i	c_i	α_i	β_i	γ_i	ε_i	a_i^{o}	θ_i^{o}
1	0.38856823203161	1	0						8.37304456	
2	2.938547594274	1	0.75						−3.70454304	
3	−5.5867188534934	1	1						2.5	
4	−0.76753199592477	1	2						1.99427042	3.15163
5	0.31729005580416	2	0.75						0.62105248	6.11190
6	0.54803315897767	2	2						0.41195293	6.77708
7	0.12279411220335	3	0.75						1.04028922	11.32384
8	2.165896154322	1	1.5	1					0.08327678	27.08792
9	1.5841735109724	2	1.5	1						
10	−0.23132705405503	4	2.5	1						
11	0.058116916431436	5	0	1						
12	−0.55369137205382	5	1.5	1						
13	0.48946615909422	5	2	1						
14	−0.024275739843501	6	0	1						
15	0.06249479050167800	6	1	1						
16	−0.12175860225246000	6	2	1						
17	−0.37055685270086000	1	3	2						
18	−0.016775879700426	1	6	2						
19	−0.11960736637987	4	3	2						
20	−0.045619362508778	4	6	2						
21	0.035612789270346	4	8	2						
22	−0.0074427727132052	7	6	2						
23	−0.0017395704902432	8	0	2						
24	−0.021810121289527	2	7	3						
25	0.024332166559236	3	12	3						
26	−0.037440133423463	3	16	3						
27	0.14338715756878	5	22	4						
28	−0.13491969083286	5	24	4						
29	−0.02315122505348	6	16	4						
30	0.012363125492901	7	24	4						
31	0.002105832197294	8	8	4						
32	−0.00033958519026368	10	2	4						
33	0.0055993651771592	4	28	5						
34	−0.00030335118055646	8	14	6						
35	−213.6548868832	2	1		25	325	1.16	1		
36	26641.569149272	2	0		25	300	1.19	1		
37	−24027.212204557	2	1		25	300	1.19	1		
38	−283.41603423999	3	3		15	275	1.25	1		
39	212.47284400179	3	3		10	275	1.22	1		
i	n_i	a_i	b_i	β_i	A_i	B_i	C_i	D_i		
40	−0.66642276540751	3.5	0.875	0.3		0.3	10	275		
41	0.72608632349897	3.5	0.925	0.3		0.3	10	275		
42	0.055068668612842	3	0.875	0.3		1	12.5	275		

D.2 Fluid properties of CO_2-lubricant mixtures

Measurement data of vapor pressure, liquid dynamic viscosity and liquid density for different mixtures of CO_2 and Reniso C 85 E are available by the manufacturer (Saito, 2014). The mixture heat capacity was determined experimentally at TVT (Saito, 2014). Further, contract measurements of the mixture thermal conductivity were conducted by ILK Dresden (Feja and Römer, 2008) and LTP Oldenburg (Ihmels, 2008a) (Ihmels, 2008b). Merged into a data base the experimental data was used for establishing or validating predictive material property correlations, see Sec. D.2.1-D.2.5.

D.2.1 Bubble point temperature and vapor pressure

The vapor pressure curve is shown in Fig. 2.26 as function of bubble point temperature at fixed oil mass fractions (left) and as function of oil mass fraction at fixed bubble point temperatures (right). Symbols represent manufacturer's data (Saito, 2014). Predicted values by Thome (1995) and Zhelezny (2007) are shown as solid lines (1) and (2), respectively. Since fitted to measured values of the investigated CO_2-oil mixtures, on the rough scale the correlation by Zhelezny seems to be precise over a the whole range of operating condition. Especially at oil mass fractions above 50 % predicted values by Thome yield large deviations. However, locally reached oil mass fractions during measurements were generally below 50 % (equal to 94 % in vapor quality at an oil mass fraction of 3 %).
For evaluating heat transfer coefficients from experiments the temperature difference between the tube wall and the boiling fluid is measured. Due to generally high flow boiling heat transfer coefficients of CO_2 temperature differences can be less then 1 K. Thus, a

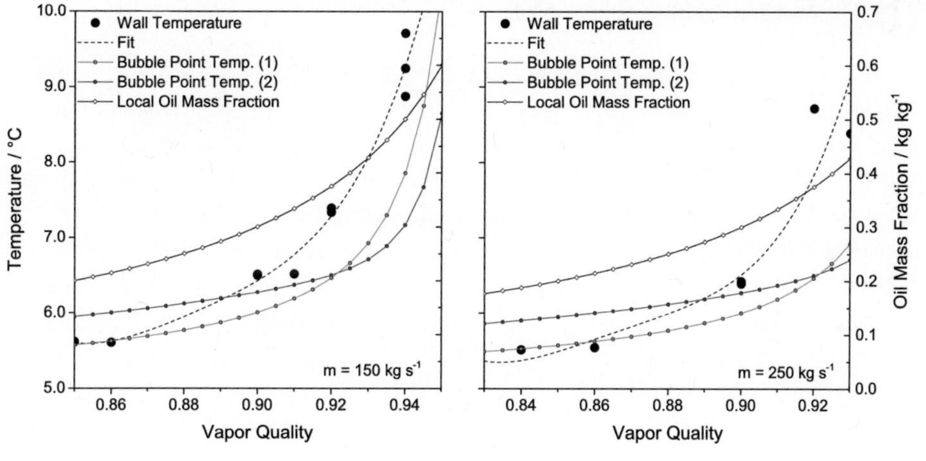

Figure D.1: Saturation Temperature

precise determination of the saturation temperature is of most significance for high quality results. For this reason measurements under near-adiabatic conditions have been made for validation purpose. Vapor quality was increased until almost all liquid was vaporized ($\dot{x} \rightarrow 1$). Measured wall temperatures (dark round symbols) and predicted bubble point temperatures of the fluid are shown in Fig. D.1. In addition, calculated local oil mass fractions are plotted as solid lines with white diamond symbols (scale on right y-axis). Although measured wall temperatures can only be treated as estimation for the bubble point temperature, predicted values by Thome (1995) obviously agree better to measurements than those by Zhelezny (2007). For the conducted heat transfer measurements, good agreement in the lower vapor quality region of Fig. D.1 is of greater importance as experimental vapor qualities are situated in this range.

D.2.2 Viscosity

The viscosity of CO_2-POE oil mixtures was calculated using Eq. 2.106 by Zhelezny (2007) in Sec. 2.5.2. The correlation was developed including manufacturer's data of the lubricant.

D.2.3 Density

The universal correlation by Medvedev et al. (2004) for the mixture density of refrigerant-lubricant mixtures was used, see Eq. 2.113 in Sec. 2.5.3. Analogous to viscosity, the model was adapted using manufacturer's data of the lubricant (Zhelezny, 2007).

D.2.4 Heat capacity

Thome (1995) suggested Eq. D.22 to be used for calculating the liquid specific heat capacity of pure lubricants ($c_{p,oil}$).

$$\frac{c_{p,oil}}{\mathrm{J\,kg^{-1}\,K^{-1}}} = 4186\left(0.388 + 0.00045\left(1.8\frac{T}{\mathrm{K}} - 459.67\right)\right)s^{-0.5} \qquad \text{(D.22)}$$

where T is the temperature, $s = \rho_{oil}/\rho_{water}$ is the lubricant's specific gravity and ρ_{oil} and ρ_{water} are the liquid densities of the lubricant and water at 15 °C.

The specific heat capacity of pure Reniso C 85 E was experimentally determined at TVT (Saito, 2014). Based on this measurement data, Eq. D.23 was determined for predicting the lubricant specific heat capacity ($c_{p,oil}$) as function of temperature (T) within the range of $-40\,°\mathrm{C} < T < 15\,°\mathrm{C}$.

$$\frac{c_{p,oil}}{\mathrm{J\,kg^{-1}\,K^{-1}}} = 1394.99 + 1.31\frac{T}{\mathrm{K}} \qquad \text{(D.23)}$$

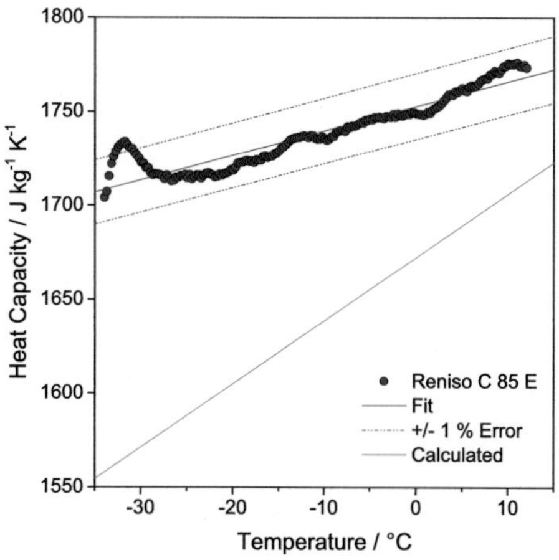

Figure D.2: Heat Capacity

In Fig. D.2 measured (symbols) and predicted values by Eq. D.23 (dark gray solid line) are plotted over the range of investigated temperatures. In comparison to Eq. D.23 predicted values by Eq. D.22 (Thome, 1995) are shown as bright gray solid line, too. This method obviously yields larger deviations, thus Eq. D.23 was used within this work.

D.2.5 Thermal conductivity

Measurements for the thermal conductivity of pure Reniso C 85 E (Ihmels, 2008a) and mixtures of CO_2 and Reniso C 85 E (Ihmels, 2008b) were done by transient hot wire method, see schematic in Fig. D.3 (left). This method is long-established for determination of the thermal conductivity of liquids (Davis et al., 1971) (Roder, 1981). A cylindrical wire rod (ohmic resistor) is placed inside the fluid. Power to the rod, regarded as infinite line heat source, is applied abruptly. Within a certain heating duration (not too short nor too long), the temperature difference between the surface of the rod and the fluid ($\Delta T = T_{rod}(t) - T_{fluid}$) follows a specific solution of Fourier's law given by Eq. D.24.

$$T_{rod}(t) - T_{fluid} = \frac{\dot{Q}}{4\pi\lambda L}\left(\ln\left(\frac{4\kappa}{r^2}t\right) - \gamma\right) \tag{D.24}$$

where $\dot{Q}$ is the electrical heating power along the length of the wire L, λ and κ are the thermal conductivity and thermal diffusivity of the fluid, respectively, r is the radius of the wire, t is the time and $\gamma = 0.57721$ is the Euler's constant.

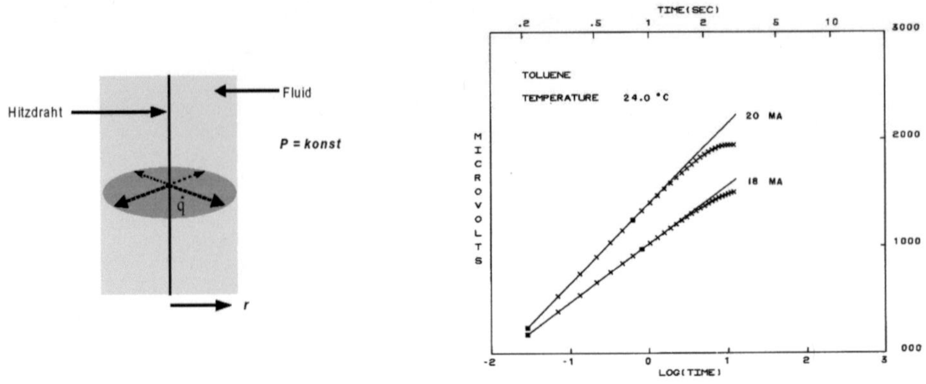

Figure D.3: Transient hot wire method for thermal conductivity determination: Schematic of applied model (taken from Ihmels (2008b), left), generic heat curve (taken from Davis et al. (1971), right)

For two measurements at t_1 and t_2 assuming constant heat rate and non-changing fluid properties, Eq. D.24 can be rewritten as Eq. D.25 or Eq. D.26, the latter from which the thermal conductivity can be calculated.

$$T_{rod}(t_2) - T_{rod}(t_1) = \frac{\dot{Q}}{4\pi\lambda L}\left(\ln(t_2) - \ln(t_1)\right) \tag{D.25}$$

$$\lambda = \frac{\dot{Q}}{4\pi\lambda L}\left(\frac{\ln(t_2) - \ln(t_1)}{T_{rod}(t_2) - T_{rod}(t_1)}\right) \tag{D.26}$$

D.2.6 Surface Tension

A literature survey for and comparison of prediction methods for the mixture surface tension was done by Kraut (2013), identifying correlations by Brock and Bird (1955), Jensen and Jackman (1984), Bell et al. (1987) and Conde (1996). Applying these to CO_2-POE oil mixtures, large variations and generally different trends were determined. As no experimental data of mixture surface tension was available and experiments could not be carried out at, the surface tension of pure CO_2 was used.

D.3 Solid material properties

Heat conductivity of insulation (Armaflex AF4):

$$\frac{\lambda_{ins}(\overline{T}_{ins})}{Wm^{-1}K^{-1}} = 6.5 \cdot 10^{-2} - 3.4 \cdot 10^{-4} \left(\frac{\overline{T}_{ins}}{K}\right) + 8 \cdot 10^{-7} \left(\frac{\overline{T}_{ins}}{K}\right)^2 \tag{D.27}$$

$$\overline{T}_{ins} = \frac{1}{2}\left(T_{ins,i} + T_{ins,o}\right) \tag{D.28}$$

Heat conductivity of measurement section tube material (nickel smooth tube, Eq. D.29, and copper micro-fin tube, Eq. D.30).

$$\frac{\lambda_{Ni}(\overline{T}_{Ni})}{Wm^{-1}K^{-1}} = 63.66 - 0.0525 \cdot \frac{\overline{T}_{Ni}}{K} \tag{D.29}$$

$$\frac{\lambda_{Cu}(\overline{T}_{Cu})}{Wm^{-1}K^{-1}} = 135.47 + 0.55 \cdot \frac{\overline{T}_{Cu}}{K} \tag{D.30}$$

where $\overline{T}_{Ni/Cu}$ is the mean nickel/copper (tube wall) temperature depending on the respective energy balance.

Appendix E

E.1 Prediction methods for flow pattern

Steiner (2002)

Based on the stratified flow model in round tubes pictured in Fig. E.1, the following geometric dimensions were defined (given here in generalized form to avoid case differentiation as stated in the VDI heat atlas chapter).

$$\tilde{h}_{\mathrm{L}} = 0.5\left(1+\cos\left(\frac{\theta_{\mathrm{strat}}}{2}\right)\right) \tag{E.1}$$

$$\tilde{U}_{\mathrm{i}} = 2\sqrt{\tilde{h}_{\mathrm{L}}\left(1-\tilde{h}_{\mathrm{L}}\right)} \tag{E.2}$$

$$\tilde{U}_{\mathrm{V}} = \cos\left(\frac{\theta_{\mathrm{strat}}}{2}\right) \tag{E.3}$$

$$\tilde{U}_{\mathrm{L}} = \pi - \tilde{U}_{\mathrm{V}} \tag{E.4}$$

$$\tilde{A}_{\mathrm{V}} = \frac{1}{8}\left(\theta_{\mathrm{strat}} - \sin\left(\frac{\theta_{\mathrm{strat}}}{2}\right)\right) \tag{E.5}$$

$$\tilde{A}_{\mathrm{L}} = \frac{\pi}{4} - \tilde{A}_{\mathrm{V}} \tag{E.6}$$

$$\tilde{h}_{\mathrm{L}} = h_{\mathrm{L}}/d \tag{E.7}$$

$$\tilde{U}_{\mathrm{L|V}} = U_{\mathrm{L|V}}/d \tag{E.8}$$

$$\tilde{A}_{\mathrm{L|V}} = A_{\mathrm{L|V}}/d^2 \tag{E.9}$$

where h_{L} is the liquid height, $U_{\mathrm{L|V}}$ is the perimeter, $A_{\mathrm{L|V}}$ is the vapor/liquid phase cross section of the vapor/liquid phase, $\tilde{U}_{\mathrm{i}}$ is the dimensionless perimeter of the vapor-liquid interface and θ_{strat} is the (stratified) dry angle defining the dry wall perimeter U_{V}.

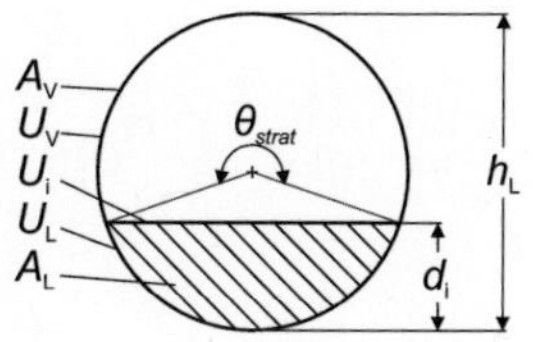

Figure E.1: Cross-sectional view and dimensions

Steiner recommended the Rouhani and Axelsson correlation (Eq. 2.3 in Sec. 2.1) for calculating the void fraction ($\varepsilon = 4\tilde{A}_V/\pi$). Then, θ_{strat} can be obtained iteratively from Eq. E.10 and Eq. E.1-E.6 be solved.

$$\theta_{strat} = 2\pi\varepsilon + \sin(\theta_{strat}) \qquad \text{(E.10)}$$

With known values for ε and θ_{strat}, no further iterations are needed. The Rouhani and Axelsson void fraction correlation, however, requires both vapor quality and mass flow velocity as input. Steiner proposed an iterative calculation method, assuming the two-fluid model to be applicable and solving the momentum equations. A condition for the Martinelli parameter was derived (see Eq. 2.2 in Sec. 2.1) that has to be solved, along with Eq. E.1-E.6, iteratively in dependency of the liquid height $\tilde{h}_L$ or dry angle θ_{strat}, respectively. The obtained value of ε from Eq. 2.3 was recommend to be used for finding a starting value of θ_{strat}.
The transition boundaries for two-phase flow in non-inclined horizontal tubes are given in dimensionless form by Eq. E.11-E.15.

$$\left(\mathrm{Re_L\,Fr'_V}\right)_{\text{S-W}} = \frac{226.3^2\,\tilde{A}_L\tilde{A}_V^2}{\pi^3} \qquad \text{(E.11)}$$

$$\left(\mathrm{Fr_{Vm}}\right)_{\text{W-I|A}} = \frac{16\tilde{A}_V^3}{\pi^2\sqrt{1-(2\tilde{h}_L-1)^2}}\left(\frac{\pi^2}{25\tilde{h}_L^2}\left(\frac{\mathrm{Fr}}{\mathrm{We}}\right)_L+1\right) \qquad \text{(E.12)}$$

$$\left(X_{tt}\right)_{\text{I-A}} = 0.34 \qquad \text{(E.13)}$$

$$\left((\mathrm{Fr\,Eu})_L\right)_{\text{I-B}} = \frac{128\tilde{A}_V\tilde{A}_L^2}{\pi^2 U_i} \qquad \text{(E.14)}$$

$$\left(\mathrm{Fr_{Vm}}\right)_{\text{A-M}} = \frac{7680\tilde{A}_V^2}{\pi^2\,\xi_h}\left(\frac{\mathrm{Fr}}{\mathrm{We}}\right)_L \qquad \text{(E.15)}$$

where

$$\left(\mathrm{Re_L\,Fr'_V}\right) = \dot{m}^3\,\dot{x}^2\,(1-\dot{x})\,\left(\rho_V\,(\rho_L-\rho_V)\,\mu_L\,g\right)^{-1} \qquad \text{(E.16)}$$

$$\mathrm{Fr_{Vm}} = \dot{m}^2\,\dot{x}^2\,\left(g\,d\,\rho_L\,\rho_V\right)^{-1} \qquad \text{(E.17)}$$

$$(\mathrm{Fr\,Eu})_L = \xi_L\,\dot{m}^2\,(1-\dot{x})^2\,\left(2d_i\,\rho_L\,(\rho_L-\rho_V)\,g\right)^{-1} \qquad \text{(E.18)}$$

$$\left(\frac{\mathrm{Fr}}{\mathrm{We}}\right)_L = \sigma\,\left(g\,d_i^2\,\rho_L\right)^{-1} \qquad \text{(E.19)}$$

$$X_{tt} = \left(\frac{1-\dot{x}}{\dot{x}}\right)^{0.875}\left(\frac{\rho_V}{\rho_L}\right)^{0.5}\left(\frac{\mu_L}{\mu_V}\right)^{0.125} \qquad \text{(E.20)}$$

The liquid Reynolds number ($\mathrm{Re_L}$), liquid (ξ_L) and phase (ξ_h) friction factors are

$$\mathrm{Re_L} = \dot{m}\,(1-\dot{x})\,d_\mathrm{i}\,\mu_\mathrm{L}^{-1} \tag{E.21}$$

$$\xi_\mathrm{L} = 0.3164\,\mathrm{Re_L^{-0.25}} \tag{E.22}$$

$$\xi_\mathrm{h} = \left(1.138 + 2\,\mathrm{lg}\left(\frac{\pi}{1.5\tilde{A}_\mathrm{L}}\right)\right)^{-2} \tag{E.23}$$

Converted into dimensionfull transition equations, Eq. E.11-E.15 yield

$$\frac{\dot{m}_\text{S-W}}{\mathrm{kg\,m^{-2}\,s^{-1}}} = \left[\frac{226.3^2\,\tilde{A}_\mathrm{L}\,\tilde{A}_\mathrm{V}^2\,\rho_\mathrm{V}\,(\rho_\mathrm{L}-\rho_\mathrm{V})\,\mu_\mathrm{L}\,g}{\dot{x}^2\,(1-\dot{x})\,\pi^3}\right]^{1/3} \tag{E.24}$$

$$\frac{\dot{m}_\text{W-I|A}}{\mathrm{kg\,m^{-2}\,s^{-1}}} = \left[\frac{16\tilde{A}_\mathrm{V}^3\,g\,d_\mathrm{i}\,\rho_\mathrm{V}\,\rho_\mathrm{L}}{\dot{x}^2\,\pi^2\,\sqrt{1-(2\tilde{h}_\mathrm{L}-1)^2}}\left(\frac{\pi^2}{25\,\tilde{h}_\mathrm{L}^2}\left(\frac{\mathrm{Fr}}{\mathrm{We}}\right)_\mathrm{L}+1\right)\right]^{0.5} \tag{E.25}$$

$$\dot{x}_\text{I-A} = \left[0.34^{1/0.875}\left(\frac{\rho_\mathrm{V}}{\rho_\mathrm{L}}\right)^{-1/1.75}\left(\frac{\mu_\mathrm{L}}{\mu_\mathrm{V}}\right)^{-1/7}+1\right]^{-1} \tag{E.26}$$

$$\frac{\dot{m}_\text{I-B}}{\mathrm{kg\,m^{-2}\,s^{-1}}} = \left[\frac{256\tilde{A}_\mathrm{V}\,\tilde{A}_\mathrm{L}^2\,d_\mathrm{i}^{1.25}\,\rho_\mathrm{L}\,(\rho_\mathrm{L}-\rho_\mathrm{V})\,g}{0.3164\,(1-\dot{x})^{1.75}\,\pi^2\,\tilde{U}_\mathrm{i}\,\mu_\mathrm{L}^{0.25}}\right]^{1/1.75} \tag{E.27}$$

$$\frac{\dot{m}_\text{A-M}}{\mathrm{kg\,m^{-2}\,s^{-1}}} = \left[\frac{7680\tilde{A}_\mathrm{L}^2\,g\,d_\mathrm{i}\,\rho_\mathrm{L}\,\rho_\mathrm{V}}{\dot{x}^2\,\pi^2\,\xi_\mathrm{h}}\left(\frac{\mathrm{Fr}}{\mathrm{We}}\right)_\mathrm{L}\right]^{0.5} \tag{E.28}$$

Saito et al. (2011)

Saito et al. recommended using the Steiner flow pattern map with modification of the wavy to intermittent or annular flow transition condition (Eq. E.12 or Eq. E.25, respectively). It follows the original definition of the critical mean vapor velocity for waves to rise by Taitel and Dukler (1976).

$$\bar{u}_\mathrm{V} \geq C_1 \left\{\left(\frac{\varepsilon\,\pi\,g\,d\,(\rho_\mathrm{L}-\rho_\mathrm{V})}{4\rho_\mathrm{V}\,\sqrt{1-(2h_\mathrm{L}/d-1)^2}}\right)\right\}^{0.5} \tag{E.29}$$

where $C_1 = 1-\tilde{h}_\mathrm{L}$ (as suggested by Taitel and Dukler) accounts for the distance from vapor-liquid interface to the crest of the tube, i.e., the distance that waves have to grow in order to yield the transition state to intermittent/ annular flow. Substituting the mean vapor phase velocity by the relative velocity of vapor to liquid flow and assuming that the liquid velocity at the interface $\bar{u}_{L,i}$ is in the range of $\bar{u}_L \leq \bar{u}_{L,i} \leq 2\bar{u}_L$, Eq. E.29 yields

$$\bar{u}_V - C_2\bar{u}_L \geq C_1 \left\{ \left(\frac{\varepsilon \pi g d (\rho_L - \rho_V)}{4\rho_V \sqrt{1-(2h_L/d-1)^2}} \right) \right\}^{0.5} \tag{E.30}$$

where $C_2 = f(\tilde{h}_L)$ accounts for the liquid velocity at the liquid-vapor interface. Combining Eq. E.30 with the original condition given by Steiner (Eq. E.12), the condition for the wavy to intermittent/annular transition yields Eq. 2.94 in Sec. 2.4. As dimensionfull expression for the mass flow velocity at the wavy to intermittent/annular transition $\dot{m}_{W\text{-}I|A}$, Eq. E.31 is obtained.

$$\frac{\dot{m}_{W\text{-}I|A}}{\mathrm{kg\,m^{-2}\,s^{-1}}} = C_1 \left(\frac{\dot{x}}{\varepsilon \rho_V} - C_2 \frac{(1-\dot{x})}{(1-\varepsilon)\rho_L} \right)^{-1} \times \left[\frac{\varepsilon \pi g d_i (\rho_L - \rho_V)}{4\rho_V \sqrt{1-(2\tilde{h}_L-1)^2}} \left(\frac{\pi^2}{25\tilde{h}_L^2} \left(\frac{\mathrm{Fr}}{\mathrm{We}} \right)_L + 1 \right) \right]^{0.5} \tag{E.31}$$

where

$$C_1 = 0.5\left(\cos\left(\pi\tilde{h}_L\right)+1\right) \tag{E.32}$$

$$C_2 = 0.5\left(\cos\left(\pi\tilde{h}_L\right)+1\right)+1 \tag{E.33}$$

Thome and El Hajal (2003)

The authors represented a modified Kattan et al. flow pattern map (which in fact is a modified Steiner map), based on according geometrical dimensions.

$$\tilde{A}_L = \frac{A}{d_i^2}(1-\varepsilon) = \frac{\pi}{4}(1-\varepsilon) \tag{E.34}$$

$$\tilde{A}_V = \frac{A}{d_i^2}\varepsilon = \frac{\pi}{4}\varepsilon \tag{E.35}$$

$$\tilde{h}_L = 0.5\left(1+\cos\left(\frac{\theta_{strat}}{2}\right)\right) \tag{E.36}$$

$$\tilde{U}_i = \sin\left(\pi - \frac{\theta_{strat}}{2}\right) \tag{E.37}$$

Herein, the void fraction ε can be predicted using the Rouhani and Axelsson drift flux model (Sec. 2.1, Eq. 2.3). The stratified dry angle θ_{strat} is determined iteratively from Eq. E.38.

$$\tilde{A}_{\mathrm{L}} = 0.125\,(2\pi - \theta_{\mathrm{strat}} - \sin(2\pi - \theta_{\mathrm{strat}})) \tag{E.38}$$

Regarding the Steiner map, the stratified to wavy and wavy to intermittent/annular flow transitions (Eq. E.24 and E.25) were modified, becoming Eq. E.39 and E.40. Equation E.40 represents the approach proposed by Kattan et al. (1998a) in the correct form published by Thome (2005). The intermittent to annular, intermittent to bubbly and annular to mist flow transitions (Eq. E.26-E.28) were adopted unchanged.

$$\frac{\dot{m}_{\mathrm{S\text{-}W}}}{\mathrm{kg\,m^{-2}\,s^{-1}}} = \left[\frac{226.3^2\,\tilde{A}_{\mathrm{L}}\tilde{A}_{\mathrm{V}}^2\,\rho_{\mathrm{V}}\,(\rho_{\mathrm{L}}-\rho_{\mathrm{V}})\,\mu_{\mathrm{L}}\,g}{\dot{x}^2\,(1-\dot{x})\,\pi^3}\right]^{1/3} + 20\dot{x} \tag{E.39}$$

$$\begin{aligned}\frac{\dot{m}_{\mathrm{W\text{-}I|A}}}{\mathrm{kg\,m^{-2}\,s^{-1}}} = & \left[\frac{16\tilde{A}_{\mathrm{V}}^3\,g\,d_{\mathrm{i}}\,\rho_{\mathrm{V}}\,\rho_{\mathrm{L}}}{\dot{x}^2\,\pi^2\sqrt{1-(2\tilde{h}_{\mathrm{L}}-1)^2}}\times\right.\\ & \left.\times\left(\frac{\pi^2}{25\,\tilde{h}_{\mathrm{L}}^2}\,(1-\dot{x})^{-F_1(\dot{q})}\left(\frac{\mathrm{Fr}}{\mathrm{We}}\right)_{\mathrm{L}}^{F_2(\dot{q})}+1\right)\right]^{0.5}\\ & + 50 - 75\,\exp\left\{-\frac{(\dot{x}^2-0.97)^2}{\dot{x}\,(1-\dot{x})}\right\}\end{aligned} \tag{E.40}$$

where $(\mathrm{Fr}/\mathrm{We})_{\mathrm{L}}$ is given by Eq. E.19. $F_1(\dot{q})$ and $F_2(\dot{q})$ are functions of the local heat flux $\dot{q}$ accounting for partial dryout.

$$F_1(\dot{q}) = 646.0\left(\frac{\dot{q}/2}{\dot{q}_{\mathrm{c}}}\right)^2 + 64.8\left(\frac{\dot{q}/2}{\dot{q}_{\mathrm{c}}}\right) \tag{E.41}$$

$$F_2(\dot{q}) = 18.8\left(\frac{\dot{q}/2}{\dot{q}_{\mathrm{c}}}\right) + 1.023 \tag{E.42}$$

The critical heat flux $\dot{q}_{\mathrm{c}}$ is given by the Eq. E.43.

$$\dot{q}_{\mathrm{c}} = 0.131\,\rho_{\mathrm{v}}^{0.5}\,\Delta h_{\mathrm{v}}\,(g\,\sigma\,(\rho_{\mathrm{L}}-\rho_{\mathrm{V}}))^{0.25} \tag{E.43}$$

Wojtan et al. (2005a)

The same geometrical dimensions as by Thome and El Hajal (2003) were considered (see Eq. E.34-E.37). To avoid any iterations, the Biberg equation is used for approximating the stratified dry angle.

$$\theta_{\text{strat}} = 2\pi - 2\left\{\left(\frac{3\pi}{2}\right)^{1/3}\left[1-2(1-\varepsilon)+(1-\varepsilon)^{1/3}-\varepsilon^{1/3}\right]+ \right. \\ \left. +\pi(1-\varepsilon)-\frac{(1-\varepsilon)\,\varepsilon}{200}\left[1-2(1-\varepsilon)\right]\left[1+4(1-\varepsilon)^2+4\varepsilon^2\right]\right\} \quad \text{(E.44)}$$

The authors proposed the following transition boundaries.

$$\frac{\dot{m}_{\text{S-W}}}{\text{kg}\,\text{m}^{-2}\,\text{s}^{-1}} = \left[\frac{226.3^2\,\tilde{A}_{\text{L}}\,\tilde{A}_{\text{V}}^2\,\rho_{\text{V}}\,(\rho_{\text{L}}-\rho_{\text{V}})\,\mu_{\text{L}}\,g}{\dot{x}^2\,(1-\dot{x})\,\pi^3}\right]^{1/3} \quad \text{(E.45)}$$

$$\frac{\dot{m}_{\text{W-I|A}}}{\text{kg}\,\text{m}^{-2}\,\text{s}^{-1}} = \left[\frac{16\tilde{A}_{\text{V}}^3\,g\,d_{\text{i}}\,\rho_{\text{V}}\,\rho_{\text{L}}}{\dot{x}^2\,\pi^2\sqrt{1-(2\tilde{h}_{\text{L}}-1)^2}}\left(\frac{\pi^2}{25\,\tilde{h}_{\text{L}}^2}\left(\frac{\text{Fr}}{\text{We}}\right)_{\text{L}}+1\right)\right]^{0.5} + 50 \quad \text{(E.46)}$$

$$\dot{x}_{\text{I-A}} = \left[0.34^{1/0.875}\left(\frac{\rho_{\text{V}}}{\rho_{\text{L}}}\right)^{-1/1.75}\left(\frac{\mu_{\text{L}}}{\mu_{\text{V}}}\right)^{-1/7}+1\right]^{-1} \quad \text{(E.47)}$$

$$\frac{\dot{m}_{\text{A-D}}}{\text{kg}\,\text{m}^{-2}\,\text{s}^{-1}} = \left[\frac{1}{0.235}\left(\ln\left(\frac{0.58}{\dot{x}}\right)+0.52\right)\left(\frac{d_{\text{i}}}{\rho_{\text{V}}\,\sigma}\right)^{-0.17} \right. \\ \left. \left(\frac{1}{g\,d_{\text{i}}\,\rho_{\text{V}}\,(\rho_{\text{L}}-\rho_{\text{V}})}\right)^{-0.37}\left(\frac{\rho_{\text{V}}}{\rho_{\text{L}}}\right)^{-0.25}\left(\frac{\dot{q}}{\dot{q}_{\text{c}}}\right)^{-0.70}\right]^{0.926} \quad \text{(E.48)}$$

$$\frac{\dot{m}_{\text{D-M}}}{\text{kg}\,\text{m}^{-2}\,\text{s}^{-1}} = \left[\frac{1}{0.0058}\left(\ln\left(\frac{0.61}{\dot{x}}\right)+0.57\right)\left(\frac{d_{\text{i}}}{\rho_{\text{V}}\,\sigma}\right)^{-0.38} \right. \\ \left. \left(\frac{1}{g\,d_{\text{i}}\,\rho_{\text{V}}\,(\rho_{\text{L}}-\rho_{\text{V}})}\right)^{-0.15}\left(\frac{\rho_{\text{V}}}{\rho_{\text{L}}}\right)^{0.09}\left(\frac{\dot{q}}{\dot{q}_{\text{c}}}\right)^{-0.27}\right]^{0.943} \quad \text{(E.49)}$$

where $(\text{Fr}/\text{We})_{\text{L}}$ is given by Eq. E.19 and the critical heat flux $\dot{q}_{\text{c}}$ by Eq. E.43. Dryout and mist flow conditions were defined by dryout inception($\dot{x}_{\text{Di}}$) and completion ($\dot{x}_{\text{De}}$) vapor qualities determined experimentally.

$$\dot{x}_{\text{Di}} = 0.58\,\exp\left\{0.52-0.235\,\text{We}_{\text{V}}^{0.17}\,\text{Fr}_{\text{V}}^{0.37}\,(\rho_{\text{V}}/\rho_{\text{V}})^{0.25}\,(\dot{q}/\dot{q}_{\text{c}})^{0.70}\right\} \quad \text{(E.50)}$$

$$\dot{x}_{\text{De}} = 0.61\,\exp\left\{0.57-0.0058\,\text{We}_{\text{V}}^{0.38}\,\text{Fr}_{\text{V}}^{0.15}\,(\rho_{\text{V}}/\rho_{\text{V}})^{-0.09}\,(\dot{q}/\dot{q}_{\text{c}})^{0.27}\right\} \quad \text{(E.51)}$$

where $\mathrm{We_V}$ and $\mathrm{Fr_V}$ are the vapor Weber and Froude numbers.

$$\mathrm{We_V} = \frac{\dot{m}^2\, d_\mathrm{i}}{\rho_\mathrm{V}\, \sigma} \tag{E.52}$$

$$\mathrm{Fr_V} = \frac{\dot{m}^2}{\rho_\mathrm{V}\,(\rho_\mathrm{L} - \rho_\mathrm{V})\, g\, d_\mathrm{i}} \tag{E.53}$$

Cheng et al. (2008a)

Cheng et al. updated the Cheng et al. (2006) flow pattern map, representing a Wojtan et al. (2005a) map with modified onset of dryout and mist flow conditions based on CO_2 data (Eq. E.54 and E.55) and including the Steiner intermittent to bubbly flow transition (Eq. E.27).

$$\dot{x}_\mathrm{Di} = 0.58 \exp\left\{0.52 - 0.236\, \mathrm{We_V^{0.17}}\, \mathrm{Fr_V^{0.17}}\, (\rho_\mathrm{V}/\rho_\mathrm{V})^{0.25}\, (\dot{q}/\dot{q}_\mathrm{c})^{0.27}\right\} \tag{E.54}$$

$$\dot{x}_\mathrm{De} = 0.57 \exp\left\{0.57 - 0.502\, \mathrm{We_V^{0.16}}\, \mathrm{Fr_V^{0.15}}\, (\rho_\mathrm{V}/\rho_\mathrm{V})^{-0.09}\, (\dot{q}/\dot{q}_\mathrm{c})^{0.72}\right\} \tag{E.55}$$

Eq. E.54 and E.55 can be converted to

$$\begin{aligned}\frac{\dot{m}_\text{A-D}}{\mathrm{kg\,m^{-2}\,s^{-1}}} = \Bigg[&\frac{1}{0.236}\left(\ln\left(\frac{0.58}{\dot{x}}\right) + 0.52\right)\left(\frac{d_\mathrm{i}}{\rho_\mathrm{V}\,\sigma}\right)^{-0.17} \\ &\left(\frac{1}{g\, d_\mathrm{i}\, \rho_\mathrm{V}\,(\rho_\mathrm{L} - \rho_\mathrm{V})}\right)^{-0.17}\left(\frac{\rho_\mathrm{V}}{\rho_\mathrm{L}}\right)^{-0.25}\left(\frac{\dot{q}}{\dot{q}_\mathrm{c}}\right)^{-0.27}\Bigg]^{1.471}\end{aligned} \tag{E.56}$$

$$\begin{aligned}\frac{\dot{m}_\text{D-M}}{\mathrm{kg\,m^{-2}\,s^{-1}}} = \Bigg[&\frac{1}{0.502}\left(\ln\left(\frac{0.61}{\dot{x}}\right) + 0.57\right)\left(\frac{d_\mathrm{i}}{\rho_\mathrm{V}\,\sigma}\right)^{-0.16} \\ &\left(\frac{1}{g\, d_\mathrm{i}\, \rho_\mathrm{V}\,(\rho_\mathrm{L} - \rho_\mathrm{V})}\right)^{-0.15}\left(\frac{\rho_\mathrm{V}}{\rho_\mathrm{L}}\right)^{0.09}\left(\frac{\dot{q}}{\dot{q}_\mathrm{c}}\right)^{-0.72}\Bigg]^{1.613}\end{aligned} \tag{E.57}$$

where $(\mathrm{Fr/We})_\mathrm{L}$ is given by Eq. E.19 and the critical heat flux $\dot{q}_\mathrm{c}$ by Eq. E.43. The flow pattern map is valid for $0.6 \geq d_i/\,\mathrm{mm} \geq 10$, $-28 \geq T/\,°\mathrm{C} \geq +25$, $50 \geq m/\,\mathrm{kg\,m^{-2}\,s^{-1}} \geq 1500$ and $1.8 \geq \dot{q}/\,\mathrm{kW\,m^{-2}} \geq 46$

Barbieri et al. (2008)

Barbieri et al. proposed a new intermittent to annular flow transition boundary transition curve, see Eq. 2.4-2.6 in Sec. 2.1. In dimensionfull expression, the intermittent to annular flow transition becomes

$$\dot{m}_{\text{I-A}} = \left[\frac{3.75\,(1-\dot{x})^{0.16}\,\rho_{\text{V}}^{1.2}\,\mu_{\text{L}}^{0.24}\,g\,d_{\text{i}}}{\dot{x}^{2.16}\,\rho_{\text{L}}^{0.2}\,\mu_{\text{V}}^{0.24}} \right]^{0.5} \tag{E.58}$$

Equation E.58 was proposed to be used in junction with the Kattan et al. (1998a) flow pattern map, substituting the original I-A transition curve. However, it is suitable for usage in any other flow pattern map as well.

Mastrullo et al. (2012a)

The flow pattern map by Mastrullo et al. was developed from adiabatic two-phase flow experiments with pure CO_2 inside a 6 mm smooth tube. The authors proposed the following transition criteria.

$$\dot{x}_{\text{Slug-I}} = 1 - a_0\,\dot{m}^{2a_1-1}\,d_{\text{i}}^{a_1-1}\,\frac{\mu_{\text{L}}}{(\rho_{\text{L}}\,\sigma)^{a_1}}\left(\frac{p_{\text{s}}}{p_{\text{c}}}\right)^{a_2} \tag{E.59}$$

$$\dot{x}_{\text{I-A}} = 1 - b_0\,\dot{m}^{2b_1-1}\,d_{\text{i}}^{b_1-1}\,\frac{\mu_{\text{L}}}{(\rho_{\text{L}}\,\sigma)^{b_1}}\left(\frac{p_{\text{s}}}{p_{\text{c}}}\right)^{b_2} \tag{E.60}$$

$$\dot{x}_{\text{A-D}} = 1 - c_0\,\dot{m}^{c_1}\,\dot{m}^{2c_2-c_1-1}\,d_{\text{i}}^{c_2-1}\,\frac{\mu_{\text{L}}}{(\rho_{\text{L}}\,\sigma)^{c_2}}\left(\frac{p_{\text{s}}}{p_{\text{c}}}\right)^{c_3} \tag{E.61}$$

where $\dot{x}_{\text{Slug-I}}$, $\dot{x}_{\text{I-A}}$ and $\dot{x}_{\text{A-D}}$ are the vapor qualities for slug to intermittent flow, intermittent to annular flow and annular flow to dryout transitions, respectively, and a_{i}, b_{i} and c_{i} are empirical parameters fitted to measurement data of pure CO_2, see Tab. E.1.

Table E.1: Constants for transition curves by Mastrullo et al. (2012a)

$a_0 = 514.7$	$a_1 = 0.6184$	$a_2 = -0.2542$	
$b_0 = 362.7$	$b_1 = 0.6280$	$b_2 = -0.3383$	
$c_0 = 20.82$	$c_1 = 0.273$	$c_2 = 1.252$	$c_3 = -0.721$

E.2 Two-phase pressure drop models

E.2.1 Pure/mixed refrigerant flow: Frictional pressure drop

Lockhart and Martinelli (1949), Chisholm and Laird (1958)

According to Lockhart and Martinelli, the two-phase frictional pressure drop is calculated from liquid/vapor single-phase pressure drop (see Eq. E.63 and E.64) multiplied by two-phase parameter of the respective phase ($\Phi^2_{L|V}$), see Eq. E.62. Equations E.63 to E.70 are given as stated by Baehr and Stephan (2006).

$$\left(\frac{\Delta p}{\Delta L}\right)_{\mathrm{tp}} = \Phi^2_{\mathrm{L|V}} \left(\frac{\Delta p}{\Delta L}\right)_{\mathrm{L|V}} \tag{E.62}$$

$$\left(\frac{\Delta p}{\Delta L}\right)_{\mathrm{L}} = 4 f_{\mathrm{L}} \frac{(\dot{m}\,(1-\dot{x}))^2\, d_{\mathrm{i}}}{2\rho_{\mathrm{L}}} \tag{E.63}$$

$$\left(\frac{\Delta p}{\Delta L}\right)_{\mathrm{V}} = 4 f_{\mathrm{V}} \frac{(\dot{m}\dot{x})^2\, d_{\mathrm{i}}}{2\rho_{\mathrm{V}}} \tag{E.64}$$

where $f_{\mathrm{L|V}} = \xi_{\mathrm{L|V}}/4$ is the Fanning friction factor given by Eq. E.65 as function of Reynolds number, see Eq. E.66 and E.67. Generally, $c = 0.00791$ and $m = 0.25$ according to Blasius (1912) are chosen, e.g., Müller-Steinhagen and Heck (1986) and Steiner (2002).

$$f_{\mathrm{L|V}} = c\,\mathrm{Re}_{\mathrm{L|V}}^{-m} \tag{E.65}$$

$$\mathrm{Re}_{\mathrm{L}} = \dot{m}\,(1-\dot{x})\, d_{\mathrm{i}}\, \mu_{\mathrm{L}}-1 \tag{E.66}$$

$$\mathrm{Re}_{\mathrm{V}} = \dot{m}\dot{x}\, d_{\mathrm{i}}\, \mu_{\mathrm{V}}^{-1} \tag{E.67}$$

Chisholm and Laird (1958) gave a correlation for $\Phi_{\mathrm{L|V}}$ as function of Martinelli parameter (X) with good agreement to listed values by Lockhart and Martinelli.

$$\Phi^2_{\mathrm{L}} = 1 + CX^{-1} + X^{-2} \tag{E.68}$$

$$\Phi^2_{\mathrm{V}} = (\Phi_{\mathrm{L}} X)^2 = X^2 + CX + 1 \tag{E.69}$$

$$X = \left((1-\dot{x})\dot{x}^{-1}\right)^{\frac{2-m}{2}} \left(\rho_{\mathrm{V}}\, \rho_{\mathrm{L}}^{-1}\right)^{0.5} \left(\mu_{\mathrm{L}}\, \mu_{\mathrm{V}}^{-1}\right)^{\frac{m}{2}} \tag{E.70}$$

where the parameter C is chosen according to flow conditions, see Tab. E.2.

Table E.2: Constant by Chisholm and Laird (1958)

flow condition	tt	lt	tl	ll
value of C	20	12	10	5

Friedel (1979)

Friedel proposed the following frictional pressure drop correlation.

$$\left(\frac{\Delta p}{\Delta L}\right)_{\text{fric}} = \Phi_{\text{LO}}^2 \left(\frac{\Delta p}{\Delta L}\right)_{\text{LO}} \quad \text{(E.71)}$$

$$\left(\frac{\Delta p}{\Delta L}\right)_{\text{LO}} = \xi_{\text{LO}} \dot{m}^2 \left(2 d_{\text{h}} \rho_{\text{L}}\right)^{-1} \quad \text{(E.72)}$$

$$\begin{aligned} \Phi_{\text{LO}}^2 = {} & A + 3.43 \dot{x}^{0.685} \left(1-\dot{x}\right)^{0.24} \left(\rho_{\text{L}} \rho_{\text{V}}^{-1}\right)^{0.8} \left(\mu_{\text{V}} \mu_{\text{L}}^{-1}\right)^{0.22} \times \\ & \times \left(1 - \mu_{\text{V}} \mu_{\text{L}}^{-1}\right)^{0.89} \text{Fr}_{\text{LO}}^{-0.047} \text{We}_{\text{LO}}^{-0.087} \end{aligned} \quad \text{(E.73)}$$

$$A = \left(1-\dot{x}\right)^2 + \dot{x}^2 \left(\rho_{\text{L}} \xi_{\text{VO}} \left(\rho_{\text{V}} \xi_{\text{LO}}\right)^{-1}\right) \quad \text{(E.74)}$$

$$\xi_{\text{LO|VO}} = \begin{cases} 64 \text{Re}_{\text{LO|VO}}^{-1} & \text{Re}_{\text{LO|VO}} \leq 1055 \\ 0.86859 \ln \left(\dfrac{\text{Re}_{\text{LO|VO}}}{1.964 \ln \text{Re}_{\text{LO|VO}} - 3.8215}\right)^{-2} & \text{Re}_{\text{LO|VO}} > 1055 \end{cases} \quad \text{(E.75)}$$

where $\text{Fr}_{\text{LO}} = \dot{m}^2 \left(g d_{\text{h}} \rho_{\text{L}}^2\right)^{-1}$ is liquid Froude number, $\text{We}_{\text{L}} = \dot{m}^2 d_{\text{h}} \left(\rho_{\text{L}} \sigma\right)^{-1}$ is liquid Weber number and $\text{Re}_{\text{LO|VO}} = \dot{m} d \left(\mu_{\text{L|V}}\right)^{-1}$ is the liquid/vapor Reynolds.

Storek and Brauer (1980)

Storek and Brauer developed a homogeneous frictional pressure drop model, additionally accounting for the two-phase flow influence due to interface deformation turbulence, see Eq. 2.7-2.10 in Sec. 2.2. The homogeneous friction factor (ξ_{h}) and correction function (ξ_{cor}) are given as follows

$$\xi_{\text{h}} = \left[\left(64 \text{Re}_{\text{h}}^{-1}\right)^2 + \left(0.3164 \text{Re}_{\text{h}}^{-0.25}\right)^2 + \left(0.036 \text{Re}_{\text{h}}^{-0.1}\right)^2\right]^{0.5} \quad \text{(E.76)}$$

$$\xi_{\text{cor}} = \frac{\left[1 + 51 \left(R_{\text{a}} d_{\text{i}}^{-1}\right)^{0.8}\right]^{\frac{1}{2}}}{\left[\left(1 + C_1 r_{\dot{x}}^{0.4}\right)^{-1} + \left(C_2 r_{\dot{x}}^{-0.8} + C_3 r_{\dot{x}}^{-2}\right)^{-2} + \left(1 + 16 r_{\dot{x}}^{-0.4}\right)^{-1}\right]^{\frac{1}{4}}} \quad \text{(E.77)}$$

where $\text{Re}_{\text{h}} = \dot{m} d_{\text{i}} \mu_{\text{h}}^{-1}$ is the homogeneous Reynolds number, R_{a} is the tube roughness, $r_{\dot{x}} = (\dot{M}_{\text{L}}/\dot{M}_{\text{V}})$ is the liquid-vapor mass flow ratio and C_1-C_3 are empirical functions of homogeneous Weber ($\text{We}_{\text{h}} = \dot{m}^2 d_{\text{i}} \left(\rho_{\text{h}} \sigma\right)^{-1}$) and Froude ($\text{Fr}_{\text{h}} = \dot{m}^2 \left(g d_{\text{i}} \rho_{\text{h}}^2\right)^{-1}$) numbers and viscosity ($r_{\mu} = \mu_{\text{L}}/\mu_{\text{V}}$) and density ($r_{\rho} = \rho_{\text{V}}/\rho_{\text{L}}$) ratios, see Eq. E.78

$$C_{\text{j}=1,2,3} = a_{\text{j}} \text{Fr}_{\text{h}}^{b_{\text{j}}} \text{We}_{\text{h}}^{c_{\text{j}}} r_{\rho}^{d_{\text{j}}} r_{\mu}^{e_{\text{j}}} \quad \text{(E.78)}$$

Table E.3: Constants for pressure drop correlation by Storek and Brauer (1980)

j	a_j	b_j	c_j	d_j	e_j
1	0.6	0.08	0	0	1
2	1.6	0.4	0	0.6	0.3
3	0.02	0.2	0.17	0.47	0

Values of a_j, b_j, c_j, d_j and e_j valid for horizontal tubes are listed in Tab. E.3.

Müller-Steinhagen and Heck (1986)

The friction pressure drop correlation of Müller-Steinhagen and Heck was presented in Sec. 2.2. Friction factors are given as

$$\xi_{\mathrm{LO|VO}} = \begin{cases} 64\,\mathrm{Re}_{\mathrm{LO|VO}}^{-1} & \mathrm{Re}_{\mathrm{LO|VO}} \leq 1187 \\ 0.3164\,\mathrm{Re}_{\mathrm{LO|VO}}^{-0.25} & \mathrm{Re}_{\mathrm{LO|VO}} > 1187 \end{cases} \tag{E.79}$$

where $\mathrm{Re}_{\mathrm{LO|VO}} = \dot{m}\,d\,\left(\mu_{\mathrm{L|V}}\right)^{-1}$ is the liquid/vapor Reynolds number.

Cavallini et al. (1997), as published by Cavallini et al. (2000)

The correlation of Cavallini et al. represents a two-phase multiplier approach using the Friedel two-phase multiplier expression (Φ_{LO}) and a liquid friction factor (f_{LO}) adapted to the micro-fin geometry.

$$\left(\frac{\Delta p}{\Delta L}\right)_{\mathrm{tp}} = \Phi_{\mathrm{LO}}^2 \left(\frac{\Delta p}{\Delta L}\right)_{\mathrm{LO}} = 2\,f_{\mathrm{LO}}\,\frac{\dot{m}^2}{\rho_{\mathrm{L}}\,d}\,\Phi_{\mathrm{LO}}^2 \tag{E.80}$$

$$f_{\mathrm{LO|VO}} = \max\{f_1, f_2\} \tag{E.81}$$

$$f_1 = \begin{cases} 16\,\mathrm{Re}_{\mathrm{LO|VO}}^{-1} & \text{(laminar)} \\ 0.079\,\mathrm{Re}_{\mathrm{LO|VO}}^{-0.25} & \text{(turbulent)} \end{cases} \tag{E.82}$$

$$f_2 = 0.25\,\left(1.74 - 2\,\lg\left(2R_{\mathrm{a}}/d\right)\right)^{-2} \tag{E.83}$$

$$(R_{\mathrm{a}}/d) = 0.18\,(h_{\mathrm{f}}/d)\,(0.1 + \cos(\beta))^{-1} \tag{E.84}$$

$$\Phi_{\mathrm{LO}}^2 = E + 3.23\,F\,H\,\mathrm{Fr}^{-0.045}\,\mathrm{We}^{-0.035} \tag{E.85}$$

$$E = (1-\dot{x})^2 + \dot{x}^2\,(\rho_{\mathrm{L}}\,f_{\mathrm{VO}})\,(\rho_{\mathrm{V}}\,f_{\mathrm{LO}})^{-1} \tag{E.86}$$

$$F = 0.224\,\dot{x}^{0.78}\,(1-\dot{x}) \tag{E.87}$$

$$H = \left(\frac{\rho_{\mathrm{L}}}{\rho_{\mathrm{V}}}\right)^{0.91} \left(\frac{\mu_{\mathrm{V}}}{\mu_{\mathrm{L}}}\right)^{0.19} \left(1 - \frac{\mu_{\mathrm{V}}}{\mu_{\mathrm{L}}}\right)^{0.7} \tag{E.88}$$

where $\bar{\rho} = \rho_V \rho_L (\dot{x}\rho_L + (1-\dot{x})\rho_V)^{-1}$ is the mean density, $Re_{LO|VO} = \dot{m} d \left(\mu_{L|V}\right)^{-1}$ is the liquid/vapor-only Reynolds number, $Fr = \dot{m}^2 \left(g d \bar{\rho}^2\right)^{-1}$ is the Froude number and $We = \dot{m}^2 d \left(\bar{\rho}\,\sigma\right)^{-1}$ is the Weber number.

Choi et al. (1999)

Another micro-fin tube correlation of the two-phase pressure drop was proposed by Choi et al., including both, frictional and acceleration pressure drop. According to the authors, it is also applicable to refrigerant-lubricant mixtures by using the mixture viscosity and regarding the oil content in the vapor fraction calculation.

$$\left(\frac{\Delta p}{\Delta L}\right)_{tot} = \left(f_N L \left(\rho_{L,out}^{-1} + \rho_{L,in}^{-1}\right) d_h^{-1} + \left(\rho_{L,out}^{-1} - \rho_{L,in}^{-1}\right)\right) \dot{m}^2 \tag{E.89}$$

where L is the axial length of segment and f_N is the newly derived friction factor as function of all liquid Reynolds number (Re_{LO}) and two-phase factor (K) and d_h is the hydraulic diameter given by Kedzierski and Goncalves (1999) (see Eq. 5.3 in Sec. 5.2.1)

$$f_N = 0.00506\, Re_{LO}^{-0.0951}\, K^{0.1554} \tag{E.90}$$

$$Re_{LO} = \dot{m} d_h \mu_L^{-1} \tag{E.91}$$

$$K = \delta \dot{x} \Delta h_v \left(L g\right)^{-1} \tag{E.92}$$

Yoon et al. (2004)

The correlation of Yoon et al. was developed for pure CO_2.

$$\left(\frac{\Delta p}{\Delta L}\right)_{fric} = \Phi_{LO}^2 \left(\frac{\Delta p}{\Delta L}\right)_{LO} \tag{E.93}$$

$$\Phi_{LO}^2 = 1 + a\left(\Gamma^2 - 1\right)\left(B\left(\dot{x}(1-\dot{x})\right)^{0.875} We_{VO}^{-1} + \dot{x}^{1.75}\right) \tag{E.94}$$

where $a = 4.2$ is a constant, $We_{VO} = \dot{m}^2 d_i \left(\rho_V \sigma\right)^{-1}$ is the Weber number and B and Γ relate to the B-coefficient method of Chisholm, given here as published by Schael (2009).

$$\Gamma = \left(\rho_L \rho_V^{-1}\right)^{0.5} \left(\mu_V \mu_L^{-1}\right)^{0.125} \tag{E.95}$$

$$B = \begin{cases} 2.364 & \Gamma < 8.9 \\ 21\Gamma^{-1} & \Gamma \geq 8.9 \end{cases} \tag{E.96}$$

Moreno Quibén and Thome (2007b)

The two-phase pressure drop model by Moreno Quibén and Thome bears on the flow pattern prediction proposed by Cheng et al. (2006). Diabatic pressure drops maybe reduced to frictional pressure drop by account for pressure drop due to acceleration.

$$\Delta p_{\mathrm{tot}} = \Delta p_{\mathrm{fric}} + \Delta p_{\mathrm{mom}} \tag{E.97}$$

$$\Delta p_{\mathrm{mom}} = \dot{m}^2 \left\{ \left[\frac{(1-\dot{x})^2}{\rho_{\mathrm{L}}\,(1-\varepsilon)} + \frac{\dot{x}^2}{\rho_{\mathrm{V}}\,\varepsilon} \right]_{\mathrm{out}} - \left[\frac{(1-\dot{x})^2}{\rho_{\mathrm{L}}\,(1-\varepsilon)} + \frac{\dot{x}^2}{\rho_{\mathrm{V}}\,\varepsilon} \right]_{\mathrm{in}} \right\} \tag{E.98}$$

where the void fraction ε can be calculated from Eq. 2.3.
(1) Pressure drop model for **annular flow**:

$$\Delta p_{\mathrm{A}} = 2\,f_{\mathrm{A}}\,L d_{\mathrm{i}}^{-1}\,\rho_{\mathrm{V}}\,u_{\mathrm{V}}^2 \tag{E.99}$$

$$f_{\mathrm{A}} = 0.67\,\left(\delta\,d_{\mathrm{i}}^{-1}\right)^{1.2}\,\left((\rho_{\mathrm{L}}-\rho_{\mathrm{V}})\,g\,\delta^2\,\sigma^{-1}\right)^{-0.4}\,\left(\mu_{\mathrm{V}}\,\mu_{\mathrm{L}}^{-1}\right)^{0.08}\,\mathrm{We}_{\mathrm{L}}^{-0.034} \tag{E.100}$$

$$u_{\mathrm{V}} = \dot{m}\dot{x}\,(\rho_{\mathrm{V}}\,\varepsilon)^{-1} \tag{E.101}$$

$$\delta = \frac{\pi\,d_{\mathrm{i}}\,(1-\varepsilon)}{2\,(2\,\pi-\theta_{\mathrm{dry}})} \quad (\theta_{\mathrm{dry}} = 0 \text{ for annular flow}) \tag{E.102}$$

$$\mathrm{We}_{\mathrm{L}} = \rho_{\mathrm{L}}\,u_{\mathrm{L}}^2\,d_{\mathrm{i}}\,\sigma^{-1} \tag{E.103}$$

$$u_{\mathrm{L}} = \frac{\dot{m}\,(1-\dot{x})}{\rho_{\mathrm{L}}\,(1-\varepsilon)} \tag{E.104}$$

(2) Pressure drop model for **slug and intermittent flow**:

$$\Delta p_{\mathrm{SLUG+I}} = \Delta p_{\mathrm{LO}}\,\left(1-\frac{\varepsilon}{\varepsilon_{\mathrm{I\text{-}A}}}\right) + \Delta p_{\mathrm{A}}\,\left(\frac{\varepsilon}{\varepsilon_{\mathrm{I\text{-}A}}}\right) \tag{E.105}$$

where Δp_{LO} is the single-phase liquid pressure drop calculated at $\dot{x} = 0$, $\varepsilon_{\mathrm{I\text{-}A}}$ is the void fraction corresponding to $\dot{x}_{\mathrm{I\text{-}A}}$ and Δp_{A} is the pressure drop calculated for the actual vapor quality $\dot{x}$ using Eq. E.99.
(3) Pressure drop model for **wavy flow**:

$$\Delta p_{\mathrm{W}} = 2\,f_{\mathrm{W}}\,L d_{\mathrm{i}}^{-1}\,\rho_{\mathrm{V}}\,u_{\mathrm{V}}^2 \tag{E.106}$$

$$f_{\mathrm{W}} = \theta_{\mathrm{dry}}\,f_{\mathrm{V}} + \left(1-\theta_{\mathrm{dry}}^{*}\right)\,f_{\mathrm{A}} \tag{E.107}$$

$$\theta_{\mathrm{dry}}^{*} = \frac{\theta_{\mathrm{strat}}}{2\,\pi}\,\left(\frac{\dot{m}_{\mathrm{S}}-\dot{m}}{\dot{m}_{\mathrm{W}}-\dot{m}_{\mathrm{S}}}\right)^{0.61} \tag{E.108}$$

$$f_{\mathrm{V}} = 0.079\,\mathrm{Re}_{\mathrm{V}}^{0.25} \tag{E.109}$$

$$\mathrm{Re}_{\mathrm{V}} = \dot{m}\dot{x}\,d_{\mathrm{i}}\,(\mu_{\mathrm{V}}\,\varepsilon)^{-1} \tag{E.110}$$

where the stratified angle θ_{strat} is calculated from Eq. E.44 in App. E.1.
(4) Pressure drop model for **slug-stratified + wavy flow**:

$$\Delta p_{SLUG+W} = \Delta p_{LO}\left(1-\frac{\varepsilon}{\varepsilon_{I\text{-}A}}\right)^{0.25} + \Delta p_{W}\left(\frac{\varepsilon}{\varepsilon_{I\text{-}A}}\right)^{0.25} \tag{E.111}$$

where calculation follows the procedure of (2) slug and intermittent flow, with Δp_W calculated by Eq. E.106.
(5) Homogeneous pressure drop model for **mist flow**:

$$\Delta p_M = 2 f_M L d_i^{-1} \dot{m}^2 \rho_h^{-1} \tag{E.112}$$

$$f_M = 0.079\,\mathrm{Re}_h^{-0.25} \tag{E.113}$$

$$\mathrm{Re}_h = \dot{m} d_i \mu_h^{-1} \tag{E.114}$$

$$\rho_h = \rho_L (1-\varepsilon_h) + \rho_V \varepsilon_h \tag{E.115}$$

$$\mu_h = \mu_L (1-\dot{x}) + \mu_V \dot{x} \tag{E.116}$$

$$\varepsilon_h = \left(1 + \frac{1-\dot{x}}{\dot{x}}\frac{\rho_V}{\rho_L}\right)^{-1} \tag{E.117}$$

(6) Pressure drop model for **dryout**:

$$\Delta p_D = \Delta p(\dot{x}_{Di}) - \frac{\dot{x}-\dot{x}_{Di}}{\dot{x}_{De}-\dot{x}_{Di}}\left[\Delta p(\dot{x}_{Di}) - \Delta p_M(\dot{x}_{De})\right] \tag{E.118}$$

where $\Delta p(\dot{x}_{Di})$ is the pressure drop calculated at the dyrout inception vapor quality $\dot{x}_{Di}$ (Eq. E.54 in App. E.1) using Eq. E.99 or E.106 and $\Delta p_M(\dot{x}_{De})$ is the pressure drop at the dryout completion vapor quality $\dot{x}_{De}$ (Eq. E.55 in App. E.1) calculated from Eq. E.112.
(7) Pressure drop model for **stratified flow**:

$$\Delta p_S = \begin{cases} 2 f_S L d_i^{-1} \rho_V u_V^2 & \text{for } \dot{x} \geq \dot{x}_{I\text{-}A} \\ \Delta p_{LO}\left(1-\dfrac{\varepsilon}{\varepsilon_{I\text{-}A}}\right) + \Delta p_{S(\dot{x}\geq\dot{x}_{I\text{-}A})}\left(\dfrac{\varepsilon}{\varepsilon_{I\text{-}A}}\right) & \text{for } \dot{x} < \dot{x}_{I\text{-}A} \end{cases} \tag{E.119}$$

$$f_S = \frac{\theta_{strat}}{2\pi} f_V + \left(1-\frac{\theta_{strat}}{2\pi}\right) f_A \tag{E.120}$$

where Δp_S is calculated from superposition analogous to the procedure of (2) slug and intermittent flow in the case of $\dot{x} < \dot{x}_{I\text{-}A}$, f_V and f_A are calculated by Eq. E.109 and E.100, respectively.

Cheng et al. (2008a)

The Cheng et al. model is an adaptation of the Moreno Quibén and Thome (2007b) model based on CO_2 data. Following, the modifications are adressed.
(1) Changes in pressure drop model for **annular flow**:

$$f_A = 3.128\,\mathrm{Re}_V^{-0.454}\,\mathrm{We}_L^{-0.0308} \tag{E.121}$$

$$\mathrm{Re}_V = \dot{m}\dot{x}d_i\,(\mu_V\,\varepsilon)^{-1} \tag{E.122}$$

(2) Changes in pressure drop model for **slug and intermittent flow**:

$$\Delta p_{SLUG+I} = \Delta p_{LO}\left(1-\frac{\varepsilon}{\varepsilon_{I\text{-}A}}\right)+\Delta p_A\left(\frac{\varepsilon}{\varepsilon_{I\text{-}A}}\right) \tag{E.123}$$

$$\Delta p_{LO} = 2f_{LO}Ld_i^{-1}\dot{m}^2\rho_L^{-1} \tag{E.124}$$

$$f_{LO} = 0.079\,\mathrm{Re}_{LO}^{0.25} \tag{E.125}$$

$$\mathrm{Re}_{LO} = \dot{m}d_i\mu_L^{-1} \tag{E.126}$$

(3) Changes in pressure drop model for **wavy flow**:

$$f_W = \theta_{dry}^{*0.02}f_V + \left(1-\theta_{dry}^*\right)^{0.02}f_A \tag{E.127}$$

(4) Changes in pressure drop model for **slug-stratified + wavy flow**:

$$\Delta p_{SLUG+W} = \Delta p_{LO}\left(1-\frac{\varepsilon}{\varepsilon_{I\text{-}A}}\right)+\Delta p_W\left(\frac{\varepsilon}{\varepsilon_{I\text{-}A}}\right) \tag{E.128}$$

(5) Changes in pressure drop model for **mist flow**:

$$f_M = 91.2\,\mathrm{Re}_h^{-0.832} \tag{E.129}$$

(6) No changes reported for the **dryout** pressure drop model.
(7) Changes in pressure drop model for **stratified flow**:

$$\Delta p_S = \begin{cases} 2f_SLd_i^{-1}\rho_V u_V^2 & \text{for } \dot{x}\geq\dot{x}_{I\text{-}A} \\ \Delta p_{LO}\left(1-\frac{\varepsilon}{\varepsilon_{I\text{-}A}}\right)+\Delta p_{S(\dot{x}\geq\dot{x}_{I\text{-}A})}\left(\frac{\varepsilon}{\varepsilon_{I\text{-}A}}\right) & \text{for } \dot{x}<\dot{x}_{I\text{-}A} \end{cases} \tag{E.130}$$

(8) New pressure drop model for **bubbly flow**:

$$\Delta p_{\mathrm{B}} = \Delta p_{\mathrm{LO}} \left(1 - \frac{\varepsilon}{\varepsilon_{\mathrm{I\text{-}A}}}\right) + \Delta p_{\mathrm{A}} \left(\frac{\varepsilon}{\varepsilon_{\mathrm{I\text{-}A}}}\right) \tag{E.131}$$

where calculation follows the procedure of (2) slug and intermittent flow.
Applicability of the prediction method yields $0.6 \geq d_i/\,\mathrm{mm} \geq 10$, $-28 \geq T/\,{}^{\circ}\mathrm{C} \geq +25$, $50 \geq m/\,\mathrm{kg\,m^{-2}\,s^{-1}} \geq 1500$ and $1.8 \geq \dot{q}/\,\mathrm{kW\,m^{-2}} \geq 46$.

Xu and Fang (2012)

The pressure drop is given by Friedel-type expression.

$$\left(\frac{\Delta p}{\Delta L}\right)_{\mathrm{fric,tp}} = \Phi_{\mathrm{LO}}^2 \left(\frac{\Delta p}{\Delta L}\right)_{\mathrm{fric,LO}} \tag{E.132}$$

The two-phase multiplier is correlated by

$$\Phi_{\mathrm{LO}}^2 = \left\{Y^2 \dot{x}^3 + (1-\dot{x})^{1/3} \left[1 + 2\dot{x}(Y^2 - 1)\right]\right\} \times \\ \times \left\{1 + 1.54\,(1-\dot{x})^{0.5}\, La^{1.47}\right\} \tag{E.133}$$

where Y^2 is the pressure drop ratio of all-vapor to all-liquid flow and La is the Laplace constant.

$$Y^2 = \left(\frac{\Delta p}{\Delta L}\right)_{\mathrm{VO}} \left(\frac{\Delta p}{\Delta L}\right)_{\mathrm{LO}}^{-1} \tag{E.134}$$

$$La = \left(\frac{\sigma}{g\,(\rho_{\mathrm{L}} - \rho_{\mathrm{V}})}\right) \tag{E.135}$$

For single-phase pressure drop, the friction factor is defined by Fang et al. (2011).

$$\left(\frac{\Delta p}{\Delta L}\right)_{\mathrm{LO|VO}} = \xi_{\mathrm{LO|VO}} \frac{\dot{m}}{\rho_{\mathrm{L|V}}\, d_{\mathrm{h}}} \tag{E.136}$$

$$\xi_{\mathrm{LO|VO}} = 0.25 \left[\lg\left(\frac{150.39}{\mathrm{Re}_{\mathrm{tp}}^{0.98865}} - \frac{152.66}{\mathrm{Re}_{\mathrm{tp}}}\right)\right]^{-2} \tag{E.137}$$

E.2.2 Acceleration pressure drop

The acceleration pressure drop assuming phase separation is described by a heterogeneous model (Moreno Quibén and Thome, 2007a), (Cheng et al., 2008a), (Katsuta et al., 2008), (Hu et al., 2009b), (Da Silva Lima et al., 2009).

$$(\Delta p)_{\text{mom}} = \dot{m} \left\{ \left(\frac{\dot{x}}{\rho_{\text{V}}\,\varepsilon} + \frac{1-\dot{x}}{\rho_{\text{L}}\,(1-\varepsilon)} \right)_{\text{i}} - \left(\frac{\dot{x}}{\rho_{\text{V}}\,\varepsilon} + \frac{1-\dot{x}}{\rho_{\text{L}}\,(1-\varepsilon)} \right)_{\text{o}} \right\} \tag{E.138}$$

where $\dot{m}$ is the total mass flow velocity, $\dot{x}$ is the vapor quality, ε is the void fraction and $\rho_{\text{L|V}}$ are the liquid/vapor densities.
Besides, a homogeneous acceleration pressure drop model can be derived assuming homogeneous flow of uniform density ρ_{h}.

$$(\Delta p)_{\text{mom}} = \dot{m}^2 \frac{\partial(\rho_{\text{h}}^{-1})}{\partial(L)} \tag{E.139}$$

$$= \dot{m}^2 \frac{\partial(\dot{x}\rho_{\text{V}}^{-1} + (1-\dot{x})\,\rho_{\text{L}}^{-1})}{\partial(L)} \tag{E.140}$$

E.2.3 Refrigerant-lubricant mixtures

Eckels et al. (1994), Eckels et al. (1998b)

The pressure drop correlation of Eckels et al. in terms of pure refrigerant pressure drop and penalty factor for correction was developed for R134a-POE oil mixtures. The Penalty factor $PF_{\Delta\text{p}}$ is given by polynomial and logarithmic empirical functions.

$$\Delta p_{\text{mix}} = PF_{\Delta\text{p}}\,\Delta p_{\text{ref}} \tag{E.141}$$

$$\begin{aligned} PF_{\Delta\text{p}} = {} & a_0 + a_1\left(w_{\text{oil,0}}\right) + a_2\left(\dot{m}'\right) + a_3\left(w_{\text{oil,0}}\dot{m}'\right) + a_4\left(w_{\text{oil,0}}^2\dot{m}'\right) \\ & + a_5\left(w_{\text{oil,0}}\dot{m}'^2\right) + a_6\left(w_{\text{oil,0}}^2\dot{m}'^2\right) + a_7\left(w_{\text{oil,0}}^2\right) + a_8\left(\dot{m}'^2\right) \end{aligned} \tag{E.142}$$

$$\begin{aligned} \ln(PF_{\Delta\text{p}}) = {} & b_0 + b_1\left(w_{\text{oil,0}}\right) + b_2\left(\dot{m}'\right) + b_3\left(w_{\text{oil,0}}\dot{m}'\right) + b_4\left(w_{\text{oil,0}}^2\dot{m}'\right) \\ & + b_5\left(w_{\text{oil,0}}\dot{m}'^2\right) + b_6\left(w_{\text{oil,0}}^2\dot{m}'^2\right) + b_7\left(w_{\text{oil,0}}^2\right) + b_8\left(\dot{m}'^2\right) \end{aligned} \tag{E.143}$$

where $\dot{m}' = (\dot{m}/250\,\text{kg}\,\text{m}^{-2}\,\text{s}^{-1})$ and $w_{\text{oil,0}}$ is the nominal oil mass fraction. The constants a_{1-8} and b_{1-8} are listed in Tab. E.4.

Table E.4: Constants by Eckels et al. (1994) and Eckels et al. (1998b)

	smooth tube			micro-fin tube		
	150 SUS	169 SUS	369 SUS	150 SUS	169 SUS	369 SUS
a_0	1	1.00	1.00	0.41	1.31	1.28
a_1	11.33	43.68	26.94	7.40	10.08	-4.07
a_2	0	24.01	-17.23	1.21	-0.22	-0.15
a_3	9.99	0	0	0	0	0
a_4	768.00	0	0	0	0	0
a_5	0	143.60	81.67	0	0	76.80
a_6	504.80	-341.80	0	36.01	0	-177.50
a_7	0	0	0	0	0	239.70
a_8	0	0	0	0.55	0	0
b_0	0	0	0	0.56	0.26	0.24
b_1	7.26	37.54	24.41	6.72	7.30	-2.46
b_2	0	-20.86	-14.46	1.12	-0.16	-0.12
b_3	0	296.20	0	0	0	0
b_4	701.40	0	0	0	0	0
b_5	5.61	0	92.97	0	0	6.09
b_6	478.50	-460.40	-87.90	57.15	0	-134.00
b_7	0	0	0	71.07	0	164.80
b_8	0	0	0	0	0.17	0

E.3 Heat transfer models

E.3.1 General established correlations

Turbulent single-phase heat transfer coefficients are commonly predicted by Nusselt correlations of Dittus and Boelter (1930), Petukhov and Popov (1963) or Gnielinski (1975).

The **Dittus and Boelter** equation, as introduced by McAdams (1942), reads

$$\mathrm{Nu_{DB}} = \frac{\alpha_{\mathrm{DB}}\, d}{\lambda} = 0.023\,\mathrm{Re}^{0.8}\,\mathrm{Pr}^{0.4} \tag{E.144}$$

Petukhov and Popov (1963) proposed a correlation that is valid in a range of $0.5 < \mathrm{Pr} < 200$ and $10^4 < \mathrm{Re} < 5 \times 10^5$.

$$\mathrm{Nu_{PP}} = \frac{\alpha_{\mathrm{PP}}\, d}{\lambda} = \frac{(\xi/8)\,\mathrm{Re}\,\mathrm{Pr}}{1 + 12.7\,(\xi/8)^{0.5}\left(\mathrm{Pr}^{2/3} - 1\right)} \tag{E.145}$$

Starting from Eq. E.145, **Gnielinski (1975)** extended its validity to $2300 < \mathrm{Re} < 10^4$.

$$\mathrm{Nu_{Gn}} = \frac{\alpha_{\mathrm{Gn}}\, d}{\lambda} = \frac{(\xi/8)\,(\mathrm{Re} - 1000)\,\mathrm{Pr}}{1 + 12.7\,(\xi/8)^{0.5}\left(\mathrm{Pr}^{2/3} - 1\right)}\left(1 + (d/L)^{2/3}\right) K(\Delta T) \tag{E.146}$$

Several predictive methods for flow boiling heat transfer emanate from superimposing heat transfer contributions by (enhanced) convection and (suppressed) nucleation. Nucleate boiling contributions are often given by pool boiling correlations of Forster and Zuber (1955), Stephan and Abdelsalam (1980) or Cooper (1984).
Equation E.147 represents the correlation **Forster and Zuber (1955)**.

$$\alpha_{\mathrm{FZ}} = 0.00122\,\lambda_L^{0.79}\,c_{p,L}^{0.45}\,\rho_L^{0.49}\,\sigma^{-0.5}\,\mu_L^{-0.29}\,\Delta h_v^{-0.24}\,\rho_V^{-0.24}\,\Delta T^{0.24}\,\Delta p^{0.75} \tag{E.147}$$

where $\Delta T = T_w - T_s(p)$ is the wall superheat and $\Delta p = p_s(T_w) - p$ is the vapor pressure difference at the wall superheat.
The correlation of **Stephan and Abdelsalam (1980)** is given by Eq. E.148

$$\mathrm{Nu}_{\mathrm{SA}} = \frac{\alpha_{\mathrm{SA}}\,d_b}{\lambda_{\mathrm{L}}} = 207\left(\frac{\dot{q}\,d_{\mathrm{b}}}{\lambda_{\mathrm{L}}\,T_{\mathrm{s}}}\right)^{0.745}\left(\frac{\rho_{\mathrm{V}}}{\rho_{\mathrm{L}}}\right)^{0.581}\mathrm{Pr}_{\mathrm{L}}^{0.533} \tag{E.148}$$

$$d_{\mathrm{b}} = 0.0146\left(\frac{\beta_0}{\mathrm{deg}}\right)\left(\frac{2\,\sigma}{g\,(\rho_{\mathrm{L}}-\rho_{\mathrm{V}})}\right)^{0.5} \tag{E.149}$$

where d_{b} is the equilibrium break-off diameter and β_0 is the contact angle. According to Stephan and Abdelsalam (1980), $\beta_0 = 35\,°$ for refrigerants.
Up to date, the **Cooper (1984)** Equation is among the most used pool boiling correlations in junction with predictive methods for flow boiling heat transfer.

$$\frac{\alpha_{\mathrm{Co}}}{\mathrm{W\,m^{-2}\,K^{-1}}} = -55\left(\frac{p_s}{p_c}\right)^{0.12}\lg\left(\frac{p_s}{p_c}\right)^{-0.55}\left(\frac{\tilde{M}}{\mathrm{g\,mol^{-1}}}\right)^{-0.5}\left(\frac{\dot{q}}{\mathrm{W\,m^{-2}}}\right)^{0.67} \tag{E.150}$$

The probably most often used or further modified predictive method for flow boiling heat transfer is the **Chen (1966)** correlation.

$$\alpha_{\mathrm{Ch}} = \alpha_{cb} + \alpha_{nb} = F\,\alpha_{DB} + S\,\alpha_{FZ} \tag{E.151}$$

$$F = (\mathrm{Re}/\mathrm{Re_L})^{0.8} = f(X_{\mathrm{tt}}) \tag{E.152}$$

$$S = \left(\Delta T_{eff}/\Delta T\right)^{0.99} = f(\mathrm{Re_L}\,F^{1.25}) \tag{E.153}$$

E.3.2 Flow boiling pure/mixed refrigerants

Steiner (2002) with proposed modifications by Schael (2009) for CO_2

The flow boiling heat transfer coefficient (α_{tp}) is calculated from cubic superposition of convective (α_{cb}) and nucleate (α_{nb}) boiling contributions, if the heat flux applied exceeds the heat flux for the onset of nucleate boiling ($\dot{q}_{onb}$). Else, nucleate boiling does not occur.

$$\alpha_{tp} = \begin{cases} \left(\alpha_{cb}^3 + \alpha_{nb}^3\right)^{1/3} & \dot{q} \geq \dot{q}_{onb} \\ \alpha_{cb} & \dot{q} < \dot{q}_{onb} \end{cases} \tag{E.154}$$

The heat flux for the onset of nucleate boiling equals

$$\dot{q}_{onb} = 2\,\sigma\,T_s\,\alpha_{LO}\,(r_c\,\rho_V\,\Delta h_v)^{-1} \tag{E.155}$$

where $r_c \approx 0.3 \times 10^{-6}$ m is the critical bubble diameter and α_{LO} the single-phase heat transfer coefficient, assuming that the liquid flows alone in the tube (see below).
For horizontal tubes, α_{cb} is equals

$$\alpha_{cb} = \alpha_{LO}\left\{(1-\dot{x})^{0.01}\left[(1-\dot{x})+1.2\dot{x}^{0.4}\left(\frac{\rho_L}{\rho_V}\right)^{0.37}\right]^{-2.2} + \right.$$
$$\left. +\dot{x}^{0.01}\left[\frac{\alpha_{VO}}{\alpha_{LO}}\left(1+(1-\dot{x})^{0.7}\left(\frac{\rho_L}{\rho_V}\right)^{0.67}\right)\right]^{-2.0}\right\}^{-0.5} \tag{E.156}$$

where $\alpha_{LO|VO}$ are single-phase heat transfer coefficients calculated by Eq. E.146 for turbulent and developed flow, assuming that the liquid/vapor phase is flowing alone inside the tube ($\mathrm{Re}_{LO|VO} = \dot{m}\,d_i\,\mu_{L|V}^{-1}$). In case of partial wetting (stratified or stratified-wavy flow), α_{cb} in Eq. E.154 is corrected as follows.

$$\alpha_{cb} = \alpha_{cb(\text{Eq. E.156})}\left(1-\frac{\theta}{2\pi}\right)+\alpha_V\,\frac{\theta}{2\pi} \tag{E.157}$$

where θ is the dry angle and α_V is the vapor single-phase heat transfer coefficient. α_V is determined using the hydraulic diameter and Reynolds number of the vapor phase.

$$d_{h,V} = d_i\left(\frac{\theta' - \sin(\theta')}{\theta' + 2\,\sin\left(\frac{\theta'}{2}\right)}\right) \tag{E.158}$$

$$\mathrm{Re}_V = \dot{m}\,d_{h,V}\,\mu_V^{-1}\,\dot{x}\,\varepsilon^{-1} \tag{E.159}$$

whereby $\theta' = 0.5\,\theta$ is the actual dry perimeter confirmed by experiments.
For nucleate boiling, the following empirical approach was proposed.

$$\alpha_{\mathrm{nb}} = \alpha_0\,\Psi\,C_{\mathrm{F}}\,(\dot{q}/\dot{q}_0)^{n(p_{\mathrm{r}})}\,F(p_{\mathrm{r}})\,F(d_{\mathrm{i}})\,F(R_{\mathrm{a}})\,F(\dot{m},\dot{x}) \tag{E.160}$$

where α_0 ($= 18.89\,\mathrm{kW\,m^{-2}\,K^{-1}}$ for CO_2) and $\dot{q}_0$ ($= 150\,\mathrm{kW\,m^{-2}}$ for CO_2) are normalization values, Ψ is a correction factor accounting for the influences of heat conductivity of the tube, boundary condition and flow pattern and C_{F} is a fluid-dependent parameter.
According to Steiner (2002) and Schael (2009) for CO_2, the particular factors are

$$C_{\mathrm{F}} = \begin{cases} \Psi\,0.789\,(\tilde{M}/\tilde{M}_{\mathrm{H_2}})^{0.11} & \text{approximation} \\ 1.57 & \mathrm{CO_2} \end{cases} \tag{E.161}$$

$$\Psi = \begin{cases} 1 & \text{intermittent or annular flow} \\ 0.86 & \text{stratified or wavy flow} \\ 1 & \mathrm{CO_2} \end{cases} \tag{E.162}$$

$$n(p_{\mathrm{r}}) = \begin{cases} 0.8 - 0.13 \times 10^{0.66\,p_{\mathrm{r}}} & \text{non-cryogenic refrigerants} \\ 0.75 - 0.027 \times 10^{1.26\,p_{\mathrm{r}}} & \mathrm{CO_2}\text{, smooth tube} \\ 0.51 & \mathrm{CO_2}\text{, micro-fin tube} \end{cases} \tag{E.163}$$

$$F(p_{\mathrm{r}}) = 2.692\,p_{\mathrm{r}}^{0.43} + \frac{1.6\,p_{\mathrm{r}}^{6.5}}{1 - p_{\mathrm{r}}^{4.4}} \tag{E.164}$$

$$F(d_{\mathrm{i}}) = (d_0/d)^0.5 \tag{E.165}$$

$$F(R_{\mathrm{a}}) = (R_{\mathrm{a}}/R_{\mathrm{a},0})^{0.133} \tag{E.166}$$

$$F(\dot{m},\dot{x}) = (\dot{m}/\dot{m}_0)^{0.25}\left(1 - p_{\mathrm{r}}^{0.1}\,(\dot{q}/\dot{q}_{\mathrm{c}}')^{0.3}\,\dot{x}\right) \tag{E.167}$$

where $d_0 = 0.01$ m, $R_{\mathrm{a},0} = 10^{-6}$ m and $\dot{m}_0 = 100\,\mathrm{kg\,m^{-2}\,s^{-1}}$ are normalization values and

$$p_{\mathrm{r}} = p/p_{\mathrm{c}} \tag{E.168}$$

$$\dot{q}_{\mathrm{c}}' = 2.79\,\dot{q}_{\mathrm{c},0}\,p_{\mathrm{r}}^{0.4}\,(1 - p_{\mathrm{r}}) \tag{E.169}$$

$$\dot{q}_{\mathrm{c},0} = 0.13\,\Delta h_{\mathrm{v},0}\,\rho_{\mathrm{V},0}^{0.5}\,(\sigma_0\,g\,(\rho_{\mathrm{L},0} - \rho_{\mathrm{V},0}))^{0.25} \tag{E.170}$$

All fluid properties in Eq. E.170 are related to a reduced pressure of $p_{\mathrm{r}} = 0.1$.

Liu and Winterton (1991)

A heat transfer correlation was presented similar to the Chen correlation.

$$\alpha_{\mathrm{tp}} = \left((F\,\alpha_{\mathrm{DB}}\,e_{\mathrm{F}})^2 + (S\,\alpha_{\mathrm{Co}}\,e_{\mathrm{S}})^2\right)^{0.5} \tag{E.171}$$

where $e_{F|S}$ are functions of liquid Froude number ($\mathrm{Fr_L} = \dot{m}^2 \left(\rho_L^2 g d_i\right)^{-1}$).

$$e_F = \begin{cases} 1 & \mathrm{Fr_L} \geq 0.05 \\ \mathrm{Fr_L}^{0.1-2\mathrm{Fr_L}} & \mathrm{Fr_L} < 0.05 \end{cases} \tag{E.172}$$

$$e_S = \begin{cases} 1 & \mathrm{Fr_L} \geq 0.05 \\ \mathrm{Fr_L}^{0.5} & \mathrm{Fr_L} < 0.05 \end{cases} \tag{E.173}$$

F and S were given by newly derived expressions of liquid Prandtl number ($\mathrm{Pr_L} = \mu_L c_{p,L} \lambda_L^{-1}$) and all-liquid Reynolds number ($\mathrm{Re_{LO}} = \dot{m} d_i \mu_L - 1$).

$$F = \left[1 + \dot{x}\mathrm{Pr_L} \left(\rho_L \rho_V^{-1} - 1\right)\right]^{0.35} \tag{E.174}$$

$$S = \left[1 + 0.055 F^{0.1} \mathrm{Re_{LO}^{0.16}}\right]^{-1} \tag{E.175}$$

Koyama et al. (1995), as published by Schael (2009)

Koyama et al. proposed the following correlation for micro-fin tubes.

$$\alpha_{tp} = \alpha_{nb} + \alpha_{cb} \tag{E.176}$$

$$\alpha_{nb} = \alpha'_{SA} S K^{0.745} \tag{E.177}$$

$$\alpha_{cb} = 0.0028 \mathrm{Re_{tp}^{0.8}} \mathrm{Pr_L^{0.4}} \lambda_L d_{fr}^{-1} \tag{E.178}$$

where $\alpha'_{SA} = 2.8 \alpha_{SA}$ and α_{SA} given by Eq. E.148. The two-phase Reynolds ($\mathrm{Re_{tp}}$) number is calculated as follows.

$$\mathrm{Re_{tp}} = F^{1.25} (1 - \dot{x}) \dot{m} d_{fr} \mu_L^{-1} \tag{E.179}$$

$$F = 1 + 2X_{tt}^{-0.88} + 0.8X_{tt}^{-1.03} \tag{E.180}$$

where X_{tt} is the Martinelli Parameter.
The nucleate boiling contribution is given as follows.

$$K^{0.745} = \left(1 + 0.875\eta + 0.158\eta^2 - 0.157\eta^3 + 0.7907\eta^4\right)^{-1} \tag{E.181}$$

$$\eta = \alpha_{cb} \left(S \alpha'_{SA}\right)^{-1} \tag{E.182}$$

$$S = (1 - \exp(-\xi)) \xi^{-1} \tag{E.183}$$

$$\xi = d_b \alpha_{cb} \lambda_L^{-1} \tag{E.184}$$

where d_b is given by Eq. E.149.

Thome et al. (1997), as published by Schael (2009)

Thome et al. proposed the following micro-fin tube correlation.

$$\alpha_{tp} = E_{mf}\left(\alpha_{Co}^3 + (E_{RB}\,\alpha_{cb})^3\right)^{1/3} \tag{E.185}$$

$$\alpha_{cb} = 0.0133\,\mathrm{Re}_{L,\delta}^{0.69}\,\mathrm{Pr}_L^{0.4}\,\lambda_L\,\delta^{-1} \tag{E.186}$$

where α_{Co} is given by Eq. E.150, $\mathrm{Pr}_L = \mu_L\,c_{p,L}\,\lambda_L^{-1}$ is the liquid Prandtl number and $\mathrm{Re}_{L,\delta}$ is the liquid film Reynolds number.

$$\mathrm{Re}_{L,\delta} = 4\dot{m}\,\delta\,(1-\dot{x})\,(\mu_L\,(1-\varepsilon))^{-1} \tag{E.187}$$

$$\delta = 0.25\,d_{fr}\,(1-\varepsilon) \tag{E.188}$$

The void fraction ε is calculated using the Rouhani and Axelsson (1970) model.

$$E_{RB} = \left\{1 + \left(2.64\,\mathrm{Re}_L^{0.036}\,\mathrm{Pr}_L^{-0.024}\left(\frac{h_f}{l_f}\right)^{0.21}\left(\frac{\beta}{90\,^\circ}\right)^{0.29}\right)^7\right\}^{1/7} \tag{E.189}$$

$$\mathrm{Re}_L = ((1-\dot{x})\,\dot{m}\,d_{fr})\,\mu_L^{-1} \tag{E.190}$$

$$E_{mf} = 1.89\,(\dot{m}/500)^2 - 3.7\,(\dot{m}/500) + 3.02 \tag{E.191}$$

where h_f is the fin height, l_f is the axial fin pitch and β is the helix angle.

Kattan et al. (1998c)

The correlation of Kattan et al. is given in Sec. 2.3. The constants in Eq. 2.54 fitted to the experimental data are $C = 0.013$ and $m = 0.69$

Cavallini et al. (1999)

The micro-fin tube correlation by Cavallini et al. was derived assuming annular flow. Nucleate and convective boiling contributions are linearly superposed.

$$\alpha_{tp} = \alpha_{nb} + \alpha_{cb} \tag{E.192}$$

$$\alpha_{nb} = \alpha_{Co}\left(\frac{d_0}{d_{ft}}\right)^c S \tag{E.193}$$

$$\alpha_{cb} = \left\{0.023\,\mathrm{Re}_L^{0.8}\,\mathrm{Pr}_L^{1/3}\,F\,\frac{\lambda_L}{d_{ft}}\right\}\left\{R_a^s\,(\mathrm{Bo\,Fr_V})^t\left(\frac{d_0}{d_{ft}}\right)^v\left(\frac{\dot{m}_0}{\dot{m}}\right)^z\right\} \tag{E.194}$$

Table E.5: Constants for heat transfer correlation by Cavallini et al. (1999)

	a	b	c	s	t	v	z
$\dot{m} < 500\,\mathrm{kg\,m^{-2}\,s^{-1}}$	1.36	0.36	0.38	2.14	-0.15	0.59	0.36
$\dot{m} \geq 500\,\mathrm{kg\,m^{-2}\,s^{-1}}$	1.36	0.36	0.38	2.14	-0.21	0.59	0.36

where d_{ft} is the fin tip diameter, the normalizing values are $\dot{m}_0 = 100\,\mathrm{kg\,m^{-2}\,s^{-1}}$ and $d_0 = 0.01$ m, $\mathrm{Bo} = g\,\rho_{\mathrm{L}}\,h_{\mathrm{f}}\,\pi\,d_{\mathrm{ft}}\,(8\,\sigma\,n_{\mathrm{f}})^{-1}$ is the boiling number, $\mathrm{Pr_L} = \mu_{\mathrm{L}}\,c_{\mathrm{p,L}}\,\lambda_{\mathrm{L}}^{-1}$ is the liquid Prandtl number and $\mathrm{Fr_V} = \dot{m}^2\,\left(\rho_{\mathrm{V}}^2\,g\,d_{\mathrm{ft}}\right)^{-1}$ is the vapor Froude number. The constants a-z are listed in Tab. E.5.

Nucleate boiling suppression factor (S) and forced convection multiplier (F) are given as

$$S = a\left[\left((1-\dot{x})\,\dot{x}^{-1}\right)^{0.9}\left(\rho_{\mathrm{V}}\,\rho_{\mathrm{L}}^{-1}\right)^{0.5}\left(\mu_{\mathrm{L}}\,\mu_{\mathrm{V}}^{-1}\right)^{0.1}\right]^{b} \tag{E.195}$$

$$F = \left[(1-\dot{x}) + 2.63\,\dot{x}\left(\rho_{\mathrm{L}}\,\rho_{\mathrm{V}}^{-1}\right)^{0.5}\right]^{0.8} \tag{E.196}$$

The tube roughness (R_{a}) is related to the grooves of the tube.

$$R_{\mathrm{a}} = \frac{1}{\cos(\beta)}\left(\frac{2\,h_{\mathrm{f}}\,n_{\mathrm{f}}\,(1-\sin(\gamma/2))}{\pi\,d_{\mathrm{ft}}\,\cos(\gamma/2)} + 1\right) \tag{E.197}$$

Yun et al. (2002)

Yun et al. suggested a model based on convective heat transfer enhanced by nucleate boiling in terms of Martinelli parameter (X_{tt}) and Boiling number ($\mathrm{Bo} = \dot{q}\,(\dot{m}\Delta h_{\mathrm{v}})^{-1}$).

$$\alpha_{\mathrm{tp}} = \alpha_{\mathrm{DB}}\,\mathrm{f}(\mathrm{Bo}, X_{\mathrm{tt}}) \tag{E.198}$$

$$\mathrm{f}(\mathrm{Bo}, X_{\mathrm{tt}}) = \left\{C_1\,\mathrm{Bo}^{C_2}\left(\frac{p_{\mathrm{s}}\,d_{\mathrm{i}}}{\sigma}\right)^{C_3} + C_4\,X_{\mathrm{tt}}^{-C_5}\left(\frac{\dot{m}\,l_{\mathrm{f}}}{\mu_{\mathrm{L}}}\right)^{C_6}\right\} \times \mathrm{Re_L}^{C_7}\,\mathrm{Pr_L}^{C_8}\left(\frac{\delta}{l_{\mathrm{f}}}\right)^{C_9} \tag{E.199}$$

where $\mathrm{Pr_L} = \mu_{\mathrm{L}}\,c_{\mathrm{p,L}}\,\lambda_{\mathrm{L}}^{-1}$ is the liquid Prandtl number and $\mathrm{Re_L} = (1-\dot{x})\,\dot{m}\,d_{\mathrm{fr}}\,\mu_{\mathrm{L}}^{-1}$ is liquid Reynolds number. α_{DB} is the convective heat transfer coefficient given by Eq. E.144. The constants C_1 to C_9 are given in Tab. E.6

Table E.6: Constants for heat transfer correlation by Yun et al. (2002)

C_1	C_2	C_3	C_4	C_5	C_6	C_7	C_8	C_9
0.009622	0.1106	0.3814	7.685	0.5100	-0.7360	0.2045	0.7452	-0.1302

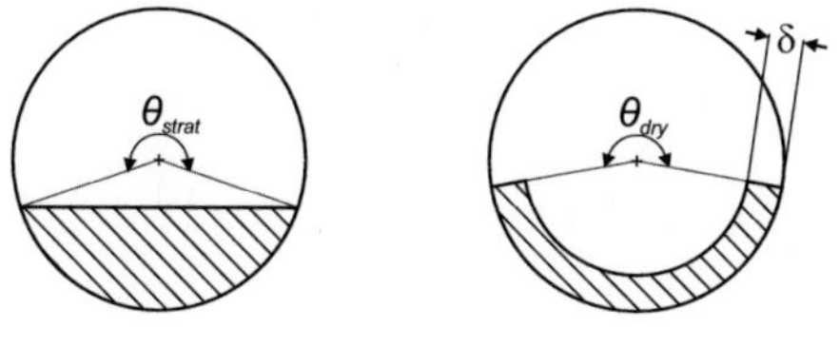

Figure E.2: Liquid distribution in stratified (left) and wavy (right) flow

Thome and El Hajal (2004)

The authors presented an updated Kattan et al. heat transfer model based on CO_2 data.

$$\alpha_{tp} = \frac{\theta_{dry}}{2\pi}\,\alpha_V + \frac{1-\theta_{dry}}{2\pi}\,\left((S\,\alpha_{nb})^3 + \alpha_{cb}^3\right)^{1/3} \tag{E.200}$$

The convective heat transfer coefficient (α_V) at the **dry perimeter** as function of vapor Reynolds ($Re_V = \dot{m}\, d_i\, \dot{x}\,(\mu_V\,\varepsilon)^{-1}$) and Prandtl numbers ($Pr_V = \mu_V\, c_{p,V}\,\lambda_V^{-1}$) yields

$$\alpha_V = 0.023\,Re_V^{0.8}\,Pr_V^{0.4}\,\lambda_V\,d_i^{-1} \tag{E.201}$$

For the **liquid wetted perimeter**, a film flow model was applied for describing forced convective heat transfer.

$$\alpha_{cb} = 0.023\,Re_{L,\delta}^{0.8}\,Pr_L^{0.4}\,\lambda_L\,\delta^{-1} \tag{E.202}$$

$$Re_{L,\delta} = 4\,\dot{m}\,\delta\,(1-\dot{x})\,(\mu_L\,(1-\varepsilon))^{-1} \tag{E.203}$$

$$\delta = \frac{\pi\, d_i\,(1-\varepsilon)}{2\,(2\pi-\theta_{dry})} \tag{E.204}$$

where $Pr_L = \mu_L\, c_{p,L}\,\lambda_L^{-1}$ is the liquid Prandtl number and θ_{dry} is the dry perimeter determined in dependency of the flow pattern.

$$\theta_{dry} = \begin{cases} \theta_{strat} & \text{stratified flow} \\ \theta_{strat}\left(\dfrac{\dot{m}_{W\text{-}I|A} - \dot{m}}{\dot{m}_{W\text{-}I|A} - \dot{m}_{S\text{-}W}}\right) & \text{wavy flow} \\ 0 & \text{annular flow} \end{cases} \tag{E.205}$$

Nucleate boiling contribution and suppression factor were substituted by Eq. E.206-E.207.

$$\alpha_{nb} = 0.71\,\alpha_{Co} + 3970\ \mathrm{W\,m^{-2}\,K^{-1}} \tag{E.206}$$

$$S = 8.264\,(1-\dot{x})^{0.5}\,Re_{L,\delta}^{-0.225} \tag{E.207}$$

Wojtan et al. (2005b)

Wojtan et al. proposed a modified Kattan et al. heat transfer model. Calculation of the dry angle (θ_{dry}) was adapted to a newly developed flow pattern map (Wojtan et al., 2005a).

$$\theta_{\mathrm{dry}} = \begin{cases} 0 & \text{slug flow} \\ \left(\dfrac{\dot{m}_{\mathrm{W\text{-}I|A}} - \dot{m}}{\dot{m}_{\mathrm{W\text{-}I|A}} - \dot{m}_{\mathrm{S\text{-}W}}}\right)^{0.61} \theta_{\mathrm{strat}} & \text{wavy flow} \\ \dfrac{\dot{x}}{\dot{x}_{\mathrm{I\text{-}A}}} \left(\dfrac{\dot{m}_{\mathrm{W\text{-}I|A}} - \dot{m}}{\dot{m}_{\mathrm{W\text{-}I|A}} - \dot{m}_{\mathrm{S\text{-}W}}}\right)^{0.61} \theta_{\mathrm{strat}} & \text{strat. slug + wavy flow} \end{cases} \tag{E.208}$$

Further modifications included:

- calculation of the film thickness δ (see Eq. E.217)
- calculation of θ_{strat} using the Biberg equation (see Eq. E.44)
- introducing a nucleate boiling suppression factor of $S = 0.8$

New correlations for the heat transfer coefficient in the mist flow (α_{M}) and dryout (α_{D}) regions were proposed.

$$\alpha_{\mathrm{M}} = 0.0117\,\mathrm{Re}_{\mathrm{h}}^{0.79}\,\mathrm{Pr}_{\mathrm{V}}^{1.06}\,Y^{-1.83}\left(\lambda_{\mathrm{V}}\,d_{\mathrm{i}}^{-1}\right) \tag{E.209}$$

$$\alpha_{\mathrm{D}} = \alpha_{\mathrm{tp}}(\dot{x}_{\mathrm{Di}}) - \frac{\dot{x} - \dot{x}_{\mathrm{Di}}}{\dot{x}_{\mathrm{De}} - \dot{x}_{\mathrm{Di}}}\left(\alpha_{\mathrm{tp}}(\dot{x}_{\mathrm{Di}}) - \alpha_{\mathrm{M}}(\dot{x}_{\mathrm{De}})\right) \tag{E.210}$$

where Re_{h} is the homogeneous Reynolds number, Pr_{V} is the vapor Prandtl number and Y is a multiplying factor.

$$\mathrm{Re}_{\mathrm{h}} = \dot{m}\,d_{\mathrm{i}}\,\mu_{\mathrm{V}}^{-1}\left(\dot{x} + \frac{\rho_{\mathrm{V}}}{\rho_{\mathrm{L}}}(1-\dot{x})\right) \tag{E.211}$$

$$\mathrm{Pr}_{\mathrm{V}} = \mu_{\mathrm{V}}\,c_{\mathrm{p,V}}\,\lambda_{\mathrm{V}}^{-1} \tag{E.212}$$

$$Y = 1 - 0.1\left(\left(\frac{\rho_{\mathrm{L}}}{\rho_{\mathrm{V}}} - 1\right)(1-\dot{x})\right)^{0.4} \tag{E.213}$$

Cheng et al. (2008b)

A further modification of the Kattan et al. (1998c) model for CO_2 was proposed, among others, introducing models for mist and bubbly flow. It is related to the updated Cheng et al. (2008a) flow pattern map, see Sec. E.1. The general formulation of the model (Eq. E.200), the convective heat transfer at the dry perimeter (Eq. E.201) and the forced convective boiling contribution at the liquid wetter perimeter (Eq. E.202-E.204) were adopted unaltered.

The superposition expression of the dry perimeter θ_{dry} for wavy flow (Eq. E.205) was modified.

$$\theta_{\mathrm{dry}} = \theta_{\mathrm{strat}} \left(\frac{\dot{m}_{\mathrm{W\text{-}I|A}} - \dot{m}}{\dot{m}_{\mathrm{W\text{-}I|A}} - \dot{m}_{\mathrm{S\text{-}W}}} \right)^{0.61} \tag{E.214}$$

The **nucleate boiling** contribution is given as modified Cooper equation.

$$\alpha_{\mathrm{nb}} = 131\, p_{\mathrm{r}}^{-0.0063} \left(-\lg(p_{\mathrm{r}})\right)^{-0.55} \left(\frac{\tilde{M}}{\mathrm{g\,mol^{-1}}} \right)^{-0.5} \left(\frac{\dot{q}}{\mathrm{W\,m^{-2}}} \right)^{0.58} \tag{E.215}$$

$$S = \begin{cases} 1 & \text{for } \dot{x} \leq \dot{x}_{\mathrm{I\text{-}A}} \\ 1 - 1.14 \left(\frac{d}{d_0} \right)^2 \left(1 - \frac{\delta}{\delta_{\mathrm{I\text{-}A}}} \right)^{2.2} & \text{for } \dot{x} > \dot{x}_{\mathrm{I\text{-}A}} \end{cases} \tag{E.216}$$

$$\delta = \min \left\{ \frac{d_{\mathrm{i}}}{2} - \left[\left(\frac{d_{\mathrm{i}}}{2} \right)^2 - \frac{2\tilde{A}_{\mathrm{L}}}{2\pi - \theta_{\mathrm{dry}}} \right]^{0.5} ; \frac{d_{\mathrm{i}}}{2} \right\} \tag{E.217}$$

where $d_0 = 7.53$ mm and $\delta_{\mathrm{I\text{-}A}} = \delta(\dot{x} = \dot{x}_{\mathrm{I\text{-}A}})$ is the liquid film thickness at the onset of annular flow ($\dot{x} = \dot{x}_{\mathrm{I\text{-}A}}$). For diameters greater d_0, $d_{\mathrm{i}} = d_0$ is used.
Heat transfer in **mist flow** was correlated based on a homogeneous model.

$$\alpha_{\mathrm{M}} = 2 \times 10^{-8}\, \mathrm{Re}_{\mathrm{h}}^{1.97}\, \mathrm{Pr}_{\mathrm{V}}^{1.06}\, Y^{-1.83} \frac{\lambda_{\mathrm{V}}}{d_{\mathrm{i}}} \tag{E.218}$$

$$\mathrm{Re}_{\mathrm{h}} = \dot{m}\, d_{\mathrm{i}}\, \mu_{\mathrm{V}}^{-1} \left[\dot{x} + \frac{\rho_{\mathrm{V}}}{\rho_{\mathrm{L}}} (1 - \dot{x}) \right] \tag{E.219}$$

$$Y = 1 - 0.1 \left[\left(\frac{\rho_{\mathrm{V}}}{\rho_{\mathrm{L}}} - 1 \right) (1 - \dot{x}) \right]^{0.4} \tag{E.220}$$

Finally, the heat transfer coefficient in the **dryout region** is given as superposition of fully wetted to mist flow.

$$\alpha_{\mathrm{D}} = \alpha_{\mathrm{tp}}(\dot{x} = \dot{x}_{\mathrm{Di}}) - \frac{\dot{x} - \dot{x}_{\mathrm{Di}}}{\dot{x}_{\mathrm{De}} - \dot{x}_{\mathrm{Di}}} \left[\alpha_{\mathrm{tp}}(\dot{x} = \dot{x}_{\mathrm{Di}}) - \alpha_{\mathrm{M}}(\dot{x} = \dot{x}_{\mathrm{De}}) \right] \tag{E.221}$$

Hu et al. (2008a)

The micro-fin tube correlation of Hu et al. is given in Sec. 2.3.

Fang (2013)

The heat transfer correlation by Fang reads

$$\mathrm{Nu} = \alpha \frac{\lambda_\mathrm{L}}{d_\mathrm{h}} = \frac{0.00061 \left(41000\,\mathrm{Bo}^{1.13} - 0.275 + \left(\frac{\dot{x}}{1-\dot{x}}\right)^a \left(\frac{\rho_\mathrm{L}}{\rho_\mathrm{V}}\right)^{0.4}\right)}{\ln\left(1.024\,\mu_\mathrm{L,s}/\mu_\mathrm{L,w}\right)} \tag{E.222}$$

where $\mu_\mathrm{L,w}$ and $\mu_\mathrm{L,s}$ are the liquid viscosities calculated at wall temperature and at saturation temperature, respectively and a is given in dependency of $\left(\mathrm{Re}_\mathrm{L}\,Fa^{0.11}\right)$.

$$a = \begin{cases} \text{if } \left(\mathrm{Re}_\mathrm{L}\,Fa^{0.11}\right) < 600 \text{ then} \\ 0.48 + 0.00524\left(\mathrm{Re}_\mathrm{L}\,Fa^{0.11}\right)^{0.85} - 5.9\times10^{-6}\left(\mathrm{Re}_\mathrm{L}\,Fa^{0.11}\right)^{1.85} \\ \text{if } \left(600 \geq \mathrm{Re}_\mathrm{L}\,Fa^{0.11}\right) \geq 6000 \text{ then} \\ 0.87 \\ \text{if } \left(\mathrm{Re}_\mathrm{L}\,Fa^{0.11}\right) > 6000 \text{ then} \\ 160.8\left(\mathrm{Re}_\mathrm{L}\,Fa^{0.11}\right)^{-0.6} \end{cases} \tag{E.223}$$

where $\mathrm{Re}_\mathrm{L} = (1-\dot{x})\,\dot{m}\,d_\mathrm{h}\,\mu_\mathrm{L}^{-1}$ is the liquid Reynolds number, $\mathrm{Pr}_\mathrm{L} = \mu_\mathrm{L}\,c_\mathrm{p,L}\,\lambda_\mathrm{L}^{-1}$ is the liquid Prandtl number, $\mathrm{Bo} = \dot{q}\,(\dot{m}\,\Delta h_\mathrm{v})^{-1}$ is the boiling number and $Fa = ((\rho_\mathrm{L} - \rho_\mathrm{V})\,\sigma)\,(\dot{m}^2\,d_\mathrm{h})^{-1}$ is a newly introduced dimensionless number.

E.3.3 Flow boiling refrigerant-lubricant mixtures

Tichy et al. (1986), as published by Shen and Groll (2003)

The correlation was developed based on R12-lubricant mixtures flow boiling in a 9.58 mm tube. The viscosity of the pure naphtenic oil was equal to 300SUS.

$$\alpha_\mathrm{tp}\,d\,\lambda_\mathrm{L}^{-1} = \mathrm{Nu}_\mathrm{DB}\,C\,(X_\mathrm{tt};\ \mathrm{Ja};\ w_\mathrm{oil}) \tag{E.224}$$

$$C = 10^{K_1\,(\lg(X_\mathrm{tt} - K_2)^{-1}} \tag{E.225}$$

$$K_1 = K_{1\mathrm{a}}\,(w_\mathrm{oil}) + K_{1\mathrm{b}}\,(w_\mathrm{oil})\,\lg(\mathrm{Ja}) \tag{E.226}$$

$$K_2 = K_{2\mathrm{a}}\,(w_\mathrm{oil}) - 2\lg(\mathrm{Ja}) \tag{E.227}$$

where Nu_DB is the Nusselt single-phase correlation by Dittus and Boelter (1930), Ja is the Jakob number, X_tt is the Martinelli parameter, C is an oil-dependent two-phase multiplier and K_xy are empirical constants, see Tab. E.7. All fluid properties correspond pure refrigerant liquid properties.

Table E.7: Constants for heat transfer correlation by Tichy et al. (1986)

w_{oil}	K_{1a}	K_{1b}	K_{2a}
0.00	3.79	-0.21	-9.0
0.02	3.56	-1.34	-11.0
0.05	2.17	-3.01	-13.0

Eckels et al. (1994), Eckels et al. (1998b)

Two-phase flow boiling heat transfer coefficients of R134a-POE oil mixtures were defined based on the oil contamination approach.

$$\alpha_{tp,mix} = EF_\alpha \, \alpha_{tp,ref} \tag{E.228}$$

$$\begin{aligned} EF_\alpha = {} & a_0 + a_1\left(w_{oil,0}\right) + a_2\left(\dot{m}'\right) + a_3\left(w_{oil,0}\dot{m}'\right) + a_4\left(w_{oil,0}^2\dot{m}'\right) + \\ & + a_5\left(w_{oil,0}\dot{m}'^2\right) + a_6\left(w_{oil,0}^2\dot{m}'^2\right) + a_7\left(w_{oil,0}^2\right) + a_8\left(\dot{m}'^2\right) \end{aligned} \tag{E.229}$$

$$\begin{aligned} \ln EF_\alpha = {} & b_0 + b_1\left(w_{oil,0}\right) + b_2\left(\dot{m}'\right) + b_3\left(w_{oil,0}\dot{m}'\right) + b_4\left(w_{oil,0}^2\dot{m}'\right) + \\ & + b_5\left(w_{oil,0}\dot{m}'^2\right) + b_6\left(w_{oil,0}^2\dot{m}'^2\right) + b_7\left(w_{oil,0}^2\right) + b_8\left(\dot{m}'^2\right) \end{aligned} \tag{E.230}$$

where $\dot{m}' = \dot{m}/250\,\mathrm{kg\,m^{-2}\,s^{-1}}$ and a_{1-8} and b_{1-8} listed in Tab. E.8.

Table E.8: Constants for heat transfer correlation by Eckels et al. (1994) and Eckels et al. (1998b)

		smooth tube			micro-fin tube	
	150 SUS	169 SUS	369 SUS	150 SUS	169 SUS	369 SUS
a_0	1	1.00	1.00	2.45	2.69	2.29
a_1	3.96	6.92	-7.05	29.95	17.27	-1.23
a_2	0	0	0	1.01	-1.43	-0.99
a_3	0	-572.10	-31.11	14.87	0	0.76
a_4	411.00	0	0	0	0	0
a_5	0	304.90	0	0	3.73	0
a_6	202.10	0	63.55	50.18	0	0
a_7	0	0	0	194.70	-508.20	0
a_8	0	0	0	0.24	0.39	0
b_0	0	0	0	0.98	1.02	0.91
b_1	4.95	9.66	-7.10	12.78	10.42	-1.67
b_2	0	0	0	0.63	-0.70	-0.60
b_3	0	-753.60	-49.49	6.20	0	1.68
b_4	498.00	-2.78	0	143.60	0	-2.64
b_5	0	449.30	0	0	1.87	0
b_6	245.20	0	57.90	78.37	0	0
b_7	0	0	0	26.56	-298.30	1.99
b_8	0	0	0	0.17	0.17	0.14

Gao et al. (2008a), as published by Li et al. (2014)

According to Li et al., Gao et al. proposed a modified Cheng et al. (2008b) type correlation, redefining the suppression factor in the original model.

$$S' = \begin{cases} S\left(-0.5\dot{x}+0.35\right) & \dot{x} \leq 0.7 \\ 0 & \dot{x} > 0.7 \end{cases} \tag{E.231}$$

where S' is the redefined and S is the original suppression factor.

Katsuta et al. (2008)

The correlation of Katsuta et al. is given in Sec. 2.4.

Li et al. (2014)

The correlation of Li et al. is given in Sec. 2.4.

Appendix F

F.1 Smooth tube measurement data

F.1.1 Flow patterns

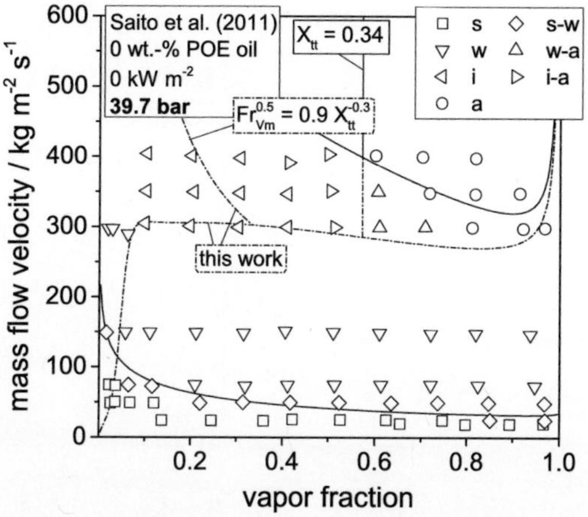

Figure F.1: Adiabatic flow patterns of CO_2 at 39.7 bar in the smooth tube plotted in Steiner flow pattern map modified by Saito et al. (2011); dot-dashed lines represent new transition curves

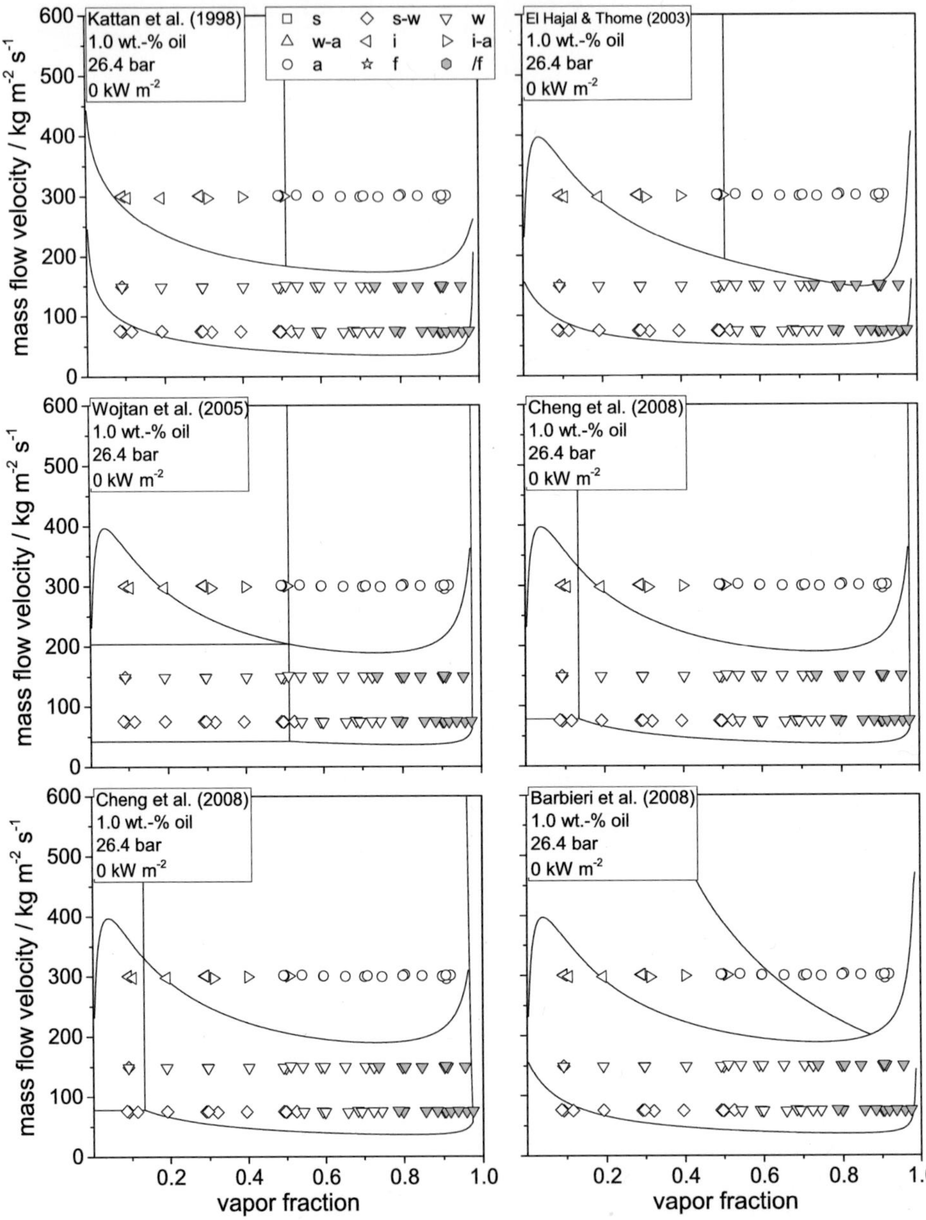

Figure F.2: Adiabatic flow patterns of CO_2-POE oil mixtures ($w_{oil,0} = 1.0$ wt.-%, $p = 26.4$ bar) in the smooth tube plotted in six established flow pattern maps

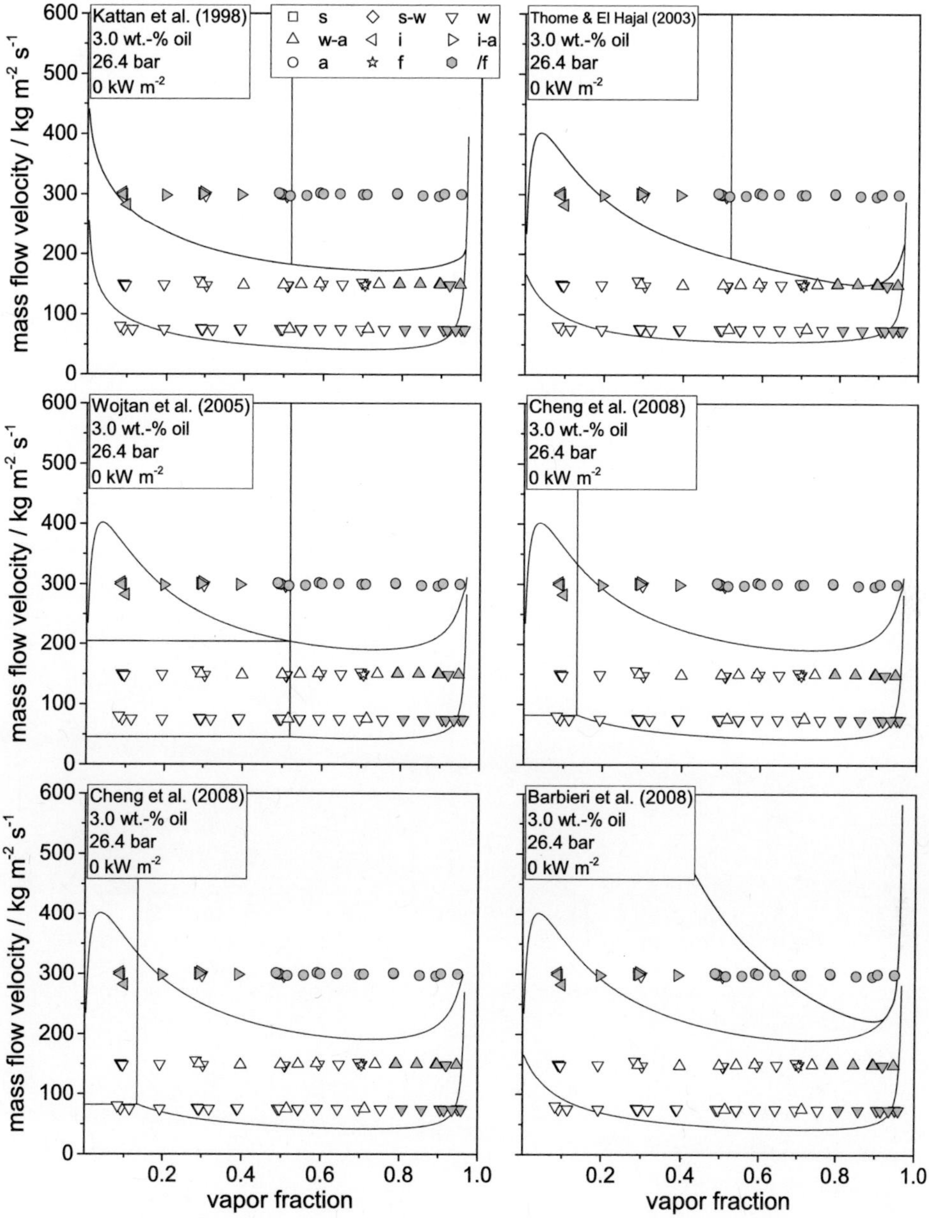

Figure F.3: Adiabatic flow patterns of CO_2-POE oil mixtures ($w_{oil,0} = 3.0$ wt.-%, $p = 26.4$ bar) in the smooth tube plotted in six established flow pattern maps

F.1.2 Pressure drop

Frictional pressure drop

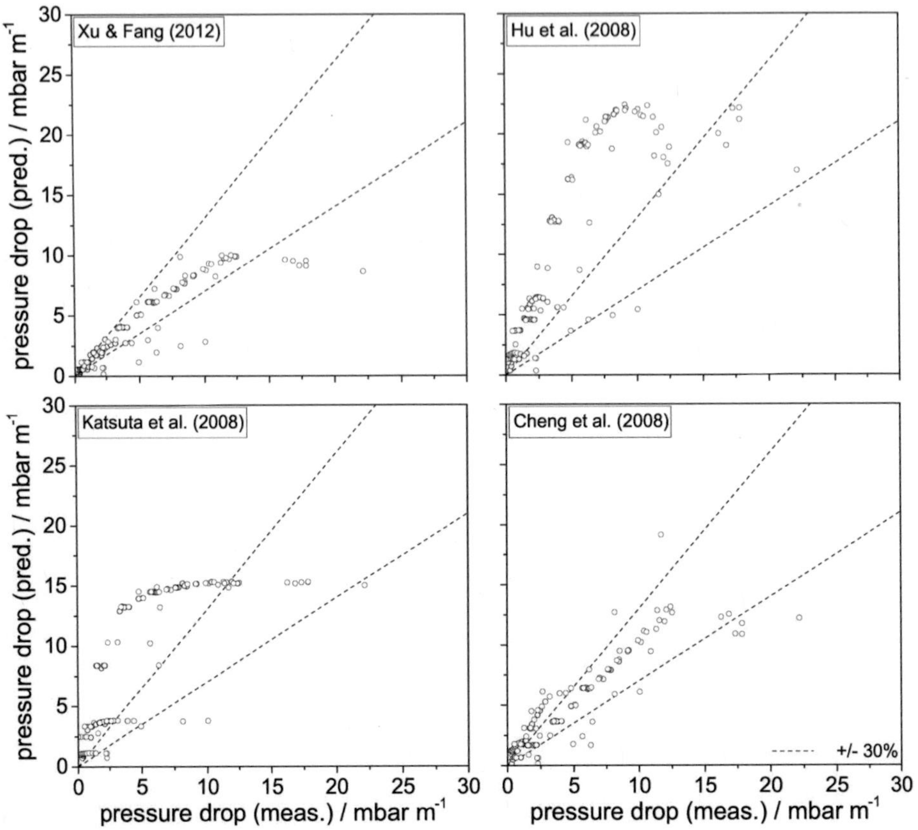

Figure F.4: Comparison of predicted to measured frictional pressure drops at 26.4 bar for CO_2-POE oil mixtures containing nominal oil mass fractions of 0 – 3 wt.-%

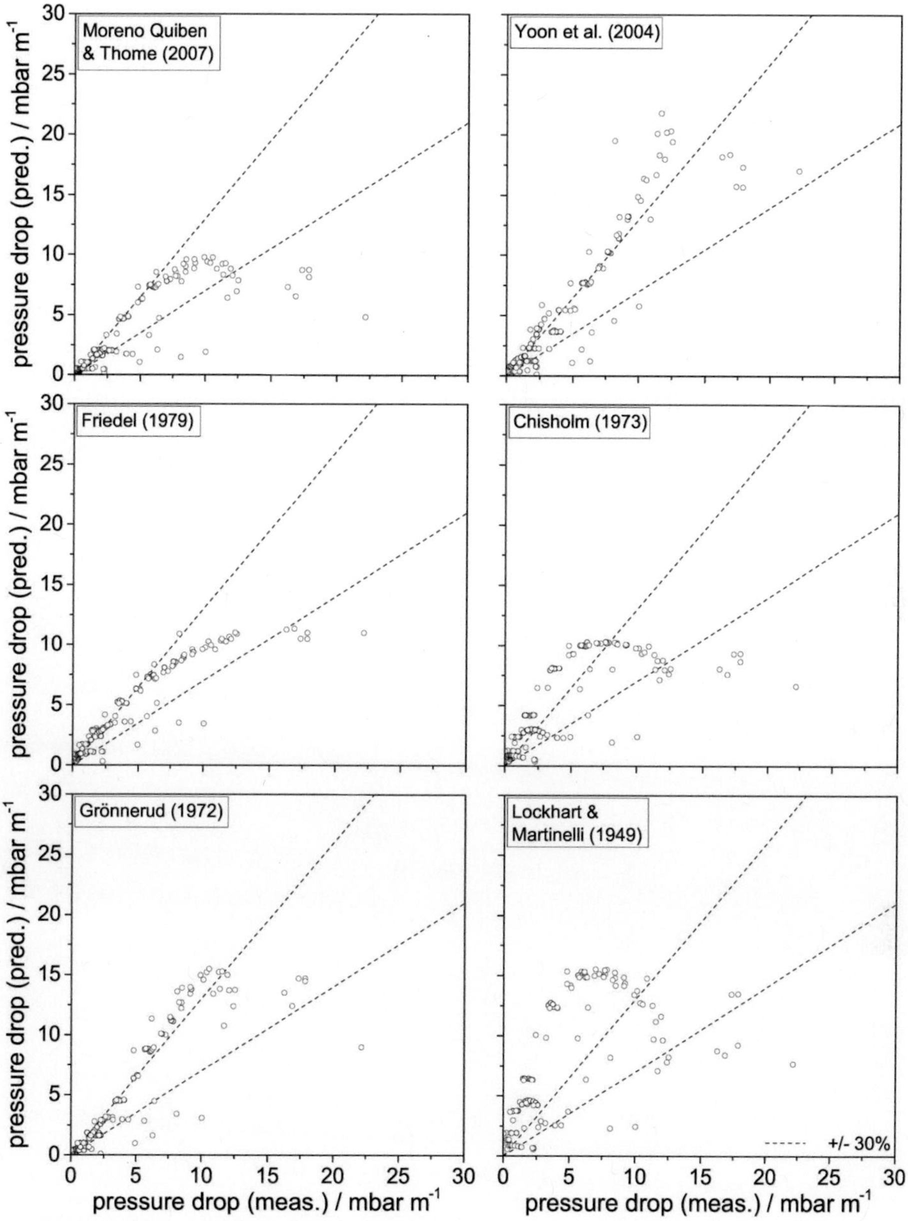

Figure F.5: Comparison of predicted to measured frictional pressure drops at 26.4 bar for CO_2-POE oil mixtures containing nominal oil mass fractions of 0 – 3 wt.-%

Total pressure drop

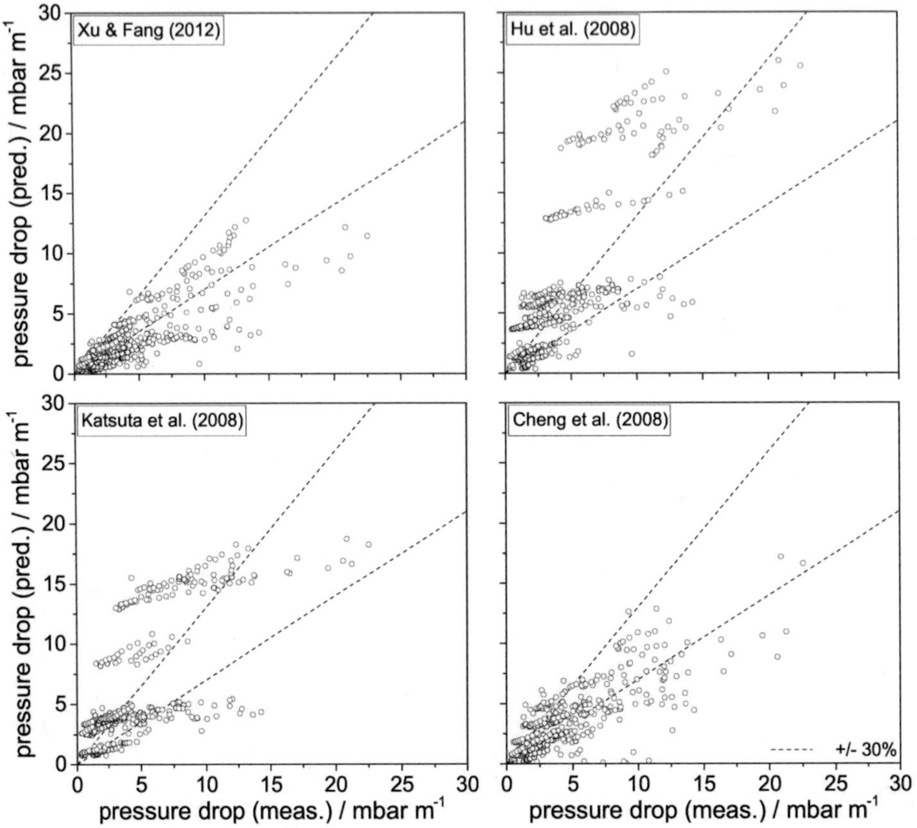

Figure F.6: Comparison of predicted to measured total pressure drops at 26.4 bar for CO_2-POE oil mixtures containing nominal oil mass fractions of 0 – 3 wt.-%

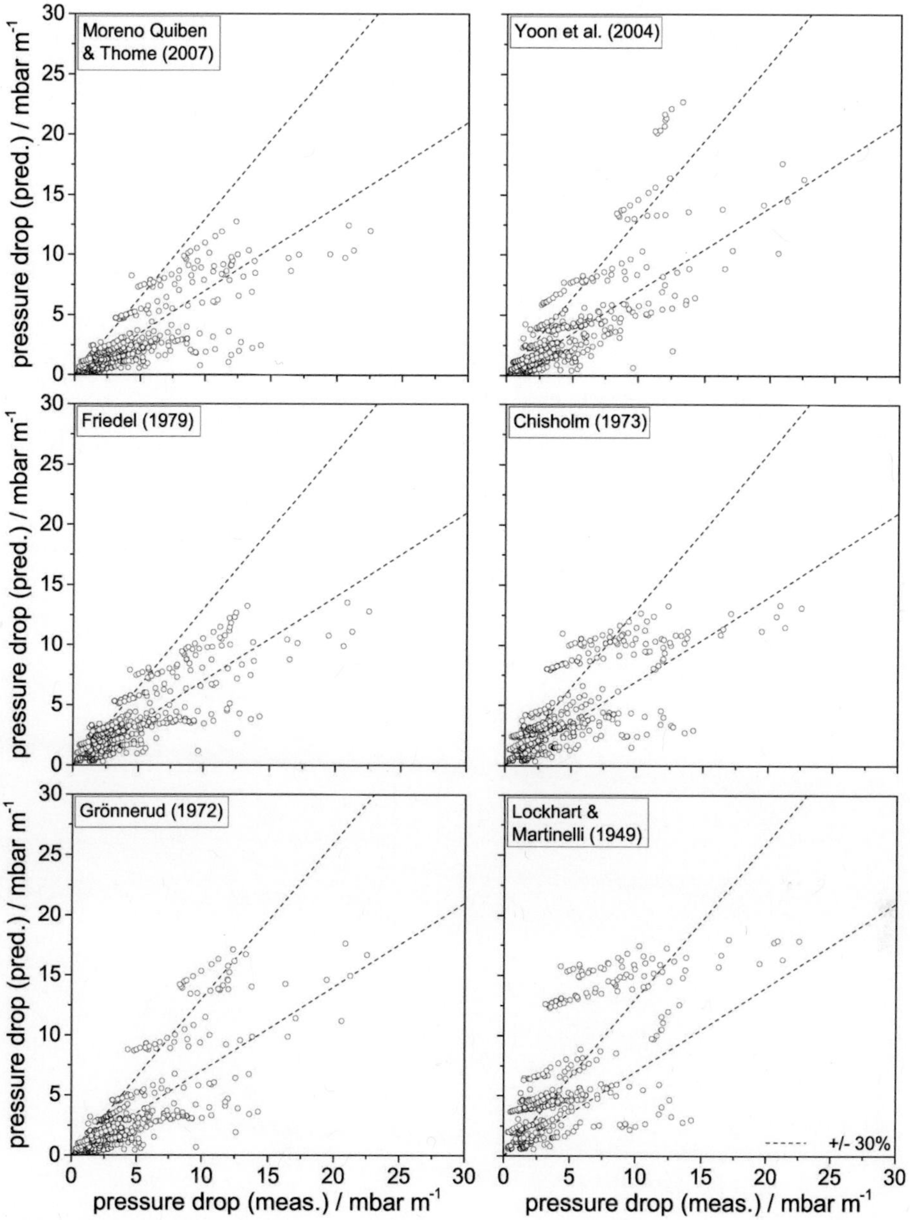

Figure F.7: Comparison of predicted to measured total pressure drops at 26.4 bar for CO_2-POE oil mixtures containing nominal oil mass fractions of 0 – 3 wt.-%

F.1.3 Heat transfer

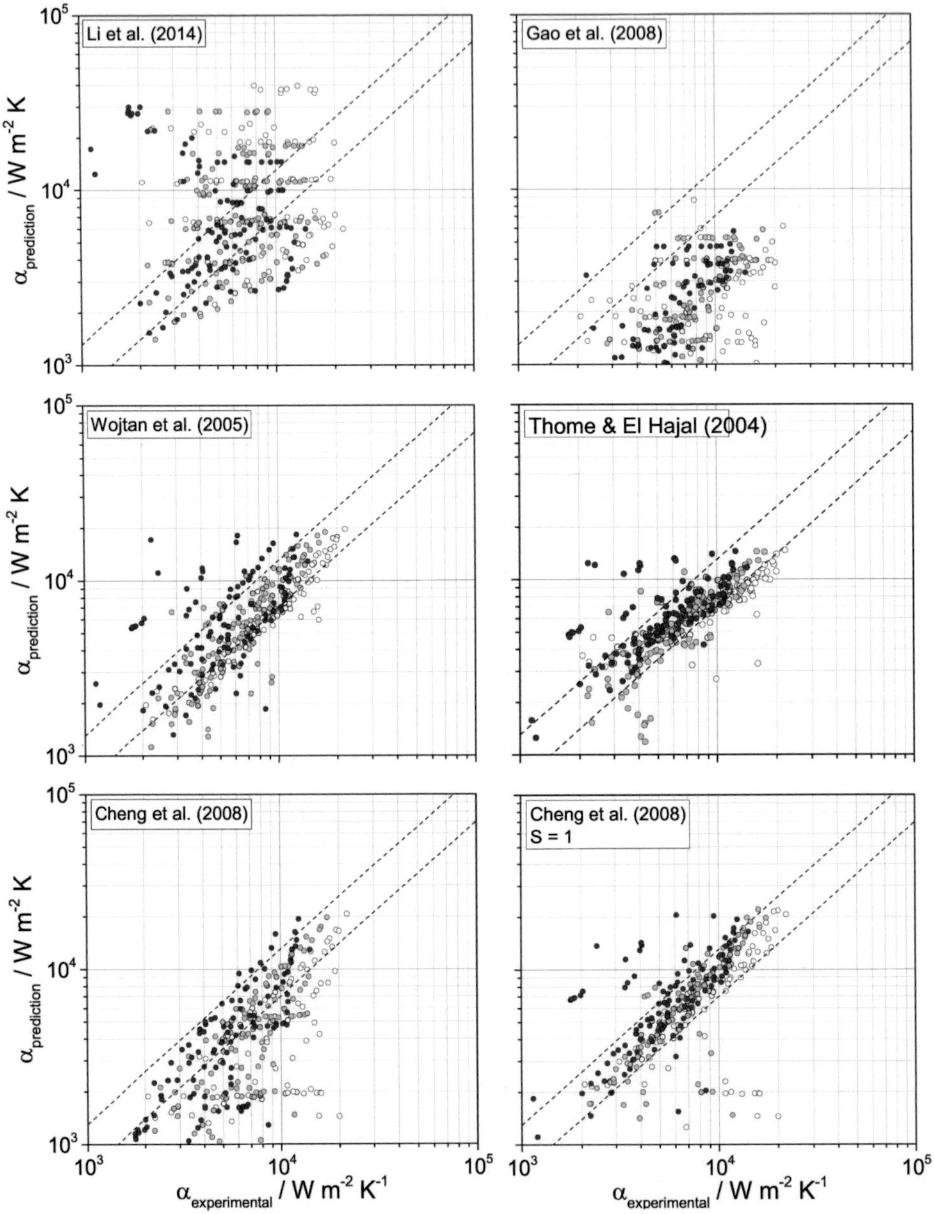

Figure F.8: Comparison of predicted to measured heat transfer coefficients inside the smooth tube tube

F.2 Micro-fin tube measurement data

F.2.1 Flow patterns

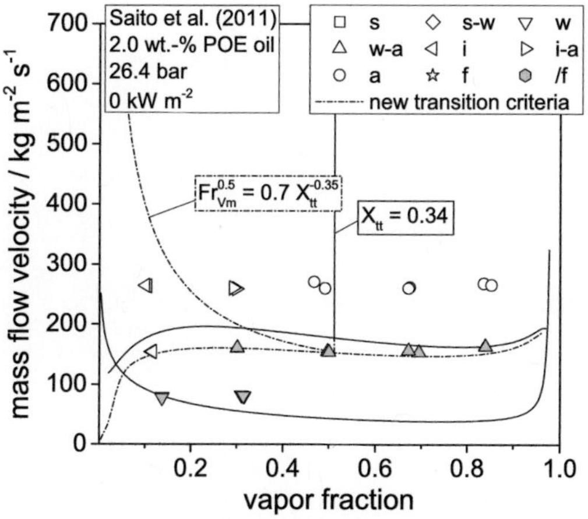

Figure F.9: Adiabatic flow patterns of CO_2-POE oil mixtures ($w_{oil,0}$ = 2.0 wt.-%, p = 26.4 bar) in the micro-fin tube plotted in newly modified Steiner flow pattern map

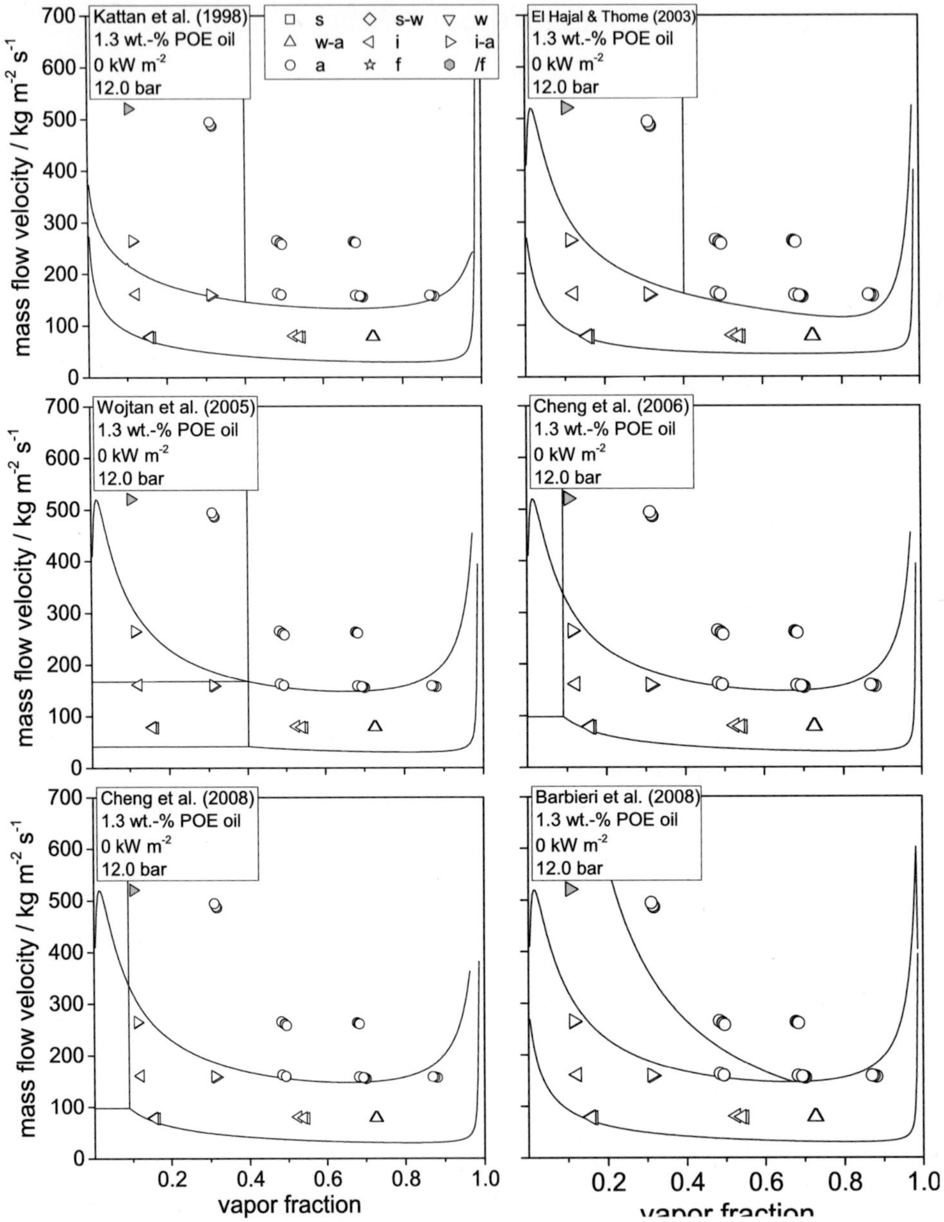

Figure F.10: Adiabatic flow patterns of CO_2-POE oil mixtures ($w_{oil,0}$ = 1.3 wt.-%, p = 12.0 bar) in the micro-fin tube plotted in six established flow pattern maps

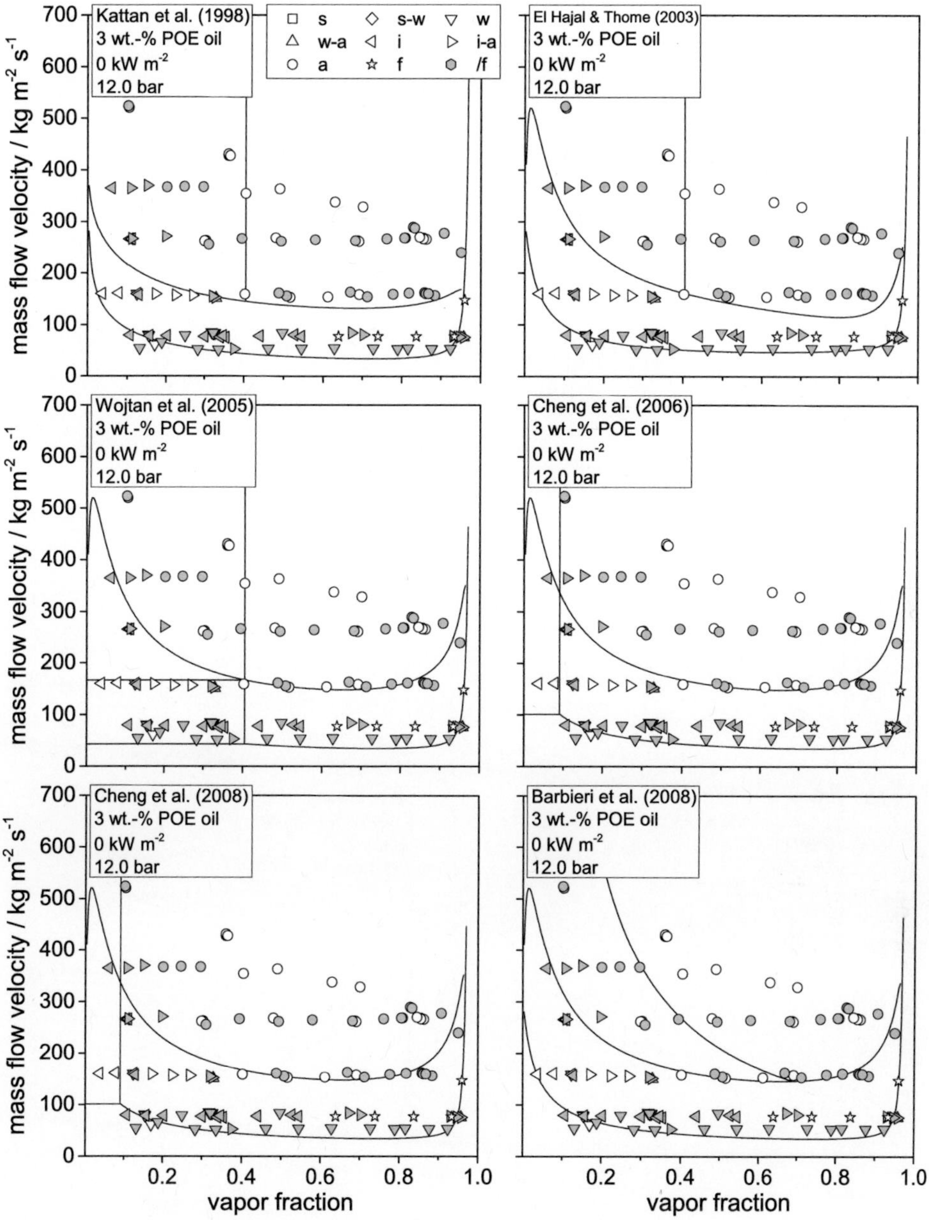

Figure F.11: Adiabatic flow patterns of CO_2-POE oil mixtures ($w_{oil,0}$ = 3.0 wt.-%, p = 12.0 bar) in the micro-fin tube plotted in six established flow pattern maps

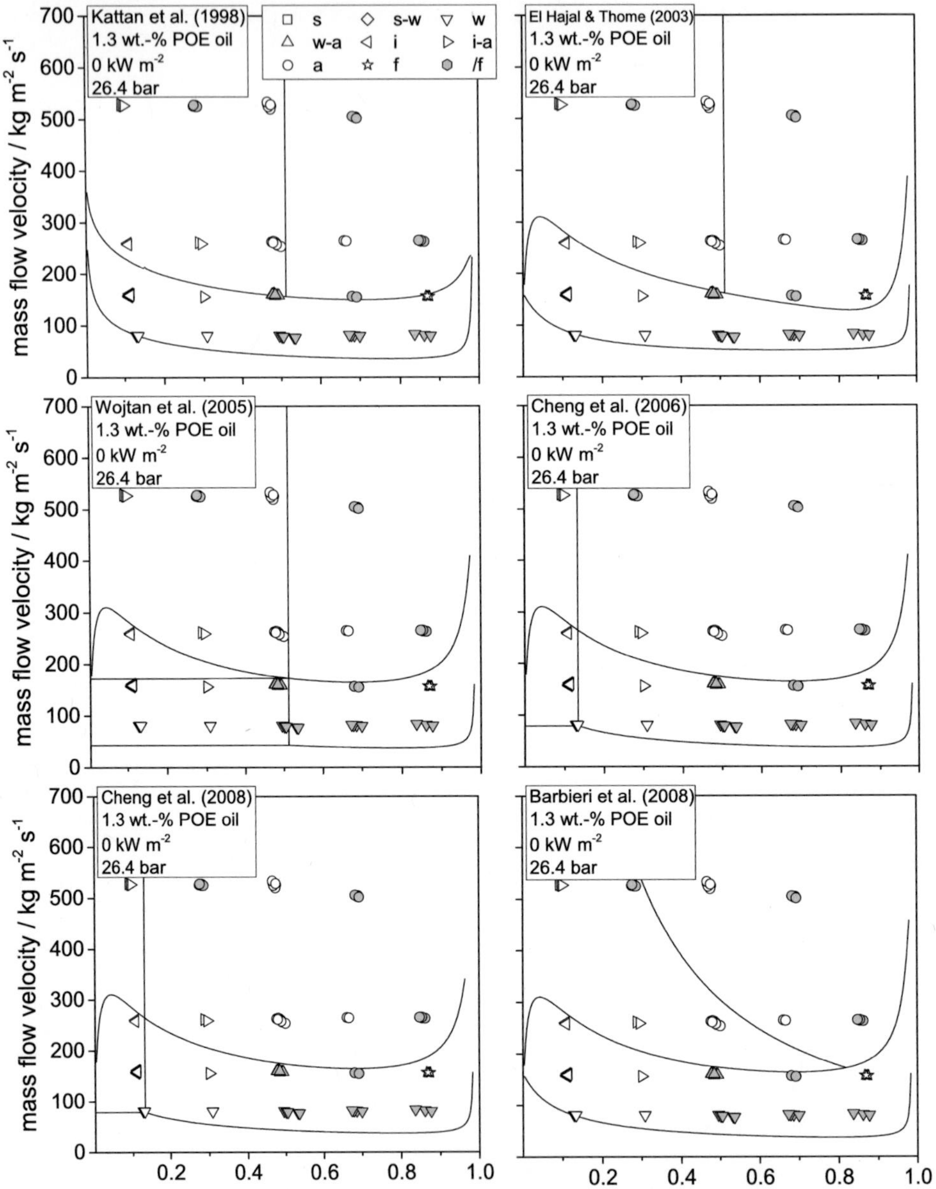

Figure F.12: Adiabatic flow patterns of CO_2-POE oil mixtures ($w_{oil,0}$ = 1.3 wt.-%, p = 26.4 bar) in the micro-fin tube plotted in six established flow pattern maps

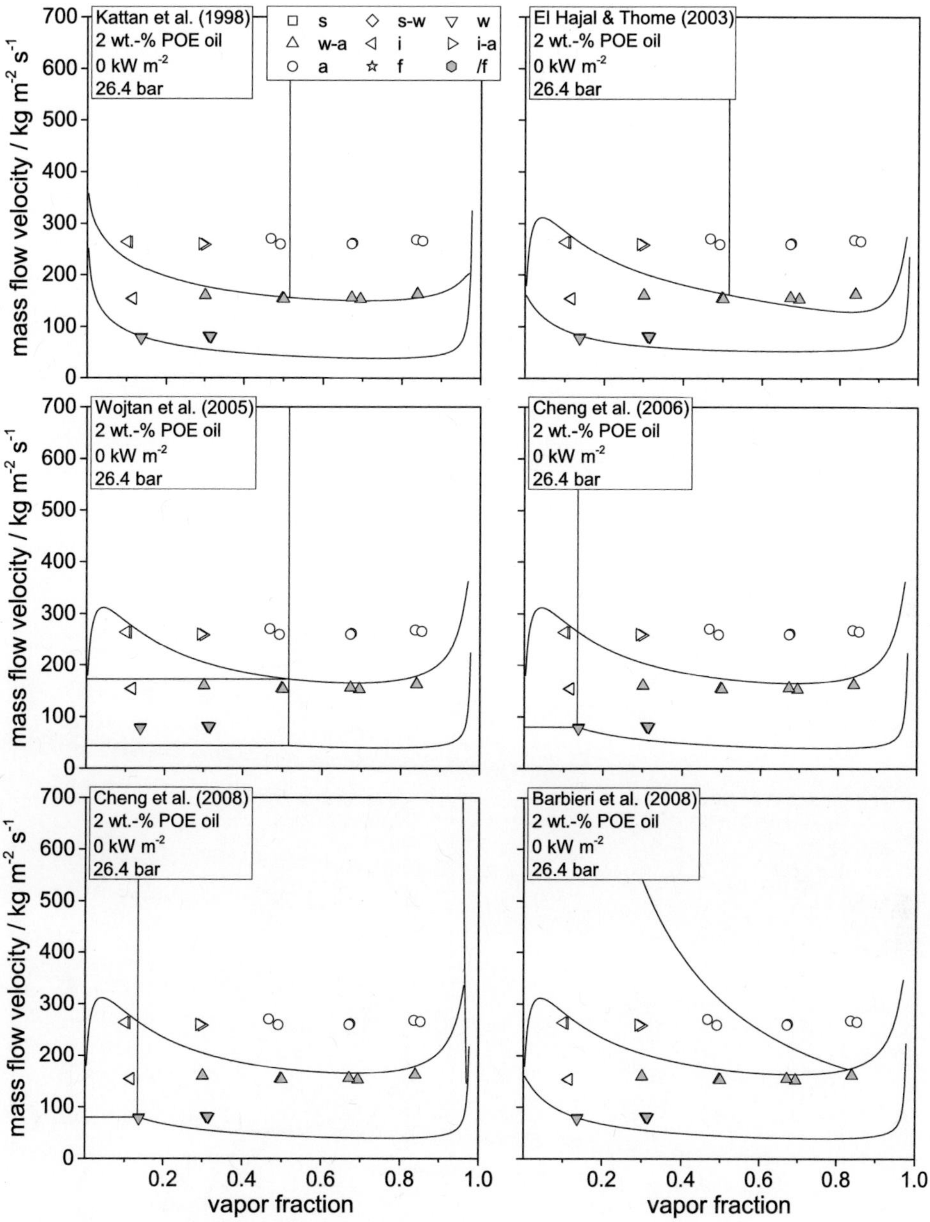

Figure F.13: Adiabatic flow patterns of CO_2-POE oil mixtures ($w_{oil,0} = 2.0$ wt.-%, $p = 26.4$ bar) in the micro-fin tube plotted in six established flow pattern maps

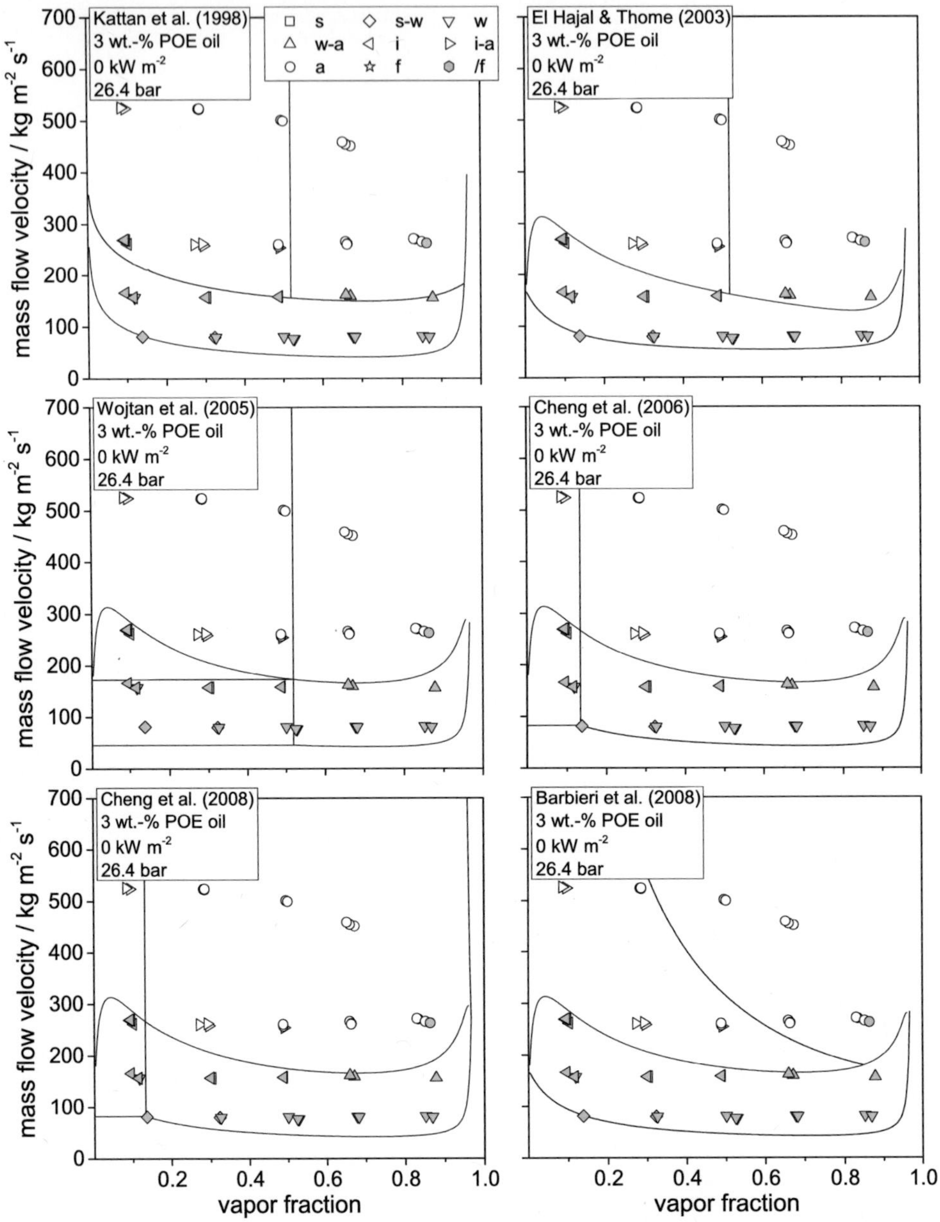

Figure F.14: Adiabatic flow patterns of CO_2-POE oil mixtures ($w_{oil,0}$ = 3.0 wt.-%, p = 26.4 bar) in the micro-fin tube plotted in six established flow pattern maps

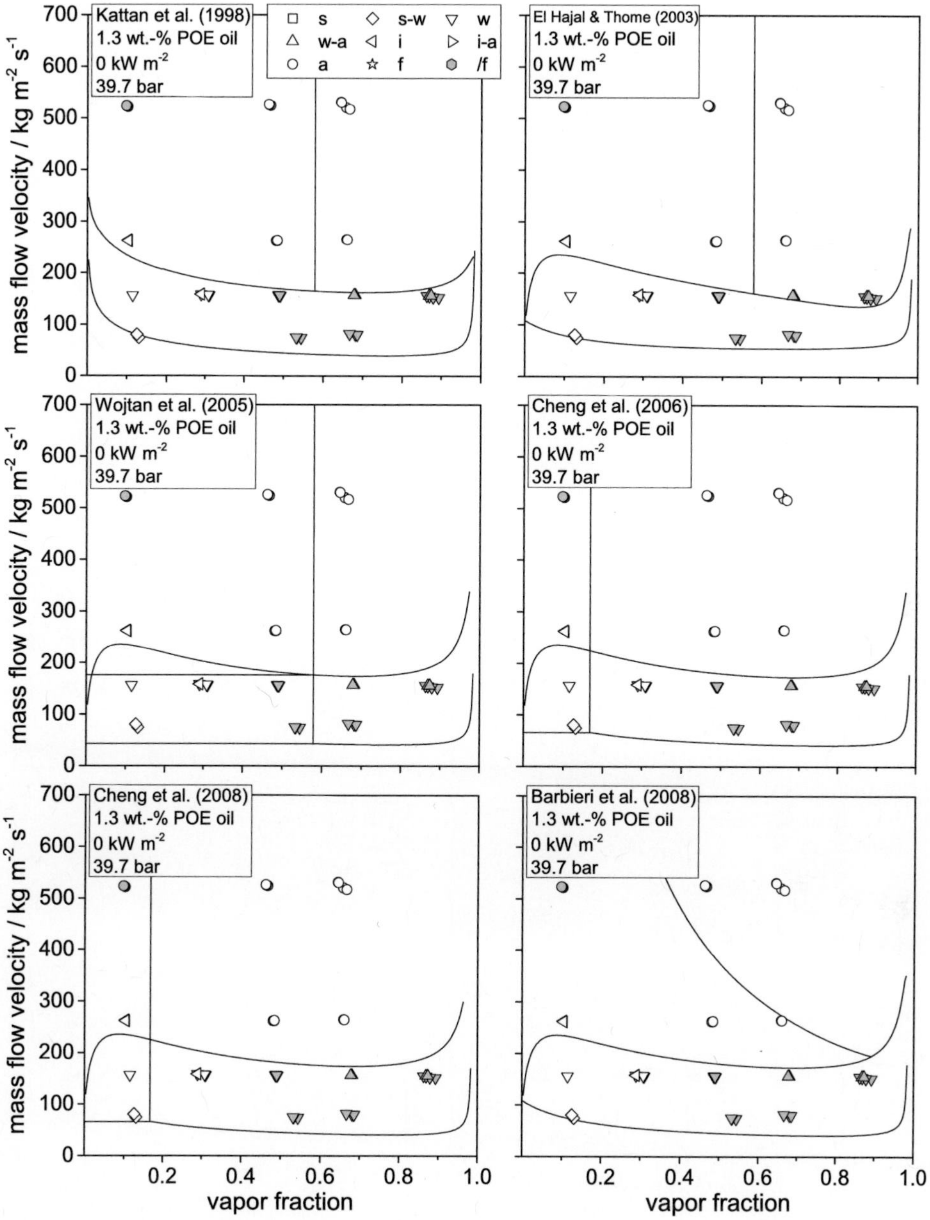

Figure F.15: Adiabatic flow patterns of CO_2-POE oil mixtures ($w_{oil,0}$ = 1.3 wt.-%, p = 39.7 bar) in the micro-fin tube plotted in six established flow pattern maps

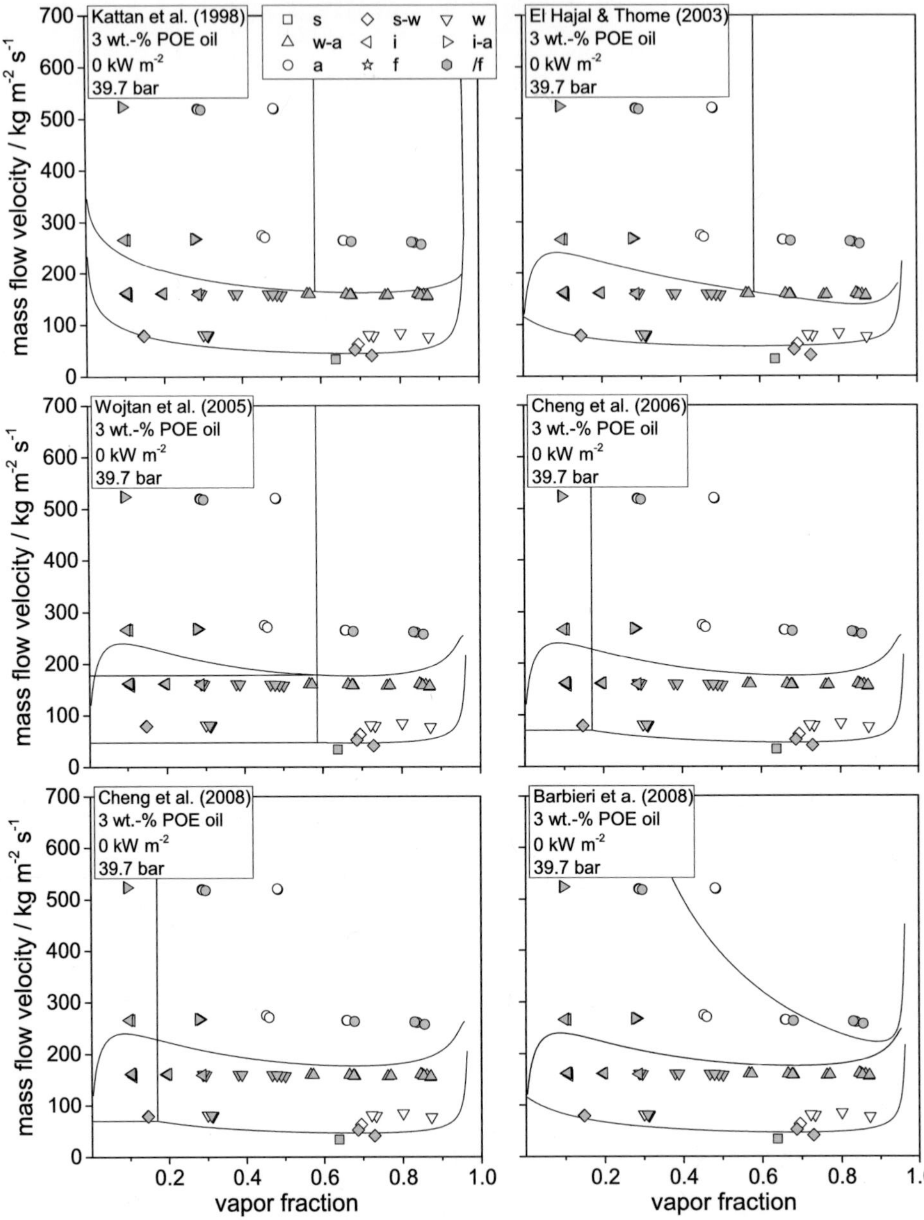

Figure F.16: Adiabatic flow patterns of CO_2-POE oil mixtures ($w_{oil,0} = 3.0$ wt.-%, $p = 39.7$ bar) in the micro-fin tube plotted in six established flow pattern maps

F.2.2 Pressure drop

Total pressure drop

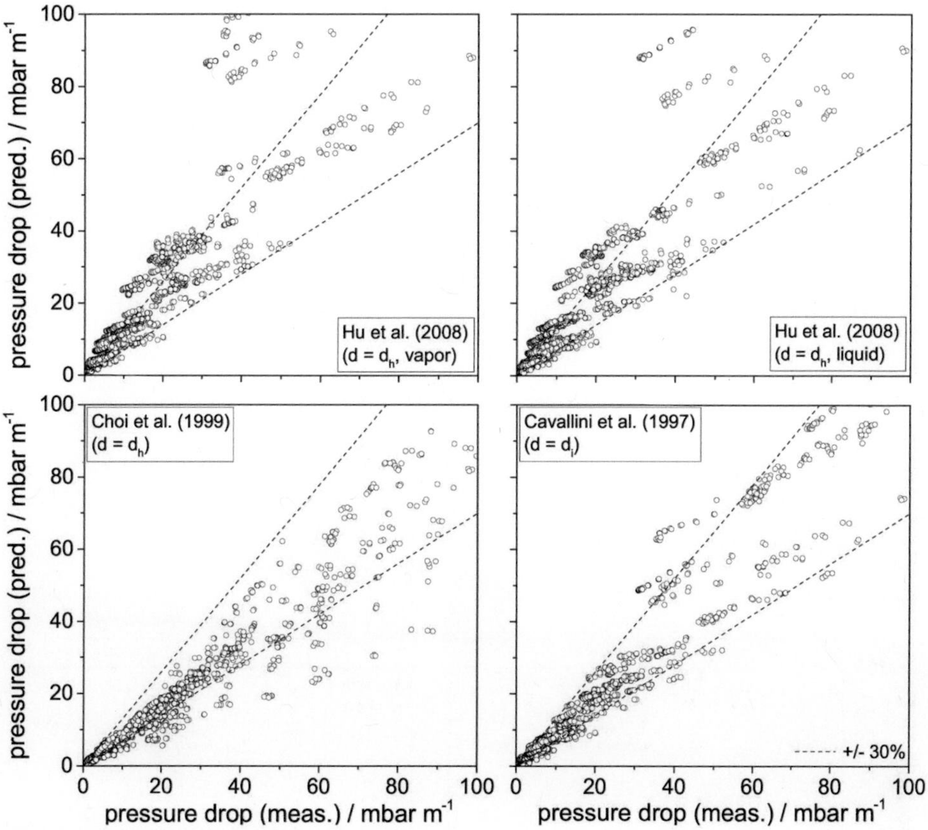

Figure F.17: Comparison of predicted to measured total pressure drops for CO_2-POE oil mixtures containing a nominal oil mass fraction of 1.3 wt.-%

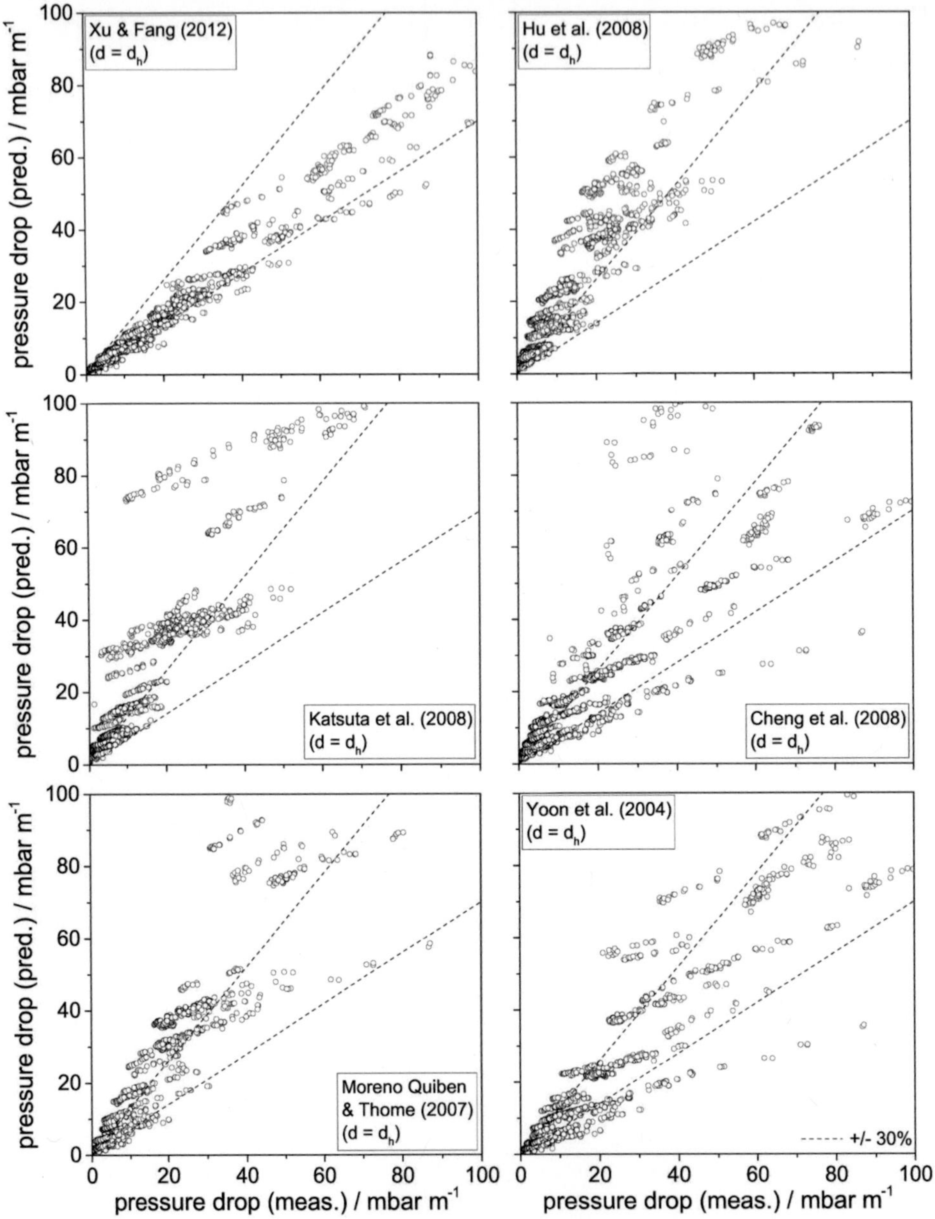

Figure F.18: Comparison of predicted to measured total pressure drops for CO_2-POE oil mixtures containing a nominal oil mass fraction of 1.3 wt.-%

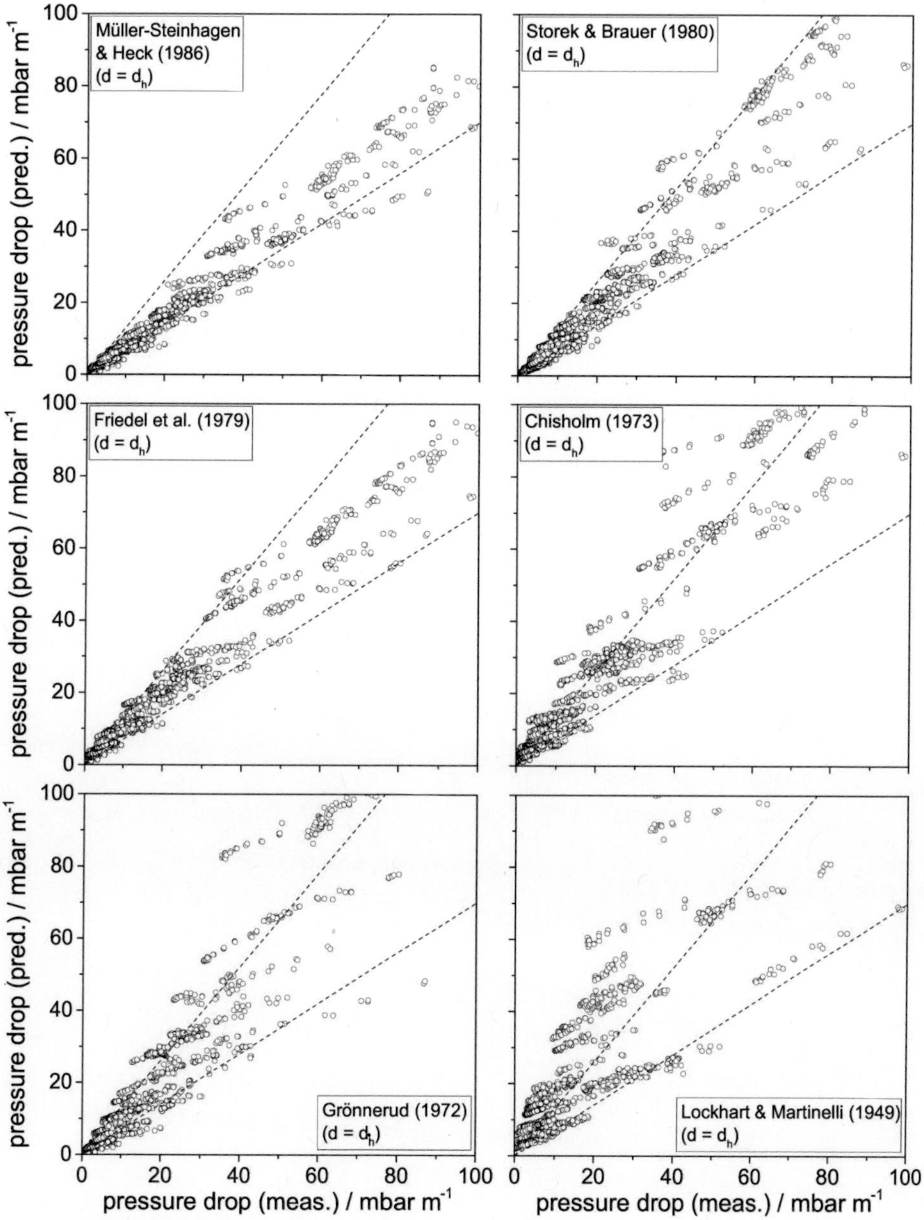

Figure F.19: Comparison of predicted to measured total pressure drops for CO_2-POE oil mixtures containing a nominal oil mass fraction of 1.3 wt.-%

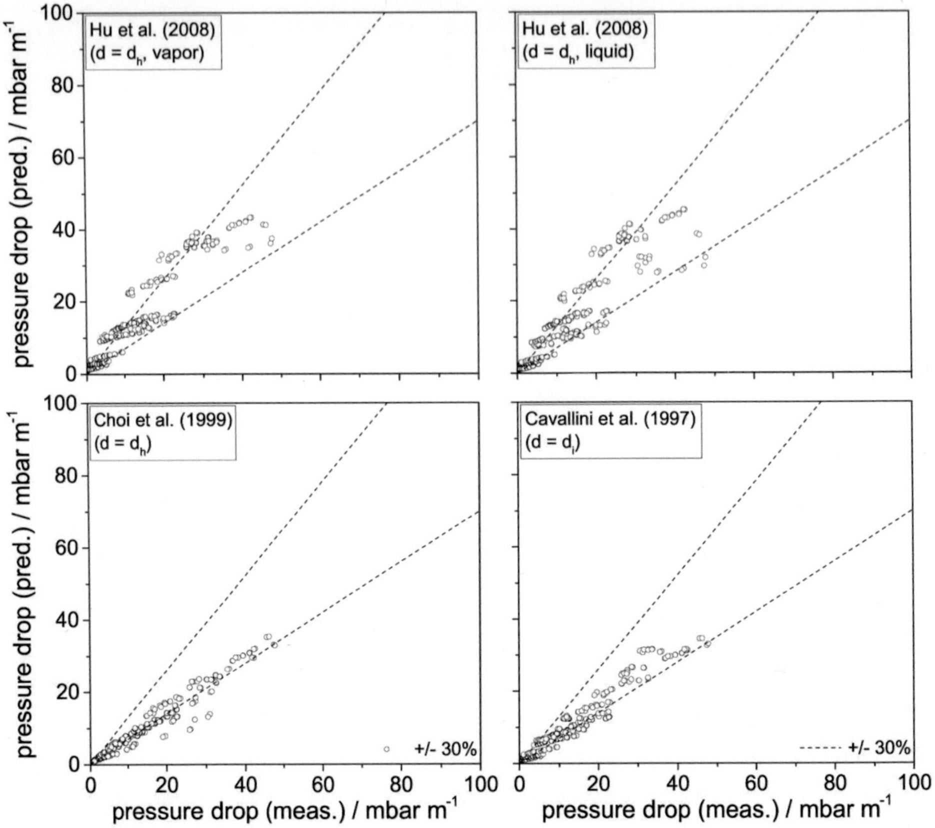

Figure F.20: Comparison of predicted to measured total pressure drops for CO_2-POE oil mixtures containing a nominal oil mass fraction of 2.0 wt.-%

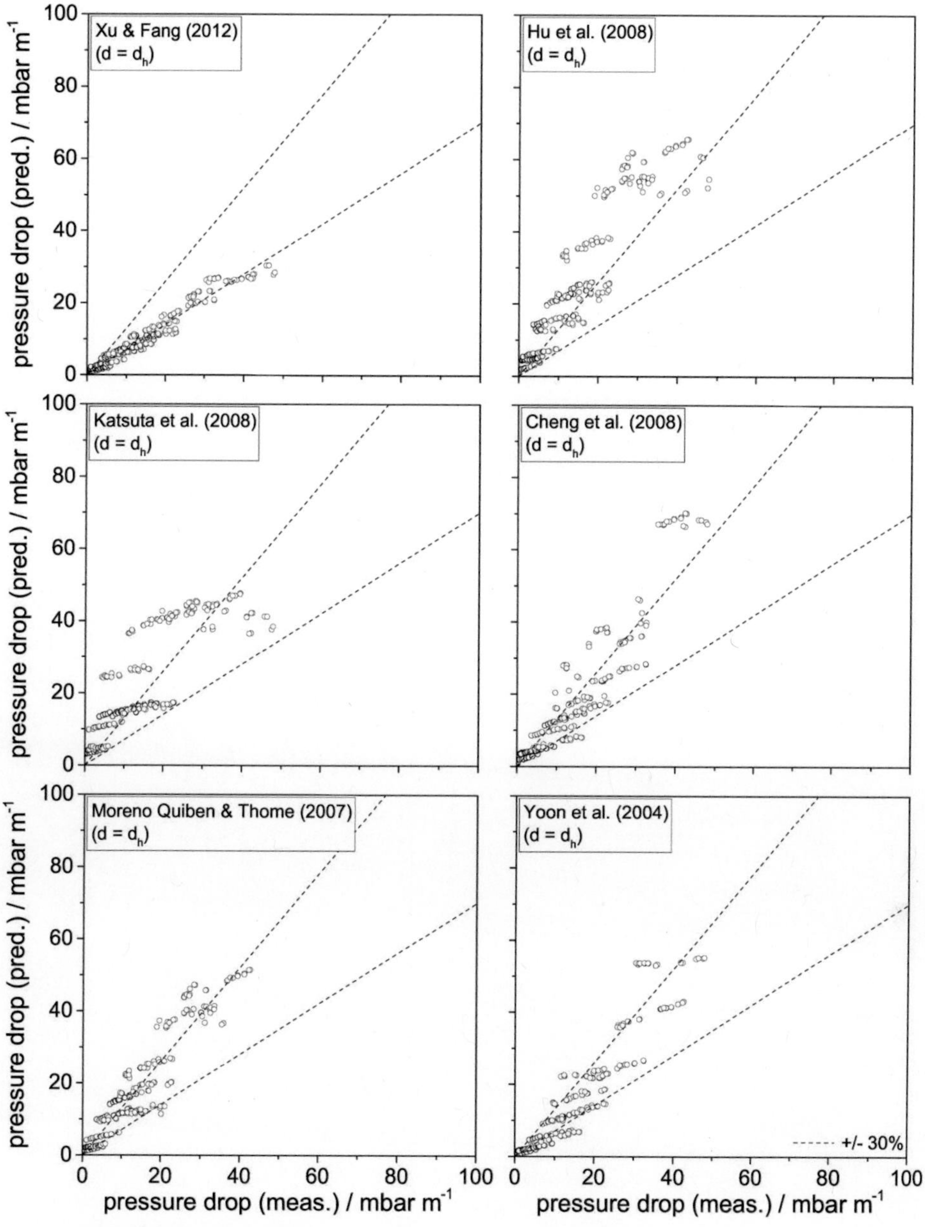

Figure F.21: Comparison of predicted to measured total pressure drops for CO_2-POE oil mixtures containing a nominal oil mass fraction of 2.0 wt.-%

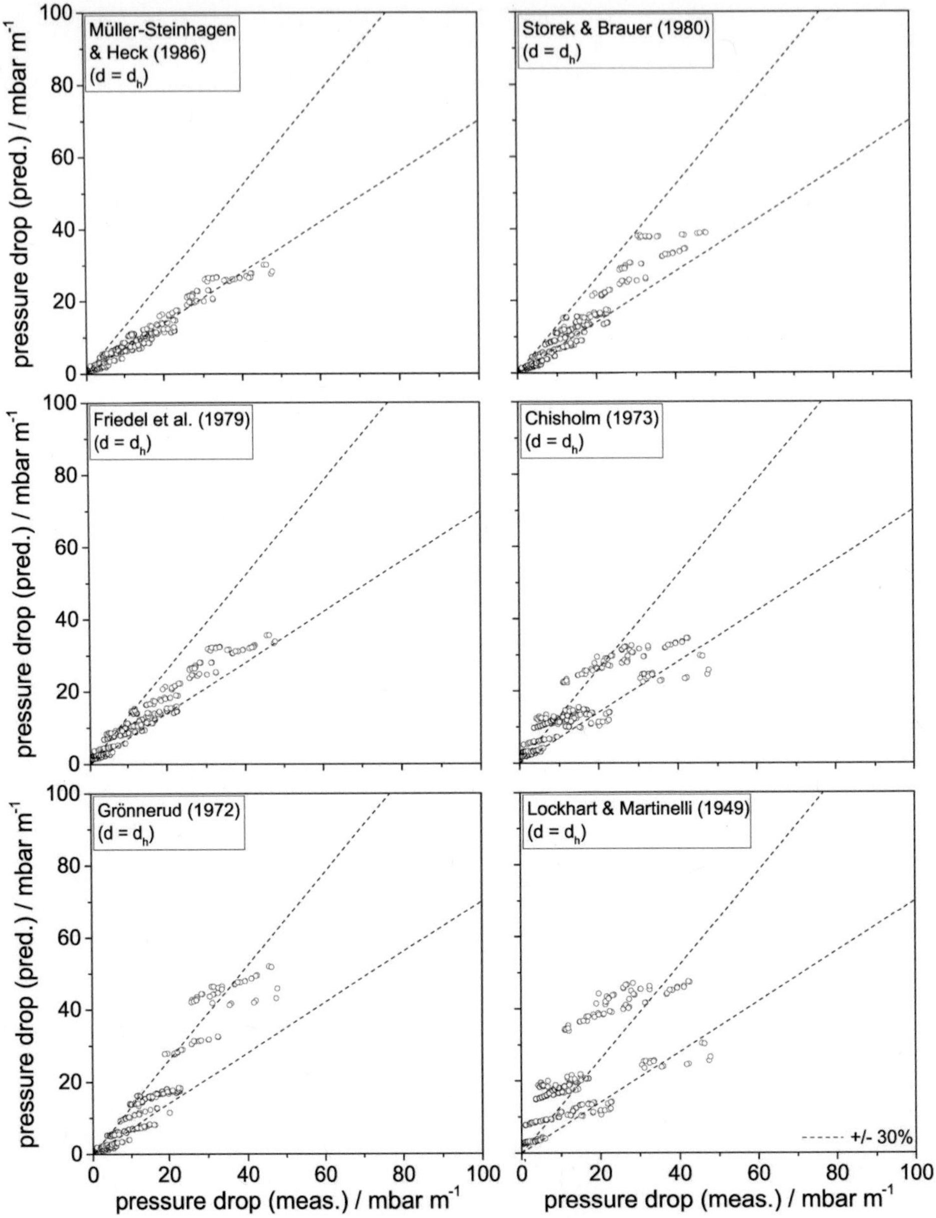

Figure F.22: Comparison of predicted to measured total pressure drops for CO_2-POE oil mixtures containing a nominal oil mass fraction of 2.0 wt.-%

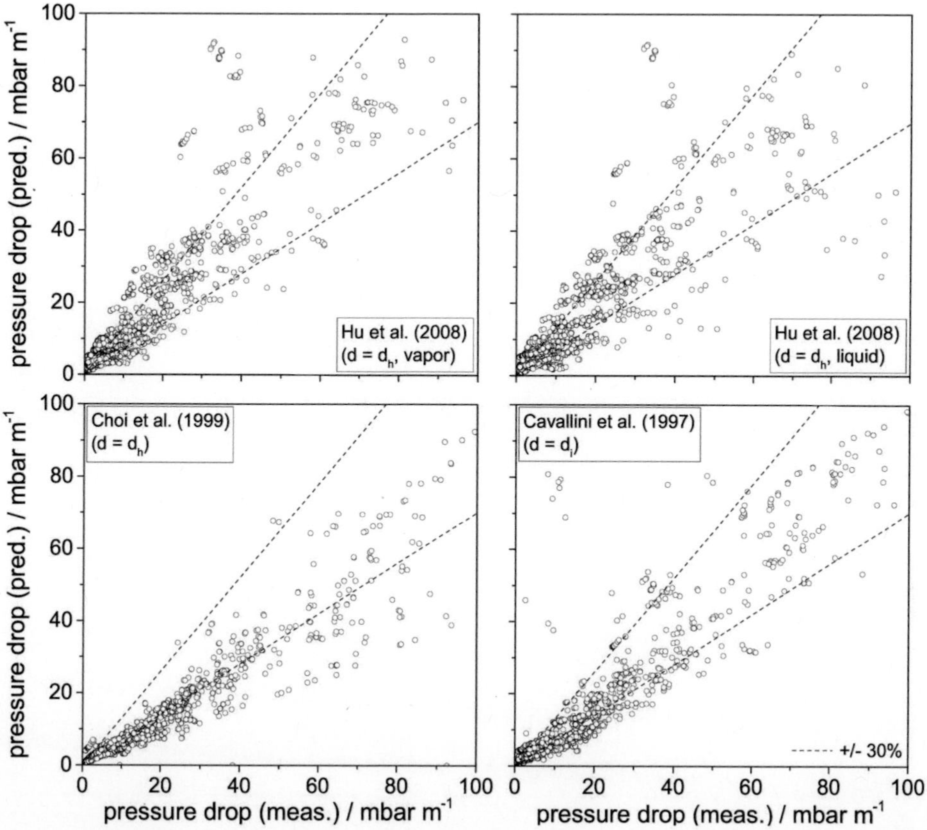

Figure F.23: Comparison of predicted to measured total pressure drops for CO_2-POE oil mixtures containing a nominal oil mass fraction of 3.0 wt.-%

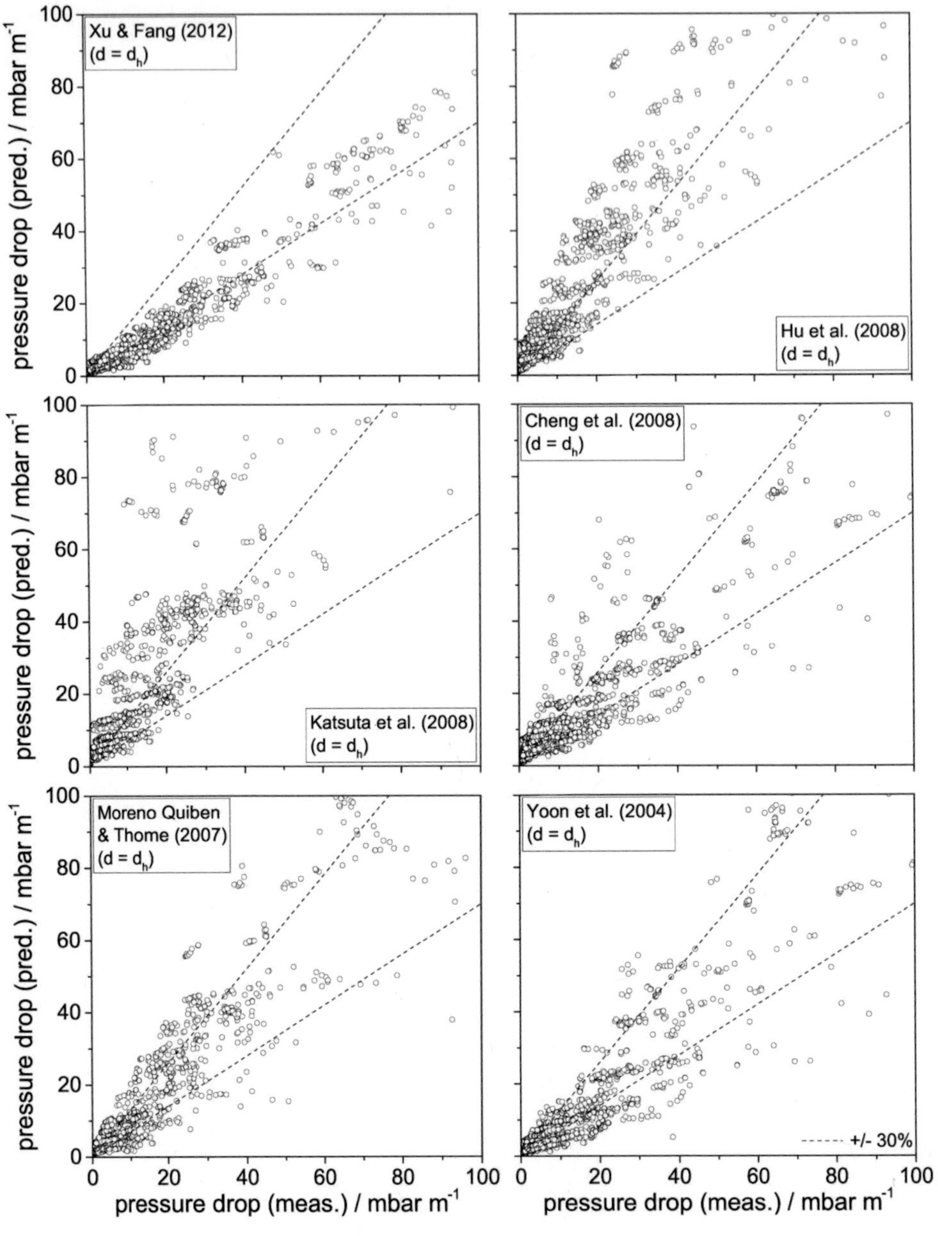

Figure F.24: Comparison of predicted to measured total pressure drops for CO_2-POE oil mixtures containing a nominal oil mass fraction of 3.0 wt.-%

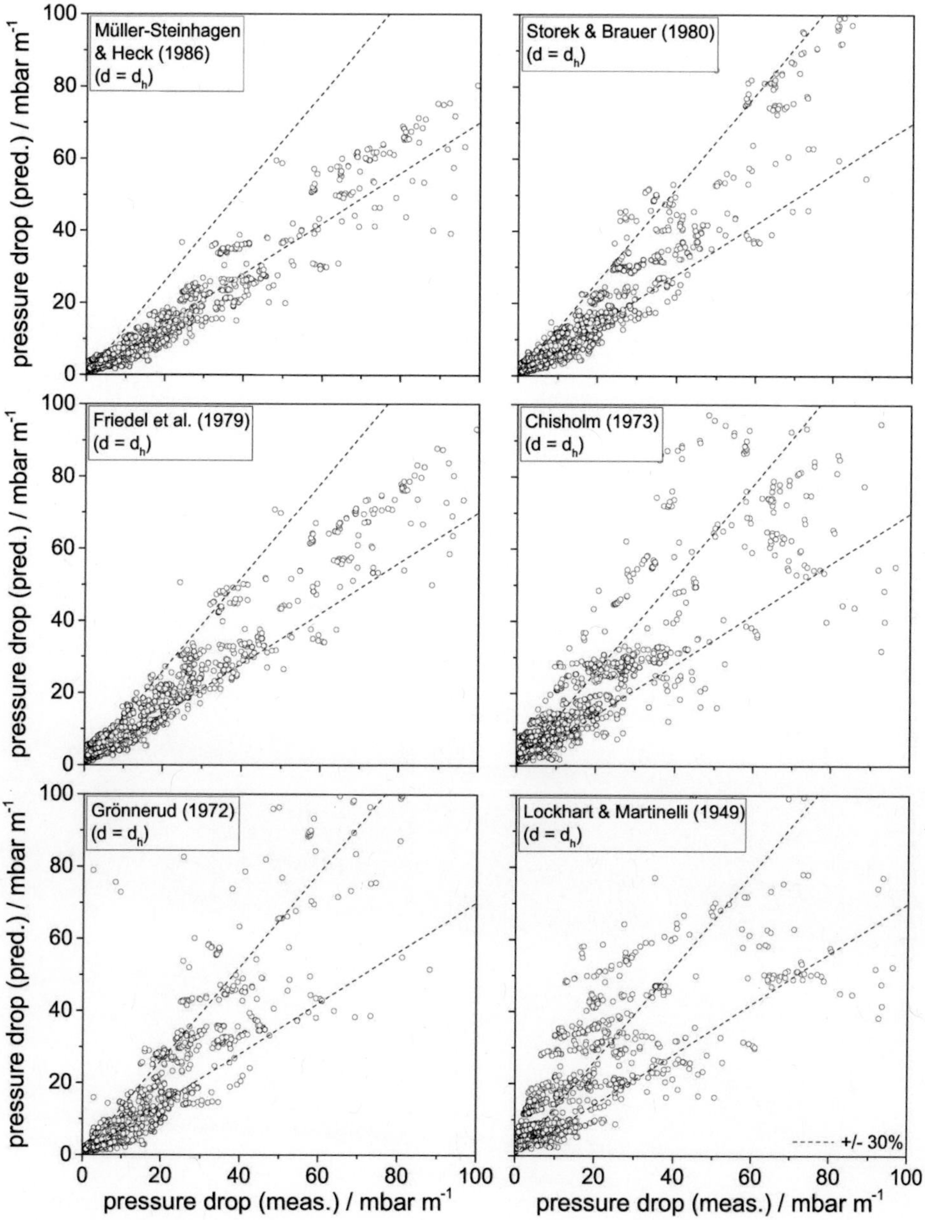

Figure F.25: Comparison of predicted to measured total pressure drops for CO_2-POE oil mixtures containing a nominal oil mass fraction of 3.0 wt.-%

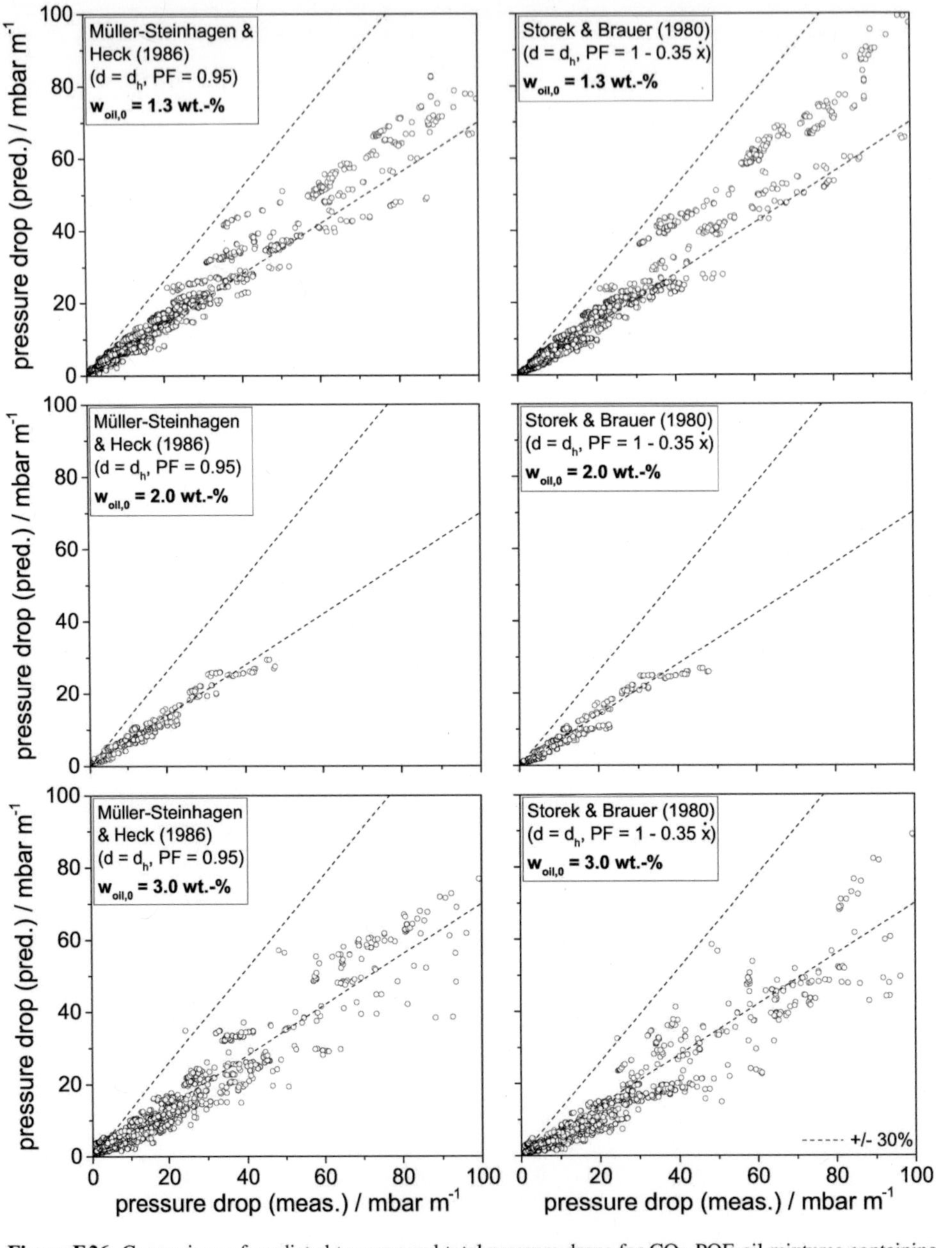

Figure F.26: Comparison of predicted to measured total pressure drops for CO_2-POE oil mixtures containing nominal oil mass fractions of 1.3 wt.-% (top), 2.0 wt.-% (middle) and 3.0 wt.-% (bottom)

F.2.3 Heat transfer

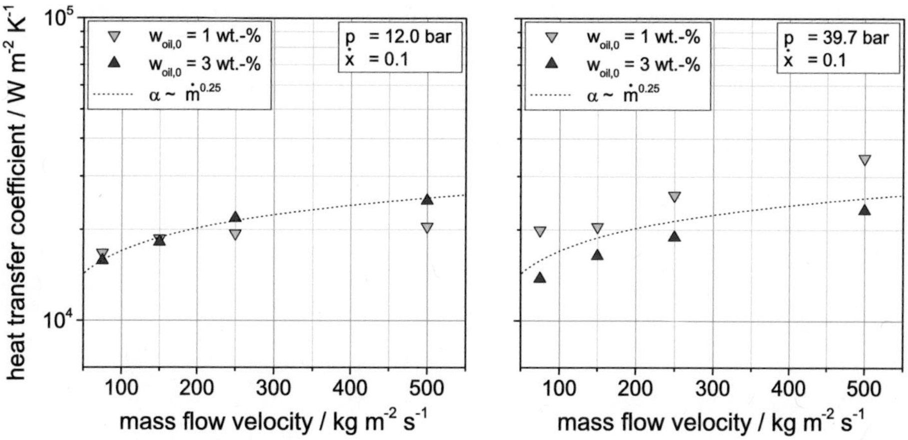

Figure F.27: Influence of the mass flow velocity on the circumference-averaged heat transfer coefficient in the micro-fin tube at a pressure of 12.0 bar

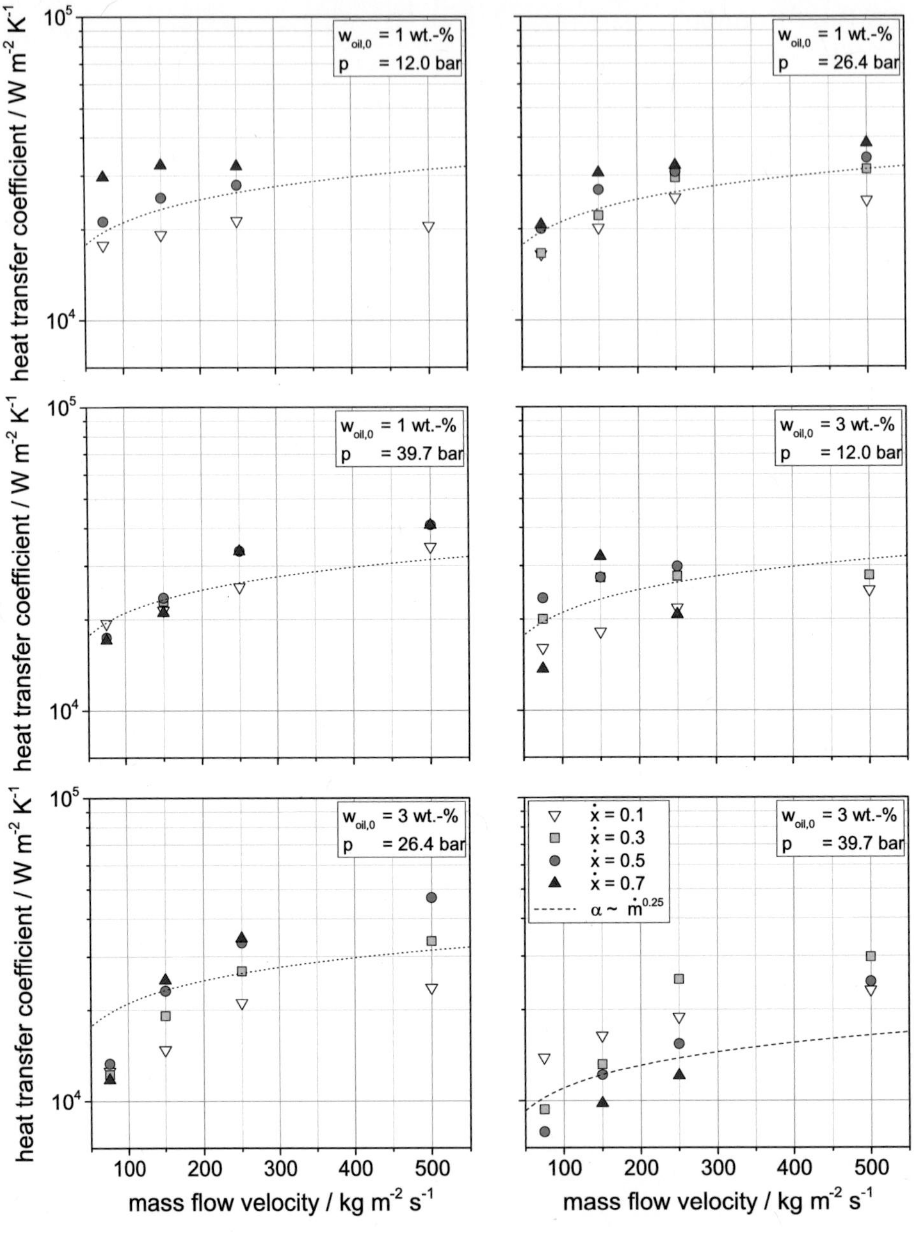

Figure F.28: Influence of mass flow velocity on the circumference-averaged heat transfer coefficient in the micro-fin tube

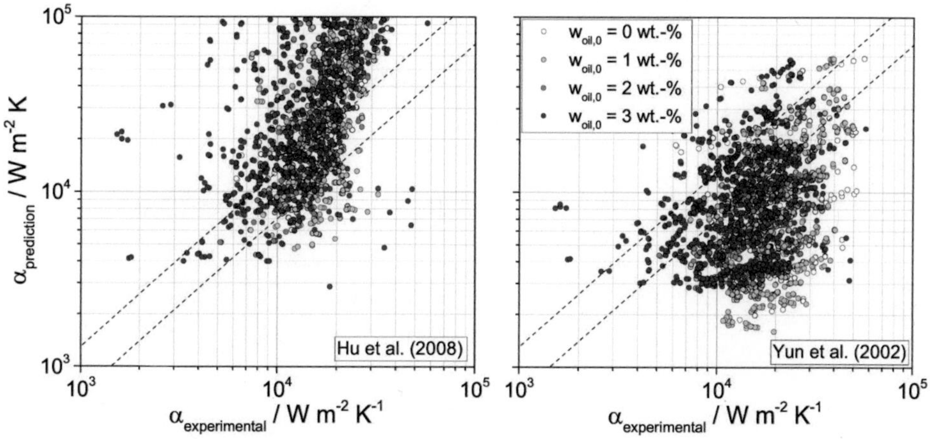

Figure F.29: Predicted to measured heat transfer coefficients inside the micro-fin tube (micro-fin tube correlations)

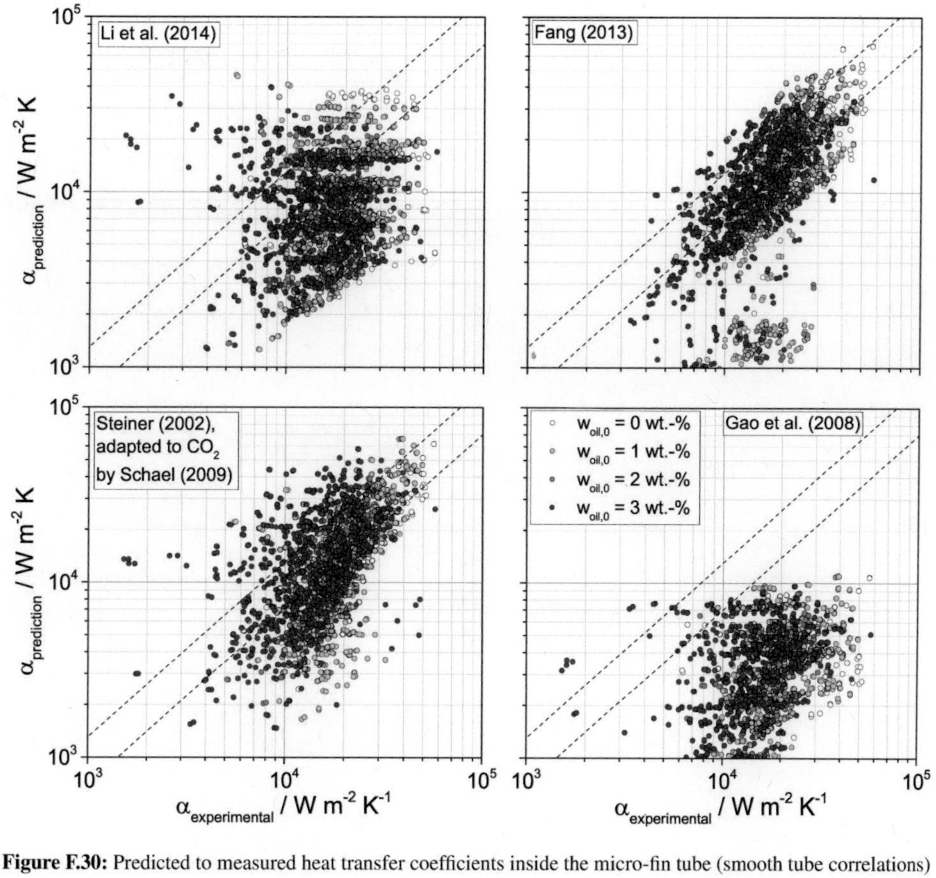

Figure F.30: Predicted to measured heat transfer coefficients inside the micro-fin tube (smooth tube correlations)

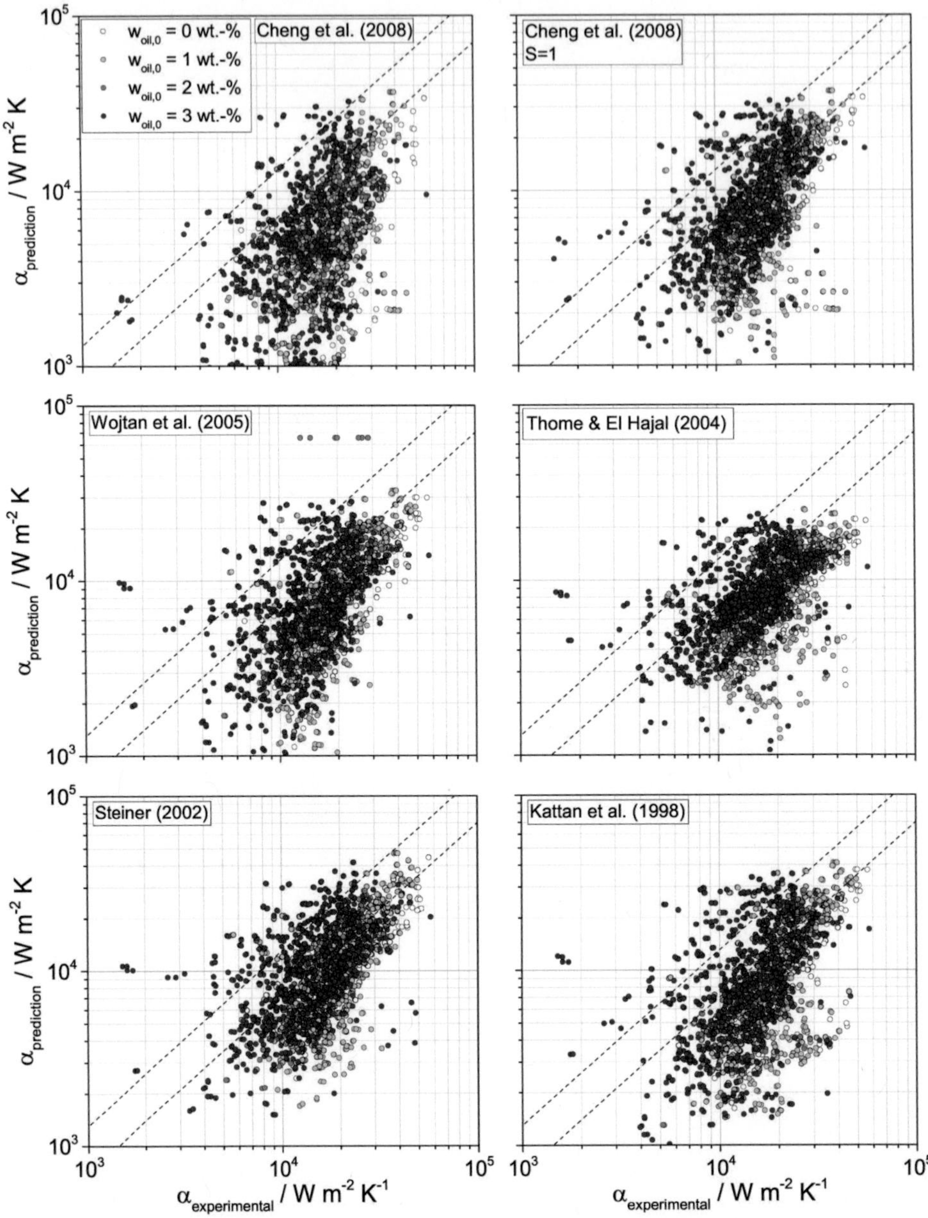

Figure F.31: Predicted to measured heat transfer coefficients inside the micro-fin tube (smooth tube correlations)

Appendix G

G.1 Thermal conductivity measurement data

Table G.1: Measurement data of thermal conductivity by Ihmels (2008b)

w_{oil} [kg kg^{-1}]	$\bar{T}$ [K]	$\bar{p}$ [bar]	λ [W m^{-1} K^{-1}]	w_{oil} [kg kg^{-1}]	$\bar{T}$ [K]	$\bar{p}$ [bar]	λ [W m^{-1} K^{-1}]
0.186	232.85	60.9	0.1698	0.186	232.85	80.8	0.1700
0.186	232.85	100.6	0.1732	0.186	232.85	120.5	0.1744
0.186	232.85	140.5	0.1767	0.186	232.85	160.4	0.1772
0.186	242.55	59.9	0.1604	0.186	242.65	80.7	0.1619
0.186	242.65	100.7	0.1642	0.186	242.65	120.6	0.1651
0.186	242.55	140.5	0.1671	0.186	242.55	160.5	0.1689
0.186	252.65	60.0	0.1506	0.186	252.65	80.8	0.1523
0.186	252.65	100.7	0.1544	0.186	252.65	120.6	0.1550
0.186	252.65	140.6	0.1576	0.186	252.65	160.5	0.1593
0.186	262.75	60.0	0.1397	0.186	262.75	80.8	0.1423
0.186	262.75	100.7	0.1430	0.186	262.75	120.6	0.1460
0.186	262.75	140.6	0.1460	0.186	262.75	160.5	0.1491
0.186	272.65	60.0	0.1301	0.186	272.65	80.8	0.1319
0.186	272.65	100.7	0.1340	0.186	272.65	120.6	0.1362
0.186	272.65	140.6	0.1370	0.186	272.65	160.5	0.1390
0.186	282.65	60.0	0.1213	0.186	282.65	80.8	0.1230
0.186	282.65	100.7	0.1240	0.186	282.65	120.6	0.1258
0.186	282.65	140.6	0.1271	0.186	282.65	160.5	0.1289
0.394	233.05	60.7	0.1615	0.394	233.05	80.5	0.1633
0.394	233.05	100.4	0.1651	0.394	233.05	120.3	0.1673
0.394	233.05	140.3	0.1681	0.394	233.05	160.2	0.1704
0.394	242.95	60.8	0.1556	0.394	242.95	80.5	0.1568
0.394	242.95	100.4	0.1568	0.394	242.95	120.4	0.1590
0.394	242.95	140.3	0.1611	0.394	242.95	160.2	0.1619
0.394	253.05	60.8	0.1474	0.394	253.05	80.5	0.1483
0.394	253.05	100.4	0.1508	0.394	253.05	120.3	0.1512
0.394	253.05	140.3	0.1530	0.394	253.05	160.2	0.1540
0.394	263.05	60.8	0.1389	0.394	263.05	80.5	0.1411
0.394	263.05	100.4	0.1425	0.394	263.05	120.4	0.1438
0.394	263.05	140.3	0.1450	0.394	263.05	160.2	0.1455
0.394	273.15	60.9	0.1328	0.394	273.15	80.5	0.1342
0.394	273.15	100.5	0.1342	0.394	273.15	120.4	0.1358
0.394	273.15	140.3	0.1370	0.394	273.15	160.2	0.1382

continued on next page

w_{oil} [$kg\,kg^{-1}$]	$\bar{T}$ [K]	$\bar{p}$ [bar]	λ [$W\,m^{-1}\,K^{-1}$]	w_{oil} [$kg\,kg^{-1}$]	$\bar{T}$ [K]	$\bar{p}$ [bar]	λ [$W\,m^{-1}\,K^{-1}$]
0.394	283.15	60.9	0.1241	0.394	283.15	80.5	0.1253
0.394	283.15	100.4	0.1276	0.394	283.15	120.4	0.1290
0.394	283.15	140.3	0.1302	0.394	283.15	160.2	0.1315
0.617	232.65	60.3	0.1553	0.617	232.55	80.5	0.1563
0.617	232.55	100.4	0.1575	0.617	232.55	120.4	0.1587
0.617	232.55	140.3	0.1599	0.617	232.55	160.2	0.1606
0.617	242.55	60.8	0.1510	0.617	242.95	80.1	0.1516
0.617	243.05	100.5	0.1528	0.617	243.05	120.4	0.1525
0.617	243.05	140.3	0.1543	0.617	243.05	160.3	0.1559
0.617	253.05	59.9	0.1460	0.617	253.05	80.0	0.1453
0.617	253.05	100.5	0.1463	0.617	253.05	120.4	0.1495
0.617	253.05	140.3	0.1492	0.617	253.05	160.3	0.1501
0.617	263.05	60.0	0.1406	0.617	263.05	80.0	0.1405
0.617	263.05	100.5	0.1422	0.617	263.05	120.4	0.1426
0.617	263.05	140.3	0.1440	0.617	263.05	160.3	0.1449
0.617	273.15	60.0	0.1347	0.617	273.15	80.0	0.1357
0.617	273.15	100.5	0.1368	0.617	273.15	120.4	0.1377
0.617	273.15	140.3	0.1385	0.617	273.15	160.2	0.1400
0.617	283.15	59.9	0.1300	0.617	283.15	79.9	0.1314
0.617	283.15	100.4	0.1326	0.617	283.15	120.4	0.1341
0.617	283.15	140.3	0.1344	0.617	283.15	160.2	0.1352
0.806	233.35	60.1	0.1493	0.806	233.35	79.9	0.1497
0.806	233.35	100.4	0.1505	0.806	233.35	120.4	0.1514
0.806	233.35	140.3	0.1525	0.806	233.35	160.2	0.1528
0.806	243.25	60.0	0.1460	0.806	243.25	79.9	0.1464
0.806	243.25	100.4	0.1475	0.806	243.25	120.4	0.1483
0.806	243.25	140.3	0.1494	0.806	243.25	160.2	0.1496
0.806	253.15	60.1	0.1432	0.806	253.15	79.9	0.1443
0.806	253.25	100.4	0.1447	0.806	253.15	120.4	0.1457
0.806	253.25	140.3	0.1463	0.806	253.15	160.2	0.1469
0.806	263.25	59.9	0.1412	0.806	263.25	79.9	0.1414
0.806	263.25	100.5	0.1431	0.806	263.25	120.4	0.1440
0.806	263.25	140.3	0.1434	0.806	263.25	160.3	0.1447
0.806	273.15	59.9	0.1386	0.806	273.15	79.9	0.1399
0.806	273.15	100.5	0.1392	0.806	273.25	120.4	0.1408
0.806	273.25	140.3	0.1415	0.806	273.25	160.3	0.1420
0.806	283.15	59.9	0.1350	0.806	283.15	80.0	0.1361
0.806	283.15	100.5	0.1358	0.806	283.15	120.4	0.1370
0.806	283.15	140.4	0.1367	0.806	283.15	160.3	0.1386
1.000	232.95	5.1	0.1420	1.000	232.95	40.2	0.1430
1.000	232.95	79.8	0.1437	1.000	232.95	119.4	0.1447
1.000	232.95	159.7	0.1457	1.000	242.95	5.4	0.1419
1.000	242.95	39.5	0.1426	1.000	243.05	80.0	0.1437
1.000	243.05	120.0	0.1442	1.000	242.95	160.1	0.1455
1.000	252.75	5.4	0.1404	1.000	252.85	39.6	0.1420
1.000	252.85	79.1	0.1425	1.000	252.85	119.9	0.1436

continued on next page

w_{oil} [kg kg^{-1}]	$\bar{T}$ [K]	$\bar{p}$ [bar]	λ [W m^{-1} K^{-1}]	w_{oil} [kg kg^{-1}]	$\bar{T}$ [K]	$\bar{p}$ [bar]	λ [W m^{-1} K^{-1}]
1.000	252.75	159.9	0.1447	1.000	262.85	5.4	0.1397
1.000	262.85	39.7	0.1413	1.000	262.85	79.1	0.1427
1.000	262.85	119.9	0.1431	1.000	262.85	159.9	0.1436
1.000	272.65	5.4	0.1392	1.000	272.65	39.7	0.1401
1.000	272.65	79.2	0.1413	1.000	272.65	120.0	0.1416
1.000	272.65	160.0	0.1431	1.000	282.65	5.4	0.1390
1.000	282.65	39.7	0.1393	1.000	282.65	79.2	0.1401
1.000	282.65	120.0	0.1417	1.000	282.75	160.0	0.1419
1.000	292.75	5.4	0.1381	1.000	292.75	39.7	0.1387
1.000	292.75	79.2	0.1403	1.000	292.75	120.0	0.1415
1.000	292.75	160.0	0.1415	1.000	302.75	5.4	0.1374
1.000	302.75	39.7	0.1377	1.000	302.75	79.2	0.1384
1.000	302.75	120.0	0.1396	1.000	302.75	160.0	0.1412

Table G.2: Measurement data of thermal conductivity by Feja and Römer (2008)

Fluid	$P/\Delta T$ [W K^{-1}]	$\bar{T}$ [K]	λ [W m^{-1} K^{-1}]	λ_0 [W m^{-1} K^{-1}]
CO_2	0.5472	273.88	0.1282	
	0.5792	263.92	0.1368	
	0.6328	254.03	0.1513	
	0.6651	244.14	0.1600	
	0.6974	234.32	0.1688	0.4317
Reniso C 85 E	0.5785	234.94	0.1366	
	0.5706	254.45	0.1345	
	0.5633	274.26	0.1325	
	0.5556	294.14	0.1304	0.1611
CO_2-oil mixture	0.5806	234.71	0.1372	
(CO_2 + 89.6 wt.-% Reniso C 85 E)	0.5729	254.39	0.1351	
	0.5676	274.21	0.1337	
	0.5467	294.08	0.1280	

G.2 Oil mass fraction measurement data

Table G.3: Oil mass fraction measurements (smooth tube)

Date	**No**	**sample weight**			**oil fraction**
		vessle	vessle/oil/CO_2	vessle/oil	
14/Jan/2009	1	576.404 g	585.190 g	576.514 g	0.012 g g^{-1}
	2	576.406 g	586.492 g	576.508 g	0.010 g g^{-1}
	3	576.404 g	586.084 g	576.500 g	0.010 g g^{-1}
24/Nov/2008	1	576.472 g	585.772 g	576.680 g	0.023 g g^{-1}

continued on next page

Date	No	sample weight			oil fraction
		vessle	vessle/oil/CO_2	vessle/oil	
	2	576.466 g	587.728 g	576.698 g	0.021 g g^{-1}
	3	576.464 g	589.382 g	576.718 g	0.020 g g^{-1}
	4	576.464 g	587.290 g	576.684 g	0.020 g g^{-1}
26/Nov/2008	5	576.468 g	589.166 g	576.724 g	0.020 g g^{-1}
	6	576.468 g	587.502 g	576.720 g	0.023 g g^{-1}
	7	576.462 g	587.446 g	576.674 g	0.019 g g^{-1}
	8	576.462 g	587.032 g	576.670 g	0.020 g g^{-1}
28/Nov/2008	9	576.466 g	586.486 g	576.674 g	0.021 g g^{-1}
	10	576.464 g	586.034 g	576.662 g	0.021 g g^{-1}
	11	576.462 g	586.236 g	576.668 g	0.021 g g^{-1}
29/Nov/2008	12	576.468 g	589.532 g	576.746 g	0.021 g g^{-1}
	13	576.468 g	589.484 g	576.736 g	0.020 g g^{-1}
08/Dec/2008	1	576.454 g	588.972 g	576.820 g	0.029 g g^{-1}
	2	576.454 g	587.682 g	576.780 g	0.029 g g^{-1}
	3	576.454 g	587.850 g	576.804 g	0.031 g g^{-1}

Table G.4: Oil mass fraction measurements (micro-fin tube)

Date	No	sample weight			oil fraction
		vessle	vessle/oil/CO_2	vessle/oil	
12/Aug/2014	1	719.905 g	731.775 g	720.030 g	0.011 g g^{-1}
	2	719.899 g	731.525 g	720.062 g	0.014 g g^{-1}
	3	719.892 g	732.162 g	720.038 g	0.012 g g^{-1}
	4	719.891 g	732.577 g	720.052 g	0.013 g g^{-1}
	5	719.886 g	733.566 g	720.072 g	0.014 g g^{-1}
	6	719.886 g	735.440 g	720.095 g	0.013 g g^{-1}
	7	719.882 g	728.324 g	719.975 g	0.011 g g^{-1}
	8	719.883 g	728.584 g	719.981 g	0.011 g g^{-1}
	9	719.884 g	728.403 g	719.983 g	0.012 g g^{-1}
	10	719.884 g	727.649 g	719.975 g	0.012 g g^{-1}
13/Aug/2014	11	719.879 g	743.785 g	720.193 g	0.013 g g^{-1}
	12	719.888 g	741.779 g	720.177 g	0.013 g g^{-1}
	13	719.885 g	738.323 g	720.131 g	0.013 g g^{-1}
	14	719.883 g	739.645 g	720.150 g	0.014 g g^{-1}
	15	719.886 g	728.619 g	720.016 g	0.015 g g^{-1}
	16	719.883 g	737.997 g	720.122 g	0.013 g g^{-1}
	17	719.885 g	733.880 g	720.074 g	0.014 g g^{-1}
	18	719.883 g	729.658 g	720.027 g	0.015 g g^{-1}
	19	719.882 g	733.364 g	720.066 g	0.014 g g^{-1}
	11	719.884 g	734.315 g	720.082 g	0.014 g g^{-1}
17/Apr/2014	1	719.918 g	737.225 g	720.282 g	0.021 g g^{-1}
	2	719.887 g	735.920 g	720.246 g	0.022 g g^{-1}
	3	719.888 g	729.430 g	720.103 g	0.023 g g^{-1}
	4	719.887 g	735.800 g	720.237 g	0.022 g g^{-1}

continued on next page

Date	No	sample weight			oil fraction
		vessle	vessle/oil/CO_2	vessle/oil	
	5	719.880 g	735.276 g	720.234 g	0.023 g g^{-1}
	6	719.894 g	733.744 g	720.204 g	0.022 g g^{-1}
	7	719.896 g	729.704 g	720.112 g	0.022 g g^{-1}
22/Apr/2014	8	719.932 g	735.221 g	720.241 g	0.020 g g^{-1}
	9	719.944 g	730.391 g	720.158 g	0.020 g g^{-1}
19/May/2014	10	719.972 g	735.312 g	720.248 g	0.018 g g^{-1}
	11	719.905 g	734.850 g	720.231 g	0.022 g g^{-1}
	12	719.895 g	741.519 g	720.398 g	0.023 g g^{-1}
	13	719.927 g	741.295 g	720.402 g	0.022 g g^{-1}
	14	719.891 g	736.185 g	720.280 g	0.024 g g^{-1}
20/May/2014	15	719.862 g	735.923 g	720.235 g	0.023 g g^{-1}
	16	719.968 g	727.767 g	720.108 g	0.018 g g^{-1}
	17	719.905 g	736.798 g	720.258 g	0.021 g g^{-1}
	18	719.906 g	735.404 g	720.220 g	0.020 g g^{-1}
21/May/2014	19	719.877 g	741.229 g	720.328 g	0.021 g g^{-1}
	20	719.918 g	739.185 g	720.310 g	0.020 g g^{-1}
	21	719.885 g	736.289 g	720.268 g	0.023 g g^{-1}
	22	719.948 g	736.178 g	720.318 g	0.023 g g^{-1}
23/May/2014	23	719.881 g	739.715 g	720.289 g	0.021 g g^{-1}
	24	719.962 g	739.311 g	720.309 g	0.018 g g^{-1}
	25	719.923 g	737.355 g	720.280 g	0.020 g g^{-1}
	26	719.903 g	739.904 g	720.340 g	0.022 g g^{-1}
	27	719.898 g	734.448 g	720.198 g	0.021 g g^{-1}
	28	719.906 g	738.212 g	720.262 g	0.019 g g^{-1}
	29	719.910 g	729.337 g	720.117 g	0.022 g g^{-1}
13/Feb/2014	1	719.905 g	730.382 g	720.243 g	0.032 g g^{-1}
	2	719.934 g	728.811 g	720.200 g	0.030 g g^{-1}
	3	719.913 g	727.750 g	720.157 g	0.031 g g^{-1}
	4	720.157 g	727.066 g	720.400 g	0.035 g g^{-1}
	5	719.935 g	727.153 g	720.148 g	0.030 g g^{-1}
	6	719.931 g	727.832 g	720.163 g	0.029 g g^{-1}
11/Mar/2014	7	719.908 g	728.473 g	720.184 g	0.032 g g^{-1}
	8	719.908 g	743.780 g	720.654 g	0.031 g g^{-1}
	9	719.922 g	731.248 g	720.308 g	0.034 g g^{-1}
25/May/2014	10	719.854 g	729.292 g	720.163 g	0.033 g g^{-1}
	11	719.857 g	729.709 g	720.184 g	0.033 g g^{-1}
	12	719.859 g	730.191 g	720.209 g	0.034 g g^{-1}
	13	719.884 g	729.547 g	720.203 g	0.033 g g^{-1}
	14	719.867 g	731.160 g	720.240 g	0.033 g g^{-1}
	15	719.906 g	729.675 g	720.218 g	0.032 g g^{-1}
	16	719.878 g	731.855 g	720.298 g	0.035 g g^{-1}

G.3 Flow boiling experimental data

The following tables list the measurement data from flow boiling experiments. Listed below, the nomenclature used inside the tables is described.

Label	Description	Unit
$w_{\mathrm{oil},0}$	Nominal oil mass fraction	–
p	System pressure	bar
T	Saturation/bubble point temperature	°C
$\dot{m}$	Total mass flow velocity	$\mathrm{kg\,m^{-2}\,s^{-1}}$
$\dot{x}$	Vapor quality	–
$\dot{q}$	Mean heat flux	$\mathrm{kW\,m^{-2}}$
α	Mean heat transfer coefficient	$\mathrm{kW\,m^{-2}\,K^{-1}}$
$\Delta p/\Delta L$	Pressure drop	$\mathrm{mbar\,m^{-1}}$
FP	Flow pattern	–
$\dot{q}_{\mathrm{j}}$	Local heat flux at segment $j = 1\ldots6$	$\mathrm{kW\,m^{-2}}$
α_{j}	Local heat transfer coefficient at segment $j = 1\ldots6$	$\mathrm{kW\,m^{-2}\,K^{-1}}$

Table G.5: Smooth tube: Heat transfer reproducibility measurements

$w_{oil,0}$	p	T	$\dot{m}$	$\dot{x}$	$\dot{q}$	α	$\Delta p/\Delta L$	FP
0	26.41	-10.1	149.1	0.11	5.4	4	0.22	s/p
0	26.42	-10.09	151.7	0.11	7.8	5	0.17	s/p
0	26.42	-10.08	143.4	0.11	11.7	6	0.33	s/p
0	26.36	-10.16	146.7	0.11	15.1	8	0.39	s/p
0	26.39	-10.13	149.8	0.11	22.4	9	0.58	s/p
0	26.35	-10.18	149.1	0.12	32.2	12	0.75	s/p
0	26.39	-10.13	147.2	0.12	44.5	14	1	s/p
0	26.41	-10.1	147.6	0.12	54.3	16	1.18	s/p
0	26.43	-10.07	149.9	0.12	73.7	19	1.68	s/p
0	26.43	-10.07	141.2	0.33	3.5	3	0.59	s/p
0	26.43	-10.07	151	0.31	3.5	3	0.67	s/p
0	26.44	-10.06	142.2	0.33	6.6	4	0.67	s/p
0	26.45	-10.05	143.4	0.32	10.4	6	0.84	s/p
0	26.39	-10.13	146.6	0.32	13.8	7	1	s/p
0	26.4	-10.11	147.2	0.32	19.6	9	1.18	s/p
0	26.36	-10.17	145.7	0.32	30.2	11	1.58	s/p
0	26.39	-10.12	147.6	0.32	40.6	13	1.93	s/p
0	26.36	-10.16	149	0.32	52.7	15	2.27	s/p
0	26.4	-10.11	148.5	0.32	70.7	18	2.61	s/p
0	26.43	-10.08	152.4	0.5	2.5	2	1.76	a
0	26.43	-10.07	146.5	0.53	4.8	3	1.84	m
0	26.38	-10.14	149.3	0.52	8.8	5	1.93	m
0	26.39	-10.13	149.8	0.51	11.5	6	2.1	m
0	26.4	-10.11	149.4	0.51	17.7	7	2.25	m
0	26.42	-10.08	151.8	0.51	25.2	9	2.51	a
0	26.38	-10.14	148.1	0.52	36.1	12	2.84	m
0	26.4	-10.11	150.1	0.5	47.5	14	3.24	a
0	26.37	-10.15	148.6	0.52	64.2	16	3.7	m
0	26.4	-10.12	148.3	0.72	1.8	2	2.27	m
0	26.4	-10.11	149.4	0.71	4.2	3	2.43	m
0	26.41	-10.09	146.6	0.73	7.9	5	2.5	m
0	26.42	-10.08	147.7	0.72	10.5	5	2.6	m
0	26.43	-10.07	148.4	0.72	16	7	2.85	m
0	26.38	-10.13	142	0.75	22.5	8	3.02	m
0	26.39	-10.12	148.7	0.71	31.9	10	3.37	m
0	26.38	-10.14	148.3	0.73	41.4	12	3.71	m
0	26.41	-10.1	148.3	0.72	58.3	14	4.21	m
0	26.41	-10.09	148.9	0.92	1.7	3	2.77	m
0	26.42	-10.08	148.6	0.92	3.5	3	2.77	m
0	26.44	-10.07	150.5	0.91	6.5	4	2.99	m
0	26.43	-10.08	146.3	0.93	8.3	4	2.94	s/p
0	26.43	-10.07	150.6	0.91	8.6	5	3.04	m
0	26.39	-10.13	150.1	0.91	12.9	6	3.2	m
0	26.4	-10.11	149.5	0.92	19.8	7	3.47	m
0	26.44	-10.06	148.4	0.93	27.4	8	3.71	s/p

$\dot{q}_1$	α_1	$\dot{q}_2$	α_2	$\dot{q}_3$	α_3	$\dot{q}_4$	α_4	$\dot{q}_5$	α_5	$\dot{q}_6$	α_6
0.3	0.22	4.51	3.77	10.45	8.02	2.11	1.65	10.53	8.01	4.28	3.57
0.3	0.17	5.44	3.51	15.36	10.18	5.05	3.35	15.52	10.33	5.02	3.22
0.3	0.15	6.09	3.33	23.7	13.14	9.69	5.36	24.3	13.5	6.04	3.3
0.3	0.13	7.74	3.85	29.99	14.91	13.94	6.88	31.06	15.38	7.7	3.84
0.3	0.11	11.36	4.72	43.1	18.26	23.89	9.97	44.26	18.69	11.73	4.88
0.3	0.1	14.87	5.35	62.67	22.85	37.95	13.83	62.24	22.62	15.33	5.52
0.3	0.09	18.13	5.88	87.94	27.75	56.13	17.87	84.32	26.62	20.06	6.53
0.3	0.08	22.45	6.52	105.44	30.77	74.57	21.51	95.71	28.39	27.16	7.95
0.4	0.08	27.93	7.54	145.57	37.11	111.06	27.72	129.98	33.71	27.37	7.38
0.3	0.24	1.91	1.57	8.05	6.56	2.21	1.8	8	6.46	0.65	0.53
0.3	0.26	1.92	1.71	8.03	6.65	2.23	1.86	7.98	6.63	0.65	0.58
0.3	0.2	2.87	1.99	14.99	9.98	4.91	3.29	14.89	9.9	1.73	1.19
0.3	0.17	2.71	1.59	22.78	12.83	9.72	5.49	23.41	13.13	3.36	1.98
0.3	0.15	3.84	1.91	29.4	14.76	14.31	7.17	29.99	15.01	4.94	2.46
0.3	0.13	4.97	2.19	40.7	17.75	23.9	10.3	40.75	17.73	7.06	3.14
0.3	0.11	6.35	2.26	63.84	23.15	37.81	13.96	62.09	22.49	10.61	3.82
0.3	0.11	6.82	2.31	85.72	27.18	56.69	18.09	81.45	25.85	12.85	4.42
0.3	0.09	9.37	2.71	109.28	31.21	75.8	21.69	104.15	29.43	17.23	5.07
0.4	0.08	11.81	2.94	145.17	36.78	111.74	27.76	136.24	34	18.82	4.74
0.3	0.28	0.64	0.61	5.78	5.29	2.25	2.13	5.47	4.93	0.59	0.56
0.3	0.21	0.63	0.46	11.46	7.89	4.99	3.42	10.96	7.42	0.59	0.42
0.3	0.17	0.93	0.54	21.09	11.6	9.65	5.32	20.19	11.04	0.68	0.4
0.3	0.16	1.56	0.8	26.55	13.19	14.16	7.01	25.62	12.69	0.69	0.35
0.3	0.12	3.62	1.43	39.39	16.65	24.02	10.13	37.41	15.69	1.57	0.62
0.3	0.12	4.31	1.67	55.04	20.4	37.45	13.81	51.18	18.88	2.74	1.06
0.3	0.1	5.22	1.71	76.76	24.53	56.56	17.99	73.26	23.34	4.62	1.51
0.3	0.09	6.74	1.96	100.89	28.78	75.69	21.48	94.22	26.88	7.22	2.1
0.3	0.08	8.69	2.07	133.21	33.57	108.54	26.86	125.57	31.24	8.99	2.14
0.3	0.39	0.57	0.72	3.84	4.74	2.15	2.68	3.55	4.49	0.29	0.37
0.3	0.23	0.52	0.4	10.15	7.36	5.11	3.71	8.64	6.46	0.29	0.22
0.3	0.17	0.65	0.37	18.99	10.88	10.12	5.82	17.31	9.96	0.29	0.16
1.4	0.61	0.33	0.16	24.08	12.31	14.26	7.25	22.43	11.4	0.29	0.14
1.3	0.53	0.36	0.15	36.7	15.96	23.65	10.3	33.91	14.73	0.29	0.12
1.3	0.45	0.3	0.1	50.28	18.7	37.8	14.02	44.9	16.66	0.3	0.1
1.4	0.44	0.85	0.3	70.8	22.96	56.11	18.06	61.74	19.87	0.31	0.11
1.4	0.39	0.94	0.28	94.13	26.78	77.38	22.04	74.24	21.44	0.32	0.09
1.3	0.33	1.15	0.28	125.77	31.1	112.48	27.54	108.56	26.56	0.49	0.12
0.3	0.48	0.3	0.48	3.68	5.45	2.19	3.22	3.55	5.12	0.3	0.48
0.3	0.25	0.29	0.24	8.05	6.32	4.95	3.89	7.33	5.82	0.29	0.24
0.3	0.18	0.29	0.17	14.64	8.71	9.62	5.75	13.76	8.17	0.29	0.17
0.3	0.15	0.29	0.14	19.15	9.92	14.1	7.35	15.96	8.42	0.29	0.14
0.3	0.16	0.29	0.15	19.09	10.18	14.15	7.47	17.59	9.34	0.29	0.15
0.3	0.13	0.29	0.12	27.77	12.38	23.95	10.56	24.89	11.07	0.29	0.12
0.3	0.1	0.29	0.09	43.03	15.43	38	14.04	36.9	13.21	0.29	0.09
0.3	0.08	0.29	0.07	57.77	17.5	57.15	18.28	48.41	14.41	0.29	0.07

Table G.5

$w_{oil,0}$	p	T	$\dot{m}$	$\dot{x}$	$\dot{q}$	α	$\Delta p/\Delta L$	FP
0	26.44	-10.06	151.8	0.9	32.8	10	4.04	m
0	26.41	-10.1	149.5	0.92	42.6	11	4.3	m
0	26.38	-10.13	298.1	0.31	2.5	4	3.54	a
0	26.39	-10.12	298.5	0.31	6.5	5	3.88	a
0	26.41	-10.1	298.9	0.31	11.9	7	4.24	a
0	26.42	-10.09	301.2	0.3	15.8	8	4.54	a
0	26.38	-10.14	294.7	0.31	25.1	11	5.01	a
0	26.4	-10.11	297.2	0.31	34.6	13	5.56	a
0	26.37	-10.15	300.7	0.3	48.3	16	6.27	a
0	26.41	-10.1	304.2	0.3	61.2	17	6.98	a
0	26.38	-10.14	297.4	0.31	80.9	20	7.95	a
0	26.38	-10.14	297.3	0.51	2.8	5	5.66	m
0	26.39	-10.13	297.5	0.51	6.6	5	5.98	m
0	26.4	-10.11	301.4	0.5	11.6	7	6.41	a
0	26.35	-10.18	301.5	0.5	16.6	9	6.79	a
0	26.38	-10.13	303.8	0.5	23.9	11	7.36	a
0	26.39	-10.12	300.6	0.51	33.6	13	7.97	a
0	26.38	-10.15	296.8	0.51	48.2	15	8.7	m
0	26.4	-10.11	297.8	0.51	58.3	17	9.38	m
0	26.4	-10.11	296.9	0.51	78.6	19	10.28	m
0	26.38	-10.14	296.5	0.71	3.3	8	8.43	m
0	26.4	-10.11	298.6	0.71	6.8	7	8.32	m
0	26.42	-10.09	300.3	0.71	12.7	8	8.65	m
0	26.42	-10.09	298.3	0.71	16.7	9	8.83	m
0	26.39	-10.13	297.7	0.71	22.5	11	9.33	m
0	26.41	-10.1	299.4	0.71	32.9	13	9.95	m
0	26.39	-10.13	298.1	0.71	45.3	15	10.74	m
0	26.41	-10.11	298.3	0.72	54.9	16	11.25	m
0	26.42	-10.09	303.7	0.7	75.5	19	12.36	m
0	26.41	-10.1	301.7	0.1	5.8	5	1.99	a
0	26.37	-10.16	297.8	0.11	11.2	7	2.29	a
0	26.39	-10.13	298.2	0.11	16.3	9	2.62	a
0	26.41	-10.11	300.3	0.11	21.8	11	2.97	a
0	26.42	-10.08	300.7	0.11	29.8	13	3.25	a
0	26.44	-10.06	296.4	0.11	41	15	3.55	a
0	26.37	-10.16	299.1	0.11	54.9	18	4.15	a
0	26.38	-10.14	297.8	0.11	66.8	19	4.73	a
0	26.41	-10.1	301.5	0.11	94.7	23	5.84	a
0	26.42	-10.09	299.8	0.91	5.9	8	11.62	m
0	26.42	-10.09	302.4	0.9	13.9	11	11.95	m
0	26.4	-10.11	302.3	0.91	18.6	11	11.99	m
0	26.4	-10.12	299.9	0.91	26.1	12	12.09	m
0	26.4	-10.11	296.8	0.92	34.2	13	11.97	m
0	26.4	-10.12	299.3	0.91	42.8	14	12.46	m
0	26.41	-10.1	301.6	0.9	54.2	16	13.32	m
0	26.43	-10.07	298.9	0.91	65.4	16	13.52	m

Table G.5

$\dot{q}_1$	α_1	$\dot{q}_2$	α_2	$\dot{q}_3$	α_3	$\dot{q}_4$	α_4	$\dot{q}_5$	α_5	$\dot{q}_6$	α_6
0.3	0.08	0.3	0.07	67.37	20.12	73	21.42	55.84	16.72	0.3	0.07
0.3	0.06	0.28	0.06	79.56	20.74	110.63	27.27	64.49	16.68	0.28	0.06
0.5	0.87	2.99	5.26	3.29	4.85	2.15	3.3	3.04	4.45	3.04	5.35
0.8	0.67	7.42	5.95	9.46	7.36	4.96	3.88	9.25	7.11	7.16	5.73
1.6	0.88	12.34	6.79	17.69	10.65	9.61	5.77	17.86	10.61	12.13	6.67
1.5	0.84	15.03	8.13	24.68	13.1	14.2	7.51	24.75	13.02	14.44	7.8
1.6	0.74	21.04	9.54	41.96	18.07	23.77	10.45	41.11	17.58	21.33	9.68
3.3	1.1	26.78	9.78	56.6	21.38	37.65	14.2	55.95	20.79	27.08	9.89
4.4	1.75	32.05	11.58	80.8	26.2	56.78	18.34	81.65	25.57	34.17	12.41
0.3	0.09	44.13	12.18	109.2	30.97	75.43	21.87	94.62	27.52	43.35	12.04
0.3	0.08	50.62	12.73	134.38	34.67	112.9	27.75	135.24	33.25	51.7	13.02
0.5	0.92	4.18	7.57	3.12	5.57	2.15	4.06	2.72	4.93	3.95	7.14
0.5	0.45	8.76	7.26	8.87	7.37	4.89	4.16	8.97	7.21	7.7	6.34
1.1	0.7	12.47	7.93	17.17	10.73	9.56	5.92	17.58	10.72	12	7.62
1.5	0.83	16.46	9.08	25.13	13.37	14.47	7.67	25.05	13.16	17.3	9.57
1.8	0.84	19.48	9.23	40.4	17.53	23.78	10.47	38.44	16.74	19.65	9.31
3.5	1.24	25.88	9.98	55.37	20.86	37.94	14.08	54.29	20.14	24.74	9.52
5	1.51	31.61	10.04	82.44	26.23	56.36	17.96	81.08	25.08	32.72	10.41
7	1.75	38.98	10.22	96.96	28.59	75.59	21.64	93.51	26.92	38	9.95
3.4	0.94	46.75	10.81	133.66	34.24	113.05	27.54	134.78	32.71	40.28	9.22
5.2	12.59	3.65	9.4	2.56	6.58	2.11	5.31	2.37	5.92	3.69	9.51
3.7	3.88	9.5	9.47	7.04	7.08	5.07	5.13	6.3	6.32	9.25	9.21
1.9	1.32	16.62	11.04	17.03	11.19	9.64	6.42	15.41	10.2	15.37	10.15
1.2	0.61	19.84	11.08	23.08	13.13	14	7.9	22.95	12.81	19.42	10.83
0.3	0.22	20.27	11.19	36.29	16.88	23.42	10.77	34.56	16.03	19.84	10.94
0.8	0.3	27.63	10.61	55.02	21.33	36.76	14.26	51.42	20	25.98	9.94
2.7	0.85	27.52	9.97	81.67	26.39	57.32	18.53	75.39	24.33	27.48	9.96
2.6	0.69	31.01	8.84	100.61	29.22	75.33	21.74	91.11	26.52	28.61	8.12
0.3	0.1	32.74	9.03	147.18	36.18	112.83	27.63	129.88	31.91	29.78	8.17
5.1	4.88	8.47	7.98	5	4.47	2.19	2.02	5	4.36	8.94	8.44
9.9	5.77	15.55	9.98	10.77	7.23	4.98	3.34	10.78	7.13	15.17	9.72
9.9	5.94	22.78	12.47	16.67	9.72	9.93	5.63	18.19	10.28	20.53	11.16
9.7	5.47	27.82	14.23	26.68	13.28	14.15	7.13	27.29	13.42	25.1	12.72
11.9	4.71	34	15.01	37.27	16.39	23.48	10.3	36.9	16.16	35	15.49
8.2	2.46	45.81	15.85	55.77	20.91	37.57	14.08	53.56	20.08	45.24	15.63
0.3	0.12	56.5	17.91	77.97	25.2	56.6	18.1	76.05	24.26	61.82	19.82
0.3	0.09	58.53	17.6	101.74	29.05	75.64	21.49	95.94	27.33	68.47	20.97
0.4	0.07	82.65	20.08	149.86	36.95	113.7	27.88	134.74	33.26	86.74	21.22
7.3	9.91	6.64	9.09	5.41	7.33	4.99	6.67	5.13	6.85	6.12	8.35
12	9.6	17.12	13.2	14.38	10.83	9.57	7.26	13.31	9.98	17.05	13.14
11.5	7.57	22.81	13.94	22.32	13.41	14.32	8.59	20.53	12.47	20.28	12.27
10.7	3.41	27.93	11.31	33.09	16.1	23.45	10.96	32.97	15.8	28.66	11.63
4.5	1.71	28.05	10.54	54.85	21.02	37.86	14.55	52.77	20.04	27.14	10.17
2.2	0.86	27.59	9.97	75.94	25.1	56.27	18.53	70.37	23.47	24.64	8.84
1.6	0.52	30.58	8.93	99.75	29.09	74.82	21.8	90.59	26.52	27.88	8.1
0.3	0.09	25.03	6.35	122.83	29.34	111.89	26.3	110.32	26.25	22.12	5.59

Table G.5

Table G.6: Smooth tube: Heat transfer measurements

$w_{oil,0}$	p	T	$\dot{m}$	$\dot{x}$	$\Delta p/\Delta L$	FP	$\dot{q}$	α
0	26.43	-10.07	149.9	0.12	1.68	w	73.6	18.5
0	26.41	-10.1	147.6	0.12	1.18	w	54.1	15
0	26.39	-10.13	147.2	0.12	1	w	44.3	13.7
0	26.35	-10.18	149.1	0.12	0.75	w	32.1	11.2
0	26.39	-10.13	149.8	0.11	0.58	w	22.3	9.2
0	26.36	-10.16	146.7	0.11	0.39	w	15	7.3
0	26.42	-10.08	143.4	0.11	0.33	w	11.6	6.2
0	26.42	-10.09	151.7	0.11	0.17	w	7.7	4.8
0	26.41	-10.1	149.1	0.11	0.22	w	5.3	4.1
0	27.97	-10.12	148.1	0.11	0.02	w	0.3	
0	26.4	-10.11	148.5	0.32	2.61	w	70.5	17.6
0	26.36	-10.16	149	0.32	2.27	w	52.6	14.7
0	26.39	-10.12	147.6	0.32	1.93	w	40.5	13
0	26.36	-10.17	145.7	0.32	1.58	w	30.1	10.9
0	26.4	-10.11	147.2	0.32	1.18	w	19.5	8.4
0	26.39	-10.13	146.6	0.32	1	w	13.7	6.7
0	26.45	-10.05	143.4	0.32	0.84	w	10.3	5.9
0	26.44	-10.06	142.2	0.33	0.67	w	6.5	4.5
0	26.43	-10.07	141.2	0.33	0.59	w	3.4	2.7
0	26.43	-10.07	151	0.31	0.67	w	3.4	2.9
0	27.99	-10.09	148.9	0.31	0.5	w	0.3	
0	26.37	-10.15	148.6	0.52	3.7	w	64.1	15.8
0	26.4	-10.11	150.1	0.5	3.24	w	47.4	13.5
0	26.38	-10.14	148.1	0.52	2.84	w	36	11.5
0	26.42	-10.08	151.8	0.51	2.51	w	25.1	9.3
0	26.4	-10.11	149.4	0.51	2.25	w	17.6	7.3
0	26.39	-10.13	149.8	0.51	2.1	w	11.4	5.9
0	26.38	-10.14	149.3	0.52	1.93	w	8.7	5
0	26.43	-10.07	146.5	0.53	1.84	w	4.7	3.4
0	26.43	-10.08	152.4	0.5	1.76	w	2.4	2.2
0	28.03	-10.08	151.6	0.51	1.71	w	0.3	
0	26.41	-10.1	148.3	0.72	4.21	w	58.2	14.2
0	26.38	-10.14	148.3	0.73	3.71	w	41.3	12
0	26.39	-10.12	148.7	0.71	3.37	w	31.8	10.5
0	26.38	-10.13	142	0.75	3.02	w	22.4	7.9
0	26.43	-10.07	148.4	0.72	2.85	w	16	6.7
0	26.42	-10.08	147.7	0.72	2.6	w	10.4	5
0	26.41	-10.09	146.6	0.73	2.5	w	7.9	4.6
0	26.4	-10.11	149.4	0.71	2.43	w	4.1	3.1
0	26.4	-10.12	148.3	0.72	2.27	w	1.7	2.1
0	28	-10.13	150.4	0.71	2.27	w	0.3	
0	26.41	-10.1	149.5	0.92	4.3	w	42.5	9.9
0	26.44	-10.06	151.8	0.9	4.04	w	32.8	9
0	26.44	-10.06	148.4	0.93	3.71	w	27.3	7.4
0	26.4	-10.11	149.5	0.92	3.47	w	19.7	6.8

$\dot{q}_1$	α_1	$\dot{q}_2$	α_2	$\dot{q}_3$	α_3	$\dot{q}_4$	α_4	$\dot{q}_5$	α_5	$\dot{q}_6$	α_6
0.2	0.1	27.8	7	145.4	36.6	110.9	27.9	129.8	32.7	27.2	6.9
0.2	0.1	22.3	6.2	105.3	29.2	74.4	20.7	95.6	26.5	27	7.5
0.2	0.1	18	5.6	87.8	27.2	56	17.3	84.2	26.1	19.9	6.2
0.2	0.1	14.7	5.1	62.5	21.7	37.9	13.1	62.1	21.6	15.2	5.3
0.2	0.1	11.3	4.6	43	17.7	23.8	9.8	44.1	18.2	11.6	4.8
0.2	0.1	7.6	3.7	29.9	14.5	13.8	6.7	31	15	7.6	3.7
0.2	0.1	6	3.2	23.6	12.5	9.6	5.1	24.2	12.9	5.9	3.2
0.2	0.1	5.3	3.3	15.3	9.6	5	3.1	15.4	9.7	4.9	3.1
0.2	0.2	4.4	3.4	10.4	8	2	1.6	10.4	8	4.2	3.2
0.2	0.1	11.7	2.9	145	36.1	111.6	27.8	136.1	33.9	18.7	4.7
0.2	0.1	9.2	2.6	109.1	30.6	75.6	21.2	104	29.2	17.1	4.8
0.2	0.1	6.7	2.1	85.6	27.4	56.5	18.1	81.3	26	12.7	4.1
0.2	0.1	6.2	2.2	63.7	23	37.7	13.6	62	22.4	10.5	3.8
0.2	0.1	4.8	2.1	40.6	17.5	23.8	10.3	40.7	17.6	7	3
0.2	0.1	3.8	1.8	29.3	14.4	14.2	7	29.9	14.7	4.8	2.4
0.2	0.1	2.6	1.5	22.7	13	9.6	5.5	23.3	13.3	3.3	1.9
0.2	0.1	2.8	1.9	14.9	10.2	4.8	3.3	14.8	10.1	1.6	1.1
0.2	0.2	1.8	1.4	8	6.3	2.1	1.7	7.9	6.2	0.5	0.4
0.2	0.2	1.8	1.6	7.9	6.8	2.1	1.8	7.9	6.8	0.5	0.5
0.2	0.1	8.6	2.1	133.1	32.9	108.4	26.8	125.4	31	8.9	2.2
0.2	0.1	6.6	1.9	100.7	28.7	75.6	21.5	94.1	26.8	7.1	2
0.2	0.1	5.1	1.6	76.7	24.4	56.5	18	73.2	23.3	4.5	1.4
0.2	0.1	4.2	1.6	54.9	20.5	37.4	13.9	51.1	19	2.7	1
0.2	0.1	3.5	1.5	39.3	16.3	23.9	9.9	37.3	15.5	1.5	0.6
0.2	0.1	1.4	0.7	26.5	13.7	14.1	7.3	25.5	13.2	0.6	0.3
0.2	0.1	0.8	0.5	21	12.1	9.5	5.5	20.1	11.6	0.6	0.4
0.2	0.1	0.5	0.4	11.4	8.3	4.9	3.6	10.8	7.9	0.5	0.3
0.2	0.2	0.5	0.5	5.7	5.3	2.2	2	5.4	5	0.5	0.4
1.2	0.3	1	0.2	125.6	30.7	112.3	27.4	108.5	26.5	0.3	0.1
1.2	0.4	0.8	0.2	94	27.3	77.3	22.5	74.1	21.5	0.2	0.1
1.2	0.4	0.8	0.2	70.7	23.4	56	18.5	61.7	20.4	0.2	0.1
1.2	0.4	0.2	0.1	50.2	17.7	37.7	13.3	44.8	15.8	0.2	0.1
1.2	0.5	0.3	0.1	36.6	15.4	23.6	10	33.8	14.3	0.2	0.1
1.3	0.6	0.3	0.1	24	11.5	14.2	6.8	22.4	10.7	0.2	0.1
0.2	0.1	0.5	0.3	18.9	11.1	10	5.9	17.3	10.2	0.2	0.1
0.2	0.2	0.4	0.3	10.1	7.7	5	3.9	8.5	6.5	0.2	0.2
0.2	0.3	0.5	0.6	3.8	4.6	2	2.5	3.5	4.3	0.2	0.3
0.2	0	0.2	0	79.5	18.5	110.6	25.7	64.4	15	0.2	0
0.2	0.1	0.2	0.1	67.3	18.4	72.9	19.9	55.7	15.2	0.2	0.1
0.2	0.1	0.2	0.1	57.7	15.8	57.1	15.6	48.3	13.2	0.2	0.1
0.2	0.1	0.2	0.1	43	14.8	37.9	13	36.8	12.7	0.2	0.1

Table G.6

$w_{oil,0}$	p	T	$\dot{m}$	$\dot{x}$	$\Delta p/\Delta L$	FP	$\dot{q}$	α
0	26.39	-10.13	150.1	0.91	3.2	w	12.8	5.5
0	26.43	-10.08	146.3	0.93	2.94	w	8.3	4.2
0	26.43	-10.07	150.6	0.91	3.04	w	8.5	4.6
0	26.44	-10.07	150.5	0.91	2.99	w	6.4	3.8
0	26.42	-10.08	148.6	0.92	2.77	w	3.4	2.9
0	26.41	-10.09	148.9	0.92	2.77	w	1.6	2.4
0	28.02	-10.1	148.7	0.92	2.64	w	0.3	
0	26.41	-10.1	301.5	0.11	5.84	i	94.5	22
0	26.38	-10.14	297.8	0.11	4.73	i	66.7	19.4
0	26.37	-10.16	299.1	0.11	4.15	i	54.7	17.9
0	26.44	-10.06	296.4	0.11	3.55	i	40.9	14.3
0	26.42	-10.08	300.7	0.11	3.25	i	29.7	13
0	26.41	-10.11	300.3	0.11	2.97	i	21.7	11.4
0	26.39	-10.13	298.2	0.11	2.62	i	16.2	9.4
0	26.37	-10.16	297.8	0.11	2.29	i	11.1	7.1
0	26.41	-10.1	301.7	0.1	1.99	i	5.7	5.2
0	27.98	-10.11	301.1	0.1	1.51	i	0.3	
0	26.38	-10.14	297.4	0.31	7.95	a	80.7	20
0	26.41	-10.1	304.2	0.3	6.98	a	61.1	17.4
0	26.37	-10.15	300.7	0.3	6.27	a	48.2	16.9
0	26.4	-10.11	297.2	0.31	5.56	a	34.5	12.7
0	26.38	-10.14	294.7	0.31	5.01	a	25	11.2
0	26.42	-10.09	301.2	0.3	4.54	a	15.7	8.3
0	26.41	-10.1	298.9	0.31	4.24	a	11.8	6.9
0	26.39	-10.12	298.5	0.31	3.88	a	6.4	5.2
0	26.38	-10.13	298.1	0.31	3.54	a	2.4	3.8
0	27.9	-10.15	299.2	0.31	3.47	a	0.3	
0	26.4	-10.11	296.9	0.51	10.28	a	78.5	19.1
0	26.4	-10.11	297.8	0.51	9.38	a	58.2	16.1
0	26.38	-10.15	296.8	0.51	8.7	a	48.1	15.3
0	26.39	-10.12	300.6	0.51	7.97	a	33.5	12.3
0	26.38	-10.13	303.8	0.5	7.36	a	23.8	10.7
0	26.35	-10.18	301.5	0.5	6.79	a	16.6	8.8
0	26.4	-10.11	301.4	0.5	6.41	a	11.6	7.2
0	26.39	-10.13	297.5	0.51	5.98	a	6.5	5.3
0	26.38	-10.14	297.3	0.51	5.66	a	2.7	5
0	27.9	-10.15	300.8	0.51	5.73	a	0.3	
0	26.42	-10.09	303.7	0.7	12.36	a	75.3	19.9
0	26.41	-10.11	298.3	0.72	11.25	a	54.8	15.6
0	26.39	-10.13	298.1	0.71	10.74	a	45.2	14.9
0	26.41	-10.1	299.4	0.71	9.95	a	32.8	12.6
0	26.39	-10.13	297.7	0.71	9.33	a	22.4	11.6
0	26.42	-10.09	298.3	0.71	8.83	a	16.7	9.3
0	26.42	-10.09	300.3	0.71	8.65	a	12.6	8.5
0	26.4	-10.11	298.6	0.71	8.32	a	6.7	6.6
0	26.38	-10.14	296.5	0.71	8.43	a	3.2	7.3
0	27.9	-10.13	298.8	0.71	8.51	a	0.3	

Table G.6

$\dot{q}_1$	α_1	$\dot{q}_2$	α_2	$\dot{q}_3$	α_3	$\dot{q}_4$	α_4	$\dot{q}_5$	α_5	$\dot{q}_6$	α_6
0.2	0.1	0.2	0.1	27.7	11.9	23.9	10.3	24.8	10.7	0.2	0.1
0.2	0.1	0.2	0.1	19	9.6	14	7.1	15.9	8	0.2	0.1
0.2	0.1	0.2	0.1	19	10.2	14.1	7.5	17.5	9.4	0.2	0.1
0.2	0.1	0.2	0.1	14.5	8.7	9.5	5.7	13.7	8.2	0.2	0.1
0.2	0.2	0.2	0.2	8	6.8	4.8	4.1	7.2	6.1	0.2	0.2
0.2	0.3	0.2	0.3	3.6	5.2	2.1	3	3.5	5	0.2	0.3
0.2	0	82.5	19.2	149.7	34.8	113.6	26.4	134.6	31.3	86.6	20.1
0.2	0.1	58.4	17	101.6	29.5	75.5	21.9	95.8	27.8	68.3	19.9
0.2	0.1	56.3	18.4	77.8	25.5	56.5	18.5	75.9	24.8	61.7	20.2
8.1	2.8	45.7	16	55.7	19.5	37.4	13.1	53.5	18.7	45.2	15.8
11.8	5.2	33.9	14.9	37.2	16.3	23.4	10.3	36.8	16.1	34.9	15.3
9.6	5	27.7	14.5	26.6	14	14.1	7.4	27.2	14.3	25	13.1
9.8	5.6	22.7	13.1	16.6	9.6	9.8	5.7	18.1	10.4	20.5	11.8
9.8	6.3	15.5	9.9	10.7	6.9	4.9	3.1	10.7	6.9	15.1	9.7
5	4.5	8.4	7.6	4.9	4.5	2.1	1.9	4.9	4.5	8.9	8.1
0.2	0.1	50.5	12.5	134.2	33.2	112.7	27.9	135.1	33.5	51.6	12.8
0.2	0.1	44	12.6	109.1	31.2	75.3	21.5	94.5	27	43.2	12.4
4.2	1.5	31.9	11.2	80.7	28.3	56.7	19.9	81.5	28.6	34	11.9
3.1	1.2	26.7	9.8	56.5	20.8	37.6	13.9	55.9	20.6	27	10
1.5	0.7	20.9	9.3	41.9	18.7	23.7	10.6	41	18.3	21.2	9.5
1.4	0.8	14.9	7.9	24.6	13	14.1	7.5	24.7	13.1	14.4	7.6
1.6	0.9	12.3	7.2	17.6	10.3	9.5	5.6	17.8	10.5	12.1	7.1
0.8	0.6	7.4	6	9.3	7.6	4.8	4	9.1	7.5	7.1	5.8
0.4	0.6	2.9	4.5	3.2	5	2	3.2	2.9	4.6	2.9	4.6
3.3	0.8	46.7	11.3	133.6	32.5	113	27.5	134.6	32.7	40.2	9.8
6.8	1.9	38.9	10.8	96.9	26.8	75.5	20.9	93.4	25.8	37.9	10.5
4.8	1.5	31.5	10	82.3	26.2	56.3	17.9	81	25.7	32.6	10.4
3.4	1.3	25.8	9.5	55.2	20.3	37.9	13.9	54.2	19.9	24.6	9
1.8	0.8	19.4	8.7	40.3	18	23.7	10.6	38.3	17.2	19.6	8.8
1.4	0.8	16.4	8.7	25	13.3	14.4	7.7	25	13.3	17.2	9.2
1	0.6	12.4	7.7	17.1	10.6	9.5	5.9	17.5	10.9	11.9	7.4
0.5	0.4	8.7	7	8.8	7.1	4.8	3.9	8.9	7.2	7.6	6.2
0.5	0.9	4.1	7.5	3.1	5.7	2	3.8	2.7	4.9	3.9	7.2
0.2	0.1	32.6	8.6	147.1	38.8	112.7	29.8	129.7	34.3	29.7	7.8
2.5	0.7	30.9	8.8	100.5	28.7	75.2	21.5	91	26	28.5	8.1
2.5	0.8	27.4	9	81.6	26.9	57.2	18.9	75.3	24.8	27.4	9
0.7	0.3	27.6	10.6	54.9	21.1	36.6	14.1	51.3	19.7	25.9	9.9
0.3	0.1	20.2	10.5	36.2	18.8	23.3	12.1	34.4	17.8	19.7	10.2
1.1	0.6	19.8	11.1	23	12.9	13.9	7.8	22.9	12.8	19.3	10.8
1.8	1.2	16.5	11.1	16.9	11.4	9.5	6.4	15.3	10.3	15.3	10.3
3.5	3.5	9.4	9.3	7	6.9	5	4.9	6.2	6.1	9.1	9
5.2	11.9	3.5	8.1	2.5	5.6	2	4.7	2.3	5.3	3.6	8.3

Table G.6

$w_{oil,0}$	p	T	$\dot{m}$	$\dot{x}$	$\Delta p/\Delta L$	FP	$\dot{q}$	α
0	26.43	-10.07	298.9	0.91	13.52	a	65.3	16
0	26.41	-10.1	301.6	0.9	13.32	a	54.1	15.9
0	26.4	-10.12	299.3	0.91	12.46	a	42.7	15.2
0	26.4	-10.11	296.8	0.92	11.97	a	34.1	13
0	26.4	-10.12	299.9	0.91	12.09	a	26.1	10.8
0	26.4	-10.11	302.3	0.91	11.99	a	18.5	11.5
0	26.42	-10.09	302.4	0.9	11.95	a	13.8	10.8
0	26.42	-10.09	299.8	0.91	11.62	a	5.8	8.5
0	26.41	-10.1	298.6	0.92	11.24	a	2.4	7.8
0	27.89	-10.13	300.7	0.91	11.38	a	0.3	
0.03	26.41	-10.29	149.6	0.12	3.04	w	72.8	9
0.03	26.43	-10.25	150.9	0.12	2.18	w	51.5	7.9
0.03	26.43	-10.26	151.9	0.11	1.77	w	39.9	7.1
0.03	26.4	-10.3	150.9	0.11	1.26	w	28.2	6
0.03	26.39	-10.31	151	0.11	1.06	w	18.8	5
0.03	26.38	-10.33	149.6	0.11	1.13	w	13.1	4.4
0.03	26.43	-10.25	148.5	0.11	1.02	w	10.4	3.8
0.03	26.45	-10.23	149.9	0.11	0.87	w	6.5	3.1
0.03	26.44	-10.24	149.5	0.11	0.59	w	3.2	2.6
0.03	28.22	-10.25	149.3	0.11	0.6	w	0.3	
0.03	26.4	-10.25	150.6	0.31	4.49	w	61.4	6.7
0.03	26.41	-10.23	150.1	0.31	3.65	w	45.5	6.2
0.03	26.38	-10.28	149	0.31	3.28	w	37.7	6.2
0.03	26.39	-10.26	150.6	0.31	3.27	w	31.9	7.3
0.03	26.42	-10.22	150	0.31	2.93	w	24.6	6.2
0.03	26.39	-10.26	149.3	0.31	2.28	w	17.4	5.5
0.03	26.37	-10.29	149.2	0.31	2.03	w	13.7	5.2
0.03	26.38	-10.28	149.1	0.31	1.7	w	9.6	4.6
0.03	26.43	-10.21	148.4	0.31	1.37	w	5	3.5
0.03	28.46	-10.24	148.4	0.32	1.1	w	0.3	
0.03	26.38	-10.19	151.3	0.51	8.6	w	81	4.1
0.03	26.37	-10.2	151.1	0.51	8.24	w	76.2	4.1
0.03	26.38	-10.18	148.6	0.52	7.81	w	69.4	4
0.03	26.43	-10.12	148.4	0.52	5.15	w	32.5	5
0.03	26.39	-10.18	152.8	0.5	3.8	w	24.6	5.5
0.03	26.38	-10.2	152.6	0.5	3.56	w	19.1	5.5
0.03	26.43	-10.12	151.9	0.5	3.56	w	15.9	5.5
0.03	26.37	-10.22	150.1	0.48	3.01	w	11.3	5.1
0.03	26.4	-10.17	149.5	0.51	2.79	w	6.9	4.7
0.03	28.42	-10.2	149.5	0.51	1.7	w	0.3	
0.03	26.44	-9.89	147.9	0.73	11.8	w	74.8	2.4
0.03	26.43	-9.92	150.5	0.72	10.6	w	52.8	3.4
0.03	26.33	-10.03	146.5	0.73	8.57	w	26.5	5.2
0.03	26.37	-10	150.8	0.71	7.16	w	17.7	6.7
0.03	26.36	-10	149.1	0.72	5.93	w	13.5	6.6
0.03	26.43	-9.94	150.2	0.7	5.6	w	11.3	6.4
0.03	26.45	-9.9	149.8	0.72	4.97	w	8.4	6.3

Table G.6

$\dot{q}_1$	α_1	$\dot{q}_2$	α_2	$\dot{q}_3$	α_3	$\dot{q}_4$	α_4	$\dot{q}_5$	α_5	$\dot{q}_6$	α_6
0.2	0.1	24.9	6.1	122.7	30.2	111.7	27.5	110.2	27.1	22	5.4
1.5	0.4	30.4	8.9	99.6	29.3	74.7	21.9	90.4	26.6	27.8	8.2
2	0.7	27.5	9.8	75.8	26.9	56.1	19.9	70.3	24.9	24.5	8.7
4.4	1.7	28	10.7	54.8	21	37.7	14.4	52.7	20.1	27	10.3
10.6	4.4	27.8	11.5	33	13.7	23.4	9.7	32.9	13.6	28.6	11.8
11.4	7.1	22.7	14.1	22.2	13.8	14.3	8.9	20.5	12.7	20.2	12.5
11.9	9.3	17.1	13.3	14.3	11.1	9.5	7.4	13.2	10.3	17	13.2
7.2	10.5	6.5	9.5	5.3	7.7	4.9	7.1	5	7.3	6	8.7
2.7	8.7	2.7	9	2.3	7.6	1.9	6.3	2	6.7	2.7	8.7
14.7	1.8	58.3	7.2	103.5	12.8	112	13.9	90.4	11.2	57.8	7.2
6.5	1	47.1	7.2	71.5	10.9	75.8	11.6	59.7	9.1	48.4	7.4
4.2	0.7	38.9	6.9	54.9	9.7	56.7	10	47.5	8.4	37.2	6.6
2	0.4	30.1	6.4	37.9	8.1	37.8	8.1	32.9	7	28.3	6
1	0.3	22	5.8	24.6	6.5	23.3	6.2	21.7	5.7	20.5	5.4
0.5	0.2	16.6	5.5	16.2	5.4	14.2	4.7	14.7	4.9	16.5	5.5
0.2	0.1	15.1	5.5	12.5	4.6	9.8	3.6	11	4	13.9	5.1
0.2	0.1	9.7	4.6	7.5	3.6	4.8	2.3	6.6	3.1	10.2	4.8
0.2	0.2	5.7	4.7	2.9	2.4	2	1.7	2.3	1.9	5.9	4.8
14.5	1.6	45.8	5	80.5	8.8	112.3	12.3	69.9	7.6	45.6	5
9.3	1.3	38.5	5.2	59.5	8.1	75	10.2	53.3	7.3	37.4	5.1
5.7	0.9	33.1	5.4	51.5	8.4	55.8	9.1	46.3	7.6	33.6	5.5
6.1	1.4	31.6	7.2	43.9	10	38.1	8.7	39.9	9.1	31.9	7.3
4.8	1.2	28.4	7.2	32.3	8.2	23.5	6	30.4	7.7	28	7.1
0.2	0.1	19.7	6.2	26	8.2	14.3	4.5	24.7	7.8	19.7	6.2
0.2	0.1	15.5	5.9	20.8	8	9.7	3.7	19.6	7.5	16.1	6.2
0.2	0.1	11.6	5.5	14.5	6.9	4.9	2.3	14.2	6.8	12.2	5.8
0.2	0.1	6.7	4.7	7.4	5.2	2	1.4	7	4.9	6.9	4.8
0.9	0	76.6	3.8	123.1	6.2	91.9	4.6	118.3	5.9	75.2	3.8
0.9	0	73.8	4	119.1	6.4	74.8	4	116.3	6.2	72.6	3.9
0.9	0.1	66.6	3.8	115.3	6.6	55.8	3.2	112.1	6.5	65.5	3.8
0.2	0	29.3	4.5	53.4	8.2	37.7	5.8	49	7.5	25.5	3.9
0.2	0	23.6	5.3	41.3	9.3	23.7	5.3	38.3	8.6	20.6	4.6
0.2	0.1	19.2	5.5	32.7	9.4	14.2	4.1	31.2	9	17	4.9
0.2	0.1	16.4	5.7	27.7	9.5	9.8	3.4	26.4	9.1	15.1	5.2
0.2	0.1	11.8	5.4	19.5	8.9	4.9	2.2	19.6	8.9	11.9	5.4
0.2	0.1	7	4.8	12.2	8.4	2	1.4	12.4	8.5	7.2	5
11.7	0.4	43.4	1.4	140.5	4.5	75.2	2.4	137.9	4.4	40.2	1.3
3.5	0.2	36.4	2.3	98	6.3	56	3.6	91.6	5.9	31.5	2
1.6	0.3	19.2	3.8	48.4	9.5	37.9	7.5	40.2	7.9	11.7	2.3
0.8	0.3	15.5	5.9	33.4	12.7	23.9	9.1	24.1	9.2	8.3	3.1
0.8	0.4	10.9	5.3	26.2	12.8	14.3	7	21.4	10.5	7.6	3.7
0.8	0.4	10	5.7	22.6	12.8	9.7	5.5	18.4	10.4	6.1	3.4
0.4	0.3	7	5.3	17.3	13	4.8	3.6	15.6	11.7	5.5	4.1

Table G.6

$w_{oil,0}$	p	T	$\dot{m}$	$\dot{x}$	$\Delta p/\Delta L$	FP	$\dot{q}$	α
0.03	26.41	-9.94	149.1	0.72	4.11	w	5.3	6
0.03	28.2	-9.99	149.5	0.72	2.44	w	0.3	
0.03	26.35	-9.33	149.1	0.91	5.84	w	10.2	0.6
0.03	26.44	-9.05	147.3	0.92	9.27	w	6.2	8.6
0.03	26.41	-9.14	149	0.92	9.29	w	0	0
0.03	26.38	-9.18	148.8	0.92	6.9	w	0	0
0.03	28.18	-9.15	148.6	0.92	4.39	w	0.3	
0.03	26.35	-10.36	305.6	0.11	7.37	i	85.1	12.4
0.03	26.4	-10.3	300.7	0.11	6	i	64.1	11.9
0.03	26.4	-10.3	295.2	0.11	5.61	i	54.4	12.1
0.03	26.37	-10.34	303.1	0.11	5.12	i	42.8	11.5
0.03	26.39	-10.31	300	0.11	4.9	i	33.9	11.3
0.03	26.35	-10.37	298.9	0.1	4.65	i	25.7	10.9
0.03	26.4	-10.31	298.6	0.1	4.26	i	20.4	10.3
0.03	26.46	-10.23	297.7	0.1	3.33	i	12.6	7.8
0.03	26.44	-10.24	297	0.1	2.48	i	6	5.9
0.03	28.2	-10.27	297.1	0.11	1.54	i	0.3	
0.03	26.4	-10.25	303	0.3	12.55	a	71.6	9.4
0.03	26.36	-10.3	302.3	0.3	10.61	a	54.1	12
0.03	26.35	-10.32	302.4	0.3	10.75	a	50.1	14.2
0.03	26.39	-10.27	302.6	0.31	8.66	a	38.6	11.4
0.03	26.36	-10.31	302.2	0.31	7.43	a	29.6	10.5
0.03	26.4	-10.25	302.1	0.31	6.4	a	21.5	9.5
0.03	26.41	-10.24	301.8	0.3	5.62	a	16.8	8.6
0.03	26.37	-10.29	301.7	0.3	4.64	a	10.4	6.9
0.03	26.39	-10.27	300.7	0.3	3.72	a	3.5	4.5
0.03	28.13	-10.28	300.6	0.31	3.56	a	0.3	
0.03	26.43	-10.12	300.2	0.51	20.6	a	72.6	6.1
0.03	26.37	-10.21	298.4	0.51	16.51	a	43.3	10.8
0.03	26.4	-10.16	298.5	0.51	13.82	a	35.4	10.9
0.03	26.38	-10.19	298.4	0.51	12.9	a	25.7	10.8
0.03	26.38	-10.2	298.9	0.51	11.85	a	17.5	10.4
0.03	26.43	-10.12	297.5	0.51	9.89	a	17.1	9.5
0.03	26.44	-10.11	297.9	0.5	8.99	a	13.7	9.2
0.03	26.47	-10.07	297.7	0.51	7.61	a	7.6	7.2
0.03	26.46	-10.08	297.4	0.51	6.51	a	3.3	4.9
0.03	28.18	-10.09	297.2	0.52	5.9	a	0.3	
0.03	26.42	-10.27	73.9	0.15	2.05	s	66.7	8.3
0.03	26.36	-10.34	75.2	0.13	1.77	s	49.9	7.4
0.03	26.41	-10.28	74.7	0.13	1.71	s	39.9	6.6
0.03	26.38	-10.32	74.6	0.12	1.56	s	29.7	5.6
0.03	26.43	-10.26	74.6	0.12	1.52	s	19.8	4.6
0.03	26.41	-10.29	74.3	0.12	1.64	s	13.9	3.8
0.03	26.39	-10.31	74.1	0.12	1.46	s	10.7	3.3
0.03	26.37	-10.34	74.1	0.12	1.3	s	7.1	2.7
0.03	26.38	-10.33	75.9	0.12	1.28	s	4.5	2.2
0.03	28.22	-10.37	75.5	0.11		s	0.3	

Table G.6

$\dot{q}_1$	α_1	$\dot{q}_2$	α_2	$\dot{q}_3$	α_3	$\dot{q}_4$	α_4	$\dot{q}_5$	α_5	$\dot{q}_6$	α_6
0.3	0.3	4.1	4.7	11	12.6	2.1	2.4	10.2	11.7	4	4.5
6.3	0.4	6.5	0.4	13	0.7	14.1	0.8	12.8	0.7	8.7	0.5
0.9	1.2	2.9	4	11.7	16.1	9.6	13.3	9.4	13	2.7	3.8
0	0	0	0	0	0	0	0	0	0	0	0
0	0	0	0	0	0	0	0	0	0	0	0
39.9	5.8	66.6	9.7	116.6	17	113.3	16.6	106.7	15.6	67.3	9.8
37.8	7	56.8	10.6	82.5	15.4	75.6	14.1	75.4	14	56.1	10.4
38.9	8.7	55.4	12.3	63	14	56.1	12.5	58	12.9	55	12.3
36.1	9.7	47.6	12.8	45.5	12.2	37.6	10.1	42.5	11.4	47.3	12.7
32.3	10.8	42	14	32.7	10.9	23.7	7.9	30.6	10.2	42.1	14.1
27.4	11.7	34.2	14.6	22.5	9.6	14.2	6	20.7	8.8	35.1	14.9
23.3	11.8	28.9	14.6	15.9	8	9.8	4.9	15.3	7.7	29.4	14.8
14.7	9.2	17.6	11	10.6	6.6	4.9	3.1	10	6.2	17.5	10.9
7.8	7.7	7.9	7.7	5.3	5.2	2.1	2.1	5.7	5.5	7.1	6.9
26.4	3.5	72.9	9.6	78.9	10.4	112.1	14.8	69.8	9.2	69.2	9.1
19.4	4.3	58.7	13	61.5	13.6	75	16.6	55.4	12.2	54.8	12.1
27.4	7.7	54.6	15.5	57.9	16.4	56.6	16	52	14.7	52.3	14.8
10.2	3	47.1	13.9	47.2	13.9	37.7	11.1	43.5	12.9	45.6	13.5
4.2	1.5	39.7	14	35.7	12.6	23.9	8.4	34.7	12.3	39.6	14
1.1	0.5	31.8	14	25.2	11.1	14	6.2	24.6	10.8	32.3	14.3
1	0.5	25.4	13	19	9.7	9.5	4.9	19.1	9.8	26.8	13.7
0.5	0.4	16	10.6	12.4	8.2	4.9	3.3	12.3	8.1	16.4	10.9
0.2	0.3	4.9	6.2	4.4	5.6	2	2.6	4.4	5.6	5.4	6.9
8.4	0.7	57.1	4.8	109.5	9.2	112.1	9.4	96.6	8.1	52.2	4.4
3.5	0.9	26.5	6.6	69.3	17.4	75.6	18.9	62.3	15.6	22.6	5.7
3.1	0.9	20.8	6.4	58.5	18	56.1	17.3	54	16.6	20.2	6.2
2	0.8	16.6	7	43.6	18.3	37.7	15.8	39.8	16.7	14.9	6.2
1.5	0.9	11.7	6.9	29.9	17.8	23.7	14.1	27.6	16.4	10.5	6.3
1.3	0.7	18.8	10.4	25.5	14.1	14.3	7.9	24.5	13.5	18.3	10.1
0.8	0.5	16.4	11	20.3	13.6	9.8	6.5	19.5	13	15.6	10.4
0.8	0.7	8.7	8.3	11.9	11.2	4.8	4.6	11.5	10.9	8	7.6
0.7	1	3.8	5.7	5.3	8	2	3.1	4.8	7.2	3	4.5
0.2	0	58.5	7.3	92.8	11.5	108.9	13.5	74.2	9.2	65.5	8.1
0.2	0	44.7	6.6	70.6	10.5	75.5	11.2	56.8	8.4	51.7	7.7
0.2	0	35.8	5.9	56.8	9.4	56.4	9.3	46.6	7.7	43.8	7.2
0.2	0	27.2	5.1	44.3	8.3	37.6	7.1	37.7	7.1	31.2	5.9
0.2	0	19.1	4.4	29.3	6.8	23.6	5.4	25.4	5.8	21.2	4.9
0.2	0.1	13.3	3.6	22.2	6	14.3	3.9	19.2	5.2	14.5	3.9
0.2	0.1	10.7	3.4	17.1	5.3	9.5	3	14.9	4.7	11.6	3.6
0.2	0.1	7.9	3	11.3	4.3	4.9	1.9	10	3.8	8.2	3.1
0.2	0.1	5.3	2.6	7.2	3.6	2.1	1.1	6.5	3.2	5.6	2.8

Table G.6

$w_{oil,0}$	p	T	$\dot{m}$	$\dot{x}$	$\Delta p/\Delta L$	FP	$\dot{q}$	α
0.03	26.42	-10.22	74.1	0.34	3.05	s-w	56.5	5.7
0.03	26.39	-10.26	75.5	0.33	2.82	s-w	39.1	5.3
0.03	26.42	-10.21	73.6	0.33	2.46	s-w	31	5
0.03	26.4	-10.25	74	0.32	2.11	s-w	23.1	4.5
0.03	26.38	-10.27	75.3	0.32	1.78	s-w	16.7	4
0.03	26.39	-10.26	74.5	0.32	1.51	s-w	12.2	3.5
0.03	26.39	-10.26	74.4	0.32	2.02	s-w	12.2	2.9
0.03	26.38	-10.27	73.9	0.32	1.19	s-w	7.7	2.4
0.03	26.38	-10.28	74.3	0.32	1.27	s-w	4.8	2
0.03	28.29	-10.3	75.1	0.31		s-w	0.3	
0.03	26.45	-10.07	75.4	0.54	3.47	w	83	3.7
0.03	26.43	-10.1	74.3	0.54	2.89	w	72.2	3.5
0.03	26.44	-10.09	75	0.54	2.55	w	63.9	3.3
0.03	26.38	-10.17	74	0.53	2.23	w	26.8	4
0.03	26.37	-10.2	74.5	0.52	1.78	w	19.3	4
0.03	26.42	-10.13	74.5	0.52	1.63	w	14.3	3.8
0.03	26.42	-10.13	75.2	0.52	1.44	w	11.9	3.7
0.03	26.4	-10.16	74.3	0.52	1.27	w	8.4	3.3
0.03	26.4	-10.16	74	0.52	1.09	w	5.6	2.8
0.03	28.31	-10.18	75.4	0.51		w	0.3	
0.03	26.43	-9.9	75.2	0.73	4.66	w	70.3	2
0.03	26.45	-9.86	74.8	0.73	4.09	w	64.4	2
0.03	26.45	-9.83	73.3	0.75	3.91	w	62.5	1.8
0.03	26.45	-9.87	75.2	0.73	3.79	w	60.6	1.8
0.03	26.45	-9.86	74.5	0.74	3.75	w	59.8	1.8
0.03	26.44	-9.89	75.6	0.72	3.71	w	58.4	1.8
0.03	26.44	-9.87	74.7	0.73	3.66	w	58	1.8
0.03	28.33	-9.97	75.7	0.7		w	0.3	
0.03	26.4	-8.96	75.4	0.93	3.51	s-w	36.8	1
0.03	26.4	-9.17	74.7	0.91	3.32	s-w	34.9	1.1
0.03	26.4	-8.95	74.3	0.93	3.11	s-w	31.2	1.2
0.03	28.28	-8.89	74.1	0.93		s-w	0.3	
0.03	26.41	-9.9	301.2	0.74	20.88	a	82.7	6.2
0.03	26.45	-9.89	298.6	0.72	22.53	a	75.9	2.2
0.03	26.38	-9.99	299.2	0.71	21.25	a	36.2	7.2
0.03	26.41	-9.96	299.9	0.71	19.46	a	28.5	10.2
0.03	26.37	-10.01	299.2	0.71	16.31	a	20.8	10.8
0.03	26.41	-9.96	298.8	0.71	13.76	a	15	9.5
0.03	26.39	-9.98	298.9	0.71	12.02	a	11.2	8.5
0.03	26.39	-9.98	298.2	0.71	11.37	a	8.1	7.4
0.03	26.38	-9.99	298.8	0.71	10.65	a	4	5.5
0.03	26.37	-10.01	298.7	0.71	9.66	a	1.7	5.1
0.03	28.21	-10.02	299.7	0.71	9.13	a	0.3	
0.01	26.38	-10.44	298.1	0.11	17.62	i	89.1	17.4
0.01	26.44	-10.36	297	0.11	17.42	i	71.9	16.2
0.01	26.42	-10.39	300.5	0.11	16.91	i	56.8	13.6
0.01	26.36	-10.47	299.7	0.11	14.59	i	38.5	11.5

Table G.6

$\dot{q}_1$	α_1	$\dot{q}_2$	α_2	$\dot{q}_3$	α_3	$\dot{q}_4$	α_4	$\dot{q}_5$	α_5	$\dot{q}_6$	α_6
0.5	0	38.3	3.8	75.6	7.6	111.9	11.2	72.4	7.3	40.4	4.1
0.3	0	25.4	3.4	54.4	7.4	75.5	10.2	51.6	7	27.5	3.7
0.3	0	19.8	3.2	45.2	7.3	56.2	9	43.8	7	20.5	3.3
0.2	0	15.4	3	35	6.8	37.6	7.3	34.4	6.7	15.7	3
0.2	0	12.1	2.9	26.9	6.4	23.5	5.6	26.1	6.2	11.7	2.8
0.2	0.1	9.5	2.7	20.5	5.9	14.1	4	20.3	5.8	9	2.6
0.2	0	9.5	2.3	22.2	5.3	9.6	2.3	22.6	5.4	8.8	2.1
0.2	0.1	5.5	1.7	15.5	4.8	4.8	1.5	15.3	4.8	5.2	1.6
0.2	0.1	3.3	1.4	10	4.1	2.1	0.9	10.2	4.2	3.1	1.3
2.6	0.1	49.1	2.2	146.4	6.6	111.9	5	136.2	6.1	51.5	2.3
1.6	0.1	43.8	2.1	137.4	6.6	75.3	3.6	132.9	6.4	42.2	2
1.6	0.1	37.7	2	127.6	6.7	56.3	2.9	123	6.4	37	1.9
0.2	0	15.3	2.3	47.3	7.1	38.1	5.7	45.4	6.8	14.3	2.1
0.2	0	11.7	2.4	35.1	7.3	23.4	4.9	34.6	7.2	10.8	2.3
0.2	0.1	9.2	2.5	26.9	7.2	14.1	3.8	26.9	7.2	8.4	2.3
0.2	0.1	8.1	2.5	22.9	7.1	9.6	3	23.1	7.2	7.3	2.3
0.2	0.1	5.7	2.2	17.2	6.7	4.8	1.9	17.4	6.8	5.3	2.1
0.2	0.1	3.3	1.7	12.3	6.3	2	1	12.6	6.5	2.9	1.5
6.8	0.2	35.1	1	147.6	4.3	56.9	1.6	141.1	4.1	34	1
6.8	0.2	29.1	0.9	144.8	4.5	37.6	1.2	136.9	4.2	31.1	1
6.1	0.2	27.4	0.8	147.6	4.2	23.7	0.7	141.5	4	29.1	0.8
6.5	0.2	26	0.8	148.9	4.5	13.8	0.4	141.3	4.3	27.1	0.8
5.3	0.2	25.9	0.8	150.2	4.4	9.3	0.3	140.9	4.2	27.5	0.8
7.2	0.2	23.7	0.7	149.4	4.7	4.7	0.1	139.4	4.3	26	0.8
6.8	0.2	22.9	0.7	149.6	4.5	2.1	0.1	141	4.3	25.9	0.8
8.3	0.2	12	0.3	92.6	2.4	9.5	0.3	84.8	2.2	13.6	0.4
7.9	0.3	9	0.3	91.5	3	4.8	0.2	84.6	2.8	11.7	0.4
7.6	0.3	7.8	0.3	83.8	3.2	2	0.1	77.1	3	8.7	0.3
5.3	0.4	43.9	3.3	149.3	11.2	112.1	8.4	142.8	10.7	42.8	3.2
22.2	0.6	18.8	0.6	148.2	4.3	112.3	3.3	132.3	3.9	21.6	0.6
5.3	1.1	9.6	1.9	63.6	12.6	75.6	15	53.8	10.7	8.9	1.8
3.1	1.1	9.8	3.5	50.1	17.9	56.2	20.1	43.2	15.5	8.5	3.1
2.2	1.1	8.7	4.5	37.2	19.2	37.5	19.4	31.4	16.3	7.8	4.1
1.4	0.9	7.7	4.8	28.7	18.1	23.5	14.8	23.2	14.6	5.7	3.6
1	0.7	7	5.3	21.4	16.4	14.3	10.9	18.7	14.3	4.6	3.5
0.8	0.7	5.5	5	15.6	14.1	9.5	8.6	13.8	12.5	3.7	3.3
0.7	0.9	2.8	3.8	7.7	10.6	4.8	6.7	6.2	8.5	1.8	2.4
0.5	1.4	1.4	4	2.8	8.2	2	6	2.3	6.8	1.4	4
34.5	6.7	80.6	15.7	121.4	23.7	112.2	21.9	108.5	21.2	77.3	15.1
28.6	6.4	73.2	16.5	98.6	22.2	75.2	16.9	86.1	19.4	70.1	15.8
24.7	5.9	55.2	13.2	79.3	19	56.8	13.6	70.8	17	54.3	13
16.4	4.9	36.8	11	55.7	16.6	37.8	11.3	49.5	14.8	34.7	10.3

Table G.6

$w_{oil,0}$	p	T	$\dot{m}$	$\dot{x}$	$\Delta p/\Delta L$	FP	$\dot{q}$	α
0.01	26.38	-10.44	298.6	0.11	14.07	i	25.4	9.3
0.01	26.43	-10.38	296.9	0.11	13.27	i	19.2	8.5
0.01	26.4	-10.42	298.3	0.1	13.11	i	13.2	6.9
0.01	26.4	-10.42	298.7	0.11	13.54	i	8.4	5.6
0.01	26.39	-10.44	297.3	0.11	13.32	i	5.2	5.1
0.01	0	-10.47	297.4	0.11		i	0.3	
0.01	26.39	-10.42	302.8	0.31	12.68	a	83	15.9
0.01	26.4	-10.4	300.7	0.3	11.37	a	62.6	14.6
0.01	26.45	-10.33	297.7	0.31	12.11	a	50.7	13.3
0.01	26.42	-10.38	300.1	0.3	18.27	a	34.6	10.6
0.01	26.38	-10.43	300	0.3	15.83	a	23	8.4
0.01	26.43	-10.36	298.6	0.31	15.46	a	14.8	7.2
0.01	26.41	-10.39	298.2	0.3	15.29	a	11	6.5
0.01	26.4	-10.4	298	0.31	15.86	a	6.6	5.5
0.01	26.39	-10.42	295.4	0.31	15.19	a	3.1	5
0.01	0	-10.43	296.9	0.31		a	0.3	
0.01	26.41	-10.36	303.8	0.51	23.4	a	68.2	14
0.01	26.37	-10.41	302.8	0.5	23.06	a	54	13.4
0.01	26.36	-10.42	303	0.5	22.98	a	46	12.6
0.01	26.4	-10.37	302.4	0.5	20.78	a	35.9	11.6
0.01	26.37	-10.42	302.1	0.5	21.8	a	26.5	10.4
0.01	26.42	-10.35	299.5	0.51	21.08	a	17.9	9.1
0.01	26.4	-10.37	300.8	0.51	20.95	a	12.4	7.8
0.01	26.39	-10.38	300.2	0.51	19.43	a	7	6.3
0.01	26.37	-10.41	300.5	0.51	19.11	a	2.7	6.3
0.01	0	-10.42	300	0.51		a	0.3	
0.01	26.42	-10.27	300.2	0.72	31.8	a	62.6	11.7
0.01	26.4	-10.29	300.9	0.71	26.5	a	44.8	13.7
0.01	26.38	-10.33	299.7	0.71	26.52	a	37.4	13.3
0.01	26.43	-10.26	301.6	0.71	23.9	a	29.2	12.5
0.01	26.39	-10.31	300.9	0.71	24.1	a	21.4	11.3
0.01	26.37	-10.34	302.4	0.7	23.69	a	14.7	9.7
0.01	26.43	-10.26	301.6	0.71	21.14	a	10.2	8.9
0.01	26.38	-10.33	301.1	0.7	22.75	a	5.2	6.5
0.01	26.38	-10.33	301.1	0.7	22.21	a	2.2	7
0.01	0	-10.32	300.5	0.71		a	0.3	
0.01	26.45	-9.84	300.2	0.91	38.98	a	48.3	2.8
0.01	26.4	-9.92	301.6	0.91	37.65	a	36	4.1
0.01	26.36	-9.99	301.5	0.91	36.54	a	25.4	8.1
0.01	26.39	-9.92	302	0.91	35.71	a	18.4	9.2
0.01	26.38	-10.01	303.4	0.9	33.86	a	11.2	9.1
0.01	26.42	-9.96	304.1	0.9	30.21	a	6.9	8
0.01	26.41	-9.97	303.7	0.9	29.83	a	5	7.2
0.01	26.38	-9.97	300.6	0.91	31.09	a	3	5
0.01	26.4	-9.85	295.9	0.92	31.72	a	1.4	5.2
0.01	0	-9.95	301.1	0.91		a	0.3	
0.01	26.41	-10.4	148.7	0.13	1.08	w	79.6	12.7

$\dot{q}_1$	α_1	$\dot{q}_2$	α_2	$\dot{q}_3$	α_3	$\dot{q}_4$	α_4	$\dot{q}_5$	α_5	$\dot{q}_6$	α_6
13.9	5.1	24.1	8.8	36.4	13.4	23.4	8.6	32.3	11.8	22.4	8.2
7.2	3.2	25.4	11.2	23.6	10.4	14.4	6.3	21.2	9.4	23.3	10.3
8.5	4.5	15.2	8	16.7	8.8	9.5	5	14.5	7.6	14.6	7.7
7.6	5.1	9.8	6.5	9.8	6.5	4.9	3.3	8.4	5.6	9.6	6.4
6.2	6.1	6.8	6.6	5	4.9	2.7	2.6	4.2	4.1	6.5	6.4
6.5	1.2	80.9	15.5	116.2	22.3	111.8	21.4	104.2	20	78.6	15.1
4.2	1	65.1	15.1	86.9	20.2	75	17.5	79.7	18.5	64.5	15
7.4	1.9	52.5	13.7	71.8	18.8	56	14.6	66.6	17.4	50.1	13.1
3.5	1.1	31.4	9.6	53.7	16.4	37.7	11.5	50.1	15.3	31.2	9.5
1.8	0.6	20.5	7.5	37.4	13.7	23.3	8.5	34.7	12.7	20.5	7.5
1.4	0.7	13.8	6.7	23.6	11.5	14.1	6.9	22.5	10.9	13.1	6.4
1.2	0.7	10.9	6.5	17.9	10.6	9.5	5.6	15.7	9.3	10.6	6.3
1	0.9	8.2	6.8	9.3	7.7	4.8	4	8.2	6.8	8	6.7
0.8	1.3	4.8	7.7	3.2	5.2	2	3.2	2.7	4.3	5.2	8.4
0.2	0	45.5	9.3	107	21.9	111.4	22.8	100	20.5	44.9	9.2
0.2	0.1	38.3	9.5	87.7	21.8	74.5	18.5	83.6	20.7	39.6	9.8
0.2	0.1	34	9.3	75.8	20.8	56.5	15.5	72.8	20	36.6	10
0.2	0.1	28.9	9.3	60.6	19.6	37.4	12.1	58	18.7	30.6	9.9
0.5	0.2	25	9.8	43.1	17	23.6	9.3	40.8	16.1	26.1	10.3
0.8	0.4	18.3	9.3	28.4	14.4	14.3	7.2	26.9	13.7	18.7	9.5
1.4	0.9	13.5	8.5	18.9	11.9	9.3	5.9	17.6	11.1	13.8	8.7
1.8	1.6	8.3	7.5	9.7	8.7	5	4.5	8.7	7.9	8.6	7.8
1.6	3.7	3.4	7.9	3.1	7.3	2	4.8	2.6	6	3.4	7.9
4.3	0.8	19.6	3.7	114.5	21.4	112.6	21	105.5	19.7	19	3.6
3.3	1	17.5	5.4	79.5	24.3	75.6	23.1	75.3	23	17.8	5.4
2.5	0.9	16	5.7	67.7	24.1	56.3	20.1	64.7	23	17.1	6.1
1.6	0.7	13.8	5.9	54	23.1	37.8	16.1	52.2	22.3	15.7	6.7
1.6	0.9	13.4	7.1	38.5	20.4	23.6	12.5	36.3	19.2	15.1	8
3.6	2.4	11.3	7.5	24	15.8	14.3	9.4	22.2	14.7	12.8	8.5
3.6	3.2	8.4	7.3	16	14	9.5	8.3	14.7	12.9	8.9	7.7
2.9	3.6	4.4	5.5	7.9	9.8	4.9	6.1	6.8	8.4	4.4	5.4
1.6	5.1	2.1	6.8	2.8	9	2.1	6.8	2.3	7.2	2.1	6.8
11.3	0.7	4	0.2	87.5	5.1	112.1	6.5	69.6	4.1	5.3	0.3
9.5	1.1	3.2	0.4	66.9	7.6	76	8.6	56.2	6.4	4.4	0.5
4.1	1.3	2.8	0.9	46.3	14.8	56.5	18	39.4	12.5	3.5	1.1
2.6	1.3	4.2	2.1	34.7	17.4	37.8	19	26.8	13.4	4.4	2.2
2	1.7	3.1	2.6	20.7	16.8	23.6	19.2	14.5	11.8	3.2	2.6
2	2.3	2.7	3	11	12.7	14.3	16.4	8.8	10.1	2.8	3.2
1.8	2.7	2.7	3.9	7	10.2	9.5	13.8	5.8	8.4	3.1	4.4
1.4	2.3	2.3	3.7	4.1	6.8	4.9	8.1	3.2	5.3	2.3	3.7
0.8	3.1	0.9	3.3	2.1	8	2	7.7	1.4	5.2	1	3.9
0.2	0	65.3	10.4	124.5	19.8	113	18	113.4	18	61.3	9.8

Table G.6

$w_{oil,0}$	p	T	$\dot{m}$	$\dot{x}$	$\Delta p/\Delta L$	FP	$\dot{q}$	α
0.01	26.43	-10.37	151	0.12	2.47	w	59.8	11.4
0.01	26.41	-10.4	148.7	0.12	3.77	w	47.4	10.1
0.01	26.37	-10.46	149.9	0.11	2.74	w	35.1	8.7
0.01	26.39	-10.43	149.2	0.11	9.61	w	24.6	7.4
0.01	26.37	-10.46	149	0.11	5.34	w	16.7	5.9
0.01	26.36	-10.47	149	0.11	2.68	w	12.3	5
0.01	26.43	-10.38	150.3	0.11	4.04	w	8	4.1
0.01	26.4	-10.42	150	0.11	4.89	w	4.2	3
0.01	0	-10.36	149.7	0.15		w	0.3	
0.01	26.4	-10.4	148.4	0.33	3.87	w	65.7	10.1
0.01	26.46	-10.33	149	0.32	4.2	w	47.9	9.3
0.01	26.44	-10.35	149.9	0.31	6.37	w	39.3	8.6
0.01	26.4	-10.4	149	0.31	5.15	w	29.2	7.7
0.01	26.37	-10.45	150.6	0.31	4.64	w	20.6	6.8
0.01	26.35	-10.46	149.9	0.31	3.74	w	14.7	5.5
0.01	26.37	-10.44	149.6	0.31	2.7	w	10.9	4.8
0.01	26.4	-10.4	149.3	0.31	5.21	w	6.6	3.9
0.01	26.39	-10.41	149.5	0.31	4.37	w	3.5	2.9
0.01	0	-10.42	148.9	0.31		w	0.3	
0.01	26.44	-10.35	149.2	0.33	12.6	w	65.6	10.2
0.01	26.39	-10.38	150.6	0.51	1.99	w	61.6	8.8
0.01	26.44	-10.32	149	0.52	1.28	w	44.5	8.2
0.01	26.4	-10.37	148.5	0.52	0.5	w	36.6	8
0.01	26.37	-10.4	148.5	0.52	3.02	w	29.1	7.6
0.01	26.41	-10.35	148.4	0.52	0.18	w	22.1	6.8
0.01	26.39	-10.38	148.1	0.52	5.23	w	15.8	5.9
0.01	26.38	-10.39	150.7	0.51	1.43	w	11.7	5.1
0.01	26.38	-10.4	151	0.51	1.38	w	6.9	4
0.01	26.36	-10.42	150.4	0.52	0.49	w	3	2.9
0.01	0	-10.44	150.8	0.51		w	0.3	
0.01	26.44	-10.24	149.3	0.72	7.91	w	67.1	6.8
0.01	26.42	-10.27	148.7	0.72	5.62	w	47.1	7.4
0.01	26.38	-10.31	149.5	0.72	6.26	w	36.8	7.8
0.01	26.43	-10.25	149.3	0.71	4.21	w	29.1	7.8
0.01	26.4	-10.28	148.9	0.72	6.07	w	23.2	7.8
0.01	26.4	-10.3	149	0.71	2.97	w	17.9	7.2
0.01	26.38	-10.32	148.9	0.72	1.41	w	14.9	7.1
0.01	26.37	-10.32	148.7	0.73	2.63	w	10.1	6.2
0.01	26.37	-10.33	148.8	0.72	0.2	w	4.9	4.4
0.01	0	-10.29	149	0.72		w	0.3	
0.01	26.46	-9.74	148.4	0.93	12	w	70.6	2.3
0.01	26.4	-9.92	150.5	0.91	11.8	w	44.6	5.1
0.01	26.44	-9.9	150.9	0.91	12.74	w	34.4	6.8
0.01	26.41	-9.86	149.7	0.92	14.25	w	24.1	7.4
0.01	26.37	-9.91	149.1	0.92	13.6	w	14.8	7.8
0.01	26.45	-9.89	150.2	0.91	10.77	w	9.5	9.3
0.01	26.42	-9.96	151.3	0.9	8.87	w	7.1	9.2

$\dot{q}_1$	α_1	$\dot{q}_2$	α_2	$\dot{q}_3$	α_3	$\dot{q}_4$	α_4	$\dot{q}_5$	α_5	$\dot{q}_6$	α_6
0.2	0	52.9	10	92.2	17.5	74.9	14.2	83.6	15.9	54.8	10.4
0.2	0	37.4	8	76.4	16.3	56.5	12.1	69.2	14.8	44.7	9.5
0.2	0.1	29.1	7.2	59.2	14.6	37.9	9.4	52.6	13	31.8	7.9
0.2	0.1	20.9	6.3	42	12.7	23.5	7.1	38.6	11.6	22.6	6.8
0.2	0.1	14.3	5	28.7	10.1	14.1	5	26.1	9.2	16.6	5.9
0.2	0.1	11	4.5	21.3	8.7	9.5	3.9	19.2	7.8	12.4	5.1
0.2	0.1	8.7	4.4	14.1	7.2	4.8	2.4	12.8	6.5	7.8	4
0.2	0.1	5.9	4.2	5.8	4.1	2	1.5	5.3	3.8	5.9	4.2
0.2	0	39.4	6.1	109	16.8	111.5	17.1	94.7	14.6	39.2	6
0.2	0	28.6	5.6	82.4	16.1	75.1	14.7	72.6	14.2	28.6	5.6
0.2	0	24.6	5.4	68.7	15.1	56.2	12.4	60.9	13.4	25	5.5
0.2	0.1	18.8	5	51.4	13.5	37.5	9.9	47.6	12.5	19.6	5.2
0.2	0.1	13.8	4.5	37	12.1	23.6	7.7	34.9	11.4	14.3	4.7
0.3	0.1	10.8	4	26.4	9.9	14.4	5.4	25.5	9.6	11	4.1
0.2	0.1	8.3	3.7	19.8	8.9	9.6	4.3	18.5	8.2	8.7	3.9
0.2	0.1	5.6	3.3	12.4	7.3	4.8	2.9	11	6.5	5.3	3.1
0.2	0.2	3.5	2.9	6.3	5.2	2.1	1.7	6.2	5.1	2.9	2.4
0.2	0	38.9	6	107.2	16.6	112.9	17.5	95.2	14.8	39	6.1
0.2	0	40.4	5.8	92.3	13.2	112.2	16	84.6	12.1	39.9	5.7
0.2	0	26.8	4.9	71.6	13.2	75.8	13.9	65.8	12.1	26.6	4.9
0.2	0	21.3	4.6	62.3	13.6	56.5	12.3	58.1	12.7	21.1	4.6
0.2	0.1	16.1	4.2	53.3	13.9	37.8	9.9	50.8	13.3	16.6	4.3
0.2	0.1	12.1	3.7	42.3	12.9	23.9	7.3	41.9	12.8	12.2	3.7
0.2	0.1	7.8	2.9	32.1	11.9	14.3	5.3	32.3	12	8.3	3.1
0.2	0.1	4.8	2.1	25	10.8	9.7	4.2	25.4	11	5.1	2.2
0.4	0.2	2.7	1.5	16	9.3	4.8	2.8	15.6	9	2.1	1.2
0.4	0.4	1.1	1	7.1	6.8	2.1	2	6.5	6.2	1	1
0.2	0	32.5	3.3	117.4	12	112.5	11.5	110	11.2	29.9	3
0.3	0.1	22.2	3.5	85.4	13.4	75.2	11.8	79.3	12.5	19.9	3.1
0.4	0.1	17.5	3.7	70.4	15	56.4	12	61.3	13	14.7	3.1
0.3	0.1	15.8	4.2	60	16	37.7	10.1	49.3	13.2	11.3	3
0.3	0.1	13.8	4.7	49.2	16.5	23.8	8	41.8	14.1	10.2	3.4
0.3	0.1	11.3	4.5	38.3	15.4	14.3	5.7	34.2	13.8	9.2	3.7
0.3	0.1	9.4	4.5	31.8	15.1	9.6	4.6	29.7	14.1	8.4	4
0.2	0.1	6.2	3.8	22.2	13.7	4.9	3	21.1	13.1	5.8	3.6
0.3	0.2	1.8	1.6	12.2	10.9	2.1	1.9	11.5	10.2	1.6	1.4
6.2	0.2	13.8	0.4	146.4	4.8	111.5	3.6	131.2	4.3	14.4	0.5
3.5	0.4	9.2	1.1	94.1	10.8	74.5	8.6	77.8	9	8.7	1
2.9	0.6	9.8	1.9	74.3	14.8	56.5	11.2	56	11.1	7	1.4
2	0.6	13.4	4.1	50.9	15.7	37.9	11.7	34.9	10.7	5.6	1.7
1.2	0.6	8.7	4.6	28.4	14.9	23.7	12.5	21.6	11.3	5.3	2.8
1.2	1.1	5.5	5.3	17.1	16.7	13.9	13.6	13.7	13.4	5.5	5.3
0.6	0.8	3.8	4.9	13.2	17.1	9.4	12.2	11.5	14.8	4.2	5.4

Table G.6

$w_{oil,0}$	p	T	$\dot{m}$	$\dot{x}$	$\Delta p/\Delta L$	FP	$\dot{q}$	α
0.01	26.43	-9.93	150.8	0.9	7.29	w	4	7.2
0.01	26.4	-9.92	150.8	0.91	9.04	w	1.8	3.8
0.01	0	-9.98	150.5	0.9		w	0.3	
0.014	14.88	-28.71	153.1	0.1	7.8	w	91.8	17.1
0.014	14.81	-28.85	153.5	0.1	6.51	w	63.8	13
0.014	14.82	-28.83	153.8	0.1	5.67	w	48.5	10.6
0.014	14.78	-28.92	153.2	0.09	5.41	w	28.3	6.9
0.014	14.81	-28.86	153.7	0.09	4.4	w	18.5	5.4
0.014	14.81	-28.86	153.6	0.09	4.5	w	12.6	4
0.014	14.81	-28.85	153.9	0.09	4.01	w	8.8	3.4
0.014	14.8	-28.87	154	0.09	4	w	5.6	2.7
0.014	14.81	-28.86	153.7	0.09	3.98	w	3.3	2.4
0.014	15.4	-28.88	153.8	0.09	3.8	w	0.4	
0.014	14.84	-28.78	79	0.11	4.87	s	59.8	11.4
0.014	14.8	-28.86	79.4	0.1	3.05	s	44	8.9
0.014	14.8	-28.87	80.1	0.1	3.28	s	30.5	7.2
0.014	14.83	-28.8	78.6	0.1	3.19	s	20.2	5.1
0.014	14.84	-28.79	79.1	0.09	3.84	s	13.4	3.9
0.014	14.83	-28.8	79.4	0.09	4.12	s	9.7	3.1
0.014	14.8	-28.86	79.7	0.09	4.01	s	6	2.2
0.014	14.84	-28.78	78.4	0.09	3.98	s	2.6	1.5
0.014	15.4	-28.77	78.5	0.09	4.04	s	0.4	
0.014	19.65	-20	76.3	0.11	3.69	s	53.4	11.2
0.014	19.65	-20	76.1	0.11	3.16	s	39.9	8.7
0.014	19.64	-20.01	76.3	0.11	2.77	s	28.3	6.9
0.014	19.62	-20.04	76.3	0.1	2.65	s	20.2	5.6
0.014	19.64	-20.02	76.6	0.1	2.73	s	14	4.4
0.014	19.64	-20.01	76	0.1	2.84	s	9.6	3.4
0.014	19.63	-20.04	76	0.09	2.54	s	5.4	2.6
0.014	19.64	-20.01	76.4	0.09	2.21	s	3.1	2
0.014	0	-20.04	76.2	0.09		s	0.3	
0.014	14.86	-28.69	77.7	0.49	8.15	w	63.5	12.4
0.014	14.82	-28.75	77.5	0.49	5.44	w	50.6	10.6
0.014	14.83	-28.74	78	0.48	4.4	w	45.3	9.5
0.014	14.8	-28.8	78	0.48	3.7	w	36.2	8.4
0.014	14.87	-28.67	77.5	0.47	4.71	w	27.2	7.2
0.014	14.85	-28.71	77.7	0.48	4.33	w	20.6	6.2
0.014	14.82	-28.77	78	0.47	3.49	w	15.4	5
0.014	14.81	-28.78	77.4	0.48	3.17	w	7.7	3.1
0.014	14.84	-28.73	77.4	0.48	3.18	w	3	1.8
0.014	15.5	-28.73	77.6	0.48	3	w	0.4	
0.014	14.81	-28.36	77.2	0.88	6.34	w	35.7	4.3
0.014	14.85	-28.25	76.7	0.89	5.35	w	18.3	10
0.014	14.82	-28.31	76.8	0.89	4.71	w	12.6	9.2
0.014	14.84	-28.3	77.2	0.88	4.41	w	9.7	7.8
0.014	14.8	-28.37	76.8	0.88	3.48	w	6.3	6.2
0.014	14.8	-28.39	77.2	0.88	3.4	w	3.9	4.8

Table G.6

$\dot{q}_1$	α_1	$\dot{q}_2$	α_2	$\dot{q}_3$	α_3	$\dot{q}_4$	α_4	$\dot{q}_5$	α_5	$\dot{q}_6$	α_6
0.2	0.4	1.8	3.3	8.5	15.2	4.9	8.8	7	12.5	1.7	3.1
0.2	0.4	0.4	0.8	4.4	9	2	4.2	3.5	7.2	0.4	0.8
28.6	5.3	76	14.2	122.6	22.8	112.5	21	128.4	23.9	83	15.5
19.8	4	53.8	11	85.3	17.4	75.4	15.4	87	17.7	61.5	12.5
16	3.5	41.5	9	64	13.9	56.6	12.3	67.9	14.8	45	9.8
10.4	2.6	20.7	5.1	38.6	9.5	37.5	9.2	40.2	9.9	22.4	5.5
8.9	2.6	12.6	3.7	24	7	23.3	6.8	28.4	8.3	13.9	4.1
8.2	2.6	11.6	3.7	13.6	4.4	14.2	4.6	16.8	5.4	11.2	3.6
7	2.7	7	2.7	9.4	3.6	9.5	3.6	11.5	4.4	8.3	3.2
4.5	2.1	7	3.3	4.7	2.2	4.8	2.3	5.5	2.6	7.2	3.4
3.1	2.2	5.2	3.7	1.7	1.2	2	1.4	2.9	2	5.2	3.7
2.8	0.5	41.6	7.9	93.8	17.9	75.2	14.4	95.7	18.3	49.4	9.4
1.8	0.4	29.1	5.9	69.6	14.2	56.3	11.5	70.8	14.4	36.2	7.4
1.9	0.5	23.1	5.5	46.5	11	37.9	9	47.1	11.1	26.3	6.2
1.9	0.5	16.8	4.3	28.4	7.2	23.6	6	30.1	7.6	20.3	5.1
0.2	0.1	14.2	4.1	16.8	4.9	14.1	4.1	19.3	5.6	16.1	4.7
0.2	0.1	11.3	3.6	11.2	3.5	9.5	3	11.5	3.7	14.5	4.6
0.2	0.1	8	3	5.8	2.1	4.7	1.7	6.9	2.5	10.6	3.9
0.2	0.1	4.2	2.4	1.6	0.9	2	1.2	2.4	1.4	4.9	2.8
0.2	0	38.1	8	79.3	16.6	75.2	15.7	83.5	17.5	43.9	9.2
0.2	0	28.3	6.2	60	13.1	56.3	12.3	61	13.4	33.8	7.4
0.2	0.1	21.7	5.3	42.3	10.4	37.7	9.2	43.4	10.6	24.6	6
0.2	0.1	17.9	4.9	28.9	8	23.6	6.5	29.5	8.2	21.1	5.8
0.2	0.1	15.2	4.8	17.4	5.4	14.3	4.5	18.8	5.9	18.4	5.8
0.2	0.1	11.8	4.2	10.4	3.7	9.4	3.4	11.7	4.2	14.2	5.1
0.2	0.1	8.1	3.8	4.9	2.3	4.8	2.3	5.7	2.7	8.7	4.1
0.2	0.1	5.5	3.4	1.7	1.1	2	1.3	2.7	1.7	6.8	4.3
0.2	0	52.2	10.2	76.9	15	112.2	21.9	76.7	15	62.8	12.3
0.2	0	37.8	7.9	72.2	15.1	75.2	15.7	72.2	15.1	46	9.6
0.2	0	32.3	6.8	71.5	15	56.3	11.8	71.3	15	40.2	8.5
0.2	0	23.1	5.3	62.3	14.4	37.7	8.7	63.6	14.7	29.9	6.9
0.2	0.1	15.5	4.1	50.6	13.4	23.5	6.2	52.5	13.9	21	5.5
0.2	0.1	10.5	3.2	39.7	11.9	14.3	4.3	44.2	13.3	14.7	4.4
0.2	0.1	7.8	2.5	30.6	9.9	9.5	3.1	34.3	11.1	10.2	3.3
0.2	0.1	3.3	1.3	15.6	6.3	4.8	1.9	18.6	7.5	3.7	1.5
0.2	0.1	1	0.6	6.5	4	2	1.2	7.3	4.4	1	0.6
2.2	0.3	22.9	2.8	62.5	7.5	37.6	4.5	61.6	7.4	27.6	3.3
0.2	0.1	13.4	7.4	26.7	14.6	23.7	13	28.6	15.7	17.2	9.4
0.2	0.1	9.4	6.8	18.8	13.6	14.3	10.3	21.2	15.4	12	8.7
0.2	0.2	6.7	5.4	15.3	12.4	9.5	7.6	17.9	14.5	8.3	6.7
0.2	0.2	2.9	2.8	11.3	11.1	4.8	4.7	14.5	14.1	4.1	4
0.2	0.2	1.6	2	7.2	8.7	2.4	2.9	10	12.1	2.3	2.7

Table G.6

$w_{oil,0}$	p	T	$\dot{m}$	$\dot{x}$	$\Delta p/\Delta L$	FP	$\dot{q}$	α
0.014	15.9	-28.41	76.9	0.88	3.57	w	0.4	
0.014	14.84	-28.64	77.8	0.68	7.64	w	49.2	12.1
0.014	14.8	-28.72	78.1	0.67	6.77	w	35.2	10.9
0.014	14.9	-28.51	77.4	0.69	6.02	w	28.7	9.8
0.014	14.79	-28.72	77.1	0.69	4.11	w	22.7	8.9
0.014	14.8	-28.72	77.5	0.68	4.67	w	19.2	7.9
0.014	14.79	-28.72	77.6	0.68	4.62	w	11.6	5.7
0.014	14.81	-28.68	77.3	0.69	3.52	w	4.1	2.9
0.014	15.4	-28.69	76.2	0.7	3.44	w	0.4	
0.014	19.6	-20.02	78.1	0.47	4.44	w	57	9
0.014	19.66	-19.93	77.2	0.47	5.27	w	41.2	8
0.014	19.64	-19.96	77.7	0.47	4.93	w	34.3	7.3
0.014	19.64	-19.95	78.1	0.46	4.87	w	27.3	6.4
0.014	19.64	-19.96	78.3	0.46	4.26	w	21.4	5.7
0.014	19.67	-19.92	78.2	0.46	3.75	w	16.3	4.9
0.014	19.67	-19.91	78.3	0.46	3.73	w	12.1	4.1
0.014	19.62	-20	78.2	0.46	2.48	w	7.7	3.2
0.014	19.65	-19.95	78.3	0.46	3.44	w	4.1	2.2
0.014	20.4	-19.98	78.5	0.46		w	0.3	
0.014	19.64	-19.75	79.3	0.81	5.64	w	23.1	9.7
0.014	19.67	-19.7	79.5	0.8	5.09	w	18.2	8.9
0.014	19.63	-19.74	79.1	0.81	4.43	w	14.8	8.3
0.014	19.66	-19.72	79.6	0.81	4.23	w	10.2	6.2
0.014	19.62	-19.78	79.4	0.81	4.38	w	6.7	5.1
0.014	20.3	-19.74	79.4	0.81		w	0.3	
0.014	14.85	-28.74	151.8	0.3	9.8	w	88.1	18.6
0.014	14.85	-28.74	152.2	0.29	7.76	w	64.7	14.9
0.014	14.85	-28.73	152.1	0.29	7.07	w	50.2	12.2
0.014	14.83	-28.78	152	0.29	5.67	w	32	8.5
0.014	14.85	-28.73	151.9	0.29	5.08	w	20.3	6.5
0.014	14.82	-28.8	151.9	0.29	4.62	w	10.6	4.3
0.014	14.83	-28.78	152.5	0.29	4.13	w	7.6	3.5
0.014	14.86	-28.72	152.3	0.29	4.03	w	4.3	2.5
0.014	14.85	-28.74	152.3	0.29	4.03	w	2	2.5
0.014	15.5	-28.69	151.8	0.29	4.13	w	0.4	
0.014	19.63	-20	292.7	0.3	16.85	a	93.1	27.3
0.014	19.62	-20.02	294.6	0.3	14.49	a	75.7	25.8
0.014	19.67	-19.94	297.3	0.3	13.29	a	66	24
0.014	19.64	-19.99	298.3	0.29	11.12	a	50.7	18.8
0.014	19.62	-20.02	297.2	0.3	8.38	a	0	0
0.014	19.69	-19.9	297.1	0.3	7.47	a	14.7	7.3
0.014	19.65	-19.97	297.4	0.3	7.18	a	10.6	6.3
0.014	19.64	-19.98	296.1	0.3	6.8	a	6.7	5.7
0.014	19.63	-20	296.1	0.3	6.61	a	2.4	4.7
0.014	20.2	-19.98	296.3	0.3	6.44	a	0.3	
0.014	19.62	-19.98	299	0.5	19.54	a	73.2	19.6
0.014	19.65	-19.93	298.1	0.5	16.05	a	55.7	22.5

Table G.6

$\dot{q}_1$	α_1	$\dot{q}_2$	α_2	$\dot{q}_3$	α_3	$\dot{q}_4$	α_4	$\dot{q}_5$	α_5	$\dot{q}_6$	α_6
0.2	0.1	43.4	10.7	72.9	18	56.3	13.9	74.5	18.4	48.1	11.9
0.2	0.1	26.7	8.3	56.5	17.5	37.7	11.7	58.7	18.2	31.4	9.7
0.2	0.1	20.5	7	50	17.1	23.7	8.1	53.1	18.1	24.5	8.4
0.2	0.1	14.1	5.5	44.1	17.4	14.1	5.6	46.5	18.3	17.1	6.7
0.2	0.1	11.5	4.7	38.1	15.6	9.5	3.9	41.9	17.2	14.3	5.9
0.2	0.1	4.9	2.4	24.5	12	4.8	2.3	29.6	14.5	5.5	2.7
0.2	0.1	0.2	0.1	9.1	6.5	1.9	1.4	12.8	9.1	0.2	0.1
0.2	0	34.9	5.5	76.1	12	112.5	17.7	73.8	11.6	44.4	7
0.2	0	21.6	4.2	58.7	11.4	75.1	14.6	63.1	12.2	28.2	5.5
0.2	0	17.7	3.8	52.4	11.2	56.5	12	56.2	12	22.6	4.8
0.2	0	13.9	3.2	45.1	10.5	37.6	8.8	49.5	11.5	17.8	4.2
0.2	0.1	10.8	2.9	37.9	10	23.7	6.2	42.2	11.1	13.8	3.6
0.2	0.1	8.1	2.4	30.9	9.2	14.1	4.2	34.3	10.2	10.5	3.1
0.2	0.1	5.8	2	23.6	8	9.5	3.3	25.5	8.7	7.8	2.7
0.2	0.1	3.8	1.5	15	6.2	4.9	2	17.3	7.1	4.7	1.9
0.2	0.1	1.6	0.9	8.4	4.5	2	1.1	9.8	5.2	2.8	1.5
0.2	0.1	15.5	6.5	39.7	16.6	23.6	9.9	41.3	17.3	18.3	7.6
0.2	0.1	12.7	6.2	32.9	16.2	14.3	7	34.7	17	14.5	7.1
0.2	0.1	10	5.6	28.3	15.8	9.5	5.3	29.4	16.4	11.5	6.4
0.2	0.1	5.9	3.5	21.1	12.7	4.9	3	22.4	13.5	7	4.2
0.2	0.2	2.7	2.1	15.4	11.7	2.1	1.6	15.9	12.1	3.7	2.8
10.6	2.2	91.7	19.4	109.7	23.2	112.2	23.7	111.2	23.5	93.3	19.7
0.2	0	69.4	16	82.5	19	75.6	17.4	84.4	19.5	75.8	17.5
0.2	0	52.5	12.7	65.7	15.9	56.6	13.7	66.9	16.2	59.3	14.4
0.2	0.1	24.3	6.4	44.9	11.9	37.4	9.9	48.1	12.7	36.8	9.8
0.2	0.1	12.6	4	31.9	10.2	23.8	7.6	37.3	12	16.3	5.2
0.2	0.1	7.6	3	15	6	14.1	5.7	18.1	7.3	8.7	3.5
0.2	0.1	6.2	2.9	9.9	4.6	9.5	4.4	13.1	6	6.6	3.1
0.2	0.1	4.5	2.6	5.1	3	4.8	2.8	6.3	3.7	5.1	3
0.2	0.2	3	3.6	1.8	2.2	2.1	2.5	2.7	3.2	2.5	2.9
31.6	9.3	98.7	29	107	31.4	111.6	32.8	106.3	31.2	103.5	30.4
27.9	9.5	87.9	30	83.2	28.4	75.1	25.6	87.9	30	92.2	31.4
27.9	10.1	80.3	29.2	68.8	25	56.4	20.5	76.9	28	85.7	31.2
26.1	9.7	60.1	22.3	51	18.9	37.9	14.1	62.8	23.3	66.1	24.5
0	0	0	0	0	0	0	0	0	0	0	0
5.9	3	14.9	7.4	14.7	7.3	14.2	7.1	19.9	9.9	18.7	9.3
5.9	3.5	11.8	7	10.5	6.3	9.4	5.6	13.1	7.8	13.1	7.8
5.9	4.9	8	6.8	5.7	4.8	4.8	4	7.8	6.6	8.2	6.9
3.8	7.5	3.2	6.3	1	1.9	2	4	2.4	4.7	2	4
1.6	0.4	55.5	14.9	94.9	25.5	112.5	30.2	109.1	29.3	65.3	17.5
3.5	1.4	45.5	18.4	71.6	28.9	75.2	30.4	83.4	33.7	55.2	22.3

Table G.6

$w_{oil,0}$	p	T	$\dot{m}$	$\dot{x}$	$\Delta p/\Delta L$	FP	$\dot{q}$	α
0.014	19.64	-19.94	298.7	0.5	14.98	a	45.4	19.9
0.014	19.64	-19.95	299	0.49	11.86	a	33.2	15.1
0.014	19.68	-19.87	299	0.5	11.25	a	27.1	14.1
0.014	19.64	-19.95	298.1	0.5	10.87	a	18.3	11.5
0.014	19.62	-19.97	297.7	0.5	10.5	a	10.6	8.1
0.014	19.62	-19.97	297.8	0.5	9.39	a	4.6	5.6
0.014	19.63	-19.96	297.7	0.5	8.96	a	1.8	5.8
0.014	20.3	-19.93	297.6	0.5	8.94	a	0.3	
0.014	19.68	-19.95	300.4	0.1	11.73	i	100.6	24.3
0.014	19.63	-20.02	302.8	0.09	10.19	i	77.8	21.6
0.014	19.62	-20.04	302.8	0.09	9.22	i	64.1	19.9
0.014	19.64	-20.01	302.7	0.09	8.08	i	47.4	16.4
0.014	19.6	-20.08	302.8	0.09	6.89	i	30.2	12.3
0.014	19.65	-19.99	299.6	0.09	5.51	i	16.2	7.5
0.014	19.65	-19.99	302	0.09	5.22	i	10.5	5.9
0.014	19.6	-20.07	301.2	0.09	4.68	i	6.1	4.8
0.014	19.61	-20.06	300.5	0.09	4.5	i	3.3	4.5
0.014	20.3	-19.97	301.3	0.09	4.17	i	0.3	
0.014	19.66	-19.35	150.1	0.9	13.37	w	42.6	3.2
0.014	19.62	-19.42	149.6	0.9	11.87	w	30	6.1
0.014	19.6	-19.44	149.8	0.9	9.87	w	17.9	8.5
0.014	19.66	-19.34	150.1	0.9	8.11	w	9.9	7.5
0.014	19.63	-19.39	150.3	0.9	6.18	w	5.9	5.1
0.014	19.64	-19.35	150.2	0.9	5.37	w	3	3.5
0.014	19.6	-19.46	150.4	0.9	4.87	w	1.2	2.9
0.014	20.4	-19.43	150.7	0.9		w	0.3	
0.03	14.88	-28.61	304.3	0.09	18.92	i	94.5	0
0.03	14.84	-28.69	304.6	0.09	16.11	i	77.3	0.1
0.03	14.83	-28.71	304.6	0.09	14.83	i	67.7	0
0.03	14.83	-28.71	304.6	0.09	13.24	i	55.1	0
0.03	14.87	-28.64	302.8	0.09	13.18	i	51.5	0
0.03	14.83	-28.72	303	0.09	11.72	i	38.1	0
0.03	14.82	-28.73	303.1	0.09	10.67	i	30	0
0.03	14.84	-28.69	302.3	0.09	5.98	i	11.2	0
0.03	14.84	-28.68	302.4	0.09	4.95	i	5.1	0
0.03	15.7	-28.69	302	0.11	4.62	i	0.4	
0.03	14.81	-28.7	302.6	0.3	38.17	a	129	7.5
0.03	14.84	-28.65	303.5	0.29	26.96	a	72.3	27.3
0.03	14.84	-28.64	300.7	0.29	23.06	a	50.9	22.6
0.03	14.83	-28.66	300.7	0.29	15.27	a	35.7	18.1
0.03	14.84	-28.65	300.9	0.29	12.24	a	28.5	15.4
0.03	14.82	-28.68	300.8	0.29	10.62	a	20	11.8
0.03	14.8	-28.72	300.2	0.3	10.42	a	14.2	9.3
0.03	14.79	-28.74	300.3	0.3	9.57	a	7	6.1
0.03	14.82	-28.67	300.3	0.29	8.96	a	2.1	4.4
0.03	15.7	-28.69	300	0.31	8.9	a	0.4	
0.03	19.65	-19.9	302.6	0.1	15.7	i	89.1	18.7

$\dot q_1$	α_1	$\dot q_2$	α_2	$\dot q_3$	α_3	$\dot q_4$	α_4	$\dot q_5$	α_5	$\dot q_6$	α_6
2	0.9	38.3	16.7	59.1	25.8	56.5	24.7	69	30.2	47.6	20.8
1.1	0.5	27.6	12.6	44.9	20.4	37.7	17.2	49.1	22.4	38.6	17.6
4.5	2.3	26.1	13.6	32.9	17.1	23.7	12.4	39.8	20.7	35.9	18.7
9.3	5.8	16.4	10.2	19.9	12.5	14.2	8.9	24	15	26.1	16.3
7.1	5.4	9.5	7.2	11.6	8.8	9.5	7.2	13.4	10.2	12.8	9.7
4.4	5.4	3.2	3.9	4.8	5.9	4.8	5.9	6.5	7.9	3.6	4.4
2.2	7	1.3	4.2	1.6	5.2	1.9	6.1	1.8	5.7	2	6.6
80.5	19.5	101.2	24.5	104.6	25.3	112.3	27.2	105.8	25.6	99.1	24
70.4	19.5	87.9	24.4	74	20.6	75	20.8	74	20.6	85.5	23.8
64.3	20	79.3	24.6	55	17.1	56.4	17.5	54.9	17	74.5	23.1
51.4	17.8	64.5	22.3	34.2	11.8	37.8	13.1	35	12.1	61.6	21.3
35.1	14.2	39.2	15.9	20.6	8.4	23.7	9.6	22.7	9.2	40.2	16.3
18.8	8.7	19.2	8.8	11.5	5.3	14.1	6.5	14.4	6.6	19	8.7
12.4	7	12	6.8	7.6	4.3	9.5	5.4	9.1	5.2	12.3	7
8	6.4	7.9	6.3	3.7	2.9	4.8	3.8	4.5	3.6	7.5	6
5.1	6.9	4.9	6.7	1.1	1.5	2.1	2.9	2.1	2.9	4.5	6.1
0.2	0	12.7	0.9	72.9	5.4	56.4	4.2	88.9	6.6	24.6	1.8
0.2	0	8.2	1.7	48.8	10	37.6	7.7	66	13.5	19.6	4
0.2	0.1	6.3	3	28.3	13.4	23.6	11.2	35.7	16.9	13.4	6.3
0.2	0.2	4.1	3.1	17.8	13.5	14.1	10.7	16.6	12.6	6.4	4.9
0.2	0.2	1.6	1.4	10.1	8.7	9.5	8.2	10.4	9	3.3	2.8
0.2	0.2	0.2	0.2	5.5	6.5	4.9	5.8	5.5	6.6	1.4	1.7
0.2	0.5	0.2	0.5	1.6	3.8	1.9	4.7	2.3	5.7	0.8	2
86.6	0	85.7	0	101.4	0	112.1	0	99.8	0	81.7	0
78.6	0.1	83.2	0.1	73.3	0.1	75.1	0.1	73.3	0.1	80.5	0.1
75.6	0	79.5	0	58.3	0	56.5	0	58.8	0	77.3	0
69	0	72.3	0	40	0	37.5	0	41.6	0	70	0
66.9	0	68.8	0	38.3	0	23.5	0	41.9	0	69.4	0
50.6	0	57.6	0	23.3	0	14.3	0	25	0	58.1	0
41.3	0	45.9	0	16.2	0	9.5	0	18.3	0	48.4	0.1
12.6	0	14.2	0	9.5	0	5	0	10.2	0	15.5	0
6.3	0	6.7	0	3.9	0	2.3	0	4.6	0	7	0
123.7	7.2	128.4	7.5	131.3	7.6	112.7	6.5	147.1	8.5	130.6	7.6
63.2	23.8	71.8	27.1	66.6	25.1	75.4	28.4	80.7	30.4	76.1	28.7
29.9	13.3	56.7	25.2	43.6	19.4	56.5	25.1	57.1	25.4	61.6	27.4
13	6.6	43.4	22	30.1	15.3	37.8	19.2	40.7	20.6	49	24.9
10.2	5.5	34.1	18.4	27.2	14.7	23.7	12.8	33.3	18	42.4	22.9
9.3	5.5	22.2	13.1	19.7	11.7	14.3	8.4	23.7	14.1	30.8	18.3
8	5.2	14.8	9.7	15	9.8	9.5	6.3	18.2	11.9	19.9	13.1
6.4	5.6	6.8	5.9	6.6	5.7	5	4.3	9.9	8.6	7.4	6.5
2.3	4.8	2	4.1	1.5	3.1	2	4.1	3.5	7.4	1.4	3
81.2	17	74.2	15.6	98.6	20.7	111.8	23.4	96.4	20.2	72.4	15.2

Table G.6

$w_{oil,0}$	p	T	$\dot{m}$	$\dot{x}$	$\Delta p/\Delta L$	FP	$\dot{q}$	α
0.03	19.63	-19.93	302.6	0.1	13.55	i	70	18.4
0.03	19.63	-19.95	302.3	0.1	12.44	i	59.7	17.5
0.03	19.66	-19.89	300.8	0.1	11.49	i	49.3	16.7
0.03	19.66	-19.89	299.1	0.1	10.63	i	38.9	15.8
0.03	19.63	-19.94	301.4	0.09	9.87	i	30.1	15
0.03	19.64	-19.92	301.7	0.09	9.08	i	22.8	13.5
0.03	19.64	-19.92	300.4	0.09	6.44	i	11.7	7.8
0.03	19.63	-19.93	299.9	0.1	5.93	i	4.9	5.4
0.03	20	-19.98	301	0.11	4.66	i	0.3	
0.03	19.69	-19.79	301.6	0.29	30.88	a	96.7	10.2
0.03	19.68	-19.81	299	0.29	25.1	a	59.4	26.7
0.03	19.63	-19.89	300.2	0.29	22.18	a	50.5	24
0.03	19.67	-19.84	300.4	0.29	15.28	a	39.9	20.5
0.03	19.64	-19.88	300.2	0.29	12.7	a	33.3	17.6
0.03	19.61	-19.93	299.9	0.3	10.88	a	27.1	15.6
0.03	19.61	-19.92	300.4	0.29	9.18	a	20.7	12.6
0.03	19.63	-19.89	299.7	0.29	7.32	a	12.5	8.9
0.03	19.68	-19.81	299.8	0.29	7.44	a	3.9	5.4
0.03	20	-19.82	299.6	0.31	7.31	a	0.3	
0.03	26.92	-9.25	301.6	0.1	10.32	i	80.5	13.4
0.03	26.95	-9.22	301.8	0.09	8.83	i	60.7	14.3
0.03	26.95	-9.22	302.3	0.09	8.24	i	49.3	13.9
0.03	26.94	-9.23	302.2	0.08	7.86	i	39.4	13.3
0.03	26.89	-9.3	301.7	0.09	7.52	i	30.7	12.6
0.03	26.9	-9.29	299.5	0.09	7.1	i	24.3	12
0.03	26.92	-9.25	303.1	0.09	6.45	i	19.7	11.6
0.03	26.91	-9.27	301.7	0.09	4.82	i	13	9.2
0.03	26.9	-9.28	301.9	0.09	4.18	i	7.4	6.6
0.03	27.4	-9.25	302.5	0.11	3.63	i	0.2	
0.03	26.95	-9.16	297.5	0.3	15.32	a	84.7	8.3
0.03	26.94	-9.17	297.2	0.3	12.71	a	60.5	11.5
0.03	26.96	-9.15	297	0.3	11.51	a	51.4	15
0.03	26.91	-9.21	296.4	0.3	8.69	a	39.3	12.3
0.03	26.94	-9.19	298.2	0.3	7.18	a	29.7	10
0.03	26.92	-9.21	298.4	0.3	6.09	a	21.5	8
0.03	26.9	-9.23	297.6	0.3	5.69	a	13.9	6.1
0.03	26.94	-9.18	297.8	0.3	5.37	a	7.2	3.7
0.03	26.93	-9.19	297.7	0.3	4.87	a	3.7	2.5
0.03	27.7	-9.16	297.3	0.32	5.13	a	0.2	
0.03	19.67	-19.87	150.1	0.12	5.96	w	76.3	12.2
0.03	19.66	-19.88	150.1	0.11	4.76	w	56.6	10.2
0.03	19.67	-19.87	150	0.11	4.13	w	43.1	8.6
0.03	19.61	-19.97	150.2	0.11	3.78	w	30.4	6.7
0.03	19.59	-20	150.3	0.11	3.76	w	22.1	5.4
0.03	19.62	-19.96	150.6	0.1	3.77	s-w	16.6	4.6
0.03	19.62	-19.95	150.8	0.1	3.73	w	12.4	4
0.03	19.64	-19.92	150.6	0.1	3.2	w	8	3

Table G.6

$\dot{q}_1$	α_1	$\dot{q}_2$	α_2	$\dot{q}_3$	α_3	$\dot{q}_4$	α_4	$\dot{q}_5$	α_5	$\dot{q}_6$	α_6
69.4	18.2	70.1	18.4	69.4	18.2	75.8	19.9	67.3	17.7	68.1	17.9
64.3	18.8	65.7	19.2	54	15.8	56.4	16.5	54.2	15.9	63.7	18.7
58.2	19.7	61.3	20.7	38.7	13.1	37.8	12.8	40.2	13.6	59.9	20.2
50.7	20.7	54	22	26.1	10.6	23.5	9.6	27.1	11	51.7	21.1
40.2	20.1	45.8	22.8	16.9	8.4	14.3	7.1	19.2	9.6	44.3	22.1
31.9	18.9	33.7	19.9	12.8	7.6	9.5	5.7	13	7.7	36.1	21.4
14.1	9.5	16.3	10.9	8	5.4	4.8	3.2	8.6	5.8	18.1	12.2
6.3	7	6.5	7.3	3.2	3.6	2.1	2.3	4	4.4	7.1	7.9
93	9.8	80.2	8.4	95.7	10.1	112.5	11.8	117	12.3	81.9	8.6
54.7	24.6	54.6	24.5	53.2	23.9	75.4	33.9	62.1	27.9	56.4	25.4
38.7	18.4	53.4	25.4	46.8	22.3	56.3	26.8	53.6	25.5	54.1	25.8
21.3	11	47.6	24.4	38.1	19.6	36.6	18.8	47.3	24.3	48.4	24.9
16.8	8.9	42.4	22.4	31.3	16.5	23.6	12.5	39.9	21.1	46	24.3
12.5	7.2	37	21.2	23.6	13.6	14.3	8.2	31.9	18.3	43.3	24.9
7.7	4.7	27.6	16.8	18.9	11.6	9.4	5.8	23.3	14.3	37.2	22.7
1	0.7	16.5	11.7	12.1	8.6	4.9	3.5	15.2	10.8	25.4	18
1.8	2.5	4.5	6.2	3.4	4.7	2.1	2.9	5.7	7.9	5.7	7.9
54.2	9	57.5	9.6	100.3	16.7	112	18.7	100.9	16.8	58.3	9.7
44.9	10.6	51.6	12.1	69.9	16.4	75.8	17.8	69.8	16.4	52.4	12.3
39.6	11.1	46.1	13	53.2	15	56.5	15.9	52.7	14.8	47.4	13.3
36	12.1	42.2	14.2	39	13.1	37.9	12.8	38.4	12.9	42.8	14.4
32.5	13.3	37.2	15.2	26.3	10.8	23.9	9.8	27.1	11.1	37.2	15.2
28.5	14.1	33.5	16.5	16.7	8.3	14.2	7	17.9	8.9	35	17.3
25	14.8	27.3	16.1	12.8	7.6	9.5	5.6	13.1	7.7	30.3	17.9
16.6	11.7	19.8	14	8	5.7	4.8	3.4	8.9	6.3	20	14.2
10.3	9.2	10.5	9.4	4.8	4.3	2	1.8	5.7	5.1	10.7	9.6
35	3.5	77.3	7.6	90.4	8.9	112.3	11.1	123.6	12.2	69.4	6.8
28.9	5.5	52.9	10.1	62.2	11.9	75.3	14.3	90.4	17.2	53.4	10.2
33.5	9.8	51.5	15	54.4	15.9	56.3	16.4	61.4	17.9	51	14.9
18.3	5.8	42.1	13.2	43.3	13.6	37.9	11.9	49.2	15.4	45.2	14.2
5.4	1.8	37	12.5	32.6	11	23.9	8.1	35.9	12.1	43.5	14.7
2	0.7	26	9.7	23.5	8.8	14.1	5.3	26.1	9.7	37.2	13.9
1.4	0.6	15.7	6.8	15.1	6.6	9.5	4.1	17.1	7.4	24.8	10.8
0.2	0.1	8.7	4.4	6.7	3.4	4.8	2.5	8.9	4.5	14	7.1
0.2	0.1	5.2	3.5	3.2	2.2	2.2	1.5	4.6	3.2	6.8	4.7
26.3	4.2	59.5	9.5	98.4	15.8	112.5	18	102.7	16.4	58.3	9.3
18.6	3.4	46.9	8.5	76.3	13.7	75.6	13.6	74.4	13.4	48	8.7
13.2	2.6	37	7.3	57.1	11.3	56.4	11.2	56.3	11.2	38.7	7.7
11.2	2.5	27.4	6	38.9	8.6	37.7	8.3	39.2	8.6	28.1	6.2
11.3	2.8	21.2	5.2	26.7	6.6	23.6	5.8	27.4	6.7	22.4	5.5
10.2	2.8	18.5	5.1	18.8	5.2	14.3	3.9	19.2	5.3	18.6	5.1
9.8	3.1	15.3	4.9	11.5	3.7	9.5	3.1	11.9	3.8	16.2	5.2
5.9	2.2	11.2	4.2	6.8	2.5	4.9	1.8	7.2	2.7	11.9	4.4

Table G.6

$w_{oil,0}$	p	T	$\dot{m}$	$\dot{x}$	$\Delta p/\Delta L$	FP	$\dot{q}$	α
0.03	19.68	-19.86	150.5	0.1	2.99	w	4.3	2
0.03	21	-19.89	150.9	0.11	1.22	w	0	
0.03	26.94	-9.22	148.1	0.11	2.48	w	71.6	9.1
0.03	26.94	-9.22	148.5	0.1	1.94	w	51.6	8.4
0.03	26.95	-9.21	148.2	0.1	5.54	w	41.1	7.5
0.03	26.93	-9.24	148.2	0.1	1.36	w	28.7	6.4
0.03	26.93	-9.24	148.5	0.1	1.3	w	19.8	5
0.03	26.95	-9.21	148.5	0.1	1.27	w	14.5	4.2
0.03	27.01	-9.14	147.5	0.1	1.2	w	11.4	3.7
0.03	26.99	-9.17	148.4	0.1	0.76	w	7	2.8
0.03	26.96	-9.2	148.3	0.09	0.86	w	4	2.1
0.03	28	-9.22	148.2	0.11	0.98	w	0.2	
0.03	26.93	-9.1	146.5	0.52	7.65	w	68.6	2.7
0.03	26.97	-9.06	147.4	0.51	6.6	w	61.3	2.9
0.03	26.93	-9.1	147.7	0.51	5.89	w	54.4	3.3
0.03	26.95	-9.09	147.5	0.51	3.44	w	23.8	6.4
0.03	26.95	-9.09	147.8	0.51	2.84	w	17.7	6
0.03	26.97	-9.06	147.9	0.51	2.61	w	14.6	5.4
0.03	26.95	-9.08	147.4	0.51	1.91	w	9.9	4.7
0.03	26.95	-9.09	147.7	0.51	2.32	w	5.8	3.4
0.03	28	-9.15	147.9	0.52	2.13	w	0.2	
0.03	26.97	-9	152.8	0.59	7.86	w	56.9	2.7
0.03	26.95	-9.05	152.6	0.56	6.02	w	31.2	6.2
0.03	26.96	-8.91	152.4	0.7	5.16	w	13.9	7.1
0.03	26.98	-8.88	152.4	0.7	3.71	w	12.3	6
0.03	26.94	-8.93	152.5	0.7	3.04	w	8.9	5.3
0.03	26.95	-8.92	152.5	0.7	2.69	w	5.6	4.8
0.03	0	-8.94	152.7	0.7	1.79	w	0.2	
0.03	26.92	-9.21	156.9	0.28	4.82	w	64.8	7.1
0.03	26.93	-9.2	156.5	0.28	4.12	w	49.1	8.5
0.03	26.94	-9.19	156.6	0.28	3.69	w	41.3	8.3
0.03	26.96	-9.16	156.3	0.28	3.38	w	35.7	8.2
0.03	26.9	-9.24	156.4	0.28	3.13	w	28.4	7.7
0.03	26.91	-9.23	154.9	0.29	2.06	w	19.5	5.9
0.03	26.91	-9.23	155.2	0.29	1.66	w	13.7	4.4
0.03	26.91	-9.22	155.9	0.28	1.59	w	10.1	3.9
0.03	26.91	-9.22	155.8	0.28	1.2	w	5.1	2.6

Table G.6

$\dot{q}_1$	α_1	$\dot{q}_2$	α_2	$\dot{q}_3$	α_3	$\dot{q}_4$	α_4	$\dot{q}_5$	α_5	$\dot{q}_6$	α_6
3.3	1.5	6.3	3	3.3	1.5	2	0.9	3.7	1.7	7.4	3.5
20.9	2.7	50.3	6.4	96.7	12.3	112.4	14.3	97.9	12.5	51.6	6.6
11	1.8	41.2	6.7	69.7	11.3	75.5	12.2	69.8	11.3	42.6	6.9
8.5	1.6	34.4	6.3	55.5	10.2	56.4	10.4	55.5	10.2	35.9	6.6
5.7	1.3	27.8	6.2	35.9	8	37.8	8.4	36.2	8.1	28.7	6.4
6	1.5	21.5	5.4	22.7	5.7	23.4	5.9	22.7	5.7	22.3	5.6
5.6	1.6	18.5	5.4	14.4	4.2	14.1	4.1	15.8	4.6	18.6	5.4
5.4	1.8	16.2	5.3	10.4	3.4	9.5	3.1	10.5	3.4	16.2	5.3
2.9	1.2	12.8	5.1	4.2	1.7	4.8	1.9	4.5	1.8	13	5.2
0.2	0.1	9.1	4.7	1.3	0.7	2.3	1.2	1.8	1	9.5	4.9
2	0.1	64.5	2.5	93	3.7	75.4	3	99.7	3.9	77.1	3
1.5	0.1	59.3	2.8	85.1	4.1	56.4	2.7	91.7	4.4	73.6	3.5
0.2	0	55.1	3.3	78.4	4.7	37.6	2.3	87.9	5.3	67.4	4
0.2	0.1	19.6	5.3	36.2	9.7	23.7	6.4	38.9	10.4	24.1	6.5
0.2	0.1	15.4	5.3	28.6	9.8	14.2	4.8	29.5	10.1	18.1	6.2
0.2	0.1	13.4	5	24.1	9	9.5	3.5	24.8	9.2	15.3	5.7
0.2	0.1	9.1	4.3	16.6	7.9	4.8	2.3	17.5	8.3	11.2	5.3
0.2	0.1	5.1	3	8.9	5.2	2	1.2	11.4	6.6	7.3	4.2
0.2	0	46.7	2.2	95.2	4.5	37.7	1.8	99.8	4.7	61.9	2.9
0.2	0	19.5	3.9	48.6	9.7	23.7	4.7	57.2	11.4	38	7.6
0.2	0.1	9.3	4.8	20.8	10.6	14.1	7.2	24.3	12.4	14.7	7.5
0.2	0.1	9.1	4.5	20.3	10	9.3	4.6	22.9	11.3	11.7	5.8
0.2	0.1	7	4.2	15.3	9.1	4.8	2.8	16.4	9.8	10	5.9
0.2	0.2	4.8	4.1	8.5	7.3	2	1.7	11.1	9.5	7	6
30.8	3.4	49.2	5.4	76.1	8.4	112.4	12.4	72	7.9	48.2	5.3
25.7	4.5	42.6	7.4	57	9.9	75.2	13	54	9.4	40	6.9
24.8	5	39.9	8	45.8	9.2	56.5	11.4	44.5	9	36.6	7.4
22.4	5.2	36.4	8.4	42.6	9.8	37.4	8.6	39.8	9.2	35.5	8.2
18.1	4.9	34	9.2	31.4	8.5	23.9	6.4	32.1	8.6	31.1	8.4
4.6	1.4	20.3	6.2	26.9	8.2	14.1	4.3	28.6	8.7	22.4	6.8
0.2	0.1	15.3	5	19.8	6.4	9.5	3.1	18.3	5.9	19.1	6.2
0.2	0.1	11.6	4.5	15	5.8	4.8	1.8	14.7	5.7	14.3	5.5
0.2	0.1	6.8	3.4	5.7	2.9	2.1	1.1	6.4	3.2	9.6	4.8

Table G.6

Table G.7: Smooth tube: Diabatic pressure drop measurements

$w_{\text{oil},0}$	p	T	$\dot{m}$	$\dot{x}$	$\Delta p/\Delta L$	FP	$\dot{q}$	α
0.03	28.53	-10.21	150	0.12	2.82	w	80.2	
0.03	28.58	-10.15	150	0.12	2.28	w	57.2	
0.03	28.55	-10.2	150	0.12	1.85	w	45.9	
0.03	28.59	-10.14	150	0.11	1.43	w	33.6	
0.03	28.55	-10.19	150	0.11	1.03	w	23	
0.03	28.59	-10.13	150	0.11	0.85	w	16.2	
0.03	28.58	-10.13	150	0.11	0.68	w	12.1	
0.03	28.58	-10.14	150	0.11	0.66	w	7.5	
0.03	28.62	-10.09	150	0.11	0.35	w	3.8	
0.03	28.61	-10.06	150	0.11	0.21	w	0	
0.03	28.62	-10.13	150	0.32	3.45	w	57.9	
0.03	28.58	-10.17	150	0.31	2.78	w	43.5	
0.03	28.54	-10.21	150	0.31	2.08	w	35.7	
0.03	28.6	-10.13	150	0.31	1.94	w	28.3	
0.03	28.58	-10.16	150	0.31	1.78	w	21.4	
0.03	28.57	-10.17	150	0.31	1.53	w	16	
0.03	28.57	-10.18	150	0.31	1.36	w	12.7	
0.03	28.6	-10.15	150	0.31	1.11	w	8.6	
0.03	28.6	-10.13	150	0.31	0.94	w	4.5	
0.03	28.59	-10.09	150	0.31	0.72	w	0	
0.03	28.64	-10.11	150	0.52	7.42	w	84.9	
0.03	28.64	-10.13	150	0.51	5.79	w	64.5	
0.03	28.67	-10.1	150	0.51	4.93	w	41.5	
0.03	28.63	-10.15	150	0.51	3.88	w	30.5	
0.03	28.59	-10.2	150	0.51	3.53	w	23.5	
0.03	28.64	-10.13	150	0.51	3.49	w	18.6	
0.03	28.63	-10.15	150	0.51	3.3	w	15.1	
0.03	28.64	-10.14	150	0.51	3.11	w	11.3	
0.03	28.62	-10.15	150	0.51	2.63	w	7.1	
0.03	28.66	-10.08	150	0.51	1.85	w	0	
0.03	28.67	-10.01	149.6	0.71	9.57	w	66.2	
0.03	28.64	-10.07	149.2	0.72	8.16	w	41.9	
0.03	28.64	-10.09	149.9	0.71	7.14	w	30.5	
0.03	28.63	-10.1	148.2	0.72	5.84	w	21.2	
0.03	28.62	-10.11	148.8	0.72	5.18	w	15.8	
0.03	28.62	-10.11	148.9	0.72	4.37	w	12.8	
0.03	28.6	-10.14	150	0.71	4.16	w	9.5	
0.03	28.59	-10.15	148.9	0.72	3.56	w	5.1	
0.03	28.64	-10.08	149.3	0.72	2.29	w	0	
0.014	20.6	-21.48	300.6	0.71	23.39	a	70.2	
0.014	20.6	-21.47	300.1	0.71	21.3	a	55	
0.014	20.5	-21.58	302.7	0.7	20.58	a	47.3	
0.014	20.5	-21.47	306	0.7	17.35	a	32.4	
0.014	20.5	-21.52	303.4	0.7	15.15	a	21.3	
0.014	20.6	-21.52	303.1	0.72	14.3	a	11	

$w_{oil,0}$	p	T	$\dot{m}$	$\dot{x}$	$\Delta p/\Delta L$	FP	$\dot{q}$	α
0.014	20.5	-21.43	302.6	0.72	13.91	a	7.4	
0.014	20.6	-21.47	301.1	0.73	13.1	a	3.1	
0.014	20.6	-21.4	302.4	0.72	12.75	a	1	
0.014	20.5	-21.5	300.2	0.73	12.67	a	0	
0.03	20	-21.44	300.8	0.51	24.44	a	80.8	
0.03	20	-21.44	300.9	0.51	23.68	a	60.3	
0.03	19.9	-21.49	300.3	0.51	19.29	a	37.2	
0.03	20.2	-21.43	300.4	0.51	15.41	a	27.4	
0.03	20.3	-21.38	300	0.51	12.9	a	20.7	
0.03	20.3	-21.4	299.5	0.51	11.73	a	15.4	
0.03	20.2	-21.37	298.4	0.51	11.8	a	10.7	
0.03	20.2	-21.44	297	0.51	10.6	a	4.7	
0.03	20.2	-21.45	298.5	0.51	10.17	a	1.2	
0.03	20.2	-21.43	298.8	0.51	10.02	a	0	
0.03	27.8	-9.87	297.2	0.52	18.84	a	80.5	
0.03	27.8	-9.93	297.1	0.52	12.35	a	37.6	
0.03	27.8	-9.86	296.6	0.52	10.66	a	29	
0.03	27.8	-9.91	294.1	0.52	9.66	a	21.7	
0.03	27.8	-9.96	295.7	0.52	7.48	a	17	
0.03	27.8	-9.91	295.6	0.52	7.49	a	13.5	
0.03	27.8	-9.99	295.9	0.52	7.01	a	7	
0.03	27.8	-9.97	296.5	0.52	6.72	a	3.2	
0.03	27.8	-9.99	296.9	0.52	6.54	a	0	
0.03	21	-20.74	150	0.52	16.1	w	99.6	
0.03	21.1	-20.71	149.4	0.52	14.54	w	81.9	
0.03	21.1	-20.68	150	0.51	12.2	w	67	
0.03	21.1	-20.61	150.4	0.51	8.98	w	35	
0.03	21.1	-20.6	150.4	0.51	7.01	w	26.6	
0.03	21.1	-20.6	150.5	0.51	6.27	w	23.4	
0.03	21.1	-20.59	150.5	0.51	5.36	w	19.1	
0.03	21.1	-20.68	150.4	0.51	4.98	w	14.1	
0.03	21.1	-20.58	150.2	0.51	4.52	w	6.3	
0.03	21.1	-20.59	150.1	0.51	3.8	w	0	
0.03	21	-20.72	149.1	0.72	14.61	a	65.5	
0.03	21.2	-20.64	149.7	0.72	14.67	a	64.5	
0.03	21.2	-20.55	149.8	0.72	13.25	a	50.3	
0.03	21.1	-20.7	147.9	0.72	9.64	a	20	
0.03	21.2	-20.6	149.7	0.71	7.26	a	13.4	
0.03	21.2	-20.63	149.9	0.71	5.46	a	9.8	
0.03	21.2	-20.57	149.6	0.71	4.58	a	6.7	
0.03	21.2	-20.62	149.5	0.71	4.9	a	2.7	
0.03	21.2	-20.56	149.6	0.71	4.46	a	0	
0.03	0	-9.96	74.9	0.52	1.74	w	44.6	
0.03	0	-9.97	75.1	0.51	2.19	w	38.5	
0.03	0	-9.94	75.9	0.54	1.85	w	56.2	
0.03	0	-9.99	75.8	0.52	1.34	w	44.8	
0.03	0	-9.96	75.3	0.51	2.13	w	17.2	

Table G.7

$w_{oil,0}$	p	T	$\dot{m}$	$\dot{x}$	$\Delta p/\Delta L$	FP	$\dot{q}$	α
0.03	0	-9.93	75.4	0.52	1.84	w	11.9	
0.03	0	-9.93	75.8	0.5	1.48	w	9.1	
0.03	28	-9.84	80.3	0.13	1.36	s	73.5	
0.03	28	-9.86	80.5	0.12	0.55	s	48.1	
0.03	28	-9.8	80.5	0.12	0.39	s	38	
0.03	28	-9.87	80.5	0.11	0.35	s	26	
0.03	28	-9.85	80.4	0.1	0.61	s	16.9	
0.03	28.3	-9.89	80.4	0.1	0.59	s	11.2	
0.03	28.3	-9.89	80.4	0.1	0.32	s	7.9	
0.03	28.3	-9.96	80.5	0.1	0.18	s	4.6	
0.03	28.3	-9.91	80.5	0.1	0.01	s	2.1	
0.03	28.3	-9.91	80.4	0.1	2.26	s	0	
0.03	28	-9.85	74.7	0.33	1.7	s-w	51.7	
0.03	28	-9.81	74.3	0.32	1.53	s-w	35.9	
0.03	28	-9.87	75.1	0.3	1.28	s-w	26.7	
0.03	28	-9.79	75.4	0.29	1.29	s-w	18.6	
0.03	28	-9.79	74.9	0.31	1.03	s-w	12.3	
0.03	28	-9.88	75	0.3	0.93	s-w	7.9	
0.03	28	-9.82	75.6	0.31	0.75	s-w	5.3	
0.03	28	-9.82	75.5	0.31	0.05	s-w	3.4	
0.03	28	-9.79	75.9	0.31	0.06	s-w	2	
0.03	26	-9.83	75.7	0.31		s-w	0	
0.03	26	-9.91	75.1	0.54	2.42	w	65.3	
0.03	26	-9.92	75.2	0.51	1.53	w	17.2	
0.03	26	-9.94	75.3	0.51	1.32	w	11.3	
0.03	26	-9.91	75.5	0.51	0.91	w	8	
0.03	26	-9.84	75.3	0.51	1.41	w	5.4	
0.03	26	-9.76	75.4	0.51		w	0	
0.03	26	-9.76	75.4	0.51		w	0	

Table G.7

Table G.8: Smooth tube: Adiabatic flow pattern measurements

$w_{\text{oil},0}$	p	T	$\dot{m}$	$\dot{x}$	$\Delta p/\Delta L$	FP	$\dot{q}$	α
0.01	26.4	-10	74.5	0.09	5.09	s-w		
0.01	26.4	-10	74.9	0.09	4.36	s-w		
0.01	26.4	-10	75.1	0.19	4.37	s-w		
0.01	26.4	-10	74.7	0.29	4.7	s-w		
0.01	26.4	-10	74.8	0.29	4.72	s-w		
0.01	26.4	-10	74.5	0.4	4.9	s-w		
0.01	26.4	-10	74.1	0.5	4.73	s-w		
0.01	26.4	-10	75	0.49	4.27	s-w		
0.01	26.4	-10	74.5	0.54	5.23	w		
0.01	26.4	-10	74.4	0.6	5.22	w		
0.01	26.4	-10	75.1	0.59	4.39	w		
0.01	26.4	-10	73.9	0.66	5.65	w		
0.01	26.4	-10	75.9	0.68	5.74	w		
0.01	26.4	-10	75	0.74	4.47	w		
0.01	26.4	-10	75.3	0.8	6.06	w/f		
0.01	26.4	-10	74.5	0.8	4.85	w/f		
0.01	26.4	-10	75	0.86	5.99	w/f		
0.01	26.4	-10	74.6	0.9	5.79	w/f		
0.01	26.4	-10	74.5	0.91	4.98	w/f		
0.01	26.4	-10	74.6	0.96	4.89	w/f		
0.01	26.4	-10	74.8	0.11	7.97	s-w		
0.01	26.4	-10	73.9	0.32	3.12	s-w		
0.01	26.4	-10	74.7	0.52	5.45	s-w		
0.01	26.4	-10	75.3	0.72	4.11	w		
0.01	26.4	-10	74.5	0.92	4.72	w/f		
0.01	26.4	-10	76.1	0.09	1.51	s-w		
0.01	26.4	-10	75	0.3	0.15	s-w		
0.01	26.4	-10	74.3	0.5	0.45	s-w		
0.01	26.4	-10	74.3	0.6	1.16	w		
0.01	26.4	-10	74.6	0.69	3.16	w		
0.01	26.4	-10	75.9	0.79	-0.35	w/f		
0.01	26.4	-10	75.5	0.88	-0.35	w/f		
0.01	26.4	-10	75.4	0.94	0.94	w/f		
0.01	26.4	-10	75.4	0.98	-0.35	w/f		
0.01	26.4	-10	149.2	0.09	4.82	s-w		
0.01	26.4	-10	148.8	0.09	5.65	w		
0.01	26.4	-10	148.6	0.19	5.23	w		
0.01	26.4	-10	148.9	0.3	5.4	w		
0.01	26.4	-10	148.5	0.3	5.83	w		
0.01	26.4	-10	148.5	0.4	5.53	w		
0.01	26.4	-10	148.2	0.5	6.12	w		
0.01	26.4	-10	148.8	0.49	6.26	w		
0.01	26.4	-10	150	0.54	6.35	w		
0.01	26.4	-10	149.9	0.59	6.62	w		
0.01	26.4	-10	149.7	0.6	6.83	w		

$w_{\text{oil},0}$	p	T	$\dot{m}$	$\dot{x}$	$\Delta p/\Delta L$	FP	$\dot{q}$	α
0.01	26.4	-10	149.5	0.65	6.83	w		
0.01	26.4	-10	149.6	0.7	7.05	w		
0.01	26.4	-10	149.5	0.74	6.9	w/f		
0.01	26.4	-10	149.1	0.8	7.01	w/f		
0.01	26.4	-10	149.3	0.8	7.32	w/f		
0.01	26.4	-10	149.1	0.85	7.39	w/f		
0.01	26.4	-10	148.6	0.9	7.8	w/f		
0.01	26.4	-10	149.3	0.91	7.37	w/f		
0.01	26.4	-10	149.3	0.95	8.27	w/f		
0.01	26.4	-10	150.8	0.51	0.89	w		
0.01	26.4	-10	149	0.72	-0.35	w		
0.01	26.4	-10	150.5	0.9	9.68	w/f		
0.01	26.4	-10	300.6	0.09	6.28	i		
0.01	26.4	-10	299.8	0.09	2.06	i		
0.01	26.4	-10	297.4	0.19	5.64	i		
0.01	26.4	-10	300.7	0.29	6.41	i		
0.01	26.4	-10	300.9	0.29	3.79	i		
0.01	26.4	-10	298.9	0.4	5.07	i-a		
0.01	26.4	-10	300.7	0.49	5.92	a		
0.01	26.4	-10	301.3	0.49	5.65	a		
0.01	26.4	-10	302.1	0.54	7.27	a		
0.01	26.4	-10	300.2	0.59	7.7	a		
0.01	26.4	-10	300.3	0.6	6.18	a		
0.01	26.4	-10	299.6	0.65	8.47	a		
0.01	26.4	-10	299.3	0.7	9.22	a		
0.01	26.4	-10	298.9	0.75	10.13	a		
0.01	26.4	-10	302.2	0.8	11.3	a		
0.01	26.4	-10	300.4	0.8	10.36	a		
0.01	26.4	-10	300.6	0.85	11.94	a		
0.01	26.4	-10	295.3	0.91	12.53	a		
0.01	26.4	-10	299.5	0.9	8.14	a		
0.01	26.4	-10	301	0.92	12.4	a		
0.01	26.4	-10	297.4	0.11	12.51	i		
0.01	26.4	-10	296.9	0.31	15.12	i-a		
0.01	26.4	-10	300	0.51	18.53	i-a		
0.01	26.4	-10	300.5	0.71	19.14	a		
0.01	26.4	-10	301.1	0.91	30.2	a		
0.03	26.4	-10	148.7	0.09	0.37	w		
0.03	26.4	-10	147.7	0.3	0.87	w		
0.03	26.4	-10	147.1	0.51	1.84	w		
0.03	26.4	-10	149.3	0.7	2.28	w		
0.03	26.4	-10	148.2	0.1	0.32	w		
0.03	26.4	-10	155.8	0.29	0.13	w		
0.03	26.4	-10	147.9	0.51	1.41	w		
0.03	26.4	-10	302.8	0.3	3.71	i-a/f		
0.03	26.4	-10	302.5	0.09	1.21	i		

Table G.8

Table G.9: Micro-fin tube: Heat transfer measurements

$w_{oil,0}$	p	T	$\dot{m}$	$\dot{x}$	$\Delta p/\Delta L$	FP	$\dot{q}$	α
0	18.59	-21.85	77.7	0.77	3.17	w	0.3	
0	18.59	-21.86	77.9	0.77	3.18	w	0.3	
0	18.59	-21.86	77.9	0.78	3.35	w	3.9	11.8
0	18.6	-21.85	78.8	0.77	3.46	w	7	13.1
0	18.6	-21.85	75.8	0.81	3.63	w	10.7	14.2
0	18.6	-21.84	78.9	0.77	3.79	w	15.2	15.7
0	18.61	-21.83	79.1	0.76	4.23	w	25.3	19.2
0	18.61	-21.82	78.9	0.76	5.11	w	39.9	21.8
0	18.62	-21.81	78.9	0.75	6.07	w	59.6	24.7
0	18.62	-21.81	80.4	0.73	7.17	w	79.4	27.8
0	18.62	-21.82	79.3	0.74	7.19	w	79.4	27.7
0	18.66	-21.75	266.8	0.83	37.38	a	0.3	
0	18.66	-21.75	263.2	0.84	37.47	a	0.3	
0	18.67	-21.72	267	0.84	37.66	a	4	17
0	18.67	-21.72	265.2	0.85	37.8	a	4	17.1
0	18.66	-21.74	265.8	0.84	37.87	a	7	19.7
0	18.66	-21.73	266.2	0.84	37.97	a	7	19.7
0	18.67	-21.72	266.1	0.85	38.08	a	10	21.4
0	18.68	-21.71	264.3	0.85	38.27	a	9.9	21.2
0	18.68	-21.7	266.2	0.86	38.57	a	15.5	23.8
0	18.69	-21.69	264.9	0.87	38.16	a	15.6	23.9
0	18.69	-21.69	263.2	0.85	40.02	a	25	25.1
0	18.69	-21.69	266.2	0.84	40.21	a	24.8	25
0	18.69	-21.68	266.1	0.84	42.65	a	39.4	27.6
0	18.68	-21.71	266.6	0.84	42.15	a	39.3	27.8
0	18.69	-21.69	264	0.84	45.59	a	59.9	32
0	18.69	-21.68	266.3	0.83	45.99	a	60	31.9
0	18.71	-21.66	265.7	0.83	48.67	a	78.5	35.9
0	18.7	-21.67	265.5	0.83	48.63	a	78.5	35.9
0	18.71	-21.66	269	0.82	55.41	a	120.2	44
0	18.71	-21.66	274.3	0.79	56.72	a	120.3	43.7
0	19.48	-20.35	526.6	0.09	18.54	i-a	0.2	
0	19.48	-20.35	526.9	0.09	18.42	i-a	0.2	
0	19.49	-20.34	522.6	0.09	18.53	i-a	4	6.2
0	19.49	-20.34	528.9	0.09	19.03	i-a	4	6.2
0	19.5	-20.33	525	0.09	19.25	i-a	8.2	8.4
0	19.49	-20.33	527.3	0.09	19.93	i-a	8.2	8.3
0	19.5	-20.32	526.2	0.09	20.11	i-a	10.9	9.4
0	19.5	-20.32	522.6	0.09	19.77	i-a	10.8	9.5
0	19.51	-20.31	528.7	0.09	20.7	i-a	15.1	11.3
0	19.5	-20.32	527.5	0.09	20.88	i-a	15.1	11.2
0	19.42	-20.46	534.2	0.09	25.82	i-a	39.8	19.7
0	19.41	-20.47	537.4	0.09	26.06	i-a	39.8	19.6
0	19.52	-20.3	527.9	0.09	22.34	i-a	25.2	15
0	19.51	-20.3	529.6	0.09	22.4	i-a	25.2	15

$\dot{q}_1$	α_1	$\dot{q}_2$	α_2	$\dot{q}_3$	α_3	$\dot{q}_4$	α_4	$\dot{q}_5$	α_5	$\dot{q}_6$	α_6
0.3		0.3		0.3		0.3		0.3		0.3	
0.3		0.3		0.3		0.3		0.3		0.3	
6	17.8	7.5	22.1	1.8	5.5	1.4	4.4	5.2	14.6	1.8	6.4
9.6	17.8	12.3	21.8	4.4	8.4	3	5.9	8.2	15	4.5	9.7
15.8	20.3	16.5	20.5	8.6	11.3	5	7.1	11.2	15	7.1	10.7
22	22	22.2	21.4	12.2	12.8	8.1	9	14.9	15.8	11.7	13.3
37.5	27.7	36.2	24.7	26.1	19.6	15.7	12.6	24.6	19.9	11.6	10.6
33.7	21.3	67.6	34	53.4	27.8	37.9	20.9	38.1	21.1	8.8	5.8
18.3	10.3	96.4	39.8	100.4	39	74.2	30.5	60	24.8	8.5	4.1
8.5	4.6	111.7	41.8	147.6	48.3	117.2	39.4	83.3	29.2	7.8	3.4
8.5	4.6	111.7	41.9	147.5	48.3	117.2	39.4	83.4	29	7.8	3.3
0.3		0.3		0.3		0.3		0.3		0.3	
0.3		0.3		0.3		0.3		0.3		0.3	
4.7	19.4	4.9	19.4	2.3	12.4	3.7	17.5	4.3	16.6	3.9	16.4
4.8	19.6	4.9	19.5	2.3	12.5	3.8	17.6	4.3	16.6	4	16.5
8.5	22.8	7.9	21.1	5.5	17.9	6.1	18.8	6.9	18.3	7	19.3
8.5	22.7	7.9	21.1	5.5	18.1	6.1	18.8	6.9	18.4	7	19.2
11.9	24.2	11.1	22.5	8.3	20.4	9.5	21.6	9.4	19.3	9.6	20.2
11.8	24.4	11	22.4	8.3	20	9.4	21.3	9.4	19.1	9.6	20.2
17.8	26.4	17.3	25.2	14	24	13.7	22.2	15.8	22.9	14.5	21.9
17.7	26.5	17.4	25.5	14	24	13.8	22.3	15.9	23.1	14.6	22.2
30.4	29	25.5	24.7	25.2	27.3	21.3	22.8	24.1	23.7	23.4	23
30.3	29	25.4	24.6	25	27.1	21.2	22.7	23.9	23.6	23.2	23
44.7	30.8	40.4	27.1	43.1	31	35.5	25.5	36.5	25.7	36	25.7
44.7	30.9	40.4	27.3	43	31.1	35.5	25.7	36.5	25.8	35.9	25.8
71.1	37.6	60.3	30.8	65.2	35.3	54.5	29.1	56.9	30.5	51.5	28.9
71.2	37.6	60.4	30.7	65.3	35.1	54.6	29	57	30.4	51.6	28.8
93.2	42.8	76.3	33.6	87.6	39.7	72	32.2	74.6	34.2	67.3	32.6
93.2	42.7	76.4	33.7	87.7	39.8	72	32.3	74.6	34.2	67.3	32.6
145.7	54.8	114.5	40.7	137.2	48.6	108.8	38.4	112.9	41.4	102.1	40.3
145.8	54.2	114.6	40.4	137.3	48.5	109	38.2	113	41.2	102.2	40
0.2		0.2		0.2		0.2		0.2		0.2	
0.2		0.2		0.2		0.2		0.2		0.2	
7.6	11.3	3.2	5.1	0.2	0.3	3	5	4.7	7.2	5.6	8.5
7.6	11.2	3.2	5	0.2	0.3	2.9	4.9	4.7	7.2	5.6	8.5
12.2	12.5	5.1	5.3	9.2	8.3	5.1	5.5	8.1	8.5	9.4	10
12.2	12.4	5.2	5.3	9.1	8.2	5.1	5.5	8.1	8.4	9.4	9.9
16.3	14.1	7.8	6.9	9.3	7.5	7.5	6.8	11.3	9.9	12.8	11.3
16.3	14.3	7.8	6.9	9.3	7.6	7.4	6.9	11.2	9.9	12.7	11.5
24.6	17.9	11.4	8.6	9.8	7.2	10.8	8.4	16.2	12	18.1	13.6
24.5	17.7	11.5	8.5	9.8	7.1	10.8	8.4	16.2	11.9	18.1	13.4
63.4	31.2	34.8	17	34.4	16.4	26.5	13.1	37.4	18.8	42.2	21.9
63.4	30.8	34.8	16.9	34.3	16.3	26.5	13	37.2	18.6	42.3	21.9
39.8	23.3	20.9	12.4	20	11.6	17.5	10.6	24.9	14.9	28.2	17.3
39.8	23.4	20.9	12.3	20	11.6	17.5	10.6	24.8	14.9	28.2	17.3

Table G.9

$w_{oil,0}$	p	T	$\dot{m}$	$\dot{x}$	$\Delta p/\Delta L$	FP	$\dot{q}$	α
0	19.43	-20.43	534.4	0.09	29.65	i-a	61	25.4
0	19.44	-20.43	535.5	0.09	29.56	i-a	61	25.5
0	19.45	-20.41	535.8	0.09	32	i-a	78.6	29.7
0	19.45	-20.41	534.9	0.09	31.73	i-a	78.5	29.6
0	19.47	-20.37	528.9	0.09	37.97	i-a	117.9	37.4
0	19.47	-20.38	531.3	0.09	37.98	i-a	117.6	37.5
0	19.48	-20.35	525.7	0.28	47.76	a	0.2	
0	19.48	-20.36	526.3	0.28	48.44	a	4.6	6.6
0	19.48	-20.36	527	0.28	47.75	a	6.9	8.1
0	19.49	-20.33	525.6	0.29	49.11	a	10.1	9.3
0	19.48	-20.35	526.7	0.28	50.27	a	14.9	11.2
0	19.44	-20.43	524.3	0.29	52.68	a	25.1	14.4
0	19.45	-20.4	526	0.29	56.03	a	41	18.9
0	19.47	-20.37	523.1	0.28	57.15	a	59.6	23.6
0	19.49	-20.35	521.1	0.29	62.06	a	79.8	27.7
0	19.55	-20.24	525.2	0.27	64.3	a	114.6	34
0	25.93	-10.74	80.2	0.3	0.62	w	0.2	
0	25.94	-10.72	81.2	0.3	0.62	w	0.2	
0	25.94	-10.73	80.9	0.3	0.74	w	4.2	11.4
0	25.94	-10.73	81	0.3	0.75	w	4.2	11.4
0	25.94	-10.73	81.2	0.3	0.94	i	7.5	11.5
0	25.94	-10.73	81.5	0.29	0.91	s-w	10.3	12.8
0	25.95	-10.72	80.7	0.3	1.05	s-w	15.1	15.1
0	25.95	-10.71	80.6	0.3	1.06	s-w	15	14.8
0	25.96	-10.7	80.7	0.3	1.5	w	25.8	18.9
0	25.96	-10.7	81.1	0.29	1.88	w	40.6	23.4
0	25.97	-10.69	81.4	0.29	1.94	w	40.7	23.1
0	25.99	-10.67	80.9	0.3	2.47	w	60.9	28.1
0	25.99	-10.66	80.7	0.3	2.5	w	60.8	28
0	26	-10.65	80.6	0.3	2.88	w	79.5	30.9
0	25.97	-10.69	80.2	0.65	2.07	w	0.2	
0	25.98	-10.68	79.9	0.66	2.05	w	0.2	
0	25.98	-10.67	80.4	0.67	2.23	w	3.9	12.6
0	25.98	-10.68	80.4	0.66	2.4	w	7.6	15.1
0	25.98	-10.67	79.7	0.67	2.34	w	7.6	15
0	25.99	-10.66	82.2	0.65	2.45	w	10.2	16.4
0	25.99	-10.66	82.3	0.62	2.51	w	15.4	18.7
0	25.99	-10.66	83.1	0.63	3.01	w	25.8	21.5
0	26	-10.65	80.5	0.65	2.98	w	25.8	21.3
0	26.01	-10.63	81.5	0.66	3.69	w	42.1	25
0	26.01	-10.63	81.2	0.65	4.35	w	61.3	29
0	26	-10.64	81.3	0.64	4.95	w	78	31.2
0	26.01	-10.64	80.7	0.65	4.94	w	78	31.3
0	26.74	-9.67	263	0.47	16.88	a	0.2	
0	26.76	-9.64	265.5	0.47	16.68	a	3.8	9
0	26.76	-9.64	263.6	0.46	16.26	a	7.1	11.9
0	26.76	-9.63	261.9	0.48	16.46	a	10.2	14

$\dot{q}_1$	α_1	$\dot{q}_2$	α_2	$\dot{q}_3$	α_3	$\dot{q}_4$	α_4	$\dot{q}_5$	α_5	$\dot{q}_6$	α_6
94.3	39.7	56.1	23.1	57.2	22.4	41.3	16.6	56.6	23.8	60.5	27.1
94.4	39.8	56.1	23.1	57.2	22.5	41.4	16.7	56.5	23.9	60.6	27.2
119.2	46.1	71.3	26.7	77.1	26.8	54.8	19.4	71.7	27.4	77.4	31.7
119.1	45.9	71.3	26.7	77	26.9	54.8	19.4	71.6	27.3	77.4	31.6
175.9	58.3	110.9	34.9	118.5	34	84	24.3	105	33.6	113.1	39.5
175.7	58.5	110.6	34.9	118.3	34.2	83.8	24.4	104.5	33.5	112.9	39.6
0.2		0.2		0.2		0.2		0.2		0.2	
5.4	7.5	3.9	5.8	3.5	4.5	4.2	6.4	5.8	8.5	4.6	6.7
8.3	9.3	5.5	6.6	5	5.6	5.9	7.4	9.3	10.9	7.6	8.7
11.6	10.5	7.7	7.2	7.7	6.8	10.3	9.8	13.2	12.2	10.1	9.5
17.8	13.2	11.4	8.7	12	8.8	14.3	11.1	19.4	14.5	14.6	11.1
32.1	18.3	18.7	10.8	24.1	13.4	23.6	13.8	29.1	16.8	22.8	13.5
53.4	24.6	33.1	15.3	41.2	18.1	36.5	16.7	46.8	21.6	34.9	16.9
80.4	32.2	50.8	20.1	62.3	23	51.4	19.8	63.7	25.5	48.8	20.7
109.7	39.1	70.1	24.4	86.5	27.5	68.9	22.8	81.1	28.7	62.7	23.9
159.4	49.2	102.7	30.7	121.6	32.8	99.6	27.5	113.8	34.2	90.5	29.6
0.2		0.2		0.2		0.2		0.2		0.2	
0.2		0.2		0.2		0.2		0.2		0.2	
7	19.1	6.1	15.8	4.2	10.7	1.3	3.4	3.3	8.9	3.6	10.4
6.9	19.1	6	15.8	4.2	10.7	1.3	3.4	3.4	9	3.5	10.5
10	14.8	11.5	18.2	7.3	11.2	3.6	5.5	6.9	10.7	5.4	8.4
10	12.1	16.6	21.1	11.1	13.7	4.9	6.1	11	13.5	8.2	10.4
10	10.1	22.8	23.5	17.9	17.7	10	9.6	18.2	17.7	11.6	11.8
10	9.7	22.8	23.1	17.8	17.5	10	9.5	18.2	17.5	11.5	11.6
10	7.2	35.3	26.2	35.1	25.9	19.6	14	34.9	25.1	19.7	14.8
9.6	5.5	49.9	29.9	58.3	34.3	37.8	20.8	59.5	33.3	28.5	16.7
9.7	5.2	50	28.9	58.5	34.1	37.9	20.7	59.6	33.1	28.6	16.7
8	3.5	70.6	33.9	88.6	42.7	70.1	30.3	89.6	40.5	38.4	17.8
8	3.5	70.5	33.6	88.4	42.6	70	30.2	89.4	40.3	38.4	17.7
8	2.8	85.4	35.3	115.7	48.1	100.6	36.3	116.2	43.9	50.9	19.1
0.2		0.2		0.2		0.2		0.2		0.2	
0.2		0.2		0.2		0.2		0.2		0.2	
6.1	18.6	4.6	15.7	0.2	0.8	3.3	11.1	3.9	12.3	5.5	17.2
10.7	20.5	8.4	16.6	6.2	12.7	4.8	9.5	6.1	12.4	9.4	18.9
10.7	20.5	8.4	16.5	6.2	12.4	4.8	9.4	6.1	12.3	9.4	18.9
15.7	23.8	10.5	17.4	6.3	11.1	7.4	12.1	7.6	12.6	13.5	21.5
21.1	24.6	21.3	25.1	10.9	14.3	11.8	14.3	11.9	14.7	15.4	19.1
14.3	14.5	44.4	36.6	32.4	25.7	27.6	21.5	25.4	21	10.7	9.8
14.3	14.2	44.5	36.5	32.4	25.7	27.6	21.4	25.4	20.8	10.7	9.5
7.4	5.8	66.8	43.2	57.7	33.8	66.5	35.2	47.5	27.5	7	4.5
5	3.3	74.5	42.9	89.3	42.8	108.5	44.5	84.1	37.4	6.1	3.2
4.9	2.4	83	41.2	118.2	49.5	142.8	50.1	113.6	42.1	5.4	2.2
4.9	2.4	82.9	41.2	118.2	49.4	142.8	50.2	113.5	42.3	5.4	2.2
0.2		0.2		0.2		0.2		0.2		0.2	
5.9	13.3	3.1	7.7	1.7	3.3	2.8	6.6	5.4	13	4.2	9.8
10.3	16.8	5.2	9.1	5.9	8.4	4.1	7.2	9.6	16.7	7.5	13
14.7	19.6	7.8	11	6.6	8.2	6.6	9.6	14	19.7	11.4	15.8

Table G.9

$w_{oil,0}$	p	T	$\dot{m}$	$\dot{x}$	$\Delta p/\Delta L$	FP	$\dot{q}$	α
0	26.76	-9.64	264.7	0.47	16.86	a	15.8	17.2
0	26.77	-9.63	263.2	0.47	17.61	a	26	22.1
0	26.77	-9.62	263.3	0.46	18.48	a	39.6	28
0	26.79	-9.6	263.1	0.47	20.18	a	58.8	34.8
0	26.79	-9.59	262.4	0.47	21.3	a	80.4	41.4
0	26.8	-9.59	265.3	0.46	23.9	a	119.1	50.1
0	26.82	-9.56	263.1	0.65	24.2	a	0.2	
0	26.81	-9.57	263.2	0.66	24.59	a	3.9	9.7
0	26.81	-9.57	262.9	0.65	24.02	a	7.1	12.6
0	26.81	-9.57	262.2	0.66	23.9	a	10.1	14.1
0	26.81	-9.58	261.9	0.66	23.35	a	15.1	16.8
0	26.81	-9.57	262.2	0.65	24.02	a	25.3	20.9
0	26.82	-9.56	266	0.65	25.48	a	39.8	26.5
0	26.83	-9.55	265.1	0.65	27.35	a	60.7	33.4
0	26.84	-9.54	263.5	0.65	28.29	a	79.7	38.4
0	26.83	-9.55	266	0.63	31.12	a	116.5	46.6
0	25.93	-10.74	261.1	0.66	23.12	a	0.2	
0	25.94	-10.72	259.7	0.67	23.07	a	0.2	
0	25.96	-10.7	260.6	0.68	23.72	a	4.1	14.8
0	25.96	-10.71	260.1	0.68	23.77	a	4.1	14.8
0	25.95	-10.71	262.9	0.65	23.4	a	7	16.4
0	25.94	-10.72	265	0.66	23.79	a	10.1	18.1
0	25.96	-10.71	263.5	0.65	24.11	a	15.3	20.6
0	25.96	-10.7	262.7	0.66	24.34	a	15.4	20.6
0	25.91	-10.77	265.6	0.66	25.32	a	25.4	25.1
0	25.92	-10.75	264.6	0.66	27	a	39.6	31
0	25.94	-10.73	266.5	0.65	29.42	a	61	38.2
0	25.95	-10.71	265.8	0.64	30.25	a	79.8	43.6
0	25.96	-10.71	265.3	0.65	33.42	a	116.8	50.3
0	25.96	-10.7	268.4	0.63	31.9	a	116.8	51.7
0	25.97	-10.69	421.6	0.81	58.17	a	0.2	
0	25.96	-10.7	418.4	0.82	58.21	a	0.2	
0	25.96	-10.71	416.3	0.8	58.13	a	4.3	16.2
0	25.94	-10.73	413.5	0.81	57.48	a	4.3	16.4
0	25.94	-10.74	422.4	0.75	57.69	a	7.4	16
0	25.93	-10.74	417.2	0.77	57.03	a	7.3	16
0	25.95	-10.72	421.4	0.76	58.29	a	10	16.4
0	25.96	-10.71	420.5	0.81	59.37	a	15.1	19
0	25.99	-10.67	425.8	0.8	61.36	a	25.2	21.8
0	25.97	-10.69	423	0.8	62.72	a	39.6	26.1
0	25.99	-10.66	418.1	0.82	65.05	a	60.9	31.8
0	26.02	-10.63	424.3	0.81	68.34	a	80.1	36.1
0	26.02	-10.62	429.9	0.79	75.89	a	121.8	43.4
0	39.31	4.62	83.5	0.29	0.46	s-w	0.1	
0	39.32	4.63	81.9	0.29	0.46	s-w	0.1	
0	39.31	4.62	84.1	0.31	0.53	s-w	3.9	9.4
0	39.32	4.63	82.6	0.3	0.57	s-w	6.9	12.5

$\dot{q}_1$	α_1	$\dot{q}_2$	α_2	$\dot{q}_3$	α_3	$\dot{q}_4$	α_4	$\dot{q}_5$	α_5	$\dot{q}_6$	α_6
21.7	23	12.1	13.2	12.2	12.3	10.8	12.4	20.6	23.2	17.6	19.5
36	29.7	20.6	17.2	22.3	17.6	19.8	17.3	32.6	28.6	24.9	22.3
53.2	37	32	22	36.9	24	31.9	22.7	47.3	34.9	36.5	27.6
81.4	47.5	48.6	28	56	30.6	48.6	28.6	69.5	42.5	48.6	31.7
107.6	55.6	67.5	33.7	83.1	38.3	68.3	34.3	91.1	48.7	65	37.6
161	69.2	100.9	41.8	124.5	48.5	102.8	41.6	132.9	55.7	92.2	43.6
0.2		0.2		0.2		0.2		0.2		0.2	
4.9	11.3	3.2	8.8	1.6	3.8	3.7	10	5.7	14.3	4	9.7
9.4	15.8	5.5	10.3	3.7	6.3	6.6	12.6	9.4	17.1	7.8	13.4
12.9	17.3	7.8	11.3	6.5	8.8	9.3	13.6	12.9	18.4	11.3	15.5
19.3	20.8	11.6	13.2	12.1	12.9	13.6	15.5	18.8	21.1	15.5	17.2
32.9	26.9	19.6	16.4	22.7	18.1	23.3	19.1	29.7	24.8	23.5	20.1
53.4	35.3	32.2	21.5	39.7	25.2	36.1	23.4	43.2	29.2	34	24.1
82.2	45.6	50.2	27.8	63.9	33.2	56.6	29.6	63.1	35.2	48.4	29.1
107.5	52.7	67.9	32.8	85.7	38.8	74.9	34	79.3	38.8	63.1	33.5
158.2	64.8	102.4	40.7	127.8	48.4	107.9	40.4	112.5	45.1	90.3	40
0.2		0.2		0.2		0.2		0.2		0.2	
0.2		0.2		0.2		0.2		0.2		0.2	
5.7	17.9	3.6	12.9	1.7	7	4.1	16.5	5.6	20.4	4	14.4
5.7	17.7	3.6	13	1.7	7	4.1	16.6	5.6	20.3	4	14.3
9.2	19.6	5.5	13.1	3.4	9.2	6.8	17.1	9.5	21.6	7.7	17.8
13	21.7	7.6	14	6.3	12.7	9.3	17.8	12.6	22.1	11.6	20.1
19.5	24.8	11.8	15.9	12.1	17.5	14.1	19.7	18.8	24.8	15.7	21.1
19.6	24.8	11.8	15.9	12.1	17.5	14.1	19.5	18.8	24.7	15.7	21
32.6	30.7	20	19.7	22.7	23.7	22.7	22.8	29.3	28.4	25.2	25.4
53.2	39.8	31.3	24.1	39.4	31.7	35.9	27.9	43.4	33.6	34.2	28.9
82.4	50.4	50.2	30.9	63.8	39.7	56.9	34.3	63.6	39.4	48.9	34.3
108.1	58.3	67.7	36.3	85.9	45.8	74.9	39.1	79.1	43.1	62.9	39
158.4	68.8	102.6	43.7	128.3	53	108.5	43.8	112.9	48	90	44.3
158.5	70.4	102.7	44.6	128.4	54.8	108.5	45.1	113	49.4	90	45.7
0.2		0.2		0.2		0.2		0.2		0.2	
0.2		0.2		0.2		0.2		0.2		0.2	
8.6	30.7	2.5	10.8	3.3	12.5	8.3	29.7	1.8	7.3	1.3	6
8.7	31.4	2.6	11.4	3.4	12.8	8.4	30.1	1.8	7.2	1.2	5.5
13.1	27.9	3.7	9.2	7.1	15.3	12.7	25.9	5.7	12.8	1.9	4.7
13.1	28	3.7	9.2	7	15.3	12.6	26	5.7	12.8	2	4.9
15.9	25.3	4.9	8.9	10.4	17.1	15.6	24.2	8.4	13.7	5.1	8.9
21.2	26.5	8.2	11.7	16.8	21.2	20.3	24.4	13.5	16.6	10.4	13.7
35.2	29.8	14.7	13.8	29	25.3	27.6	23.4	23.8	20.1	20.6	18.2
55.3	35.5	23.9	16.9	46	30.6	38.6	24.9	39.7	25.4	34.4	23.3
84	43.9	39.8	22.5	68.8	36.2	55.7	28.1	63.9	31.7	53	28.7
114.1	51.4	57.3	27.5	89.4	39.7	73.5	31.6	76.1	33.3	70.1	33.1
173.8	63.3	96	35.9	139.9	47.1	113.7	37.6	107.3	37.8	100.4	38.9
0.1		0.1		0.1		0.1		0.1		0.1	
0.1		0.1		0.1		0.1		0.1		0.1	
3.7	7.9	6.8	16.5	2.8	8	4.5	11.3	3.9	9.3	1.3	3.2
1.8	3	12.3	21.7	7.9	15.3	8.2	14.4	8.6	15.3	2.9	5.4

Table G.9

$w_{oil,0}$	p	T	$\dot{m}$	$\dot{x}$	$\Delta p/\Delta L$	FP	$\dot{q}$	α
0	39.31	4.62	83.2	0.29	0.65	s-w	9.8	15
0	39.31	4.62	82.9	0.3	0.78	s-w	15.1	18.5
0	39.33	4.64	82.3	0.3	0.96	s-w	24.7	23.4
0	39.33	4.64	81.5	0.29	1.29	w	39.6	28.7
0	39.34	4.65	82.2	0.29	1.27	w	39.6	28.7
0	39.34	4.65	82.4	0.28	1.66	w	60.1	34.4
0	39.33	4.64	83.3	0.27	1.66	w	60.1	34.4
0	39.31	4.62	81.9	0.28	1.95	w	79.1	39.8
0	39.35	4.66	81.6	0.28	1.95	w	79.1	39.7
0	39.25	4.56	81.4	0.7	1.31	w	0.1	
0	39.24	4.56	81.4	0.7	1.32	w	0.1	
0	39.25	4.56	80.7	0.69	1.38	w	4.1	13.5
0	39.24	4.55	79.2	0.72	1.47	w	7	15.7
0	39.25	4.56	81.4	0.67	1.49	w	9.7	17.1
0	39.25	4.56	81.5	0.68	1.68	w	15	19.2
0	39.25	4.56	81.1	0.69	1.92	w	25.3	22.1
0	39.24	4.55	81.9	0.67	2.34	w	39.7	24.7
0	39.25	4.56	82.3	0.67	2.34	w	39.7	24.6
0	39.24	4.55	81.2	0.68	2.71	w	59.5	27
0	39.24	4.55	81.1	0.68	3.19	w	77.4	29.7
0	39.29	4.6	260.3	0.47	9.09	i-a	0.1	
0	39.28	4.59	262.9	0.45	9.07	i-a	0.1	
0	39.28	4.59	261.3	0.46	9.22	i-a	4	15.5
0	39.28	4.59	263.7	0.44	9.2	i-a	7.1	18.7
0	39.24	4.55	265	0.44	9.44	i-a	9.8	21.5
0	39.29	4.6	262.1	0.44	9.65	i-a	15.4	26.5
0	39.3	4.61	266.1	0.44	10.39	i-a	25.4	33.7
0	39.28	4.59	265	0.46	11.33	i-a	39.6	41.3
0	39.31	4.62	265.5	0.44	12.26	i-a	60	49.2
0	39.31	4.62	265.7	0.43	13.4	i-a	80.1	49.4
0	39.31	4.62	265.4	0.44	13.46	i-a	80.1	50.9
0	39.31	4.62	265.5	0.43	14.47	a	109.9	49
0	39.31	4.62	266.3	0.43	14.56	a	109.8	50
0	39.35	4.66	525.2	0.06	9.58	i-a	0.1	
0	39.34	4.65	525.3	0.06	9.41	i-a	0.1	
0	39.34	4.65	519.9	0.08	10.23	i-a	3.9	10.4
0	39.34	4.66	521.7	0.07	10.01	i-a	7	13.4
0	39.32	4.63	524	0.06	10.33	i-a	10	16.4
0	39.23	4.54	520.8	0.07	11.07	i-a	14.9	20.4
0	39.22	4.53	524.4	0.06	11.79	i-a	24.8	27.2
0	39.22	4.53	523.3	0.07	13.56	i-a	40.3	35.5
0	39.24	4.55	519.9	0.07	15.22	i-a	59.5	43
0	39.25	4.56	524.5	0.06	17.48	i-a	80.9	49.6
0	39.25	4.56	523.6	0.06	17.39	i-a	80.9	49.6
0	39.23	4.54	521.8	0.06	20.49	i-a	109.7	56.4
0	39.23	4.54	521	0.07	20.45	i-a	109.7	56.3
0.013	11.58	-36.06	77.7	0.17	0.6	i	0.3	

$\dot{q}_1$	α_1	$\dot{q}_2$	α_2	$\dot{q}_3$	α_3	$\dot{q}_4$	α_4	$\dot{q}_5$	α_5	$\dot{q}_6$	α_6
0.1	0.2	15.6	24	13.6	21.5	11.1	16.5	13.1	19.9	5	8.2
0.1	0.1	21.5	26.4	22.2	28.3	16.7	19.9	21.2	25.5	8.7	11
0.1	0.1	32.1	29.9	40.8	39.3	27.8	25.6	35	33.1	12.7	12.7
0.1	0.1	47.4	33.7	68.5	50.5	46.7	33.1	56.6	41.3	18.5	13.5
0.1	0.1	47.4	33.6	68.5	50.4	46.7	33	56.7	41.4	18.5	13.8
0.1	0.1	69.3	38.2	104.9	61.1	72.8	41.1	84.1	49.1	29.3	16.8
0.1	0.1	69.3	38	104.8	61.1	72.9	41.1	84.2	49.1	29.4	16.7
0.1	0.1	85	45.6	143	70.9	98.7	47	115	57.3	32.6	18.3
0.1	0.1	85	45.3	143.1	70.8	98.6	46.9	115.1	57.2	32.7	18.2
0.1		0.1		0.1		0.1		0.1		0.1	
0.1		0.1		0.1		0.1		0.1		0.1	
6.2	21.7	3.6	12.5	6	17	4.1	11.7	3.8	13.1	1.1	4.8
8.9	21.9	7.2	16.9	10.5	21.2	6.7	13.6	7.1	16.9	1.3	3.8
5.6	14.3	12.8	24.4	17.1	25.5	13.4	20.1	9.5	17.9	0.1	0.3
4.3	8.3	19	28.7	27.6	31.7	22	24.5	17.1	21.5	0.1	0.2
2.2	2.4	26.4	28.8	51.7	42.8	41	32.4	30.6	26.1	0.1	0.1
0.1	0.1	35.1	23	85.4	53.3	68	40.8	49.5	31.2	0.1	0.1
0.1	0.1	35.1	23.2	85.5	52.8	68	40.5	49.4	31	0.1	0.1
0.1	0	26.1	12.4	138.3	62.2	115.8	51	76.8	36.3	0.1	0
0.1	0	36.2	13.6	176.8	68.3	149.4	56.2	101.4	39.8	0.1	0
0.1		0.1		0.1		0.1		0.1		0.1	
0.1		0.1		0.1		0.1		0.1		0.1	
6.4	24.6	2.7	11	3.3	12	3.6	12.8	4.2	15.3	4	17.3
10.2	27	6.1	15.9	6	15.2	5.6	13.8	7.5	19.1	7.2	20.9
14.3	31.3	8.5	18.5	9	18.9	6.9	14.4	10.5	22.6	9.7	23.4
21.5	37.3	13.4	23	14.8	24.1	11.7	18.8	16.1	27.5	15	28.7
37.6	49.8	21.4	29	24.9	31.2	19.1	23.6	26	34.6	23.2	34.2
57	60.2	34.5	36.8	40.4	38.8	31.2	29.7	39.3	41.2	35.2	40.9
90.5	74.6	51.8	44.6	61.4	46.9	47.4	35.8	57	46.8	52	46.6
119.7	73	81.8	54.5	88.2	54.5	72.5	42.8	55.8	35.2	62.5	36.5
119.7	77.4	81.8	54.9	88.2	54	72.5	43.4	56	36.8	62.6	39.1
125.9	58.3	147.8	68	167	68.3	132.3	54.6	66.1	34.4	20	10.5
125.9	62.1	147.7	69.2	166.9	68.5	132.2	55	66.2	34.4	20	10.7
0.1		0.1		0.1		0.1		0.1		0.1	
0.1		0.1		0.1		0.1		0.1		0.1	
7.7	20.3	1.4	3.8	3.1	8.6	4	10.6	3.2	8	4	11
12.2	23.8	4.1	7.9	6.5	12.3	6.5	12	6.2	11.2	6.6	13.5
16.7	28	7.2	11.6	9.7	15.3	9.2	14.1	8.2	12.9	9.2	16.3
24.6	34.8	12.3	16.4	14.7	19.1	12.4	15.9	12	16	13.2	20
41.4	46.8	21.7	23.3	24.2	24.9	20.6	20.6	19.5	20.9	21.5	26.5
65.9	60.8	37.6	32.4	41.5	33.3	32.3	25.6	31.1	27.1	33.3	33.8
97.6	74.2	54.5	38.9	62.6	40.8	48.4	31	45.2	32.3	48.7	41
130.4	84.7	75.9	45.8	85.2	47.1	66.2	35.7	60.9	36.7	66.7	47.5
130.5	84.6	75.9	45.8	85.2	47.2	66.3	35.7	60.9	36.7	66.7	47.5
176.2	96.6	100.2	51.3	120.3	55	91.2	40.8	81.5	41.2	89	53.4
176	96.4	100.2	51.3	120.2	55	91.2	40.8	81.4	41.1	89	53.3
0.3		0.3		0.3		0.3		0.3		0.3	

Table G.9

$w_{oil,0}$	p	T	$\dot{m}$	$\dot{x}$	$\Delta p/\Delta L$	FP	$\dot{q}$	α
0.013	11.58	-36.07	78	0.17	0.73	i	0.3	
0.013	11.59	-36.05	77.1	0.18	0.9	i	0.3	
0.013	11.58	-36.08	77.5	0.17	0.96	i	0.3	
0.013	11.57	-36.08	78.6	0.17	0.72	i	0.3	
0.013	11.58	-36.07	78	0.17	0.97	i	0.3	
0.013	11.54	-36.15	77.7	0.17	1.3	i	7.9	10.8
0.013	11.54	-36.17	78.2	0.17	0.98	i	7.7	10.6
0.013	11.53	-36.19	78.3	0.17	1.21	i	13.4	12.1
0.013	11.54	-36.17	78.2	0.17	1.48	i	13.5	11.9
0.013	11.54	-36.16	77.9	0.17	1.66	i	20.2	13.3
0.013	11.55	-36.15	77.6	0.17	1.62	i	20.2	13.3
0.013	11.55	-36.14	78	0.17	1.96	i	24.2	13.9
0.013	11.55	-36.14	77.5	0.17	2.29	i	24.1	13.9
0.013	11.55	-36.14	78	0.17	2.06	i	24.2	14
0.013	11.55	-36.14	78.4	0.17	2.81	i	34.3	15.1
0.013	11.56	-36.12	78.8	0.17	2.44	i	34.1	14.8
0.013	11.55	-36.14	77.8	0.17	2.77	i	34.3	15
0.013	11.57	-36.1	78.4	0.18	3.98	i	43.3	16.9
0.013	11.55	-36.15	78.6	0.17	4	i	43.4	16.4
0.013	11.56	-36.13	78	0.17	6.47	i	61.8	19.8
0.013	11.51	-36.23	79.3	0.17	5.31	i	61.8	19.4
0.013	11.54	-36.16	78	0.2	7.8	i (f)	78.4	22.8
0.013	11.57	-36.1	76.9	0.2	7.88	i (f)	78.2	22.9
0.013	11.54	-36.17	78.4	0.2	7.4	i (f)	78.4	22.6
0.013	11.58	-36.08	76.8	0.25	9.12	i (f)	96.9	25.8
0.013	11.54	-36.16	77.3	0.25	9.52	i (f)	97	25.5
0.013	11.57	-36.09	77.6	0.58	4.04	i	0.3	
0.013	11.58	-36.06	80.6	0.56	4.13	i	0.3	
0.013	11.58	-36.06	78.5	0.57	4.05	i	0.3	
0.013	11.58	-36.06	76.5	0.59	4.53	i	8.6	11.8
0.013	11.6	-36.02	77.6	0.58	4.31	i	8.4	11.8
0.013	11.6	-36.02	80.2	0.56	4.91	i (f)	12.6	12.8
0.013	11.57	-36.09	77.6	0.58	4.77	i (f)	12.6	12.7
0.013	11.59	-36.05	80.7	0.55	5.25	i (f)	20	15
0.013	11.59	-36.05	77	0.58	5.24	i (f)	20	15
0.013	11.59	-36.04	77.5	0.58	5.7	i (f)	25.8	15.8
0.013	11.59	-36.04	76.9	0.59	5.73	i (f)	25.7	16
0.013	11.6	-36.02	76.1	0.6	5.9	i (f)	25.7	15.9
0.013	11.6	-36.02	79.3	0.57	6.42	w-a (f)	32.3	16
0.013	11.59	-36.05	77.6	0.57	6.51	w-a (f)	32.3	15.7
0.013	11.6	-36.01	77.6	0.57	7.77	w-a (f)	42.1	17
0.013	11.59	-36.05	76.3	0.58	7.61	w-a (f)	42.1	16.8
0.013	11.59	-36.04	78.3	0.57	7.76	w-a (f)	42.1	17.1
0.013	11.59	-36.04	78	0.57	9.59	w-a (f)	54.9	20
0.013	11.6	-36.03	77.6	0.58	9.55	w-a (f)	54.9	19.9
0.013	11.59	-36.04	75.5	0.6	12.61	w-a (f)	71.5	22.2
0.013	11.61	-35.99	77.7	0.58	12.41	w-a (f)	71.5	23.5

$\dot{q}_1$	α_1	$\dot{q}_2$	α_2	$\dot{q}_3$	α_3	$\dot{q}_4$	α_4	$\dot{q}_5$	α_5	$\dot{q}_6$	α_6
0.3		0.3		0.3		0.3		0.3		0.3	
0.3		0.3		0.3		0.3		0.3		0.3	
0.3		0.3		0.3		0.3		0.3		0.3	
0.3		0.3		0.3		0.3		0.3		0.3	
0.3		0.3		0.3		0.3		0.3		0.3	
14.1	19.1	10.7	14.6	4	5.7	1.9	2.8	4.9	6.8	11.7	16
13.9	18.6	10.5	14.5	3.8	5.4	1.7	2.4	4.8	6.6	11.6	15.9
16.5	15.4	21.3	19.1	8.2	7.3	4.6	4.2	8.8	7.9	21.1	18.9
16.6	15.1	21.4	18.8	8.3	7.2	4.7	4.2	8.8	7.7	21.2	18.6
18	12	34.4	22.5	12.2	8.1	9.3	6.3	12.6	8.4	34.9	22.6
18	11.9	34.3	22.3	12.2	8.1	9.3	6.2	12.5	8.4	35	22.6
14.8	8.7	42.7	24.3	16.6	9.7	13.8	8	19.2	11	38.3	21.9
14.7	8.8	42.5	24	16.5	9.5	13.7	7.9	19	10.9	38.2	22.1
14.8	8.8	42.6	24.5	16.6	9.7	13.8	8	19.1	11	38.3	22.1
17.4	7.6	49.8	21.9	36.2	16.3	22.6	10.4	30.9	13.8	48.6	20.8
17.3	7.3	49.7	21.6	35.9	16.1	22.5	10.3	30.9	13.6	48.4	20.3
17.5	7.6	49.8	21.8	36.3	16.2	22.6	10.3	31	13.7	48.6	20.3
17.4	7.1	56.9	22	55.2	21.1	36.1	14	47.3	18.3	46.8	18.9
17.5	6.5	57.1	21.2	55.2	20.8	36.3	13.9	47.5	17.9	46.9	17.9
33.8	10.8	75.3	23.6	86.1	27.1	54.1	17.5	71.3	22.8	50	17.2
33.8	10.4	75.4	23.3	86.2	27	54.1	17.3	71.3	22.6	50.1	16
40.5	11.9	106.4	30.3	103.2	29.9	71.9	21.1	97	28.3	51.5	15.7
40.3	12.2	106.3	30.3	103	29.7	71.7	21	96.7	28.1	51.3	16
40.6	11.9	106.4	30	103.2	29.6	71.9	20.9	96.9	27.8	51.6	15.1
46.8	12.4	135.2	36	130.1	34.5	107.4	28.2	113.4	30.2	48.4	13.2
46.9	12.3	135.2	35.5	130.1	34.5	107.5	28.2	113.5	29.9	48.5	12.8
0.3		0.3		0.3		0.3		0.3		0.3	
0.3		0.3		0.3		0.3		0.3		0.3	
0.3		0.3		0.3		0.3		0.3		0.3	
10.4	14.4	8.5	11.8	8.6	11.6	2.4	3.3	8.6	11.8	13.2	17.8
10.2	14.4	8.3	11.8	8.5	11.7	2.2	3.1	8.4	11.7	13	17.9
16.4	16.4	12.3	12.6	9.8	10.2	4.8	4.9	13.5	13.6	18.9	18.8
16.3	16.5	12.2	12.6	9.8	9.9	4.8	4.9	13.5	13.5	18.8	18.8
26.9	20.2	20.3	15.1	15.9	12.1	9	6.8	19.8	14.9	28.3	21.2
26.9	20.1	20.3	15.2	15.9	12	9	6.8	19.8	14.9	28.3	21.2
40.4	24.5	25.4	15.8	23.6	14.4	13.7	8.5	25.2	15.4	26.5	16.3
40.1	24.7	25.4	15.7	23.5	14.4	13.6	8.4	25.2	15.6	26.4	17
40.2	24.5	25.4	15.7	23.5	14.3	13.7	8.4	25.2	15.5	26.4	16.8
56.9	28.3	39.6	19.5	31.2	15.2	22.6	11.2	35.8	17.8	7.7	4
57.1	27.4	39.6	19.3	30.9	15.1	22.6	11.1	35.8	17.4	7.6	3.9
66.3	26.6	56.6	23.2	44.4	17.7	36.1	14.8	46	18.6	3.1	1.3
66.3	25.6	56.6	22.7	44.5	18	36.1	14.9	46	18.2	3.1	1.3
66.3	27	56.6	23	44.4	17.7	36.1	14.8	45.9	18.6	3.2	1.4
87.9	32	71.3	26.2	59.7	21.4	54.2	20	50.4	18.2	6	2.2
87.9	32.3	71.2	26	59.8	20.9	54.2	19.5	50.4	18.3	6	2.2
109.6	30.6	91.1	28.7	86.3	29.9	71.9	23.8	54.3	16.1	15.8	4.4
109.6	36.2	91.1	30	86.4	27.6	71.9	24	54.3	18	15.8	5.3

Table G.9

$w_{oil,0}$	p	T	$\dot{m}$	$\dot{x}$	$\Delta p/\Delta L$	FP	$\dot{q}$	α
0.013	11.62	-35.97	77.2	0.57	12.28	w-a (f)	73.2	23.9
0.013	11.65	-35.89	77.7	0.58	14.76	w-a (f)	92.1	26.8
0.013	11.6	-36.02	78.1	0.58	15.07	w-a (f)	92.3	26.9
0.013	11.58	-36.07	77.8	0.77	5.5	w-a	0.3	
0.013	11.58	-36.06	78.4	0.77	5.52	w-a	0.3	
0.013	11.6	-36.03	78.2	0.77	5.32	w-a	0.3	
0.013	11.61	-36	78	0.77	6.11	w-a	3.7	15
0.013	11.59	-36.05	77.8	0.77	5.95	w-a	3.8	15
0.013	11.58	-36.06	77.9	0.77	6.45	w-a (f)	8.1	15.6
0.013	11.58	-36.06	78.3	0.77	6.3	w-a (f)	8.2	15.4
0.013	11.6	-36.03	78.1	0.77	6.18	w-a (f)	8.2	15.3
0.013	11.6	-36.01	78.9	0.76	6.83	w-a (f)	13.5	15.7
0.013	11.59	-36.04	77.8	0.77	6.97	w-a (f)	13.5	15.5
0.013	11.6	-36.02	78.5	0.77	7.39	w-a (f)	20.2	17.2
0.013	11.61	-36	77.4	0.78	7.54	w-a (f)	20.3	17.1
0.013	11.61	-36	77.7	0.78	7.76	w-a (f)	20.2	16.6
0.013	11.6	-36.02	77.7	0.78	8.82	w-a (f)	29.8	19.5
0.013	11.61	-35.99	78.3	0.77	8.86	w-a (f)	29.8	18.8
0.013	11.61	-35.99	78	0.78	11.13	w-a (f)	41.8	22.1
0.013	11.62	-35.97	79.5	0.76	11.14	w-a (f)	41.8	21.4
0.013	11.62	-35.96	83.4	0.72	14.21	w-a (f)	58.4	24.8
0.013	11.63	-35.95	82.5	0.73	13.68	w-a (f)	58.2	23.6
0.013	11.62	-35.96	82.8	0.73	15.73	w-a (f)	76	23.2
0.013	11.64	-35.92	83	0.73	17.17	w-a (f)	76	24.6
0.013	11.64	-35.91	84	0.72	16.86	w-a (f)	75.8	20.4
0.013	11.63	-35.94	80.3	0.75	15.99	w-a (f)	76	26.1
0.013	11.61	-36	155.3	0.02			0.3	
0.013	11.63	-35.96	159.1	0.13	3.16		0.3	
0.013	11.64	-35.93	159.7	0.13	3.25		0.3	
0.013	11.63	-35.95	160.4	0.13	3.54		8.6	10.8
0.013	11.65	-35.9	160	0.13	3.57		8.5	10.9
0.013	11.65	-35.91	160.3	0.13	5.91		27.3	14.3
0.013	11.66	-35.88	158.9	0.13	6.29		27.3	14.4
0.013	11.64	-35.93	159.7	0.14	4.89		19	12.4
0.013	11.64	-35.92	153.6	0.13	5.11		25.6	14.9
0.013	11.64	-35.92	154.3	0.14	5.96		25.6	14.8
0.013	11.67	-35.84	159.5	0.14	7.24		31.5	14.9
0.013	11.63	-35.94	160.4	0.14	7.38		31.2	14.2
0.013	11.65	-35.89	154.2	0.14	6.29		32.9	16
0.013	11.66	-35.88	153	0.14	7.24		32.9	16
0.013	11.67	-35.86	153.5	0.14	7.4		37.9	16.7
0.013	11.64	-35.91	153.2	0.14	5.63		37.8	16.8
0.013	11.64	-35.92	159.3	0.14	10.11		47.7	18
0.013	11.65	-35.91	157.5	0.14	10.48		47.8	18.5
0.013	11.66	-35.87	165	0.13	9.21		45.5	17.7
0.013	11.65	-35.89	152.9	0.14	8.42		45.3	17.1
0.013	11.68	-35.82	157.5	0.14	8.48		56.9	19.5

Table G.9

$\dot{q}_1$	α_1	$\dot{q}_2$	α_2	$\dot{q}_3$	α_3	$\dot{q}_4$	α_4	$\dot{q}_5$	α_5	$\dot{q}_6$	α_6
112.6	37.4	94.1	30.8	89.2	27.9	72.4	23.9	55.3	18.2	15.8	5.2
138.6	38.6	119.1	35.1	109.4	32.7	107	31.6	58.1	16.8	20.7	5.7
138.8	40.2	119.2	35.1	109.6	32.1	107.1	31.2	58.2	16.8	21	6
0.3		0.3		0.3		0.3		0.3		0.3	
0.3		0.3		0.3		0.3		0.3		0.3	
0.3		0.3		0.3		0.3		0.3		0.3	
2.9	11.3	3.1	12.4	4.6	19.9	2	8.2	4.9	19.5	4.6	18.7
3.2	12.5	3.3	12.6	4.6	19.3	2.1	8.5	4.9	18.7	4.6	18.3
9.9	18.5	7.6	14.4	8.8	16.8	4.8	9.5	8.1	15.8	9.6	18.8
10.1	18.4	7.6	14.2	8.9	17	4.9	9.4	8.1	15.4	9.7	18.2
10	18.2	7.6	14.1	8.9	16.8	4.9	9.2	8.1	15.3	9.7	18.2
15.7	18.2	14.2	16.7	13.5	15.7	9.3	10.9	12.4	14.3	15.9	18.6
15.7	18	14.2	16.4	13.4	15.4	9.3	10.8	12.3	14.1	15.8	18.3
24.4	20.6	21.9	18.7	20.6	17.4	13.7	11.8	18.5	15.6	22	18.9
24.5	20.6	21.9	18.7	20.7	17.2	13.7	11.7	18.6	15.5	22.1	18.8
24.4	19.9	21.9	18.1	20.6	16.6	13.7	11.4	18.6	15.1	22.1	18.4
40.2	25.6	32.2	20.8	29.5	19.5	22.7	15.3	24.4	16.1	29.8	19.7
40.3	24.5	32.2	20.5	29.5	18.9	22.7	14.7	24.4	15.4	29.8	18.7
63.3	32.6	43.1	23	44.5	24	36.2	19.8	27.3	14.4	36.6	18.9
63.3	31.6	43.1	22.4	44.4	23.1	36.2	18.9	27.2	13.9	36.6	18.4
86.7	37.3	62.9	26.4	69.7	29.3	54.2	23	38.4	15.9	38.2	16.5
86.6	35.2	62.7	25.2	69.6	28.3	54.1	22.4	38.4	15.3	38.1	15.4
107.3	32	84.4	25.9	103.6	33.7	71.8	22.3	42	12.3	47	13
107.1	31.5	84.3	27.4	103.5	36.5	71.7	24.6	42.1	13.5	47	14
107	26.5	84.1	22.9	103.4	30.7	71.6	20.3	41.9	11	46.8	11.2
107.2	34.8	84.3	29.2	103.6	40.6	71.7	25.6	42.2	13	47	13.5
0.3		0.3		0.3		0.3		0.3		0.3	
0.3		0.3		0.3		0.3		0.3		0.3	
0.3		0.3		0.3		0.3		0.3		0.3	
12.9	16.1	11.4	14.4	6.4	8.2	1.4	1.7	8.5	10.2	11.2	14.2
12.9	16.6	11.2	14.4	6.4	8.3	1.2	1.4	8.3	10.1	11.2	14.4
31.5	17	39.3	20.8	29.1	13.9	5.1	2.6	19.6	10.4	39.3	20.9
31.5	17.3	39.2	20.9	29.1	14.1	5.1	2.6	19.6	10.5	39.3	21.2
27.1	17.9	29.2	19	10.9	7.2	5.1	3.2	14.3	9.1	27.5	18
30.4	17.5	38.9	22.6	21.4	12.6	9.5	5.5	22	12.7	31.4	18.3
30.4	17.6	38.9	22.4	21.4	12.6	9.5	5.6	22	12.6	31.4	18.3
38	18.3	44.9	21.5	28.9	13.1	8.3	3.8	29.5	13.7	39.2	18.8
37.1	16.4	45.1	20.9	29.1	13.3	6.8	3.1	29.7	13.6	39.4	18
30.4	14.9	49.5	24	32.3	15.9	13.6	6.7	33.5	15.9	37.8	18.6
30.4	14.9	49.5	23.9	32.3	16	13.7	6.7	33.5	16	37.8	18.7
40.3	17.6	52.7	23.3	35.8	16.1	24	10.4	36.6	16.1	37.8	17.1
40.3	17.9	52.7	24	35.8	16	23.9	10.3	36.6	16.1	37.8	16.6
31.8	11.6	65.7	24.3	47.8	19.1	36.5	14.1	51.5	19.4	52.8	19.2
31.9	12.5	65.8	25.2	47.9	19.2	36.6	14.1	51.6	19.8	52.8	20.2
34.9	13.8	67.6	26.6	35.9	15.2	36.3	14	44.7	17.1	53.5	19.7
34.9	12.8	67.5	25.1	35.2	14.8	36.3	13.9	44.7	16.8	53.4	19.2
33.8	11.9	77.1	26.7	63.4	22.5	54.4	18.2	70.8	23.5	41.7	14

Table G.9

$w_{oil,0}$	p	T	$\dot{m}$	$\dot{x}$	$\Delta p/\Delta L$	FP	$\dot{q}$	α
0.013	11.64	-35.93	158.4	0.14	10.66		57.1	19.8
0.013	11.69	-35.8	164.1	0.17	11.33		52.4	18.7
0.013	11.67	-35.84	164.6	0.13	12.43		55.2	18.6
0.013	11.67	-35.86	163.3	0.13	9.55		55.1	18
0.013	11.66	-35.88	158.2	0.13	13.49		69.4	22.3
0.013	11.65	-35.9	158.3	0.14	13.56		69.5	21.7
0.013	11.67	-35.85	170.1	0.13	18.61		75.2	23.1
0.013	11.65	-35.9	170.4	0.13	15.43		74.6	23.5
0.013	11.55	-36.13	160.7	0.13	3.89	i	0.3	
0.013	11.55	-36.13	161.2	0.13	3.68	i	0.3	
0.013	11.55	-36.13	161.1	0.13	3.3	i	0.3	
0.013	11.56	-36.12	162.1	0.13	4.18	i	6.5	12.6
0.013	11.55	-36.13	161.5	0.13	4.19	i	6.5	12.5
0.013	11.53	-36.19	160.9	0.14	4.08	i	9.9	13
0.013	11.53	-36.19	160.7	0.14	5.24	i	9.9	13.1
0.013	11.51	-36.23	160.6	0.14	5.01	i	16.2	14
0.013	11.52	-36.21	160.5	0.14	5.15	i	16.2	13.8
0.013	11.53	-36.19	160.6	0.14	5.33	i	23.3	15.6
0.013	11.53	-36.18	160.6	0.14	5.55	i	23.2	15.5
0.013	11.56	-36.13	161.5	0.14	6.39	i	23.3	15.5
0.013	11.54	-36.17	159.2	0.14	7.06	i	34.7	17.3
0.013	11.56	-36.12	160	0.14	7.39	i	34.8	17.6
0.013	11.54	-36.17	159.7	0.14	9.06	i	45.8	19.1
0.013	11.55	-36.15	160.5	0.13	9.66	i	45.8	19.2
0.013	11.55	-36.14	160.8	0.13	12.94	i (f)	60.2	21.1
0.013	11.57	-36.09	160.3	0.13	15.04	i (f)	60.2	21.2
0.013	11.56	-36.11	160.9	0.13	11.73	i (f)	60.3	20.8
0.013	11.57	-36.09	159.8	0.13	11	i (f)	76	23
0.013	11.58	-36.08	159.9	0.13	14.89	i (f)	75.9	23.7
0.013	11.6	-36.03	162.8	0.13	20.11	i (f)	99.4	27.6
0.013	11.6	-36.03	162.4	0.14	20.02	i (f)	99.4	27.5
0.013	11.59	-36.05	159.1	0.14	16.64	i (f)	99.3	27.2
0.013	11.53	-36.19	158.2	0.34	11.24	i-a	0.3	
0.013	11.54	-36.17	159.8	0.33	11.09	i-a	0.3	
0.013	11.55	-36.14	159.9	0.33	10.91	i-a	0.3	
0.013	11.55	-36.13	160.5	0.33	11.51	i-a	5.8	15.1
0.013	11.53	-36.19	160.2	0.33	11.58	i-a	5.8	15.2
0.013	11.53	-36.18	159.2	0.33	11.36	i-a	9.7	14.8
0.013	11.55	-36.15	160	0.33	11.64	i-a	9.7	14.7
0.013	11.55	-36.14	159.8	0.34	12.1	i-a	14.7	15.6
0.013	11.56	-36.12	161.1	0.33	12.31	i-a	14.8	15.5
0.013	11.56	-36.12	160.3	0.33	11.35	i-a	14.8	15.6
0.013	11.56	-36.12	161	0.33	12.84	i-a	20.5	16.6
0.013	11.55	-36.13	159.7	0.33	12.82	i-a	20.4	16.6
0.013	11.53	-36.18	159.9	0.33	13.73	i-a	30.7	18.6
0.013	11.56	-36.1	161.2	0.33	13.89	i-a	30.8	18.6
0.013	11.57	-36.1	160	0.33	16.07	i-a	45	21.4

$\dot{q}_1$	α_1	$\dot{q}_2$	α_2	$\dot{q}_3$	α_3	$\dot{q}_4$	α_4	$\dot{q}_5$	α_5	$\dot{q}_6$	α_6
34.1	12.1	77.3	26.9	63.6	22.6	54.7	18.4	71	23.9	41.9	14.9
46.2	16.5	87.1	30.1	37.5	14.5	53.9	18.5	53.1	18.9	36.6	13.8
46.2	15	87.1	27.9	37.4	14.4	53.9	18.1	57.4	19.5	49	16.7
46.1	14.4	87	27.7	37.3	14.2	53.8	17.8	57.3	19	48.9	15.1
64.5	21	84.1	27.5	74	24.1	72.9	22.4	73.3	23.6	47.5	15.4
64.7	19.5	84.2	26.6	74.1	24.1	73	22.3	73.6	23.2	47.6	14.7
63.2	20.2	96	29.6	71.1	22.4	107.6	29.9	69.3	21.9	43.9	14.8
55.3	19.1	101.8	32.2	71.2	22.5	106.3	29.9	69.2	22.1	44	15.1
0.3		0.3		0.3		0.3		0.3		0.3	
0.3		0.3		0.3		0.3		0.3		0.3	
0.3		0.3		0.3		0.3		0.3		0.3	
10	19.3	7.2	14	4.6	8.8	2.6	5.1	4.7	9.2	9.6	19.4
10	19.1	7.3	13.9	4.6	8.6	2.7	5.2	4.6	9	9.7	19.1
12.8	17.4	12.5	16.4	6.2	7.9	4.8	6.2	6.8	8.9	16.2	21.2
12.8	17.4	12.5	16.4	6.2	8.2	4.7	6.3	6.7	8.9	16.1	21.3
24.4	20.8	18.3	15.9	10.8	9.2	9.2	8	11.6	10	23.1	19.9
24.4	20.4	18.3	15.8	10.8	9.2	9.2	7.9	11.6	9.9	23.1	19.4
30.2	20.4	27.6	18.7	17.6	11.6	13.9	9.3	17.7	11.7	32.8	21.8
30.1	20.2	27.5	18.7	17.5	11.5	13.8	9.2	17.7	11.8	32.7	21.9
30.2	20.4	27.6	18.6	17.6	11.6	13.8	9.3	17.8	11.8	32.8	21.5
40.3	19.7	38.9	19.8	34.5	17.2	22.8	11.5	33.3	16.5	38.3	19.1
40.4	20.3	39	20.1	34.6	17.3	22.9	11.6	33.4	16.7	38.4	19.7
44.1	18.5	51.3	21.5	53.2	21.7	36.2	15.1	47.9	19.9	42.1	17.9
44.1	18.2	51.1	21.5	53.2	22	36.2	15.3	48	19.8	42.1	18.1
48.8	17.1	67.9	23.7	76.2	26.3	54.5	19.1	70.2	24.2	43.8	15.8
48.9	17.3	68	23.3	76.3	26.6	54	19.1	70.4	24.4	43.9	16.8
48.9	16.9	68.1	23.4	76.3	26.3	54	19	70.5	24	44	15.5
67.8	19.5	87.5	27.3	95.4	29.4	71.8	22.5	83.9	25.1	49.6	13.9
67.7	21.3	87.5	27.4	95.3	29.4	71.7	22.6	83.8	26	49.6	15.5
81.2	22.2	123.1	34.2	121.8	34.3	108.1	30.1	98.4	27.1	63.5	17.7
81.2	22.1	123.2	33.9	122	33.9	108.2	29.8	98.5	27.2	63.6	18
81.1	22	123	34.1	121.8	33.9	108	29.7	98.4	26.7	63.4	16.8
0.3		0.3		0.3		0.3		0.3		0.3	
0.3		0.3		0.3		0.3		0.3		0.3	
0.3		0.3		0.3		0.3		0.3		0.3	
8.9	23	3.4	8.9	5.1	12.4	2.5	6.3	6.5	17.2	8.4	22.6
8.9	23	3.5	9	5.1	12.3	2.5	6.2	6.6	17.7	8.4	23.2
12.5	19.2	8	12.1	7.5	11.4	4.8	7.4	9.6	14.8	15.9	24
12.5	19	7.9	12	7.5	11.2	4.8	7.4	9.5	14.8	15.9	24
20.2	21.3	10.9	11.7	11.8	12.5	9.2	9.7	15.3	16.1	21.2	22.2
20.2	21.2	10.9	11.7	11.8	12.4	9.2	9.7	15.3	15.9	21.2	22
20.3	21.3	11	11.9	11.8	12.5	9.2	9.8	15.3	15.9	21.2	22
28	22.4	15.9	12.9	17.7	14.4	13.7	11.3	20.2	16.4	27.8	22.2
27.8	22.4	15.6	12.8	17.6	14.4	13.6	11.3	19.9	16.3	27.7	22.3
38	23.2	23	14	32.7	19.2	22.8	13.8	31	18.7	37	22.5
38	23.2	23	14	32.7	19.3	22.8	13.8	31	18.7	37	22.5
52.7	25.4	34.4	16.4	50	23.4	36.2	17.4	43.5	20.7	53.3	25.2

Table G.9

$w_{oil,0}$	p	T	$\dot{m}$	$\dot{x}$	$\Delta p/\Delta L$	FP	$\dot{q}$	α
0.013	11.58	-36.07	160.1	0.33	16.58	i-a	45	21.3
0.013	11.57	-36.09	160.6	0.33	15.93	i-a	44.9	21.4
0.013	11.58	-36.06	159.8	0.33	19.84	i-a (f)	63.5	25.2
0.013	11.61	-36	159.2	0.34	20.38	i-a (f)	63.6	25
0.013	11.57	-36.09	159.6	0.33	22.25	i-a (f)	77.5	27.7
0.013	11.59	-36.05	160.7	0.33	21.63	i-a (f)	77.5	27.7
0.013	11.57	-36.08	159.4	0.34	27.03	i-a (f)	107	32.2
0.013	11.6	-36.01	158.9	0.34	27.52	i-a (f)	106.9	32.7
0.013	11.61	-36	160.1	0.33	27.13	i-a (f)	106.9	32.1
0.013	11.72	-35.73	154.4	0.54	18.95		0.3	
0.013	11.73	-35.7	153.4	0.54	18.46		0.3	
0.013	11.76	-35.64	154.7	0.54	18.64		1.9	
0.013	11.72	-35.73	154.4	0.54	18.85		3.4	18.6
0.013	11.71	-35.76	156.3	0.53	18.47		8.8	17.2
0.013	11.73	-35.7	157.4	0.53	18.94		8.8	16.8
0.013	11.75	-35.67	157.4	0.53	17.78		12.7	17.1
0.013	11.74	-35.67	157.8	0.53	19.06		12.7	16.8
0.013	11.75	-35.65	158.4	0.53	20.05		18	17.1
0.013	11.77	-35.6	157.9	0.54	20.64		17.2	16.8
0.013	11.73	-35.7	154.8	0.54	19.73		26.5	18.3
0.013	11.75	-35.66	159	0.52	19.59		26.3	18.1
0.013	11.81	-35.5	160.3	0.53	25.6		46.3	21.2
0.013	11.74	-35.68	161.5	0.52	24.88		46.4	21.7
0.013	11.74	-35.67	160.4	0.53	29.02		66.9	25.8
0.013	11.76	-35.64	160.8	0.52	28.62		67.1	25.9
0.013	11.74	-35.67	161.4	0.53	31.4		78.2	27.9
0.013	11.77	-35.61	161.5	0.52	31.76		78.2	27.6
0.013	11.79	-35.55	159.9	0.53	33.64		92.4	29.5
0.013	11.76	-35.63	161.6	0.5	33.74		92.7	30.1
0.013	11.66	-35.88	156.6	0.54	17.67		0.3	
0.013	11.69	-35.81	155.4	0.54	17.95		0.3	
0.013	11.57	-36.1	155.2	0.54	18.1		0.3	
0.013	11.6	-36.01	155.4	0.54	17.78		0.3	
0.013	11.58	-36.06	161.7	0.03			0.3	
0.013	11.6	-36.02	155.8	0.54	17.67		3.3	
0.013	11.57	-36.09	155.5	0.54	18.35		3.4	20.8
0.013	11.58	-36.07	155	0.54	18.13		6.6	18.5
0.013	11.6	-36.03	155.9	0.54	19.16		6.6	18.4
0.013	11.58	-36.06	156	0.54	18.12		12.8	17.6
0.013	11.6	-36.01	155.6	0.53	18.17		12.9	17.5
0.013	11.61	-36	156	0.54	18.67		12.9	17.2
0.013	11.61	-35.99	155.2	0.54	17.79		18.3	17.8
0.013	11.6	-36.01	155.9	0.54	20.03		18.4	17.2
0.013	11.6	-36.02	155.6	0.54	21.87		29.8	18.5
0.013	11.61	-36	156.2	0.54	21.92		29.8	18.3
0.013	11.6	-36.01	156.7	0.54	23.56		44	20.7
0.013	11.63	-35.94	155.8	0.54	23.78		44	20.7

Table G.9

$\dot{q}_1$	α_1	$\dot{q}_2$	α_2	$\dot{q}_3$	α_3	$\dot{q}_4$	α_4	$\dot{q}_5$	α_5	$\dot{q}_6$	α_6
52.7	25.3	34.4	16.3	49.9	23.4	36.1	17.3	43.4	20.5	53.3	25.1
52.7	25.4	34.3	16.4	49.9	23.4	36.1	17.3	43.4	20.7	53.3	25.2
76.7	30.2	48.3	19.1	69.5	27.8	54	21.6	61.8	24.5	70.4	27.9
76.8	30.1	48.4	19	69.6	27.8	54.1	21.5	61.9	24.2	70.7	27.6
100.4	35.7	63.7	22.4	82.4	29.6	72.2	25.8	68.9	24.6	77.5	28
100.5	35.3	63.7	22.4	82.4	29.8	72.2	25.8	68.9	24.7	77.6	28.2
144.3	44.1	91.7	27.5	126.1	37.3	108	32.5	73.5	21.9	98.5	29.9
144	44.1	91.6	27.4	125.9	37.2	107.8	32.9	73.5	23.1	98.4	31.7
144.1	43.9	91.7	27.4	125.9	37.3	107.8	32.5	73.7	21.9	98.4	29.9
0.3		0.3		0.3		0.3		0.3		0.3	
0.3		0.3		0.3		0.3		0.3		0.3	
0.3		3.6		0.6		1.5		4.9		0.5	
2.2	12.1	3.3	17.5	4.8	27.5	1.7	8.1	5.8	31.3	2.5	15.1
10.2	19.7	6.9	13.7	8.6	17.8	5.3	10.1	10.3	20.1	11.1	21.6
10.4	19.3	7	13.3	8.6	16.9	5.4	9.9	10.5	19.9	11.2	21.2
14.3	19	11.4	15.3	11	16	9	12.1	15.4	20.2	15	19.9
14.3	18.9	11.4	15.1	11	15.5	9	11.7	15.5	19.9	14.9	19.7
20.9	19.6	18.4	17	15.7	15.5	13.9	13.3	18.8	18.1	20.2	19.4
20.8	20.1	14.4	14.2	15.5	15.7	13.8	13.3	18.7	18.2	20.2	19.6
29.4	20	25.3	17.6	22.6	16.4	22.7	15.7	27	18.3	31.9	21.4
29.4	20	25.3	17.5	22.6	16.4	22.7	15.7	27	18.4	30.6	20.6
53.5	24.4	44.1	20.1	45.1	21.1	36	16.6	45.2	20.7	53.7	24.4
53.7	24.9	44.2	20.6	45.3	21.6	36.1	17	45.3	21.1	53.9	25
79.5	30.5	60.7	23.3	65.6	25.8	54.8	21.1	62.3	24	78.6	30
79.4	30.5	60.6	23.3	67	26.4	54.5	21	62.3	24	78.5	30.1
92.5	32.7	71	25.2	65.7	25	72.1	25.2	74.9	26.5	92.8	32.9
92.5	32.8	71	25.2	65.7	24.7	72.1	24.8	74.9	26	92.8	32.4
96.2	32.8	90	28.9	86.3	28.6	107.3	30.7	92.6	28	82.2	28
96.7	33.1	90.3	28.7	86.7	28.7	107.6	31.6	92.7	29.3	82.3	29
0.3		0.3		0.3		0.3		0.3		0.3	
0.3		0.3		0.3		0.3		0.3		0.3	
0.3		0.3		0.3		0.3		0.3		0.3	
0.3		0.3		0.3		0.3		0.3		0.3	
0.3		0.3		0.3		0.3		0.3		0.3	
3		0.3		4.7		1.9		5.7		4.4	
3.2	18.6	0.3	1.9	4.8	30.6	2	12.3	5.8	34.1	4.5	26.9
8.2	22.2	4.1	11.1	6.2	17.5	4.8	13.5	8.3	23.6	7.9	22.7
8.1	22.1	4.3	11.3	6.1	17.4	4.8	13.4	8.2	23.3	7.9	22.8
15.4	20.9	8.1	11.3	12.8	17.6	9.3	13.1	14	19.5	17.2	23.3
15.5	20.9	8.2	11.3	12.8	17.4	9.3	12.9	14.1	19.3	17.2	23
15.5	20.3	8.2	10.9	12.9	17.1	9.4	12.8	14.1	19.1	17.2	22.8
23.6	22.1	13.6	13.1	17.8	17.5	14	14.1	17.7	17.7	22.9	22
23.7	21.7	13.6	12.6	18	16.5	14.1	13.4	17.8	17.2	23	21.6
36.6	22.5	21.4	13.4	31.1	19.1	22.9	14.5	30.5	19	36.6	22.5
36.5	22.3	21.3	13.2	31.1	18.9	22.8	14.4	30.5	18.8	36.6	22.2
53.9	25	36.4	16.9	43.4	20.7	36.1	17.4	42.5	20.2	51.5	24.3
53.9	25.1	36.4	17.1	43.5	20.5	36.2	17.2	42.5	20.1	51.5	24.1

$w_{oil,0}$	p	T	$\dot{m}$	$\dot{x}$	$\Delta p/\Delta L$	FP	$\dot{q}$	α
0.013	11.61	-36	156.5	0.54	28.21		66.9	26.2
0.013	11.64	-35.93	156.2	0.54	28.7		66.4	25.7
0.013	11.63	-35.93	157.2	0.53	28.77		66.4	25.5
0.013	11.62	-35.96	157.7	0.53	34.25		95.1	30.9
0.013	11.64	-35.91	157.5	0.53	33.38		95.1	28.5
0.013	11.71	-35.76	154	0.54	38.76		124.4	36.3
0.013	11.67	-35.85	156	0.54	40		124.5	36.8
0.013	11.71	-35.76	156.4	0.54	39.88		124.4	35.9
0.013	11.59	-36.03	159.6	0.53	18.22	a	0.3	
0.013	11.59	-36.03	162.6	0.52	18.24	a	0.3	
0.013	11.57	-36.09	159.8	0.53	17.41	a	0.3	
0.013	11.56	-36.1	160	0.52	18.84	a	4.9	17.2
0.013	11.59	-36.05	160.5	0.52	18.75	a	4.9	16.8
0.013	11.6	-36.03	160	0.52	19.14	a	4.9	16.5
0.013	11.57	-36.08	159.9	0.53	20.44	a	9	16.5
0.013	11.59	-36.05	159.2	0.53	18.76	a	9	16.8
0.013	11.59	-36.03	160.3	0.52	19.25	a	9	16.6
0.013	11.57	-36.09	160	0.52	18.97	a	15	17.1
0.013	11.6	-36.02	159.6	0.52	19.89	a	14.3	16.7
0.013	11.6	-36.01	160.8	0.52	20.18	a	15	16.5
0.013	11.6	-36.01	159.7	0.52	19.38	a	19.8	17.5
0.013	11.6	-36.01	160.5	0.52	20.54	a	19.8	17.2
0.013	11.64	-35.93	159.7	0.53	20.4	a	19.6	17
0.013	11.58	-36.08	160.7	0.52	22.08	a (f)	30.3	18.7
0.013	11.6	-36.03	160.6	0.52	21.9	a (f)	30.3	18.6
0.013	11.61	-36	160	0.52	22.88	a (f)	30.4	18.4
0.013	11.59	-36.04	160.6	0.49	23.47	a (f)	44.2	20.5
0.013	11.61	-35.99	162.6	0.48	23.13	a (f)	44.1	20.4
0.013	11.59	-36.05	161.6	0.49	21.94	a (f)	43.9	20.7
0.013	11.58	-36.06	160.7	0.52	25.69	a (f)	47.2	20.3
0.013	11.59	-36.03	161.4	0.52	25.31	a (f)	47.2	20.5
0.013	11.62	-35.98	160.9	0.52	24.88	a (f)	47.2	20.3
0.013	11.63	-35.94	160.8	0.52	29.42	i-a (f)	67.4	24.6
0.013	11.62	-35.97	160.2	0.52	29.95	i-a (f)	67.5	25.2
0.013	11.64	-35.93	160	0.53	34.17	i-a (f)	90.7	29.8
0.013	11.64	-35.92	160.9	0.52	33.34	i-a (f)	90.6	29.7
0.013	11.62	-35.97	161	0.53	40.83	i-a (f)	121.7	36.4
0.013	11.64	-35.93	161.4	0.53	40.78	i-a (f)	121.8	36.4
0.013	11.65	-35.9	161.1	0.53	41.62	i-a (f)	121.6	36.3
0.013	11.57	-36.09	155	0.74	23.52	a	0.3	
0.013	11.6	-36.01	157.5	0.73	22.56	a	0.3	
0.013	11.61	-35.98	158.8	0.72	23.8	a	0.3	
0.013	11.61	-35.99	156.5	0.74	24.04	a	1.3	
0.013	11.58	-36.06	157.3	0.73	25.18	a	1.3	
0.013	11.58	-36.07	156.9	0.73	23	i-a (f)	5.3	22.6
0.013	11.59	-36.04	158.5	0.73	22.77	i-a (f)	5.3	22.7
0.013	11.6	-36.01	156	0.74	23.83	i-a (f)	8.3	21.1

Table G.9

$\dot{q}_1$	α_1	$\dot{q}_2$	α_2	$\dot{q}_3$	α_3	$\dot{q}_4$	α_4	$\dot{q}_5$	α_5	$\dot{q}_6$	α_6
79.2	30.7	53	20.5	72.6	28.5	54.3	21.8	62.7	24.9	79.8	31.1
76.7	29.9	51.8	20.1	72.6	28	54.3	21.4	62.7	24.4	79.9	30.6
76.7	29.6	52	20.1	72.6	27.4	54.3	21	62.7	24.1	79.9	30.4
110.3	36	81.9	26	104.1	34.2	72.3	23.9	86.5	28.2	115.7	37.2
110.3	34.2	81.9	25.1	104.1	30.9	72.3	21.6	86.5	25.5	115.7	33.5
142.7	42.5	106.1	30.3	140.8	42.8	107.5	30.9	110.3	31.2	138.8	40
142.9	42.3	106.3	30.5	140.8	44	107.7	31.5	110.6	31.5	138.8	40.7
142.7	42.6	106	30.5	140.9	42.6	107.5	30.1	110.6	30.3	138.8	39.3
0.3		0.3		0.3		0.3		0.3		0.3	
0.3		0.3		0.3		0.3		0.3		0.3	
0.3		0.3		0.3		0.3		0.3		0.3	
5.9	20.2	1.6	6	6.5	24.2	1.9	7.2	5.8	21	7.4	24.6
5.9	19.6	1.7	5.9	6.5	23.3	2.1	7.6	5.8	20.3	7.4	24.1
6	19.4	1.8	6.1	6.5	22.3	2	7	5.8	20.2	7.4	24
10.5	19.6	6.4	11.6	10	17.7	4.6	8.4	11.5	21	10.9	20.5
10.5	19.9	6.4	11.9	10	18.6	4.6	8.7	11.5	21.3	10.9	20.6
10.6	19.7	6.5	11.8	10	18.2	4.6	8.6	11.5	21	10.9	20.5
18.7	21.3	9.7	11.2	17.7	19.5	9.2	10.7	16.5	18.9	18.2	20.9
18.8	21.6	9.7	11.6	13.5	16.2	9.4	11.1	16.5	18.9	18.1	20.8
18.8	20.9	9.6	10.7	17.6	18.2	9.3	10.3	16.5	18.3	18.1	20.3
23.6	20.9	15.9	14.2	20.2	17.7	13.8	12.5	20.1	17.8	25.1	21.9
23.6	20.7	15.8	13.8	20.2	17.3	13.7	12.1	20.1	17.5	25.1	21.8
23.4	20.5	15.6	13.5	20	16.9	13.6	12	19.8	17.4	24.9	21.5
38	23.2	24.8	15.5	30.1	18.6	22.7	14.3	29.7	18.4	36.2	22.2
38	23.2	24.8	15.4	30	18.5	22.7	14.2	29.7	18.2	36.3	22.1
38.2	23	25	15.2	30.2	18	22.9	13.9	29.9	18.1	36.4	22
50.6	23.5	39.6	18.1	44	20.1	36.6	17.1	42.1	19.7	52.1	24.4
50.5	23.6	39.6	18.3	44	20	36.2	16.8	42.1	19.6	51.9	24.2
50.4	23.9	39.5	18.4	43.8	20.5	36.1	17.1	42.1	19.8	51.7	24.4
56.5	24.2	39.3	16.8	45.9	20.1	36.1	15.8	45.1	19.5	60.3	25.5
56.5	24.3	39.3	16.9	45.9	20.3	36.1	16	45.2	19.6	60.3	25.7
56.4	24.3	39.3	16.9	45.9	20.3	36.1	15.8	45.2	19.2	60.3	25.3
79.1	29.2	55.2	20.7	70.6	25.7	54.2	19.8	65.4	23.5	80	28.9
79.2	29.5	55.3	20.6	70.7	26.5	54.2	20.5	65.5	24.5	80	29.9
104.8	34.7	71.9	24	100.5	32.9	71.9	23.5	90.3	29.1	105	34.5
104.6	34.9	71.8	24	100.3	33	71.8	23.7	90.1	28.8	104.8	33.9
138.9	41.9	93.6	28.1	136.7	43	107.5	31.9	112.3	32.3	141.3	41
139	41.7	93.7	28	136.8	41.8	107.7	31.8	112.4	33.2	141.4	42.2
138.8	42	93.5	27.9	136.6	41.9	107.5	31.6	112.2	32.8	141.3	41.9
0.3		0.3		0.3		0.3		0.3		0.3	
0.3		0.3		0.3		0.3		0.3		0.3	
0.3		0.3		0.3		0.3		0.3		0.3	
0.3		0.3		1.8		2.2		2		1.5	
0.3		0.3		1.7		2.4		2.1		1.3	
4.5	18.1	3	12.4	7.2	31.8	4.9	21.1	6.3	27.7	5.8	24.6
4.5	17.4	3.1	12.7	7.2	32.8	4.5	20.4	6.3	28	5.9	24.7
7.8	19.9	5	13.2	9.3	24.6	9.4	22.5	9.7	24.2	8.6	22.4

Table G.9

$w_{oil,0}$	p	T	$\dot{m}$	$\dot{x}$	$\Delta p/\Delta L$	FP	$\dot{q}$	α
0.013	11.62	-35.98	156.4	0.73	22.39	i-a (f)	8.3	21.4
0.013	11.59	-36.03	154.1	0.75	23.34	i-a (f)	13.4	21.4
0.013	11.61	-36	155.6	0.74	23.63	i-a (f)	13.5	21.1
0.013	11.62	-35.98	155	0.74	23.46	i-a (f)	13.5	21
0.013	11.61	-35.99	154.9	0.74	24.5	i-a (f)	25	21.5
0.013	11.62	-35.96	153.6	0.75	25.7	i-a (f)	25	21.1
0.013	11.63	-35.95	155	0.74	29.04	i-a (f)	38.4	22.9
0.013	11.61	-36	153.5	0.75	30.1	i-a (f)	38.3	23.1
0.013	11.61	-35.98	153.1	0.75	34.19	i-a (f)	58.7	27
0.013	11.64	-35.91	155	0.74	31.8	i-a (f)	58.7	25.7
0.013	11.65	-35.89	154.4	0.75	34.12	i-a (f)	58.6	26.1
0.013	11.63	-35.95	153.7	0.74	40.45	i-a (f)	80.7	31.1
0.013	11.65	-35.89	153.5	0.74	38.68	i-a (f)	80.7	30.7
0.013	11.63	-35.95	152.6	0.76	49.32	i-a (f)	122.6	34.7
0.013	11.66	-35.87	153.5	0.75	47.57	i-a (f)	122.6	35.5
0.013	11.58	-36.07	155.6	0.74	25.41		0.3	
0.013	11.58	-36.07	155.8	0.74	24.01		0.3	
0.013	11.6	-36.03	156.4	0.73	23.52		0.3	
0.013	11.6	-36.02	155.8	0.74	24.59		0.3	
0.013	11.6	-36.03	155	0.74	25.19		1.5	
0.013	11.6	-36.02	154	0.75	23.42		1.5	
0.013	11.6	-36.03	155.6	0.74	25.13		1.5	
0.013	11.57	-36.09	156.3	0.73	25.69		4.8	22.6
0.013	11.6	-36.03	157.2	0.73	22.6		4.8	22.9
0.013	11.6	-36.02	156.4	0.73	25		4.7	22.4
0.013	11.58	-36.07	157.5	0.73	24.9		9	21.7
0.013	11.6	-36.03	157	0.73	25.48		9	21.4
0.013	11.61	-35.99	157.1	0.73	24.44		9	21.5
0.013	11.58	-36.08	155.8	0.74	26.18		13.4	21.4
0.013	11.6	-36.03	156.1	0.73	25.44		13.4	21.4
0.013	11.61	-36	156.7	0.73	25.77		13.4	21.3
0.013	11.63	-35.96	156.9	0.73	29.01		22.7	20.9
0.013	11.58	-36.06	157.6	0.73	27.93		22.8	21.4
0.013	11.61	-35.99	158.3	0.73	27.57		22.8	21.5
0.013	11.6	-36.03	157	0.73	30.95		35.5	22.8
0.013	11.62	-35.98	155.6	0.74	30.05		35.5	22.6
0.013	11.62	-35.97	154.1	0.75	30.39		35.5	23.1
0.013	11.63	-35.95	158.5	0.73	36.2		56.5	26.5
0.013	11.63	-35.94	156.9	0.73	35.5		56.6	26.9
0.013	11.62	-35.98	157.4	0.73	36.35		56.6	27.5
0.013	11.61	-36.01	158.2	0.73	34.85		56.5	27.2
0.013	11.63	-35.95	159.2	0.72	38.92		80.1	28.7
0.013	11.64	-35.92	158.5	0.73	41.3		80.1	28.7
0.013	11.65	-35.91	160.8	0.71	40.06		79.9	30.6
0.013	11.66	-35.86	160.6	0.72	49.74		119.3	34.2
0.013	11.61	-36	159.5	0.72	46.96		119.3	30.9
0.013	11.65	-35.89	159.6	0.72	52.08		119.4	38.1

$\dot{q}_1$	α_1	$\dot{q}_2$	α_2	$\dot{q}_3$	α_3	$\dot{q}_4$	α_4	$\dot{q}_5$	α_5	$\dot{q}_6$	α_6
7.8	20.2	5	13.3	9.3	25.3	9.4	23.1	9.6	24.2	8.5	22.6
13.1	21	8.7	14.3	16.2	25.7	13.8	21.6	14.2	22.4	14.5	23.4
13.2	20.7	8.8	14	16.3	25.4	13.9	21.4	14.2	22.1	14.6	23.2
13.1	20.7	8.7	14	16.2	25	13.9	21.1	14.2	22	14.6	22.9
26.2	22.5	17.8	15.4	32.9	27.5	22.8	20	25.1	21.5	25.3	22
26.2	22.4	17.8	15.2	32.9	26.7	22.9	19.4	25.1	21.1	25.3	21.5
40.4	23.7	30.8	18.2	45.7	27.8	36.5	22.4	36.3	21.6	40.8	23.8
40.4	24.2	30.7	18.6	45.7	27.7	36.4	22.3	36.1	21.8	40.6	24.2
62.5	28.5	46.3	21.6	72.2	33.4	54	25.1	53.1	24.4	63.9	29.1
62.6	27.7	46.4	20.6	72.3	31.6	54	23.8	53.1	23.1	63.9	27.5
62.4	27.6	46.3	20.4	72.1	32	53.9	24.2	53	24	63.8	28.5
87.3	33.7	63.3	24.9	99.1	39.5	72.1	27.5	73.4	27.6	89	33.3
87.4	34	63.3	24.7	99	37.8	72.1	27.3	73.4	27.5	89	33.2
139.1	38.6	105.3	29.4	153.5	47.1	107.7	29.4	105	28.9	124.8	34.5
139.1	41.2	105.3	30.9	153.5	49.5	107.7	29.1	105	28.4	124.9	34.1
0.3		0.3		0.3		0.3		0.3		0.3	
0.3		0.3		0.3		0.3		0.3		0.3	
0.3		0.3		0.3		0.3		0.3		0.3	
0.3		0.3		0.3		0.3		0.3		0.3	
0.3		0.3		3.1		1.9		1.3		2	
0.3		0.3		3.1		1.8		1.4		2	
0.3		0.3		3.2		1.9		1.5		1.9	
5.4	24	2.8	13.2	5.7	30.4	4.8	23.3	4.9	21.5	4.9	23.3
5.3	23.3	2.9	13.8	5.7	32.1	4.8	23.9	4.9	21.4	4.9	22.8
5.3	23	2.8	13	5.7	30.7	4.8	23.4	4.8	21.5	4.9	22.8
9.4	22.2	6.7	16.3	9.4	24.3	9.4	22.8	8.2	19.1	10.9	25.6
9.3	22	6.7	16.1	9.4	24.3	9.4	22.4	8.3	18.8	10.9	24.9
9.3	21.7	6.7	16.1	9.4	24.4	9.4	22.8	8.3	19	10.9	24.8
14	22.2	10.3	16.7	13.8	23.1	13.8	22.3	12.7	19.7	15.6	24.5
14	22.2	10.4	16.9	13.8	23.4	13.9	22.3	12.7	19.6	15.6	24.2
14.1	22.4	10.5	16.9	13.9	23	13.9	22	12.8	19.5	15.6	24.1
21.3	20.1	20.2	18.4	25.6	23.2	22.6	21	19.1	17.7	27.4	25
21.4	20.4	20.3	18.8	25.7	23.9	22.7	21.7	19.2	18.1	27.5	25.4
21.3	20.7	20.3	19	25.6	23.9	22.6	21.6	19.2	18.2	27.5	25.6
38.7	24.6	30.8	19.6	39	25.1	36.1	23.4	29.6	19.1	38.9	24.9
38.6	24.4	30.7	19.5	38.9	25.1	36.1	23.4	29.5	18.9	38.8	24.5
38.7	24.9	30.8	19.9	39	25.6	36.1	23.8	29.6	19.3	38.9	25
62.8	29.5	47.1	22.1	65	30.3	54.2	25.8	46.8	22	63.3	29.5
62.8	29.8	47.1	22.3	65.1	30.9	54.3	26.1	46.8	22.2	63.3	29.9
62.9	30.6	47.1	23	65.2	31.4	54.3	26.6	46.8	22.8	63.3	30.4
62.8	30.1	47.1	22.6	65.1	31.2	54.3	26.4	46.7	22.5	63.3	30.2
89.7	32.6	67.8	24.1	94.9	34.3	71.9	25.7	66.8	23.8	89.2	31.5
89.7	32	67.8	24.1	94.9	33.3	72	26	66.9	24.3	89.1	32.1
89.4	34.4	67.7	25.9	94.6	36.3	71.9	27.8	66.7	25.6	88.9	33.7
137.9	39.1	97.9	28.2	142.8	44	107.4	29.7	104.1	28.9	125.6	35.3
137.8	37.4	98	26.1	142.8	37.9	107.6	26.8	104	25.8	125.7	31.5
137.9	44	98.1	30.9	142.9	47.9	107.7	33.7	104	32.6	125.7	39.8

Table G.9

$w_{oil,0}$	p	T	$\dot{m}$	$\dot{x}$	$\Delta p/\Delta L$	FP	$\dot{q}$	α
0.013	11.62	-35.98	156.7	0.93		a	0.3	
0.013	11.64	-35.92	158	0.92		a	0.3	
0.013	11.65	-35.9	158.8	0.92		a	0.3	
0.013	11.61	-35.99	159.7	0.92	23.27	a	1.8	12.3
0.013	11.64	-35.92	160.3	0.91	22.72	a	1.8	12.7
0.013	11.62	-35.97	158.9	0.92	20.85	a (f)	4.2	15.8
0.013	11.64	-35.91	158.5	0.92	24.16	a (f)	4.1	16.3
0.013	11.62	-35.97	158.2	0.93	22.46	i-a (f)	9.6	16.4
0.013	11.66	-35.88	159.5	0.92	23.61	a (f)	9.6	15
0.013	11.66	-35.87	158.2	0.93	24.42	a (f)	9.6	15.5
0.013	11.62	-35.96	158.1	0.93	23.91	i-a (f)	14.5	17.9
0.013	11.65	-35.89	158	0.93	24.61	i-a (f)	14.4	16
0.013	11.64	-35.93	158.8	0.92	29.54	i-a (f)	22.1	19.4
0.013	11.66	-35.86	158.9	0.92	28.44	i-a (f)	22.1	19.7
0.013	11.64	-35.93	158.1	0.93	33.66	i-a (f)	35.1	20.2
0.013	11.67	-35.86	157.7	0.93	31.79	i-a (f)	35.1	20.4
0.013	11.67	-35.85	158	0.93	33.29	i-a (f)	35	16.9
0.013	11.65	-35.9	158.3	0.93	42.73	i-a (f)	56.2	20.6
0.013	11.67	-35.84	159.7	0.92	39.49	i-a (f)	56.1	20.5
0.013	11.55	-36.14	264.8	0.13	10.09	i-a	0.3	
0.013	11.56	-36.12	263.8	0.13	10.23	i-a	0.3	
0.013	11.56	-36.11	265	0.13	10.49	i-a	0.3	
0.013	11.57	-36.09	265.3	0.12	10.86	i-a	5.2	11.1
0.013	11.54	-36.15	266	0.13	10.68	i-a	5.3	11.2
0.013	11.55	-36.15	264.8	0.12	10.98	i-a	8.3	11.5
0.013	11.56	-36.12	265.8	0.12	10.91	i-a	8.3	11.4
0.013	11.56	-36.11	265.1	0.12	11.57	i-a	8.3	11.5
0.013	11.57	-36.09	265.2	0.12	11.73	i-a	12.2	12.4
0.013	11.57	-36.08	266.6	0.12	11.3	i-a	12	12.4
0.013	11.51	-36.23	265.4	0.13	12.8	i-a	18.4	14.1
0.013	11.52	-36.21	265.6	0.12	12.13	i-a	18.4	14.1
0.013	11.53	-36.2	265.6	0.13	14.05	i-a (f)	29.2	16.8
0.013	11.53	-36.18	263.8	0.13	14.35	i-a (f)	29.3	16.8
0.013	11.54	-36.17	265.3	0.13	13.88	i-a (f)	29.3	16.8
0.013	11.55	-36.15	266.8	0.12	18.06	i-a (f)	46.8	20.4
0.013	11.55	-36.14	265.8	0.12	17.16	i-a (f)	46.6	20.4
0.013	11.54	-36.16	263.9	0.13	22.4	i-a (f)	66.4	23.9
0.013	11.54	-36.17	264.9	0.13	22.8	i-a (f)	66.4	23.8
0.013	11.54	-36.15	264.9	0.13	25.76	i-a (f)	80.9	25.9
0.013	11.56	-36.11	266.6	0.13	24.47	i-a (f)	80.7	25.9
0.013	11.56	-36.1	263.4	0.13	25.56	i-a (f)	80.8	26.1
0.013	11.58	-36.08	264.4	0.13	30.38	i-a (f)	104.5	28.4
0.013	11.57	-36.09	263.8	0.13	29.92	i-a (f)	104.6	28.6
0.013	11.58	-36.06	262.8	0.52	46.9	a	0.3	
0.013	11.54	-36.16	263.1	0.52	47.92	a	0.3	
0.013	11.58	-36.06	265.3	0.51	46.58	a	0.3	
0.013	11.53	-36.18	263.6	0.52	47.59	a	3.2	32.4

$\dot{q}_1$	α_1	$\dot{q}_2$	α_2	$\dot{q}_3$	α_3	$\dot{q}_4$	α_4	$\dot{q}_5$	α_5	$\dot{q}_6$	α_6
0.3		0.3		0.3		0.3		0.3		0.3	
0.3		0.3		0.3		0.3		0.3		0.3	
0.3		0.3		0.3		0.3		0.3		0.3	
0.3	1.9	0.3	1.9	4	26.3	2.5	16.5	2.4	17.1	1.4	9.9
0.3	1.9	0.3	2	4	27.5	2.2	15.1	2.4	18.1	1.6	11.4
1.8	6.5	3.2	11.5	6	23.5	4.9	18.6	3.9	15.3	5.3	19.2
1.6	5.7	3.1	11.5	6.1	24.8	4.9	19.3	3.9	16.3	5.4	20.1
6.7	11.4	7.3	12.5	13.9	23.6	9.2	16.3	9.6	16.4	10.7	17.8
6.7	10.8	7.3	11.6	13.9	21.3	9.3	14.7	9.6	15	10.7	16.4
6.7	11	7.2	11.9	13.9	22.4	9.3	15.1	9.6	15.6	10.7	17.1
15.3	18.2	12.3	15.3	18.6	23.8	13.8	17.9	15.2	18.4	11.6	14
15.2	16	12.2	13.6	18.5	21.2	13.7	16.1	15.2	16.5	11.6	12.5
19.9	16.6	18	15.5	31	28.9	22.7	20.9	20.5	17.5	20.7	16.8
19.9	16.7	18	15.6	31	28.8	22.7	21.1	20.5	18.2	20.7	17.7
32.4	17.6	25.1	14	50.5	30.7	36.2	21.8	35.2	19.9	30.9	16.9
32.4	18.1	25.1	14.2	50.5	32.7	36.2	21.7	35.3	19.3	30.9	16.2
32.4	15.3	25	11.9	50.4	25.5	36.1	17.4	35.3	16.6	31	14.4
59	20.1	45	15.7	75.5	29.2	54.3	20	53	20	50.2	18.5
58.9	20.7	45	15.8	75.4	29.2	54.3	19.3	53	19.5	50.1	18.3
0.3		0.3		0.3		0.3		0.3		0.3	
0.3		0.3		0.3		0.3		0.3		0.3	
0.3		0.3		0.3		0.3		0.3		0.3	
6.8	14.3	3.5	7.6	4.8	10.3	1.8	4	3.4	7.6	10.8	22.8
6.9	14.3	3.7	7.8	4.8	10.4	2	4.4	3.5	7.8	10.9	22.8
10.6	14.7	6.1	8.4	7.3	9.9	4.9	6.7	6.3	8.8	14.4	20.2
10.6	14.8	6.1	8.4	7.3	9.8	4.9	6.7	6.3	8.8	14.4	20
10.6	14.9	6.1	8.4	7.3	9.8	4.9	6.8	6.4	8.9	14.4	20.1
15.8	16.2	10.1	10.3	9.3	9.3	9	9	9.9	10.1	18.8	19.4
15.6	16.2	10	10.3	9.2	9.3	8.9	9	9.8	10	18.7	19.5
24.3	18.8	16.6	12.7	14.3	10.9	13.7	10.5	15	11.6	26.2	20.3
24.4	18.8	16.6	12.7	14.4	10.8	13.7	10.4	15	11.5	26.2	20.2
36.9	21.3	27.1	15.6	25.5	14.6	22.5	12.9	24.4	14.1	38.9	22.4
37	21.3	27.2	15.6	25.6	14.6	22.5	13	24.4	14	38.9	22.3
37	21.3	27.2	15.6	25.6	14.6	22.5	12.9	24.4	14	38.9	22.3
59.4	25.5	42.1	18.4	46.9	20.4	36.3	16	40.4	17.7	55.5	24.3
59.2	25.8	42	18.4	46.9	20.4	36.2	16	40.1	17.6	55.3	24.4
84.8	30.4	62.1	22.3	70.9	25.5	54.1	19.7	59.7	21.2	66.8	24
84.8	30.1	62.2	22.2	70.9	25.4	54.1	19.6	59.7	21.3	66.9	24
101.3	32.6	79.8	25.5	88.6	28.2	72.1	23.3	70.1	22.3	73.4	23.6
101.1	32.8	79.6	25.6	88.5	28.2	71.9	23.3	69.9	22.3	73.2	23.5
101.2	33.3	79.7	25.7	88.5	28.3	72	23.4	70	22.3	73.3	23.5
134.7	36.2	107.7	29.6	117.6	32.5	108	29.7	81.2	22	77.6	20.7
134.9	36.4	107.9	29.8	117.9	32.6	108.2	29.8	81.2	22.2	77.6	20.9
0.3		0.3		0.3		0.3		0.3		0.3	
0.3		0.3		0.3		0.3		0.3		0.3	
0.3		0.3		0.3		0.3		0.3		0.3	
1.7	18.7	0.3	2.8	7	69.5	1.8	21.7	3.8	41.4	4.2	40.5

Table G.9

$w_{oil,0}$	p	T	$\dot{m}$	$\dot{x}$	$\Delta p/\Delta L$	FP	$\dot{q}$	α
0.013	11.57	-36.1	263.2	0.52	48.41	a	3.2	
0.013	11.55	-36.15	260.1	0.53	49.64	a	5.3	23.1
0.013	11.58	-36.07	260.8	0.52	47.6	a	5.3	23.3
0.013	11.56	-36.1	260.2	0.52	50.58	a	9.9	19.8
0.013	11.58	-36.06	262.8	0.52	49.34	a	9.9	20.1
0.013	11.56	-36.13	264.4	0.51	49.05	a	13.6	20
0.013	11.59	-36.05	264	0.51	50.64	a	13.6	19.3
0.013	11.59	-36.05	263.5	0.51	48.85	a	13.6	19.8
0.013	11.64	-35.92	262.6	0.52	51.33	a	22.8	20.2
0.013	11.59	-36.05	263.9	0.51	52.67	a	22.9	20
0.013	11.56	-36.11	264.2	0.51	55.12	a	37.1	21.6
0.013	11.6	-36.02	263.3	0.52	54.1	a	37.2	21.6
0.013	11.62	-35.96	263.2	0.52	63.61	a	56.7	24.3
0.013	11.59	-36.04	265.9	0.51	59.55	a	56.7	24.8
0.013	11.61	-35.99	263.4	0.52	59.96	a	56.7	24.7
0.013	11.63	-35.95	259.4	0.53	68.14	a	78.4	28.1
0.013	11.63	-35.95	260.4	0.52	66.81	a	78.3	28.2
0.013	11.7	-35.78	262.9	0.52	77.81	a (f)	118.8	34.7
0.013	11.62	-35.97	263.1	0.52	78.36	a (f)	118.7	34.4
0.013	11.64	-35.92	260.8	0.52	77.84	a (f)	118.8	34.5
0.013	11.54	-36.17	261	0.52	47.37	a	0.3	
0.013	11.56	-36.12	260.4	0.52	46.04	a	0.3	
0.013	11.58	-36.07	257.8	0.53	47.14	a	0.3	
0.013	11.57	-36.09	258.4	0.53	47.52	a	2.8	
0.013	11.53	-36.18	257	0.53	47.93	a	2.9	
0.013	11.54	-36.18	255.8	0.53	49.65	a	4.9	17.7
0.013	11.57	-36.1	260.2	0.52	48.33	a	4.8	17.6
0.013	11.54	-36.17	256.3	0.53	49.46	a	9.7	17.5
0.013	11.57	-36.1	257.4	0.53	49.36	a	9.7	17.4
0.013	11.54	-36.17	258.2	0.53	50.51	a	14.8	18.1
0.013	11.56	-36.1	258.3	0.53	48.36	a	14.9	18.1
0.013	11.54	-36.16	260.9	0.52	52.54	a	23.4	18.7
0.013	11.58	-36.07	261.4	0.52	51.27	a	23.3	18.8
0.013	11.59	-36.03	256.3	0.53	52.16	a	23.3	18.6
0.013	11.56	-36.11	258.5	0.53	55.06	a	38.4	20.7
0.013	11.58	-36.06	260.1	0.52	55.21	a	38.3	20.7
0.013	11.56	-36.11	262.4	0.52	59.74	a	58.2	24.2
0.013	11.6	-36.02	261	0.52	61.45	a	58.2	24
0.013	11.58	-36.08	260.8	0.52	65.29	a	79.2	27.8
0.013	11.6	-36.02	259.2	0.53	68.28	a	79.2	27.4
0.013	11.62	-35.97	260	0.52	68.33	a	79.1	27.4
0.013	11.66	-35.88	268.8	0.51	78.67	a (f)	121.8	34.3
0.013	11.63	-35.95	267.3	0.51	80.39	a (f)	121.6	34.1
0.013	11.65	-35.9	268.1	0.51	79.13	a (f)	121.7	34.4
0.013	11.58	-36.07	263.1	0.72	62.51	a	0.3	
0.013	11.52	-36.21	261.8	0.72	62.35	a	0.3	
0.013	11.57	-36.1	261	0.72	61.42	a	0.3	

$\dot{q}_1$	α_1	$\dot{q}_2$	α_2	$\dot{q}_3$	α_3	$\dot{q}_4$	α_4	$\dot{q}_5$	α_5	$\dot{q}_6$	α_6
1.9		0.3		7.2		1.9		3.9		4.3	
4.9	21.6	1.8	7.6	8.9	36.2	4.8	21.8	5.1	23.7	6.2	27.8
4.9	21.8	1.8	7.6	8.9	37	4.9	21.9	5.1	23.9	6.2	27.7
10.5	21.1	4.9	9.5	12.9	26.2	9.3	19	10.7	21.5	10.7	21.7
10.6	21.2	4.9	9.7	13	26.5	9.4	19.2	10.7	21.8	10.8	21.9
13.7	20.4	7.8	11.4	16.7	24.9	13.7	20.2	14.9	21.6	14.6	21.7
13.7	19.5	7.9	10.9	16.7	23.9	13.7	19.6	14.8	21.1	14.5	21
13.7	20.1	7.9	11.3	16.7	24.5	13.8	19.9	14.9	21.3	14.6	21.3
24.9	22	15.8	14.1	25.5	22.8	22.9	20.5	22.1	19.5	25.4	22.2
25	21.7	16	13.8	25.6	22.5	23	20.2	22.3	19.5	25.7	22.1
40.1	23.5	27.3	16	44.7	25.5	36.1	21.1	35.1	20.3	39.4	22.9
40.2	23.5	27.4	16	44.7	25.5	36.2	21.2	35.1	20.4	39.4	22.9
63.1	27.3	45	19.4	68.3	29	54.4	23.4	48.8	21	60.7	26
63	27.8	45	19.9	68.3	29.6	54.4	23.9	48.8	21.4	60.7	26.4
63.1	27.7	45	19.8	68.3	29.4	54.4	23.7	48.8	21.3	60.7	26.3
91.7	32.6	62.3	22.4	93.8	33.6	72.1	26	66.8	24	83.6	29.7
91.6	32.8	62.3	22.5	93.8	33.8	72.1	26.3	66.7	24.1	83.5	30
138.6	40.7	96.9	28.3	144.4	41.9	107.8	31.5	100.5	29.3	124.5	36.4
138.5	40.5	96.9	28.1	144.3	41.6	107.7	31.2	100.5	29	124.5	36
138.6	40.4	96.8	28.1	144.3	41.5	107.7	31.4	100.5	29.3	124.5	36.4
0.3		0.3		0.3		0.3		0.3		0.3	
0.3		0.3		0.3		0.3		0.3		0.3	
0.3		0.3		0.3		0.3		0.3		0.3	
0.3		0.3		7		1.9		2.9		4.6	
0.3		0.3		7.1		2		2.9		4.6	
2.5	9.3	1.7	6.4	8.3	30.3	4.9	17.7	4.1	15.6	7.5	27.2
2.3	8.5	1.6	6	8.3	30.9	4.8	17.5	4.1	15.5	7.4	26.9
8.4	15	5.5	9.9	13.8	24.3	9.1	16.4	9.8	17.8	11.6	21.3
8.4	14.9	5.6	9.8	13.9	24.3	9.2	16.4	9.8	17.6	11.6	21.3
13.5	16.6	10.4	12.6	18.6	22.7	13.8	17	14.9	18.1	17.8	21.7
13.5	16.6	10.4	12.6	18.6	22.7	13.8	17	15	18.1	17.8	21.7
23.6	18.8	18.2	14.3	27.6	21.7	23.1	18.6	20.9	16.9	26.9	21.6
23.6	18.9	18.2	14.5	27.6	21.9	22.8	18.6	20.9	17	26.9	21.7
23.6	18.7	18.2	14.3	27.6	21.7	22.8	18.5	20.9	16.9	26.8	21.6
43.3	23.1	31	16.8	43.9	23.5	36.5	19.9	33.7	18.3	41.9	22.6
43.3	23	31	16.7	43.8	23.4	36.5	20	33.7	18.3	41.8	22.6
62.9	26.4	48.6	20.4	72.2	29.3	54.2	22.6	47.9	20	63.3	26.3
63	26.2	48.7	20.2	72.2	29.1	54.2	22.5	48	19.9	63.3	26.2
90.3	31.7	66.9	23.6	95.4	33.2	71.8	25.5	64.7	22.8	86	30.1
90.3	31.2	66.8	23.2	95.4	32.7	71.8	25	64.7	22.5	86	29.8
90.3	31.2	66.8	23.2	95.3	32.7	71.7	25.1	64.7	22.5	86	29.8
143.1	40.4	103.9	29.3	146.5	40.7	108	30.5	102.1	28.7	127.3	36.1
143	40.2	103.7	29.1	146.3	40.5	107.7	30.4	101.9	28.6	127.2	36
143.1	40.6	103.8	29.3	146.4	40.8	107.9	30.7	101.9	28.8	127.2	36.3
0.3		0.3		0.3		0.3		0.3		0.3	
0.3		0.3		0.3		0.3		0.3		0.3	
0.3		0.3		0.3		0.3		0.3		0.3	

Table G.9

$w_{oil,0}$	p	T	$\dot{m}$	$\dot{x}$	$\Delta p/\Delta L$	FP	$\dot{q}$	α
0.013	11.53	-36.2	259.3	0.73	61.14	a	3.8	26.4
0.013	11.56	-36.12	260.3	0.73	61.24	a	3.7	22.6
0.013	11.52	-36.2	258.7	0.73	62.55	a	5.3	20.5
0.013	11.56	-36.11	258.3	0.73	61.36	a	5.4	20.6
0.013	11.53	-36.19	260	0.73	63.03	a	10.4	20.8
0.013	11.57	-36.08	261	0.72	63.08	a	10.4	20.7
0.013	11.53	-36.18	265	0.71	65.14	a	14.9	21
0.013	11.56	-36.11	261	0.72	64.77	a	14.9	20.9
0.013	11.58	-36.06	261.2	0.72	66.07	a	14.9	20.7
0.013	11.54	-36.16	261.5	0.73	68.36	a	24.5	21.2
0.013	11.57	-36.08	263	0.72	67.29	a	24.4	21.3
0.013	11.55	-36.13	264.3	0.72	71.28	a	36.7	22.9
0.013	11.57	-36.08	265.3	0.71	71.04	a	36.7	23
0.013	11.55	-36.13	266.6	0.71	75.89	a	58.4	26.5
0.013	11.6	-36.02	267.7	0.71	77.86	a	58.3	26.2
0.013	11.6	-36.01	265.4	0.71	78.47	a	58.3	26.6
0.013	11.56	-36.11	267.1	0.71	82.97	a	82.4	31.5
0.013	11.6	-36.02	267.6	0.71	84.64	a	82.4	30.9
0.013	11.63	-35.95	269.5	0.7	97.57	a (f)	130.8	40
0.013	11.58	-36.06	267.8	0.71	98.59	a (f)	130.7	40.1
0.013	11.63	-35.95	267.9	0.7	97.83	a (f)	130.8	40.3
0.013	11.54	-36.17	520.8	0.12	35.09	i-a (f)	0.3	
0.013	11.54	-36.16	521.1	0.12	34.64	i-a (f)	0.3	
0.013	11.54	-36.15	520.4	0.12	34.06	i-a (f)	0.3	
0.013	11.58	-36.07	520.6	0.12	34.96	i-a (f)	2.7	16.7
0.013	11.53	-36.2	519.9	0.12	34.36	i-a (f)	2.8	16.5
0.013	11.55	-36.15	522.3	0.12	36.09	i-a (f)	7.2	13.2
0.013	11.55	-36.14	522.9	0.12	36.44	i-a (f)	7.1	13.4
0.013	11.55	-36.14	521.6	0.12	36.68	i-a (f)	10.8	13.8
0.013	11.58	-36.06	521.7	0.11	37.4	i-a (f)	10.6	13.4
0.013	11.52	-36.21	518	0.12	36.78	i-a (f)	10.7	13.4
0.013	11.55	-36.13	518.8	0.12	39.68	i-a (f)	15.6	13.8
0.013	11.53	-36.18	522	0.12	39.46	i-a (f)	15.7	13.9
0.013	11.53	-36.19	529.3	0.12	43.47	i-a (f)	26.5	15.8
0.013	11.56	-36.13	522.2	0.12	43.27	i-a (f)	26.5	15.7
0.013	11.53	-36.18	520.8	0.12	51.5	i-a (f)	45.6	19.6
0.013	11.56	-36.11	519.5	0.12	50.81	i-a (f)	45.7	19.6
0.013	11.57	-36.09	521.8	0.12	50.37	i-a (f)	45.6	19.6
0.013	11.55	-36.15	517.4	0.12	63.86	i-a (f)	66.7	23.7
0.013	11.54	-36.16	519.4	0.12	61.8	i-a (f)	66.8	23.8
0.013	11.54	-36.16	527.3	0.12	71	i-a (f)	95	28.2
0.013	11.55	-36.13	525	0.12	72.68	i-a (f)	94.9	28.1
0.013	11.57	-36.1	525.2	0.12	72.73	i-a (f)	94.8	28.2
0.013	11.57	-36.09	524.9	0.12	86.68	i-a (f)	139.1	34.7
0.013	11.61	-36	525	0.12	87.02	i-a (f)	139.2	34.8
0.013	11.58	-36.07	486.3	0.34	88.72	a	0.3	
0.013	11.57	-36.09	488.9	0.34	87.47	a	0.3	

Table G.9

$\dot{q}_1$	α_1	$\dot{q}_2$	α_2	$\dot{q}_3$	α_3	$\dot{q}_4$	α_4	$\dot{q}_5$	α_5	$\dot{q}_6$	α_6
0.3	2.4	2.6	18.4	8.1	57.5	2.5	16.9	4.4	28.5	4.7	34.4
0.3	2	2.6	15.5	8.1	50.2	2.5	15.1	4.4	25	4.6	27.9
1.5	6.3	3.2	13.4	10.3	38.6	4.8	18.4	4.8	18.8	7	27.4
1.9	7.6	3.5	13.9	10.3	38.4	4.8	18	4.8	18.4	7	27.2
7.5	15.5	8.2	16.6	16.3	31.5	9.3	18.5	9.4	18.8	11.7	23.9
7.5	15.4	8.2	16.5	16.3	31.7	9.3	18.5	9.4	18.8	11.7	23.4
12.9	18.5	11.9	17.1	21	29.1	14	19.8	13.2	18.6	16.3	23.1
13	18.5	12	17.1	21	28.8	14	19.7	13.2	18.5	16.3	23.1
12.9	18.4	11.9	16.9	21	28.5	13.9	19.3	13.2	18.2	16.3	22.6
23.9	20.9	19.9	17.4	33.1	27.6	23	20.2	20.4	18	26.3	22.9
23.9	21	19.9	17.5	33.1	27.7	23	20.4	20.4	18.1	26.3	23.1
39.1	24.3	32.2	20.2	42.1	26.3	36.3	22.7	31.9	19.9	38.4	24.1
39.2	24.5	32.2	20.2	42.1	26.3	36.3	22.7	31.9	20	38.5	24.3
64.1	29	54.8	24.6	64.4	29.4	54.3	24.9	48.3	22.1	64.3	29.2
64.1	28.5	54.8	24.2	64.4	29	54.3	24.7	48.2	22	64.3	29.1
64.1	29.2	54.8	24.7	64.3	29.4	54.4	24.9	48.2	22.1	64.3	29.2
92.3	35.4	71.8	27.5	95.4	36.1	72.1	27.7	69.7	26.6	93.1	35.4
92.4	34.9	71.8	27.1	95.4	35.5	72.1	27.1	69.7	26.1	93.1	34.7
163.9	49.8	113.9	34.5	145.1	44.4	107.8	33.3	108	33.3	146.2	44.7
163.8	49.9	113.7	34.6	145	44.4	107.6	33.3	107.9	33.5	146.4	45.1
163.9	50.4	113.9	34.9	145.1	44.7	107.7	33.5	108	33.4	146.4	44.9
0.3		0.3		0.3		0.3		0.3		0.3	
0.3		0.3		0.3		0.3		0.3		0.3	
0.3		0.3		0.3		0.3		0.3		0.3	
2.1	12.5	0.3	1.9	3.5	21.7	1.9	11.3	3.7	23.4	4.7	29.2
2.1	12	0.3	1.8	3.5	20.8	2.2	12.5	3.8	23	4.8	29.1
9.1	16.7	2.9	5.3	7.4	13.6	4.7	8.8	9.4	17.2	9.4	18
9	16.9	2.8	5.3	7.3	13.6	4.7	8.8	9.4	17.5	9.4	18.3
11.6	15	7	8.7	9.7	12.3	9.1	11.4	11.5	14.6	16.2	20.8
11.3	14.5	6.8	8.4	9.5	11.9	8.9	11.2	11.4	14.4	15.9	20.2
11.5	14.4	6.9	8.4	9.6	11.9	9	11.3	11.5	14.3	15.9	20
17.9	15.9	10.7	9.4	13.2	11.7	13.7	12.2	16.5	14.7	21.6	19.1
18	16	10.8	9.5	13.3	11.8	13.8	12.2	16.5	14.6	21.7	19.2
31.2	18.5	17.2	10.3	25	14.9	22.8	13.6	27.8	16.5	35.2	20.8
31.2	18.4	17.2	10.2	25.1	14.9	22.9	13.6	27.7	16.4	35.2	20.8
55.8	23.8	30.8	13.2	47	20.1	36	15.6	46.7	20.1	57.5	24.8
55.8	23.9	30.9	13.3	47	20.1	36.1	15.6	46.7	20.1	57.5	24.7
55.8	23.9	30.8	13.3	47	20.2	36	15.6	46.6	20.1	57.4	24.7
82.1	29.2	46.1	16.5	70.1	24.8	53.9	19.2	63.9	22.8	84.2	29.8
82.2	29.3	46.1	16.5	70.1	24.8	54	19.2	64	22.8	84.2	29.9
124	36.7	69.9	20.8	97.3	28.8	72	21.5	90.1	26.8	116.7	34.6
123.9	36.5	69.9	20.8	97.1	28.8	71.9	21.5	90	26.8	116.5	34.5
123.7	36.6	69.8	20.8	97	28.8	71.9	21.6	90	26.8	116.5	34.6
177.8	44.7	111.6	27.7	136.4	34	107.6	26.8	130.7	32.5	170.2	42.3
178.2	44.9	111.7	27.8	136.5	34.1	107.7	26.9	131	32.7	170.1	42.5
0.3		0.3		0.3		0.3		0.3		0.3	
0.3		0.3		0.3		0.3		0.3		0.3	

Table G.9

$w_{oil,0}$	p	T	$\dot{m}$	$\dot{x}$	$\Delta p/\Delta L$	FP	$\dot{q}$	α
0.013	11.6	-36.01	494.4	0.33	83.35	a	0.3	
0.013	11.6	-36.01	482.7	0.34	87.52	a	2.4	22.5
0.013	11.56	-36.11	485.1	0.34	87.85	a	2.4	22.3
0.013	11.58	-36.06	486.2	0.34	87.62	a	4	17.4
0.013	11.57	-36.09	489.9	0.34	87.34	a	4	17.3
0.013	11.6	-36.02	491.3	0.34	89.45	a	4	16.7
0.013	11.6	-36.02	495.8	0.33	89.99	a (f)	8.1	15.7
0.013	11.62	-35.96	496.7	0.33	87.8	a (f)	8	15.9
0.013	11.56	-36.1	492.7	0.33	90.23	a (f)	11.3	15.9
0.013	11.58	-36.06	494.5	0.33	90.98	a (f)	11.2	15.5
0.013	11.52	-36.2	493.3	0.33	88.77	a (f)	11.3	15.9
0.013	11.58	-36.08	506.3	0.32	93.91	a (f)	19.7	17.2
0.013	11.65	-35.9	515.7	0.32	89.79	a (f)	19.6	17.1
0.013	11.58	-36.07	530.1	0.31	94.08	a (f)	33.4	19.7
0.013	11.59	-36.05	523.5	0.31	96.88	a (f)	33.3	19.7
0.013	11.59	-36.05	518.6	0.32	98.28	a (f)	33.2	19.2
0.013	11.55	-36.13	507	0.32	99.68	a (f)	50.2	22.3
0.013	11.59	-36.05	503.9	0.33	102.48	a (f)	50.2	22.2
0.013	11.62	-35.98	526.6	0.31	104.2	a (f)	73.7	25.8
0.013	11.61	-36	507.8	0.33	114.42	a (f)	73.6	24.3
0.013	11.55	-36.13	523.4	0.31	107.94	a (f)	73.7	24.1
0.013	11.58	-36.06	519.9	0.32	121.26	a (f)	111.2	29.1
0.013	11.62	-35.97	523.1	0.32	124.16	a (f)	111.3	29
0.013	11.63	-35.94	527.6	0.31	119.92	a (f)	111.2	31.1
0.013	25.95	-10.7	81.1	0.14	0.86	w	0.2	
0.013	25.97	-10.67	81.6	0.14	0.9	w	0.2	
0.013	25.95	-10.7	80.9	0.15	0.84	w	0.2	
0.013	25.96	-10.69	81.1	0.15	0.97	w	4.8	8.6
0.013	25.96	-10.69	80.5	0.14	0.96	w	4.7	8.5
0.013	25.96	-10.68	81.2	0.15	1.1	w	9.2	9.7
0.013	25.97	-10.68	82.3	0.14	1.11	w	9.2	9.8
0.013	25.96	-10.69	80.9	0.14	1.19	w	11.7	9.4
0.013	25.96	-10.68	81.1	0.15	1.19	w	11.6	9.9
0.013	26	-10.63	81.5	0.15	1.37	w	16.3	10.9
0.013	25.95	-10.69	80.2	0.15	1.39	w	16.2	11
0.013	25.97	-10.67	79.7	0.15	1.8	w	27.4	13.1
0.013	25.98	-10.66	80.5	0.15	1.76	w	27.4	13.1
0.013	25.98	-10.67	80.2	0.15	2.23	w	40.3	15.6
0.013	25.96	-10.69	80	0.15	2.2	w	40.3	15.5
0.013	25.98	-10.66	80.7	0.15	2.21	w	40.3	15.6
0.013	25.99	-10.65	80	0.15	2.84	w (f)	59.1	18.3
0.013	25.99	-10.64	79.9	0.15	2.87	w (f)	59.2	18.4
0.013	26	-10.63	81.1	0.15	2.84	w (f)	59.1	18.5
0.013	26.01	-10.63	80.2	0.2	3.63	w (f)	74.7	18.9
0.013	25.98	-10.67	82.9	0.18	3.71	w (f)	74.8	19.1
0.013	26.02	-10.61	79.5	0.19	3.62	w (f)	74.8	19.3
0.013	26.03	-10.59	79.6	0.24	5	w (f)	94.8	18.4

$\dot{q}_1$	α_1	$\dot{q}_2$	α_2	$\dot{q}_3$	α_3	$\dot{q}_4$	α_4	$\dot{q}_5$	α_5	$\dot{q}_6$	α_6
0.3		0.3		0.3		0.3		0.3		0.3	
0.3	2.8	0.3	2.6	4.3	45.1	2.3	24.1	2.9	26.6	4.3	33.6
0.3	2.7	0.3	2.5	4.3	46.2	2	21.6	2.9	27.6	4.4	33.3
2.6	11.6	0.3	1.4	6.7	28.9	4.9	21.3	4.6	19.9	4.8	21.1
2.8	12.5	0.3	1.4	6.7	28	4.9	20.8	4.6	20	4.8	20.8
2.7	11.8	0.3	1.4	6.7	27.2	5	20.1	4.6	19.3	4.8	20.4
8.2	15.7	4.2	8	10.2	19.7	9.4	18.4	7.7	15.1	8.9	17.4
8.1	15.8	4.1	7.9	10.2	20.2	9.3	18.7	7.6	15.3	8.9	17.6
13.2	17.7	5.7	7.9	13.7	19.2	13.8	19.6	9.7	14	12	16.9
13.1	17.2	5.6	7.7	13.6	18.8	13.7	19.2	9.6	13.6	11.8	16.3
13.3	18	5.7	8	13.7	19.3	13.8	19.6	9.6	14	11.8	16.6
21.3	18.3	12.7	11.1	23.4	20.5	22.7	20.1	18.6	16.3	19.4	17
21.2	18.1	12.6	11.1	23.2	20.4	22.7	19.9	18.5	16.1	19.3	16.7
31.7	18.9	28	16.4	39.2	23.1	36.2	21.6	31.6	18.6	33.6	19.8
31.5	18.6	28	16.3	39.1	23.3	36.2	21.7	31.5	18.6	33.7	19.8
31.3	18.3	27.9	15.9	39	22.5	36.1	21	31.5	18.1	33.6	19.3
50.4	22.7	40	18	60.1	26.4	54.4	23.9	47.5	20.9	48.6	21.7
50.3	22.7	40	18	60.1	26.4	54.4	23.8	47.5	20.8	48.6	21.6
81	28.3	61.5	21.5	88.3	30.7	71.9	25.3	67.5	23.7	71.8	25.4
80.9	26.7	61.3	20.2	88.2	29.1	71.8	24	67.4	22.3	71.7	23.9
81	26.2	61.5	20.1	88.3	28.8	72	23.8	67.5	22.2	71.8	23.6
124	32.1	98.5	25.6	133.5	34.9	107.5	28.3	96.2	25.3	107.7	28.3
124.1	32.3	98.5	25.7	133.7	34.8	107.6	28.1	96.2	25.1	107.7	28
124	35.1	98.4	27.3	133.4	37.3	107.4	30	96.3	26.9	107.7	30.3
0.2		0.2		0.2		0.2		0.2		0.2	
0.2		0.2		0.2		0.2		0.2		0.2	
0.2		0.2		0.2		0.2		0.2		0.2	
3.9	7.1	9.7	17.5	3.9	6.7	2	3.7	3.5	6.2	5.7	10.1
3.7	6.7	9.6	17.3	3.9	6.7	2.1	3.9	3.5	6.2	5.6	10.2
0.2	0.2	21.1	22.4	7.1	7.3	4.9	5.1	10.6	11	11.4	12.3
0.2	0.2	21	22.8	7.1	7.4	4.8	5.1	10.6	11.1	11.3	12.4
0.2	0.1	23.5	18.8	10.6	8.5	9	7.2	14.6	11.7	12.1	9.8
0.2	0.2	23.4	20.3	10.6	8.8	9	7.4	14.5	12.1	12.1	10.6
0.2	0.1	29.4	20	17.6	11.5	13.5	8.9	21	13.9	16	10.9
0.2	0.1	29.4	20.4	17.5	11.6	13.5	8.9	20.9	14.1	15.8	11
0.2	0.1	55.5	26.3	32.9	15.6	22.6	10.9	32.7	15.7	20.6	9.9
0.2	0.1	55.5	26.4	32.9	15.7	22.6	10.9	32.8	15.7	20.7	9.9
0.2	0.1	76.1	29.3	53.2	20.3	36.2	14	43.8	17.1	32.3	12.5
0.2	0.1	76.1	29.3	53.3	20.3	36.2	14	43.9	17.1	32.3	12.5
0.2	0.1	76.1	29.5	53.3	20.4	36.2	14	43.8	17.1	32.3	12.6
0.2	0.1	98.6	30.5	85.4	26	54.1	16.7	61.7	19.3	55	17.2
0.2	0.1	98.5	30.7	85.4	26.1	54.1	16.8	61.7	19.5	55	17.4
0.2	0.1	98.3	30.7	85.3	26.3	54.2	16.9	61.8	19.5	55.1	17.4
0.2	0	115.7	29.1	110.4	28.1	72	18.3	80.6	20.4	69.6	17.4
0.2	0	115.7	29.3	110.5	28.3	72	18.3	80.7	20.7	69.6	17.7
0.2	0	115.8	29.7	110.7	28.7	72.1	18.5	80.6	20.9	69.7	18.1
13.2	2.5	138.1	26.4	127.2	25.8	106.9	20.2	114.6	21.9	69	13.4

Table G.9

$w_{oil,0}$	p	T	$\dot{m}$	$\dot{x}$	$\Delta p/\Delta L$	FP	$\dot{q}$	α
0.013	26.02	-10.61	80.6	0.24	4.96	w (f)	94.8	18.6
0.013	26.04	-10.58	81.3	0.23	5	w (f)	94.8	18.9
0.013	25.98	-10.66	81.1	0.33	0.69	w	0.2	
0.013	25.99	-10.64	80.4	0.33	0.68	w	0.2	
0.013	26.01	-10.62	78.2	0.35	0.92	w	5.4	10.1
0.013	26.02	-10.61	80.1	0.34	0.89	w	4.9	9.7
0.013	26	-10.62	78.9	0.35	1.31	w (f)	11.9	10.2
0.013	26.02	-10.6	80.5	0.34	1.06	w (f)	11.9	9.1
0.013	26.02	-10.61	80.1	0.34	1.33	w (f)	11.4	10.3
0.013	26.03	-10.59	80.4	0.34	1.42	w (f)	16	10.2
0.013	26.02	-10.6	80	0.34	1.45	w (f)	15.9	10.7
0.013	26.02	-10.6	80.9	0.34	1.64	w (f)	20.7	11.5
0.013	26	-10.64	80.5	0.34	1.94	w (f)	20.5	11.5
0.013	26.03	-10.58	79.9	0.34	2.86	w (f)	32	13.5
0.013	26.04	-10.58	80.7	0.33	2.5	w (f)	31.8	13
0.013	26.04	-10.58	82.3	0.33	3.03	w (f)	42	14.6
0.013	26.06	-10.55	78.4	0.35	3.05	w (f)	42	14.5
0.013	26.04	-10.57	83	0.33	4.2	f	53.1	15.7
0.013	26.09	-10.51	81.1	0.34	3.91	f	53.1	16.1
0.013	26.04	-10.58	80.9	0.34	3.82	f	63.5	16.5
0.013	26.06	-10.55	82.8	0.33	4.31	f	63.5	16.6
0.013	26.06	-10.55	75.4	0.36	4.74	f	82.5	17
0.013	26.05	-10.56	82.3	0.33	4.82	f	82.6	17.8
0.013	25.65	-11.09	81.3	0.53	1.5	w (f)	0.2	
0.013	25.66	-11.08	79.3	0.53	1.47	w (f)	0.2	
0.013	25.67	-11.07	78.9	0.54	1.51	w (f)	0.2	
0.013	25.67	-11.07	79.1	0.54	1.64	w (f)	3.8	9.8
0.013	25.69	-11.04	79.1	0.55	1.63	w (f)	3.9	10
0.013	25.72	-11	79.8	0.54	1.94	w (f)	10.1	11.4
0.013	25.66	-11.08	79.1	0.54	1.9	w (f)	10.1	11.3
0.013	25.68	-11.06	78.7	0.55	2.15	w (f)	13.2	11
0.013	25.69	-11.04	79.6	0.54	2.15	w (f)	13.1	11
0.013	25.68	-11.06	79.2	0.54	2.34	w (f)	16.8	11.6
0.013	25.65	-11.1	79.8	0.54	2.34	w (f)	16.8	11.5
0.013	25.66	-11.08	79	0.55	3.1	f	29.6	15
0.013	25.7	-11.03	80.5	0.53	3.08	f	29.7	15.2
0.013	25.7	-11.02	79.8	0.54	3.69	f	43.4	17.7
0.013	25.7	-11.03	78.5	0.55	3.75	f	43.4	17.6
0.013	25.69	-11.03	79.7	0.54	4.56	f	55.8	18.6
0.013	25.69	-11.05	80.7	0.53	4.56	f	55.8	18.5
0.013	25.71	-11.02	79.9	0.53	5.1	f	66.3	18
0.013	25.7	-11.03	79.3	0.54	5.2	f	66.1	18.6
0.013	26.01	-10.6	77.6	0.56	1.65	w (f)	0.2	
0.013	26.03	-10.58	77	0.57	1.64	w (f)	0.2	
0.013	26.04	-10.56	76.5	0.6	1.92	w (f)	7.8	13.2
0.013	26.03	-10.58	76.9	0.57	1.84	w (f)	7.9	13
0.013	26.04	-10.57	76.3	0.57	1.81	w (f)	7.8	12.9

$\dot{q}_1$	α_1	$\dot{q}_2$	α_2	$\dot{q}_3$	α_3	$\dot{q}_4$	α_4	$\dot{q}_5$	α_5	$\dot{q}_6$	α_6
13.3	2.5	138.2	26.6	127.1	26.3	106.9	20.4	114.5	22.2	69	13.6
13.2	2.6	138.2	27.1	127.2	26.5	106.9	20.8	114.5	22.4	69	13.8
0.2		0.2		0.2		0.2		0.2		0.2	
0.2		0.2		0.2		0.2		0.2		0.2	
7.9	14.9	11.8	21.3	3.6	6.7	1.9	3.5	3	5.5	4.4	8.4
7.8	15.1	11.8	23	0.3	0.8	2	4.3	3	6.2	4.4	9
3.8	3.1	25.5	21.4	10.8	9.3	4.7	4.1	15.8	13.7	10.7	9.4
3.8	2.7	25.5	20.1	10.8	8.6	4.7	3.8	15.8	11.8	10.7	7.7
3.8	3.3	25.6	22.6	10.1	9.1	4.7	4.4	13.6	12.5	10.8	9.9
0.2	0.1	30.7	19.8	18.5	11.9	9.3	6	25.2	15.7	11.8	7.6
0.2	0.1	30.7	20.2	18.5	12.4	8.9	6	25.1	17.1	11.8	8.5
2.9	1.6	32.1	17.8	28.9	15.8	13.7	7.6	32.3	18.1	14.3	8.1
2.9	1.6	32	17.8	28.8	15.8	12.5	7.1	32.3	18.3	14.2	8.4
0.2	0.1	44.5	19.1	51.6	21	23.5	9.9	52.7	22.1	19.6	8.7
0.2	0.1	44.5	18.2	51.5	20.8	22.5	9.4	52.7	21.6	19.6	8.1
2.5	0.9	51.5	17.9	72.2	24.6	36	12.6	66	23	23.6	8.3
2.3	0.8	51.6	17.8	72.3	24.6	36.1	12.6	66	22.8	23.7	8.2
4.1	1.2	64.1	18.7	87.2	25.7	53.9	16	80.9	23.9	28.2	8.6
4.1	1.3	64.1	19.7	87.2	26.4	54	16.2	80.9	24.4	28.1	8.7
5.3	1.4	81.1	20.7	95.6	25.8	72.1	18.7	94.2	24.3	32.7	8.3
5.3	1.4	81	20.7	95.6	25.7	72.1	18.8	94.1	24.5	32.7	8.6
14.6	3	95.6	19.9	121.5	25.2	108.2	21.6	119.4	24.6	35.6	7.5
14.6	3.2	95.8	21.1	121.6	26.5	108.3	22.3	119.6	25.5	35.7	8
0.2		0.2		0.2		0.2		0.2		0.2	
0.2		0.2		0.2		0.2		0.2		0.2	
0.2		0.2		0.2		0.2		0.2		0.2	
4.8	12	4.3	11.4	3	8.5	1.6	4	2.7	6.6	6.4	16.2
4.8	12.3	4.4	11.7	2.8	7.9	1.9	4.8	2.9	7.1	6.5	16.4
15.2	17.1	17.1	19.5	10.5	11.9	5.1	5.7	8.4	9.3	4.5	5
15.2	16.9	17.2	19.5	10.5	11.8	5.1	5.6	8.4	9.1	4.5	4.9
7.8	6.5	30.1	25.1	15.4	12.9	9.2	7.7	14.9	12.3	1.9	1.5
7.5	6.4	30	25.4	15.3	12.9	9.2	7.7	14.8	12.3	1.8	1.6
4.1	2.8	37.7	26	21.4	15	13.7	9.4	20	13.7	3.9	2.7
4.1	2.8	37.6	25.9	21.4	14.9	13.7	9.4	20	13.6	3.8	2.6
10.2	5.2	58.3	29.7	44.9	22.8	22.4	11.3	33.6	16.9	8.4	4.2
10.3	5.4	58.5	30.1	45	22.8	22.6	11.5	33.6	17.1	8.4	4.3
17.8	7.4	78.7	31.7	65.9	27.1	36.3	14.6	46.5	19	15.3	6.2
17.8	7.4	78.7	31.7	65.9	26.9	36.3	14.5	46.6	18.9	15.3	6.2
20.8	7	95	31.3	80.2	27.8	54.1	17.7	65.1	21.3	19.5	6.4
20.9	7	95	31.2	80.2	27.7	54.2	17.5	65.2	21.3	19.6	6.4
20.9	5.3	106	27.6	96.2	28.8	72.1	19.4	82.7	21.5	20.2	5.1
19.5	5.5	106	29.2	96.1	29.5	72	19.7	82.6	22.2	20.2	5.5
0.2		0.2		0.2		0.2		0.2		0.2	
0.2		0.2		0.2		0.2		0.2		0.2	
14.7	25	10.8	18.1	7.8	12.4	1.9	3.3	5.3	9	6.6	11.5
14.8	24.7	10.8	17.8	7.8	12.2	2	3.3	5.3	8.8	6.6	11.1
14.5	24.5	10.6	17.7	7.7	12.1	2.1	3.5	5.2	8.7	6.5	11.1

Table G.9

$w_{oil,0}$	p	T	$\dot{m}$	$\dot{x}$	$\Delta p/\Delta L$	FP	$\dot{q}$	α
0.013	26.02	-10.59	77.6	0.57	1.97	w (f)	10.7	11.2
0.013	26.02	-10.59	76.9	0.57	1.94	w (f)	10.7	11.1
0.013	26.04	-10.57	78.4	0.56	1.95	w (f)	10.7	11.2
0.013	26.04	-10.56	77.6	0.56	2.23	w (f)	14.8	11.2
0.013	26.04	-10.57	76.2	0.57	2.21	w (f)	14.8	11
0.013	26.06	-10.54	77.5	0.56	2.21	w (f)	14.8	11.1
0.013	26.06	-10.54	77.2	0.55	2.48	w (f)	19.8	12.1
0.013	26.08	-10.52	78	0.54	2.43	w (f)	19.8	12.1
0.013	26.11	-10.48	77.5	0.56	2.44	w (f)	19.9	12.2
0.013	26.05	-10.56	76.2	0.57	3.36	f	36	15.8
0.013	26.08	-10.52	79.2	0.55	3.29	f	36	15.7
0.013	26.08	-10.51	73.2	0.6	3.33	f	36	15.8
0.013	26.07	-10.52	77.1	0.57	4.15	f	49	17.3
0.013	26.07	-10.52	76.9	0.57	4.13	f	49	17.1
0.013	26.06	-10.54	76.1	0.58	4.98	f	60.3	17.7
0.013	26.07	-10.53	77.6	0.57	4.88	f	60.3	17.3
0.013	26.06	-10.53	77.3	0.57	4.93	f	60.2	17.5
0.013	26.1	-10.49	77.4	0.56	5.18	f	66.6	16.9
0.013	26.1	-10.49	77.9	0.55	5.03	f	66.6	16.8
0.013	26.1	-10.48	77.6	0.56	5.27	f	66.4	17
0.013	26.1	-10.49	78.9	0.55	5.22	f	66.6	16.9
0.013	26.07	-10.52	75.7	0.57	5.43	f	66.6	17.1
0.013	25.97	-10.65	81.4	0.73	2.9	w (f)	0.2	
0.013	25.97	-10.65	82.1	0.72	2.92	w (f)	0.2	
0.013	25.95	-10.66	80	0.74	2.87	w (f)	0.2	
0.013	25.96	-10.66	82.2	0.71	2.83	w (f)	0.2	
0.013	25.97	-10.64	80.9	0.73	3.13	w (f)	4.3	13.1
0.013	25.96	-10.66	80.9	0.73	3.15	w (f)	4.2	12.7
0.013	25.96	-10.66	81.3	0.72	3.13	w (f)	4.2	12.8
0.013	25.97	-10.64	83.1	0.71	3.09	w (f)	4.2	12.7
0.013	25.95	-10.67	81.9	0.71	3.35	w (f)	9.1	14.4
0.013	25.96	-10.66	81.4	0.72	3.3	w (f)	9.2	14
0.013	25.97	-10.64	81.2	0.71	3.31	w (f)	9.1	13.8
0.013	25.97	-10.65	81	0.72	3.53	w (f)	13.6	11.7
0.013	25.97	-10.64	80.8	0.73	3.5	w (f)	13.6	11.7
0.013	25.97	-10.64	82.4	0.71	3.57	w (f)	13.5	11.9
0.013	26	-10.6	78.5	0.75	3.72	f	17.7	12
0.013	25.99	-10.62	81.3	0.72	3.75	f	17.7	11.9
0.013	25.97	-10.64	81.8	0.72	3.79	f	17.7	12
0.013	25.98	-10.63	81.4	0.72	4.48	f	27.5	13.7
0.013	25.98	-10.63	81.7	0.72	4.45	f	27.5	13.5
0.013	25.99	-10.61	79.9	0.74	4.44	f	27.5	13.6
0.013	25.99	-10.61	81.8	0.72	5.65	f	40	14.8
0.013	26	-10.61	80.5	0.73	5.57	f	39.9	14.8
0.013	26	-10.6	80.2	0.74	5.56	f	39.9	15
0.013	26.02	-10.58	82.2	0.72	6.78	f	49.7	15
0.013	26	-10.6	80.7	0.74	6.71	f	49.6	15.3

Table G.9

$\dot{q}_1$	α_1	$\dot{q}_2$	α_2	$\dot{q}_3$	α_3	$\dot{q}_4$	α_4	$\dot{q}_5$	α_5	$\dot{q}_6$	α_6
14.6	15.4	22.4	24	12.5	12.3	4.6	5	9.7	10.5	0.2	0.2
14.6	15.1	22.4	23.8	12.5	12.3	4.6	4.9	9.7	10.4	0.2	0.2
14.7	15.3	22.5	23.9	12.5	12.3	4.6	5	9.7	10.5	0.2	0.2
3.8	2.9	34.7	26.5	19.5	14.1	9.2	7.1	18.1	13.8	3.4	2.6
3.9	2.9	34.7	25.9	19.5	13.9	9.3	7	18	13.4	3.6	2.6
4	3	34.7	26.2	19.5	14	9.3	7.1	18	13.6	3.4	2.5
2.8	1.7	42.4	26.6	29.4	17.2	13.9	8.5	25.9	15.8	4.6	2.9
2.7	1.7	42.5	26.5	29.4	17.3	13.6	8.4	26	15.9	4.7	2.9
3	1.9	42.5	26.7	29.5	17.4	13.6	8.4	26	16	4.7	2.9
17.2	7.7	62.6	28.2	60.7	25.4	22.4	9.9	43.8	19.2	9.1	4
17.3	7.6	62.6	28	60.6	25.4	22.4	9.9	43.8	19.1	9.1	4
17.3	7.9	62.7	28.3	60.6	25.5	22.4	9.9	43.9	19.2	9.1	4
24.5	8.9	82	29.1	82.2	28.1	35.9	12.7	57.9	20.6	11.4	4.1
24.4	8.9	82	29	82.2	27.8	35.9	12.6	58	20.4	11.4	4.1
23.7	7.3	98.9	29.2	95.1	27.8	53.6	15.5	76.3	22.4	14	4.2
23.7	6.9	98.8	28.4	95.1	27.2	53.7	15.4	76.3	21.8	14	4
23.6	7.1	98.8	28.7	95	27.3	53.6	15.5	76.3	22.4	13.9	4.1
21.3	5.3	113.7	28.3	89.6	24	71.9	17.9	87.7	21.9	15.7	3.9
21.3	5.1	113.7	27.6	89.6	24.7	71.9	17.9	87.6	21.6	15.7	3.8
21.1	5.4	113.3	28.8	89.3	23.7	71.7	17.9	87.3	22.1	15.5	4
21.3	5.2	113.7	28	89.5	24.8	71.8	17.9	87.5	21.7	15.6	3.8
21.3	5.6	113.7	28.8	89.5	23.2	71.8	18.1	87.7	22.8	15.7	4.1
0.2		0.2		0.2		0.2		0.2		0.2	
0.2		0.2		0.2		0.2		0.2		0.2	
0.2		0.2		0.2		0.2		0.2		0.2	
0.2		0.2		0.2		0.2		0.2		0.2	
5.4	17.2	6.8	20.5	3.7	10.5	2.1	6.5	3.3	10.2	4.3	13.7
5.3	16.7	6.9	20.3	3.5	9.8	2.1	6.2	3.3	10	4.2	13.3
5.3	16.6	6.8	20.1	3.6	10.2	2	6.2	3.4	10.3	4.3	13.6
5.4	16.7	6.7	19.9	3.7	10.4	2.1	6.3	3.2	9.7	4.1	13
15.6	24.7	12.4	19.7	8.1	12.7	4.9	7.6	5.9	9.2	7.9	12.6
15.6	23.9	12.5	19.5	8.2	12.5	5	7.5	5.9	8.8	8	12
15.6	23.6	12.4	19.1	8.1	12.4	4.8	7.3	6	8.9	7.9	11.8
14.4	12.3	30.5	26.3	15	12.8	9.1	8	12.5	10.7	0.2	0.2
14.4	12.4	30.5	26.5	15	12.8	9.1	8	12.4	10.6	0.2	0.2
14.3	12.9	30.3	26.7	14.9	12.8	9	8	12.4	10.9	0.2	0.2
8.2	5.6	43.5	29.5	21.9	14.6	13.8	9.4	19	12.6	0.2	0.1
8.1	5.6	43.6	29.5	21.8	14.6	13.7	9.4	18.7	12.4	0.2	0.1
8.1	5.6	43.5	29.7	21.8	14.7	13.7	9.5	18.8	12.7	0.2	0.1
13.5	6.8	61.4	30.7	37.7	18.7	22.9	11.5	29.5	14.8	0.2	0.1
13.5	6.5	61.3	30	37.5	18.4	22.8	11.3	29.4	14.4	0.2	0.1
13.5	6.7	61.3	30.4	37.5	18.6	22.8	11.4	29.4	14.6	0.2	0.1
19	7	79.2	29.2	59.7	22.1	36.1	13.5	44.1	16.5	1.7	0.6
18.9	7	79.1	29.4	59.6	21.9	36	13.4	44	16.4	1.7	0.6
18.9	7.2	79.1	29.8	59.6	22.2	36	13.6	44	16.7	1.8	0.7
21.2	6.5	89.3	27.4	71.7	21.2	54.1	16.2	53	16.2	8.7	2.6
21.2	6.6	89.2	27.8	71.7	21.6	54	16.4	52.9	16.6	8.7	2.7

Table G.9

$w_{oil,0}$	p	T	$\dot{m}$	$\dot{x}$	$\Delta p/\Delta L$	FP	$\dot{q}$	α
0.013	26	-10.47	81.8	0.91	3.13	w (f)	0.2	
0.013	26.01	-10.51	83.7	0.88	3.08	w (f)	0.2	
0.013	26.01	-10.4	80.5	0.93	3.12	w (f)	0.2	
0.013	26.01	-10.43	80.8	0.92	3.51	w (f)	5	19.6
0.013	26.01	-10.5	82.8	0.9	3.34	w (f)	5	15.1
0.013	26.03	-10.46	82.5	0.9	3.37	w (f)	5	16.2
0.013	26.02	-10.47	82.6	0.9	3.4	w (f)	5	15.8
0.013	26.01	-10.48	81.7	0.9	3.68	w (f)	9.4	14.2
0.013	26.02	-10.48	82.2	0.9	3.56	w (f)	9.4	14.5
0.013	26	-10.49	81.9	0.9	3.73	w (f)	9.4	14.9
0.013	26.01	-10.5	82.7	0.89	3.58	w (f)	9.5	14
0.013	26.01	-10.46	81.8	0.91	4.1	f	15.5	11.9
0.013	26.02	-10.48	83.3	0.9	4.22	f	15.6	12
0.013	26.03	-10.43	82.3	0.91	4.17	f	15.6	11.5
0.013	26.02	-10.47	82.7	0.9	4.09	f	15.6	11.4
0.013	26.01	-10.52	81.5	0.88	4.12	f	18.5	11.6
0.013	26	-10.55	84	0.86	4.06	f	18.5	11.4
0.013	26.05	-10.47	82	0.88	4.14	f	18.5	11.6
0.013	26	-10.5	81.9	0.9	5.13	f	26.6	11.4
0.013	26	-10.52	83.2	0.89	5.14	f	26.6	11.2
0.013	26.01	-10.5	82	0.89	5.02	f	26.7	11
0.013	26.03	-10.46	82.1	0.9	4.92	f	26.6	10.8
0.013	25.64	-11.12	156.3	0.12	1.15	i	0.2	
0.013	25.65	-11.11	155.4	0.13	1.49	i	6.6	11.4
0.013	25.63	-11.14	155.1	0.13	1.56	i	6.7	11.4
0.013	25.65	-11.1	155.6	0.13	2.17	i	14.5	15.1
0.013	25.65	-11.11	155.7	0.13	2.22	i	14.6	15
0.013	25.65	-11.11	155.2	0.13	2.75	i	22.9	17.4
0.013	25.66	-11.09	156	0.12	2.95	i	22.9	17.6
0.013	25.66	-11.09	155.8	0.13	3.37	i	28.7	18.4
0.013	25.63	-11.13	155.6	0.13	3.49	i	28.9	18.8
0.013	25.66	-11.09	155.7	0.12	4.32	i (f)	38.6	19.7
0.013	25.66	-11.09	154	0.13	4.27	i (f)	38.6	20
0.013	25.68	-11.07	156.4	0.13	5.39	i (f)	50.2	21.8
0.013	25.63	-11.13	155.3	0.13	4.88	i (f)	50.2	22.2
0.013	25.66	-11.09	155.9	0.12	5.47	i (f)	63.8	23.6
0.013	25.66	-11.09	156.1	0.13	5.79	i (f)	63.8	23.7
0.013	25.69	-11.05	155.3	0.13	5.7	i (f)	63.8	23.7
0.013	25.7	-11.04	155.5	0.13	6.42	i (f)	76.3	25.1
0.013	25.68	-11.07	155.2	0.13	6.07	i (f)	76.4	24.8
0.013	25.67	-11.08	155.3	0.14	7.61	i-a (f)	109.7	28.2
0.013	25.68	-11.07	155.1	0.13	7.33	i-a (f)	109.7	28.6
0.013	26.03	-10.6	162	0.12	2.7	i	0.2	
0.013	25.98	-10.66	160.8	0.13	2.56	i	0.2	
0.013	26	-10.63	162.7	0.13	2.99	i	8.2	12.1
0.013	26	-10.63	161.8	0.13	2.92	i	8.1	12
0.013	26.01	-10.62	163.2	0.12	3.35	i	15.4	15.2

$\dot{q}_1$	α_1	$\dot{q}_2$	α_2	$\dot{q}_3$	α_3	$\dot{q}_4$	α_4	$\dot{q}_5$	α_5	$\dot{q}_6$	α_6
0.2		0.2		0.2		0.2		0.2		0.2	
0.2		0.2		0.2		0.2		0.2		0.2	
0.2		0.2		0.2		0.2		0.2		0.2	
6	24.7	7.2	30.4	7.2	23.5	2	7.8	3.7	15.5	3.7	15.8
6	19.1	7.2	22.8	7.2	19.1	2	6	3.7	11.7	3.7	11.9
6	20.6	7.2	24.6	7.2	19.9	2	6.5	3.7	12.6	3.7	12.8
6.1	19.6	7.2	24.2	7.2	19.7	2.1	6.8	3.7	12.1	3.7	12.1
13.3	20	15.8	24.1	9.2	13.9	4.4	6.8	6.5	9.8	7.1	10.8
13.3	20.3	15.8	24.1	9.3	14.3	4.5	7	6.5	10	7.1	11.2
13.3	20.8	15.8	25.3	9.3	14.5	4.5	7.3	6.5	10.3	7.1	11.3
13.4	19.5	15.9	23.7	9.4	13.9	4.6	6.9	6.5	9.7	7.2	10.6
13.4	10.6	37.6	28.9	19.1	14.3	9.1	7.2	12	9.1	2.1	1.6
13.5	10.7	37.6	29	19.1	14.2	9.1	7.2	12	9.3	2.1	1.7
13.4	10.2	37.6	27.6	19.1	13.7	9.1	6.9	12.1	9	2	1.5
13.4	10	37.6	27.3	19.1	13.7	9.1	6.9	12.2	8.9	2.1	1.5
10.4	6.9	49.1	30.6	22.3	13.8	13.5	8.6	15.4	9.7	0.2	0.1
10.4	6.8	49	30.2	22.3	13.5	13.6	8.4	15.5	9.5	0.2	0.1
10.4	7	49.1	30.6	22.3	13.8	13.6	8.5	15.4	9.7	0.2	0.1
14	6	59.6	25.3	41.8	18	22.3	9.7	21.9	9.3	0.2	0.1
14	5.9	59.6	24.9	41.8	17.6	22.3	9.7	21.9	9.3	0.2	0.1
14	5.8	59.7	24.6	41.8	17.2	22.3	9.4	21.9	9.1	0.2	0.1
14	5.8	59.6	24.1	41.8	16.6	22.3	9.2	21.9	9	0.2	0.1
0.2		0.2		0.2		0.2		0.2		0.2	
8.8	15.5	9.6	16.7	4.6	8	3.3	5.6	3.6	6	9.7	16.7
9.6	16.4	9.6	16.4	4.6	7.9	3.4	5.6	3.6	6	9.6	16.1
20.3	21.3	18.7	19.7	12.8	13.3	4.5	4.7	9.6	9.9	21.3	21.7
20.4	21	18.7	19.4	13	13.4	4.5	4.7	9.6	9.9	21.3	21.7
30.5	23.1	29.3	22.5	23.5	17.7	9.1	6.9	19.8	14.8	25.4	19.4
30.5	23.4	29.3	22.6	23.4	17.6	9.1	6.8	19.9	15	25.4	20
35	22	35.7	23.2	31.6	20.1	14	8.9	27.1	17.2	29	19
35.1	22.9	35.9	23.7	31.7	20.2	14	8.9	27.3	17.5	29.1	19.5
35.2	17.7	48.8	24.9	50.1	25.4	22.7	11.7	35.5	18.4	39.4	20.3
35.2	18.4	48.8	25.5	50	25.3	22.7	11.7	35.5	18.5	39.4	20.3
32.3	14.5	65.7	28.1	71.1	30.1	36.3	15.5	51.6	22.5	43.9	19.9
32.3	15.3	65.8	29.2	71.2	30.3	36.4	15.6	51.7	22.7	44	20.2
38.5	14.9	81.5	30.3	92.1	33.7	54.3	19.7	62.1	23	54.2	20.1
38.6	15	81.5	30.3	92.1	33.8	54.4	19.8	62.1	23.2	54.1	20.3
38.6	15.2	81.6	30.5	92.1	33.8	54.4	19.8	62	23	54	20.2
44	14.9	91.6	30.9	108.6	36	75.6	24	70.3	22.9	67.5	22.2
44	13.7	97.1	31.1	104.2	35	75.5	24.1	70.3	22.9	67.6	21.9
68.6	17.1	140.4	36	153.9	40.8	107.3	27	96.3	24.7	91.8	23.9
68.6	17.4	140.5	36.5	153.9	41.6	107.2	27.1	96.2	24.8	91.8	24.2
0.2		0.2		0.2		0.2		0.2		0.2	
0.2		0.2		0.2		0.2		0.2		0.2	
13.2	19.6	9.9	15.1	6.3	8.8	2.6	3.7	6.1	8.9	11	16.6
13	19.4	9.8	15.2	6.2	8.7	2.6	3.8	6.1	8.7	10.9	16.3
22.3	22.3	19.2	19.3	13.1	12.3	4.8	4.7	12.3	11.9	20.9	20.6

Table G.9

$w_{oil,0}$	p	T	$\dot{m}$	$\dot{x}$	$\Delta p/\Delta L$	FP	$\dot{q}$	α
0.013	26	-10.63	162.2	0.13	3.42	i	15.5	15
0.013	25.97	-10.68	163.2	0.13	3.88	i	22.5	17.1
0.013	25.98	-10.66	162.4	0.12	3.85	i	22.5	16.7
0.013	26.04	-10.58	163.4	0.12	4.17	i (f)	27.4	17.2
0.013	26.02	-10.61	162.3	0.13	4.14	i (f)	27.3	17.7
0.013	26.03	-10.59	160.9	0.13	5.43	i (f)	36.5	18.9
0.013	26.07	-10.53	161.8	0.13	5.34	i (f)	37.2	18.4
0.013	26.03	-10.59	162.1	0.12	6.49	i (f)	51.1	20.9
0.013	26.06	-10.56	162	0.12	6.02	i (f)	51.1	21.1
0.013	26.06	-10.56	163.3	0.12	7.11	i (f)	65.1	22.4
0.013	26.05	-10.57	164.3	0.12	7.09	i (f)	64.8	22.6
0.013	26.06	-10.56	157.2	0.13	7.53	i (f)	78.7	23.6
0.013	26.1	-10.49	156.4	0.13	7.55	i (f)	78.8	23.6
0.013	25.99	-10.64	157.2	0.13	8.18	i-a (f)	109.3	26.2
0.013	25.94	-10.72	157.2	0.13	8.35	i-a (f)	109.3	26.4
0.013	26	-10.64	158.5	0.12	2.4	i	0.2	
0.013	25.99	-10.64	158	0.12	2.18	i	0.2	
0.013	25.97	-10.67	158.6	0.13	2.34	i	0.2	
0.013	26.03	-10.6	156.9	0.13	2.34	i	5.8	11.9
0.013	25.97	-10.67	157.2	0.12	2.41	i	5.9	11.8
0.013	25.99	-10.65	157.4	0.12	2.37	i	5.8	11.8
0.013	26	-10.63	157.8	0.12	2.44	i	5.8	11.7
0.013	26.01	-10.62	158.2	0.12	2.4	i	8.4	12.9
0.013	26.01	-10.62	158.1	0.12	2.59	i	8.4	12.8
0.013	25.98	-10.66	158.3	0.12	2.58	i	8.3	12.8
0.013	25.99	-10.65	157.6	0.13	3.44	i	20	16.8
0.013	26.01	-10.62	157.5	0.12	3.46	i	20.1	16.7
0.013	26	-10.63	158.3	0.12	3.2	i	20.1	16.5
0.013	26.01	-10.62	158	0.12	4.01	i	26.2	17.7
0.013	26.02	-10.61	156.6	0.12	3.84	i	26.1	17.9
0.013	26	-10.64	156.7	0.13	3.78	i	26.1	18.2
0.013	26.01	-10.62	158	0.12	4.48	i (f)	31.8	18.4
0.013	26	-10.64	156.8	0.13	4.39	i (f)	31.7	18.5
0.013	25.98	-10.65	157.7	0.12	4.46	i (f)	31.7	18.7
0.013	26.01	-10.61	156.8	0.13	5.35	i (f)	43.8	20.6
0.013	26.02	-10.61	157.7	0.12	5.14	i (f)	43.8	20.2
0.013	26	-10.64	158.3	0.12	5.66	i (f)	43.8	20.6
0.013	26.03	-10.59	157	0.13	6.78	i (f)	59.5	23.1
0.013	26.02	-10.61	156.7	0.13	6.5	i (f)	59.4	23.2
0.013	26.07	-10.53	157	0.13	7.18	i (f)	76.1	24.8
0.013	26.01	-10.62	157.7	0.12	6.96	i (f)	76.1	24.5
0.013	26.04	-10.58	156.6	0.13	6.76	i (f)	76.1	24.5
0.013	26.04	-10.58	157.7	0.13	8	i-a (f)	108.5	27.6
0.013	26.05	-10.56	156.9	0.13	8.1	i-a (f)	108.5	27.8
0.013	26.04	-10.58	157.7	0.13	7.98	i-a (f)	108.5	27.8
0.013	25.95	-10.69	155.6	0.32	3.59	i-a	0.2	
0.013	25.95	-10.69	155.9	0.32	3.62	i-a	0.2	

Table G.9

$\dot{q}_1$	α_1	$\dot{q}_2$	α_2	$\dot{q}_3$	α_3	$\dot{q}_4$	α_4	$\dot{q}_5$	α_5	$\dot{q}_6$	α_6
22.3	21.7	19.2	19	13.1	12.3	4.9	4.7	12.3	11.8	21	20.5
27.4	21.2	30.8	23.2	18.8	14	8.8	6.6	19.6	14.8	29.4	23.2
27.4	20.5	30.8	23.1	18.8	13.9	8.8	6.5	19.6	14.5	29.4	21.9
32.9	20.4	36.2	23.2	23.8	15.4	13.5	8.6	24.8	15.7	32.9	19.9
32.9	21.1	36.2	23.8	23.8	15.6	13.5	8.7	24.7	15.9	32.9	20.9
37	18.9	47.1	23.8	44.1	22	23	11.8	35.6	18.8	32.4	17.8
37.4	18	47.2	23	44.2	21.8	23.1	11.7	37.1	18.7	34.3	17.1
39	16	62.6	25.6	67.3	27.3	36.7	15.2	51.5	21.2	49.5	20.2
39	16.7	62.7	26.2	67.3	27.3	36.7	15.2	51.5	21.1	49.4	19.8
43.4	14.4	82	27.7	89	30.8	54.4	19.2	63.5	22.3	58	20.2
44.8	15.9	79.2	27.9	88.9	30.8	54.4	19	63.4	22.2	57.9	19.7
39.3	12.2	98.8	29.8	111.8	33.5	72.9	21.9	70.5	21.4	78.9	23.1
39.5	12.1	98.8	29.7	112	33.6	73	22	70.6	21.4	79.1	22.8
58.2	14.1	145.5	34.6	149.1	36.6	106.8	25.2	91.8	21.8	104.3	24.9
58.3	14.2	145.6	34.9	149.1	36.9	106.9	25.5	91.7	22	104.3	24.9
0.2		0.2		0.2		0.2		0.2		0.2	
0.2		0.2		0.2		0.2		0.2		0.2	
0.2		0.2		0.2		0.2		0.2		0.2	
9.2	18.1	8.6	17.7	3.4	6.9	2.2	4.6	4.1	8.2	7.6	15.7
9.2	18.3	8.6	17.5	3.4	6.8	2.2	4.5	4.1	8	7.7	15.5
9.3	18.9	8.7	17.7	3.3	6.7	2.3	4.6	3.9	7.7	7.4	15
9.2	18.8	8.7	17.7	3.3	6.7	2.3	4.8	3.7	7.4	7.3	14.9
12.6	19.6	10.9	17.5	5.4	8.1	4.4	6.6	5.9	8.7	10.9	16.9
12.6	19.5	10.9	17.2	5.4	8.1	4.4	6.7	5.9	8.7	11	16.8
12.5	19.2	10.8	17	5.3	8.1	4.4	6.6	5.8	8.7	11	17.1
22.6	19.2	27.8	23.4	16.3	13.6	8.9	7.5	17	14.1	27.6	23
22.6	19.1	27.8	23.2	16.3	13.7	9.1	7.7	16.9	14	27.7	22.8
22.6	18.6	27.8	23.4	16.3	13.7	9.1	7.7	17	13.9	27.7	22
21.2	14.9	39	25.8	25	16.8	13.5	9.3	26.1	17.6	32.4	22
21.2	15.3	38.9	26.3	24.9	16.8	13.5	9.3	26	17.6	32.3	22.3
21.1	15.7	38.8	26.9	24.9	16.9	13.5	9.3	25.9	17.7	32.3	22.6
24.1	14.4	44.9	25.6	32.1	18.6	22.6	13	31.7	18.3	35.6	20.5
24.1	14.5	44.9	25.9	32	18.6	22.5	13	31.4	18.3	35.4	20.7
24.1	14.8	44.8	26	32.1	18.5	22.5	13	31.6	18.4	35.5	21.7
25.3	12.8	60.6	28.1	52.3	23.9	36.1	16.7	45.4	21.3	43.4	20.9
25.3	12.3	60.6	27.5	52.2	23.7	36.1	16.6	45.4	21	43.4	20
25.2	12.8	60.5	28	52.2	23.7	36.1	16.6	45.3	21.3	43.3	21
31	12.8	79.4	30.4	77	29.4	54.2	20.9	59.3	23.2	56.3	22.1
30.8	12.9	79.3	30.8	76.8	29.3	54.1	20.8	59.2	23.2	56.2	22.2
38.9	13.2	103	33.3	103	33	72	23.6	67.3	22.4	72.4	23.2
38.9	12.5	103	32.3	103.1	33.2	72	23.7	67.4	22.3	72.5	23.1
38.9	12.7	103	32.4	103.1	33.2	71.9	23.6	67.3	22.2	72.5	22.6
65.9	16.5	130	33.6	153.3	39.6	107.3	27	103.1	25.6	91.5	23.2
65.8	16.8	130	33.8	153.1	40	107.2	27.1	103.2	25.7	91.5	23.3
65.9	16.6	130	33.7	153.2	39.8	107.4	27.1	103.2	25.8	91.5	23.5
0.2		0.2		0.2		0.2		0.2		0.2	
0.2		0.2		0.2		0.2		0.2		0.2	

Table G.9

$w_{oil,0}$	p	T	$\dot{m}$	$\dot{x}$	$\Delta p/\Delta L$	FP	$\dot{q}$	α
0.013	25.96	-10.69	156	0.32	3.76	i-a	4.2	9.9
0.013	26	-10.63	156.3	0.32	3.42	i-a	4.2	10
0.013	25.96	-10.69	158	0.31	4.16	i-a (f)	13.8	14.8
0.013	25.98	-10.66	156.5	0.32	4.1	i-a (f)	13.8	14.8
0.013	25.97	-10.67	155.4	0.33	4.53	i-a (f)	21.8	17.2
0.013	25.98	-10.66	156.9	0.33	4.63	i-a (f)	21.8	17.1
0.013	25.97	-10.67	156.6	0.33	5.1	i-a (f)	26.1	17.8
0.013	26	-10.63	156.5	0.33	5.11	i-a (f)	26.1	18
0.013	25.98	-10.65	156.9	0.32	5.03	i-a (f)	26	17.8
0.013	26.03	-10.59	156	0.33	5.91	i-a (f)	36.3	19.7
0.013	25.95	-10.7	156.3	0.32	5.99	i-a (f)	36.3	19.4
0.013	26.02	-10.61	157.1	0.33	6.93	i-a (f)	47.1	21.7
0.013	25.99	-10.64	158	0.33	6.89	i-a (f)	47.1	21
0.013	26.01	-10.62	154.8	0.33	7.41	w-a (f)	55.4	22.1
0.013	25.99	-10.65	156.2	0.33	7.48	w-a (f)	55.4	22.6
0.013	26	-10.63	155	0.34	8.7	w-a (f)	68	22.6
0.013	26.02	-10.61	154.1	0.34	9.1	w-a (f)	68	23.1
0.013	25.99	-10.64	155.2	0.33	9.57	w-a (f)	80.5	22.7
0.013	26.02	-10.6	156.6	0.33	10.14	w-a (f)	80.5	22.7
0.013	25.58	-11.19	161.8	0.5	6.87		0.2	
0.013	25.61	-11.15	162.3	0.5	6.87		0.2	
0.013	25.6	-11.16	151.8	0.53	6.46		2.3	15.1
0.013	25.6	-11.17	151.7	0.54	6.43		2.2	14.7
0.013	25.6	-11.16	153.2	0.54	6.81		9.3	15.7
0.013	25.61	-11.15	154.1	0.53	6.72		9.3	15.7
0.013	25.61	-11.15	153	0.54	6.99		19.3	18.8
0.013	25.6	-11.17	153.8	0.54	6.82		19.4	18.9
0.013	25.63	-11.12	153.8	0.55	7.32		22.6	20.2
0.013	25.6	-11.16	154.8	0.54	7.52		22.7	19.9
0.013	25.62	-11.14	155.1	0.54	8.62		36.2	22.2
0.013	25.61	-11.15	154.5	0.54	8.54		36.4	22.4
0.013	25.61	-11.14	154.6	0.54	9.5		46.2	22.7
0.013	25.65	-11.1	154.8	0.53	9.24		46.2	22.6
0.013	25.64	-11.11	154.4	0.53	10.67		58.6	23.5
0.013	25.65	-11.09	155	0.53	10.97		58.6	23.3
0.013	25.66	-11.09	153.5	0.55	12.86		72	23.2
0.013	25.63	-11.12	154.3	0.55	12.73		72.1	22.9
0.013	25.63	-11.12	154	0.54	14.59		84.7	23.8
0.013	25.68	-11.06	153.7	0.55	14.34		84.7	23.8
0.013	25.59	-11.18	157.1	0.51	7.01	w-a (f)	0.2	
0.013	25.59	-11.17	160	0.51	7.05	w-a (f)	0.2	
0.013	25.59	-11.18	159.5	0.51	7.13	w-a (f)	0.2	
0.013	25.65	-11.1	161.9	0.51	7.26	w-a (f)	6.9	14
0.013	25.59	-11.18	159.7	0.52	7.31	w-a (f)	6.9	14
0.013	25.61	-11.16	160.3	0.51	7.2	w-a (f)	10.3	15.1
0.013	25.6	-11.17	161.4	0.5	7.17	w-a (f)	10.3	15.1
0.013	25.55	-11.23	161	0.51	7.63	w-a (f)	19.8	18.4

Table G.9

$\dot{q}_1$	α_1	$\dot{q}_2$	α_2	$\dot{q}_3$	α_3	$\dot{q}_4$	α_4	$\dot{q}_5$	α_5	$\dot{q}_6$	α_6
5.8	13.6	6.1	14.3	2.3	5.7	2	4.6	4.2	9.7	4.9	11.7
5.9	13.8	6.1	14.5	2.2	5.7	2	4.8	4.1	9.7	4.9	11.8
21.6	22.7	17.1	18.2	10.5	11.5	4.7	5.2	13.7	14.7	15	16.6
21.7	22.8	17.1	18.2	10.5	11.5	4.8	5.2	13.7	14.7	15	16.5
26	21.1	32.6	25.1	20	15.5	9.3	7.3	24.9	19.3	18	14.8
26	21.1	32.6	25	20.1	15.6	9.4	7.3	24.9	19	18	14.5
27.8	19.4	41.4	27.6	26	17.4	13.7	9.3	29.6	20.1	18.2	12.8
27.8	19.8	41.4	27.7	26	17.5	13.7	9.4	29.6	20.3	18.2	13.3
27.7	19.7	41.3	27.7	25.9	17.4	13.6	9.3	29.5	20	18.2	12.8
34.4	18.9	61.2	32.5	36.9	20.2	22.8	12.5	42.8	23.2	19.8	11.1
34.3	18.2	61.3	32	36.9	20.1	22.8	12.4	42.8	23	19.8	11
34	16.3	82.5	36.9	49.7	22.9	36.1	16.5	53.3	24.6	26.7	12.9
34	15.2	82.6	35.9	49.7	22.7	36.1	16.3	53.4	23.9	26.8	12
34.2	14.1	92.4	36.6	59.6	24.1	53.9	20.9	61.4	24.4	31	12.4
34.2	14.8	92.4	37.5	59.6	24.4	53.9	21.1	61.3	24.7	31	12.9
39	12.9	111.9	36.8	72.6	25.6	72	23.7	75.7	24.7	36.7	11.8
39	13.6	111.9	37.6	72.7	25.9	72	23.8	75.7	25.4	36.7	12.4
36.6	10.5	125.5	35.9	84.5	25.5	107.9	29	79.9	22.5	48.7	13
36.6	10.4	125.6	35.5	84.4	25.5	107.8	29.3	79.9	22.5	48.7	13.2
0.2		0.2		0.2		0.2		0.2		0.2	
0.2		0.2		0.2		0.2		0.2		0.2	
2	14	3.5	22.5	1.6	12.5	1.5	9	1.8	11.5	3.4	21.2
2	13.7	3.3	21.6	1.6	12.2	1.5	9.3	1.7	10.9	3.3	20.6
15.3	25.4	10.9	18.5	5.7	10.1	4.3	7.3	8.8	14.8	11	18.3
15.3	25.3	11	18.5	5.7	10.1	4.3	7.3	8.8	14.8	11	18.3
27.4	26.8	24.4	23.8	16.1	15.7	8.9	8.7	17.4	16.8	21.7	21
27.6	26.9	24.4	23.8	16.1	15.7	8.9	8.8	17.4	16.9	21.8	21.1
33.7	29.8	29.2	25.9	17	15.4	12.4	10.9	20.1	17.8	23.4	21.1
33.6	29.3	29.1	25.6	16.9	15.2	12.4	10.9	19.9	17.3	24.2	21.3
55.7	33.7	49	30.5	38.9	23.8	22.4	13.8	28.3	17.4	22.9	14.1
56	34.2	49.2	30.7	39.1	24	22.6	13.9	28.5	17.5	23.1	14
60.6	29	74.8	36.6	50.1	25.3	36.6	18	41.7	20.5	13.5	6.5
60.6	28.6	74.8	36.6	50.1	25.5	36.6	18	41.6	20.2	13.5	6.4
54.2	21.7	101.5	40	72.5	29.5	54.3	21.7	59.2	24.1	9.8	4.1
54.2	21.5	101.6	39.6	72.3	29.3	54.4	21.6	59.2	24	9.9	4
56.2	17.7	113.9	36.5	96	31.7	72.8	23.5	77.4	24.7	15.9	5
56.3	17.1	113.9	35.9	96.1	31.9	72.9	23.3	77.5	24.4	16	4.9
56.1	16.2	129.9	36.6	106.7	30.7	107.3	29.1	92.2	25.9	16	4.6
56.1	16	129.9	36.6	106.7	30.8	107.3	29.1	92.3	26	16	4.6
0.2		0.2		0.2		0.2		0.2		0.2	
0.2		0.2		0.2		0.2		0.2		0.2	
0.2		0.2		0.2		0.2		0.2		0.2	
9.7	20	6.5	13.4	6.6	13	2.8	5.6	5.9	11.8	9.9	20.1
9.7	20	6.5	13.4	6.6	12.9	2.9	5.7	5.9	11.7	9.9	20.1
14.6	21.8	11	16.2	9	12.7	4.6	6.6	8.7	12.7	13.8	20.5
14.6	21.8	11	16.2	9	12.8	4.5	6.7	8.7	12.7	13.8	20.5
28.3	26.5	22.4	21.1	16.8	15.4	9.3	8.6	17.7	16.3	24	22.4

Table G.9

$w_{oil,0}$	p	T	$\dot{m}$	$\dot{x}$	$\Delta p/\Delta L$	FP	$\dot{q}$	α
0.013	25.55	-11.24	162.5	0.5	7.56	w-a (f)	19.8	18.4
0.013	25.55	-11.23	161.7	0.51	8.49	w-a (f)	28.3	21
0.013	25.51	-11.29	162.1	0.51	8.27	w-a (f)	28.4	21
0.013	25.5	-11.3	161.2	0.52	9.35	w-a (f)	40.7	22.6
0.013	25.55	-11.23	160.3	0.52	9.43	w-a (f)	40.6	22.7
0.013	25.56	-11.21	161.3	0.51	10.53	w-a (f)	50.8	23.4
0.013	25.57	-11.21	161.3	0.52	10.49	w-a (f)	50.8	23.3
0.013	25.58	-11.19	161.3	0.51	11.83	w-a (f)	62.5	23.8
0.013	25.63	-11.12	160.5	0.52	11.76	w-a (f)	62.5	24.1
0.013	25.58	-11.19	160.2	0.52	13.48	w-a (f)	74.2	23.9
0.013	25.6	-11.17	161.6	0.51	13.26	w-a (f)	74.2	24.2
0.013	25.6	-11.16	161.3	0.51	15.43	f	91	24.7
0.013	25.58	-11.2	161.2	0.52	15.26	f	91.1	24.4
0.013	25.55	-11.24	159.9	0.51	7.42	w-a (f)	0.2	
0.013	25.56	-11.23	159.5	0.51	7.4	w-a (f)	0.2	
0.013	25.58	-11.19	159.8	0.52	7.45	w-a (f)	6.8	13.3
0.013	25.59	-11.18	159.2	0.52	7.72	w-a (f)	6.8	13.2
0.013	25.58	-11.19	160	0.51	7.43	w-a (f)	10.2	14.5
0.013	25.55	-11.24	160.3	0.51	7.56	w-a (f)	10.4	14.5
0.013	25.6	-11.16	160.9	0.51	7.75	w-a (f)	19.9	17.6
0.013	25.58	-11.19	160.7	0.52	8.07	w-a (f)	19.9	17.6
0.013	25.59	-11.18	161	0.51	8.46	w-a (f)	28.9	20.1
0.013	25.58	-11.2	160.4	0.52	8.5	w-a (f)	28.9	20
0.013	25.61	-11.15	161.1	0.51	9.79	w-a (f)	41	21.4
0.013	25.62	-11.14	160.2	0.52	9.68	w-a (f)	40.8	21.3
0.013	25.57	-11.21	161	0.52	10.81	w-a (f)	51.2	21.7
0.013	25.61	-11.15	160.5	0.52	10.49	w-a (f)	51.1	21.6
0.013	25.55	-11.24	160.3	0.52	11.95	w-a (f)	62.4	21.9
0.013	25.57	-11.21	160	0.52	12.26	w-a (f)	62.4	22.2
0.013	25.51	-11.29	159.2	0.53	13.85	w-a (f)	74.8	21.9
0.013	25.57	-11.21	161.2	0.52	13.74	w-a (f)	74.6	22.1
0.013	25.51	-11.28	153.6	0.53	14.48	w-a (f)	85.5	20.8
0.013	25.56	-11.22	158.6	0.52	15.11	w-a (f)	85.5	22.6
0.013	25.58	-11.2	160.1	0.52	15.43	w-a (f)	91.1	22.7
0.013	25.51	-11.29	160.9	0.52	16.1	f	91.2	22.5
0.013	25.95	-10.7	158.4	0.52		w-a (f)	0.2	
0.013	25.93	-10.71	158	0.52		w-a (f)	0.2	
0.013	25.95	-10.7	157.8	0.52	5.71	w-a (f)	3.4	12.7
0.013	25.93	-10.72	158.2	0.52	5.81	w-a (f)	3.4	12.7
0.013	25.95	-10.69	158.1	0.52	5.65	w-a (f)	3.4	12.5
0.013	25.93	-10.71	158.4	0.52	6.15	w-a (f)	8.3	14.9
0.013	25.96	-10.68	157.6	0.53	6.03	w-a (f)	8.4	14.9
0.013	25.95	-10.69	158	0.52	6.02	w-a (f)	8.3	14.8
0.013	26	-10.62	158.5	0.51	6.29	w-a (f)	16.5	17.5
0.013	25.93	-10.72	158.6	0.52	6.74	w-a (f)	16.6	17.4
0.013	25.95	-10.7	158	0.51	6.46	w-a (f)	16.5	17.4
0.013	25.97	-10.66	159	0.51	7.48	w-a (f)	28.3	20.5

$\dot{q}_1$	α_1	$\dot{q}_2$	α_2	$\dot{q}_3$	α_3	$\dot{q}_4$	α_4	$\dot{q}_5$	α_5	$\dot{q}_6$	α_6
28.4	26.4	22.5	21.1	16.8	15.5	9.3	8.6	17.8	16.4	24	22.4
43.1	31.9	33.8	25.2	25.2	18.6	13.8	10.3	22.9	17.1	31	23.2
43.2	31.9	33.9	25.3	25.3	18.6	13.9	10.3	23.1	17	31.1	22.9
57.6	32.3	59.1	32.7	47.8	25.6	22.9	12.8	32.3	18.3	24.6	13.7
57.4	32.6	59	32.7	47.7	25.6	22.8	12.8	32.3	18.4	24.4	13.9
59	27.8	78.3	36.3	62.3	28.1	36.6	16.7	47.7	21.9	21	9.7
59	27.4	78.4	36.2	62.3	28.2	36.6	16.7	47.7	21.8	21	9.5
60.8	23	98.4	37.5	83.6	31.3	54.8	20.8	59.5	23.1	18.2	7
60.7	23.8	98.4	37.9	83.6	31.6	54.8	20.9	59.5	23.5	18.1	7.2
59.6	18.9	114.4	36.8	101.7	33.1	72.3	23.4	76.6	24.7	20.5	6.6
59.6	19.5	114.5	37.6	101.9	33.1	72.3	23.5	76.6	25.1	20.4	6.6
64.5	18.2	129.2	35.7	127.7	33.6	107.3	28.2	92.7	25.8	24.7	6.7
64.6	17.2	129.2	34.7	127.7	34.1	107.5	28.3	92.8	25.6	24.8	6.7
0.2		0.2		0.2		0.2		0.2		0.2	
0.2		0.2		0.2		0.2		0.2		0.2	
9.6	19.6	6.4	13.1	6.5	11.2	2.6	5.1	5.6	11	10	19.8
9.5	19.6	6.4	13	6.5	11.2	2.6	5	5.5	10.9	10	19.8
14.5	21.5	10.8	15.8	8.9	11.3	4.6	6.3	8.7	12.4	13.8	19.9
14.7	21.5	11	15.8	9.1	11.3	4.7	6.4	8.9	12.3	13.9	19.8
28.4	25.8	22.6	20.7	16.9	13.8	9.5	8.3	17.8	15.5	24.4	21.5
28.4	25.7	22.6	20.5	16.9	13.9	9.5	8.3	17.9	15.6	24.4	21.5
43.7	30.8	34.6	24.5	26.1	17.3	14.1	9.8	23.8	16.5	31.3	21.6
43.6	30.7	34.6	24.5	26.1	17.2	14.2	9.9	23.7	16.4	31.2	21.6
57.5	30.7	59.8	31.2	47.9	23.4	23.5	12.5	32.7	17.6	24.7	13.1
57.2	30.5	59.8	31.1	47.7	23.5	22.8	12.2	32.6	17.6	24.6	12.9
59.5	25.1	78.4	33.7	62.7	26	36.9	16	48.3	20.5	21.1	8.8
59.4	25	78.4	33.7	62.7	26.1	36.9	16	48.2	20.4	21.1	8.5
60.5	20.8	98.3	34.7	83.8	28.7	54.8	19.6	59.1	21.2	18.1	6.2
60.5	21.4	98.3	35	83.7	28.8	54.8	19.9	59.1	21.6	18	6.5
60.3	16.9	115.1	33.8	102.8	30.1	72.7	22	77.1	22.9	20.7	5.9
60.1	17.3	114.9	34.2	102.6	30.1	72.5	22.1	76.9	23.2	20.6	5.9
64.9	13.7	129.4	30.9	93.1	25.9	107.8	26.7	93.3	22	24.5	5.3
64.9	16.4	129.6	34.2	92.9	26.5	107.9	27.8	93.2	24.4	24.6	6.3
64.8	16.2	129.3	32.7	127.1	30.6	107.7	26.8	93.1	24	24.4	6.1
65.1	15.5	129.5	31.9	127.2	31.5	107.9	26.8	93.2	23.5	24.6	6
0.2		0.2		0.2		0.2		0.2		0.2	
0.2		0.2		0.2		0.2		0.2		0.2	
4.2	15.2	4.4	17	2.3	9.3	1.9	7.1	3.3	12.1	4.3	15.5
4.2	15	4.3	16.7	2.2	9.1	2	7.6	3.3	12.2	4.3	15.4
4.2	14.9	4.3	16.5	2.3	9.2	1.9	7.1	3.3	12	4.3	15.3
12	21.4	10.5	18.9	5.8	10.3	4.8	8.5	8.3	14.6	8.7	15.8
12	21.3	10.5	18.8	5.8	10.4	4.8	8.5	8.3	14.6	8.6	15.7
11.9	21.3	10.5	18.7	5.8	10.3	4.8	8.4	8.3	14.6	8.5	15.4
23.7	24.9	21.2	22.5	12.1	12.9	9.1	9.7	15.1	16	18	19.1
23.8	24.9	21.2	22.3	12.1	12.8	9.1	9.7	15.1	15.9	18	19
23.8	24.8	21.2	22.3	12	12.7	9.1	9.6	15.1	15.9	18	19
44.4	31.8	37.7	27	26.2	18.7	13.7	9.9	26.9	19.4	21.1	15.8

Table G.9

$w_{oil,0}$	p	T	$\dot{m}$	$\dot{x}$	$\Delta p/\Delta L$	FP	$\dot{q}$	α
0.013	25.97	-10.67	157.6	0.52	7.44	w-a (f)	28.3	20.2
0.013	25.96	-10.68	158.6	0.51	7.45	w-a (f)	28.3	20.2
0.013	25.96	-10.68	157.2	0.53	8.47	w-a (f)	37.6	20.9
0.013	25.99	-10.64	158.1	0.53	8.57	w-a (f)	37.6	20.9
0.013	25.95	-10.69	156.5	0.54	9.37	w-a (f)	48	21.7
0.013	25.98	-10.65	158.5	0.53	9.55	w-a (f)	48	21.7
0.013	25.99	-10.64	157.5	0.53	9.66	w-a (f)	48.1	21.4
0.013	26.01	-10.61	158	0.52	10.95	w-a (f)	61.6	22.1
0.013	25.99	-10.63	157.9	0.53	10.86	w-a (f)	61.6	22
0.013	25.98	-10.65	158	0.51	10.97	w-a (f)	61.6	22.2
0.013	26	-10.62	157	0.52	12.41	w-a (f)	74.7	22.6
0.013	26	-10.62	157.8	0.51	12.41	w-a (f)	74.7	22.5
0.013	26.01	-10.61	157.4	0.52	14.35	f	90.8	22.3
0.013	26.02	-10.6	156.4	0.53	14.36	f	90.7	22.4
0.013	26.01	-10.61	157.8	0.52	14.47	f	90.5	22.6
0.013	25.94	-10.7	161.6	0.51		w-a (f)	0.2	
0.013	25.91	-10.74	160.7	0.51		w-a (f)	0.2	
0.013	25.96	-10.68	160.3	0.51		w-a (f)	0.2	
0.013	25.98	-10.65	159.8	0.51		w-a (f)	0.2	
0.013	25.97	-10.67	161	0.51	6.52	w-a (f)	3.2	10.8
0.013	25.96	-10.68	160.1	0.51	6.56	w-a (f)	3.1	10.5
0.013	25.95	-10.69	160.2	0.51	6.92	w-a (f)	8	13.2
0.013	26.01	-10.61	162	0.51	6.99	w-a (f)	8	13.2
0.013	25.94	-10.71	160.5	0.51	7.05	w-a (f)	8	13.2
0.013	25.97	-10.66	160.3	0.51	7.25	w-a (f)	16	16.1
0.013	25.97	-10.66	158.8	0.52	7.37	w-a (f)	15.8	16.1
0.013	25.97	-10.67	159.9	0.51	7.33	w-a (f)	15.7	16.1
0.013	25.97	-10.67	160.3	0.51	8.15	w-a (f)	25.5	19.1
0.013	25.98	-10.65	160.5	0.51	8.17	w-a (f)	25.5	19
0.013	25.95	-10.69	161	0.51	7.75	w-a (f)	25.8	19
0.013	25.97	-10.67	161.4	0.51	8.44	w-a (f)	35	20.4
0.013	25.99	-10.64	160.7	0.51	8.57	w-a (f)	35.7	20.4
0.013	26.01	-10.61	160.5	0.51	8.27	w-a (f)	36	20.4
0.013	26	-10.62	160.6	0.51	9.89	w-a (f)	46.4	22.7
0.013	25.98	-10.65	161	0.51	9.53	w-a (f)	46.5	22
0.013	25.98	-10.65	161.6	0.5	9.69	w-a (f)	46.5	21.7
0.013	25.98	-10.64	161.1	0.52	11.37	w-a (f)	61.9	23.4
0.013	26.02	-10.6	160.4	0.51	11.63	w-a (f)	61.8	23.2
0.013	25.99	-10.63	161.1	0.51	11.3	w-a (f)	61.8	22.8
0.013	26.03	-10.59	162.1	0.51	12.63	w-a (f)	74.2	24
0.013	26.03	-10.58	161.9	0.51	12.51	w-a (f)	74.2	24.4
0.013	25.98	-10.65	161.8	0.51	14.08	f	89.8	23.1
0.013	26.02	-10.6	163.9	0.51	14.61	f	89.8	23.6
0.013	26.01	-10.61	161.5	0.51	14.42	f	89.8	24.2
0.013	26.02	-10.59	161.6	0.51	14.61	f	89.8	24.2
0.013	25.99	-10.62	155.3	0.73	8.24	a (f)	0.2	
0.013	25.99	-10.62	156.5	0.72	8.14	a (f)	0.2	

Table G.9

$\dot{q}_1$	α_1	$\dot{q}_2$	α_2	$\dot{q}_3$	α_3	$\dot{q}_4$	α_4	$\dot{q}_5$	α_5	$\dot{q}_6$	α_6
44.3	31.4	37.7	26.8	26.1	18.7	13.7	9.9	26.9	19.1	21	15.5
44.3	31.4	37.6	26.8	26.1	18.5	13.7	9.9	26.9	19.1	21	15.6
50.3	28.3	60.8	33.6	34.4	19.6	22.6	12.6	34.3	18.8	23.2	12.7
50.4	28.2	60.9	33.6	34.4	19.7	22.6	12.6	34.3	18.9	23.2	12.7
53.3	25	80.5	36.1	55.2	24.2	36.2	16.2	46.7	21.3	16.3	7.3
53.3	24.9	80.5	36	55.2	24.3	36.1	16.2	46.7	21.4	16.3	7.4
53.4	24	80.6	35.6	55.3	24	36.2	16.2	46.8	21.1	16.4	7.4
50.7	18.6	101.4	36.3	82.5	28.5	54.1	19.4	64.1	23.4	17	6.2
50.6	18.6	101.3	36.3	82.6	28.5	54	19.4	64	23.3	16.9	6.1
50.7	18.8	101.4	36.5	82.6	28.6	54.1	19.5	64.1	23.5	17	6.3
53.3	16.1	117.5	35.7	96.4	29.1	72.2	21.8	86.4	25.9	22.6	6.7
53.3	15.8	117.5	35.5	96.3	29.1	72.1	21.9	86.5	26	22.5	6.7
59.6	14.6	139.3	34.1	113.1	28.1	107.6	26.2	101.2	25.1	23.9	5.8
59.6	14.6	139.1	34.1	113.1	28	107.5	26.3	101.2	25.6	23.9	5.9
59.4	14.7	138.9	34.4	112.9	28.3	107.2	26.4	101	25.8	23.7	6
0.2		0.2		0.2		0.2		0.2		0.2	
0.2		0.2		0.2		0.2		0.2		0.2	
0.2		0.2		0.2		0.2		0.2		0.2	
0.2		0.2		0.2		0.2		0.2		0.2	
3.7	12.6	4.2	14.2	3.1	10.5	1.8	6.3	2.3	8	3.9	13.1
3.3	10.9	4.2	14.1	3.1	10.3	1.8	6.2	2.4	8.1	3.9	13.3
10.7	17.8	10.1	16.7	5	8.3	4.6	7.6	6.9	11.5	10.5	17.4
10.7	17.7	10.1	16.5	5.1	8.3	4.6	7.6	6.9	11.5	10.6	17.5
10.7	17.8	10.2	16.6	5	8.3	4.6	7.6	6.9	11.5	10.6	17.6
21.4	21.6	20.7	20.8	10.6	10.8	9	9.1	14.3	14.4	20.1	20
21.5	21.9	20.8	20.9	10.7	10.9	9.1	9.2	14.3	14.6	18.8	19.2
21.5	22.1	20	20.4	10.7	10.9	9.1	9.3	14.3	14.7	18.7	19.4
39	29	32.1	23.7	21.6	15.8	13.6	10.2	23.3	17.6	23.5	18.3
38.9	28.8	32.1	23.7	21.5	15.8	13.5	10.2	23.3	17.6	23.7	18.3
39	28.5	32.1	23.7	21.5	15.8	13.6	10.2	23.3	17.3	24.9	18.6
53.9	30.8	52.3	30.2	30.3	18	22.6	13.5	30	17.7	20.9	12.1
53.6	30.2	52.2	29.7	35	19.9	22.4	13.1	29.9	17.5	20.8	12
53.9	30.1	52.4	29.8	35.8	20.3	22.6	13.1	30.1	17.3	20.9	11.8
54.8	28	72.7	35.1	49.6	23.8	36.2	17.3	48	23.2	17.3	8.9
54.8	26.6	72.8	34.5	49.7	23.5	36.2	16.9	48	22.2	17.3	8.1
54.9	26	72.8	34.2	49.8	23.3	36.2	16.8	48.1	22	17.3	8.1
55.4	21.6	100.3	37.6	80.1	29.7	54.4	20.7	67.1	25.4	13.9	5.4
55.3	21.2	100.3	37.3	80	29.4	54.3	20.4	67	25.2	13.8	5.3
55.3	20.4	100.3	36.9	80	29.5	54.3	20.5	67	24.7	13.8	5.1
57.6	19.2	116.6	37.8	98.7	31.7	71.7	23.2	84.7	27.2	15.7	5.2
57.5	19.5	116.8	38.3	98.7	31.9	71.8	23.6	84.8	28	15.7	5.3
57.5	14.6	135	34.7	116.1	30.3	107.6	27.6	100.9	25.9	21.6	5.4
57.5	15.2	135	35.5	116.3	30.7	107.7	28	100.9	26.4	21.6	5.7
57.5	16.1	134.9	36.5	116.1	30.9	107.7	28.4	100.8	27.3	21.6	5.9
57.5	15.9	135.1	36.4	116.1	31.5	107.8	28.3	100.8	27.3	21.6	5.9
0.2		0.2		0.2		0.2		0.2		0.2	
0.2		0.2		0.2		0.2		0.2		0.2	

Table G.9

$w_{oil,0}$	p	T	$\dot{m}$	$\dot{x}$	$\Delta p/\Delta L$	FP	$\dot{q}$	α
0.013	25.98	-10.63	155.2	0.73	8.1	a (f)	0.2	
0.013	25.99	-10.61	154.9	0.74	8.46	a (f)	2.4	16.1
0.013	25.98	-10.63	156.6	0.73	8.32	a (f)	2.4	15.5
0.013	25.99	-10.62	155.3	0.73	8.47	a (f)	2.6	16.6
0.013	25.99	-10.62	156.5	0.72	8.63	a (f)	6.3	16.8
0.013	25.98	-10.63	156.9	0.72	8.5	a (f)	6.2	16.7
0.013	25.99	-10.61	155.6	0.73	8.75	a (f)	6.2	16.6
0.013	25.98	-10.63	156.8	0.72	9.53	a (f)	11.9	18.3
0.013	25.98	-10.63	155.8	0.73	9.67	a (f)	11.8	18.4
0.013	25.98	-10.63	156.5	0.72	9.47	a (f)	11.8	18.3
0.013	26	-10.6	155.7	0.73	11.04	a (f)	27.7	22.1
0.013	26	-10.6	156.4	0.72	11.37	a (f)	27.7	22.1
0.013	26	-10.61	156.5	0.73	11	a (f)	27.6	22.1
0.013	26.01	-10.58	156.3	0.72	12.19	a (f)	38.3	23.1
0.013	26.01	-10.59	156.5	0.72	12.29	a (f)	38.2	22.9
0.013	26.01	-10.59	156.3	0.73	12.9	a (f)	38.2	22.3
0.013	26.03	-10.57	156.7	0.73	13.54	a (f)	47.1	22.5
0.013	26.01	-10.58	155.7	0.74	13.36	a (f)	47	21.9
0.013	26.02	-10.58	155.7	0.73	13.32	a (f)	47.1	22.1
0.013	26	-10.44	156.5	0.92	10.42	f	0.2	
0.013	26.02	-10.43	157.3	0.91	10.79	f	0.2	
0.013	26.01	-10.41	156.6	0.92	11.23	f	0.2	
0.013	26.04	-10.4	156.6	0.92	10.96	f	0.2	
0.013	26.03	-10.39	156.6	0.92	11.52	f	1.7	
0.013	26.04	-10.35	156.5	0.93	11.26	f	1.7	
0.013	26.01	-10.41	156.5	0.92	11.33	f	1.6	
0.013	26.04	-10.33	155.6	0.93	11.19	f	1.7	
0.013	26.04	-10.36	155.8	0.93	12.63	f	7.4	24.7
0.013	26.01	-10.36	155.1	0.93	12.88	f	7.5	28.7
0.013	26.04	-10.36	155.9	0.93	12.02	f	7.5	23.5
0.013	26.01	-10.39	156.2	0.93	12.61	f	7.5	24
0.013	26	-10.39	155.9	0.93	13.69	f	14	22.6
0.013	26.02	-10.41	156.3	0.92	13.92	f	14	21.6
0.013	26.04	-10.34	156.1	0.93	13.37	f	14	22.2
0.013	26.05	-10.34	156.2	0.93	17.92	f	45.5	19.2
0.013	26.04	-10.34	155.6	0.93	17.39	f	45.5	19.6
0.013	26.04	-10.39	156.8	0.92	18.24	f	45.5	18.1
0.013	26.01	-10.38	155.9	0.93	13.59	f	21.5	18.8
0.013	26.04	-10.37	157.2	0.92	14.36	f	21.5	18.9
0.013	26.03	-10.39	156.6	0.92	13.73	f	21.6	18.8
0.013	26.03	-10.37	156.3	0.93	14.74	f	21.6	19.1
0.013	26	-10.42	156.6	0.93	16.3	f	49.8	15.8
0.013	26.05	-10.36	156.2	0.92	17.3	f	50	15.9
0.013	26.05	-10.34	156.5	0.93	18.75	f	50	18.2
0.013	26.06	-10.37	156.9	0.92	17.27	f	49.9	15.6
0.013	26.04	-10.37	156.7	0.92	16.9	f	54.4	11.4
0.013	26	-10.4	155.7	0.93	18.18	f	54.4	15.5

$\dot{q}_1$	α_1	$\dot{q}_2$	α_2	$\dot{q}_3$	α_3	$\dot{q}_4$	α_4	$\dot{q}_5$	α_5	$\dot{q}_6$	α_6
0.2		0.2		0.2		0.2		0.2		0.2	
2.3	15.8	2.7	18.4	3.5	20.8	2.3	15.6	1.8	12.4	1.9	13.7
2.3	14.6	2.9	19.3	3.5	20.1	2	13	2	13.4	1.9	12.8
2.7	17.9	2.9	19.1	3.5	20.3	2.2	14.7	2	13.8	2	13.9
7.1	18.9	7.9	21.5	4.3	11.9	4.6	12.9	5.4	14.6	8.3	21.2
6.8	18.4	7.9	21.5	4.3	11.8	4.6	12.8	5.4	14.4	8.3	21.1
6.8	18.4	7.9	21.6	4.3	11.8	4.6	12.8	5.4	14.2	8.2	21.1
16.6	25.3	13.5	21	7.9	12.2	8.8	13.6	10.8	16.9	13.6	21
16.6	25.3	13.5	21	7.9	12.2	8.8	13.7	10.7	16.9	13.5	21.1
16.6	25.2	13.5	20.9	7.9	12.3	8.8	13.7	10.7	16.7	13.5	20.9
40.9	32.8	33.8	27.3	24.8	19.2	13.6	11.1	22.8	18.3	30.2	24
41	32.7	33.8	27.1	24.8	19	13.6	11.1	22.8	18.3	30.2	24.1
40.8	32.8	33.7	27.3	24.7	19.3	13.6	11.1	22.7	18.2	30.1	23.9
53.2	32.2	54.2	32.5	44.7	26.7	22.6	13.6	23.9	14.9	31.1	19
53.2	32	54.2	32.2	44.6	26.7	22.5	13.4	23.8	14.5	31.1	18.5
53.1	31.6	54.1	31.1	44.5	25.1	22.4	12.8	23.8	14.5	31	18.9
59.8	28.1	72.9	34.8	53.1	27.6	35.8	16.6	29.8	13.8	31.1	14
59.8	27.8	72.8	34	53.1	26.2	35.6	16	29.9	13.5	31.1	13.7
59.8	27.7	72.8	34.3	53	27.5	35.6	16.4	30	13.5	31.1	13.5
0.2		0.2		0.2		0.2		0.2		0.2	
0.2		0.2		0.2		0.2		0.2		0.2	
0.2		0.2		0.2		0.2		0.2		0.2	
0.2		0.2		0.2		0.2		0.2		0.2	
1.3		2.2		2.5		1.8		2.3		0.2	
1.4		2		2.5		1.6		2.3		0.2	
0.8		2		2.5		1.9		2.3		0.2	
1.1		2.2		2.7		1.8		2.3		0.2	
9.6	31.6	7.4	25.1	9.6	29.5	4.5	15.7	8.3	27.7	5.2	18.4
9.7	37.4	7.5	29.1	9.7	33.7	4.5	18	8.4	32.2	5.2	21.7
9.7	30.3	7.5	23.8	9.8	28.5	4.5	14.8	8.4	26.1	5.2	17.3
9.7	30.9	7.5	24.5	9.8	28.9	4.5	15.1	8.4	26.9	5.2	17.8
18.5	29.2	15	24.2	14.9	23.9	9.2	15.4	13.9	22.6	12.4	20.4
18.5	28.1	15	23.1	14.9	22.8	9.2	14.6	13.9	21.5	12.3	19.4
18.6	29	15	23.9	14.9	23.3	9.2	15.1	13.9	21.9	12.3	19.8
61.6	24.8	59.4	23.5	86.8	40.4	13.6	5.3	26.9	10.9	24.8	10
61.6	25.1	59.4	24.3	86.9	42	13.5	5.4	26.9	10.9	24.8	10
61.6	23.6	59.4	22.6	86.9	37.4	13.5	5.2	26.9	10.5	24.8	9.6
28.6	24.7	25.2	22	20.5	18.1	13.6	12.3	20.2	17.5	21	18.2
28.6	24.6	25.2	21.8	20.6	18	13.7	12.4	20.2	18	20.9	18.6
28.7	24.3	25.2	21.6	20.6	18	13.7	12.6	20.3	18.1	21	18.5
28.7	24.8	25.2	22	20.6	18.3	13.8	12.7	20.2	18.1	21	18.6
67.8	20.6	76.5	23.2	88.3	31.1	23.1	7.1	23.1	7.2	20.2	6
67.9	20.4	76.7	23.3	88.5	31	23.2	7.2	23.2	7.2	20.3	6.1
67.8	22.4	76.6	26.1	88.5	37.6	23.3	8.1	23.2	8	20.4	6.7
67.8	19.5	76.6	23.2	88.4	31.5	23.2	7	23.2	6.9	20.3	5.8
65.8	12.6	87.8	17.6	99.6	23.7	36.1	7.3	27.3	5.4	9.4	1.8
65.9	17.8	87.9	23.9	99.7	31.4	36.1	9.8	27.4	7.6	9.4	2.6

Table G.9

$w_{oil,0}$	p	T	$\dot{m}$	$\dot{x}$	$\Delta p/\Delta L$	FP	$\dot{q}$	α
0.013	26.04	-10.39	156.2	0.92	18.54	f	54.2	16
0.013	26.05	-10.35	155.5	0.93	17.73	f	54.1	17.2
0.013	25.85	-10.84	260.4	0.12	4.92	i	0.2	
0.013	25.86	-10.83	257.9	0.12	4.86	i	0.2	
0.013	25.86	-10.82	261.1	0.11	5.11	i	4.2	9.9
0.013	25.86	-10.82	260.2	0.12	5.09	i	4.2	10
0.013	25.84	-10.86	261.4	0.12	5.7	i	10.8	13.3
0.013	25.84	-10.85	260.9	0.12	5.7	i	10.8	13.3
0.013	25.85	-10.84	262.9	0.11	5.76	i	10.7	13.3
0.013	25.87	-10.82	258.6	0.12	6.49	i	18.7	16.9
0.013	25.86	-10.82	261.7	0.12	6.52	i	18.7	16.8
0.013	25.86	-10.82	263.1	0.12	7.22	i	26.5	19.8
0.013	25.88	-10.8	261.2	0.12	7.27	i	26.5	19.8
0.013	25.88	-10.8	260.9	0.11	7.18	i	26.4	19.7
0.013	25.88	-10.79	262.5	0.11	8.81	i (f)	40.8	24.2
0.013	25.92	-10.74	260.7	0.12	8.75	i (f)	40.8	24.3
0.013	25.88	-10.8	261	0.12	10.79	i (f)	59.6	28.2
0.013	25.89	-10.78	261.9	0.11	10.88	i (f)	59.6	28.1
0.013	25.89	-10.78	262.4	0.12	12.72	i-a (f)	76.6	30.5
0.013	25.92	-10.75	260.5	0.12	12.83	i-a (f)	76.8	30.5
0.013	25.9	-10.76	261.9	0.12	12.64	i-a (f)	76.8	30.8
0.013	25.89	-10.79	263.1	0.12	14.07	i-a (f)	91.4	31.7
0.013	25.91	-10.76	261	0.12	14.14	i-a (f)	91.4	31.5
0.013	25.91	-10.76	262.6	0.11	14.14	i-a (f)	91.4	31.5
0.013	25.9	-10.77	262.5	0.12	16.86	i-a (f)	118.9	33.2
0.013	25.92	-10.74	263	0.11	16.67	i-a (f)	119	33.3
0.013	25.93	-10.73	262.3	0.12	16.67	i-a (f)	119	32.9
0.013	26.08	-10.53	261.4	0.31	11.29	i-a	0.2	
0.013	26.06	-10.54	258.9	0.32	11.08	i-a	0.2	
0.013	26.05	-10.57	261.8	0.31	10.94	i-a	2.7	10.5
0.013	26.09	-10.51	263.2	0.31	10.95	i-a	2.8	10.6
0.013	26.11	-10.48	260.4	0.32	11.34	i-a	5.3	12
0.013	26.09	-10.51	260.6	0.31	11.22	i-a	5.4	12
0.013	26.15	-10.43	265.2	0.31	11.95	i-a (f)	17.8	17.4
0.013	26.11	-10.49	265.5	0.31	12.13	i-a (f)	17.9	17.3
0.013	26.12	-10.46	265.2	0.31	12.56	i-a (f)	25.6	20.2
0.013	26.11	-10.48	265.8	0.3	12.83	i-a (f)	25.6	20.2
0.013	26.15	-10.43	264	0.3	14.01	i-a (f)	45.5	26.1
0.013	26.12	-10.47	266	0.31	14.54	i-a (f)	44.7	26.1
0.013	26.15	-10.43	265.2	0.31	16.2	i-a (f)	65	29
0.013	26.19	-10.37	264.7	0.31	16.34	i-a (f)	64.8	29.1
0.013	26.13	-10.45	263.8	0.31	16.35	i-a (f)	65	29.1
0.013	26.12	-10.47	264.9	0.31	16.26	i-a (f)	65	29.1
0.013	26.11	-10.49	265.6	0.31	17.8	i-a (f)	78	29.9
0.013	26.17	-10.4	265.3	0.31	17.86	i-a (f)	78	29.7
0.013	26.17	-10.41	263.1	0.3	19.09	a (f)	91.8	30.6
0.013	26.17	-10.41	265.3	0.29	19.25	a (f)	91.7	30.9

$\dot{q}_1$	α_1	$\dot{q}_2$	α_2	$\dot{q}_3$	α_3	$\dot{q}_4$	α_4	$\dot{q}_5$	α_5	$\dot{q}_6$	α_6
65.7	18.1	87.7	24.2	99.4	33	36	10.2	27.4	7.7	9.1	2.5
65.6	19.5	87.5	26.1	99.2	35.8	36	10.9	27.3	8.3	9	2.7
0.2		0.2		0.2		0.2		0.2		0.2	
0.2		0.2		0.2		0.2		0.2		0.2	
6.3	14.6	6.1	14.3	2.3	5.5	2.2	5.3	2.6	6.1	5.8	13.7
6.3	14.6	6.1	14.4	2.2	5.4	2.1	5.1	2.8	6.6	5.8	13.6
14.9	18.4	13.7	16.9	8.3	10.2	4.5	5.6	9.4	11.3	13.9	17.3
14.9	18.6	13.6	16.9	8.3	10.2	4.5	5.6	9.4	11.3	13.9	17.2
14.9	18.5	13.6	16.9	8.2	10.2	4.5	5.6	9.4	11.3	13.9	17.1
27.3	24.3	22.3	20.2	15.9	14.2	9.1	8.3	15.1	13.5	22.7	20.6
27.4	24.4	22.3	20.2	15.9	14.3	9.2	8.3	15	13.4	22.7	20.4
35	26.3	33.2	24.7	22.7	16.9	13.4	10.1	20.9	15.6	33.5	24.9
34.9	26.4	33.2	24.7	22.7	16.9	13.6	10.2	20.8	15.6	33.5	25.1
35	26.2	33.2	24.6	22.7	16.9	13.4	10	20.9	15.6	33.3	24.8
55.9	33.1	48.8	29	36.3	21.7	22.6	13.5	31.6	18.9	49.4	29.3
56	33.2	48.9	29	36.2	21.6	22.6	13.5	31.6	18.9	49.4	29.3
77.5	36.5	72.3	33.8	58.4	27.7	36.3	17.3	48.2	22.9	65	30.7
77.5	36.2	72.2	33.9	58.4	27.7	36.3	17.3	48.2	22.9	64.9	30.5
92.3	36.6	88.9	35.8	82.8	33	53.9	21.6	64.7	25.8	77.2	30.5
92.6	36.5	89.1	35.6	83	33	54	21.5	64.9	25.8	77.3	30.7
92.6	37.6	89.1	36.1	83	33	54	21.6	64.8	25.7	77.3	30.5
99.8	35.3	107.1	37	106.4	36.5	71.9	24.8	76.5	26.6	86.9	30
99.8	34.4	106.9	36.7	106.3	36.6	71.9	24.8	76.5	26.7	86.8	30.1
99.8	34.5	107.1	36.6	106.2	36.4	71.9	24.8	76.5	26.7	86.9	30.1
117.8	32.7	136.6	37.8	148.6	42.2	107.2	29.9	104.4	29.2	98.8	27.3
117.9	33.3	136.7	37.9	148.6	42.3	107.1	29.8	104.6	29.1	99	27.6
117.8	32.4	136.7	37.3	148.6	42	107.1	29.6	104.6	28.8	98.9	27.3
0.2		0.2		0.2		0.2		0.2		0.2	
0.2		0.2		0.2		0.2		0.2		0.2	
3.8	13.9	3.5	13.2	1.7	6.9	2.4	9.3	1.5	6.1	3.4	13.5
3.9	14.2	3.5	12.8	1.8	7	2.4	9	1.6	6.2	3.7	14.4
7.6	16.8	6.8	15	1.4	3.4	4.4	9.7	3.6	8.2	8.3	18.7
7.6	16.8	6.9	15.1	1.5	3.6	4.4	9.8	3.5	8	8.4	18.8
25.7	25	19.8	19.1	13.6	13.3	9.2	8.9	14.7	14.4	23.9	23.4
25.8	25	19.9	19.1	13.6	13.3	9.2	8.9	14.8	14.4	23.9	23.3
37.1	29.3	26.4	20.9	21.2	16.8	13.4	10.6	21.3	16.7	34	26.6
37.2	29.3	26.4	21	21.3	16.9	13.4	10.6	21.4	16.8	34.1	26.6
67.3	38.6	50.1	28.6	42.7	24.4	22.5	13.2	37.2	21.5	53.1	30.5
67.3	39.1	47.6	27.7	42.7	24.5	22.5	13.3	34.9	20.7	53	31.1
94.3	42.7	74.1	33.3	81.6	34.2	36.2	16.5	52.4	23.9	51.4	23.3
94.1	42.6	74	33.4	81.3	34.3	36.1	16.5	52.3	24	51.2	23.7
94.3	42.6	74.1	33.3	81.5	34.4	36.1	16.5	52.4	24.1	51.3	23.8
94.2	42.7	74	33.3	81.5	34.4	36.1	16.5	52.4	24	51.3	23.6
95.5	37.9	101.3	38.7	99.1	36.3	54.3	20.9	65.9	25.6	52.2	20.2
95.6	37.4	101.3	38.6	99.1	36.2	54.1	20.9	66	25.4	52.2	19.8
97.9	33.5	119.6	40	121.5	39	72.1	24	85.2	28.7	54.6	18.6
97.9	33.8	119.5	40	121.4	39.2	72.1	24.1	85.1	29.1	54.5	18.9

Table G.9

$w_{oil,0}$	p	T	$\dot{m}$	$\dot{x}$	$\Delta p/\Delta L$	FP	$\dot{q}$	α
0.013	26.18	-10.4	268	0.31	22.1	a (f)	111.8	30.9
0.013	26.17	-10.4	268.2	0.31	22.42	a (f)	111.9	30.9
0.013	26.15	-10.42	264.3	0.31	21.85	a (f)	111.8	31.1
0.013	25.68	-11.05	257.4	0.52	18.96	a	0.2	
0.013	25.68	-11.06	256.5	0.52	18.71	a	0.2	
0.013	25.64	-11.11	253.9	0.53	19.09	a	0.2	
0.013	25.69	-11.04	257.7	0.52	19.22	a	2.6	12
0.013	25.64	-11.11	256.2	0.52	19.33	a	2.7	14.2
0.013	25.7	-11.03	257.3	0.52	19.71	a	7.4	14.5
0.013	25.69	-11.05	257.2	0.52	19.7	a	7.4	14.4
0.013	25.69	-11.04	259.8	0.51	20.01	a	11.9	16.1
0.013	25.65	-11.1	258.2	0.52	19.97	a	12	16
0.013	25.7	-11.04	263.4	0.51	19.91	a (f)	20.6	19.5
0.013	25.74	-10.98	260.7	0.52	19.61	a (f)	20.4	19.5
0.013	25.66	-11.09	264.8	0.51	20.46	a (f)	34.1	25.3
0.013	25.71	-11.02	264.8	0.51	19.83	a (f)	32.5	24.4
0.013	25.75	-10.96	266.7	0.5	24.53	a (f)	65.6	34
0.013	25.72	-11	267.8	0.5	24.47	a (f)	65.5	34.1
0.013	25.7	-11.02	266	0.5	26.56	a (f)	87.1	36.4
0.013	25.74	-10.97	264.5	0.51	26.98	a (f)	87.2	36.2
0.013	25.76	-10.95	264.4	0.51	29.3	a (f)	100.2	34.8
0.013	25.75	-10.97	264.9	0.51	28.48	a (f)	100.3	34.1
0.013	25.76	-10.95	265.7	0.51	30.59	a (f)	109.2	32.8
0.013	25.69	-11.05	264.9	0.51	30	a (f)	109.4	32.3
0.013	25.95	-10.7	260.3	0.52	21.27	a	0.2	
0.013	25.95	-10.68	263	0.51	21.41	a	0.2	
0.013	25.98	-10.65	260.2	0.52	21.15	a	0.2	
0.013	25.95	-10.69	261	0.51	20.88	a	3.5	11.5
0.013	25.96	-10.67	261.6	0.51	21.17	a	3.5	11.4
0.013	25.96	-10.68	262.9	0.51	20.9	a	7.6	14.1
0.013	25.96	-10.67	263.7	0.51	20.81	a	7.6	14
0.013	25.97	-10.66	264.5	0.51	20.93	a	11.5	15.9
0.013	25.96	-10.68	264.8	0.51	20.56	a	11.7	15.9
0.013	25.93	-10.71	268.3	0.5	20.93	a (f)	18.9	18.4
0.013	25.97	-10.66	267.3	0.5	20.46	a (f)	18.9	18.4
0.013	25.99	-10.64	261.2	0.52	21.07	a (f)	38	25.3
0.013	25.99	-10.64	261.6	0.52	20.85	a (f)	38.1	25
0.013	26.03	-10.58	260.1	0.52	25.2	a (f)	71	33.1
0.013	26.03	-10.59	261.4	0.52	25.25	a (f)	70.8	32.8
0.013	25.97	-10.66	260.7	0.53	27.5	a (f)	86.8	33.8
0.013	26.04	-10.57	259.7	0.53	27.84	a (f)	86.9	33.6
0.013	26.03	-10.58	272.9	0.5	29.9	a (f)	98.1	33.3
0.013	26.04	-10.57	273.8	0.49	29.24	a (f)	98.2	33
0.013	26.03	-10.58	258.6	0.52	30.4	a (f)	107.4	28.6
0.013	26.05	-10.56	257.8	0.53	30.16	a (f)	107.4	28.3
0.013	25.98	-10.64	259.5	0.53	30.85	a (f)	111	28.6
0.013	26.05	-10.56	260.2	0.53	30.86	a (f)	110.9	28.8

$\dot{q}_1$	α_1	$\dot{q}_2$	α_2	$\dot{q}_3$	α_3	$\dot{q}_4$	α_4	$\dot{q}_5$	α_5	$\dot{q}_6$	α_6
102	28.5	154.1	42.3	141	39.2	107.4	29.3	101.1	28.3	65.1	18
102	27.8	154.2	42	141.2	39.4	107.4	29.5	101.2	28.6	65.2	18.1
102	28.3	154.1	42.1	141.1	39.9	107.3	29.6	101.1	28.3	65.1	18.1
0.2		0.2		0.2		0.2		0.2		0.2	
0.2		0.2		0.2		0.2		0.2		0.2	
0.2		0.2		0.2		0.2		0.2		0.2	
0.2	0.9	2.7	12	4.5	20	1.4	6.4	2.6	12	4.2	20.5
0.2	1.1	2.7	13.3	4.6	23.8	1.5	7.4	2.8	14.5	4.3	25
8.4	16.3	5.2	10.2	8.1	15.8	5.2	10.2	6.8	13.4	10.9	21.1
8.2	16.2	5.2	10.2	8.1	15.6	5.2	10.2	6.7	13.3	10.8	21.1
13.8	18.6	9.6	12.9	9.7	13.7	9.8	13.2	12	15.9	16.6	22.1
13.7	18.3	9.6	12.9	9.7	13.7	9.8	13.1	12.2	16.1	16.7	22.2
23.5	22.5	15.2	14.5	23.5	21.9	13.8	13.2	19.5	18.6	28.2	26.5
23.7	22.9	15.4	14.8	22	20.8	14	13.4	19.7	18.7	27.7	26.2
43.4	32	29.5	21.9	34.5	25.8	22.6	17	31.3	23.3	43.2	31.9
41	30.8	28.1	21.1	33.3	24.7	22.6	16.9	30.3	22.7	39.8	30.1
88.1	45.8	58.4	30.8	77.4	38.3	36.6	19.4	55	29	77.9	40.6
88	46	58.4	30.8	77.3	38.4	36.6	19.4	54.9	29.1	77.9	40.7
115	48	90	37.1	114.9	49.2	55.1	22.8	67.6	28.2	79.9	33.4
115.5	47.4	89.9	37.1	115.4	49.8	55.1	22.7	67.7	27.7	79.9	32.6
111.7	39.3	120	40.6	144.9	51.8	73	25	80.3	27.5	71	24.2
111.9	38.6	120.1	40.5	145.1	51.1	73	24.8	80.4	26.7	71	23.2
107.7	33.9	141.7	42.2	153.3	46.4	106	30.7	84.3	25.4	61.9	18.4
108	33	141.9	41.8	153.5	46	106.2	30.6	84.7	24.7	62	17.5
0.2		0.2		0.2		0.2		0.2		0.2	
0.2		0.2		0.2		0.2		0.2		0.2	
0.2		0.2		0.2		0.2		0.2		0.2	
2.4	7.9	4.3	14.4	3.5	11.1	2.1	6.9	4.1	13.8	4.4	14.8
2.6	8.2	4.3	14.2	3.5	11	2	6.7	4.1	13.5	4.4	14.6
8.7	16	7.7	14.3	6.1	11.1	5.1	9.6	8.7	16.1	9.2	17.6
8.7	15.8	7.7	14.3	6.1	11.1	5.1	9.5	8.7	15.9	9.2	17.3
12.9	17.8	11.4	15.7	8.8	12	8.3	11.5	12.3	17	15.3	21.2
13.1	17.9	11.6	15.7	9	12.1	8.5	11.5	12.5	16.9	15.3	20.9
23.2	22.6	17	16.7	15.5	15.1	14.2	13.8	19	18.6	24.3	23.8
23.2	22.5	17	16.7	15.5	15	14.2	13.7	19	18.6	24.4	23.9
50.7	33.9	38.2	25.4	34.5	22.4	22.9	15.4	36.8	24.7	45.2	30.3
50.8	33.4	38.2	25	34.6	21.9	22.9	15.1	37	24.5	45	29.9
96.8	45.2	73.5	34.3	79.5	37	41.4	19.4	58.7	27.2	76.3	35.4
96.7	44.9	73.2	34	79.3	37	41.3	18.8	58.5	26.8	76.1	35
116.3	44.6	96.5	37.1	112.9	46.3	54.7	21.1	66.7	25.7	73.9	28.2
116.3	44	96.5	37	113	46.1	54.8	21	66.7	25.5	74	27.7
120.9	41.2	117.9	39.4	133.2	45.9	72.3	24.2	79.9	27	64.3	22
121.1	40.9	118.1	39.4	133.3	45.6	72.3	24.1	80	26.6	64.3	21.5
118.5	30.9	144.4	38.6	144.4	39.8	107.5	28.4	91.2	24	38.1	9.9
118.5	30.5	144.4	38.4	144.5	38.8	107.6	28.3	91.4	23.8	38.2	9.7
116.2	29.8	153.4	39.2	155.4	40.8	107.5	27.5	101.4	26.1	32.1	8.2
116.1	30.3	153.3	39.5	155.2	41	107.4	27.6	101.3	26.2	32.1	8.2

Table G.9

$w_{oil,0}$	p	T	$\dot{m}$	$\dot{x}$	$\Delta p/\Delta L$	FP	$\dot{q}$	α
0.013	26	-10.62	262.8	0.52		a	0.2	
0.013	26	-10.62	262.1	0.52		a	0.2	
0.013	26.01	-10.61	263.9	0.51		a	0.2	
0.013	25.99	-10.64	263.5	0.51	16.67	a	2.1	11.3
0.013	26	-10.63	262.7	0.51	16.64	a	2.2	12.1
0.013	26	-10.63	264.4	0.51	16.94	a	2.2	12.1
0.013	25.99	-10.64	264.5	0.5	17.33	a	6.7	14.9
0.013	26.01	-10.62	263.4	0.51	17.11	a	6.7	14.8
0.013	26.02	-10.59	263.1	0.52	17.26	a	12.4	17.1
0.013	25.95	-10.68	262.5	0.52	17.82	a	12.4	17
0.013	26	-10.62	263.3	0.51	17.53	a	12.4	17.1
0.013	26.01	-10.61	264	0.51	17.91	a (f)	18.4	19.2
0.013	26.02	-10.6	264	0.5	18.07	a (f)	18.4	19.2
0.013	26.03	-10.59	265.8	0.51	19.73	a (f)	31.7	23.8
0.013	25.98	-10.65	263.2	0.52	19.78	a (f)	31.7	23.7
0.013	26.04	-10.58	265.4	0.5	22.52	a (f)	56	30.5
0.013	26.04	-10.57	265.5	0.51	22.81	a (f)	55.7	30.5
0.013	26.05	-10.56	265.7	0.51	22.87	a (f)	55.7	30.5
0.013	26.06	-10.54	265.2	0.51	26.78	a (f)	88.4	33.1
0.013	26.07	-10.54	264.3	0.51	26.76	a (f)	88.3	33.6
0.013	26.08	-10.52	264.8	0.51	26.71	a (f)	88.3	33.9
0.013	26.07	-10.52	264.9	0.51	28.32	a (f)	97.5	30
0.013	26.07	-10.53	264.7	0.51	28.2	a (f)	97.4	29.4
0.013	26.01	-10.61	266.2	0.5	28.15	a (f)	99.1	31.2
0.013	26.07	-10.53	265.3	0.51	30.59	a (f)	117.3	31.1
0.013	26.07	-10.53	264.9	0.51	30.45	a (f)	117.1	30.7
0.013	26.07	-10.53	265.8	0.5	30.53	a (f)	117	30.6
0.013	25.98	-10.65	260.7	0.51		a	0.2	
0.013	25.98	-10.65	258	0.52		a	0.2	
0.013	25.98	-10.64	261.6	0.51		a	0.2	
0.013	26	-10.63	261.3	0.51	18.16	a	2.9	12.4
0.013	25.99	-10.63	260.5	0.52	18.57	a	2.8	12.4
0.013	25.99	-10.64	260.5	0.51	18.88	a	7.7	14.3
0.013	25.97	-10.66	260.9	0.52	18.92	a	7.8	14.3
0.013	25.98	-10.65	260.8	0.51	18.85	a	7.8	14.2
0.013	25.97	-10.66	260.3	0.51	19.39	a	14	16.2
0.013	25.99	-10.63	261.6	0.51	19.17	a	14.1	16.3
0.013	26	-10.62	262.3	0.51	19.68	a (f)	21.9	19
0.013	25.99	-10.64	263	0.51	19.87	a (f)	21.9	19
0.013	26.05	-10.55	262.4	0.51	19.75	a (f)	21.8	19
0.013	26	-10.63	261.6	0.52	22	a (f)	38.6	24.9
0.013	26	-10.63	262.7	0.52	21.79	a (f)	38.8	24.9
0.013	26.01	-10.61	262.3	0.51	24.13	a (f)	59.7	30.6
0.013	26.02	-10.6	262.1	0.51	23.85	a (f)	59.7	30.7
0.013	26	-10.62	262.2	0.51	24.13	a (f)	59.8	30.7
0.013	26.03	-10.58	262.8	0.51	27.7	a (f)	90.9	34.6
0.013	26.02	-10.6	262.6	0.52	27.65	a (f)	90.9	34.7

Table G.9

$\dot{q}_1$	α_1	$\dot{q}_2$	α_2	$\dot{q}_3$	α_3	$\dot{q}_4$	α_4	$\dot{q}_5$	α_5	$\dot{q}_6$	α_6
0.2		0.2		0.2		0.2		0.2		0.2	
0.2		0.2		0.2		0.2		0.2		0.2	
0.2		0.2		0.2		0.2		0.2		0.2	
0.2	1	3	17.3	2.6	13.8	1.8	9.8	2.1	11.2	2.9	14.9
0.2	1.1	3	18.3	2.7	14.6	1.9	10.2	2.4	13.3	2.8	15
0.2	1.1	3	17.9	2.7	14.6	1.8	9.7	2.6	14.4	2.8	15
6.5	14.9	8	17.6	5.4	12.3	4.5	10.2	8	17.4	7.7	17.1
6.5	14.5	8	17.6	5.4	12.3	4.5	10.2	8	17.4	7.7	16.9
12.4	17.8	14	19	11.7	15.4	9.3	12.7	13.7	18.9	13.1	18.7
12.5	17.8	13.9	18.7	11.6	15.2	9.5	12.7	13.8	18.8	13.3	18.7
12.5	17.9	13.9	18.7	11.6	15.3	9.4	12.8	13.8	19	13.3	18.8
21.3	22.3	20.3	20.9	13.8	14.8	13.7	14.4	20.3	20.9	20.8	21.7
21.4	22.4	20.3	20.8	13.9	14.9	13.7	14.4	20.3	20.9	20.8	21.7
39.9	29.9	32.9	24.8	27.3	20.4	22.3	16.8	31.8	23.7	36.2	27.1
39.9	29.8	32.9	24.7	27.3	20.3	22.4	16.8	31.8	23.6	36.2	27
71.7	39.4	63.2	34.1	50.6	27.2	36.5	20	50.3	27.5	63.9	34.8
71.8	39.5	61.5	33.5	50.6	27.4	36	19.9	50.2	27.7	63.9	35
71.8	39.5	61.5	33.4	50.6	27.3	36	19.9	50.3	27.7	63.9	35
116.8	43.7	96	35.8	130.2	48.3	54	20.2	73.8	27.7	59.3	22.7
116.6	44.7	95.9	36.3	130.2	48.7	54	20.4	73.8	28.3	59.2	23.4
116.8	45.3	96	36.5	130.3	49	54	20.6	73.8	28.5	59.2	23.7
115.4	35.7	120.2	37.2	148.4	44.8	72.1	22.6	82.7	25.7	46.4	14
115.3	34.6	119.9	36.7	148.2	44.1	72.1	22.3	82.5	25.1	46.4	13.6
117.9	37.2	120.8	38.2	147.8	46.2	71.8	23	82.3	26.1	54	16.6
120.5	31.9	154.8	40.8	165.7	44.6	107.6	28.7	98.9	26.5	56	14.3
120.3	31.1	154.6	40.3	165.6	44.2	107.4	28.5	98.7	26.2	55.8	13.8
120.2	31.2	154.5	40.3	165.5	44.2	107.4	28.4	98.7	26	55.9	13.6
0.2		0.2		0.2		0.2		0.2		0.2	
0.2		0.2		0.2		0.2		0.2		0.2	
0.2		0.2		0.2		0.2		0.2		0.2	
1.4	6	3.7	15.4	4	16	1.6	6.8	2.9	12.8	3.9	17.1
0.9	3.8	3.7	15.9	3.9	16.4	1.7	7.4	3	13.4	3.9	17.5
9.4	17.3	5.9	11.2	7.1	13.2	4.5	8.5	8.7	16.1	10.7	19.7
9.4	17.1	6	11.3	7.4	13.4	4.5	8.5	8.7	15.8	10.9	19.8
9.4	16.9	6	11.3	7.4	13.2	4.6	8.4	8.6	15.7	10.9	19.6
17.2	20	12.2	14.2	11.1	12.8	9.1	10.6	14.8	17.1	19.6	22.6
17.3	20.1	12.2	14.3	11.2	12.8	9.2	10.6	14.9	17.2	19.7	22.7
29.7	25.7	20.1	17.5	16.9	14.7	13.7	12	22	19.2	28.7	25.1
29.7	25.5	20.1	17.4	16.9	14.7	13.7	12	22.1	19.2	28.7	25
29.7	25.5	20	17.4	16.9	14.6	13.7	12	22	19.2	28.7	25.1
52.8	34.2	39	25.1	32.5	20.7	22.5	14.7	36.2	23.4	48.5	31.4
53	34.1	39.1	25	32.7	20.6	22.7	14.6	36.4	23.5	48.8	31.4
85	43.5	61.3	31.6	52.7	26.8	36.1	18.5	52.2	26.9	70.7	36.4
85	43.6	61.4	31.7	52.8	26.9	36.2	18.6	52.1	26.9	70.7	36.6
85	43.5	61.5	31.6	52.8	26.8	36.4	18.6	52.3	27	70.7	36.5
123.5	46.7	96.9	36.9	118.3	45.9	54	20.4	72.4	27.4	80.3	30.4
123.5	47	96.9	37	118.3	46	54.1	20.5	72.4	27.4	80.3	30.4

Table G.9

$w_{oil,0}$	p	T	$\dot{m}$	$\dot{x}$	$\Delta p/\Delta L$	FP	$\dot{q}$	α
0.013	26.03	-10.59	263.2	0.51	27.86	a (f)	90.7	34.8
0.013	25.97	-10.66	262.9	0.51	29.18	a (f)	102.2	33.4
0.013	25.93	-10.72	263	0.51	29.51	a (f)	102.2	33.8
0.013	25.99	-10.64	262.1	0.51	31.8	a (f)	116.7	31.2
0.013	25.93	-10.72	260.8	0.51	31.32	a (f)	116.6	31.2
0.013	25.98	-10.65	263.1	0.51	31.64	a (f)	116.5	31.6
0.013	25.94	-10.68	264.2	0.7	23.62	a	0.2	
0.013	25.95	-10.67	265.2	0.7	23.27	a	0.2	
0.013	25.96	-10.66	264.5	0.71	23.31	a	0.2	
0.013	25.96	-10.66	265.5	0.7	23.68	a	2.2	14.8
0.013	25.95	-10.67	264.7	0.7	23.48	a	2.3	14.6
0.013	25.96	-10.66	265.6	0.7	24	a	5.1	15.8
0.013	26.03	-10.56	263.6	0.71	23.8	a	5	15.8
0.013	25.96	-10.66	264.9	0.7	23.88	a	10	17.4
0.013	25.91	-10.72	263.4	0.71	24.44	a	10	17.3
0.013	25.96	-10.66	265.8	0.7	24.34	a	10	17.2
0.013	25.96	-10.66	266.1	0.7	25	a (f)	16.6	19.6
0.013	25.97	-10.65	267.5	0.69	25.1	a (f)	16.5	19.3
0.013	26.02	-10.58	265	0.7	27.39	a (f)	31.2	24.7
0.013	26.01	-10.59	262.6	0.71	27.53	a (f)	31.1	24.6
0.013	26	-10.6	263.9	0.7	26.36	a (f)	27	23.4
0.013	26.01	-10.58	260.2	0.72	35.49	a (f)	87.8	36.9
0.013	26.03	-10.56	260.9	0.72	35.71	a (f)	87.8	37.1
0.013	26.02	-10.58	261.4	0.71	35.79	a (f)	87.7	37.2
0.013	26	-10.6	260.4	0.72	36.93	f	96.1	31.8
0.013	26.01	-10.59	260.2	0.72	36.75	f	96.1	31.7
0.013	26.03	-10.56	260.6	0.72	36.79	f	96.1	32.2
0.013	26.02	-10.57	261	0.73	37.24	f	101.2	28.1
0.013	26.04	-10.55	261.3	0.72	37.61	f	101.1	29.1
0.013	26.02	-10.57	259.5	0.73	37.88	f	101.2	28.7
0.013	26.03	-10.57	259.7	0.72	38.44	f	109	26
0.013	26.01	-10.59	259.2	0.72	37.8	f	109	25.9
0.013	26.01	-10.48	264	0.9	26.94	a (f)	0.2	
0.013	26.01	-10.48	264.7	0.9	26.19	a (f)	0.2	
0.013	26	-10.49	263.6	0.9	26.53	a (f)	0.2	
0.013	26.01	-10.46	263.2	0.91	26.62	a (f)	0.5	
0.013	25.99	-10.51	265.6	0.9	26.44	a (f)	0.6	9.4
0.013	26.02	-10.48	265.4	0.89	26.47	a (f)	0.6	
0.013	26.03	-10.45	263.5	0.9	27.88	a (f)	4.3	17.4
0.013	26.03	-10.43	262.8	0.91	27.42	a (f)	4.4	18.3
0.013	26.01	-10.47	263	0.91	27.79	a (f)	4.4	17.5
0.013	26.04	-10.42	262.2	0.91	29.25	a (f)	11.2	19.6
0.013	26.04	-10.45	264.1	0.9	28.55	a (f)	11.1	18.7
0.013	26.03	-10.43	260.8	0.91	29.73	a (f)	11.1	19.7
0.013	26.03	-10.42	261	0.92	30.35	a (f)	15.9	20.1
0.013	26.03	-10.44	263	0.9	30.44	a (f)	15.8	19.5
0.013	26.04	-10.44	263.1	0.91	30.42	a (f)	15.9	19.6

Table G.9

$\dot{q}_1$	α_1	$\dot{q}_2$	α_2	$\dot{q}_3$	α_3	$\dot{q}_4$	α_4	$\dot{q}_5$	α_5	$\dot{q}_6$	α_6
123.2	47	96.5	36.9	118.1	46.2	53.9	20.5	72.3	27.5	80.2	30.7
125.8	40.9	123.3	39.8	141.1	47.3	71.7	23.4	81	26.4	70.2	22.4
126	41.8	123.4	40.1	141.1	47.9	71.6	23.6	81	26.5	70.2	22.7
122.4	33.3	153.4	40.3	167.6	44.6	107.4	28.5	104.2	28	45.3	12.4
122.1	33.8	153.4	40.7	167.3	44.4	107.3	28.5	104.4	27.8	45.3	12.2
121.9	34.1	153.3	40.8	167.2	44.8	107.2	28.7	104.3	28.3	45.3	12.7
0.2		0.2		0.2		0.2		0.2		0.2	
0.2		0.2		0.2		0.2		0.2		0.2	
0.2		0.2		0.2		0.2		0.2		0.2	
0.2	1.4	3.5	21.8	3.4	24.6	1.9	13.5	2.8	16.9	1.5	10.5
0.2	1.4	3.6	21.6	3.4	23.8	2	13.7	2.8	16.7	1.5	10.4
5.2	16.5	4.6	14.3	6.6	20.3	4.8	14.9	4.5	14	4.7	14.9
5.2	16.5	4.5	14.3	6.5	20.2	4.7	14.9	4.5	14.2	4.7	15
11	19	9.3	16	10.2	18.2	9	15.9	9.9	17	10.7	18.4
11	18.9	9.3	15.9	10.3	18	9	15.7	9.9	16.8	10.6	18.1
11	18.9	9.3	15.9	10.2	18.2	9	15.6	9.9	16.7	10.7	18.1
17.5	21.1	15.1	17.8	17.3	20.7	13.8	16.6	16.2	19	19.5	22.8
17.5	20.7	15.1	17.5	17.3	20.4	13.8	16.3	16.2	18.7	19.4	22.4
39	30.6	28.3	22.2	28.5	22.9	24.1	19.4	30	23.7	37.1	29.2
39	30.5	28.2	22.2	28.5	22.8	24	19.3	30	23.7	37.1	29.2
32.5	28	23.6	20.6	23.6	21.2	22.6	19.6	26.9	22.9	32.7	27.9
121.3	45.9	84.3	32.2	136.9	72.6	36.1	13.9	75.3	28.6	72.9	28.2
121.3	46.2	84.3	32.3	136.9	72.6	36.1	14	75.3	29	72.9	28.6
121.3	46.1	84.2	32.3	136.9	73.3	36.1	14.1	75.2	28.8	72.8	28.4
129.5	39.8	111.2	34	146.1	58.1	54	16.6	80.5	24.9	55.4	17.4
129.4	39.6	111.2	33.9	146.3	58.3	53.9	16.6	80.5	24.8	55.4	17.3
129.4	39.8	111.1	34.1	146.3	60.5	54	16.8	80.5	24.7	55.4	17.2
129.5	33.1	133	34.2	150.1	51.3	72	18.6	82.5	21.2	40.3	10.3
129.4	34.6	133	35.3	149.9	51.5	71.9	19.5	82.5	22.7	40.2	11.1
129.4	33.7	133	35	149.9	50.8	72	19.4	82.5	22.3	40.3	10.8
127.9	28.8	149.7	33.9	159	43.8	99.6	22.6	93.3	21.3	24.6	5.6
127.9	29	149.7	34	158.9	42.6	99.6	22.5	93.3	21.6	24.5	5.7
0.2		0.2		0.2		0.2		0.2		0.2	
0.2		0.2		0.2		0.2		0.2		0.2	
0.2		0.2		0.2		0.2		0.2		0.2	
0.2		0.2		0.9		1.7		0.2		0.2	
0.2	2	0.2	3.5	1.1	16.8	1.8	29.2	0.2	2.4	0.2	2.7
0.2		0.2		1.1		1.7		0.2		0.2	
2.4	9.3	5	21.1	4.9	21.8	4.6	17.8	4.5	16.5	4.6	18
2.4	9.7	5.1	22.4	5	22.7	4.6	18.6	4.6	17.3	4.7	18.9
2.3	9.1	5.2	21.3	5	20.2	4.6	18.8	4.6	16.9	4.7	18.9
11.6	20	11.2	19.8	13.8	24.3	9.3	16.5	10.8	18.5	10.3	18.3
11.5	19.1	11.1	19.1	13.7	23.4	9.3	15.8	10.7	17.8	10	17.2
11.5	20.1	11.1	19.9	13.7	24.6	9.3	16.7	10.7	18.8	10	18.2
16.7	21.3	15.9	20.5	18.3	23.5	13.5	17	17.5	21.1	13.2	17.2
16.7	20.7	15.9	20	18.3	22.8	13.3	16.3	17.5	20.6	13.2	16.8
16.7	20.7	16	20	18.3	23.1	13.3	16.4	17.6	20.7	13.3	16.8

Table G.9

$w_{oil,0}$	p	T	$\dot{m}$	$\dot{x}$	$\Delta p/\Delta L$	FP	$\dot{q}$	α
0.013	26.02	-10.45	262.3	0.91	34.18	f	26	22
0.013	26.04	-10.44	262.9	0.9	33.98	f	26.1	21.7
0.013	26.03	-10.43	261.6	0.91	33.83	f	26.1	22.1
0.013	26.05	-10.36	258.7	0.92	39.29	f	42.4	26.7
0.013	26.06	-10.33	257.5	0.93	38.51	f	42.3	25.9
0.013	26.06	-10.34	258.5	0.93	39.41	f	42.4	26.8
0.013	25.99	-10.41	259.3	0.93	40.72	f	66.1	25.9
0.013	26.07	-10.31	259.3	0.93	42.18	f	66	28
0.013	26.06	-10.36	260.5	0.92	40.93	f	66	25.8
0.013	25.88	-10.8	526.4	0.11	18.61	i-a	0.2	
0.013	25.9	-10.77	529.7	0.1	18.57	i-a	0.2	
0.013	25.92	-10.74	528.8	0.1	18.52	i-a	0.2	
0.013	25.92	-10.74	527.4	0.11	18.28	i-a	0.2	
0.013	25.91	-10.75	526.8	0.11	18.31	i-a	0.2	
0.013	25.9	-10.76	526.4	0.11	18.5	i-a	2.3	10.6
0.013	25.91	-10.76	529.1	0.1	18.54	i-a	2.3	10.6
0.013	25.92	-10.75	528.5	0.11	18.69	i-a	5.4	11.5
0.013	25.93	-10.73	527	0.1	18.83	i-a	5.5	11.4
0.013	25.92	-10.75	526.9	0.1	19.95	i-a	11.7	14.6
0.013	25.93	-10.72	529.4	0.1	19.93	i-a	11.7	14.7
0.013	25.89	-10.78	529.2	0.1	21.3	i-a	19.8	18.3
0.013	25.95	-10.7	526.8	0.11	21.25	i-a	19.8	18.3
0.013	25.94	-10.72	526.5	0.11	21.17	i-a	19.8	18.3
0.013	25.94	-10.71	525.6	0.11	23.44	i-a (f)	33.9	23.9
0.013	25.96	-10.69	526.2	0.11	23.31	i-a (f)	33.9	23.8
0.013	25.92	-10.75	526.6	0.11	27.62	i-a (f)	56.6	30.7
0.013	25.95	-10.71	529.6	0.11	27.56	i-a (f)	56.7	30.7
0.013	25.96	-10.68	526.4	0.11	27.28	i-a (f)	56.8	30.7
0.013	25.96	-10.69	528.9	0.1	32.42	i-a (f)	85.6	37.3
0.013	25.99	-10.65	524.9	0.11	32.27	i-a (f)	85.7	37.3
0.013	25.94	-10.71	531.7	0.11	36.26	i-a (f)	107.4	40.7
0.013	25.98	-10.66	533.4	0.1	36.38	i-a (f)	107.3	40.7
0.013	26	-10.64	530.9	0.11	43	i-a (f)	147.9	44.6
0.013	26.02	-10.61	528.1	0.11	42.86	i-a (f)	147.8	44.6
0.013	25.99	-10.64	531.9	0.1	42.8	i-a (f)	147.7	44.6
0.013	25.97	-10.67	524.9	0.3	37.04	a	0.2	
0.013	26	-10.63	526.3	0.3	37.42	a	0.2	
0.013	26	-10.64	527.7	0.3	37.23	a (f)	1.5	12.8
0.013	26.02	-10.61	525	0.31	36.61	a (f)	1.4	12.5
0.013	25.98	-10.66	527.8	0.3	37.27	a (f)	1.4	12.7
0.013	25.98	-10.66	524.3	0.31	37.52	a (f)	4	12.3
0.013	25.97	-10.67	526.4	0.31	37.15	a (f)	4	12.5
0.013	25.99	-10.65	524.1	0.31	37.93	a (f)	4	12.5
0.013	26.02	-10.61	526.8	0.3	38.16	a (f)	7.6	13.5
0.013	26.01	-10.62	526.8	0.31	37.95	a (f)	7.6	13.6
0.013	25.97	-10.68	525	0.3	39.79	a (f)	13.2	15.7
0.013	26	-10.63	525.2	0.31	39.77	a (f)	13.2	15.7

$\dot{q}_1$	α_1	$\dot{q}_2$	α_2	$\dot{q}_3$	α_3	$\dot{q}_4$	α_4	$\dot{q}_5$	α_5	$\dot{q}_6$	α_6
27.1	23	22.9	19.5	33	28.1	22.5	19.2	27.6	22.6	23.2	19.7
27.2	22.6	22.9	19.2	33	27.9	22.5	19	27.7	22.3	23.3	19.4
27.2	23.2	22.9	19.7	33	28.2	22.5	19.3	27.6	22.7	23.2	19.9
43.4	26.3	33.6	20.9	53.3	35.9	36.1	23.2	42.3	26.6	45.6	27.4
43.4	25.2	33.5	20.1	53.2	34.7	36	22.5	42.3	26	45.6	26.8
43.4	26.4	33.5	20.8	53.3	35	36.1	23.4	42.3	27.1	45.6	28.1
84.7	30.6	57.5	20.5	72.1	41.9	46.7	15.7	64.1	21.8	71.3	24.8
84.6	33	57.5	22.5	72.1	45.2	46.6	16.8	64.1	23.3	71.3	26.9
84.6	30.4	57.5	20.6	72	41.9	46.6	15.3	64	21.6	71.3	24.8
0.2		0.2		0.2		0.2		0.2		0.2	
0.2		0.2		0.2		0.2		0.2		0.2	
0.2		0.2		0.2		0.2		0.2		0.2	
0.2		0.2		0.2		0.2		0.2		0.2	
0.2		0.2		0.2		0.2		0.2		0.2	
2.1	9.4	2.4	11.2	1.4	7.6	2.2	10.9	1.6	7.1	3.9	17.4
2.2	9.7	2.6	11.6	1.5	7.5	2.2	10.6	1.6	7	4	17.1
7.2	15.1	5.2	10.8	3.6	7.9	4.6	9.9	4.2	8.8	7.6	16.4
7.3	15.1	5.2	10.8	3.7	7.9	4.6	9.8	4.3	8.8	7.7	16.3
16.1	20.1	10.4	13.1	8.6	11	9	11.2	9.2	11.2	16.8	21.1
16.1	20.1	10.4	13.1	8.6	11	9.1	11.3	9.2	11.3	17.1	21.2
27.9	25.7	18.3	17	16.3	15.3	14	13	14.9	13.8	27.3	25.3
27.9	25.7	18.3	16.9	16.3	15.2	14	13	14.9	13.8	27.3	25.3
27.9	25.7	18.3	16.9	16.2	15.2	14	12.9	14.9	13.7	27.3	25.2
46.4	32.7	33.5	23.4	29.6	21	22.6	15.9	25.3	17.9	46.3	32.3
46.4	32.6	33.5	23.4	29.5	21	22.6	15.9	25.3	17.9	46.2	32.2
81	43.8	53.7	29.3	53.8	29.1	35.5	19.4	44.2	24.1	71.3	38.6
81.1	43.8	53.8	29.3	53.9	29	35.8	19.5	44.2	24	71.4	38.6
81.1	43.8	53.8	29.2	53.8	28.9	36.5	19.7	44.2	24	71.3	38.6
124.9	54.2	85.4	37.1	81.8	35.5	53.8	23.6	66.8	29.3	101	44.3
125	54.2	85.4	37.1	81.8	35.5	53.9	23.6	66.9	29.3	101.1	44.2
154.9	58.3	111.4	42.2	103.3	39.4	72	27.5	80.5	30.9	122	45.9
154.9	58.3	111.4	42.2	103.1	39.3	72	27.5	80.2	30.9	121.9	45.9
176.9	55.4	166	49.8	161.2	47.6	107.6	32.2	114.6	34.5	160.8	48.2
176.9	55.3	165.9	49.8	161	47.8	107.5	32.3	114.6	34.5	160.9	47.8
176.8	55.2	165.9	49.9	161	48	107.4	32.4	114.5	34.5	160.7	47.9
0.2		0.2		0.2		0.2		0.2		0.2	
0.2		0.2		0.2		0.2		0.2		0.2	
0.2	1.3	2.3	22.3	1.5	13.7	1.8	16.6	1.1	9.5	1.8	13.6
0.2	1.3	2.3	22.4	1.5	13.4	1.6	14.8	1.2	10.4	1.7	13
0.2	1.3	2.3	23.2	1.5	13.9	1.8	15.8	1.3	11.9	1.3	10.4
2.2	6.5	5.3	17.1	2.6	8.5	4.8	14.8	4.1	12.5	4.9	14.6
2.1	6.4	5.3	17.3	2.7	8.8	4.8	14.9	4.1	12.6	4.9	14.8
2.3	7	5.3	17.3	2.5	8.3	4.8	15	4.1	12.6	4.9	14.8
6	10.6	9.6	17	4.5	8.3	9	15.8	8.5	14.8	7.9	14.3
6	10.7	9.6	17.1	4.6	8.4	9.1	15.9	8.5	14.9	7.9	14.3
12.3	14.6	14.1	16.8	9.8	11.8	13.6	16.1	14.7	17.3	14.9	17.8
12.3	14.6	14	16.7	9.8	11.8	13.6	16.1	14.7	17.3	14.7	17.7

Table G.9

$w_{oil,0}$	p	T	$\dot{m}$	$\dot{x}$	$\Delta p/\Delta L$	FP	$\dot{q}$	α
0.013	26.02	-10.6	529.4	0.29	39.31	a (f)	25	20.7
0.013	26.02	-10.6	526.4	0.3	40.05	a (f)	25	20.6
0.013	26.03	-10.59	528.6	0.3	40.76	a (f)	25.1	20.5
0.013	25.98	-10.65	525	0.31	43.34	a (f)	46.8	27.7
0.013	26.04	-10.58	527	0.3	43.15	a (f)	46.9	27.7
0.013	26.02	-10.6	527.6	0.3	48.14	a (f)	76	35.4
0.013	26.08	-10.52	528.3	0.3	47.28	a (f)	76.1	35.6
0.013	26.04	-10.58	533.7	0.28	53.72	a (f)	109.1	41.5
0.013	26.09	-10.51	531.9	0.3	54.38	a (f)	108.7	41.7
0.013	26.08	-10.52	531.3	0.3	54.64	a (f)	108.7	41.8
0.013	26.1	-10.49	533.1	0.3	62.44	a (f)	151.7	46
0.013	26.1	-10.5	532.4	0.29	63.07	a (f)	152.4	46.4
0.013	25.76	-10.95	528	0.5	60.04	a	0.2	
0.013	25.74	-10.97	526.1	0.5	59.82	a	0.2	
0.013	25.75	-10.96	526	0.51	59.73	a	1.2	11.3
0.013	25.76	-10.95	529.7	0.5	60.61	a	1.3	11.2
0.013	25.71	-11.01	528.8	0.5	60.69	a	4	12.7
0.013	25.76	-10.95	528.8	0.5	60.84	a	4.3	12.7
0.013	25.71	-11.01	515.9	0.51	60.01	a	8.5	14
0.013	25.73	-10.98	516.9	0.51	60.61	a	8.5	14.1
0.013	25.72	-11	516.1	0.51	61.86	a	14.1	16.1
0.013	25.76	-10.95	522.5	0.5	60.66	a	12.4	15.6
0.013	25.72	-11	499.1	0.53	61.4	a (f)	23.5	19.9
0.013	25.78	-10.92	520.7	0.51	59.81	a (f)	23.3	19.6
0.013	25.74	-10.98	515.2	0.5	62.99	a (f)	46	27.4
0.013	25.79	-10.9	514.2	0.52	63.57	a (f)	46.1	27.2
0.013	25.82	-10.87	511.6	0.52	68.79	a (f)	70.5	33.3
0.013	25.81	-10.88	519.7	0.51	67.66	a (f)	70.6	34.2
0.013	25.81	-10.88	527.8	0.5	73.64	a (f)	103.1	41.5
0.013	25.84	-10.85	530	0.49	72.95	a (f)	99.8	41.4
0.013	25.84	-10.84	529.4	0.5	73.08	a (f)	99.9	41.2
0.013	25.85	-10.83	531.1	0.5	81.79	a (f)	148.8	49.2
0.013	25.77	-10.93	529.5	0.5	83.39	a (f)	148.8	49.4
0.013	26.16	-10.41	520.1	0.51		a	0.2	
0.013	26.13	-10.45	524.6	0.5		a	0.2	
0.013	26.12	-10.46	523.9	0.49	58.49	a	1.5	11.1
0.013	26.16	-10.41	520.5	0.5	58.95	a	1.6	11.6
0.013	26.11	-10.47	517.2	0.5	57.28	a	2.8	12
0.013	26.15	-10.42	517.5	0.51	57.21	a	2.7	11.8
0.013	26.16	-10.41	519.7	0.5	56.99	a	2.7	11.8
0.013	26.15	-10.43	514.2	0.52	59.48	a	7.6	13.8
0.013	26.17	-10.4	513.2	0.52	58.42	a	7.5	13.7
0.013	26.16	-10.41	517.4	0.51	59.12	a	7.6	13.8
0.013	26.13	-10.45	522.6	0.5	58.94	a	11.5	15.1
0.013	26.17	-10.4	523.4	0.5	58.54	a	11.5	15
0.013	26.19	-10.37	526	0.5	60.4	a (f)	24.9	19.6
0.013	26.18	-10.38	527.6	0.5	62.7	a (f)	24.8	19.6

$\dot{q}_1$	α_1	$\dot{q}_2$	α_2	$\dot{q}_3$	α_3	$\dot{q}_4$	α_4	$\dot{q}_5$	α_5	$\dot{q}_6$	α_6
26.5	21.9	25.7	21.1	20.3	16.8	22.7	18.7	24.8	20.5	30.2	24.9
26.5	21.8	25.7	21.1	20.3	16.8	22.7	18.7	24.8	20.4	30.2	24.8
26.6	21.8	25.8	21	20.3	16.8	22.7	18.6	24.8	20.3	30.2	24.7
56.8	33.7	45.7	27.2	43.4	25.2	36.1	21.5	43.1	25.6	55.9	33
56.8	33.8	45.8	27.2	43.4	25.2	36.2	21.6	43.1	25.6	55.9	33
97.8	45.8	74.9	35.1	74.8	33.8	54.1	25.6	65.9	31.1	88.6	41.2
97.9	46	75	35.2	74.9	33.8	54.1	25.7	66	31.2	88.6	41.5
142.6	54.9	106.8	40.9	118.2	42.5	73.8	28.6	92.8	35.7	120.6	46.4
142.6	55.2	106.7	41.2	118.1	42.8	71.7	28.3	92.5	36	120.4	46.6
142.6	55.3	106.7	41.3	118.1	43	71.7	28.3	92.6	36.1	120.5	46.7
198.5	60.7	151.1	46	176.9	52.2	106.4	32.6	121.7	37	155.9	47.3
200.1	61.2	151.1	46.4	177.8	52.8	107.7	33.3	121.8	37.3	156	47.4
0.2		0.2		0.2		0.2		0.2		0.2	
0.2		0.2		0.2		0.2		0.2		0.2	
1.5	10	1.8	12.5	1.9	24.2	1.2	12.1	0.6	5.7	0.4	3.1
1.7	11	1.8	12.5	2	23.7	1.1	10.9	0.6	5	0.5	4
2.1	6.8	3.4	10.6	5.9	19.4	5	14.9	3.7	11.2	4.2	13.2
3.4	9.7	3.1	9.1	5.9	18.5	4.9	14.2	3.6	10.4	5	14.2
7.5	12.5	6.3	10.4	11.5	19.2	8.9	14.6	8.1	13.2	8.5	14.1
7.6	12.5	6.3	10.5	11.6	19.5	9	14.7	8.1	13.3	8.5	14
13.7	15.8	10	11.6	19.4	21.9	13.9	15.9	13.2	15.1	14.3	16.3
12.7	15.8	10.1	12.7	13.6	18.1	14.1	17.4	12.4	15.3	11.3	14.2
23.2	19.4	18.6	15.8	28	24.4	22.8	19.1	21.3	17.8	27.2	22.5
23.1	19.3	19.5	16.3	27.8	23.6	22.7	19.1	21	17.9	25.7	21.5
49	29.4	38.9	23.4	56	33.2	36	21.5	42.3	25.2	53.7	31.8
49.3	29.4	40	23.7	55.9	32.8	36	21.2	42.3	24.9	53.4	31.3
79.6	37.9	62.4	29.7	83.2	38.5	53.8	25.4	63	29.9	81.1	38.2
80.8	39.6	62.3	30.4	83	39.4	53.7	26	62.8	30.5	81	39.2
141	56.8	93.1	37.8	110.1	43.6	72.6	29.3	85.5	34.5	116.4	46.9
133.9	55.5	92.3	38.3	104.9	43.3	72.3	30.1	80.1	33.5	115.4	47.8
134	55.3	92.3	38.1	105.1	43	72.4	29.7	80.1	33.3	115.5	47.6
199.2	64.6	142.2	46.4	168.5	58.1	107.5	35.5	116.8	38.7	158.8	52
199.3	65.3	142.1	46.8	168.4	56.7	107.6	35.8	116.8	39	158.8	52.6
0.2		0.2		0.2		0.2		0.2		0.2	
0.2		0.2		0.2		0.2		0.2		0.2	
0.2	1.2	3.3	24.6	1.7	12.7	1.8	13.2	1.9	13.2	0.2	1.4
0.2	1.2	3.5	25.8	1.7	13	1.9	13.8	2.1	14.6	0.2	1.3
0.2	0.8	5.5	24.4	1.7	7.6	4.6	18.8	2.2	9.8	2.3	10.3
0.2	0.8	5.5	24.4	1.6	7.5	4.6	19	1.9	8.4	2.3	10.8
0.2	0.8	5.5	24.3	1.7	7.5	4.6	18.9	2.1	9.5	2.2	9.8
6	10.8	10.1	18.5	5.8	10.9	9.1	16.5	7	12.7	7.7	13.7
5.9	10.6	9.7	17.9	5.7	11	9.1	16.5	7	12.7	7.6	13.8
6	10.7	9.7	18	5.8	11	9.1	16.5	7.1	12.7	7.7	13.8
10.1	13.1	14.5	18.7	9.5	12.7	13.4	17.4	10.9	14.3	10.8	14.3
10	13	14.5	18.7	9.4	12.7	13.4	17.4	10.9	14.2	10.7	14.1
28.2	22.1	25.2	19.9	23.9	19.2	22.6	17.9	24.5	19	24.8	19.7
28.2	22.1	25.1	19.9	23.9	19.2	22.6	17.8	24.4	19	24.8	19.6

Table G.9

$w_{oil,0}$	p	T	$\dot{m}$	$\dot{x}$	$\Delta p/\Delta L$	FP	$\dot{q}$	α
0.013	26.13	-10.45	526.2	0.51	64.26	a (f)	50	27.8
0.013	26.19	-10.37	527.8	0.51	63.33	a (f)	49.9	27.7
0.013	26.19	-10.37	526.3	0.51	64.01	a (f)	49.9	27.6
0.013	26.21	-10.34	533.8	0.49	67.6	a (f)	76.3	34.6
0.013	26.23	-10.31	536.4	0.49	66.14	a (f)	76.4	34.5
0.013	26.23	-10.32	533.4	0.49	65.45	a (f)	76.4	34.6
0.013	26.18	-10.38	531.5	0.49	73.39	a (f)	105.3	40.2
0.013	26.23	-10.31	532.4	0.49	74.35	a (f)	105.3	40.2
0.013	26.22	-10.32	534.5	0.49	78.89	a (f)	133.9	43.9
0.013	26.22	-10.32	538.1	0.48	78.64	a (f)	141	44.9
0.013	26.24	-10.31	535.9	0.48	80.71	a (f)	141	45.1
0.013	25.97	-10.67	533.6	0.5		a	0.2	
0.013	25.95	-10.69	528.5	0.51		a	0.2	
0.013	25.95	-10.7	532.6	0.5	60.06	a	3.1	14.4
0.013	25.89	-10.77	531.8	0.51	60.27	a	3.1	14.3
0.013	25.94	-10.71	533.3	0.5	60.17	a	4.6	14.5
0.013	25.97	-10.66	532.2	0.5	60.41	a	4.6	14.3
0.013	25.98	-10.65	525.5	0.5	60.53	a	8.7	15
0.013	25.97	-10.66	523.7	0.51	60.28	a	8.6	15
0.013	25.96	-10.67	514.8	0.52	60.76	a	13.7	17
0.013	25.91	-10.75	516.6	0.52	60.81	a	13.7	16.8
0.013	26	-10.62	522.4	0.51	62.01	a (f)	24	20.9
0.013	25.93	-10.71	521.6	0.51	62.74	a (f)	24.2	20.9
0.013	25.95	-10.69	523.6	0.51	62.81	a (f)	46.3	28.3
0.013	26	-10.62	523.9	0.51	61.33	a (f)	46.4	28.4
0.013	26.04	-10.57	517.3	0.51	62.53	a (f)	46.2	28.2
0.013	25.92	-10.73	510.9	0.52	63.23	a (f)	46.3	28.1
0.013	25.99	-10.64	515.6	0.52	67.06	a (f)	71.3	35.2
0.013	26.04	-10.57	515.9	0.51	66.32	a (f)	71.3	35.2
0.013	26.01	-10.6	516.7	0.52	72.29	a (f)	99.8	41.1
0.013	26.04	-10.58	517.8	0.51	71.89	a (f)	99.6	41.1
0.013	25.97	-10.67	523.9	0.52	76.89	a (f)	131.9	45.7
0.013	25.96	-10.67	524.2	0.51	76.41	a (f)	131.7	46.1
0.013	25.93	-10.71	516	0.53	76.61	a (f)	135.7	46.3
0.013	25.99	-10.64	518.1	0.52	77.9	a (f)	135.7	46.5
0.013	25.89	-10.77	516.3	0.52	79.07	a (f)	135.6	46.6
0.013	25.98	-10.65	513.8	0.52	79.44	a (f)	135.6	46.6
0.013	25.93	-10.72	508.8	0.53	80.06	a (f)	135.5	46.6
0.013	25.98	-10.65	512.8	0.52	79.93	a (f)	136	46.7
0.013	26.09	-10.48	504.3	0.73	74.25	a (f)	0.2	
0.013	26.09	-10.48	504.7	0.73	74	a (f)	0.2	
0.013	26.1	-10.47	502.6	0.73	73.67	a (f)	0.2	
0.013	26.11	-10.45	502	0.73	74.34	a (f)	0.2	
0.013	26.12	-10.44	502.3	0.73	74.3	a (f)	1.4	15.3
0.013	26.1	-10.46	505.7	0.72	74.54	a (f)	1.4	14.4
0.013	26.09	-10.48	501.7	0.73	74.45	a (f)	1.4	14.5
0.013	26.09	-10.48	501.4	0.73	74.51	a (f)	4	16.4

$\dot{q}_1$	α_1	$\dot{q}_2$	α_2	$\dot{q}_3$	α_3	$\dot{q}_4$	α_4	$\dot{q}_5$	α_5	$\dot{q}_6$	α_6
62.3	34.8	46.4	26.1	57.4	30.6	36.1	20.4	42.9	24.1	54.7	30.6
62.2	34.8	46.5	26.2	57.5	30.5	35.7	20.2	42.8	24.1	54.8	30.6
62.3	34.7	46.6	26	57.5	30.4	35.7	20.1	42.9	24.1	54.7	30.5
97.8	44.9	75.3	34.2	82.5	36	54.2	24.9	65.2	29.8	82.9	37.9
97.9	44.8	75.4	34.1	82.5	35.9	54.3	24.9	65.3	29.8	82.9	37.9
97.8	44.8	75.4	34.2	82.5	35.8	54.2	24.8	65.3	29.9	82.9	38.1
141.9	54.4	102.3	39.2	115.3	42.4	72	27.9	87.1	33.6	113.1	43.5
141.9	54.5	102.4	39.3	115.3	42.1	72	27.9	87.1	33.6	113.1	43.6
172.1	57.5	130	43	142.7	47.1	106.9	33.9	116.6	37.2	135	44.9
186.2	59.3	142.8	45.2	146.6	47.6	107.5	33.8	116.6	36.7	146.1	46.5
186.2	59.9	142.9	45.6	146.6	47.4	107.5	33.9	116.6	37	146.1	46.9
0.2		0.2		0.2		0.2		0.2		0.2	
0.2		0.2		0.2		0.2		0.2		0.2	
3.3	13.7	2.9	13.5	3.4	17.4	3	15.4	2.5	11.7	3.3	14.9
2.6	10.7	3	13.8	3.8	18.9	3.1	15.7	2.5	11.6	3.4	15.3
4.3	13.2	4.6	14.6	4.5	15.3	4.8	15.6	4.8	14.5	4.4	13.7
4.3	13	4.6	14.4	4.5	15.1	4.8	15.3	5	14.7	4.3	13.3
9	15	8.6	14.7	9	15.8	9.1	16.2	7.3	12.8	9.2	15.8
8.9	14.9	8.5	14.7	8.9	15.7	9	16.1	7.2	12.7	9.1	15.8
13.4	16.6	13.3	16.4	13.8	17.3	14.6	18.3	12.3	15.3	14.6	18.2
13.5	16.6	13	15.9	13.9	17.1	14.9	18.2	12.4	15.2	14.6	17.9
27	23.4	22.5	19.5	25.1	21.8	22.6	19.8	21.6	18.8	25.5	22.2
27.2	23.3	22.7	19.4	25.2	21.8	22.7	19.7	21.7	18.7	25.6	22.1
58	35.4	42.5	26.1	50.6	30.3	36.4	22.7	39.4	24.4	50.8	31.1
58.1	35.5	42.6	26.1	50.6	30.4	36.4	22.8	39.5	24.5	50.9	31.3
57.8	35.3	42.4	25.9	50.5	30.1	36.2	22.5	39.4	24.3	51	31.2
58.1	35.3	42.6	25.9	50.7	30.1	36.4	22.4	39.6	24.2	50.7	30.9
93.4	46.2	68.1	33.6	73.6	35.8	54.2	27	60.1	29.9	78	38.7
93.5	46.2	68.1	33.6	73.6	35.7	54.2	27.1	60.2	29.9	78.1	38.7
132.3	55	95.6	39.4	106.6	42.5	73.4	30.5	83.1	34.5	107.9	44.8
132	55	95.4	39.5	106.2	42.5	73.2	30.5	82.9	34.5	107.7	44.8
169.3	58.2	125.9	43.2	140	51.6	107.3	36.4	112.6	38	136.3	46.6
169	58.6	125.7	43.5	139.8	52.5	107.1	36.7	112.4	38.2	136.3	46.9
183.5	61.8	133.3	45.2	139.7	48.6	107.1	36.5	111.9	38.1	138.5	47.4
183.4	62	133.4	45.4	139.7	49.1	107.2	36.7	112	38.2	138.5	47.5
183.3	62.2	133.1	45.4	139.5	49.9	107.2	36.4	112	38.2	138.4	47.6
183.2	62.4	133.1	45.5	139.5	49.6	107.2	36.2	112	38	138.5	47.6
183.1	62.9	133.1	45.7	139.4	47.9	107.2	36.4	112	38.4	138.4	48.2
183.8	63.1	133.5	45.9	139.9	48	107.5	36.5	112.5	38.5	138.9	48.3
0.2		0.2		0.2		0.2		0.2		0.2	
0.2		0.2		0.2		0.2		0.2		0.2	
0.2		0.2		0.2		0.2		0.2		0.2	
0.2		0.2		0.2		0.2		0.2		0.2	
0.2	1.8	2.4	24.7	2.4	26.9	1.6	20.8	1.4	15.6	0.2	2.2
0.2	1.7	2.5	24	2.2	24	1.7	19.8	1.4	14.7	0.2	2
0.2	1.7	2.7	26.7	2.2	22.1	1.7	21	1.2	13.4	0.2	2
2.5	10	4.3	17.9	5.3	21.6	4.8	20.1	3.3	13.5	3.9	15.5

Table G.9

$w_{oil,0}$	p	T	$\dot{m}$	$\dot{x}$	$\Delta p/\Delta L$	FP	$\dot{q}$	α
0.013	26.09	-10.47	501.9	0.74	74.85	a (f)	4	16.6
0.013	26.1	-10.47	499.5	0.74	74.99	a (f)	4	16.7
0.013	26.09	-10.48	503.8	0.73	75.36	a (f)	7.7	17.5
0.013	26.11	-10.45	503.1	0.73	75.06	a (f)	7.7	17.4
0.013	26.1	-10.46	501.7	0.74	75.27	a (f)	7.7	17.5
0.013	26.09	-10.49	506.5	0.72	76.22	a (f)	11.9	18.3
0.013	26.11	-10.46	502.6	0.73	76.23	a (f)	11.9	18.3
0.013	26.1	-10.47	504.2	0.73	75.55	a (f)	11.8	18.2
0.013	26.1	-10.46	502.7	0.73	77.38	a (f)	22.4	21.9
0.013	26.11	-10.45	506.2	0.72	77.64	a (f)	22.4	21.9
0.013	26.13	-10.43	506.8	0.72	80.02	a (f)	39	28.2
0.013	26.12	-10.43	507.5	0.72	80.56	a (f)	39	28.2
0.013	26.12	-10.44	504.2	0.73	80.34	a (f)	38.9	28.1
0.013	26.14	-10.41	508.3	0.72	85.77	a (f)	63.1	34.8
0.013	26.13	-10.42	505.7	0.73	85.25	a (f)	63.1	34.8
0.013	26.14	-10.42	508.9	0.72	85.78	a (f)	63	34.9
0.013	26.15	-10.41	510.7	0.72	88.21	a (f)	85.6	39.2
0.013	26.15	-10.4	507.1	0.73	88.21	a (f)	85.6	39.3
0.013	26.15	-10.4	511.2	0.72	88.21	a (f)	85.6	39.2
0.013	26.18	-10.36	512.7	0.71	88.21	a (f)	131	45.2
0.013	26.18	-10.36	509.5	0.72	88.2	a (f)	131	45.1
0.013	26.19	-10.35	511.3	0.72	88.21	a (f)	130.9	45.1
0.013	39.37	4.71	74.3	0.15	0.21	s-w	0.1	
0.013	39.26	4.61	80.3	0.14	0.25	s-w	0.1	
0.013	39.27	4.62	80.4	0.14	0.24	s-w	0.1	
0.013	39.23	4.58	79.2	0.14	0.29	s-w	2.4	7.3
0.013	39.26	4.6	80.2	0.14	0.29	s-w	2.4	7.4
0.013	39.29	4.64	80.4	0.14	0.35	s-w	4.7	9.2
0.013	39.3	4.65	79.6	0.14	0.34	s-w	4.6	9.1
0.013	39.31	4.66	79.9	0.14	0.44	s-w	7.6	10.6
0.013	39.31	4.66	80.2	0.14	0.44	s-w	7.6	10.6
0.013	39.27	4.62	79.5	0.14	0.57	s-w (f)	10.7	11.9
0.013	39.27	4.62	78.6	0.15	0.56	s-w (f)	10.7	11.9
0.013	39.27	4.62	80.6	0.14	0.55	s-w (f)	10.6	11.8
0.013	39.3	4.65	80.4	0.13	0.75	s-w (f)	18	14.5
0.013	39.34	4.69	80.5	0.14	0.75	s-w (f)	18	14.4
0.013	39.27	4.62	80.5	0.13	1.03	s-w (f)	29.1	17.1
0.013	39.35	4.7	79.4	0.13	1	s-w (f)	29.1	17
0.013	39.25	4.6	74.7	0.14	1.27	w (f)	43	18.6
0.013	39.37	4.72	74.2	0.15	1.25	w (f)	43	18.7
0.013	39.37	4.72	80.3	0.13	1.31	w (f)	43.1	19.1
0.013	39.27	4.62	74.4	0.17	1.76	w (f)	62.1	20.3
0.013	39.31	4.66	74	0.17	1.72	w (f)	61.9	20.3
0.013	39.41	4.76	74.6	0.17	1.73	w (f)	62	20.3
0.013	39.39	4.75	74.3	0.35	3.34	f	119.7	10.3
0.013	39.35	4.71	74.5	0.35	3.4	f	119.8	10.1
0.013	39.29	4.68	74	0.58	1.1	w (f)	0.1	

$\dot{q}_1$	α_1	$\dot{q}_2$	α_2	$\dot{q}_3$	α_3	$\dot{q}_4$	α_4	$\dot{q}_5$	α_5	$\dot{q}_6$	α_6
2.5	10.1	4.3	18.2	5.3	21.8	4.7	20.4	3.3	13.7	3.9	15.6
2.7	10.8	4.3	18.2	5.3	21.9	4.7	20.2	3.3	13.6	3.9	15.7
6	13.9	8.4	19	9	20.3	9.1	20.6	6.3	14.2	7.3	16.7
6	13.8	8.4	18.9	9	20.5	9.1	20.6	6.2	14.1	7.3	16.6
6	13.8	8.4	19	9.1	20.6	9.1	20.7	6.3	14.2	7.3	16.8
10.6	16.5	11.9	18.3	14.6	21.9	13.4	20.5	9.4	14.7	11.3	17.8
10.6	16.6	11.9	18.3	14.5	22	13.3	20.6	9.4	14.7	11.3	17.8
10.6	16.6	11.9	18.2	14.5	21.8	13.3	20.4	9.4	14.7	11.3	17.8
23.2	22.5	21.5	20.9	24.8	24.6	22.7	22.3	20.8	20.1	21.3	20.9
23.2	22.7	21.5	21	24.8	24.6	22.7	22.2	20.8	20	21.3	21
43.7	31.7	34.6	25.3	44.3	32	35.9	26	34	24.6	41.5	29.9
43.7	31.7	34.6	25.3	44.2	32.2	35.9	26.1	33.9	24.4	41.5	29.8
43.6	31.5	34.6	25.2	44.1	32	35.9	26	33.9	24.4	41.5	29.7
75.8	41.5	57.3	31.7	69.5	38.3	54	30.1	55.2	30.4	66.6	36.6
75.9	41.7	57.2	31.6	69.5	38.3	54	30.1	55.2	30.5	66.6	36.6
75.8	41.9	57.1	31.7	69.5	38.5	53.9	30.2	55.1	30.5	66.6	36.7
109.5	49.8	80	36.5	90.8	41.5	72.2	33.3	72.2	33.3	89	40.9
109.5	49.9	80	36.5	90.7	41.7	72.1	33.4	72.3	33.4	89	40.9
109.5	49.8	80.1	36.5	90.7	41.5	72.1	33.3	72.3	33.3	89	40.9
172.6	58.9	121.3	41.8	138.4	48.3	108.1	37.8	112	38.5	133.4	45.9
172.5	58.6	121.3	41.8	138.5	48.4	108	37.8	112.1	38.5	133.4	45.8
172.6	58.7	121.3	41.8	138.4	48.2	107.9	37.7	112	38.6	133.3	46
0.1		0.1		0.1		0.1		0.1		0.1	
0.1		0.1		0.1		0.1		0.1		0.1	
0.1		0.1		0.1		0.1		0.1		0.1	
0.1	0.4	4.5	14.3	3.7	10.9	2.3	6.9	3.5	10.9	0.1	0.4
0.1	0.4	4.4	14.2	3.7	11	2.3	7.2	3.5	10.9	0.1	0.4
0.1	0.2	7.2	14.4	6.8	12.9	4.4	8.8	6.4	12.6	3.1	6.1
0.1	0.2	7.2	14.5	6.7	12.8	4.4	8.7	6.3	12.5	3	5.8
0.1	0.2	10.4	14.6	10.9	14.9	9.2	12.9	10.4	14.6	4.7	6.6
0.1	0.2	10.3	14.5	10.8	14.9	9.2	12.8	10.4	14.5	4.7	6.6
0.1	0.1	14.2	15.7	15.2	16.9	13.7	15.2	15	16.8	6.1	6.8
0.1	0.1	14.2	15.7	15.1	17	13.6	15.2	15	16.8	6	6.8
0.1	0.1	14.1	15.5	15.1	16.7	13.6	15.1	15	16.7	6	6.7
0.1	0.1	26.4	21.2	25	20	22.5	18.1	24.9	20	9.1	7.4
0.1	0.1	26.5	21.1	25.1	19.9	22.5	18	24.9	19.9	9.1	7.3
0.1	0.1	46.1	27.1	38.8	22.8	36	21.2	38.9	22.9	14.5	8.7
0.1	0.1	46.2	26.9	38.8	22.6	36	21.1	39	22.8	14.5	8.6
0.1	0	61.8	26.8	61.4	26.4	54.2	23.5	56.1	24.3	24.5	10.7
0.1	0	61.9	27	61.4	26.4	54.2	23.5	56.1	24.5	24.4	10.9
0.1	0	62.1	28.1	61.4	26.6	54.2	23.8	56.1	25	24.5	11.3
6.3	2.1	77.4	25.2	94.2	30.6	71.9	23.6	78.5	25.7	44.4	14.6
6.2	2.1	77.3	25.3	93.8	30.5	71.7	23.6	78.2	25.8	44.1	14.7
6.3	2.1	77.4	25.2	94.1	30.4	71.8	23.5	78.4	25.8	44.2	14.7
0.1	0	177.4	15	177	15.4	107.6	9	177.8	15.7	78.6	6.7
0.1	0	177.6	14.7	177.1	15	107.7	8.8	177.6	15.3	78.6	6.6
0.1		0.1		0.1		0.1		0.1		0.1	

Table G.9

$w_{oil,0}$	p	T	$\dot{m}$	$\dot{x}$	$\Delta p/\Delta L$	FP	$\dot{q}$	α
0.013	39.21	4.6	74.9	0.57	1.1	w (f)	0.1	
0.013	39.27	4.65	75.2	0.57	1.11	w (f)	0.1	
0.013	39.31	4.7	73.8	0.58	1.23	w (f)	4	11.1
0.013	39.3	4.69	76.1	0.56	1.23	w (f)	4	11
0.013	39.39	4.77	74.6	0.57	1.25	w (f)	5.1	10.3
0.013	39.34	4.73	74.1	0.58	1.25	w (f)	5.1	10.4
0.013	39.37	4.75	80.2	0.52	1.42	w (f)	8.4	11.2
0.013	39.4	4.77	82.4	0.51	1.4	w (f)	8.5	11
0.013	39.36	4.73	81.3	0.52	1.41	w (f)	8.5	11
0.013	39.35	4.72	81.1	0.51	1.52	w (f)	11.1	11.6
0.013	39.37	4.75	80	0.52	1.52	w (f)	11.1	11.6
0.013	39.38	4.76	79.8	0.53	1.74	w (f)	16.5	12.5
0.013	39.34	4.72	80.4	0.53	1.7	w (f)	16.5	12.5
0.013	39.38	4.76	80.5	0.52	2.09	w (f)	26.8	13.4
0.013	39.38	4.76	80.6	0.52	2.11	w (f)	26.8	13.4
0.013	39.29	4.68	80.5	0.53	2.1	w (f)	26.9	13.6
0.013	39.28	4.66	79.9	0.53	2.75	w (f)	42.9	13.7
0.013	39.25	4.63	79.7	0.53	2.74	w (f)	42.9	13.7
0.013	39.3	4.68	79.9	0.53	3.28	f	62.6	12.4
0.013	39.32	4.7	80.8	0.52	3.26	f	62.4	12.4
0.013	39.29	4.67	80.2	0.53	4.22	f	100.4	9.1
0.013	39.34	4.72	80.5	0.52	4.27	f	100.6	9.1
0.013	39.21	4.64	80.4	0.72	1.53	w (f)	0.1	
0.013	39.32	4.75	80.1	0.73	1.51	w (f)	0.1	
0.013	39.32	4.75	81.8	0.71	1.51	w (f)	0.1	
0.013	39.32	4.76	80.2	0.73	1.65	w (f)	4.2	12.7
0.013	39.28	4.72	79.7	0.73	1.63	w (f)	4.1	12.3
0.013	39.31	4.74	79.6	0.73	1.66	w (f)	5.3	11
0.013	39.31	4.74	80.9	0.72	1.66	w (f)	5.3	10.8
0.013	39.32	4.75	82.2	0.71	1.66	w (f)	8	10.8
0.013	39.33	4.76	80.4	0.73	1.64	w (f)	8	10.8
0.013	39.34	4.77	81.4	0.72	1.64	w (f)	8	10.8
0.013	39.25	4.68	80.4	0.72	1.8	w (f)	12	11.7
0.013	39.25	4.69	79.4	0.73	1.79	w (f)	12.1	11.8
0.013	39.32	4.74	81.8	0.71	2.12	w (f)	18.3	12.7
0.013	39.31	4.74	81.2	0.71	2.09	w (f)	18.2	12.6
0.013	39.31	4.74	80	0.72	2.69	f	30.3	13
0.013	39.34	4.77	79.9	0.73	2.66	f	30.3	12.9
0.013	39.35	4.78	79.7	0.73	2.66	f	30.3	12.9
0.013	39.33	4.76	79.8	0.73	3.32	f	45	12
0.013	39.33	4.76	80.4	0.72	3.28	f	45	12
0.013	39.26	4.7	80.5	0.72	4.15	w-a	69.9	7.6
0.013	39.32	4.76	80.1	0.73	4.46	w-a	69.9	7.7
0.013	39.32	4.76	80.2	0.74	5.17	w-a	89.1	5.6
0.013	39.35	4.78	80.4	0.73	5.12	w-a	88.9	5.6
0.013	39.42	4.77	156.2	0.13	0.81	w	0.1	
0.013	39.4	4.75	156.3	0.13	0.81	w	0.1	

$\dot{q}_1$	α_1	$\dot{q}_2$	α_2	$\dot{q}_3$	α_3	$\dot{q}_4$	α_4	$\dot{q}_5$	α_5	$\dot{q}_6$	α_6
0.1		0.1		0.1		0.1		0.1		0.1	
0.1		0.1		0.1		0.1		0.1		0.1	
4.1	11.9	10.3	27.7	3.8	10.4	2	5.5	3.9	10.9	0.1	0.3
4	11.6	10.3	27.5	3.6	9.9	2.1	6.1	3.9	10.7	0.1	0.3
1.7	3.6	13.8	27.6	4.9	9.9	4.5	9.2	5.5	11.1	0.1	0.2
1.9	4	13.8	27.6	4.9	9.8	4.5	9.2	5.6	11.3	0.1	0.2
0.1	0.2	19.2	26	11	14.3	9	12.2	10.7	14.6	0.1	0.1
0.1	0.1	19.4	25.3	11.2	14.2	9.2	12.1	10.9	14.3	0.1	0.1
0.1	0.1	19.4	25.4	11.2	14.3	9.2	12.1	10.9	14.3	0.1	0.1
0.1	0.1	22	23.1	15	15.7	13.7	14.3	15.6	16.2	0.1	0.1
0.1	0.1	22	22.9	15	15.7	13.7	14.4	15.6	16.2	0.1	0.1
0.1	0.1	24.6	19.5	27.2	20.3	22.7	17.1	24.3	18.3	0.1	0.1
0.1	0.1	24.7	19.4	27.2	20.2	22.8	17	24.3	18.2	0.1	0.1
0.1	0	34.4	17.3	49.7	24.6	36	18.2	39.3	19.6	1.2	0.6
0.1	0.1	34.5	17.3	49.8	24.6	36	18.2	39.4	19.6	1.2	0.6
0.1	0.1	34.5	17.8	49.8	24.9	36	18.4	39.5	20.1	1.3	0.7
0.1	0	45.7	14.8	88.6	28	54.1	17.3	62.8	20	6.1	2
0.1	0	45.7	14.7	88.5	28.1	54.1	17.3	62.8	20	6	1.9
0.1	0	65.3	12.6	126.7	25.6	71.8	14	100.3	19.7	11.7	2.3
0.1	0	64.2	12.5	126.7	25.9	71.8	14.1	100.3	19.8	11.6	2.3
0.1	0	98.8	8.6	178.3	16.4	107.6	9.4	178.6	16.8	39.1	3.5
0.1	0	99.1	8.6	178.4	16.4	107.8	9.4	178.9	16.9	39.4	3.5
0.1		0.1		0.1		0.1		0.1		0.1	
0.1		0.1		0.1		0.1		0.1		0.1	
0.1		0.1		0.1		0.1		0.1		0.1	
5.4	17	9.3	27.8	4.5	13.4	1.9	5.7	3.4	10.6	0.5	1.5
5.5	16.7	9.3	27.4	4.5	13.2	1.7	5.2	3.4	10.2	0.4	1.1
1.6	3.5	14.7	30.4	5.5	11.4	4.5	9.4	5.3	11.1	0.1	0.2
1.7	3.6	14.6	29.6	5.5	11.3	4.5	9.2	5.3	10.9	0.1	0.2
0.1	0.2	20.2	27	9.8	13.4	9	12.2	8.6	11.6	0.1	0.2
0.1	0.2	20.2	27	9.9	13.4	9.1	12.2	8.6	11.6	0.1	0.2
0.1	0.2	20.3	27.1	9.9	13.4	9	12.2	8.6	11.6	0.1	0.2
0.1	0.1	27.1	26.2	17	16.9	13.7	13.5	14	13.6	0.1	0.1
0.1	0.1	27.2	26.4	17.1	17	13.8	13.6	14	13.6	0.1	0.1
1	0.7	34.5	23.8	31.8	21.5	22.5	15.6	20.3	14.2	0.1	0.1
0.8	0.6	34.3	23.9	31.5	21.4	22.3	15.5	20.1	14.2	0.1	0.1
1.9	0.8	45.2	19.4	60.6	26	36.1	15.6	37.8	16.3	0.1	0
1.7	0.7	45.2	19.1	60.7	25.8	36.1	15.5	37.8	16.1	0.1	0
1.8	0.8	45.2	19.3	60.7	25.9	36.1	15.5	37.8	16.1	0.1	0
0.1	0	56.2	14.8	94.2	25.4	53.9	14.1	65.7	17.5	0.1	0
0.1	0	56.2	14.9	94.2	25.4	53.9	14.2	65.7	17.4	0.1	0
0.1	0	58.4	6.3	145.8	16.2	71.8	7.7	143.3	15.6	0.1	0
0.1	0	58.3	6.4	145.9	16.3	71.8	7.8	143.2	15.7	0.1	0
0	0	64.8	4	177.1	10.8	106.2	6.5	178.8	11.6	7.6	0.5
0	0	64.6	4	176.7	11	105.8	6.6	178.4	11.8	7.5	0.5
0.1		0.1		0.1		0.1		0.1		0.1	
0.1		0.1		0.1		0.1		0.1		0.1	

Table G.9

$w_{oil,0}$	p	T	$\dot{m}$	$\dot{x}$	$\Delta p/\Delta L$	FP	$\dot{q}$	α
0.013	39.51	4.85	156.4	0.13	0.96	w	5	12.7
0.013	39.43	4.78	157	0.13	0.96	w	4.9	12.8
0.013	39.45	4.8	154.8	0.13	0.96	w (f)	5	12.8
0.013	39.47	4.82	156.2	0.13	1.12	w (f)	8.6	13.9
0.013	39.48	4.83	156.2	0.13	1.1	w (f)	8.6	13.7
0.013	39.48	4.82	156.6	0.13	1.33	w (f)	13.8	14.4
0.013	39.46	4.81	156.3	0.13	1.39	w (f)	13.7	14.6
0.013	39.48	4.83	155.9	0.12	1.4	w (f)	13.8	14.7
0.013	39.43	4.77	155.7	0.13	1.68	w (f)	19	15.6
0.013	39.41	4.76	156.7	0.13	1.66	w (f)	19	15.8
0.013	39.43	4.77	156.5	0.13	2.02	w (f)	25.7	17.4
0.013	39.43	4.78	155.3	0.13	2.08	w (f)	25.9	17.3
0.013	39.47	4.82	156.8	0.12	2.72	w (f)	38.3	19.5
0.013	39.45	4.8	157.2	0.12	2.76	w (f)	38.3	19.5
0.013	39.42	4.77	157	0.13	3.33	w (f)	57.1	22
0.013	39.41	4.76	156.2	0.13	3.35	w (f)	57	21.9
0.013	39.47	4.82	156.3	0.13	3.34	w (f)	56.9	22
0.013	39.56	4.91	156.9	0.12	3.87	w (f)	71.8	23.3
0.013	39.45	4.8	156.8	0.13	3.84	w (f)	71.7	23.3
0.013	39.5	4.84	156.7	0.13	3.82	w (f)	71.7	23.2
0.013	39.48	4.82	157.1	0.13	5.26	f	106.6	23.1
0.013	39.54	4.89	157.3	0.13	5.15	f	106.6	23.1
0.013	39.48	4.84	158	0.31	3	w	0.1	
0.013	39.49	4.84	158.3	0.31	2.99	i	0.1	
0.013	39.48	4.84	158.2	0.31	2.87	i	0.1	
0.013	39.49	4.85	158.9	0.31	3.15	i (f)	9.2	15.1
0.013	39.47	4.83	157.8	0.32	3.28	i (f)	9.2	14.8
0.013	39.49	4.84	159.1	0.31	3.4	i (f)	12	15.1
0.013	39.49	4.85	158.9	0.31	3.36	i (f)	12	15.1
0.013	39.48	4.84	158.2	0.3	3.8	w (f)	20.6	15.3
0.013	39.48	4.84	159.1	0.31	3.79	i (f)	20.5	15.7
0.013	39.51	4.87	158.4	0.32	4.27	i (f)	23.9	15.2
0.013	39.5	4.86	159.3	0.31	3.96	i (f)	24	15.4
0.013	39.49	4.85	157.3	0.31	3.42	i (f)	26.6	18.5
0.013	39.49	4.85	157.2	0.31	3.41	i (f)	26.6	18.6
0.013	39.52	4.88	158.2	0.32	4.57	i (f)	33.6	15.1
0.013	39.51	4.87	156.8	0.32	4.51	i (f)	33.7	14.8
0.013	39.46	4.82	158.3	0.31	4.27	i (f)	40.4	19.8
0.013	39.4	4.76	158.3	0.32	4.19	i (f)	40.5	19.6
0.013	39.45	4.81	158.7	0.32	4.3	i (f)	40.5	19.8
0.013	39.43	4.78	156.8	0.32	4.28	i (f)	40.5	19.9
0.013	39.47	4.82	156.9	0.31	4.54	i (f)	50.8	20.5
0.013	39.47	4.83	158.1	0.31	4.72	i (f)	50.7	20.6
0.013	39.45	4.81	157.9	0.31	4.59	i (f)	50.7	20.1
0.013	39.43	4.78	158.5	0.31	4.66	i (f)	50.7	20.5
0.013	39.46	4.82	157.4	0.31	5.34	f	65.4	21.4
0.013	39.45	4.81	159.2	0.31	5.1	f	65.3	21

Table G.9

$\dot{q}_1$	α_1	$\dot{q}_2$	α_2	$\dot{q}_3$	α_3	$\dot{q}_4$	α_4	$\dot{q}_5$	α_5	$\dot{q}_6$	α_6
9.4	24.5	6.1	15.9	4.3	10.5	1.7	4.3	4	10.3	4.3	11
9.4	24.8	6.1	16	4.3	10.6	1.7	4.4	4.1	10.3	4.1	10.5
9.4	24.5	6.1	15.9	4.3	10.6	1.7	4.3	4.1	10.5	4.2	11
9.4	16.1	14.7	23.6	7.6	11.5	4.8	7.6	8.3	13.3	6.9	11.4
9.4	15.7	14.7	23.1	7.6	11.4	4.8	7.6	8.3	13.1	6.9	11
4.7	5	23.8	24.9	15.3	15.1	9.3	9.8	14.6	15.5	15.2	16
4.6	5.1	23.8	25	15.2	15.1	9.2	9.8	14.5	15.7	15.1	16.6
4.7	5.4	23.8	25.1	15.2	15.1	9.2	9.8	14.5	15.9	15.1	16.9
4.8	3.9	29.5	24.3	21.4	17.2	13.3	11.3	22	18.3	23.1	18.6
4.6	3.9	29.4	24.6	21.4	17.2	13.3	11.2	22	18.6	23.1	19.4
5.4	4	38.3	26.1	28.5	18.8	22.4	14.9	33.7	22.4	26.2	18
5.4	3.8	38.4	25.8	28.6	18.8	22.5	14.9	33.8	22.4	26.3	18.2
8.4	4.4	55.8	28.2	43.5	21.9	36.2	18.4	49.5	25.3	36.3	18.8
8.5	4.6	55.9	28.5	43.6	21.9	36.2	18.3	49.7	25.2	36.4	18.8
22.5	8.6	78.6	30.4	69	26.5	53.8	21	74.6	28.6	44.4	17.2
22.4	8.6	78.5	30.3	68.8	26.4	53.7	20.9	74.5	28.5	44.1	17
22.3	8.5	78.4	30.2	68.7	26.4	53.6	20.9	74.4	28.6	44.1	17.2
35.8	11.5	89.9	29.2	91.7	29.2	71.8	23.4	86.3	28.3	55.3	18.1
35.7	11.4	89.9	29.1	91.7	29.2	71.7	23.3	86.2	28.3	55.1	18.2
35.7	11.5	90	29.3	91.8	29.2	71.8	23.4	85.9	28.1	55	18
66.4	14.3	125.9	27.2	142.5	31.1	107.7	23.3	117.7	25.5	79.4	17.3
66.3	14.3	125.9	27.2	142.4	31.1	107.7	23.3	117.9	25.5	79.4	17.3
0.1		0.1		0.1		0.1		0.1		0.1	
0.1		0.1		0.1		0.1		0.1		0.1	
0.1		0.1		0.1		0.1		0.1		0.1	
15.9	26.8	11.6	18.3	8.2	13	1.9	3.1	8.8	14.4	8.7	14.8
15.9	26.3	11.6	18	8.3	12.9	1.9	3	8.8	14.2	8.8	14.7
15.8	20.6	17.5	21.9	10.2	12.7	4.6	5.7	9.7	12.4	14.1	17.1
15.9	20.7	17.5	22	10.3	12.8	4.6	5.7	9.7	12.3	14.1	17
13.6	10.5	39.5	28.9	20.8	15.2	9.2	6.8	26.5	19.7	14.3	10.8
13.6	11.2	39.4	29.8	20.6	15.3	9.1	6.8	26.4	19.9	14.1	11.1
12.9	8.4	43.5	27.8	25.4	15.9	13.3	8.2	31	19.5	17.2	11.2
12.9	8.8	43.9	28.7	25.4	16	13.6	8.4	30.9	19.6	17.2	11
17.2	11.8	42.3	29.1	28.6	20	22.4	15.7	35.4	24.6	13.5	9.5
17.2	12	42.3	29.3	28.6	20	22.4	15.8	35.4	24.7	13.5	9.7
17.3	7.9	63.1	28.1	37	16.4	22.4	10	42.5	19.3	19.3	8.8
17.4	7.4	63.3	27.2	37.1	16.4	22.5	10	42.6	19.1	19.4	8.5
19.4	9.8	56.4	28.1	54.5	26.2	36	18	52.4	25.4	23.6	11.4
19.5	9.5	56.5	27.5	54.6	26.2	36.1	17.9	52.4	25	23.7	11.2
19.6	9.8	56.5	28.2	54.6	26.2	36.1	18	52.5	25.4	23.7	11.4
19.5	9.9	56.5	28.2	54.6	26.5	36	18.1	52.4	25.4	23.7	11.4
22	9	64.7	26.4	71.4	28.3	53.9	21.9	63.6	25.9	29.2	11.7
21.9	9.2	64.6	26.6	71.3	28.4	53.8	21.9	63.5	25.8	29.2	11.9
21.9	8.7	64.6	25.6	71.3	28.1	53.9	21.6	63.5	25.2	29.2	11.2
21.9	9.3	64.6	26.6	71.3	28.4	53.8	21.8	63.5	25.5	29.2	11.6
24.5	8.5	80.5	26.3	96.1	31	71.7	23.5	89.3	29.1	30.2	10.3
24.4	8.2	80.4	25.8	95.9	31	71.6	23.2	89.2	28.3	30.2	9.7

Table G.9

$w_{oil,0}$	p	T	$\dot{m}$	$\dot{x}$	$\Delta p/\Delta L$	FP	$\dot{q}$	α
0.013	39.46	4.82	159.3	0.3	5.28	f	65.3	20.9
0.013	39.42	4.78	158.7	0.31	5.04	f	65.2	20.9
0.013	39.53	4.88	157.8	0.3	6.23	f	90.9	19.8
0.013	39.47	4.83	158	0.31	6.23	f	90.8	19.8
0.013	39.48	4.84	157.7	0.31	6.42	f	90.7	20
0.013	39.5	4.86	156.7	0.31	6.38	f	90.7	19.5
0.013	39.55	4.91	156.2	0.33	2.6	w	0.1	
0.013	39.51	4.86	156.1	0.33	2.64	w	0.1	
0.013	39.49	4.85	157.4	0.33	2.57	w	0.1	
0.013	39.42	4.78	155.6	0.32	2.73	w	5.5	14.2
0.013	39.43	4.79	155.5	0.32	2.73	i	5.5	14.2
0.013	39.43	4.78	155.9	0.32	2.76	w	5.4	14
0.013	39.39	4.75	156.8	0.32	2.88	w (f)	9.6	16.2
0.013	39.39	4.75	155.8	0.32	2.91	i (f)	9.6	16.1
0.013	39.44	4.8	156.6	0.32	2.91	i (f)	9.6	16.2
0.013	39.43	4.79	156.8	0.33	3.11	i (f)	13.6	17.2
0.013	39.42	4.78	155.8	0.33	3.06	i (f)	13.6	17.3
0.013	39.42	4.78	156.1	0.33	3.19	i (f)	13.6	17.3
0.013	39.41	4.77	156.2	0.33	3.36	i (f)	18.7	17.7
0.013	39.4	4.76	156.8	0.32	3.42	i (f)	18.6	17.8
0.013	39.44	4.79	156	0.33	3.31	i (f)	18.6	17.8
0.013	39.43	4.79	156.4	0.33	3.76	i (f)	26.3	19.1
0.013	39.44	4.8	155.6	0.33	3.78	i (f)	26.4	19
0.013	39.45	4.81	156.8	0.33	3.73	i (f)	26.3	19.1
0.013	39.46	4.82	156.9	0.33	4.38	i (f)	39.5	20.7
0.013	39.47	4.83	156.3	0.33	4.32	i (f)	39.5	20.5
0.013	39.46	4.82	155.9	0.33	4.39	i (f)	39.6	20.6
0.013	39.5	4.85	156.1	0.33	4.9	i (f)	52.2	21
0.013	39.5	4.86	156.3	0.33	4.91	i (f)	52.2	20.5
0.013	39.47	4.83	156.1	0.33	4.94	i (f)	52.3	21.1
0.013	39.43	4.79	155.4	0.33	5.77	f	67.5	19.9
0.013	39.46	4.82	155.7	0.33	5.84	f	67.5	20.1
0.013	39.45	4.81	155.2	0.33	5.68	f	67.4	19.9
0.013	39.48	4.84	156.5	0.33	6.94	f	91.1	18.1
0.013	39.47	4.83	155.1	0.33	6.96	f	91	18
0.013	39.48	4.84	157.1	0.32	7.08	f	91.1	18
0.013	39.49	4.86	156.4	0.52	3.13	w (f)	0.1	
0.013	39.47	4.85	155.7	0.52	3.13	w (f)	0.1	
0.013	39.46	4.84	155.7	0.52	3.12	w (f)	0.1	
0.013	39.45	4.83	156.3	0.51	3.08	w (f)	0.1	
0.013	39.46	4.84	155.5	0.52	3.35	w (f)	4.6	14.3
0.013	39.47	4.85	154.9	0.53	3.35	w (f)	4.5	14.3
0.013	39.5	4.88	155.4	0.52	3.36	w (f)	4.6	14.1
0.013	39.44	4.82	153	0.54	3.37	w (f)	4.6	14.2
0.013	39.46	4.84	155.5	0.52	3.65	w (f)	9.8	16.5
0.013	39.46	4.84	156.7	0.52	3.66	w (f)	9.9	16.5
0.013	39.46	4.84	157.4	0.52	3.63	w (f)	9.8	16.4

$\dot{q}_1$	α_1	$\dot{q}_2$	α_2	$\dot{q}_3$	α_3	$\dot{q}_4$	α_4	$\dot{q}_5$	α_5	$\dot{q}_6$	α_6
24.4	8.1	80.4	25.6	95.9	30.8	71.6	23.2	89.1	28.2	30.2	9.7
24.3	8	80.4	25.7	95.8	30.8	71.5	23.1	89.1	28.1	30.1	9.6
24.3	5.4	102.6	22.2	136.2	30	107.4	23.3	121.2	26.4	53.4	11.6
24.4	5.4	102.5	22.2	136.1	30.1	107.2	23.4	121.1	26.5	53.4	11.6
24.4	5.5	102.4	22.4	136	30.1	107.2	23.4	121	26.6	53.4	11.6
24.4	5.3	102.3	21.6	136	29.5	107.1	23.2	121	26.3	53.4	11.4
0.1		0.1		0.1		0.1		0.1		0.1	
0.1		0.1		0.1		0.1		0.1		0.1	
0.1		0.1		0.1		0.1		0.1		0.1	
9.4	24.2	6.1	15.7	4.4	11.2	2.3	6	4.8	12.5	6	15.6
9.4	24.3	6.1	15.6	4.4	11.2	2.4	6.2	4.8	12.3	6	15.4
9.4	24.1	6	15.5	4.4	11.1	2.2	5.7	4.7	12.3	5.9	15.4
15.2	25.8	10.5	18.1	8.5	14.1	4.8	8.1	9.7	16.1	9	14.9
15.2	25.5	10.6	18	8.5	13.9	4.8	8.1	9.7	16.1	9	15.1
15.2	25.7	10.6	18.1	8.5	14	4.7	8	9.7	16.2	9.1	15.1
17.6	22.4	19.2	24.7	10.4	13.3	9.2	11.7	14.6	18.4	10.4	12.7
17.6	22.5	19.2	24.7	10.4	13.3	9.2	11.7	14.7	18.6	10.5	12.7
17.7	22.5	19.1	24.5	10.4	13.3	9.3	11.8	14.7	18.5	10.4	12.9
18.8	17.4	31.6	29.5	16.4	15.7	13.5	13.1	23.7	23	8	7.7
18.8	17.8	31.6	29.7	16.4	15.7	13.4	13	23.7	23	8	7.8
18.8	17.7	31.6	29.7	16.5	15.7	13.5	13	23.6	22.9	7.9	7.6
19	13.6	43.8	31.6	25	18.3	22.4	16.3	36.5	26.3	11.3	8.2
19.1	13.8	43.8	31.8	25	18.2	22.5	16.2	36.5	26	11.3	8.1
19	14	43.9	31.9	25	18.3	22.4	16.2	36.5	26	11.2	8.1
20.5	11.3	63.3	33.1	46.8	23.8	36.1	18.9	51.9	27.4	18.6	9.8
20.5	11	63.2	32.6	46.7	23.6	36	18.8	51.8	27	18.6	9.6
20.5	11.1	63.3	32.6	46.9	23.7	36.2	18.9	52	27.3	18.7	9.8
26.1	10.9	72.9	29.4	66.2	26.6	54	21.6	72.8	29	21.2	8.7
26.1	10.2	72.9	28.4	66.2	26.4	54	21.4	72.7	28.4	21.1	8.3
26.2	11	72.9	29.7	66.3	26.6	54.2	21.6	72.8	29	21.3	8.7
30.4	8.8	82.5	23.9	96.4	29.1	71.7	21.6	92.8	27	31	8.9
30.4	9	82.5	24.2	96.4	29.2	71.8	21.7	92.8	27.4	31	9
30.3	8.7	82.4	24	96.4	28.9	71.7	21.7	92.8	27.3	31	8.9
25	5	102.4	20.2	138.7	27.5	107.8	21.3	125	25.1	47.5	9.5
25	5	102.3	20	138.7	27.4	107.7	21.4	124.9	25	47.5	9.4
25	5.1	102.4	20.2	138.6	27.4	107.8	21.3	125.1	25	47.5	9.3
0.1		0.1		0.1		0.1		0.1		0.1	
0.1		0.1		0.1		0.1		0.1		0.1	
0.1		0.1		0.1		0.1		0.1		0.1	
0.1		0.1		0.1		0.1		0.1		0.1	
9.1	28.6	3.9	12.7	4.4	13.2	1.7	5.4	3.9	12.4	4.3	13.5
9.1	28.5	3.9	12.8	4.4	13.3	1.6	5.2	3.9	12.4	4.3	13.5
9.1	27.9	4	12.7	4.4	12.9	1.7	5.3	3.9	12.3	4.3	13.4
9.1	27.7	4	12.7	4.4	13.1	1.8	5.8	3.9	12.3	4.3	13.4
16.9	28.2	10.5	18	9	14.6	4.6	7.7	10.6	17.6	7.4	12.8
17	28.3	10.5	17.9	9.1	14.6	4.6	7.7	10.6	17.5	7.4	12.8
16.9	28.2	10.5	17.9	9	14.6	4.6	7.7	10.6	17.6	7.4	12.8

Table G.9

$w_{oil,0}$	p	T	$\dot{m}$	$\dot{x}$	$\Delta p/\Delta L$	FP	$\dot{q}$	α
0.013	39.48	4.85	158.5	0.51	3.68	w (f)	9.9	16.4
0.013	39.49	4.87	156.5	0.52	3.81	w (f)	13.7	17.6
0.013	39.51	4.89	155.6	0.52	3.78	w (f)	13.8	17.6
0.013	39.46	4.84	155.8	0.52	3.79	w (f)	13.7	17.6
0.013	39.44	4.82	155	0.53	3.79	w (f)	13.8	17.6
0.013	39.44	4.82	155.6	0.52	3.96	w-a (f)	17.3	17.7
0.013	39.47	4.85	155.3	0.52	4.01	w-a (f)	17.3	17.7
0.013	39.46	4.84	154	0.53	4	w-a (f)	17.2	17.7
0.013	39.45	4.83	156	0.52	3.99	w-a (f)	17.3	17.7
0.013	39.48	4.86	158.2	0.51	4.35	w-a (f)	23.8	17.5
0.013	39.48	4.86	156.9	0.52	4.33	w-a (f)	23.8	17.6
0.013	39.46	4.84	155.6	0.52	4.36	w-a (f)	23.7	17.6
0.013	39.49	4.87	156.6	0.52	4.31	w-a (f)	23.7	17.5
0.013	39.48	4.86	156.8	0.52	5.23	w-a (f)	35.4	19.2
0.013	39.48	4.86	154.9	0.53	4.95	w-a (f)	35.3	17.6
0.013	39.41	4.79	156.7	0.52	4.99	w-a (f)	35.3	17.6
0.013	39.48	4.86	156.3	0.52	4.96	w-a (f)	35.4	17.6
0.013	39.43	4.81	155.6	0.53	5.66	f	46.5	18.5
0.013	39.5	4.88	155.2	0.52	5.68	f	46.5	18.5
0.013	39.48	4.86	155.9	0.52	5.64	f	46.5	18.4
0.013	39.5	4.88	156.3	0.52	5.61	f	46.5	18.3
0.013	39.43	4.81	155.9	0.52	6.38	f	58.6	18.3
0.013	39.4	4.78	156.3	0.52	6.34	f	58.6	18.5
0.013	39.44	4.82	154.9	0.53	6.44	f	58.7	18.7
0.013	39.52	4.9	156.1	0.52	6.22	f	58.7	18.6
0.013	39.44	4.81	156.7	0.52	7.71	f	82.7	16.7
0.013	39.41	4.79	156.2	0.52	7.77	f	82.6	16.9
0.013	39.43	4.81	155.9	0.52	7.8	f	82.6	16.9
0.013	39.39	4.77	156.3	0.52	7.84	f	82.4	17
0.013	39.22	4.6	157	0.52	3.4	w (f)	0.1	
0.013	39.25	4.63	156	0.52	3.39	w (f)	0.1	
0.013	39.26	4.64	156.2	0.52	3.39	w (f)	0.1	
0.013	39.22	4.6	153.7	0.53	3.67	w (f)	4.9	15.5
0.013	39.26	4.64	156.5	0.52	3.64	w (f)	4.8	15.4
0.013	39.29	4.67	154.8	0.52	3.62	w (f)	4.9	15.4
0.013	39.27	4.65	156.7	0.52	3.81	w (f)	8.6	17
0.013	39.28	4.66	155.7	0.52	3.79	w (f)	8.5	16.7
0.013	39.32	4.71	154.7	0.53	3.8	w (f)	8.6	16.7
0.013	39.22	4.6	155.9	0.52	4.04	w (f)	14.4	18.3
0.013	39.23	4.61	156	0.52	4.09	w (f)	14.6	18.2
0.013	39.27	4.65	155.4	0.52	4.09	w (f)	14.6	18.3
0.013	39.26	4.64	155.7	0.53	4.41	w-a (f)	18.6	18.1
0.013	39.27	4.65	156.2	0.52	4.28	w-a (f)	18.6	18
0.013	39.28	4.66	155.8	0.52	4.32	w-a (f)	18.5	18.4
0.013	39.28	4.66	156.7	0.52	4.63	w-a (f)	25.4	18.2
0.013	39.26	4.64	155.8	0.52	4.71	w-a (f)	25.3	18.5
0.013	39.23	4.61	156.2	0.52	4.64	w-a (f)	25.3	18.2

Table G.9

$\dot{q}_1$	α_1	$\dot{q}_2$	α_2	$\dot{q}_3$	α_3	$\dot{q}_4$	α_4	$\dot{q}_5$	α_5	$\dot{q}_6$	α_6
16.9	28.2	10.5	17.9	9	14.5	4.6	7.7	10.6	17.5	7.5	12.7
20.9	26.8	18.2	23.6	11.4	14.3	8.9	11.2	14.1	18	8.9	11.7
20.9	26.7	18.2	23.5	11.4	14.4	8.9	11.2	14.1	18	8.9	11.8
20.9	26.8	18.2	23.4	11.4	14.3	8.9	11.2	14.1	18	8.9	11.8
21	26.8	18.3	23.4	11.5	14.4	9	11.2	14.2	18	8.9	11.7
24.3	24.8	27	27.7	13.7	14.3	13.6	13.7	16.9	17.5	8	8.3
24.3	24.7	27	27.7	13.7	14.2	13.6	13.8	16.9	17.4	8	8.3
24.3	24.8	27	27.6	13.7	14.2	13.5	13.7	16.9	17.5	8.1	8.3
24.3	24.7	27	27.7	13.7	14.3	13.6	13.8	16.9	17.4	8	8.2
25	18.2	42.2	31.1	22.9	17	22.6	16.6	26.1	19.3	4.3	3.1
24.9	18.3	42.1	31.2	22.7	16.9	22.6	16.6	26.1	19.3	4.2	3.1
24.8	18.3	42.1	31.3	22.7	16.9	22.5	16.5	26	19.3	4.2	3
24.9	18.2	42.1	31.2	22.7	16.9	22.5	16.6	26.1	19.4	4.2	3.1
23	12.7	61	33	42.7	22.6	35.9	19.6	45.4	25	4.4	2.4
23	10.9	60.9	30.5	42.7	21.9	35.9	18.3	45.2	22.1	4.2	2
23	10.9	61	30.6	42.7	21.8	35.9	18.3	45.2	22	4.1	1.9
23	10.9	60.9	30.5	42.7	21.9	35.9	18.3	45.3	22.2	4.2	2
24.7	10	71.1	28.5	70.5	27.4	54.1	21.6	58.6	23.6	0.1	0
24.8	10	71.1	28.5	70.4	27.3	54.1	21.6	58.7	23.5	0.1	0
24.7	9.9	71.1	28.3	70.4	27.3	54.1	21.5	58.7	23.4	0.1	0
24.7	9.8	71.1	28.2	70.4	27.2	54.1	21.5	58.6	23.3	0.1	0
26.8	8.3	83.5	25.8	90.9	28.7	71.8	22.7	76.8	23.9	1.7	0.5
26.8	8.5	83.5	26.2	90.9	28.8	71.8	22.7	76.8	24.2	1.8	0.6
26.9	8.7	83.5	26.5	91.1	29.2	71.9	23	76.8	24.4	1.8	0.6
26.8	8.5	83.6	26.4	91.2	29.2	71.8	22.8	76.8	24.2	1.7	0.5
33.4	6.7	107	21.5	132	27	107.6	21.6	113.6	22.9	2.4	0.5
33.4	6.9	106.7	21.7	131.9	27.2	107.5	21.9	113.6	23.1	2.7	0.6
33.3	6.9	106.8	21.8	131.9	27.2	107.6	22	113.7	23.1	2.3	0.5
33.3	7	106.4	21.9	131.7	27.3	107.3	22	113.5	23.4	2.3	0.5
0.1		0.1		0.1		0.1		0.1		0.1	
0.1		0.1		0.1		0.1		0.1		0.1	
0.1		0.1		0.1		0.1		0.1		0.1	
8.6	27.3	4.4	14.1	4.7	14.8	2	6.3	4.2	13.4	5.4	17.1
8.6	27.1	4.4	14.1	4.6	14.8	1.9	6	4.2	13.3	5.4	17.1
8.6	27.2	4.4	14.1	4.6	14.5	2	6.1	4.2	13.2	5.5	17.3
13.2	26.6	9.7	19.1	7.5	14.7	4.6	9.1	8.4	16.4	8	16.1
13.1	26	9.7	18.9	7.4	14.5	4.6	8.9	8.3	16.2	7.9	15.7
13.2	25.9	9.8	18.8	7.5	14.5	4.6	8.9	8.4	16.1	8	15.7
21.4	27	17.3	22.3	13.6	17.3	9	11.4	14.2	17.8	10.8	13.8
21.6	26.8	17.5	22.2	13.8	17.3	9.2	11.4	14.4	17.7	11	13.8
21.6	27	17.5	22.2	13.8	17.3	9.1	11.4	14.4	17.9	11	13.9
25.1	24.1	27.5	27	19.5	19	13.6	13.4	19	18.4	7.1	6.9
25.1	23.9	27.5	26.8	19.5	19	13.6	13.3	19	18.3	7.1	6.8
25	24.7	27.4	27.1	19.4	19.1	13.5	13.5	18.9	18.8	6.9	7
24.1	17.3	41.9	30.2	29.1	20.9	22.5	16.3	31.3	22.2	3.5	2.5
23.9	17.9	41.8	30.7	28.9	21	22.4	16.3	31.2	22.3	3.5	2.6
24	17.1	41.8	30.1	28.9	21	22.4	16.2	31.3	22	3.5	2.5

Table G.9

$w_{oil,0}$	p	T	$\dot{m}$	$\dot{x}$	$\Delta p/\Delta L$	FP	$\dot{q}$	α
0.013	39.34	4.72	155.9	0.53	5.15	w-a (f)	33.6	18.5
0.013	39.25	4.63	158.7	0.52	5.14	w-a (f)	33.6	18.5
0.013	39.26	4.64	156.4	0.52	5.08	w-a (f)	33.5	18.5
0.013	39.28	4.67	156.5	0.53	6.63	f	46.2	21
0.013	39.27	4.65	157.7	0.52	6.01	f	46.2	20
0.013	39.37	4.75	155.5	0.53	6.57	f	58.8	18.9
0.013	39.23	4.62	155.3	0.53	6.61	f	58.8	19
0.013	39.24	4.63	156.2	0.53	8.12	f	83	17.1
0.013	39.3	4.68	156.7	0.53	8.02	f	83	17
0.013	39.41	4.84	156.6	0.72	5.3	w-a (f)	0.1	
0.013	39.44	4.87	156.7	0.72	5.34	w-a (f)	0.1	
0.013	39.43	4.86	156.8	0.72	5.79	w-a (f)	5.1	17.1
0.013	39.43	4.86	156.5	0.72	5.75	w-a (f)	5.1	17.2
0.013	39.47	4.9	155.4	0.73	6.11	w-a (f)	8.9	18.6
0.013	39.42	4.85	155.8	0.72	6.08	w-a (f)	8.9	18.4
0.013	39.45	4.88	156.6	0.72	6.07	w-a (f)	8.9	18.4
0.013	39.43	4.86	156.5	0.72	6.59	w-a (f)	17.5	21.4
0.013	39.44	4.87	156.5	0.72	6.56	w-a (f)	17.5	21.3
0.013	39.4	4.83	156.3	0.72	7.01	w-a (f)	24.8	20
0.013	39.46	4.88	156.4	0.72	6.95	w-a (f)	24.8	20
0.013	39.46	4.89	156.3	0.72	7.09	w-a (f)	29.2	18.5
0.013	39.47	4.9	157.4	0.72	7.18	w-a (f)	29.3	18.5
0.013	39.41	4.84	157.4	0.71	7.32	f	33.6	17.2
0.013	39.47	4.89	156.4	0.72	7.28	f	33.7	17.2
0.013	39.52	4.95	156.1	0.73	8.2	f	43.8	14.6
0.013	39.48	4.92	156.1	0.73	8.22	f	43.8	14.7
0.013	39.48	4.91	156.6	0.72	8.14	f	43.8	15
0.013	39.48	4.91	156.9	0.72	8.6	f	51	16.1
0.013	39.42	4.85	156.1	0.73	8.6	f	51	16
0.013	39.4	4.83	156.8	0.72	8.75	f	50.7	15.3
0.013	39.42	4.85	157	0.72	9.88	f	69.6	11.4
0.013	39.47	4.9	156.6	0.73	9.89	f	69.4	11.2
0.013	39.44	5.22	155.6	0.92		w-a (f)	0.1	
0.013	39.42	5.17	155.5	0.92		w-a (f)	0.1	
0.013	39.41	5.2	155.4	0.92		w-a (f)	0.1	
0.013	39.45	5.27	155.2	0.93	9.14	w-a (f)	4.7	
0.013	39.42	5.29	154.8	0.93	9.09	w-a (f)	4.7	
0.013	39.45	5.21	156.4	0.92	9.21	w-a (f)	4.7	
0.013	39.49	5.31	154.8	0.93	9.82	f	10.2	
0.013	39.4	5.21	155	0.93	10.43	f	10.3	
0.013	39.48	5.33	154.9	0.93	10.12	f	10.2	
0.013	39.44	5.27	155.2	0.93	10.62	f	25.6	16.8
0.013	39.39	5.19	155.6	0.92	10.88	f	25.5	16.6
0.013	39.42	5.3	154.5	0.93	10.45	f	27.7	14.4
0.013	39.41	5.3	154.6	0.93	10.35	f	27.8	13.6
0.013	39.39	5.25	154.8	0.93	10.31	f	27.8	14
0.013	39.48	5.44	153.6	0.94	10.29	f	29.9	12.2

$\dot{q}_1$	α_1	$\dot{q}_2$	α_2	$\dot{q}_3$	α_3	$\dot{q}_4$	α_4	$\dot{q}_5$	α_5	$\dot{q}_6$	α_6
23.5	12.7	55.5	30.7	43.4	24	36.3	20	39.4	21.6	3.7	2
23.4	12.9	55.4	30.9	43.3	23.9	36.2	20	39.2	21.5	3.7	2
23.4	12.8	55.4	30.9	43.3	24	36.2	20	39.2	21.4	3.7	2
26.9	13.3	68.8	31.7	66.8	29	53.9	23.8	57.8	26.5	2.9	1.4
26.9	12	68.8	30.3	66.8	28.6	53.9	23.1	57.8	24.8	2.7	1.2
29.5	9.4	82	25.9	92.5	30.2	71.9	23.2	75.2	24	1.6	0.5
29.5	9.5	81.9	26.1	92.6	30.2	72	23.4	75.1	24.1	1.6	0.5
31.5	6.7	106.6	21.9	136.9	28.3	107.4	22.1	110.8	22.8	4.6	1
31.4	6.5	106.7	21.7	137	28.2	107.5	22.1	110.8	22.8	4.6	1
0.1		0.1		0.1		0.1		0.1		0.1	
0.1		0.1		0.1		0.1		0.1		0.1	
9.2	30.5	4.2	14	5.8	18.7	1.7	5.8	5.3	17.9	4.6	15.7
9.2	30.5	4.2	14.2	5.9	18.9	1.8	6.3	5.3	17.9	4.5	15.6
15.6	32.3	7.4	15.7	9.3	19.1	4.6	9.8	7.8	16.7	8.7	18.2
15.6	31.6	7.4	15.5	9.2	18.8	4.6	9.7	7.8	16.5	8.7	18.1
15.6	31.8	7.4	15.6	9.2	18.8	4.6	9.7	7.8	16.4	8.7	18.1
25.6	32	22.3	26.9	23.3	27.7	9.3	11.3	14.4	17.7	10.2	13.1
25.5	31.7	22.3	26.8	23.3	27.5	9.3	11.2	14.3	17.5	10.2	13
27.2	22.6	35.6	28.5	42	33.6	13.3	10.9	22.7	18.2	7.8	6.5
27.2	22.7	35.6	28.5	42.1	33.5	13.4	11	22.7	18.1	7.8	6.5
26	16.6	43.7	27.7	50.2	32	22.6	14.4	27.3	16.6	5.8	3.6
26	16.8	43.7	27.8	50.2	31.8	22.7	14.4	27.2	16.6	5.8	3.6
24.3	12.7	53.5	27.5	58.3	29.8	36.1	18.3	29.4	15	0.1	0.1
25.1	13	53.4	27.3	58.3	29.8	36.2	18.3	29.3	14.9	0.1	0.1
16.5	5.6	66.3	22.5	86.6	28.4	53.7	18	39.9	13.3	0.1	0
16.4	5.6	66.3	22.6	86.6	28.2	53.7	18.1	39.9	13.4	0.1	0
16.4	5.7	66.3	23	86.6	29	53.7	18.5	39.9	13.7	0.1	0
14.2	4.5	80.2	25.2	85.6	27	72	22.8	53.7	16.9	0.1	0
14.2	4.5	80.3	25.2	85.5	26.8	72	22.8	53.7	16.8	0.1	0
14.2	4.3	81.6	24.6	83.2	24.9	72	21.9	53.5	16.2	0.1	0
3.6	0.6	107.2	17.7	96.1	15.7	107.7	17.5	102.6	16.9	0.1	0
3.5	0.6	107.1	17.3	95.9	15.6	107.5	17.3	102.4	16.7	0.1	0
0.1		0.1		0.1		0.1		0.1		0.1	
0.1		0.1		0.1		0.1		0.1		0.1	
0.1		0.1		0.1		0.1		0.1		0.1	
10.6		2.2		6.2		2.1		5		2.2	
10.6		2.1		6.2		2.4		4.9		2.2	
10.7		2.1		6.2		2.4		4.9		2.1	
17.2		9.8		10.4		4.5		10.2		9.1	
17.2		9.9		10.4		4.6		10.3		9.1	
17.2		9.8		10.3		4.5		10.2		9.1	
42.2	27.6	35.1	22.5	47	31.5	8.9	5.7	14	9	6.6	4.3
42.1	27.4	35	22.1	46.9	31.2	8.8	5.6	13.9	9	6.5	4.3
35.4	19	48.4	24.9	50.2	25.8	13.8	7.2	17.9	9.5	0.5	0.3
35.4	17.5	48.5	23.6	50.3	24.5	13.8	6.8	17.9	8.7	1	0.5
35.5	18.7	48.5	24	50.2	25.3	13.8	6.8	18	9	0.9	0.5
26.2	10.7	59.4	24	50	20.8	22.4	9.1	19.6	7.9	1.5	0.6

Table G.9

$w_{oil,0}$	p	T	$\dot{m}$	$\dot{x}$	$\Delta p/\Delta L$	FP	$\dot{q}$	α
0.013	39.5	5.3	156.3	0.93	10.51	f	29.8	11.7
0.013	39.54	5.37	155.2	0.93	9.89	f	30	11.6
0.013	39.56	5.4	155.4	0.93	10.29	f	36.4	9.4
0.013	39.59	5.45	155	0.93	10.41	f	36.5	9.5
0.013	39.48	5.37	154.7	0.93	10.51	f	45.2	8.4
0.013	39.53	5.4	154.8	0.93	10.44	f	45.2	8.2
0.013	39.5	5.38	155.1	0.93	10.57	f	45.2	7.9
0.013	39.57	5.37	155.1	0.92	10.82	f	50.8	8.1
0.013	39.52	5.33	154.9	0.93	10.74	f	50.8	8.1
0.013	39.54	5.27	156.9	0.91	10.8	f	50.8	8.2
0.013	39.52	5.25	155.1	0.91	11.99	w (f)	68.5	7.4
0.013	39.54	5.24	156.1	0.91	11.75	w (f)	68.6	7.4
0.013	39.24	5.22	152	0.94		w (f)	0.1	
0.013	39.29	5.04	154.7	0.92	6.27	w (f)	0.1	
0.013	39.29	5.11	152.7	0.93	6.55	w (f)	0.1	
0.013	39.29	5.01	155.4	0.91	6.27	w (f)	0.1	
0.013	39.3	5.13	153.9	0.93	6.95	f	4.8	
0.013	39.28	5.09	153.8	0.93	6.97	f	4.8	
0.013	39.28	5.13	153.4	0.93	6.96	f	4.7	
0.013	39.3	5.09	154	0.92	8.64	f	11.8	25.3
0.013	39.31	5	156.5	0.91	8.74	f	11.8	20.7
0.013	39.29	5	156.1	0.91	8.54	f	11.7	21.4
0.013	39.28	5.08	154.3	0.92	9.63	f	21.6	14.8
0.013	39.26	5.07	154.4	0.93	9.71	f	21.8	14.6
0.013	39.25	5.1	154	0.93	9.63	f	21.9	14.6
0.013	39.27	5.05	154.1	0.92	8.78	f	21.4	13
0.013	39.29	5.11	153.9	0.93	9.67	f	25.3	13.7
0.013	39.3	5.1	154.7	0.92	9.7	f	25.3	13.2
0.013	39.31	5.03	155.9	0.91	9.85	f	25.4	13.1
0.013	39.28	5.14	154.1	0.93	10.52	f	31	12
0.013	39.29	5.11	154.1	0.93	10.66	f	31.1	11.1
0.013	39.3	5.09	155	0.92	10.22	f	29.5	12.8
0.013	39.3	5.06	154.8	0.92	9.93	f	29.1	13
0.013	39.27	4.94	159.7	0.9	9.98	f	37	10
0.013	39.27	4.92	160.3	0.9	10.09	f	37	10
0.013	39.31	4.98	160.1	0.9	12.35	f	43.4	11.9
0.013	39.3	4.96	160.7	0.9	12.34	f	43.5	11.8
0.013	39.34	4.68	263.1	0.12	3.38	i	0.1	
0.013	39.4	4.75	261.6	0.12	3.33	i	0.1	
0.013	39.32	4.66	262	0.12	3.29	i	0.1	
0.013	39.34	4.69	262.3	0.12	3.57	i	4.2	14
0.013	39.36	4.71	263.5	0.12	3.57	i	4.2	13.6
0.013	39.33	4.68	262.7	0.11	3.74	i	8	16.2
0.013	39.31	4.66	261.7	0.12	3.75	i	8.1	16.2
0.013	39.43	4.77	264.5	0.11	4.17	i (f)	16.6	20.3
0.013	39.38	4.72	264.2	0.11	4.23	i (f)	16.6	20.2
0.013	39.24	4.59	261.6	0.12	4.78	i (f)	24.7	22.4

$\dot{q}_1$	α_1	$\dot{q}_2$	α_2	$\dot{q}_3$	α_3	$\dot{q}_4$	α_4	$\dot{q}_5$	α_5	$\dot{q}_6$	α_6
26.1	10.3	59.3	23.1	50	19.9	22.4	8.8	19.5	7.6	1.6	0.6
26.3	10.1	59.5	22.6	50.1	19.9	22.5	8.6	19.8	7.5	1.6	0.6
22.4	5.7	64.5	16.5	69	18.3	35.9	9.4	26.5	6.7	0.1	0
22.4	5.8	64.6	16.7	69.1	17.9	36	9.4	26.5	6.8	0.1	0
17.2	3.2	57.5	10.5	108.2	20.3	53.9	9.8	34.3	6.3	0.1	0
17.2	3	57.5	10.2	108	20.2	53.8	9.5	34.3	6.1	0.1	0
17.3	3	57.6	10	108.1	19.2	53.9	9.2	34.4	5.9	0.1	0
10.9	1.7	64.9	10	116.9	19.6	71.7	11	40.4	6.3	0.1	0
10.8	1.7	64.9	10	116.8	19.5	71.8	11	40.4	6.2	0.1	0
10.8	1.7	64.9	10.1	116.9	19.7	71.8	11.1	40.4	6.3	0.1	0
4.9	0.5	57.3	5.9	176	20.1	107.2	11	65.4	6.7	0.1	0
4.9	0.5	57.3	5.9	176.1	20.6	107.5	11	65.5	6.7	0.1	0
0.1		0.1		0.1		0.1		0.1		0.1	
0.1		0.1		0.1		0.1		0.1		0.1	
0.1		0.1		0.1		0.1		0.1		0.1	
0.1		0.1		0.1		0.1		0.1		0.1	
7.8		5.5		4.1		1.5		4.8		5.3	
7.8		5.5		4		1.5		4.8		5.3	
7.7		5.4		4		1.5		4.7		5.2	
20	41.2	13.9	29.2	11.1	25.5	4.5	10.1	10.4	22.1	10.8	23.6
20.1	34.1	13.9	24	11.1	20.6	4.5	8.2	10.4	18.1	10.7	19.2
20	35.3	13.8	24.7	11	21.2	4.4	8.5	10.3	18.9	10.6	19.9
28.5	19.8	34.4	23.7	39.1	26.4	9	6	15.3	10.5	3.4	2.4
28.8	19.4	34.6	23.5	39.3	26	9.2	6.1	15.4	10.4	3.6	2.5
28.8	19.1	34.7	23.5	39.4	26.4	9.2	6.1	15.4	10.2	3.6	2.4
28.4	17.1	35.2	21.6	40.1	24.5	8.8	5.3	11.9	7.2	4.1	2.5
26.7	14.4	45.6	24.8	43.1	23.6	13.5	7.3	18.2	9.8	4.4	2.4
26.8	13.7	45.6	23.8	43.2	23.1	13.5	7.1	18.3	9.4	4.4	2.3
26.9	13.8	45.7	23.6	43.2	22.5	13.6	7.1	18.3	9.4	4.4	2.3
27.5	10.6	58.5	22.1	54.9	21.5	22.5	8.6	19.1	7.5	3.8	1.5
27.8	10	58.5	20.6	54.9	19.8	22.5	8	19.1	7	3.8	1.4
22.5	10.1	55.6	24	53.6	22.9	22.6	9.8	19	8.4	3.7	1.7
22.3	10.3	55.5	24.6	52	23.1	22.5	10	18.8	8.5	3.6	1.7
24.3	6.6	68.5	18.5	67.2	18.5	36	9.5	22	5.9	4.1	1.1
24.3	6.8	68.5	18.8	67.2	17.8	35.9	9.6	22	6.1	4.1	1.1
28.7	7.8	75.2	20.2	72	20.2	53.9	14.7	27.9	7.6	2.6	0.7
28.8	7.8	75.3	20	72.1	20.3	53.9	14.6	28	7.6	2.9	0.8
0.1		0.1		0.1		0.1		0.1		0.1	
0.1		0.1		0.1		0.1		0.1		0.1	
0.1		0.1		0.1		0.1		0.1		0.1	
9.6	31.5	2.9	9.9	3.5	11.5	2.2	7.2	2.9	9.7	4.2	14.3
9.5	30.7	2.9	9.8	3.3	10.7	2.1	6.8	2.9	9.7	4.2	14.2
14.1	29.3	8.2	16.4	6.6	12.8	4.2	8.4	5.9	12.1	8.8	18.4
14.1	28.9	8.2	16.3	6.6	12.6	4.9	9.4	5.9	11.9	8.8	18.3
26.1	32.1	20.4	24.5	12.3	15	8.9	10.8	14.9	18.1	17	21.3
26.1	32.1	20.3	24.4	12.3	15	8.8	10.7	14.9	18	16.9	21.1
32.9	29.7	32.1	28.6	22.2	19.8	13.5	12.4	23.8	21.8	23.6	22.1

Table G.9

$w_{oil,0}$	p	T	$\dot{m}$	$\dot{x}$	$\Delta p/\Delta L$	FP	$\dot{q}$	α
0.013	39.27	4.62	261.7	0.12	4.67	i (f)	24.5	22.6
0.013	39.2	4.54	260.8	0.12	4.69	i (f)	24.6	22.3
0.013	39.2	4.55	261.4	0.12	5.49	i (f)	34.7	23.8
0.013	39.23	4.57	262.1	0.11	5.51	i (f)	34.8	23.9
0.013	39.29	4.64	263.7	0.12	6.33	i (f)	50.5	25.6
0.013	39.32	4.67	263.4	0.12	6.22	i (f)	50.5	25.9
0.013	39.26	4.6	264.6	0.11	6.58	i (f)	64.3	27
0.013	39.28	4.63	264.4	0.11	6.57	i (f)	64.3	27.1
0.013	39.24	4.59	265.4	0.11	6.6	i (f)	64.2	27.2
0.013	39.25	4.59	264.6	0.12	7.56	i (f)	78.6	27.5
0.013	39.27	4.61	264.2	0.12	7.55	i (f)	78.6	27.5
0.013	39.4	4.75	265.6	0.11	9.16	a (f)	106.9	27.4
0.013	39.38	4.73	266.8	0.11	9.49	a (f)	106.7	27.3
0.013	39.43	4.78	266.9	0.11	9.39	a (f)	105.6	27.7
0.013	39.27	4.65	262.2	0.52	9.55	a	0.1	
0.013	39.33	4.71	262.1	0.51	9.54	a	0.1	
0.013	39.32	4.7	262.8	0.52	9.58	a	0.1	
0.013	39.35	4.73	264.2	0.51	9.86	a	4.7	16.6
0.013	39.35	4.73	263.9	0.51	9.88	a	4.7	16.5
0.013	39.23	4.61	263	0.51	10.36	a	9.7	18.6
0.013	39.21	4.59	263	0.51	10.41	a	9.8	18.4
0.013	39.22	4.6	262.9	0.51	10.32	a	9.8	18.4
0.013	39.21	4.59	263.1	0.52	10.99	a (f)	15.5	21.2
0.013	39.2	4.58	261.7	0.52	11.01	a (f)	15.5	21.2
0.013	39.29	4.67	262.5	0.52	11.69	a (f)	25.8	24.8
0.013	39.2	4.58	262.5	0.51	11.78	a (f)	25.9	24.9
0.013	39.23	4.61	262.5	0.51	12.78	a (f)	41.1	26.6
0.013	39.19	4.57	262.2	0.52	12.75	a (f)	41.1	26.6
0.013	39.23	4.61	262.5	0.51	12.78	a (f)	41.1	26.7
0.013	39.26	4.64	264	0.51	13.28	a (f)	49	24.8
0.013	39.25	4.63	264.4	0.51	13.35	a (f)	49	25
0.013	39.25	4.63	263.3	0.52	13.33	a (f)	49	24.9
0.013	39.26	4.64	263.6	0.51	13.83	a (f)	57.6	23.4
0.013	39.26	4.63	265	0.51	13.83	a (f)	57.4	23.4
0.013	39.27	4.65	263.5	0.52	14.35	a (f)	65.4	22.5
0.013	39.25	4.63	263.6	0.51	14.39	a (f)	65.3	22.6
0.013	39.28	4.66	264	0.51	14.35	a (f)	65.3	22.6
0.013	39.24	4.62	263.2	0.51	15.87	a (f)	83.6	20.9
0.013	39.22	4.6	263	0.52	15.88	a (f)	83.5	21
0.013	39.25	4.67	264.6	0.7	12.81	a	0.1	
0.013	39.19	4.61	264.3	0.7	12.79	a	0.1	
0.013	39.23	4.65	263.7	0.71	13.26	a	3.9	20.4
0.013	39.25	4.67	263.9	0.7	13.2	a	3.8	20.4
0.013	39.25	4.68	263.1	0.71	13.46	a	7.5	19.4
0.013	39.25	4.68	264.2	0.71	13.46	a	7.5	19.2
0.013	39.24	4.66	264	0.71	14.01	a	12.6	20.4
0.013	39.23	4.65	264	0.7	14.1	a	12.6	20.2

Table G.9

$\dot{q}_1$	α_1	$\dot{q}_2$	α_2	$\dot{q}_3$	α_3	$\dot{q}_4$	α_4	$\dot{q}_5$	α_5	$\dot{q}_6$	α_6
32.8	30.7	31.9	28.9	22.1	19.8	13.4	12.3	23.6	21.7	23.5	22.1
32.8	29.7	32	28.8	22.2	19.6	13.4	12.2	23.7	21.6	23.5	21.9
40.1	27.3	41.9	28.9	35.4	23.9	22.7	15.7	35.3	24.1	33	22.8
40.1	27.5	41.9	29.3	35.4	24	22.7	15.8	35.4	24.2	33	22.4
48.8	24.8	65.9	32.7	51.6	26.6	36.2	18.9	54.9	27.4	45.3	22.9
48.8	25.4	65.9	33.4	51.6	26.7	36.2	18.9	54.9	27.6	45.4	23.1
52.3	22.4	77.2	32.4	74.1	30.7	53.9	22.8	66.8	28.1	61.3	25.6
52.2	22.6	77.1	32.5	74.1	30.8	53.9	22.8	66.9	28.1	61.3	25.7
52.1	22.9	77	32.8	74.2	30.8	53.9	22.8	66.8	28.2	61.3	25.9
56.3	19.9	87.3	30.6	95.1	33.1	72.2	25.2	84.4	29.3	76.5	26.6
56.3	19.7	87.3	30.5	95.1	33.3	72.2	25.3	84.3	29.3	76.4	26.6
83.3	21.2	108.1	27.6	139	35.9	107.1	27.6	112.1	28.7	91.8	23.5
83	20.7	108	26.9	138.6	35.9	107	27.7	111.9	28.8	91.5	23.6
80.4	20.8	103.3	26.8	138.9	36.5	107.2	28.1	112.2	29.4	91.7	24.4
0.1		0.1		0.1		0.1		0.1		0.1	
0.1		0.1		0.1		0.1		0.1		0.1	
0.1		0.1		0.1		0.1		0.1		0.1	
7.2	26.7	4.4	15.6	4.5	15.2	2.2	7.5	4.1	14.6	5.6	19.7
7.2	26.5	4.4	15.5	4.5	15	2.3	7.8	4.1	14.5	5.6	19.6
16.2	31.2	9.2	17.6	7.6	14.6	4.6	9	10.1	19	10.3	20.3
16.4	30.8	9.3	17.3	7.8	14.6	4.7	8.8	10.2	18.7	10.4	20
16.4	30.8	9.3	17.3	7.8	14.6	4.6	8.7	10.2	18.8	10.5	20.1
27.6	36.8	15.4	20.8	10.6	14.6	8.9	12.2	13.6	18.9	16.9	23.7
27.6	36.9	15.4	20.9	10.5	14.5	8.9	12.2	13.5	18.8	17	23.8
42.8	40.9	28.8	27.7	20.5	19.4	13.8	13.4	21.5	21	27.5	26.7
42.8	41.1	28.9	27.7	20.6	19.4	13.9	13.4	21.5	21	27.5	26.9
60	39	50.1	32.4	49.4	31.4	22.7	14.7	34	22.1	30.5	20.2
60	39.1	50.1	32.4	49.4	31.3	22.7	14.7	34	22.2	30.5	20.2
60	39.1	50.1	32.4	49.4	31.5	22.7	14.8	34	22.2	30.4	20.2
61.9	31.6	66.6	33.7	65.2	32.6	36	18.3	43.3	21.9	21.2	11
61.9	31.8	66.6	33.8	65.1	32.7	35.9	18.3	43.2	22	21.2	11.1
61.9	31.7	66.6	33.8	65.1	32.7	35.9	18.3	43.2	22	21.1	10.9
57.7	24	82.3	33.5	85.3	34.3	53.8	22	54.5	22.1	11.7	4.8
57.5	23.9	82.2	33.6	85.1	34.2	53.7	22	54.4	22.1	11.6	4.8
55.6	19.2	93.6	32.1	102.5	35.2	71.9	24.8	64.3	22.1	4.2	1.5
55.6	19.4	93.5	32.2	102.3	35.4	71.9	24.9	64.4	22.2	4.2	1.5
55.6	19.4	93.5	32.2	102.3	35.3	71.9	24.9	64.4	22.2	4.2	1.5
48	12.1	114.2	28.4	142.8	35.9	107.4	26.9	86.3	21.6	2.9	0.7
48	12.1	114.2	28.5	142.6	35.9	107.3	26.9	86.3	21.6	2.5	0.6
0.1		0.1		0.1		0.1		0.1		0.1	
0.1		0.1		0.1		0.1		0.1		0.1	
5.4	28.4	4.1	21.4	4.7	23.7	1.8	10.1	2.7	14.7	4.6	24.1
5.3	27.6	4	21.5	4.6	23.9	1.5	8.6	2.9	16.3	4.6	24.3
10.7	27.8	7.2	18.8	7.8	19.6	4.7	12.3	7.5	19.4	6.9	18.3
10.8	27.6	7.2	18.7	7.8	19.5	4.7	12.2	7.5	19.1	7	18.2
16.5	26.9	13	21	11.2	18	9	14.8	12.3	20	13.3	21.7
16.6	26.6	13.1	20.7	11.2	17.7	9.1	14.6	12.4	19.8	13.3	21.5

Table G.9

$w_{oil,0}$	p	T	$\dot{m}$	$\dot{x}$	$\Delta p/\Delta L$	FP	$\dot{q}$	α
0.013	39.24	4.66	263.6	0.71	14.05	a	12.6	20.2
0.013	39.25	4.68	263.8	0.7	15.51	a	21.5	24.1
0.013	39.25	4.68	263.7	0.71	15.45	a (f)	21.5	24
0.013	39.28	4.7	263.5	0.71	17.05	a (f)	48.6	25.1
0.013	39.27	4.69	264.4	0.7	17.13	a (f)	48.5	25.2
0.013	39.23	4.66	263.8	0.71	16.9	a (f)	50	21.4
0.013	39.27	4.69	264.7	0.7	16.96	a (f)	50	21.6
0.013	39.29	4.72	264.4	0.71	17.38	a (f)	54	20.4
0.013	39.28	4.71	262.5	0.71	17.51	a (f)	53.9	20.3
0.013	39.27	4.69	263	0.71	17.46	a (f)	54	20.5
0.013	39.23	4.65	262.6	0.71	18.08	a (f)	59.6	17.1
0.013	39.25	4.68	262.4	0.71	17.96	a (f)	59.5	17
0.013	39.25	4.68	263.2	0.7	18.69	a (f)	65.8	14.9
0.013	39.28	4.71	263	0.71	18.62	a (f)	65.9	14.9
0.013	39.27	4.69	263.6	0.7	18.64	a (f)	65.8	15
0.013	39.27	4.7	256.7	0.73	19.97	a (f)	77.5	12.3
0.013	39.22	4.65	256.1	0.73	20.14	a (f)	77.3	12.4
0.013	39.26	4.69	257	0.73	20.14	a (f)	77.2	12.4
0.013	39.32	4.66	521.9	0.12	20.31	a (f)	0.1	
0.013	39.2	4.55	523.5	0.11	20.36	a (f)	0.1	
0.013	39.3	4.65	523.7	0.11	20.28	a (f)	0.1	
0.013	39.27	4.62	523.8	0.1	20.46	a (f)	3.6	17.5
0.013	39.35	4.69	523.6	0.11	20.27	a (f)	3.7	16.8
0.013	39.22	4.57	522.9	0.11	21.4	a (f)	7.5	17.2
0.013	39.25	4.6	522.9	0.11	21.14	a (f)	7.5	17.2
0.013	39.26	4.61	522.7	0.11	21.06	a (f)	7.5	17
0.013	39.22	4.57	523.2	0.11	22.34	a (f)	17.2	22.9
0.013	39.26	4.6	522.6	0.11	22.22	a (f)	17.3	22.9
0.013	39.23	4.58	523.9	0.11	23.09	a (f)	23.6	26.4
0.013	39.26	4.61	523.7	0.11	22.95	a (f)	23.6	26.4
0.013	39.23	4.57	526.9	0.11	23.27	a (f)	42.1	33.1
0.013	39.27	4.62	526.3	0.11	23.19	a (f)	42.1	33.1
0.013	39.24	4.59	526.1	0.11	23.31	a (f)	42.1	33
0.013	39.31	4.66	524.1	0.11	23.57	a (f)	58.6	36.1
0.013	39.23	4.57	523.8	0.11	23.85	a (f)	58.7	36.2
0.013	39.3	4.65	526	0.11	24.69	a (f)	76.9	38.2
0.013	39.33	4.68	526.4	0.11	24.6	a (f)	76.9	38.2
0.013	39.33	4.68	524.6	0.11	24.57	a (f)	77	38.2
0.013	39.3	4.65	521.8	0.11	25.43	f	95.8	38.9
0.013	39.35	4.69	525.4	0.1	25.01	f	95.6	39.1
0.013	39.14	4.48	525.8	0.1	25.18	f	95.6	39.2
0.013	39.28	4.62	524.7	0.11	27.58	f	125.9	37.8
0.013	39.2	4.55	525.4	0.11	27.69	f	126	38.5
0.013	39.31	4.66	525	0.1	27.28	f	126	39
0.013	39.3	4.65	524.7	0.11	27.34	f	125.9	39.4
0.013	39.52	4.9	524.4	0.5	30.87	a	0.1	
0.013	39.52	4.9	525.6	0.49	30.87	a	0.1	

$\dot{q}_1$	α_1	$\dot{q}_2$	α_2	$\dot{q}_3$	α_3	$\dot{q}_4$	α_4	$\dot{q}_5$	α_5	$\dot{q}_6$	α_6
16.5	26.5	13	20.8	11.2	17.8	9.1	14.6	12.3	19.8	13.3	21.5
31.1	34.6	18.2	20.7	19.8	21.7	13.7	15.5	22	24.8	23.9	27.2
31.1	34.4	18.2	20.6	19.8	21.5	13.7	15.4	22.1	24.8	23.9	27.1
70.7	35.7	56.7	28.5	69.8	39.3	22.6	11.3	43.6	21.7	27.9	14.3
70.6	35.7	56.7	28.5	69.7	39.4	22.6	11.3	43.6	21.9	27.9	14.5
70.7	29.9	73.4	30.6	74	33.3	22.2	9.4	40.2	16.9	19.7	8.5
70.6	30.3	73.4	30.7	74	33.4	22.2	9.5	40.1	17.3	19.6	8.6
68.7	25.9	78.4	29.5	80	30.6	36	13.6	47.5	17.7	13.2	5
68.6	25.8	78.4	29.4	79.9	30.3	35.9	13.5	47.5	17.8	13.1	5
68.8	26	78.6	29.6	80.1	30.7	36	13.7	47.6	17.9	13.2	5
56.1	16.4	102.7	29.5	86.4	23.9	54	15.8	52.8	15.4	5.3	1.5
56	16.2	102.6	29.4	86.4	23.9	54	15.8	52.9	15.3	5.3	1.5
45.3	10.5	121.1	27.6	94.2	20.4	71.8	16.6	62.3	14.5	0.1	0
45.4	10.5	121.2	27.6	94.3	20.5	71.9	16.7	62.3	14.5	0.1	0
45.4	10.5	121.2	27.6	94.3	20.6	71.8	16.7	62.2	14.5	0.1	0
30.3	4.9	131.1	20.8	116.1	18.7	107.4	17	80.1	12.8	0.1	0
30.2	4.9	130.9	21	115.9	18.7	107.1	17.1	79.9	12.8	0.1	0
30.1	4.9	130.7	21.1	115.7	18.6	106.9	17.2	79.7	12.9	0.1	0
0.1		0.1		0.1		0.1		0.1		0.1	
0.1		0.1		0.1		0.1		0.1		0.1	
0.1		0.1		0.1		0.1		0.1		0.1	
5.4	27.6	2.4	11.2	3.6	17	1.7	8.1	3.2	14.5	5.6	26.6
5.5	25.8	2.6	11.8	3.7	15.8	1.7	7.5	3.3	14.3	5.6	25.4
12	27.9	5.1	11.3	7.6	16.1	4.4	9.9	5.4	12.7	10.7	25.2
11.9	27.8	5.1	11.4	7.6	16.1	4.4	9.9	5.4	12.6	10.8	25.2
11.9	27.6	5.1	11.3	7.6	15.9	4.3	9.7	5.4	12.6	10.8	25
30	39.9	12.4	16.6	13.5	17.9	9.1	12.2	14.4	19.2	24	32
30	39.9	12.5	16.6	13.6	17.9	9.2	12.1	14.5	19	24.1	32.1
42.8	47.3	17.3	19.5	17.4	19.5	13.4	15.1	17.2	19.5	33.6	37.4
42.8	47.1	17.3	19.5	17.4	19.5	13.4	15	17.2	19.6	33.8	37.4
75.7	58.9	35.1	27.7	34.2	26.8	22.4	17.9	32.6	25.9	52.5	41.5
75.7	59	35.1	27.7	34.2	26.7	22.4	17.9	32.6	25.7	52.5	41.4
75.7	58.8	35.1	27.7	34.2	26.7	22.4	17.8	32.6	25.8	52.5	41.5
102.2	62	53.7	33.3	50.8	31.4	36.1	22.5	48	29.7	60.6	37.7
102.4	62.3	53.7	33.2	51.8	31.7	36.1	22.4	47.9	29.6	60.6	37.7
120.5	59.4	75.1	37.3	74.9	37.1	53.4	26.7	66.4	33.1	71.1	35.4
120.4	59.3	75.1	37.4	75	37.2	53.5	26.7	66.5	33.1	71.1	35.5
120.5	59.4	75.2	37.3	75.1	37.2	53.6	26.7	66.5	33.1	71.1	35.5
136.3	55	91.3	37.3	104.7	42.5	72	29.6	86.9	35.4	83.6	33.9
136.2	55.3	91.1	37.5	104.3	42.5	71.9	29.7	86.6	35.6	83.3	34.1
136.2	55.6	91.3	37.8	104.6	42.6	71.9	29.7	86.5	35.5	83.3	34.3
154.1	46.1	119.2	35.7	153.3	45.9	107.5	32.4	120.1	36.1	101.4	30.5
154.2	47.1	119.2	36.4	153.4	46.8	107.6	33	120.1	36.7	101.3	30.9
154.2	47.6	119.2	36.9	153.4	47.5	107.6	33.5	120.2	37.1	101.4	31.2
154.3	48.2	119.1	37.4	153.3	48.1	107.4	33.8	120.1	37.5	101.3	31.4
0.1		0.1		0.1		0.1		0.1		0.1	
0.1		0.1		0.1		0.1		0.1		0.1	

Table G.9

$w_{oil,0}$	p	T	$\dot{m}$	$\dot{x}$	$\Delta p/\Delta L$	FP	$\dot{q}$	α
0.013	39.46	4.84	526.4	0.49	30.93	a	0.1	
0.013	39.54	4.91	525.7	0.49	31.19	a (f)	2.2	25.9
0.013	39.46	4.83	525.9	0.49	31.16	a (f)	2.2	26.1
0.013	39.47	4.85	526	0.49	31.01	a (f)	2.2	25.7
0.013	39.49	4.87	526.1	0.49	31.43	a (f)	5	19.6
0.013	39.51	4.89	525.5	0.49	31.44	a (f)	4.9	19.4
0.013	39.46	4.83	526.9	0.49	31.51	a (f)	4.8	19.3
0.013	39.6	4.97	526.6	0.49	31.92	a (f)	10.7	20.7
0.013	39.5	4.88	525	0.5	32.09	a (f)	10.8	20.6
0.013	39.51	4.89	525.9	0.49	32.12	a (f)	10.8	20.5
0.013	39.53	4.91	517.7	0.51	31.74	a (f)	15.8	23.3
0.013	39.52	4.89	517.1	0.5	31.83	a (f)	16	23.4
0.013	39.51	4.89	518.1	0.5	31.81	a (f)	15.9	23.4
0.013	39.6	4.97	519.4	0.5	33.36	a (f)	29.7	29.1
0.013	39.53	4.91	519.9	0.5	33.28	a (f)	29.7	29.1
0.013	39.55	4.93	519.5	0.5	33.22	a (f)	29.7	29
0.013	39.53	4.9	521.4	0.5	35.86	a (f)	49	35.1
0.013	39.53	4.9	519.8	0.5	35.98	a (f)	49.1	35.1
0.013	39.45	4.82	521.2	0.5	35.98	a (f)	49.2	35.1
0.013	39.65	5.02	521.9	0.5	38.66	a (f)	75	39.2
0.013	39.59	4.97	521.4	0.5	38.7	a (f)	75	39.3
0.013	39.55	4.93	521.6	0.5	38.77	a (f)	75	39.3
0.013	39.58	4.95	523.8	0.5	42.77	a (f)	106.3	41.5
0.013	39.59	4.97	524.9	0.5	42.54	a (f)	106.1	41.7
0.013	39.55	4.93	523.1	0.5	42.79	a (f)	103.1	40.8
0.013	39.53	4.91	525.6	0.49	43.01	a (f)	103.1	41.2
0.013	39.6	4.97	524.4	0.5	44.15	a (f)	116.6	37.8
0.013	39.61	4.99	521.9	0.5	44.28	a (f)	117	37.3
0.013	39.51	4.93	520.6	0.7	35.45	a	0.1	
0.013	39.57	4.98	520.4	0.7	35.34	a	0.1	
0.013	39.45	4.87	517.4	0.71	35.35	a	0.1	
0.013	39.49	4.9	531.3	0.68	36.08	a	0.1	
0.013	39.53	4.94	530.6	0.69	36.18	a	0.1	
0.013	39.55	4.96	521.2	0.69	35.93	a	3.3	16.7
0.013	39.52	4.94	521.2	0.7	35.72	a	3.3	16.6
0.013	39.49	4.92	516.5	0.7	36.02	a	3.3	16.7
0.013	39.54	4.96	521.2	0.7	36.2	a	7.7	16.1
0.013	39.53	4.96	519.3	0.7	36.36	a	7.7	16.2
0.013	39.54	4.95	530.2	0.68	37.57	a (f)	11.6	16
0.013	39.5	4.91	532.7	0.68	37.37	a (f)	11.6	16.1
0.013	39.49	4.91	530.6	0.68	38.8	a (f)	17.6	18.4
0.013	39.48	4.9	528.8	0.69	39.08	a (f)	17.6	18.8
0.013	39.54	4.95	530.3	0.69	41.39	a (f)	28	24.7
0.013	39.47	4.89	529.4	0.69	41.69	a (f)	28	25
0.013	39.5	4.92	529.6	0.69	45.22	a (f)	42.7	29.2
0.013	39.53	4.94	527.9	0.69	45.49	a (f)	42.6	29.5
0.013	39.59	5	539.7	0.67	50.48	a (f)	101.3	19.8

$\dot{q}_1$	α_1	$\dot{q}_2$	α_2	$\dot{q}_3$	α_3	$\dot{q}_4$	α_4	$\dot{q}_5$	α_5	$\dot{q}_6$	α_6
0.1		0.1		0.1		0.1		0.1		0.1	
2.4	29.9	2.8	33.1	2.5	31.5	1.7	18.9	1.5	16.3	2.6	26.1
2.3	27.7	2.9	35.3	2.5	32.5	1.8	20.9	1.4	15.9	2.3	24.4
2.3	28.6	2.9	33.4	2.5	32.2	1.8	21.6	1.4	15.9	2.1	22.8
5.2	20.8	6.1	23.8	5	19.9	5	19	3.8	14.7	5.1	19.6
5.2	20.7	6.1	23.7	4.9	20.3	4.2	17	3.9	15.1	5.2	19.8
5.1	20.8	6	23.5	4.9	20.3	4.2	16.8	3.7	14.9	5.1	19.6
14.4	27.3	10.1	19.6	10.5	20.4	9.1	17.7	10	19.5	10.1	19.6
14.5	27.2	10.2	19.5	10.6	20.3	9.2	17.6	10.1	19.4	10.2	19.6
14.5	27.1	10.1	19.4	10.5	20.2	9.1	17.5	10.1	19.3	10.2	19.5
20.5	30.4	13.9	21.2	15	22	13.6	19.9	14.7	21.8	17	24.8
20.7	30.4	14.1	21.2	15.2	22	13.7	19.9	14.9	21.9	17.2	25
20.7	30.4	14	21.1	15.1	22.1	13.7	19.9	14.9	21.8	17.2	25
39.6	39.4	27.3	27.3	28.6	27	22.4	21.9	28.6	28	31.8	31.4
39.6	39.3	27.3	27.2	28.6	27	22.3	21.8	28.6	27.9	31.9	31.3
39.5	39.1	27.3	27.1	28.6	26.9	22.3	21.8	28.6	27.9	31.9	31.3
70.3	49.8	48.8	35.1	43.3	30.8	36	26.2	43.7	31.6	51.8	37.1
70.5	49.9	49	35	43.5	30.8	36.2	26.2	43.7	31.6	51.9	37.2
70.5	50	49	35	43.5	30.8	36.3	26.2	43.8	31.6	52	37.3
105.6	55	74.1	38.8	81.8	41.5	54	28.5	64.2	33.9	70.4	37.4
105.7	55.3	74.1	39	81.8	41.6	54	28.6	64.1	33.9	70.3	37.6
105.6	55.2	74.1	39	81.8	41.4	54	28.6	64.1	33.9	70.4	37.7
134.9	49.8	91.8	33.9	179.2	79.6	71.9	26.4	89.2	33	70.6	26.7
134.6	49.9	91.6	33.9	178.9	79.9	71.7	26.5	89	33	70.5	26.7
141.7	51.7	109.1	39.8	120.3	63.9	71.2	26.1	82.7	29.9	93.5	33.6
141.9	52	109.2	40	120.2	64.5	71.2	26.4	82.5	30.2	93.6	33.9
140.6	42.1	123.8	36.6	178.9	71.1	107.2	31.8	99.2	29.9	49.8	15.3
141.1	41.4	124.4	36.3	179	70.3	107.7	31.5	99.7	29.3	50	14.8
0.1		0.1		0.1		0.1		0.1		0.1	
0.1		0.1		0.1		0.1		0.1		0.1	
0.1		0.1		0.1		0.1		0.1		0.1	
0.1		0.1		0.1		0.1		0.1		0.1	
0.1		0.1		0.1		0.1		0.1		0.1	
6.2	30.6	2.7	13	4.2	22.1	2.3	11.5	3.6	17.5	1	5.5
6.3	30.2	2.8	13.4	4.2	22.7	2.2	11.1	3.6	17.5	0.9	4.9
6.2	29.9	2.5	12.5	4.1	22.4	2.3	11.5	3.6	17.3	1.2	6.3
9.2	20.3	7.2	14.7	10.7	22	4.8	10.2	8	16.5	6.1	13
9.2	20	7.4	15.1	10.7	22	5.1	10.7	8	16.4	6	12.9
14.2	20.1	10.6	14.6	14.4	19.8	9	12.5	11.5	15.6	9.6	13.5
14.3	20.2	10.9	15	14.4	19.9	9.1	12.6	11.4	15.5	9.6	13.6
20.8	22	15.5	16	22.4	23.1	13.3	13.9	17	17.5	16.7	17.7
20.8	22.5	15.6	16.3	22.3	23.6	13.3	14.2	17.1	17.9	16.7	18
34.6	30.5	25.3	22.1	32.2	28.5	22.5	19.9	25.9	22.7	27.2	24.2
34.6	30.9	25.3	22.4	32.2	29	22.6	20.2	26	22.9	27.2	24.5
51.4	35.5	38.6	26.2	46.8	32	36.1	24.6	38.1	26	45.4	31
51.1	35.8	38.6	26.4	46.7	32.3	35.9	24.9	38	26.3	45.2	31.3
143.5	23.4	148.6	25	149.6	44.4	53.8	8.4	112.4	17.8	0.1	0

Table G.9

$w_{oil,0}$	p	T	$\dot{m}$	$\dot{x}$	$\Delta p/\Delta L$	FP	$\dot{q}$	α
0.013	39.54	4.96	529.9	0.68	49.83	a (f)	69.9	34.3
0.013	39.5	4.92	527.7	0.69	49.79	a (f)	69.7	34.7
0.013	39.53	4.95	528	0.69	50	a (f)	69.7	34.9
0.02	25.96	-10.67	79.6	0.15	0.16	w (f)	0.2	
0.02	25.97	-10.67	77.6	0.15	0.23	w (f)	0.2	
0.02	25.97	-10.67	77.6	0.14	0.26	w (f)	3	5.2
0.02	25.97	-10.67	78.6	0.13	0.26	w (f)	2.9	5.1
0.02	25.98	-10.65	78	0.16	0.38	w (f)	8.6	7.7
0.02	25.98	-10.65	77.7	0.15	0.47	i (f)	8.6	7.7
0.02	25.99	-10.65	78	0.15	0.68	i (f)	13.9	8.3
0.02	25.98	-10.65	77.7	0.15	0.73	i (f)	13.9	8.3
0.02	25.99	-10.63	77.4	0.15	1.2	i (f)	23.3	10.4
0.02	25.99	-10.64	76.6	0.16	1.18	i (f)	23.3	10.4
0.02	25.91	-10.75	79.3	0.16	1.6	i (f)	32.6	12
0.02	25.92	-10.73	77.9	0.15	1.58	i (f)	32.7	12
0.02	25.93	-10.71	76.8	0.16	2.36	i (f)	48	14.1
0.02	25.93	-10.72	77.1	0.16	2.37	i (f)	47.9	14.2
0.02	25.94	-10.7	76.7	0.17	2.98	i (f)	60.5	15.3
0.02	25.94	-10.7	77.7	0.17	3.04	i (f)	60.5	15.3
0.02	25.95	-10.69	78.2	0.19	3.88	f	73.1	15.6
0.02	25.96	-10.68	77.4	0.2	3.92	f	73	15.8
0.02	25.96	-10.68	75.7	0.25	5.2	f	89.8	15.7
0.02	25.96	-10.68	74.6	0.25	5.28	f	89.7	15.7
0.02	25.92	-10.73	82.1	0.33	0.78	w (f)	0.2	
0.02	25.92	-10.73	80.8	0.34	0.73	w (f)	0.2	
0.02	25.93	-10.72	81.3	0.34	0.76	w (f)	0.2	
0.02	25.92	-10.73	81.1	0.34	0.94	w (f)	6.2	8.1
0.02	25.93	-10.72	81.5	0.34	0.93	w (f)	6.3	7.9
0.02	25.93	-10.72	81.3	0.35	1.24	w (f)	10.9	8.2
0.02	25.92	-10.73	81.8	0.34	1.16	w (f)	9.5	8.2
0.02	25.93	-10.71	82.4	0.33	1.43	i-a (f)	15.7	10.3
0.02	25.92	-10.72	81.8	0.33	1.56	i-a (f)	15.7	10.1
0.02	25.93	-10.71	74	0.38	2.05	i-a (f)	24.2	10.8
0.02	25.93	-10.71	80.8	0.35	2.03	i-a (f)	24.1	11.1
0.02	25.93	-10.71	76.1	0.37	2.73	w (f)	31.1	11.6
0.02	25.93	-10.71	75.9	0.37	2.72	i-a (f)	31.1	11.6
0.02	25.95	-10.69	74.4	0.37	3.3	i-a (f)	42.4	12.8
0.02	25.95	-10.69	73.6	0.38	3.4	i-a (f)	42.5	13
0.02	25.95	-10.69	74.7	0.38	3.91	i-a (f)	51.5	13.3
0.02	25.95	-10.69	74.7	0.38	4.1	i-a (f)	51.4	13.4
0.02	25.95	-10.69	76	0.37	4.44	f	58.5	13.4
0.02	25.96	-10.68	76.8	0.37	4.66	f	58.4	13.7
0.02	25.96	-10.67	82.3	0.35	5.33	f	70.8	14.2
0.02	25.94	-10.7	82.9	0.34	5.89	f	70.6	14.3
0.02	25.91	-10.74	153.9	0.13	0.95	i	0.2	
0.02	25.91	-10.75	153.9	0.13	1.01	i	0.2	
0.02	25.93	-10.72	151.5	0.13	1.99	i (f)	11.5	12.6

$\dot{q}_1$	α_1	$\dot{q}_2$	α_2	$\dot{q}_3$	α_3	$\dot{q}_4$	α_4	$\dot{q}_5$	α_5	$\dot{q}_6$	α_6
89.4	44.2	57.9	28.4	81.3	39.3	54.8	26.9	62.6	30.9	73.3	36.3
89.3	44.8	57.9	28.8	81.3	40	53.9	26.9	62.6	31.3	73.3	36.7
89.4	45.3	57.9	28.9	81.2	40.2	53.9	27.1	62.5	31.4	73.2	36.8
0.2		0.2		0.2		0.2		0.2		0.2	
0.2		0.2		0.2		0.2		0.2		0.2	
3.2	5.3	4.9	9	1.4	2.8	1.2	2.2	2	3.6	4.9	8.2
3.3	5.3	4.9	9	1.4	2.8	1.3	2.3	1.9	3.3	5	8.1
3.5	3	17.6	16.1	6.9	6.3	4.8	4.2	9.1	8.1	9.6	8.4
3.6	3.1	17.7	15.9	6.9	6.3	4.9	4.2	9.2	8.1	9.6	8.5
0.2	0.1	29.7	17.4	18	10.4	8.9	5.2	15.3	9.3	11.6	7.1
0.2	0.1	29.7	17.5	17.5	10.2	8.9	5.3	15.2	9.3	11.6	7.2
0.2	0.1	50.7	22.6	31.2	13.5	13.7	6.1	23.7	10.8	20.1	9.1
0.2	0.1	50.8	22.4	31.2	13.6	13.8	6.2	23.7	10.8	20.1	9.1
5.5	2	67.5	24.9	42.6	15.5	22.3	8.3	30.7	11.6	27.1	10
5.4	1.9	67.5	25	42.7	15.5	22.4	8.3	30.8	11.6	27.2	10.1
15.4	4.4	86.5	25.6	62.2	18.5	36.5	10.8	46.4	13.6	40.9	11.9
15.4	4.4	86.4	25.7	62.1	18.5	36.4	10.7	46.4	13.7	40.9	11.9
21.7	5.4	104.4	26.2	72.3	19.2	54.2	13.6	58.7	14.7	51.6	12.8
21.6	5.4	104.6	26.3	72.3	19.2	54.2	13.6	58.7	14.7	51.6	12.8
32.1	6.6	119.6	25	86.2	19.5	72	15.2	71.3	15.1	57.3	12
32.1	6.7	119.5	25.4	86.1	19.7	72	15.3	71.2	15.3	57.2	12.2
31.5	5.5	148.2	25.5	86.2	16.6	108	18.3	89.3	15.5	75.6	13.1
31.5	5.5	148.2	25.5	86.2	16.5	107.9	18.3	89.2	15.6	75.5	13.1
0.2		0.2		0.2		0.2		0.2		0.2	
0.2		0.2		0.2		0.2		0.2		0.2	
0.2		0.2		0.2		0.2		0.2		0.2	
5.4	7.4	12.1	15.9	8.5	10.3	1.3	1.7	5.9	7.9	3.9	5.3
5.5	7.3	12.2	15.5	8.6	10.1	1.4	1.8	6	7.9	3.9	5.1
8.2	5.7	15.5	12.5	17.5	12.5	4.8	3.9	9.1	7.2	10.3	7.5
6.3	5.6	15.5	14.3	17.5	13.4	4.8	4.2	9	8.3	3.9	3.6
9.9	6.8	16.5	11.7	26.6	16.1	9.2	5.9	21.7	14.2	10.4	7.3
9.9	6.4	16.6	11.3	26.6	16	9.2	5.9	21.6	14	10.5	7.1
8.4	3.8	31.7	14.5	41.1	18.1	13.7	6.2	39.7	17.7	10.4	4.8
8.4	4	31.6	15	41	18.3	13.6	6.3	39.7	18.3	10.4	5
9.5	3.4	46.2	17.1	48.6	18.2	22.7	8.7	49.2	18.3	10.4	4
9.5	3.4	46.2	17	48.5	18.2	22.7	8.7	49.3	18.2	10.4	4
10.9	3.2	72.1	21.3	57.5	18	36.4	11.3	64.6	19.3	13.1	4
10.9	3.2	72.2	21.6	57.5	18.1	36.4	11.3	64.7	19.6	13.1	4
13.6	3.4	89.2	22.5	61.6	16.7	54.3	14.2	74	19	16.3	4.1
13.5	3.4	89.1	22.6	61.5	16.8	54.1	14.2	73.9	19	16.2	4.2
17.1	3.7	95.8	21.7	64.6	15.8	72.1	16.6	77.5	17.4	23.7	5.2
17.1	3.9	96	22.3	64.5	16.1	72	16.6	77.3	17.9	23.7	5.4
19.3	3.9	116.5	23.2	67.7	14.5	107.7	20.9	82.2	16.7	31.2	6.1
19.2	3.9	116.4	23.1	67.6	14.7	107.6	20.9	82.1	16.8	31.1	6.2
0.2		0.2		0.2		0.2		0.2		0.2	
0.2		0.2		0.2		0.2		0.2		0.2	
18.8	19.8	16.6	18	5.6	6.9	1.8	2.1	7	7.8	19.5	20.8

Table G.9

$w_{oil,0}$	p	T	$\dot{m}$	$\dot{x}$	$\Delta p/\Delta L$	FP	$\dot{q}$	α
0.02	25.93	-10.72	154.9	0.13	2.23	i (f)	13.6	13.1
0.02	25.94	-10.71	154.3	0.13	2.67	i (f)	22.1	14.5
0.02	25.93	-10.72	152.9	0.14	2.68	i (f)	22.2	14.4
0.02	25.94	-10.71	154	0.13	3.4	i (f)	28.8	15.1
0.02	25.95	-10.7	154.3	0.13	3.27	i (f)	28.7	15.3
0.02	25.95	-10.69	153.8	0.13	3.6	i (f)	34.1	16.4
0.02	25.95	-10.7	154.1	0.14	3.6	i (f)	34.1	16.5
0.02	25.96	-10.68	153.2	0.13	4.61	i (f)	42.1	17.3
0.02	25.95	-10.69	153.6	0.13	4.75	i (f)	42	17.5
0.02	25.96	-10.68	154.2	0.13	5.37	i (f)	55.2	18.8
0.02	25.97	-10.67	154	0.13	5.92	i (f)	55.2	18.7
0.02	25.97	-10.67	153.7	0.13	6.12	i (f)	63.8	19.6
0.02	25.97	-10.66	153	0.14	6.22	i (f)	63.7	19.7
0.02	25.98	-10.65	150.9	0.13	7.58	i (f)	77.6	20.3
0.02	25.96	-10.68	154.7	0.13	7.37	i (f)	77.6	20.4
0.02	25.98	-10.66	153.6	0.13	9.2	f	103.8	21.6
0.02	25.98	-10.66	154.8	0.13	9.41	f	103.8	21.6
0.02	25.97	-10.66	159.8	0.32	3.63	w (f)	0.2	
0.02	25.96	-10.68	160.5	0.32	3.63	w (f)	0.2	
0.02	25.97	-10.66	161.8	0.31	4.53	i (f)	13	12.8
0.02	25.98	-10.64	161.2	0.31	4.51	i (f)	13	12.8
0.02	25.98	-10.65	161.5	0.31	5.11	i (f)	21.4	15
0.02	25.98	-10.65	162	0.31	5.1	i (f)	21.4	15.1
0.02	25.99	-10.64	161.8	0.32	5.97	i (f)	28.1	16.6
0.02	26	-10.62	161.2	0.32	5.74	i (f)	28.2	16.6
0.02	25.99	-10.63	162	0.32	6.43	i (f)	33.7	17
0.02	26	-10.63	162.1	0.32	6.31	i (f)	33.7	17.2
0.02	25.93	-10.72	162.7	0.32	8.5	i (f)	45.6	19.4
0.02	25.92	-10.72	161.5	0.32	8	i (f)	45.6	19.3
0.02	25.93	-10.71	161.3	0.32	9.51	i (f)	60.8	21.1
0.02	25.94	-10.7	161.5	0.32	9.31	i (f)	60.8	21.1
0.02	25.94	-10.7	160.4	0.32	10.67	i (f)	70.8	21.4
0.02	25.95	-10.69	161.3	0.32	10.27	i (f)	70.8	21.5
0.02	25.97	-10.66	161.6	0.32	12.04	f	79.7	21.5
0.02	25.97	-10.67	161.7	0.32	11.91	f	79.8	21.6
0.02	25.98	-10.65	161.5	0.33	14.35	f	96.6	21
0.02	25.97	-10.66	161.6	0.32	13.7	f	96.7	21.1
0.02	25.92	-10.72	154.4	0.53	7.01	w-a (f)	0.2	
0.02	25.93	-10.71	153.2	0.53	6.99	w-a (f)	0.2	
0.02	25.93	-10.71	156.6	0.52	7.81	w-a (f)	7.6	10.7
0.02	25.92	-10.71	156.4	0.52	7.81	w-a (f)	7.6	10.6
0.02	25.94	-10.7	157.5	0.52	8.27	w-a (f)	18	15.2
0.02	25.93	-10.7	156.1	0.52	8.45	w-a (f)	18	15.3
0.02	25.93	-10.7	156.2	0.52	9.11	w-a (f)	26.6	17
0.02	25.94	-10.69	156.6	0.52	9.22	w-a (f)	26.6	17.1
0.02	25.94	-10.68	156.4	0.51	9.62	w-a (f)	33	17.4
0.02	25.94	-10.69	156.5	0.52	9.38	w-a (f)	33	17.4

Table G.9

$\dot{q}_1$	α_1	$\dot{q}_2$	α_2	$\dot{q}_3$	α_3	$\dot{q}_4$	α_4	$\dot{q}_5$	α_5	$\dot{q}_6$	α_6
18.9	18.4	16.6	15.8	18	15.5	1.8	1.7	7.1	7.2	19.5	19.8
26.9	17.3	30.3	20.3	21.8	14.6	4.8	3.3	19	12.5	29.8	19.2
27	17	30.3	19.9	21.9	14.7	4.9	3.3	19.2	12.5	29.9	19
31.7	16.2	38.4	20.1	35.4	18.5	9.3	5	28.1	14.9	29.7	16
31.7	16.6	38.3	20.4	35.3	18.4	9.2	5	27.8	15	29.7	16.3
30.6	15.4	43.2	20.9	49.2	22.1	13.8	6.6	33.4	16.1	34	17
30.6	15.6	43.2	21.3	49.2	22.1	13.8	6.6	33.5	16.1	34	17
34.7	14.4	51.9	21.6	57.8	23.3	21.5	8.9	44.5	18	42.2	17.4
34.7	14.8	51.8	22	57.7	23.2	21.4	8.9	44.4	18.3	42.1	17.8
42	14.5	68.8	23.5	77.8	26.2	36.5	12.4	57.5	19.5	48.4	16.4
42	14.5	68.9	23.3	77.9	26.1	36.5	12.4	57.5	19.5	48.4	16.3
49.6	15.8	72.6	22.8	90.9	27.4	54.4	16.2	60.1	18.5	55.1	16.7
49.6	15.7	72.6	23	90.7	27.4	54.4	16.3	60.1	18.6	55	17
58.8	15	101.4	26.2	90.8	25.3	71.9	18.9	67.4	17.7	75.4	18.8
58.8	15.1	101.2	26.2	90.7	25.4	71.9	18.9	67.4	17.8	75.3	19.2
88.1	17.9	144.3	29.3	99.9	22.7	108	22	82	16.9	100.7	20.6
88	17.7	144.2	29.2	99.8	23	107.9	22.2	82.1	17	100.7	20.3
0.2		0.2		0.2		0.2		0.2		0.2	
0.2		0.2		0.2		0.2		0.2		0.2	
19.6	18.9	17.2	17	11.1	10.9	2	2	14.9	14.7	13.3	13.3
19.5	19	17.1	17	11.1	10.9	2.2	2.2	14.9	14.6	13.2	13.3
25.5	17.8	31.8	22.9	22.2	15.4	4.6	3.4	24.4	17.1	20	13.5
25.5	18.1	31.8	23.1	22.2	15.5	4.6	3.4	24.4	17.2	20	13.5
31.7	18.7	39	24.3	29.8	17.6	9	5.4	35.5	20.2	23.9	13.4
31.7	18.6	39	24.3	29.8	17.6	9	5.4	35.5	20.2	24	13.4
39.9	18.8	49.8	26	37.3	19.2	14.1	7.4	40.5	20.5	20.4	10
39.9	19.3	49.8	26.5	37.3	19.3	14	7.4	40.5	20.7	20.4	9.8
51.2	20.7	63.3	27.7	54.1	23.2	22.4	9.8	52.4	22.2	30.2	12.6
51.2	20.6	63.3	27.7	54.1	23.1	22.2	9.7	52.5	22.1	30.2	12.3
57.8	20	82.4	29.1	84.1	28.6	36.4	12.9	70.2	24.3	33.7	11.8
57.8	20.2	82.5	29.1	84.1	28.6	36.4	12.9	70.3	24.2	33.7	11.6
63	18.9	94.9	28.9	94.5	28.7	54.3	16.5	80.2	24.1	37.9	11.4
63	19	94.8	29	94.4	28.7	54.3	16.6	80.1	24.1	37.9	11.4
66	17.7	103.1	27.9	105	28.6	72.9	19.6	89.9	24	41.1	11
66.2	18	103.2	28.2	105	28.7	73	19.6	90.1	24.2	41.2	11.2
78.4	16.2	120.4	25.9	119.4	27.6	107.7	23.3	105.9	23.1	48.2	10.2
78.4	16.5	120.4	26.1	119.5	27.6	107.8	23.4	105.9	22.9	48.3	10.1
0.2		0.2		0.2		0.2		0.2		0.2	
0.2		0.2		0.2		0.2		0.2		0.2	
13.2	17.3	4.3	6.4	8	11	2.5	3.7	5.5	8.1	12.3	17.4
13.3	17.4	4.3	6.4	8	11.1	2.4	3.6	5.2	7.8	12.3	17.4
28	23.1	21.7	19.2	18.6	15.1	4.6	4	15	12.9	20	16.9
28.1	23.3	21.7	19.2	18.6	15.1	4.6	4	15	13	20	17.1
43.6	26.8	39.2	25.5	25.3	16.4	9.1	6	22.6	14.6	20	12.9
43.6	27	39.1	25.6	25.3	16.5	9.1	6	22.6	14.7	20	13
48.4	24.7	54.5	28.7	35.1	18.8	13.7	7.4	30.7	16.2	15.5	8.4
48.4	24.8	54.4	28.8	35.1	18.8	13.7	7.4	30.7	16.2	15.4	8.3

Table G.9

$w_{oil,0}$	p	T	$\dot{m}$	$\dot{x}$	$\Delta p/\Delta L$	FP	$\dot{q}$	α
0.02	25.95	-10.67	157.2	0.53	10.84	w-a (f)	41.3	17.7
0.02	25.95	-10.67	156.4	0.53	10.87	w-a (f)	41.2	18
0.02	25.99	-10.63	159.4	0.52	12.03	w-a (f)	53.8	17.9
0.02	25.99	-10.62	160.1	0.52	12.32	w-a (f)	53.8	18
0.02	25.96	-10.66	157.6	0.52	12.28	w-a (f)	53.7	17.4
0.02	25.97	-10.65	156.9	0.52	12.07	w-a (f)	53.7	17.2
0.02	25.99	-10.62	159	0.52	13.75	w-a (f)	68.3	17.8
0.02	26	-10.61	158.3	0.53	13.73	w-a (f)	68.3	17.7
0.02	26.01	-10.59	159.6	0.52	15.17	f	78	17
0.02	26	-10.61	159.4	0.52	15.16	f	78.1	16.8
0.02	26.02	-10.58	159.6	0.53	16.77	f	90.8	15.7
0.02	26.01	-10.59	159.7	0.52	16.99	f	90.8	15.6
0.02	25.96	-10.62	152.8	0.73	9.62	w-a (f)	0.2	
0.02	25.95	-10.65	155.8	0.71	9.66	w-a (f)	0.2	
0.02	25.96	-10.63	154.1	0.71	10.1	w-a (f)	5.8	9.1
0.02	25.96	-10.63	154.2	0.71	10.03	w-a (f)	5.8	9.2
0.02	26	-10.58	158.3	0.72	13.18	w-a (f)	17.2	16.9
0.02	26	-10.58	159.2	0.71	13.76	w-a (f)	17.2	16.7
0.02	25.96	-10.62	152.7	0.73	13.02	w-a (f)	17.1	15.6
0.02	25.97	-10.62	153.5	0.73	12.99	w-a (f)	17	15.6
0.02	26	-10.58	160.3	0.71	15	w-a (f)	35.8	19.6
0.02	26.02	-10.55	159.9	0.71	14.79	w-a (f)	35.9	19.4
0.02	26.02	-10.56	160	0.71	15.3	w-a (f)	39.9	18.9
0.02	26.01	-10.56	159.8	0.71	15.31	w-a (f)	39.8	18.9
0.02	26.02	-10.55	159.7	0.71	17.01	w-a (f)	47.2	18.9
0.02	26.02	-10.56	159.6	0.71	16.97	w-a (f)	47.3	18.8
0.02	26.03	-10.54	159.9	0.71	18.44	w-a (f)	53.3	18
0.02	26.02	-10.55	160	0.71	18.21	w-a (f)	53.3	17.7
0.02	26.03	-10.53	155.7	0.74	16.85	i (f)	59.9	13.5
0.02	26.03	-10.54	155	0.74	17.7	i (f)	59.9	13.7
0.02	26.04	-10.51	154.9	0.74	22.01	f	68.1	16.2
0.02	26.03	-10.52	154.4	0.75	21.82	f	68.1	16.1
0.02	26.03	-10.53	155.4	0.74	22.79	f	77.5	14.4
0.02	26.03	-10.53	154.9	0.74	22.6	f	77.5	14.3
0.02	25.94	-10.49	163.1	0.89	11.99	w-a (f)	0.2	
0.02	25.95	-10.49	162.2	0.89	11.56	w-a (f)	0.2	
0.02	25.95	-10.46	161.6	0.89	12.51	w-a (f)	1.6	7.8
0.02	25.94	-10.47	161.7	0.89	12.43	w-a (f)	1.6	8.1
0.02	25.96	-10.43	162	0.9	12.42	w-a	2.9	9.8
0.02	25.95	-10.44	162.1	0.9	12.4	w-a	2.8	9.6
0.02	25.94	-10.42	160.1	0.91	15.49	w-a (f)	7	15.5
0.02	25.95	-10.42	160.9	0.9	15.12	w-a (f)	7	15.3
0.02	25.94	-10.52	160.1	0.88	18.03	w-a (f)	18.9	15.1
0.02	25.96	-10.5	160.8	0.88	18.06	w-a (f)	19	15.1
0.02	25.96	-10.43	160.6	0.9	19.87	w-a (f)	29.3	15.4
0.02	25.97	-10.33	158.1	0.91	20.15	w-a (f)	29.2	17.1
0.02	25.98	-10.42	161.2	0.9	19.99	w-a (f)	39.4	16.5

$\dot{q}_1$	α_1	$\dot{q}_2$	α_2	$\dot{q}_3$	α_3	$\dot{q}_4$	α_4	$\dot{q}_5$	α_5	$\dot{q}_6$	α_6
48.6	20.3	70.1	30.1	49.6	21.5	23	9.9	41.9	17.9	14.5	6.3
48.5	20.9	70.1	30.5	49.6	21.7	22.9	10	41.8	18.2	14.5	6.4
49.2	15.8	90.6	29.8	73.6	24.4	36.3	12.4	57.9	20	14.9	5
49.2	16	90.6	30	73.7	24.5	36.4	12.4	57.9	20	14.9	5
49.5	15.8	90.4	29.3	73.4	24	36	11.8	57.6	18.8	15.5	5
49.4	15.3	90.4	28.9	73.3	24.1	36	11.8	57.6	18.3	15.5	4.9
55.1	14.1	114.6	29.6	87.7	23	54.2	14.4	80.5	21.3	17.5	4.5
55.1	14.1	114.5	29.5	87.7	23	54.3	14.2	80.5	20.8	17.6	4.5
57.5	12.5	120.4	26.6	105.3	22.1	72	15.9	90.6	20.4	22.1	4.8
57.5	12.3	120.5	26.3	105.4	22	72	15.8	90.7	20	22.1	4.7
54.9	9.3	139.8	24.2	105.3	18.2	107.6	18.5	105	18.6	32.1	5.4
54.9	9.3	139.8	24	105.3	18.2	107.5	18.3	105.2	18.5	32.1	5.3
0.2		0.2		0.2		0.2		0.2		0.2	
0.2		0.2		0.2		0.2		0.2		0.2	
9.9	14.7	1.9	3.2	8.3	13.2	1.3	2.2	4.5	7.5	9	14
9.9	14.8	1.8	3.2	8.2	13.4	1.4	2.4	4.5	7.5	9	13.8
33	29.9	14.4	14.4	15.7	15.7	4.3	4.6	15.2	16.1	20.5	20.9
32.9	29.3	14.4	14.2	15.6	15.4	4.3	4.6	15.2	15.8	20.5	20.6
33.3	28.4	14.6	13.3	15.4	14	4.1	4.2	14.5	14.3	20.5	19.5
33.3	28.3	14.5	13.3	15.3	13.8	4.1	4.2	14.5	14.3	20.5	19.5
54.5	29.1	50.7	27.7	53.1	28.4	9.2	5.3	26.8	15.4	20.4	11.8
54.6	28.7	50.8	27.4	53.2	28.2	9.2	5.3	26.9	15.1	20.5	11.5
54.4	25.5	60.4	28.6	59.7	28.1	13.2	6.5	32.8	15.6	18.8	9.1
54.4	25.4	60.4	28.6	59.7	28.1	13.1	6.5	32.7	15.7	18.7	9.2
57.2	22.9	70.3	28.1	74.5	29.6	25.1	10.2	37.5	15	18.9	7.7
57.2	22.7	70.3	27.9	74.5	29.4	25	10.1	37.7	15	18.9	7.7
57.9	19.9	83.4	27.8	81	27.2	37.8	12.8	44.6	14.9	15	5.3
57.9	19.6	83.4	27.5	81	26.8	37.8	12.7	44.5	14.6	15	5.1
62.5	13.3	87.2	19.3	90.7	22.2	59.9	13.3	46.1	9.9	13	2.7
62.5	13.3	87.1	19.7	90.6	23	59.9	13.5	46.1	9.8	13	2.7
66.7	15.8	100.7	23.6	101.7	24.9	72.4	17.1	53.5	12.6	13.8	3.3
66.7	15.5	100.6	23.3	101.7	24.9	72.5	16.9	53.4	12.5	13.8	3.3
60.9	11.2	122.3	22.2	99.6	20.1	107.8	19.6	62.2	11.5	11.9	2.2
60.9	11	122.3	22	99.6	19.9	107.9	19.5	62.1	11.3	11.9	2.1
0.2		0.2		0.2		0.2		0.2		0.2	
0.2		0.2		0.2		0.2		0.2		0.2	
2.2	8.8	1.7	9.3	0.2	1.6	1.6	8.9	1.8	9.4	1.9	9
2.3	9.3	1.7	9.6	0.2	1.6	1.6	9.2	1.7	9.6	1.9	9.1
4.8	14.1	3.3	11.7	0.2	0.9	4.7	16.5	1.8	6.9	2.4	8.6
4.9	14.2	3.3	11.8	0.2	0.9	4.8	16.7	1.9	7.4	1.9	6.7
9.7	19.6	8.7	20	0.2	0.6	9.3	20.7	6.3	14.7	7.7	17.3
9.7	19.4	8.7	19.8	0.2	0.6	9.3	20.3	6.3	14.5	7.7	17.1
25.6	19.7	17.6	14.4	22.1	17.9	11.6	9.5	15.2	12.5	21.2	16.7
25.8	19.7	17.7	14.3	22.2	17.9	11.6	9.5	15.3	12.4	21.3	16.6
39.5	19.9	31.7	16.9	30.4	17.3	21.7	11.4	22	11.4	30.2	15.3
39.5	22	31.7	18.8	30.4	19	21.7	12.7	22	12.8	30.2	17.1
57.6	23.3	48.1	20.1	33.1	17.1	36.4	14.2	25.3	10.1	36	14.3

Table G.9

$w_{oil,0}$	p	T	$\dot{m}$	$\dot{x}$	$\Delta p/\Delta L$	FP	$\dot{q}$	α
0.02	25.97	-10.42	161.5	0.9	21.05	w-a (f)	39.4	17
0.02	25.97	-10.41	160.9	0.9	22.19	w-a (f)	51.9	12.9
0.02	25.97	-10.43	162.1	0.89	22.57	w-a (f)	51.9	14.6
0.02	25.94	-10.71	262.9	0.12	4.2	i	0.2	
0.02	25.94	-10.71	264	0.11	4.6	i	0.2	
0.02	25.94	-10.71	261.7	0.12	4.92	i (f)	2.7	6.6
0.02	25.93	-10.72	263.5	0.12	5.11	i (f)	2.8	6.8
0.02	25.95	-10.7	264	0.11	5.5	i (f)	9.4	11.4
0.02	25.94	-10.7	263.4	0.11	5.51	i (f)	9.3	11.4
0.02	25.96	-10.68	262.2	0.12	6.69	i (f)	19.9	16.1
0.02	25.96	-10.69	264.3	0.11	6.62	i (f)	20	16.1
0.02	25.97	-10.67	264.9	0.1	8.66	i (f)	34.8	20.9
0.02	25.96	-10.68	265.2	0.1	8.55	i (f)	34.8	20.8
0.02	25.99	-10.64	262.1	0.12	11.68	i (f)	59.2	25.1
0.02	25.99	-10.64	263.5	0.12	11.54	i (f)	59.2	25
0.02	25.99	-10.64	263	0.12	12.66	i (f)	65.9	26.2
0.02	26	-10.62	265	0.11	12.62	i (f)	66	26.3
0.02	26.02	-10.61	263.8	0.1	13.58	i (f)	78.5	26.5
0.02	26.01	-10.61	263.4	0.11	13.46	i (f)	78.5	26.3
0.02	26.03	-10.59	262	0.11	15.01	i (f)	91.2	26.7
0.02	26.02	-10.6	261.6	0.12	14.91	i (f)	91.1	26.7
0.02	26.03	-10.59	262.8	0.1	16.39	i (f)	99.8	26.5
0.02	26.03	-10.59	262.8	0.1	16.15	i (f)	99.7	26.9
0.02	25.93	-10.71	259.4	0.31	10.98	i-a	0.2	
0.02	25.93	-10.71	258.8	0.32	11.04	i-a	0.2	
0.02	25.92	-10.73	259.9	0.31	11.29	i-a	0.2	
0.02	25.94	-10.7	260.1	0.32	11.8	i-a	10.6	12.6
0.02	25.94	-10.7	260.4	0.32	11.9	i-a	10.6	12.6
0.02	25.94	-10.7	262	0.29	12.05	i-a (f)	14.2	14.5
0.02	25.94	-10.7	262.2	0.3	11.86	i-a (f)	14.2	14.4
0.02	25.95	-10.68	262.1	0.31	14.77	i-a (f)	39.5	22.8
0.02	25.95	-10.68	261.5	0.32	14.88	i-a (f)	39.4	22.8
0.02	25.96	-10.67	262.7	0.31	16.67	i-a (f)	49.4	23.6
0.02	25.96	-10.67	262.9	0.31	15.74	i-a (f)	49.3	22.9
0.02	25.97	-10.66	264.7	0.31	16.51	i-a (f)	56.7	23.8
0.02	25.97	-10.66	265.9	0.31	16.94	i-a (f)	56.7	23.9
0.02	25.98	-10.65	263.5	0.32	18.27	i-a (f)	67.4	25.6
0.02	25.97	-10.66	264.5	0.31	18.53	i-a (f)	67.5	25.6
0.02	25.99	-10.63	262.5	0.33	19.45	i-a (f)	76.3	25.8
0.02	25.99	-10.63	264.3	0.31	19.64	i-a (f)	76.4	26
0.02	26	-10.62	263.6	0.31	20.84	i-a (f)	86.2	25.8
0.02	26	-10.62	263.8	0.31	20.94	i-a (f)	86.2	25.8
0.02	26.01	-10.61	262.8	0.31	22.93	a (f)	99.8	25.2
0.02	26.01	-10.6	263.7	0.32	22.42	a (f)	99.7	25.1
0.02	25.94	-10.69	270.7	0.5	19.63	a	0.2	
0.02	25.94	-10.69	259.6	0.52	18.99	a	0.2	
0.02	25.97	-10.65	258.1	0.52	21.33	a	3.3	8.6

$\dot{q}_1$	α_1	$\dot{q}_2$	α_2	$\dot{q}_3$	α_3	$\dot{q}_4$	α_4	$\dot{q}_5$	α_5	$\dot{q}_6$	α_6
57.7	23.7	48.1	20.7	33.1	17.7	36.4	14.8	25.4	10.5	36	14.8
62.4	14.3	97.9	23.1	46.1	15.8	54.1	12.7	25.1	5.9	25.9	5.9
62.5	16.2	98	25.6	46.1	18.2	54.1	14.1	25.1	6.6	25.9	6.7
0.2		0.2		0.2		0.2		0.2		0.2	
0.2		0.2		0.2		0.2		0.2		0.2	
6.1	13.8	0.2	0.6	2	5.5	1.1	2.9	1.1	2.9	5.7	13.8
6.3	14.2	0.2	0.6	2	5.6	1.3	3.4	1.2	3.1	5.8	13.8
16.9	19.5	6.4	8.5	6	7.8	4.6	5.9	5.7	7.4	16.5	19.3
16.9	19.6	6.4	8.5	5.9	7.8	4.6	5.9	5.7	7.3	16.5	19.5
35.8	27.5	17.1	14.5	15	12.7	9.3	7.8	14.7	12.4	27.4	21.7
35.9	27.6	17.1	14.6	15.1	12.7	9.4	7.8	14.8	12.4	27.6	21.6
55.8	32.7	34.8	21.5	31.9	19.3	13.9	8.6	28.8	17.5	43.5	25.6
55.8	32.6	34.8	21.4	31.8	19.2	13.9	8.6	28.8	17.4	43.4	25.5
75.4	32.3	58.2	25.4	82.2	32.4	22.9	10.1	54.1	23.3	62.4	26.8
75.3	32.1	58.1	25.3	82.2	32.5	22.9	10.1	54	23.2	62.5	26.7
79	32.1	62.6	25.8	93.7	34.2	36.2	14.4	58.4	23.7	65.6	27
79.1	32.5	62.7	26.1	93.7	34.2	36.2	14.5	58.6	23.8	65.7	27
89.9	30.1	80.1	27.3	102	33.7	54.4	18.6	70.1	24.1	74.7	25.2
89.9	29.6	80.1	27.1	101.9	33.7	54.4	18.5	70	24	74.7	25
98.9	28.4	88.7	26.3	118.4	35	71.9	21.3	81.9	24.2	87.6	25.1
98.8	28.2	88.6	26.2	118.3	35	71.8	21.2	81.7	24.2	87.4	25.1
85.5	23.6	106.8	28	120.1	32.2	107.6	27	86.8	23.2	91.9	24.9
85.5	24.4	106.7	28.5	120	32.5	107.4	27.1	86.6	23.4	91.8	25.5
0.2		0.2		0.2		0.2		0.2		0.2	
0.2		0.2		0.2		0.2		0.2		0.2	
0.2		0.2		0.2		0.2		0.2		0.2	
20.5	23.8	2.6	3.3	13.5	15.1	2.6	3.3	8.3	10.6	16	19.4
20.4	23.6	2.6	3.3	13.5	15.1	2.7	3.3	8.4	10.7	16	19.4
26.4	26.1	9.5	9.9	14.8	14.2	4.6	4.9	9.2	10.3	20.7	21.5
26.4	26	9.5	9.9	14.8	14.2	4.6	4.9	9.2	10.2	20.7	21.3
71.8	40.6	41.3	24.2	42.7	23.3	9.1	5.4	37.3	21.8	34.8	21.2
71.7	40.8	41.2	24.2	42.7	23.3	9	5.4	37.2	21.9	34.7	21.3
77.8	37.4	58.5	28.5	62	27.8	13.9	6.9	50	23.9	34.1	17
77.7	35.6	58.4	28.1	61.9	27.7	13.9	6.8	50	23.1	34.1	16
85.6	34.9	67.5	29.5	69	28.5	22.5	10	54.7	23.3	40.7	16.8
85.6	35	67.5	29.6	69	28.7	22.5	10	54.7	23.5	40.7	17
85.6	33	89.2	33.6	85.8	31.3	36.4	13.9	64.6	24.7	43.1	17
85.7	33.1	89.3	33.7	85.9	31.2	36.5	14	64.7	24.5	43.2	16.9
85.7	29.4	102.6	34.5	94.2	31.6	54.2	18.1	73.8	24.7	47.1	16.2
85.9	29.9	102.8	34.8	94.5	32	54.2	18.3	73.9	24.9	47.2	16.3
90	26.7	119	35.3	99.4	31.4	72.5	21.5	82.6	24.4	53.4	15.6
90.1	26.7	119.2	35.2	99.5	31.5	72.6	21.6	82.4	24.4	53.3	15.6
93.3	22.5	141.3	35	103.3	28.9	108	26.7	90.8	23.1	62.3	15.2
93.1	22.4	141.2	35	103.1	28.6	107.9	26.7	90.7	22.8	62.3	15
0.2		0.2		0.2		0.2		0.2		0.2	
0.2		0.2		0.2		0.2		0.2		0.2	
4.1	9.7	0.2	0.5	5.6	14.6	2	5.5	1.5	4.5	6.6	16.7

Table G.9

$w_{oil,0}$	p	T	$\dot{m}$	$\dot{x}$	$\Delta p/\Delta L$	FP	$\dot{q}$	α
0.02	25.96	-10.66	267	0.5	21.68	a	3.3	8.6
0.02	25.97	-10.65	259.9	0.52	21.61	a	9.9	12.9
0.02	25.96	-10.66	260.2	0.52	21.32	a	9.9	12.9
0.02	25.97	-10.65	262.4	0.51	22.24	a	19.1	17.7
0.02	25.96	-10.66	261.6	0.51	22.17	a	19.3	17.9
0.02	25.96	-10.66	263.3	0.51	23.52	a	30.7	22.9
0.02	25.97	-10.64	263	0.51	23.16	a	30.7	22.8
0.02	25.98	-10.64	265.5	0.51	25.92	a	53.9	28.9
0.02	25.98	-10.64	266.4	0.51	25.85	a	53.9	28.9
0.02	25.99	-10.62	266.5	0.5	26.76	a (f)	70.2	28.7
0.02	26	-10.61	265.9	0.51	26.47	a (f)	70.2	28.4
0.02	26.02	-10.59	258.2	0.52	27.81	a (f)	77.3	25
0.02	26	-10.61	266.8	0.5	28.28	a (f)	77.4	26.2
0.02	26.02	-10.59	259.3	0.52	30.16	w-a (f)	81.7	23.4
0.02	26.01	-10.59	259.1	0.52	30.23	w-a (f)	81.6	23.6
0.02	26.01	-10.59	256.2	0.53	32.51	w-a (f)	93.6	20.4
0.02	26.03	-10.57	259.8	0.52	32.43	w-a (f)	93.5	20.5
0.02	25.96	-10.63	261.7	0.71	26.06	a	0.2	
0.02	25.96	-10.63	260	0.71	25.88	a	0.2	
0.02	25.96	-10.64	267.3	0.7	27.31	a	4.9	10.4
0.02	25.97	-10.62	261.8	0.71	26.75	a	4.8	8.1
0.02	25.97	-10.62	262	0.7	26.49	a	4.8	8.2
0.02	25.94	-10.66	260.5	0.72	27.33	a	7.7	11.9
0.02	25.97	-10.62	261.2	0.71	27.13	a	7.6	11.8
0.02	25.96	-10.64	269.1	0.69	28.65	a	14.3	15.4
0.02	25.96	-10.63	269.7	0.69	28.43	a	14.3	15.3
0.02	25.98	-10.6	260.7	0.72	31.43	a	27.3	22.7
0.02	25.98	-10.6	260.5	0.72	31.3	a	27.3	22.5
0.02	26.01	-10.56	261.2	0.72	36.71	a (f)	63.5	33.3
0.02	26.01	-10.56	262	0.72	36.84	a (f)	63.6	33.3
0.02	26	-10.58	262.9	0.71	38.16	a (f)	70.8	29.7
0.02	26.02	-10.55	263.1	0.71	37.54	a (f)	70.8	30.9
0.02	25.99	-10.59	264.3	0.71	39.44	a (f)	75.2	29.5
0.02	26.01	-10.56	263.8	0.71	39.41	a (f)	75.2	29.4
0.02	25.99	-10.59	264.8	0.71	41.16	w-a (f)	79.8	26.4
0.02	26.02	-10.56	264	0.72	41.1	w-a (f)	79.9	27.7
0.02	26.01	-10.56	266.3	0.71	42.55	w-a (f)	89.1	24.8
0.02	26	-10.59	266.4	0.71	42.24	w-a (f)	89.2	24.3
0.02	25.98	-10.45	268.3	0.88	30.95	a	0.2	
0.02	25.98	-10.38	265.9	0.9	30.5	a	0.2	
0.02	25.99	-10.34	264.8	0.91	31.13	a	2.2	11.1
0.02	25.97	-10.46	268.4	0.89	31.44	a	2.3	7.6
0.02	25.97	-10.45	267.5	0.89	32.59	a	4.6	10.3
0.02	25.98	-10.41	265.9	0.9	32.48	a	4.6	11.1
0.02	25.98	-10.45	267.5	0.89	33.56	a (f)	7.2	12
0.02	25.98	-10.43	266.6	0.89	33.47	a (f)	7.2	12.4
0.02	25.99	-10.33	260.7	0.91	35.85	a (f)	12.3	15.9

$\dot{q}_1$	α_1	$\dot{q}_2$	α_2	$\dot{q}_3$	α_3	$\dot{q}_4$	α_4	$\dot{q}_5$	α_5	$\dot{q}_6$	α_6
4.1	9.8	0.2	0.5	5.6	14.9	1.7	4.8	1.5	4.6	6.6	16.9
15.8	19.1	3.3	4.5	11.1	15	5.3	7.4	10.5	14.3	13.3	17.2
15.8	19.2	3.3	4.5	11.1	15	5.3	7.4	10.5	14.4	13.3	17.2
28.1	24.9	10.6	10.2	17.8	17.2	10.5	10.3	20.7	19.5	26.8	24.4
28	24.8	10.5	10.1	19.5	18.3	10.4	10.1	20.4	19.4	26.7	24.5
45.8	32.9	21.1	16.2	30.9	22.9	15.5	12.1	29.4	22.6	41.5	30.7
45.8	32.8	21.1	16.3	30.9	22.8	15.6	12	29.4	22.6	41.5	30.5
80.6	41.9	50.7	27.7	59.8	31.9	24.2	13.4	47.8	26	60.3	32.5
80.7	41.9	50.7	27.6	59.8	31.8	24.2	13.4	47.6	26	60.3	32.6
100.1	39.3	79.6	32.4	87.8	39.7	36.5	15	62.7	24.5	54.3	21.5
100	39	79.6	32.3	87.8	39.3	36.5	14.9	62.7	24.1	54.4	21
94.4	30.8	98.7	32.3	110.1	36	54.2	17.5	71.9	22.3	34.7	11.1
94.4	32.4	98.7	33.2	110.1	37.6	54.3	18.1	71.9	23.7	34.8	12.1
95.9	26.3	112.5	31.6	110.1	33.6	72.7	21.1	71.4	20.1	27.2	7.6
95.9	26.6	112.4	31.8	110	34.1	72.7	21.4	71.4	20.2	27.2	7.7
95.3	20.5	149	31.4	110.4	25.5	108.1	23.5	82.1	17.8	16.7	3.6
95.2	20.7	148.8	31.6	110.2	25.5	108	23.6	82.1	18	16.7	3.6
0.2		0.2		0.2		0.2		0.2		0.2	
0.2		0.2		0.2		0.2		0.2		0.2	
7.4	14.9	0.2	0.4	8.2	17.2	2.2	5	4.7	10.9	6.8	14.3
5.5	9.1	0.2	0.3	10.4	17.4	1.8	3.3	3.9	7	6.8	11.4
5.3	8.7	0.2	0.3	10.9	17.9	2	3.6	3.9	7	6.8	11.4
12	17.4	1.6	2.7	10.2	16	5	8	6.6	10.9	10.8	16.4
12	17.2	1.6	2.7	10.1	16	4.9	8	6.4	10.8	10.8	16.3
18.6	19.4	6.4	7.5	15.7	17.6	9.1	10	15.2	16.5	20.5	21.3
18.7	19.5	6.4	7.5	15.7	17.4	9.1	10	15.2	16.5	20.6	21.2
40	32	15.2	13.2	29.3	23.8	13.6	12	29.4	25.2	36.1	29.9
40.1	32	15.2	13.1	29.3	23.7	13.6	11.9	29.3	24.9	36.1	29.6
95.7	43.5	52.9	25.5	98.9	66.7	22.7	11	57.4	27.5	53.5	25.7
95.8	43.8	53	25.7	98.9	65.3	22.8	11.1	57.5	28	53.5	26.1
99.8	36.2	79.6	29.5	108.3	60.4	36.2	13.8	57.7	22	42.9	16.2
99.9	37.6	79.6	30.2	108.3	62.8	36.2	14.3	57.8	23.3	42.9	17.3
99.7	33.9	95.9	32.5	99.4	57	54.3	18.4	57.7	19.9	44	15.1
99.7	34	96	32.6	99.4	54.9	54.2	18.8	57.6	20.5	44	15.5
107.1	30.7	109	31.6	99.2	46.7	72.4	21.3	57.1	17.5	34.1	10.3
107.2	32.6	109.3	32.7	99.3	48.2	72.5	22.3	57.2	18.8	34.2	11.2
109.9	28.1	138.1	34.7	75.6	34.3	107.4	25.7	67.1	16.7	36.7	9.4
110.1	27.6	138.2	34.4	75.7	33.4	107.5	25.3	67.2	16.3	36.8	9.1
0.2		0.2		0.2		0.2		0.2		0.2	
0.2		0.2		0.2		0.2		0.2		0.2	
2.9	11.2	0.2	0.9	5.5	30.2	1.3	7	1.9	10	1.4	7
3.3	9	0.2	0.6	5.6	20.1	1.3	4.7	1.8	6.5	1.4	4.7
6.5	12.9	0.2	0.5	7.8	18.9	4.8	11	3.6	8.4	4.8	10.3
6.4	13.9	0.2	0.5	7.7	20.5	4.8	11.7	3.7	9.2	4.7	11
6.9	10.8	4.1	7.3	8.6	15.9	9.3	15.1	5.4	9.1	8.7	13.8
6.9	11.2	4.2	7.6	8.5	16.4	9.3	15.6	5.5	9.5	8.7	14.4
15.7	18.1	9.2	11.8	13.8	19.7	13.5	17.7	7.5	10.2	14.3	17.6

Table G.9

$w_{oil,0}$	p	T	$\dot{m}$	$\dot{x}$	$\Delta p/\Delta L$	FP	$\dot{q}$	α
0.02	26	-10.31	259.8	0.91	35.43	a (f)	12.3	16.1
0.02	26	-10.32	260.6	0.91	42.24	w-a (f)	23.1	19.6
0.02	26	-10.3	260.1	0.91	41.73	w-a (f)	23.1	20
0.02	26	-10.3	260.7	0.91	47.47	w-a (f)	40.5	26.8
0.02	26.01	-10.36	263	0.9	47.76	w-a (f)	40.5	25.6
0.02	26	-10.49	265.9	0.86	45.62	w-a (f)	64.6	28.6
0.02	26	-10.48	265.8	0.86	46.26	w-a (f)	64.5	28.7
0.03	11.63	-35.94	80.4	0.17	1.32	w (f)	0.3	
0.03	11.63	-35.94	78.5	0.17	1.1	i (f)	0.3	
0.03	11.64	-35.92	77.9	0.17	2.46	w (f)	11.7	10.5
0.03	11.64	-35.91	79.2	0.17	2.1	i (f)	14.9	10.9
0.03	11.65	-35.91	78.8	0.17	3.04	i (f)	20	10.9
0.03	11.64	-35.91	78.8	0.18	3.72	i (f)	25.3	11.5
0.03	11.65	-35.89	78.2	0.18	5.19	i (f)	36.3	14
0.03	11.66	-35.87	78.2	0.18	6.29	i (f)	45.1	15.7
0.03	11.67	-35.85	78	0.17	8.97	i (f)	61	18.7
0.03	11.67	-35.84	79.2	0.19	11.08	i (f)	71.2	21.1
0.03	11.65	-35.9	79	0.23	15.93	i (f)	92.1	23.4
0.03	11.68	-35.83	77.9	0.24	15.59	i (f)	92.2	23.3
0.03	11.55	-36.14	79.9	0.17			0.3	
0.03	11.55	-36.14	79.4	0.18			0.3	
0.03	11.56	-36.12	79.7	0.18			11.6	11
0.03	11.56	-36.12	80.7	0.18			11.6	10.4
0.03	11.56	-36.11	81.3	0.17			14.7	10.8
0.03	11.56	-36.12	81.7	0.16			14.7	11
0.03	11.56	-36.11	80	0.18			20	11.4
0.03	11.57	-36.1	79.4	0.18			20	10.7
0.03	11.58	-36.07	80.4	0.17			25.3	11.4
0.03	11.58	-36.07	80	0.18			25.3	11.3
0.03	11.58	-36.07	80.1	0.17			36.4	13.8
0.03	11.58	-36.06	81.2	0.17			36.4	13.5
0.03	11.59	-36.05	79.7	0.17			42.5	14.4
0.03	11.59	-36.04	80.4	0.17			45.7	15.2
0.03	11.59	-36.04	80.1	0.18			61.8	18.5
0.03	11.59	-36.03	80.6	0.18			61.8	18.1
0.03	11.59	-36.03	80.9	0.19			71.8	20.1
0.03	11.6	-36.03	81.2	0.19			71.8	20.3
0.03	11.6	-36.01	80	0.24			95.4	22
0.03	11.6	-36.01	80	0.24			95.4	22
0.03	11.58	-36.07	75.9	0.37	2.41	i (f)	0.3	
0.03	11.58	-36.08	75.1	0.38	2.32	i (f)	0.3	
0.03	11.58	-36.06	73.9	0.38	2.64	i (f)	8.7	10.6
0.03	11.58	-36.06	76.5	0.37	3.59	i (f)	8.7	10.5
0.03	11.59	-36.05	75.2	0.38	3.35	i (f)	15.9	12.3
0.03	11.59	-36.04	77	0.37	3.28	i (f)	16	12
0.03	11.59	-36.03	75.4	0.38	3.82	i (f)	20	12.3
0.03	11.6	-36.02	74.3	0.39	4.18	i (f)	19.8	12.4

$\dot{q}_1$	α_1	$\dot{q}_2$	α_2	$\dot{q}_3$	α_3	$\dot{q}_4$	α_4	$\dot{q}_5$	α_5	$\dot{q}_6$	α_6
15.7	18.4	9.2	12	13.8	19.9	13.5	18	7.5	10.4	14.4	18
26	21.5	14.6	12.7	30.8	26	22.5	18.8	18.6	16.3	26.2	22.2
25.9	21.9	14.6	13	30.7	26.7	22.4	19.2	18.6	16.6	26.2	22.7
50.3	31.5	32.3	20.8	42.6	30.5	36.5	24.3	34.2	23.2	47.4	30.3
50.4	30.3	32.3	20	42.6	28.8	36.4	23.1	34	22.2	47.2	29.3
84.9	31.7	78.5	29.3	49.8	47.2	53.8	19.7	51.8	19	68.5	24.5
84.9	31.4	78.4	29.2	49.8	49.1	53.7	19.4	51.7	18.7	68.5	24.2
0.3		0.3		0.3		0.3		0.3		0.3	
0.3		0.3		0.3		0.3		0.3		0.3	
14.3	12.5	22.2	19.5	11.6	10	1.4	1.3	9.8	9.1	10.9	10.5
13	9.8	24.7	18.9	15.8	11.1	4.7	3.2	20.6	14.2	10.9	8.4
14.2	7.9	28.6	16.2	24.2	12.8	8.9	4.6	27.4	14.5	16.8	9.6
16.9	7.6	39.3	18.1	27.3	12.4	13.4	5.9	31.8	14.3	22.9	10.5
18.9	7.2	67.8	26.5	35.6	13.9	23.2	8.6	43.9	16.6	28.3	11.2
29.1	10.1	81	28.7	43.3	15.1	36.8	12.2	46	16.1	34.2	11.9
59.1	18	80.8	26.2	70.6	21.3	55.9	16.4	60	18.4	39.6	12.1
68.4	20.2	90.8	27.3	89.8	26.2	72	20.5	66.5	20.1	40.1	12.1
95.3	23.2	117.3	29.2	91.6	26.5	107.7	26.6	95.8	23.7	44.8	11.1
90.7	22.2	114.2	28.2	101.6	28.7	107.8	26.5	95.6	23.8	43.4	10.7
0.3		0.3		0.3		0.3		0.3		0.3	
0.3		0.3		0.3		0.3		0.3		0.3	
13	12.4	23	21.5	12.1	10.7	1.5	1.3	9.4	8.9	10.6	10.9
13	11.4	23	20.5	12.1	10.6	1.5	1.3	9.3	8.7	10.6	10.2
12.9	9.6	24.3	18.9	16	11.3	4.5	3	20	13.6	10.5	8.3
13	10	24.3	19.1	16.1	11.4	4.4	3	20	13.9	10.5	8.6
14.2	8.4	28.9	17.9	24.4	13.3	8	4.2	27.7	14.8	16.8	10.1
14.2	7.5	28.9	16.3	24.4	13	8.1	4.1	27.7	14.3	16.9	9.4
16.5	7.2	39.1	18.2	27	12.4	13.3	5.8	32.6	14.3	23.4	10.6
16.5	7	39.1	18	27	12.4	13.3	5.8	32.6	14.1	23.4	10.5
18.9	7	68.3	26.3	37.1	14.4	23.3	8.6	42.6	15.8	28	10.5
19	6.7	68.4	25.5	37.1	14.4	23.4	8.7	42.6	15.6	28	10.2
30.7	9.7	81.7	28.6	25.5	9.7	36.5	12.2	46.4	15.4	34.1	10.9
30.7	9.8	81.8	27.8	44.7	15.4	36.6	12	46.5	15.5	34.2	11.1
60.8	17.2	81.3	25.3	72.5	22.4	56.5	16.7	60	18	39.5	11.3
60.9	16.7	81.3	24.7	72.5	22	56.5	16.4	60.1	17.5	39.6	11.1
69.4	18.1	92.1	25.7	89.5	26.6	72.7	20.7	67.3	19	39.6	10.5
69.4	18.8	92.1	26.2	89.5	26.2	72.7	20.3	67.3	19.2	39.7	11
90.3	19.7	113.8	25	122.5	31.3	106.9	24.3	96.4	22	42.6	9.5
90.3	19.7	113.8	25	122.5	31.3	106.9	24.3	96.4	22	42.6	9.5
0.3		0.3		0.3		0.3		0.3		0.3	
0.3		0.3		0.3		0.3		0.3		0.3	
13.8	15.9	10.9	13.9	8.3	10.1	1.6	1.9	4	5.3	13.7	16.5
13.8	15.7	10.9	13.5	8.4	10	1.6	1.9	4	5.3	13.8	16.6
14.4	11.4	30.7	24.3	10.2	8	4.9	3.6	17.9	13.5	17.1	13.2
14.5	11.1	30.8	24.1	10.3	7.9	5	3.6	17.9	13.1	17.2	12.6
15.1	9.5	37.8	23.7	16.4	10	9.4	5.5	21.9	13.4	19.2	11.9
14.9	9.4	37.7	23.9	16.3	9.9	9.3	5.5	21.8	13.5	19.1	11.9

Table G.9

$w_{oil,0}$	p	T	$\dot{m}$	$\dot{x}$	$\Delta p/\Delta L$	FP	$\dot{q}$	α
0.03	11.6	-36.01	75.9	0.38	5.39	i (f)	28.5	12.3
0.03	11.6	-36.02	76.4	0.38	5.03	i (f)	28.5	12.7
0.03	11.61	-36	81.2	0.37	4.44	i (f)	42.1	16.6
0.03	11.6	-36.01	81.7	0.37	3.45	i (f)	42.1	16.8
0.03	11.6	-36.01	84.2	0.34	7.02	i (f)	32.1	13.4
0.03	11.61	-35.99	80.2	0.36	5.24	i (f)	48.5	18.4
0.03	11.61	-35.99	81.6	0.36	4.99	i (f)	48.5	17.7
0.03	11.6	-36.01	81.2	0.36	7.14	i (f)	37.6	14.8
0.03	11.61	-36	81.3	0.36	7.43	i (f)	36.7	15.1
0.03	11.62	-35.98	80.2	0.36	8.29	i (f)	61.2	20.4
0.03	11.62	-35.98	78.8	0.38	8.44	i (f)	61.2	20
0.03	11.61	-35.99	79.4	0.37	8.81	i (f)	71.4	18.8
0.03	11.62	-35.98	80	0.37	9.16	i (f)	71.5	19.4
0.03	11.62	-35.97	80.1	0.37	13.11	i (f)	85.7	15.4
0.03	11.62	-35.96	79.1	0.38	13.79	i (f)	85.7	14
0.03	11.62	-35.96	83.6	0.35		w	0.3	
0.03	11.63	-35.95	85.2	0.34		w	0.3	
0.03	11.64	-35.92	84.2	0.36	0.76	i (f)	16.6	13
0.03	11.64	-35.92	85.4	0.35	0.85	w (f)	16.5	13.6
0.03	11.64	-35.92	86	0.34	0.91	i (f)	19.1	13.8
0.03	11.64	-35.92	82.4	0.35	0.89	i (f)	19.1	13.3
0.03	11.65	-35.9	85.5	0.34	1.11	w (f)	25.3	14.2
0.03	11.64	-35.91	83.5	0.34	0.84	i (f)	19.7	12.3
0.03	11.65	-35.89	78.9	0.37	1.34	w (f)	33.4	16
0.03	11.66	-35.88	80.8	0.37	1.44	i (f)	33.3	15.6
0.03	11.66	-35.88	79.3	0.37	1.5	i (f)	39.9	17.1
0.03	11.66	-35.88	79.5	0.37	1.61	i (f)	39.9	17.2
0.03	11.66	-35.88	79.9	0.37	1.77	i (f)	44.3	18.4
0.03	11.66	-35.87	78.5	0.38	1.73	i (f)	44.6	17.9
0.03	11.66	-35.87	82.6	0.36	2.34	i (f)	54.6	20
0.03	11.66	-35.87	83	0.35	2.09	i (f)	54.6	20.2
0.03	11.66	-35.87	83.6	0.35	2.43	i (f)	67	20.9
0.03	11.66	-35.86	83.7	0.35	2.84	i (f)	66.6	21.4
0.03	11.66	-35.87	79.9	0.36	2.92	i (f)	84.9	19.9
0.03	11.67	-35.86	82.3	0.35	2.66	i (f)	85	21.3
0.03	11.64	-35.93	78.2	0.39	10.47		58.5	23.2
0.03	11.65	-35.91	79.8	0.37	10.41		46.3	21.6
0.03	11.65	-35.9	80.1	0.37	14.62	i (f)	71.2	23.6
0.03	11.66	-35.86	79.9	0.37	13.12		70.3	19.8
0.03	11.65	-35.9	77.8	0.37	12.39		56.7	17.7
0.03	11.62	-35.97	79.2	0.36	11.65	i (f)	54.5	23.6
0.03	11.6	-36.02	83.4	0.35	2.67	w (f)	0.3	
0.03	11.61	-36	83.7	0.34	2.89	w (f)	0.3	
0.03	11.62	-35.97	83.7	0.35	3.82	i (f)	13.2	14.9
0.03	11.64	-35.93	81.9	0.37	4.12	i (f)	13.4	15.3
0.03	11.63	-35.93	83.8	0.35	4.54	i (f)	16.6	15.6
0.03	11.6	-36.02	85.4	0.35	4.42	i (f)	16.7	15.4

$\dot{q}_1$	α_1	$\dot{q}_2$	α_2	$\dot{q}_3$	α_3	$\dot{q}_4$	α_4	$\dot{q}_5$	α_5	$\dot{q}_6$	α_6
21.5	8.5	60.7	24.8	25.7	11.9	14	6.5	27.2	12.8	22	9.3
21.5	8.9	60.7	26	25.6	12	14	6.6	27.1	12.9	22	9.7
44.7	16	68.7	27.5	49.9	22	22.6	9.5	50.7	19.1	15.7	5.7
44.7	15.7	68.7	27.6	50	23.1	22.7	10.1	50.8	19.1	15.7	5.4
21.7	8.5	66.9	27.1	25.8	11.4	22.9	9.8	30.3	13.3	24.7	10.2
44.8	15.8	75.4	28.4	59.7	26.4	36.5	14.1	59	20.4	15.7	5.3
44.8	14.8	75.4	27.3	59.7	25.6	36.5	14	59	19.5	15.7	5
26.6	10.2	70.6	28.4	27.8	11.5	36.3	13.9	31.9	12.7	32.5	11.9
24.1	9.9	73.2	30.5	28.6	12.1	36.8	14.5	31.2	13.2	26	10.6
50.5	16.7	89.9	29.8	76.1	28.4	54.7	18.1	75.3	23	20.4	6.4
50.6	15.7	89.9	29	76.1	29	54.7	17.8	75.3	22.2	20.4	6.1
56.1	13.2	102	24.8	92.1	31.5	72.4	18.9	85.1	20	20.6	4.6
56.1	14.9	102.2	26.4	92.3	29.4	72.5	19.5	85.1	21	20.6	5.1
72	11	109.7	17.9	99.3	26.5	107.6	17.8	100.6	15.6	24.8	3.7
72.1	9.7	109.7	15.9	99.4	25	107.7	16.1	100.6	14.2	24.9	3.3
0.3		0.3		0.3		0.3		0.3		0.3	
0.3		0.3		0.3		0.3		0.3		0.3	
19.9	15.4	21.4	17.7	21.6	16.2	2.4	1.9	19.5	15.4	14.7	11.6
19.8	16.9	21.3	18.3	21.6	15.9	2.3	1.8	19.5	15.9	14.7	12.7
23.3	16.8	29.2	21.4	21.9	14.5	4.6	3.3	19.9	14.9	15.6	11.9
23.4	15.6	29.2	20.6	22	14.7	4.6	3.3	19.9	14.7	15.6	11.2
23.3	13.4	39.2	22.5	34.1	17.5	9.5	5.1	30.3	17.4	15.6	9.3
23.4	13.8	39.2	25.4	0.3	0.3	9.5	6.3	30.2	18.6	15.5	9.5
26.7	13.1	60.4	29.1	40.3	18.2	13.8	6.6	40.1	19.6	18.9	9.4
26.6	12.4	60.4	28.2	40.2	18.4	13.8	6.7	40.1	19	18.9	9
36.6	15.4	73.2	31	44	18.9	22.5	10	43.8	19.1	19.4	8.3
36.6	15.5	73.3	31.1	44	18.9	22.5	9.9	43.8	19.2	19.4	8.4
38	16.1	71	30	50.4	20.8	36.3	14.9	48.9	19.9	21.3	8.7
38.2	14.2	71.3	28.6	50.7	23.4	36.5	15.1	49.4	18.2	21.6	7.6
46.5	17.7	86.8	31.7	58.9	22.4	54.5	19.2	59.6	21.3	21.3	7.9
46.5	17.2	86.7	31.5	58.9	23.7	54.4	20	59.6	21.1	21.3	7.5
51.5	15.4	96.2	29	77.4	29.8	73.2	22.4	77	21.7	26.5	7.3
51.2	15.9	95.6	29.7	77	30.5	72.9	22.5	76.6	22.3	26.2	7.6
58.3	12.7	109.7	24.1	102	32.5	108.6	23.6	100.1	20.3	30.6	6.2
58.4	11.9	109.8	23	102.2	43.4	108.8	23.6	100.2	19.7	30.8	5.8
60	23.4	82	32.1	58.1	27.7	72.9	27.4	57.8	21.4	20	7.2
29	19.5	59.7	31.5	63.7	24.8	73.8	27	40.7	19.8	10.7	6.7
62.7	20.4	92.9	29.7	72.1	31.9	108	31.5	72.2	21.8	19.3	5.9
62.7	17.1	92.9	25.3	72.2	27.2	108	26.9	72.2	18.6	13.6	3.6
39.2	13.3	74	23	84.5	27.4	84.8	23.9	46.9	14.8	10.9	3.6
29	17.6	74	33.2	84.6	32.3	84.9	31.1	44.1	21.3	10.5	6.2
0.3		0.3		0.3		0.3		0.3		0.3	
0.3		0.3		0.3		0.3		0.3		0.3	
19.3	21.6	19.8	23.1	6.4	7.6	5.2	5.6	15	16.4	13.8	15.4
19.4	22.3	20	23.1	6.5	7.6	5.3	5.7	15.2	16.9	13.9	16.1
20.6	19.3	28.5	26.6	9.5	8.9	9.3	8.2	16	15.5	15.8	15.2
20.6	18.9	28.6	26.2	9.6	8.9	9.3	8.2	16.1	15.2	15.8	14.8

Table G.9

$w_{oil,0}$	p	T	$\dot{m}$	$\dot{x}$	$\Delta p/\Delta L$	FP	$\dot{q}$	α
0.03	11.61	-35.99	83.4	0.35	3.99	i (f)	16.7	15.4
0.03	11.61	-36	83.2	0.36	5.21	i (f)	21.8	15
0.03	11.62	-35.96	83.8	0.36	4.64	i (f)	21.9	15.4
0.03	11.62	-35.97	82.9	0.36	5.47	i (f)	21.8	15.5
0.03	11.63	-35.96	83.2	0.35	6.17	i (f)	28.7	16.3
0.03	11.61	-35.99	83.8	0.35	6.57	i (f)	28.8	15.8
0.03	11.63	-35.95	83.4	0.35	5.67	i (f)	28.8	14.8
0.03	11.66	-35.86	82.3	0.37	7.8	i (f)	37.9	18.9
0.03	11.61	-36	82.6	0.36	7.41	i (f)	39.6	18.4
0.03	11.63	-35.94	83.8	0.36	8.56	i (f)	39.5	16.9
0.03	11.62	-35.96	83	0.36	9.54	i (f)	48.2	16.5
0.03	11.64	-35.92	83.2	0.35	10.37	i (f)	48.2	17.7
0.03	11.63	-35.93	82.4	0.36	10.69	i (f)	48.1	19.7
0.03	11.64	-35.91	82.5	0.36	9.66	i (f)	61.3	20.9
0.03	11.64	-35.91	83.1	0.36	12.03	i (f)	61.4	22.6
0.03	11.64	-35.92	79	0.38	11.51	i (f)	61.3	20.2
0.03	11.66	-35.87	81.9	0.36	15.23	i (f)	71.4	20.5
0.03	11.63	-35.96	82.4	0.36	10.08	i (f)	71.3	19
0.03	11.64	-35.92	82.4	0.36	12.98	i (f)	72.3	17.3
0.03	11.59	-36.05	83.9	0.53	4.72	w (f)	0.3	
0.03	11.58	-36.06	80.2	0.56	4.91	i (f)	0.3	
0.03	11.59	-36.05	83.4	0.54	4.98	i (f)	2.3	8.9
0.03	11.59	-36.05	83.7	0.54	4.89	i (f)	2.2	9.1
0.03	11.59	-36.03	79.3	0.58	5.18	i (f)	4.2	10.6
0.03	11.59	-36.04	82	0.55	5.18	i (f)	4.2	10.5
0.03	11.6	-36.03	80.6	0.56	5.28	i (f)	7.2	11.6
0.03	11.59	-36.04	81.9	0.55	5.35	i (f)	7.1	11.7
0.03	11.59	-36.04	83.5	0.53	5.46	i (f)	7.2	11.7
0.03	11.6	-36.01	79.7	0.57	6.09	i (f)	14.2	13.7
0.03	11.6	-36.02	84.5	0.53	6.18	i (f)	14.2	13.8
0.03	11.61	-35.99	81.9	0.55	7.43	i (f)	19.7	15.2
0.03	11.61	-36.01	82.3	0.55	7.21	i (f)	19.7	15.3
0.03	11.61	-36	80.7	0.57	9.18	i (f)	23.3	16.4
0.03	11.61	-36	84.9	0.52	8.19	i (f)	24.4	16.5
0.03	11.62	-35.98	81.7	0.56	10.36	i (f)	36.6	17
0.03	11.62	-35.97	84	0.54	10.74	i (f)	36.6	19.3
0.03	11.63	-35.95	82.4	0.55	14.65	i (f)	50.5	20.8
0.03	11.67	-35.86	82.7	0.54	15.24	i (f)	59.9	20.6
0.03	11.69	-35.8	81.1	0.8	19.18	i (f)	59.9	9.8
0.03	11.67	-35.86	79.9	0.56	17.93	i (f)	73	20.5
0.03	11.68	-35.82	80.6	0.74	20.13	i (f)	72.8	6.1
0.03	11.56	-36.13	80.5	0.74	7.08	i-a (f)	0.3	
0.03	11.55	-36.13	84	0.71	6.95	i-a (f)	0.3	
0.03	11.56	-36.12	82.8	0.72	7.42	i-a (f)	3.1	8.3
0.03	11.56	-36.12	84	0.71	7.63	i-a (f)	3.1	8.1
0.03	11.56	-36.11	84	0.71	9.27	i-a (f)	8.7	12.1
0.03	11.56	-36.11	84.3	0.7	9.09	i-a (f)	8.8	12.1

$\dot{q}_1$	α_1	$\dot{q}_2$	α_2	$\dot{q}_3$	α_3	$\dot{q}_4$	α_4	$\dot{q}_5$	α_5	$\dot{q}_6$	α_6
20.6	19.4	28.6	26.5	9.6	8.7	9.3	8.1	16.2	15	15.7	14.7
23.9	15.8	38.8	26.4	14.4	10.2	13.9	9.3	25.7	18.1	14.2	10
24.1	16.9	38.8	27.3	14.5	10.3	14	9.3	25.8	18.1	14.3	10.3
23.9	16.9	38.5	27.5	14.4	10.3	13.9	9.3	25.8	18.5	14.3	10.7
32.6	17.9	50	28.3	22.3	12.4	22.8	12.4	25.9	15.5	19	11
32.6	16.9	50.1	27.6	22.3	12.4	22.8	12.3	26	15	19	10.5
32.6	14.7	50	24.9	22.3	12.6	22.8	12.6	26	14.5	19.2	9.4
32.8	19.5	60	31.6	37.1	15.9	37.1	15.4	39.2	19.2	20.9	11.6
41.7	18.5	60.2	29.2	37.4	18.7	37.4	17.2	39.4	17.9	21.2	9.2
41.7	17	60.3	26.6	37.4	17	37.3	15.7	39.4	16.5	21	8.5
41.5	15.1	69.9	24.4	60.4	21.3	54.7	17.4	48.5	16	13.9	4.9
41.5	15.2	70	25.5	60.4	24.1	54.7	19.7	48.6	17	13.8	4.9
41.5	17.5	69.9	28.7	60.4	26.4	54.8	21.3	48.5	18.8	13.8	5.6
58.7	17.9	81.5	26.4	73.4	32.9	72.8	24	63.3	19	18.1	5.3
58.8	21.5	81.6	29.7	73.4	30	72.8	25.2	63.4	22.2	18.2	6.6
58.7	17.6	81.5	25.5	73.2	32	72.7	22.5	63.2	18.4	18.1	5.1
54.2	15.5	87.4	25.5	88.6	29.7	94.4	24.8	73.6	19.3	30.2	8.2
54	13.9	87.3	23.1	88.4	29.6	94.2	23.6	73.5	17.2	30.3	6.8
66.6	14.1	87.2	19.4	88.4	29.6	94.2	20.5	73.6	15.1	23.7	4.8
0.3		0.3		0.3		0.3		0.3		0.3	
0.3		0.3		0.3		0.3		0.3		0.3	
5.1	18.6	1.7	7.7	1.6	6.9	1.7	6.4	1.8	7.6	1.6	6.1
5.2	19.5	1.7	8.1	1.6	7.2	1.7	6.4	1.7	7.2	1.6	6.3
8.3	19.8	4.6	12.8	1.6	4.4	3.9	9.5	1.8	4.9	5.1	12.5
8.3	19.6	4.5	12.5	1.6	4.5	3.9	9.5	1.8	4.8	5	12.3
13.2	20.2	7.5	13.2	4.1	7	5.1	8.2	4.6	7.9	8.4	13.3
13.2	20.4	7.4	13.1	4.1	7	5.1	8.2	4.5	7.8	8.3	13.4
13.3	20.3	7.6	13.1	4.2	7	5.2	8.2	4.7	7.9	8.4	13.4
23.5	21.6	17.6	17.7	7.7	7.8	9.7	9.3	9.8	9.8	17	16.2
23.6	21.7	17.6	17.8	7.7	7.7	9.7	9.3	9.9	9.9	17	16.3
31.8	23.5	28.3	22.7	9.7	8	14	10.7	15.4	12	18.7	14.2
31.8	23.9	28.4	22.7	9.8	7.9	14.1	10.7	15.6	12.2	18.7	14.4
33.2	22.9	32.2	23.2	15.3	11.4	23.6	16	19.7	13.9	15.8	11.2
34	22.7	36.5	25.3	15.2	10.5	23.5	15.2	19.8	13.6	17.1	11.7
48.9	21.5	52.7	25.2	27.8	14.9	36.8	17	32.7	14.4	20.7	8.8
48.8	25.4	52.6	28.4	27.7	15.4	36.8	18.5	32.7	17	20.7	10.8
54.6	22.1	76.5	31.5	44.1	21.4	54.4	22.2	48.4	18.5	24.9	9.5
52.8	17.8	110.5	35.8	40.6	17.9	75.5	25.7	54.9	18	24.9	8.2
52.5	8.5	110.6	17	40.5	8.8	75.4	11.7	55	8.7	25.1	4
60.1	16.7	128.3	34.5	40.7	16.5	107.3	28.7	68.6	18	32.8	8.7
59.9	4.8	128	10.5	40.6	4.4	107	8.7	68.7	5.7	32.8	2.5
0.3		0.3		0.3		0.3		0.3		0.3	
0.3		0.3		0.3		0.3		0.3		0.3	
6.2	14.5	1.9	6.3	1.7	5.2	2.3	6	1.9	5.7	4.7	11.9
6.2	14.4	1.9	6.1	1.7	5.1	2.3	5.8	1.9	5.4	4.8	11.8
14.6	18.6	8.3	12.8	8.5	11.8	4.8	6.6	6.6	9.6	9.7	13.1
14.7	18.4	8.4	12.7	8.7	12	4.9	6.7	6.7	9.6	9.8	13.1

Table G.9

$w_{oil,0}$	p	T	$\dot{m}$	$\dot{x}$	$\Delta p/\Delta L$	FP	$\dot{q}$	α
0.03	11.57	-36.09	84.5	0.72	10.47	i-a (f)	12	12.9
0.03	11.57	-36.08	83.9	0.72	10.21	i-a (f)	12	13
0.03	11.57	-36.09	83.7	0.69	10.8	i-a (f)	14.7	13.8
0.03	11.57	-36.09	84.4	0.69	10.87	i-a (f)	14.7	13.8
0.03	11.58	-36.07	83.9	0.72	14.26	i-a (f)	22.4	16.5
0.03	11.58	-36.06	84.6	0.71	14.74	i-a (f)	22.5	16.6
0.03	11.59	-36.04	84.4	0.72	16.51	i-a (f)	31.1	13.2
0.03	11.59	-36.05	82.8	0.73	14.57	i-a (f)	31.1	11.6
0.03	11.6	-36.01	83.4	0.72	17.86	i-a (f)	37.8	14.2
0.03	11.59	-36.04	83.3	0.73	12.78	i-a (f)	37.7	6.5
0.03	11.58	-34.21	75.5	0.95	9.59	w-a (f)	0.4	
0.03	11.59	-34.18	75.3	0.95		i-a (f)	0.4	
0.03	11.58	-29.46	77.7	0.96		i-a (f)	1.3	
0.03	11.58	-34.21	76.1	0.95		i-a (f)	1.4	
0.03	11.58	-34.21	76.7	0.95	8.24	w (f)	4.7	
0.03	11.58	-29.58	77.9	0.96	9.19	w (f)	4.7	
0.03	11.58	-26.83	77.4	0.96	10.05	w (f)	9.7	
0.03	11.59	-28.82	78.1	0.96	11.22	w (f)	9.7	
0.03	11.58	-28.87	77.5	0.96	10.72	w (f)	14.2	
0.03	11.58	-29.08	77.5	0.96	10.98	w (f)	14.2	
0.03	11.59	-26.82	76	0.96	11.72	w (f)	19.7	
0.03	11.59	-30.48	77.1	0.96	12.47	w (f)	19.8	
0.03	11.59	-36.04	86	0.87	14.99	i-a (f)	29	12.9
0.03	11.59	-36.04	85.1	0.88	14.88	i-a (f)	29	12.9
0.03	11.62	-35.97	157.5	0.14	4.25	i (f)	0.3	
0.03	11.61	-35.98	159.9	0.14	4	w (f)	0.3	
0.03	11.63	-35.95	162.1	0.14	5.52	i (f)	17.7	14.6
0.03	11.64	-35.93	159	0.14	6.43	i (f)	26	16.4
0.03	11.64	-35.91	160.9	0.14	9.2	i (f)	34.4	16.7
0.03	11.65	-35.9	160	0.14	9.75	i (f)	39.9	17.1
0.03	11.65	-35.89	163.1	0.13	10.89	i (f)	46.4	18.2
0.03	11.66	-35.88	162.4	0.14	13.9	i (f)	54.1	20.4
0.03	11.67	-35.86	162.5	0.14	14.69	i (f)	59.2	21.2
0.03	11.67	-35.85	158.3	0.14	16.28	i (f)	67	22.6
0.03	11.67	-35.84	155.9	0.14	18.42	i (f)	79.6	24.9
0.03	11.63	-35.94	152.1	0.35	10.06	i-a (f)	0.3	
0.03	11.63	-35.96	150.9	0.35	10.22	w-a (f)	0.3	
0.03	11.63	-35.94	150.9	0.35	10.24	i-a (f)	3.5	12.2
0.03	11.64	-35.92	151.3	0.35	11.2	w-a (f)	9.3	13.7
0.03	11.64	-35.92	164.7	0.32	11.49	i-a (f)	13	14.7
0.03	11.65	-35.9	148.9	0.36	14.96	i-a (f)	21.4	18.1
0.03	11.65	-35.9	150.5	0.36	16.59	i-a (f)	31.7	21
0.03	11.67	-35.85	151.2	0.36	19.98	i-a (f)	44.3	24.4
0.03	11.68	-35.83	156.5	0.34	21.01	i-a (f)	57.4	26.5
0.03	11.68	-35.82	151.8	0.36	26.39	i-a (f)	65.8	26
0.03	11.69	-35.79	152.5	0.35	29.52	i-a (f)	80.6	27.5
0.03	11.6	-36.01	155.4	0.54	18.62	a (f)	0.3	

$\dot{q}_1$	α_1	$\dot{q}_2$	α_2	$\dot{q}_3$	α_3	$\dot{q}_4$	α_4	$\dot{q}_5$	α_5	$\dot{q}_6$	α_6
18.7	18.8	11.7	13.7	11.9	12.7	9.1	9.8	7.1	8.2	13.5	14.3
18.8	19	11.7	13.8	11.9	12.7	9.2	9.9	7	8.1	13.6	14.4
19.9	17.8	17	16.8	13.7	12.6	13.7	12.6	8	8.1	16.2	15.1
19.8	17.9	17	16.9	13.7	12.5	13.7	12.5	8.1	8.1	16.2	15.2
28.2	20	27.4	20.8	16.5	13	22.7	16.5	14.5	10.9	24.8	17.8
28.4	20.3	27.5	21.1	16.7	13	22.8	16.5	14.6	11	24.8	17.9
35.1	14.4	45.1	19.5	21.1	10.1	36.2	15.1	19.1	7.9	30.1	12.1
35.2	12.8	45.1	17.1	21.2	8.8	36.2	13.1	19.1	7	30.1	10.7
39.6	14.5	55.4	20.2	23	9.4	54.1	19.9	19.3	7.6	35.2	13.4
39.5	6.9	55.3	9.9	23	4.4	54.1	9	19.2	3.3	35.1	5.8
0.4		0.4		0.4		0.4		0.4		0.4	
0.4		0.4		0.4		0.4		0.4		0.4	
3.3		0.4		0.4		1.5		0.4		2.2	
3.2		0.4		0.4		1.6		0.4		2.3	
10		1.9		4.9		4.9		1.8		4.8	
10.1		1.9		4.9		4.9		1.8		4.8	
19		7.9		9.9		9.2		2.6		9.6	
18.9		7.9		9.8		9.2		2.6		9.6	
25.8		15.2		13.3		13.9		6.9		9.9	
25.9		15.3		13.4		13.9		6.9		9.9	
32		20.3		13.8		22.7		9.8		19.6	
32.1		20.3		13.9		22.7		9.9		19.6	
46.5	19.9	35	15.8	15.5	7.4	36.1	15.8	13.2	5.9	27.8	12.3
46.5	19.9	35	16	15.5	7.5	36.1	15.7	13.3	5.9	27.8	12.3
0.3		0.3		0.3		0.3		0.3		0.3	
0.3		0.3		0.3		0.3		0.3		0.3	
33.2	26	20.3	17.7	14.9	12.4	1.7	1.5	14.8	12.5	21.2	17.6
33.1	21.7	35.1	22.7	27	15.7	4.7	2.9	27.1	16.7	28.8	18.6
37.2	18.3	50.4	24	40	18.1	9.4	4.6	40.3	19.8	28.9	15.3
40.7	16.8	62.1	26	44.3	18.8	13.7	6.2	44.6	19.6	33.8	15.2
45	16.9	77.8	29.9	49.4	19.2	22.7	9.3	49.4	20.1	33.9	13.8
52.6	20.2	91.3	33.7	54	19.3	36.3	13.3	54.1	21.1	36.1	14.6
58.5	20.8	99.2	34.7	55	19.4	54.2	18.7	55.2	20.7	33	12.7
58.4	19.8	110.9	36.3	58.4	20.7	72.3	23.5	58.7	20.4	43.3	15
75	23.2	120.9	37.1	59.8	21.4	108.1	32.5	59.7	19.2	54.1	16.1
0.3		0.3		0.3		0.3		0.3		0.3	
0.3		0.3		0.3		0.3		0.3		0.3	
11	34.1	0.3	1.3	2.2	9.7	1.5	6.1	1.7	7.9	4	14.2
18.6	26.6	4.7	7.6	8.3	12.4	4.8	7	8.2	12.5	11	15.9
24	26.3	9.6	11.5	10.9	12	9.3	10.1	10.9	12.8	13.6	15.6
33.9	27.5	20.8	18.1	18.6	16.1	13.8	12	18.8	16.3	22.5	18.9
50.8	32.9	31.2	21.2	27.3	17.6	22.8	14.7	27.5	18.6	30.7	21
73.3	38.7	47.1	26.5	35.6	20.5	36.2	20.2	32.9	18.6	40.4	22.1
90.8	40.9	76	34.5	38.7	20	54	24.9	38.6	18.1	46.4	20.8
97.2	36.5	87.7	34.8	44.6	21.2	72.2	28.6	44.7	17.3	48.6	17.6
102.2	34.7	108.7	36.8	50.5	21.1	107.8	35	50.5	17.2	63.7	20.4
0.3		0.3		0.3		0.3		0.3		0.3	

Table G.9

$w_{oil,0}$	p	T	$\dot{m}$	$\dot{x}$	$\Delta p/\Delta L$	FP	$\dot{q}$	α
0.03	11.6	-36.02	161.2	0.52	19.23	a (f)	0.3	
0.03	11.6	-36.01	156	0.53	18.38	a (f)	1.6	6.8
0.03	11.6	-36.01	157.9	0.53	18.8	a (f)	6	11.4
0.03	11.57	-36.09	157.8	0.53	19.92	i-a (f)	10.7	13.7
0.03	11.58	-36.08	160.5	0.52	20.34	i-a (f)	14.5	15.1
0.03	11.59	-36.05	159.7	0.53	23.09	i-a (f)	22.9	17.6
0.03	11.6	-36.03	157	0.54	27.16	i-a (f)	34.7	20.5
0.03	11.61	-35.99	153.5	0.55	33.68	i-a (f)	53.9	23.8
0.03	11.61	-35.98	154.1	0.55	37.47	i-a (f)	71.3	24.2
0.03	11.63	-35.95	155.2	0.55	38.7	i-a (f)	84.9	23.1
0.03	11.58	-36.06	154	0.75	24.87	a (f)	0.3	
0.03	11.57	-36.09	163	0.71	25.26	a (f)	0.3	
0.03	11.58	-36.08	159.9	0.71	25.09	a (f)	2.2	10.2
0.03	11.57	-36.09	151.9	0.76	23.87	a (f)	5.6	12.9
0.03	11.57	-36.08	151.8	0.76	24.19	i-a (f)	10.1	14.6
0.03	11.58	-36.06	157.2	0.73	26.45	i-a (f)	10.2	14.9
0.03	11.6	-36.03	152.8	0.76	26.61	i-a (f)	13.6	15.5
0.03	11.59	-36.05	154.3	0.74	27.78	i-a (f)	13.6	15.8
0.03	11.6	-36.02	156.6	0.73	29.92	i-a (f)	23	17.5
0.03	11.59	-36.05	155.3	0.75	31.04	i-a (f)	22.5	17.8
0.03	11.6	-36.01	157.5	0.73	34.35	i-a (f)	36.6	20.4
0.03	11.6	-36.01	156.3	0.74	35.47	i-a (f)	34.6	18.4
0.03	11.59	-36.03	156	0.74	36.01	i-a (f)	34.6	20.5
0.03	11.6	-36.01	157.1	0.74	34.67	i-a (f)	32.8	19.2
0.03	11.61	-36	160.8	0.72	43.03	i-a (f)	55.3	22.8
0.03	11.62	-35.98	162.1	0.72	42.54	i-a (f)	54.6	24.1
0.03	11.62	-35.98	162.5	0.71	48.63	i-a (f)	75.7	25.1
0.03	11.65	-35.91	164.5	0.7	60.46	a (f)	102	34.8
0.03	11.62	-35.96	159.6	0.73	60.94	i-a (f)	104.3	35.7
0.03	11.65	-35.89	161.6	0.72	60.96	i-a (f)	104.3	36
0.03	11.66	-35.88	166.5	0.69	59.42	i-a (f)	104.9	34.9
0.03	11.64	-35.92	167.6	0.69	58.03	i-a (f)	105.1	32
0.03	11.63	-35.95	171.6	0.68	52.18	w-a (f)	108.1	11.1
0.03	11.58	-36.07	161.9	0.91	27.18	a (f)	0.3	
0.03	11.57	-36.07	160	0.91	25.67	a (f)	0.3	
0.03	11.58	-35.76	156.6	0.93	25.67	a (f)	0.3	
0.03	11.61	-32.58	151.8	0.95	26.95	a (f)	2.2	
0.03	11.58	-35.73	156.7	0.93	27.06	a (f)	3.2	18.4
0.03	11.58	-36.06	166.4	0.88	29.76	a (f)	6.4	9
0.03	11.6	-35.3	155.4	0.94	33.43	a (f)	11.1	34.9
0.03	11.55	-35.94	158.8	0.93	38.04	a (f)	24.8	12.9
0.03	11.6	-35.92	159.1	0.92	39.87	w-a (f)	24.6	13.6
0.03	11.52	-36.22	162.8	0.9	44.15	w-a (f)	40.6	12
0.03	11.52	-35.94	157.5	0.93	41.22	w-a (f)	38.5	13.2
0.03	11.61	-35.73	157.4	0.93	50.62	w-a (f)	55	22.6
0.03	11.63	-35.96	164.8	0.89	52.64	w-a (f)	49.6	19.2
0.03	11.57	-35.77	157.1	0.93	46.46	w-a (f)	47	14.4

$\dot{q}_1$	α_1	$\dot{q}_2$	α_2	$\dot{q}_3$	α_3	$\dot{q}_4$	α_4	$\dot{q}_5$	α_5	$\dot{q}_6$	α_6
0.3		0.3		0.3		0.3		0.3		0.3	
2	7.9	1.8	8.5	0.3	1.5	3.2	12.2	0.3	1.3	2.3	9.4
11.8	20.3	3.3	7	3.1	6.7	6.3	12.4	2.2	4.4	9.5	17.4
17.2	21	7.8	10.8	7.9	10.5	10.7	13.4	7	9.3	13.5	17.3
21.3	21	12	13.2	11	12	14.1	14.6	9.5	10.2	18.9	19.3
30.8	22.8	19.5	15.7	18.2	15	22.8	17.3	18.2	14	27.6	20.8
44.4	24.6	35.6	21.4	25.4	17.5	37.1	22.2	25	14.7	40.9	22.7
65.8	27.6	59.6	26.1	45.4	23.7	56.9	24.7	45.8	19.4	50.2	21.3
83.3	26.8	88.8	29.7	60.8	26.5	72.6	23.8	61.7	19.2	60.3	19.2
90.4	24.2	106	28.2	69.3	23.9	107.7	26.8	69.8	17.8	66.1	17.5
0.3		0.3		0.3		0.3		0.3		0.3	
0.3		0.3		0.3		0.3		0.3		0.3	
4.8	18.9	0.3	1.8	2	12	1.8	8.6	1.8	9.7	2.6	10.4
8.9	18.5	3	7.8	4.5	12.5	4.7	11.1	4.7	11.3	8.1	16.4
15.9	21.8	4.9	8	9.7	15.2	9.4	13.4	9.7	14.4	10.7	14.8
16.1	22.3	5	8.2	9.8	15.5	9.6	13.7	9.9	14.8	10.9	15.2
20.9	22.4	9.6	11.7	12.7	15.6	13.6	15.1	12.6	14.6	11.8	13.3
20.9	23.3	9.6	11.8	12.7	15.4	13.6	15.2	12.7	15.3	11.9	13.9
31.5	22.4	20	15.7	20.2	17	22.9	17.6	19.6	15.1	24.1	17.4
29.2	22	19.8	15.9	20	16.9	22.8	17.9	19.6	15.8	23.7	18
49.4	26.2	33.6	18.6	29.2	18.6	36.6	20.6	32.5	18.1	38	20.4
41.3	21.1	30.2	16.4	29.2	17.4	36.6	19.3	32.4	16.7	38.2	19.1
41.3	23.7	30.2	18.5	29.1	19.6	36.6	21.6	32.5	18.7	37.9	21
41.3	23.4	30.1	17.9	27.3	17.6	36.6	20.9	27.4	16	34.3	19.3
67.6	26.9	57.5	23.9	46.8	21.1	54.5	22.1	46.1	18.9	59.2	23.8
66.9	29.1	51.9	23.2	49.1	23.9	54.7	23.5	46.1	19.9	59.1	25.1
99.7	31.2	84.2	27.8	62.4	24.7	73	23.6	61.1	19.7	73.7	23.4
128.9	42.4	120.1	39.7	80.2	33.3	107.5	34.8	79.8	26.6	95.7	32
133.6	43.3	121.2	40	80.3	35.1	108	35.1	79.9	26.7	103.1	33.9
133.5	43.8	121	40.2	80.3	34.9	107.9	35.4	80	27	103.3	34.8
135.7	43.1	121.3	38.9	80.4	34.1	107.8	33.8	80.2	26	104.2	33.7
136.2	39.2	121.7	36.1	80.5	32.4	107.8	30.6	80.1	23.3	104	30.1
147.5	14.5	128	13.1	98.5	11.6	108.1	10.9	80.4	8	85.8	8.2
0.3		0.3		0.3		0.3		0.3		0.3	
0.3		0.3		0.3		0.3		0.3		0.3	
0.3		0.3		0.3		0.3		0.3		0.3	
3.8		1.7		1.6		2.8		1.7		1.7	
5.4	24.9	1.7	14.5	1.6	15.8	5	27.7	1.7	10.8	3.7	17
10	13.1	5	7.5	4.4	7	9.3	12.8	3.7	5.4	6.2	8.4
15.9	40	9.7	33.2	7.9	38	14.4	42.8	8	26.3	10.7	29.2
26.9	13.4	16.7	9	25.7	15.2	28	14.4	25.9	13	25.4	12.3
26.6	14.2	16.7	9.6	25.5	15.9	27.6	15	26.2	13.8	25.3	13
50.3	14.3	41.9	12.3	36.3	11.7	35.7	10.8	38.2	11.2	41.2	11.9
45.9	14.9	38.5	13.1	34.8	13.2	36	12.5	34.2	11.7	41.4	13.5
73	28.4	57.7	23.4	43.9	21.6	54.4	21.7	43.6	17.7	57.6	22.8
55.6	20.7	43.8	17	43.1	18.3	54.2	21	43.7	16.9	57.1	21.3
49.8	15.2	39	12.1	43.1	14	54.2	16	44.1	13.4	51.7	15.5

Table G.9

$w_{oil,0}$	p	T	$\dot{m}$	$\dot{x}$	$\Delta p/\Delta L$	FP	$\dot{q}$	α
0.03	11.57	-36.09	265.7	0.12	9.61	i (f)	0.3	
0.03	11.58	-36.07	265.8	0.12	10.92	i (f)	0.3	
0.03	11.58	-36.07	268	0.11	11.65	i (f)	7.3	11.4
0.03	11.58	-36.07	268.6	0.12	11.01	i (f)	7.2	11.2
0.03	11.59	-36.05	253.1	0.13	15.09	i (f)	18.9	15.8
0.03	11.59	-36.04	256.4	0.13	13.94	i (f)	18.9	15.7
0.03	11.6	-36.03	254.8	0.12	17.95	i (f)	29.3	19.1
0.03	11.59	-36.04	254.7	0.13	17.28	i (f)	29.2	18.8
0.03	11.6	-36.03	256	0.13	16.4	i (f)	38.3	21
0.03	11.6	-36.02	257.6	0.12	17.84	i (f)	38.4	20.7
0.03	11.61	-35.99	270.1	0.11	22.14	i (f)	57	24.9
0.03	11.61	-36	272.6	0.11	22	i (f)	57	25.3
0.03	11.62	-35.96	269.3	0.13	27.32	i (f)	66.2	27.5
0.03	11.62	-35.97	269.3	0.12	25.86	i (f)	66.1	27.9
0.03	11.63	-35.94	268.6	0.11	30.41	i (f)	77	29.3
0.03	11.62	-35.96	268	0.12	28.42	i (f)	76.9	29.1
0.03	11.64	-35.93	270.6	0.11	32.02	i (f)	84.1	30
0.03	11.64	-35.92	266.9	0.12	30.51	i (f)	84.1	30.9
0.03	11.64	-35.91	267.2	0.11	33.13	i (f)	93	30.7
0.03	11.65	-35.9	267	0.11	34.64	i (f)	93	31.7
0.03	11.58	-36.06	263	0.32	27.89	a	0.3	
0.03	11.59	-36.05	262.1	0.32	28.37	a	0.3	
0.03	11.59	-36.05	263.3	0.32	29.39	a	4.4	12.7
0.03	11.59	-36.04	265.1	0.32	27.91	a (f)	7	14.3
0.03	11.6	-36.02	266.9	0.32	30.37	a (f)	10.7	14.4
0.03	11.61	-36.01	266.7	0.33	31.57	i-a (f)	16.3	16.3
0.03	11.61	-35.99	271.4	0.31	34.4	i-a (f)	27.7	20.4
0.03	11.62	-35.96	272.8	0.31	39.56	i-a (f)	41.9	25.1
0.03	11.63	-35.94	272	0.31	44.04	i-a (f)	55.8	28.7
0.03	11.65	-35.9	266.9	0.32	59.47	i-a (f)	72.4	31.8
0.03	11.66	-35.87	266	0.32	64.04	i-a (f)	90.9	34.8
0.03	11.65	-35.89	268.4	0.32	57.52	i-a (f)	91	32.2
0.03	11.61	-35.98	268.4	0.51	50.57	a	0.3	
0.03	11.62	-35.98	264.7	0.52	50.04	a	3.2	9.9
0.03	11.62	-35.96	257.8	0.53	49.97	a	6.3	11.7
0.03	11.62	-35.96	260.2	0.53	51.99	a	11.9	13.3
0.03	11.56	-36.13	262	0.52	54.14	a	18.3	15
0.03	11.56	-36.1	262.5	0.52	58.81	a	34.4	20.2
0.03	11.56	-36.1	259.4	0.53	58.1	a	47.3	20.5
0.03	11.58	-36.07	260.3	0.53	64.88	i-a (f)	58.9	27.3
0.03	11.59	-36.05	262.2	0.52	68.18	i-a (f)	71.2	29.1
0.03	11.59	-36.04	263.1	0.52	74.56	i-a (f)	90.7	33.3
0.03	11.59	-36.04	263.6	0.52	73.26	a	91.2	34.1
0.03	11.6	-36.03	262.8	0.73	64.63	a	0.3	
0.03	11.6	-36.03	261.4	0.73	64.59	a	0.3	
0.03	11.6	-36.01	256.9	0.75	64.35	a	2.1	5.8
0.03	11.6	-36.01	262.1	0.73	65.75	a	2.1	6.3

$\dot{q}_1$	α_1	$\dot{q}_2$	α_2	$\dot{q}_3$	α_3	$\dot{q}_4$	α_4	$\dot{q}_5$	α_5	$\dot{q}_6$	α_6
0.3		0.3		0.3		0.3		0.3		0.3	
0.3		0.3		0.3		0.3		0.3		0.3	
13.2	20	3.2	5.4	8.1	12.3	1.6	2.5	8	13.5	9.5	14.9
13	19.7	3.1	5.3	8.1	12.1	1.6	2.5	8	13.4	9.4	14.6
29.6	24.4	14.7	13.2	18.7	15.1	4.9	4.2	18.8	16.2	26.5	21.8
29.6	24.3	14.7	13.1	18.6	14.9	4.8	4.1	18.8	16.2	26.6	21.7
50.4	31.5	25.7	17.4	27.1	17.1	9.3	6.2	27.4	18.6	36.1	23.5
50.3	31.4	25.6	17.4	27.1	16.8	9.2	6.1	27.3	18.2	36	23.1
62.8	33.9	38.6	22.1	35.3	18.5	13.6	7.6	35.6	20	44.2	24
62.9	32.6	38.6	21.7	35.4	18.7	13.7	7.6	35.6	19.9	44.2	23.7
97.3	41	60.4	27.5	52.5	22.6	22.8	10.4	55.5	24.8	53.4	23.3
97.4	42.3	60.4	27.8	52.5	22.7	22.9	10.4	55.5	24.9	53.4	23.4
103.8	42.5	78.7	32.8	59.8	24.3	36.4	15.3	60.1	25.4	58.5	24.7
103.6	44	78.7	33.1	59.7	23.8	36.3	15	60	25.7	58.4	25.7
116.9	44.4	98	36.7	63.2	23.9	54.4	20.3	63.2	25	66.6	25.6
116.6	43.6	97.7	36.9	63.1	24.5	54.3	20.7	63.1	24.8	66.6	24.2
121	42	111.8	39.1	65.7	25.1	72	25.6	65.6	24.1	68.7	23.8
121	44.2	111.7	40	65.7	24.8	71.9	25.7	65.7	25.1	68.7	25.2
127.8	41.4	116.3	38.8	67.1	24.5	107.9	33.5	67.1	23.1	71.7	23.1
127.8	43.7	116.4	39.8	67.1	24.3	108.1	33.4	67.2	23.7	71.7	25.3
0.3		0.3		0.3		0.3		0.3		0.3	
0.3		0.3		0.3		0.3		0.3		0.3	
8.6	22.6	0.3	1	6.4	18.8	2.6	7.3	4.2	13.5	4.5	12.8
12.3	23.6	1.7	3.9	8.8	18.9	4.7	9.8	6.2	13.9	8	16
17.4	22.4	3.9	5.7	11	15.4	9.4	12.5	9	12.9	13.2	17.5
24.6	23.4	9.5	10	15.8	16.7	13.9	13.9	15.1	15.5	18.9	18.5
37.5	26.9	18.4	14.3	31.9	23.3	22.5	16.5	24.1	18	32.1	23.1
56.3	33.2	33.7	20.9	46.3	27.2	36	21.3	35.9	21.6	43.5	26.1
66.3	35	44.5	23.9	65.4	32.2	54.2	26.8	44	23	60.5	31.6
83.5	35.6	65.1	28.9	71.9	35.5	72	31.8	60.6	26	81.1	33.3
115.6	41.8	89.2	34.7	71.8	34.5	107.6	38.7	67.1	25	94.2	33.8
115.6	39.4	89.3	32.5	72	31.4	107.6	35.4	67.4	23	94	31.3
0.3		0.3		0.3		0.3		0.3		0.3	
5.2	14.4	0.3	1.1	3.9	14.3	3.5	10.9	1.6	5.4	4.5	13.3
10.5	18	0.3	0.7	9	18.2	4.8	9.2	4	7.8	9.1	16.2
18.1	19.1	4.2	5.1	13.7	16.1	12.3	13.8	7.1	8.4	15.7	17
26.2	20.4	9.8	8.5	21.8	18.6	16.3	13.5	15.1	12.5	20.5	16.5
45.2	26.9	19.5	12.3	54.8	29.8	22.8	13.7	27.3	16.5	36.9	21.9
60.2	25.7	32	14.5	65.6	28	36.6	16.4	36.5	16.3	52.6	22.4
73.1	33	45.4	21.6	65.6	31.2	54.3	25.1	47.6	22.3	67.3	30.5
87.8	34.9	62.7	25.6	65.7	29.3	73	29.4	58.1	23.7	80.1	31.9
109.2	39	94.5	34	65.6	28.9	107.3	37.3	65.4	24	101.9	36.3
113.2	40.5	94.3	34.1	65.6	29.9	107.1	38.4	65.5	24.6	101.6	36.8
0.3		0.3		0.3		0.3		0.3		0.3	
0.3		0.3		0.3		0.3		0.3		0.3	
4.4	11.3	0.3	1	2.1	7.2	1.5	4.8	0.3	0.9	3.6	9.8
4.5	11.8	0.3	1	2.4	8.8	1.5	5.1	0.3	1	3.6	9.9

Table G.9

$w_{oil,0}$	p	T	$\dot{m}$	$\dot{x}$	$\Delta p/\Delta L$	FP	$\dot{q}$	α
0.03	11.6	-36.03	258.5	0.74	65.51	a	4.3	9.2
0.03	11.6	-36.01	257.3	0.74	64.15	a	4.4	9.3
0.03	11.6	-36.02	258.8	0.74	67.24	a	7.2	11.2
0.03	11.61	-35.98	255.8	0.75	67.16	a	7.1	11.3
0.03	11.6	-36.02	248	0.77	65.28	a	11.8	12.9
0.03	11.61	-35.99	256.9	0.74	67.7	a	11.8	13
0.03	11.61	-36	262.7	0.73	72.85	a	27.5	16
0.03	11.62	-35.98	260.5	0.73	73.09	a	27.7	16.2
0.03	11.6	-36.01	268.6	0.72	57.8	a (f)	41.4	9.3
0.03	11.61	-36	270.2	0.71	61.96	a (f)	41.4	8.9
0.03	11.61	-35.98	266.6	0.72	64.18	i-a (f)	48.9	11.1
0.03	11.61	-35.99	267.6	0.71	63.57	i-a (f)	49.8	11.8
0.03	11.6	-36.01	274.8	0.7	64.89	i-a (f)	57.2	12.7
0.03	11.62	-35.97	278.2	0.68	71.02	i-a (f)	57.2	14.8
0.03	11.57	-36.1	268.3	0.9	68.56	a	0.3	
0.03	11.57	-36.07	266	0.91	70.89	a	0.3	
0.03	11.57	-36.1	266.8	0.9	68.78	a	1.1	3.3
0.03	11.57	-36.1	270.2	0.89	68.84	a	1.1	3.5
0.03	11.57	-36.09	268.1	0.9	72.95	a	4.5	8.9
0.03	11.57	-36.06	264.6	0.91	73.31	a	4.5	9.1
0.03	11.58	-36.06	265.6	0.91	75.43	a	8.9	12
0.03	11.57	-36.08	266.5	0.91	73.69	a	8.9	11.7
0.03	11.58	-36.02	264.9	0.91	78.08	a	11.2	13.4
0.03	11.58	-36.05	265.7	0.91	77.04	a	11.2	12.8
0.03	11.61	-35.55	257.6	0.94	82.85	i-a (f)	19.3	20.2
0.03	11.59	-35.54	257.4	0.94	85.84	i-a (f)	19.4	20.7
0.03	11.65	-35.3	256.1	0.94	93.42	i-a (f)	31.5	23.2
0.03	11.63	-35.68	259.3	0.93	93.29	i-a (f)	31.5	17.4
0.03	11.64	-35.83	263	0.92	96.12	i-a (f)	48.1	17
0.03	11.65	-35.8	262.1	0.92	91.75	i-a (f)	47.9	18.3
0.03	11.55	-36.15	520.8	0.12	33.54	a (f)	0.3	
0.03	11.55	-36.14	524.1	0.11	35.83	a (f)	0.3	
0.03	11.55	-36.14	524.5	0.11	34.57	a (f)	1.6	6.2
0.03	11.55	-36.14	521.3	0.12	35.62	a (f)	1.6	6.3
0.03	11.55	-36.13	524.8	0.12	34.32	i-a (f)	3.4	8.2
0.03	11.55	-36.13	524.4	0.12	35.19	i-a (f)	3.5	8.4
0.03	11.56	-36.11	524.8	0.12	37.76	i-a (f)	8.8	10
0.03	11.56	-36.12	529.1	0.11	36.09	i-a (f)	8.8	10.1
0.03	11.57	-36.1	518.2	0.12	41.38	i-a (f)	21.2	16.9
0.03	11.57	-36.09	519.9	0.12	41.43	i-a (f)	21.2	17
0.03	11.57	-36.09	530.6	0.12	24.31	i-a (f)	13.6	11.4
0.03	11.57	-36.09	531.8	0.11	40.34	i-a (f)	13.7	11.5
0.03	11.58	-36.08	523.7	0.12	45.92	i-a (f)	32.5	21.2
0.03	11.58	-36.07	523.2	0.12	46.04	i-a (f)	32.5	21.3
0.03	11.59	-36.05	523.3	0.12	54.71	i-a (f)	49.7	26.5
0.03	11.59	-36.04	526	0.12	54.63	i-a (f)	49.8	26.5
0.03	11.6	-36.02	520.3	0.12	73.32	i-a (f)	60.8	30.9

$\dot{q}_1$	α_1	$\dot{q}_2$	α_2	$\dot{q}_3$	α_3	$\dot{q}_4$	α_4	$\dot{q}_5$	α_5	$\dot{q}_6$	α_6
7.4	14.3	2.5	5.8	4.3	9.8	5.1	11.3	0.3	0.7	6.4	13.6
7.4	14.5	2.6	6	4.4	9.9	5.1	11.3	0.3	0.7	6.4	13.7
11.1	16.1	3.2	5.5	6.9	11.7	9.1	14.5	1.9	3	11	16.5
11	16.3	3.2	5.6	6.9	11.8	9.1	14.5	1.8	3	10.8	16.5
14.6	15.6	7.3	8.4	15.7	16.7	13.6	14.8	3.8	4.3	15.9	17.4
14.6	15.8	7.3	8.5	15.7	16.8	13.6	15	3.7	4.3	15.9	17.6
38.5	21.5	18.6	11.2	35	20.5	22.7	13.6	22.3	13	27.8	16.2
38.7	21.8	18.9	11.4	35.2	20.7	22.8	13.7	22.4	13.1	27.9	16.3
48.1	10.6	35.3	8.1	49.1	11.4	36.2	8.3	32.4	7.2	47.4	10.2
48.1	9.9	35.3	7.6	49.1	10.8	36.2	8	32.4	7.1	47.4	10.1
55.3	12.2	45.3	10.2	51.5	12.1	54.1	12.3	33	7.7	54.2	12.2
55.4	12.6	45.3	10.7	51.7	12.9	54.2	12.9	37.9	9.1	54.2	12.4
62.2	13.7	57.9	12.8	53.8	12.4	71.9	15.5	36.4	8.3	61.2	13.5
62.2	15.4	57.9	14.6	53.9	14.7	71.8	18.3	36.4	10	61.1	15.9
0.3		0.3		0.3		0.3		0.3		0.3	
0.3		0.3		0.3		0.3		0.3		0.3	
1.9	5.7	0.3	1.1	0.3	1.4	1.8	5.8	0.3	1.1	1.7	5.1
2	6	0.3	1.1	0.3	1.3	2	6.3	0.3	1.1	1.7	5.2
9.3	16.7	0.3	0.7	5	11.5	5.5	10.9	1.6	3.4	5.3	9.9
9.2	17	0.3	0.7	5	11.9	5.5	11.3	1.6	3.4	5.3	10.3
15.6	19.6	4.5	6.5	9.4	14.1	10.8	14.6	4.8	6.8	8	10.7
15.6	19	4.5	6.3	9.4	13.6	10.7	14.1	4.8	6.6	8	10.4
15.6	17.7	7.2	9.2	9.4	13	14	16.5	7.8	9.2	13.3	14.9
15.6	17	7.2	8.8	9.4	12.3	14	15.7	7.9	8.8	13.3	14.2
24.8	24.4	14.1	15.4	20.3	23.8	22.8	23.3	15.5	15.9	18.5	18.3
24.9	25	14.2	15.7	20.4	24.4	22.9	23.7	15.4	16.3	18.5	18.8
38.8	26.4	24.9	18.2	31.5	27	38.1	27.5	27.3	19.8	28.6	20.1
38.8	20.1	24.9	13.7	31.5	19.8	38.1	20.8	27.3	15	28.6	15.2
53.8	17.9	42.8	14.6	46.2	18.8	53.5	18.8	43.5	15.4	48.5	16.5
53.7	19.7	42.8	16	46.1	20.2	53.5	20.3	43.2	16.2	48.3	17.6
0.3		0.3		0.3		0.3		0.3		0.3	
0.3		0.3		0.3		0.3		0.3		0.3	
3.9	14.4	0.3	1.4	3.3	11.2	1.7	6.8	0.3	1.6	0.3	1.7
4	15.1	0.3	1.5	3.2	10.9	1.7	7.1	0.3	1.7	0.3	1.8
5.8	13.8	0.3	0.9	3.3	8	4.8	11.1	1.9	4.6	4.4	11
5.8	13.8	0.3	1	3.4	7.9	4.7	11	2	5	4.5	11.7
14.8	16.3	3	3.8	8	9.2	9.2	10.6	4.3	5.1	13.5	15.2
14.8	16.4	3.2	4	8.1	9.3	9.2	10.6	4.1	4.9	13.3	15.2
33.3	26.1	9.3	8.1	24.6	18.6	14.1	11.7	16.8	14.3	29.1	22.8
33.3	26	9.3	8.1	24.6	18.7	14.1	11.7	16.9	14.3	29.1	22.8
20.6	16.8	7.5	6.5	13.5	11.2	13.5	11.3	9.7	8.3	16.9	14.3
20.8	17	7.6	6.6	13.7	11.3	13.7	11.4	9.7	8.3	17.1	14.5
46.8	30.2	21.2	14.6	35.6	21.9	22.7	15.1	27.3	18.5	41.2	26.6
46.7	30.4	21.2	14.8	35.6	22.1	22.8	15.3	27.3	18.6	41.2	26.7
69	36.2	37.6	20.7	52.3	26.8	36	19.9	41.1	22.7	62.4	32.8
69	36.2	37.6	20.7	52.3	26.7	36	19.8	41.1	22.7	62.4	32.9
82.9	41.2	48.5	25.4	57.1	29.2	54.1	27.6	47.3	24.7	74.6	37.4

Table G.9

$w_{oil,0}$	p	T	$\dot{m}$	$\dot{x}$	$\Delta p/\Delta L$	FP	$\dot{q}$	α
0.03	11.6	-36.01	524.4	0.12	69.33	i-a (f)	60.8	30.8
0.03	11.65	-35.9	525.9	0.11	116.67	f	73.2	37.8
0.03	11.65	-35.89	525.5	0.11	116.39	f	73.2	37.7
0.03	11.63	-35.94	522	0.11	120.14	f	105	45.2
0.03	11.64	-35.93	523	0.11	119.29	f	105.2	45.5
0.03	11.56	-36.1	427.6	0.39	80.73	a	0.3	
0.03	11.57	-36.09	428.3	0.39	81.02	a	0.3	
0.03	11.57	-36.1	430.9	0.38	81.11	a	1	5.5
0.03	11.55	-36.13	427.7	0.39	80.64	a	0.9	5.7
0.03	11.57	-36.1	438.4	0.38	81.1	a	4.8	11.3
0.03	11.57	-36.1	436.1	0.38	82.47	a	4.8	11
0.03	11.57	-36.08	443.5	0.37	83.75	a (f)	9.6	13.7
0.03	11.58	-36.08	440.4	0.38	85.45	a (f)	9.6	13.5
0.03	11.57	-36.09	458.9	0.36	86.23	a (f)	15.8	15.4
0.03	11.57	-36.08	464	0.35	84.39	a (f)	15.8	15.5
0.03	11.58	-36.07	485.8	0.33	89.36	a (f)	26.8	18.5
0.03	11.58	-36.07	485.1	0.33	90.73	a (f)	26.9	18.4
0.03	11.59	-36.03	491.4	0.34	99.64	a (f)	45.7	24.5
0.03	11.59	-36.03	495.3	0.33	99.29	a (f)	45.8	24.7
0.03	11.61	-36	509.6	0.32	109.31	a (f)	67.3	30.6
0.03	11.61	-35.99	507.6	0.32	111.6	a (f)	67.5	30.8
0.03	11.62	-35.98	522.9	0.32	118.65	a (f)	83.7	34.1
0.03	11.62	-35.96	523.5	0.31	120.41	a (f)	83.7	34.4
0.03	11.65	-35.91	533.1	0.31	134.95	a (f)	110.7	39.7
0.03	11.64	-35.92	533.6	0.31	132.03	a (f)	110.8	37.4
0.03	25.99	-10.63	80.7	0.15	0.25	s-w (f)	0.2	
0.03	25.99	-10.63	81.1	0.15	0.38	s-w (f)	2.1	5.5
0.03	26	-10.62	80.9	0.15	0.5	s-w (f)	6.7	7.3
0.03	26	-10.61	79.7	0.15	0.6	i (f)	10.9	7.5
0.03	26	-10.61	81.9	0.15	0.75	i (f)	14.1	7.9
0.03	26.01	-10.6	80.3	0.14	1.27	i (f)	23.7	9.6
0.03	26.02	-10.59	80.9	0.14	1.86	i (f)	35.4	11.2
0.03	26.03	-10.57	82.5	0.15	2.69	i (f)	49.3	12.6
0.03	26.04	-10.55	80.2	0.19	3.93	i (f)	64.3	13.4
0.03	26.06	-10.52	80.2	0.26	7.09	i (f)	95.6	13.4
0.03	25.92	-10.71	79.7	0.34	0.87	s-w (f)	0.2	
0.03	25.93	-10.7	80.4	0.34	1.07	w (f)	5.4	6.8
0.03	25.93	-10.7	80.3	0.35	1.53	i (f)	10.4	8.7
0.03	25.93	-10.7	80.5	0.34	1.7	i (f)	13.6	8.6
0.03	25.93	-10.7	80.3	0.34	2.3	i (f)	19.2	9
0.03	25.94	-10.68	79.4	0.35	2.69	i (f)	25.9	9.8
0.03	25.95	-10.67	80.2	0.34	3.92	i (f)	39.4	11.1
0.03	25.95	-10.66	79.9	0.35	4.87	i (f)	51.9	11.6
0.03	25.95	-10.67	78.8	0.35	5.57	i (f)	60	11.9
0.03	25.94	-10.68	79.1	0.36	7.03	f	79	12.3
0.03	25.93	-10.7	79.8	0.35	0.75	w (f)	0.2	
0.03	25.93	-10.7	80	0.34	0.81	w (f)	2	4.5

$\dot{q}_1$	α_1	$\dot{q}_2$	α_2	$\dot{q}_3$	α_3	$\dot{q}_4$	α_4	$\dot{q}_5$	α_5	$\dot{q}_6$	α_6
82.9	40.9	48.5	25.2	57.2	29.2	54.1	27.7	47.3	24.8	74.7	37.3
96.5	48.1	61.9	32.4	67.7	35.9	72.6	37.5	56.9	30	83.8	42.7
96.4	48	61.9	32.3	67.7	35.9	72.6	37.4	56.9	30	83.8	42.6
147.7	60.9	98.7	41.6	75.9	37.7	108	45.5	81.8	35.4	118	50
147.9	61.3	98.9	41.8	76.1	38.2	108.1	45.8	82	35.6	118.1	50.3
0.3		0.3		0.3		0.3		0.3		0.3	
0.3		0.3		0.3		0.3		0.3		0.3	
2.6	13.5	0.3	2.2	0.3	2.2	1.7	10.1	0.3	2.8	0.3	2.1
2.6	13.6	0.3	2.4	0.3	2.4	1.7	10.5	0.3	3	0.3	2.2
8.7	19.4	0.3	0.9	7.3	16.9	4.8	11.8	2.1	6	5.4	12.9
8.8	18.9	0.3	0.9	7.3	16.3	4.8	11.4	2.1	5.8	5.4	12.5
14.4	20.2	3.7	5.6	15.5	20.4	9.6	13.5	6.2	10.1	8.2	12.5
14.4	19.8	3.9	5.8	15.5	20	9.6	13.4	6.1	9.8	8.2	12.2
24.4	22.8	7.8	8	23	21.3	13.4	13.4	11.6	12.4	14.5	14.5
24.4	22.8	7.7	8	23	21.5	13.4	13.5	11.6	12.5	14.5	14.6
37.7	25.5	16.4	12	37.5	24.4	22.8	15.9	20.3	14.8	26.4	18.5
37.7	25.4	16.4	12	37.5	24.4	22.8	15.8	20.3	14.7	26.4	18.4
59.2	31.6	33.6	19.1	63.1	31.9	36.2	19.4	39.2	21.4	43.1	23.5
59.3	31.8	33.7	19.2	63.2	32	36.3	19.6	39.3	21.6	43.1	23.7
87.7	38.9	54.5	25.6	79.3	35.9	54.4	25.2	56.5	26	71.6	32.3
87.8	39.1	54.6	25.7	79.5	35.9	54.5	25.2	56.6	26.2	71.7	32.5
110	43.5	70.1	29.3	89.1	37	72	29.6	63.5	26.6	97.2	38.7
110	43.7	70.2	29.4	89.2	37.4	72	29.9	63.5	26.9	97.1	39
147.7	51.4	100.3	36.1	100.5	38.9	107.8	37.5	84.3	30.4	123.4	43.6
147.7	48.1	100.3	34	101.1	36.6	107.8	35.8	84.2	28.9	123.5	41.3
0.2		0.2		0.2		0.2		0.2		0.2	
4.5	12.1	2.4	6.9	0.2	0.5	1.4	3.8	0.2	0.5	3.6	9.4
4.6	5	13.9	15.5	3	3.2	4.8	5.2	2.9	3.2	10.6	11.4
3.1	2.1	22.6	15.8	8.1	5.4	9.2	6.3	8	5.5	14.6	10
3	1.6	30.2	17	10.6	5.9	13.6	7.6	10.8	6.1	16.4	9.2
6.2	2.4	51.6	21.2	17.3	7.1	22.8	9.1	17.5	7.1	26.6	10.4
15.5	4.8	76.4	24.1	28	9.1	36.2	11.3	26.3	8.4	30	9.5
30.3	7.6	97.9	24.9	37.6	10.2	54	13.7	36.5	9.4	39.2	10
48.7	9.8	110.1	22.8	49.9	11.4	72.6	15	50.5	10.4	53.9	11
61	8.6	139.3	19.5	91.4	13.5	107.6	14.8	90.5	12.5	83.7	11.6
0.2		0.2		0.2		0.2		0.2		0.2	
8.2	10	7.7	10.7	3.6	4.6	2.4	3.1	2.9	3.6	7.7	9
8.7	7.1	17	14.3	11.3	9.1	4.7	3.9	11.8	9.9	8.8	7.6
8.9	5.4	22.4	14.2	15.7	9.9	9.2	6	16.7	10.6	8.9	5.6
10.3	4.6	27.3	12.9	26.7	12.7	13.9	6.6	26.1	12.1	11	5.1
13.7	4.8	40.9	15.3	33	12.9	22.9	8.8	32.5	12.3	12.7	4.7
19	5.1	65.4	18.1	49	14.4	40.5	11.5	48.3	13.4	14.2	4
24.8	5.3	78.5	17.6	67.5	15.8	54.5	12.3	68.7	14.9	17.7	3.8
30.9	6.1	85.9	17.2	75.2	15.3	72.4	14.1	75.3	14.6	20.3	3.9
38.9	5.9	116	17.5	89.4	15.2	107.2	16.4	89.2	13.7	33.4	5
0.2		0.2		0.2		0.2		0.2		0.2	
4	8.4	2.6	6	1.3	3.4	1.7	3.9	0.9	2	1.6	3.6

Table G.9

$w_{oil,0}$	p	T	$\dot{m}$	$\dot{x}$	$\Delta p/\Delta L$	FP	$\dot{q}$	α
0.03	25.93	-10.7	83.3	0.33	1.29	i (f)	8.9	8.8
0.03	25.94	-10.69	81.7	0.34	1.32	i (f)	13.3	8.8
0.03	25.94	-10.68	81.2	0.34	1.73	i (f)	17.1	9.2
0.03	25.95	-10.67	79.7	0.35	2.31	i (f)	23	9.9
0.03	25.96	-10.67	82.8	0.33	3.8	i (f)	37.1	11.2
0.03	25.96	-10.65	78.7	0.35	4.83	i (f)	54.4	11.9
0.03	25.96	-10.66	79.6	0.35	6.04	f	64.1	12.2
0.03	25.96	-10.66	80.1	0.35	6.83	f	78.8	12.2
0.03	25.95	-10.66	80.7	0.53	1.74	w (f)	0.2	
0.03	25.95	-10.65	80	0.54	2.13	w (f)	5.9	8
0.03	25.95	-10.65	80.7	0.53	2.57	w (f)	10.2	8.1
0.03	25.95	-10.65	81.4	0.54	2.47	w (f)	10.5	8.3
0.03	25.96	-10.64	78.1	0.57	2.46	w (f)	10.6	8.3
0.03	25.96	-10.64	81.3	0.54	2.49	w (f)	10.2	8.1
0.03	25.97	-10.63	80.3	0.54	2.89	w (f)	17.4	9.4
0.03	25.96	-10.64	81.5	0.53	2.97	w (f)	17.5	9.2
0.03	25.97	-10.63	80.9	0.53	3.38	w (f)	24	10
0.03	25.97	-10.63	79.7	0.55	4.8	f	38	10.6
0.03	25.99	-10.6	80.6	0.54	5.48	f	45	10.6
0.03	25.98	-10.61	81.2	0.54	5.97	f	50.9	10.6
0.03	25.97	-10.62	79.9	0.54	6.63	f	68.2	10
0.03	25.98	-10.61	76.6	0.56	1.41	w (f)	0.2	
0.03	25.97	-10.62	76.9	0.56	1.4	w (f)	0.2	
0.03	25.98	-10.61	76.8	0.56	1.98	w (f)	9.3	7.9
0.03	25.98	-10.61	77.6	0.55	2.15	w (f)	11.5	8.2
0.03	25.98	-10.6	78	0.55	2.22	w (f)	15	8.7
0.03	25.99	-10.59	77.8	0.56	2.63	w (f)	20.6	9.5
0.03	26	-10.59	79.8	0.54	3.46	i (f)	28.5	10.5
0.03	25.99	-10.59	80.6	0.54	3.48	i (f)	28.5	10.6
0.03	26	-10.59	78.8	0.55	3.96	i (f)	33.6	10.7
0.03	26	-10.58	80.2	0.54	4.04	i (f)	33.6	10.8
0.03	26.01	-10.58	80.6	0.54	4.94	i (f)	42.3	10.7
0.03	26.01	-10.57	79.6	0.55	5.08	i (f)	42.3	10.8
0.03	26	-10.57	79	0.55	6.16	f	49.3	10.6
0.03	26.01	-10.57	79.6	0.55	5.68	f	49.3	10.8
0.03	25.99	-10.59	78.5	0.56	6.09	f	59.3	10.4
0.03	26	-10.58	79.2	0.55	6.39	f	59.1	10.2
0.03	25.97	-10.58	81	0.72	2.52	w (f)	0.2	
0.03	25.97	-10.57	80.5	0.72	2.51	w (f)	0.2	
0.03	25.98	-10.57	80.8	0.72	2.73	w (f)	3.3	7.2
0.03	25.98	-10.57	80.4	0.71	2.57	w (f)	5.5	8.4
0.03	25.98	-10.56	81.1	0.71	3.04	w (f)	7.9	8.9
0.03	25.99	-10.53	79.7	0.75	4.12	i (f)	12.1	9.7
0.03	25.98	-10.55	80.2	0.74	3.91	i (f)	12.1	9.7
0.03	25.99	-10.55	81.5	0.72	4.4	i (f)	16.1	9.9
0.03	26	-10.53	80.4	0.74	5.03	f	22.9	9.5
0.03	26	-10.52	79.9	0.74	5.46	f	31.1	9

Table G.9

$\dot{q}_1$	α_1	$\dot{q}_2$	α_2	$\dot{q}_3$	α_3	$\dot{q}_4$	α_4	$\dot{q}_5$	α_5	$\dot{q}_6$	α_6
9.3	8.9	14.8	14.8	8.7	8.7	4.8	4.7	8.5	8.4	7.5	7.5
9.3	5.9	21	13.9	15.6	10.7	9	6.1	15.8	10.5	9.3	6
8.9	4.6	24.9	13.5	22.7	12.4	13.8	7.4	22.4	11.8	10.2	5.4
10.2	4.3	34.3	14.9	30.2	13.2	22.6	9.5	30.1	12.9	10.8	4.7
12.9	3.8	61.2	18.4	50	15.5	36.2	10.9	50	14.9	12.4	3.8
26	5.6	83.7	18.1	75.9	16.9	54	11.9	75.8	16.3	11	2.4
32.1	6	93.1	17.4	86.3	17.2	72	13.6	86.2	16.1	14.6	2.8
43.4	6.4	116.8	17.3	90.6	15.5	106.6	16.3	90.7	13.9	24.5	3.7
0.2		0.2		0.2		0.2		0.2		0.2	
11.3	14.7	10.2	14.3	3.1	4.2	3.4	4.7	3.3	4.6	3.9	5.4
11	8.5	24.1	19.2	6.6	5.3	9	7.1	6.5	5.3	4	3.2
11.1	8.4	23.9	19	8.5	6.6	9.1	7.1	6.6	5.3	4	3.1
11	8.5	24.2	19.4	8.5	6.6	9	7.1	6.6	5.3	4	3.1
11	8.4	24.2	19.4	6.6	5.3	9	7.2	6.6	5.3	4	3.1
12.6	6.6	33.8	18.7	18.9	10.3	13.8	7.5	16.6	8.9	8.5	4.5
12.7	6.3	33.9	18.1	19	10.2	13.9	7.5	16.7	8.9	8.6	4.4
12.4	4.9	47.6	20	26	11.1	22.8	9.4	24.8	10.2	10.7	4.3
15.2	4.1	81.5	21.9	40.5	11.9	37.1	10.5	40.6	11.5	13	3.5
17.4	4	90.3	20.7	46.9	11.5	54.1	12.8	47	11.1	14.4	3.3
19.4	4.1	95.6	19.7	49.9	11.1	72.4	14.8	50.6	10.5	17.5	3.6
23.5	3.3	114.4	16	69.9	12.7	108.4	15.4	70.5	9.6	22.4	3
0.2		0.2		0.2		0.2		0.2		0.2	
0.2		0.2		0.2		0.2		0.2		0.2	
9.2	7.6	17.1	14.9	12.9	11	2.4	2.1	10	8.5	4.3	3.5
9.9	6.8	22.5	16.3	15.4	11	5.1	3.7	11.1	8	5	3.4
10.9	6.1	30	17.7	19.5	11.4	9.4	5.5	14.4	8.4	5.5	3.1
12.9	5.7	45.9	21.1	24	11.3	15.1	7	19.5	9	6	2.7
13.4	4.7	60.1	21.9	29.1	11.4	25.5	9.3	35.7	12.8	7.3	2.6
13.4	4.8	60.1	22.1	29.1	11.6	25.5	9.4	35.7	12.8	7.3	2.6
15.3	4.8	69.1	22	33	11.3	35.6	11.1	40.7	12.7	8.2	2.5
15.3	4.9	69	22.1	33	11.3	35.6	11.1	40.6	12.7	8.1	2.6
19.9	4.9	84.3	21	41.7	11.2	52.6	13.2	46.2	11.7	9	2.3
19.9	5.1	84.3	21.3	41.7	11.2	52.6	13.1	46.3	11.9	9	2.3
20.2	4.3	94.7	20.1	46.1	10.7	70	14.7	54.4	11.6	10.5	2.3
20.2	4.4	94.6	20.5	46.1	10.7	70	14.8	54.4	11.9	10.5	2.3
25.9	4.4	116.2	20	34.3	7.5	108	18.4	60.8	10.4	10.4	1.8
25.8	4.4	115.8	19.8	34.2	7.1	107.7	18.1	60.6	10.4	10.3	1.8
0.2		0.2		0.2		0.2		0.2		0.2	
0.2		0.2		0.2		0.2		0.2		0.2	
9.3	19.4	3.2	7.9	1.5	3.3	1.5	3.4	1.4	3.2	2.9	6.2
10.2	14.9	10.5	16.6	3.3	4.9	2.9	4.4	3	4.6	3.2	4.9
11.8	12.7	17.1	19.6	4.7	5.4	4.8	5.4	4.6	5.3	4.4	4.9
15.9	12.2	25	20.5	8.4	7	9.5	7.6	8.2	6.7	5.7	4.5
15.9	12.1	24.9	20.4	8.4	7	9.5	7.6	8.2	6.6	5.7	4.4
18.5	10.9	35.3	21.8	11.1	7.2	13.6	8.4	12	7.5	6.2	3.8
17.8	7.1	52.2	21.5	18.7	8.2	22.7	9.4	19.5	8.2	6.5	2.6
17.7	5	69.3	19.9	28.6	8.7	35.9	10.1	28.4	8.2	6.9	2

Table G.9

$w_{oil,0}$	p	T	$\dot{m}$	$\dot{x}$	$\Delta p/\Delta L$	FP	$\dot{q}$	α
0.03	26	-10.52	80.4	0.74	7.16	i (f)	48.5	7.9
0.03	26	-10.53	81.2	0.73	7.71	w (f)	60.4	7.9
0.03	25.98	-10.12	81.8	0.9	2.56	w (f)	0.2	
0.03	25.99	-9.87	80.5	0.91	2.59	w (f)	0.2	
0.03	25.99	-9.74	80.5	0.92	8.23	w-a (f)	3.1	
0.03	25.98	-9.68	80.2	0.92	8.31	w-a (f)	3.1	
0.03	25.99	-9.73	79.9	0.92	6.56	w (f)	4.9	
0.03	26	-9.69	80.2	0.92	8.07	w (f)	9.8	47.8
0.03	26	-9.31	79.9	0.93	8.29	w (f)	13.6	37.2
0.03	26	-9.42	80.7	0.93	8.06	w (f)	13.7	18.2
0.03	26	-8.94	79.7	0.94	10.89	w (f)	22.6	48.3
0.03	25.99	-9.39	80.4	0.93	10.01	w (f)	21.8	18.4
0.03	25.99	-9.7	81.1	0.92	9.29	w (f)	31.7	8.2
0.03	26	-9.17	79.8	0.93	8.78	w (f)	31.8	10.4
0.03	25.99	-9.14	79.6	0.93	9.01	w (f)	33.1	8.7
0.03	25.99	-9.46	80.4	0.93	8.66	w (f)	40.4	4.5
0.03	26	-9.27	80.1	0.93	8.86	w (f)	41.2	5.3
0.03	25.94	-10.69	157	0.13	0.95	w (f)	0.2	
0.03	25.94	-10.7	156.4	0.13	0.96	i (f)	0.2	
0.03	25.95	-10.68	156.9	0.13	1.51	i (f)	4.6	8
0.03	25.95	-10.68	157.4	0.13	1.88	i (f)	10.9	10.9
0.03	25.96	-10.67	157.1	0.13	2.49	i (f)	18.8	12.2
0.03	25.96	-10.67	156.5	0.13	2.53	i (f)	22.6	12.1
0.03	25.96	-10.66	156.9	0.13	4.17	i (f)	33.8	13.6
0.03	25.97	-10.66	156.4	0.13	5.12	f	43.7	14.6
0.03	25.98	-10.64	155.9	0.13	6.02	f	52.9	15.3
0.03	25.98	-10.64	155.7	0.14	7.81	f	67.2	16.2
0.03	25.99	-10.62	155.5	0.13	10.03	f	91.7	17.6
0.03	25.9	-10.75	166	0.11	1.21	i (f)	0.2	
0.03	25.9	-10.75	166.2	0.11	1.75	i (f)	4.4	7.8
0.03	25.91	-10.73	165.7	0.11	2.49	i (f)	10.9	11.2
0.03	25.92	-10.72	165.3	0.11	2.87	i (f)	19	12.6
0.03	25.91	-10.74	165.9	0.1	3.31	i (f)	22.7	12.5
0.03	25.94	-10.7	166.5	0.11	4.87	i (f)	33.8	13.2
0.03	25.94	-10.7	164.8	0.11	5.19	i (f)	43.3	13.8
0.03	25.95	-10.68	163.8	0.11	6.76	f	53.1	14.3
0.03	25.96	-10.67	163.7	0.11	8.16	f	67.3	15.2
0.03	25.97	-10.66	163.7	0.11	11.11	f	91.7	16.8
0.03	25.97	-10.65	157.1	0.13	1	i (f)	0.2	
0.03	25.96	-10.67	157	0.12	1.93	i (f)	6.2	8.6
0.03	25.98	-10.64	158.1	0.13	2.3	i (f)	13.2	10.5
0.03	25.93	-10.71	154.4	0.13	3.51	i (f)	25.5	12.8
0.03	25.98	-10.64	156.2	0.13	2.7	i (f)	18.2	10.7
0.03	25.95	-10.69	154.4	0.13	3.84	i (f)	32	12.7
0.03	25.95	-10.68	153.9	0.13	4.42	i (f)	37.6	13.3
0.03	25.96	-10.67	153.6	0.13	5.91	i (f)	48.4	14.1
0.03	25.96	-10.67	164.5	0.12	7.54	f	60.6	15.2

Table G.9

$\dot{q}_1$	α_1	$\dot{q}_2$	α_2	$\dot{q}_3$	α_3	$\dot{q}_4$	α_4	$\dot{q}_5$	α_5	$\dot{q}_6$	α_6
22.7	3.7	95.9	15.5	55.3	9.2	54.2	8.9	54.4	8.9	8.6	1.4
28.8	3.7	107.8	13.7	69.4	9.7	75	9.6	69.8	8.9	11.6	1.5
0.2		0.2		0.2		0.2		0.2		0.2	
0.2		0.2		0.2		0.2		0.2		0.2	
6.4		3		1.5		2.9		1.5		3.1	
6.4		3		1.5		2.9		1.5		3.1	
10.9		5.9		2.2		4.9		1.5		3.8	
13.8	54.5	17	83.4	6.9	51.9	9.1	42.2	7	32.2	5.2	22.3
20.8	44.8	22	59.1	10	45.4	13.9	36.1	10.1	25.5	5.1	12.1
20.8	23.6	22	29.9	10	18.1	13.9	18.5	10.1	13.3	5.1	6.2
27.7	46.1	40.3	69.2	18.4	85.5	25.1	45.8	18.5	32.6	5.5	10.3
27.9	21.5	36.3	29.5	18.5	20.5	25.5	20.7	15.7	12.8	7	5.6
25.1	6.2	72.2	17.8	26	8.5	36.4	9.3	26.2	6.7	4.2	1
25.3	7.9	72.3	22.2	26.1	11	36.5	11.7	26.2	8.3	4.2	1.3
25.2	6.5	72.2	18.4	26.1	8.2	44.8	11.4	26.2	6.8	4.2	1.1
29.4	3.1	84.6	9.2	31.9	4.5	54	5.9	33.4	3.7	9.4	1
29.3	3.5	84.4	10.5	31.9	5.6	60.4	7.4	33.5	4.1	7.5	0.9
0.2		0.2		0.2		0.2		0.2		0.2	
0.2		0.2		0.2		0.2		0.2		0.2	
10.2	16.7	2.7	5	2.4	4.8	3.5	6.3	1.8	3.2	6.8	12.1
20.3	19.5	10.1	10.7	9.8	10.2	4.6	4.7	5.9	6.1	14.4	14.2
26.2	16.9	22.1	15	17.6	11.3	9.3	5.8	18.8	11.7	18.9	12.3
27.7	14.7	30	16.3	22.4	12.2	13.5	7.2	22.3	11.8	19.9	10.6
35.7	14	48.4	19.3	32.8	13.6	22.9	9.2	32.8	13.2	30.3	12.3
41.4	13.3	59.9	19.8	45	15.7	36.1	12	44.7	14.8	34.9	11.8
49.5	13.7	75.2	21.1	50.9	15.7	53.8	15.5	51.1	14.9	36.8	11
67.7	15.3	87.1	20.8	62	16.5	71.8	17.3	61.2	14.7	53.1	12.3
84.1	15.8	126.2	23.7	76.6	16.3	107.3	20.2	76.4	14.5	79.4	15
0.2		0.2		0.2		0.2		0.2		0.2	
10	16.9	2.3	4.6	2.9	5.4	3.2	5.5	1.3	2.4	6.6	12.2
20.3	20.3	10.1	10.9	9.6	9.6	4.6	4.6	6.2	6.6	14.5	15.3
26.5	17.8	22.5	15.6	17.8	11.4	9.3	5.8	19	11.9	18.7	12.7
27.6	15.5	29.9	16.9	22.1	11.9	13.7	7.1	22.4	11.9	20.3	11.6
35.7	13.9	48.6	19	32.6	12.9	22.8	8.7	32.9	12.7	30.5	12.1
41.1	12.9	60	19.1	42.3	14.2	36.4	11.4	44.9	13.9	34.9	11.2
49.7	13.3	75.2	19.7	52	14.7	54.1	14.2	51.3	13.8	36.6	10.2
67.8	15	87.8	19.6	62.2	15.1	72	16.1	61.1	13.7	52.9	11.7
84	15.5	125.9	22.7	76.9	15.2	107.4	19	76.6	13.7	79.2	14.6
0.2		0.2		0.2		0.2		0.2		0.2	
10.7	14	8.9	12.7	1.4	2.3	2.3	3.4	1.9	2.7	11.8	16.3
26.5	19.5	19.8	16.2	4.8	4.4	4.7	4	4.2	3.6	19	15.1
31.8	16.3	29.9	15.6	34.9	16.4	9.2	4.6	23.7	12	23.8	12.2
32.2	17.6	28	16.7	7.1	4.7	9.3	5.6	9.6	5.9	22.8	13.5
36	14.1	38	15.6	43.5	17	13.7	5.5	37.2	14.5	23.5	9.4
36.1	13	47.9	17.4	47.3	16.4	22.8	7.8	42	14.6	29.6	10.4
48.6	13.9	59.3	17.5	62.3	18.2	36.4	10.6	49.7	14.6	34.3	9.9
55.3	13.6	78.5	19.4	73.5	18.7	54	13.5	59.8	15	42.8	10.7

Table G.9

$w_{oil,0}$	p	T	$\dot{m}$	$\dot{x}$	$\Delta p/\Delta L$	FP	$\dot{q}$	α
0.03	25.98	-10.65	164.5	0.12	8.27	f	70.2	15.3
0.03	25.98	-10.64	164.8	0.12	10.48	f	85.4	16.4
0.03	25.98	-10.63	156.5	0.32	3.45	i (f)	0.2	
0.03	25.99	-10.63	157.1	0.32	3.43	i (f)	0.2	
0.03	25.99	-10.62	158.3	0.31	4.54	i (f)	8.2	11.3
0.03	26	-10.61	155.7	0.33	5.41	i (f)	15.2	14.3
0.03	26.01	-10.59	158.1	0.32	6.72	i (f)	25.4	15.7
0.03	26.02	-10.58	156.6	0.33	7.6	f	33.7	16.5
0.03	26.02	-10.58	157.2	0.32	7.25	f	38.7	16.8
0.03	26.03	-10.56	156.4	0.33	8.08	f	49.3	16.8
0.03	26.03	-10.56	159.7	0.32	10.29	f	59.8	17.1
0.03	26.05	-10.54	157.9	0.33	11.77	f	73.3	16.9
0.03	26.05	-10.54	155.4	0.33	12.95	f	85.4	16.5
0.03	25.94	-10.67	158.2	0.52	6.44	i (f)	0.2	
0.03	25.93	-10.68	158	0.52	6.32	i (f)	0.2	
0.03	25.94	-10.67	158.2	0.52	6.89	i (f)	4	9.9
0.03	25.95	-10.65	158.2	0.52	9.22	i (f)	14.8	15.7
0.03	25.96	-10.63	158.3	0.53	11.68	i (f)	23.8	19.1
0.03	25.97	-10.63	158.5	0.52	11.45	i (f)	24	19
0.03	25.98	-10.61	157.9	0.52	13.13	i (f)	36.5	19.1
0.03	25.98	-10.61	157.5	0.53	13.36	i (f)	36.4	19.4
0.03	25.99	-10.59	160.7	0.52	14.11	i (f)	47.6	17.7
0.03	26	-10.59	161.6	0.51	14.93	i (f)	60.8	15.9
0.03	26	-10.59	153.4	0.54	16.58	i (f)	70.4	13.5
0.03	26.01	-10.57	153.7	0.54	16.63	i (f)	70.8	13.4
0.03	26.01	-10.58	153.8	0.54	16.52	i (f)	71	13.2
0.03	26.01	-10.57	165.3	0.49	18.42	f	92	14.3
0.03	25.99	-10.6	161.1	0.5	18.62	f	92.3	14.6
0.03	25.99	-10.6	160	0.51	18.54	f	92.2	14.7
0.03	26.01	-10.58	162	0.51	17.86	f	89.6	13.7
0.03	26.01	-10.57	162.9	0.51	17.88	f	84.8	13.2
0.03	26.01	-10.57	162.3	0.51	18.54	f	86	13.2
0.03	25.98	-10.61	154.2	0.54	18.29	f	86	12.5
0.03	25.96	-10.59	159.2	0.71	9.5	w-a (f)	0.2	
0.03	25.96	-10.6	161.5	0.7	9.65	w-a (f)	0.2	
0.03	25.96	-10.59	159.1	0.7	10.4	w-a (f)	2.3	6.6
0.03	25.96	-10.59	157.9	0.71	12.14	w-a (f)	6.2	12.2
0.03	25.97	-10.57	159.5	0.71	15.33	w-a (f)	10.8	16.1
0.03	25.97	-10.58	159.8	0.7	16.89	w-a (f)	14.9	18.1
0.03	25.98	-10.56	158.9	0.72	19.41	a (f)	24.1	19.3
0.03	25.99	-10.55	160.2	0.71	20.38	a (f)	33.9	18.8
0.03	26	-10.53	158.7	0.72	21.66	a (f)	43.7	18.5
0.03	26	-10.53	158.8	0.72	22.69	f	53.6	16.4
0.03	26.04	-10.48	158.8	0.71	20.14	f	75.7	10.3
0.03	26.06	-10.46	160	0.71	20.27	a (f)	75.7	10.2
0.03	26.07	-10.44	159.5	0.71	20.22	f	75.8	10.7
0.03	26.07	-10.44	159.2	0.71	23.73	a (f)	75.6	12.5

$\dot{q}_1$	α_1	$\dot{q}_2$	α_2	$\dot{q}_3$	α_3	$\dot{q}_4$	α_4	$\dot{q}_5$	α_5	$\dot{q}_6$	α_6
66.7	13.9	98	20.6	73.7	17	72.2	16	59.9	13.4	50.9	11
75	14.4	114.2	21.5	75.6	15.6	107.7	20	65.3	12.7	74.7	14.1
0.2		0.2		0.2		0.2		0.2		0.2	
0.2		0.2		0.2		0.2		0.2		0.2	
15.2	19.8	5.3	7.6	8.7	12.4	1.7	2.4	8.5	12.2	9.5	13.2
24.3	22.1	15.2	15	14.9	14.2	4.8	4.6	15.1	14.3	16.8	15.8
38.1	22.5	31.6	19.9	27.3	17	9.2	5.8	27.2	17.1	18.9	11.9
44.2	21.3	44.6	22.2	40.8	19.7	13.6	6.8	40.7	19.8	18.1	9.5
49.8	21.5	51	22.9	45.8	19.5	22.6	9.6	45.8	19.6	17.4	7.7
50.6	16.9	76.3	25.8	56.9	19.7	36.3	12.4	56.8	19.6	19.1	6.6
55.6	15.5	84.5	24.1	67.1	20.2	54.1	15.3	67.4	18.9	29.9	8.6
64.8	14.6	99.1	22.7	86.3	20.9	71.8	16.5	86.4	19.6	31.4	7
66.7	13	109.1	21.2	96.6	19.7	107.2	20	97.3	18.5	35.6	6.9
0.2		0.2		0.2		0.2		0.2		0.2	
0.2		0.2		0.2		0.2		0.2		0.2	
8.3	18.8	1.5	4.1	3.8	10	1.5	4	3.5	9.2	5.5	13.1
28.5	29.1	13.3	14.9	14.9	15.3	5.6	6.1	13.2	14.7	13.2	14.4
41.7	32.6	26.3	21.8	20.3	16.2	13.7	10.9	19.4	15.9	21.3	17
41.9	31.9	26.5	21.8	20.5	16.5	13.8	11.1	19.6	15.8	21.5	16.7
52.9	27	49.3	26.2	38	19.8	23.5	12.3	28.8	15.5	26.3	13.8
52.9	27.6	49.3	26.5	38	20	23.5	12.4	28.7	15.8	26.2	14.1
61.5	22.2	69.4	26	50.1	19.1	36.9	13.6	43.5	16.3	24.1	9
65.1	16.9	92.9	24.5	66.8	17.6	51.4	13.5	69.7	18.2	19	4.9
55.2	10.8	108.7	20.8	85.4	15.8	73	13.8	82.3	16.1	17.7	3.4
57.4	10.8	108.8	20.4	85.4	16.5	72.3	13.7	82.2	15.6	18.8	3.6
57.6	10.7	109	20.3	85.6	15.9	72.5	13.4	82.3	15.5	19	3.5
87.9	13.4	124.1	19.2	101.7	16.4	106.4	16.4	102.1	15.8	30	4.6
86.3	13.4	123.8	19.4	102.7	16.9	106.5	16.8	102.1	16.3	32.4	5.1
86.2	13.6	123.8	19.6	102.6	16.8	106.3	16.9	102	16.3	32.4	5.2
75.2	11.1	123.9	18.7	102.8	16.2	106.4	16.2	102.1	15.7	27.4	4.1
62.4	9.5	115.4	18	102.7	16.3	106.5	16.1	102	16	19.9	3
62.3	9.6	122.2	18.7	102.7	16	106.5	15.9	102.2	15.9	19.9	3.1
62.4	8.9	122.2	17.4	102.7	15.6	106.5	15.3	102.1	14.7	20	2.8
0.2		0.2		0.2		0.2		0.2		0.2	
0.2		0.2		0.2		0.2		0.2		0.2	
3.3	8.3	0.2	0.6	3.4	10.8	2	5.7	3.1	9.1	1.8	5
9.1	16.2	3.8	7.9	6.5	13.9	4.7	9.3	6.5	12.9	6.9	13
14.9	20.4	9.7	14.8	9.4	15.8	9.2	13.8	9.5	14.7	11.9	17.3
19.6	22.2	14	17.4	12.2	16.7	13.6	16.5	12.4	15.4	17.4	20.5
31.7	23.7	25.9	20.9	17.1	16.4	22.8	18.5	17.1	13.9	30	22.6
48.3	25.3	39.8	22.2	24.8	16.2	36	19.4	24.8	13.3	29.8	16.2
63.1	25.1	61.5	25.7	26.8	14.6	54	22.2	26.5	11.1	30.3	12.7
68.6	19.9	86.5	25.8	35	14.1	71.8	21.3	34.8	10.3	24.7	7.3
50.6	6.8	143.7	19.3	71	10.2	107.8	14.4	70.8	9.6	10.3	1.4
50.6	6.7	143.7	19	71	10.2	107.7	14.3	70.7	9.5	10.3	1.4
50.8	7	143.9	19.9	71	10.6	107.8	14.9	70.7	10	10.3	1.4
50.8	8.3	143.1	22.1	70.9	14.9	107.7	17.1	70.6	10.7	10.3	1.6

Table G.9

$w_{oil,0}$	p	T	$\dot{m}$	$\dot{x}$	$\Delta p/\Delta L$	FP	$\dot{q}$	α
0.03	26	-9.65	155.9	0.92	14.48	w-a (f)	0.2	
0.03	26	-9.63	155.7	0.92	15.21	w-a (f)	0.2	
0.03	25.99	-9.1	154	0.94	25.5	w-a (f)	3.2	
0.03	26.01	-9.81	157.5	0.92	15.63	w-a (f)	4.8	
0.03	26.02	-9.71	156.5	0.92	15.29	w-a (f)	9.4	24.1
0.03	26.02	-9.7	156.5	0.92	22.69	w-a (f)	26	19.5
0.03	26.01	-9.63	158.5	0.92	20.33	w-a (f)	19	15.8
0.03	26.02	-9.82	157.4	0.91	21.92	a (f)	40	9.8
0.03	26.02	-9.85	158.8	0.91	22.51	a (f)	54.1	7
0.03	26.02	-9.67	157.5	0.92	22.22	a (f)	51.5	8.6
0.03	25.96	-10.68	260.9	0.11	3.34		0.2	
0.03	25.96	-10.67	262.1	0.11	4.88		4	8.3
0.03	25.97	-10.65	260.4	0.11	5.81		7.8	11.1
0.03	25.98	-10.64	261	0.1	6.11		15.4	13.9
0.03	25.99	-10.63	259.8	0.1	7.09		25.1	16.1
0.03	26	-10.62	259.8	0.11	9.67		40.4	17.1
0.03	26.02	-10.59	259	0.11	11.76		62	18.2
0.03	26.03	-10.58	260	0.1	13.51		80.8	18.5
0.03	26.03	-10.58	258.5	0.1	18.21		117.6	20
0.03	25.96	-10.66	261.6	0.11	4.16	i (f)	0.2	
0.03	25.98	-10.65	261.9	0.11	6.75	i (f)	11.5	14.8
0.03	25.99	-10.64	259.9	0.12	7.15	i (f)	18	16.1
0.03	25.99	-10.63	261.8	0.11	7.7	i (f)	23.9	17.5
0.03	26.01	-10.61	262.1	0.11	9.41	i (f)	37.1	18.9
0.03	26.02	-10.59	259.9	0.12	12.16	i (f)	54.6	19.6
0.03	26.01	-10.61	260	0.12	11.63	i-a	62.4	20.1
0.03	26.03	-10.57	259.2	0.12	14.18	i-a	73.2	20.1
0.03	26.06	-10.54	259.9	0.11	15.34	i-a	85.8	19.7
0.03	26.05	-10.55	258.2	0.12	18.21	i-a	103.4	19.8
0.03	25.95	-10.68	269.1	0.1	3.64	i (f)	0.2	
0.03	25.96	-10.67	266.8	0.11	5.74	i (f)	4.7	9.2
0.03	25.97	-10.66	266.8	0.11	7.41	i (f)	9.4	13
0.03	25.97	-10.65	267.9	0.11	7.8	i (f)	14.7	14.8
0.03	25.98	-10.65	267.1	0.11	8.85	i (f)	22.6	16.8
0.03	25.98	-10.64	267.6	0.11	8.43	i (f)	31.9	18.3
0.03	26	-10.61	265.7	0.12	10.57	i-a	47	19.3
0.03	26	-10.62	267.5	0.11	12.82	i-a	56.8	19.9
0.03	26.01	-10.6	266.1	0.12	13.96	i-a	68	19.8
0.03	26.01	-10.6	265.5	0.12	13.94	i-a	70.9	20.5
0.03	26.03	-10.58	265.1	0.11	19.66	i-a	110.5	19.8
0.03	25.94	-10.7	269	0.11	4.05	i (f)	0.2	
0.03	25.95	-10.69	269.8	0.11	4.31	i (f)	0.2	
0.03	25.94	-10.69	269.5	0.11	5.31	i (f)	2.8	6.8
0.03	25.96	-10.68	268.6	0.11	7	i (f)	16.3	14.4
0.03	25.96	-10.67	269.8	0.11	7.89	i (f)	26.3	16.4
0.03	25.96	-10.67	270.6	0.11	8.38	i (f)	26.3	16.4
0.03	25.97	-10.66	267.2	0.12	17.73	f	38.6	22.7

$\dot{q}_1$	α_1	$\dot{q}_2$	α_2	$\dot{q}_3$	α_3	$\dot{q}_4$	α_4	$\dot{q}_5$	α_5	$\dot{q}_6$	α_6
0.2		0.2		0.2		0.2		0.2		0.2	
0.2		0.2		0.2		0.2		0.2		0.2	
3.7		1.6		3.2		3.9		3.2		3.4	
5.6		2.9		5.3		4.8		5.2		5.2	
10.6	24.8	7.6	20.6	9	26.4	8.8	22.4	9.7	25.1	10.6	25.6
45.1	31.1	20.9	15.2	26.9	23.4	13.7	10.2	26.6	19.8	23	17.4
26.6	21.3	13.8	11.6	21.5	18.4	13.8	11.4	21.8	18.1	16.7	14.3
52.5	12.1	78.9	18.8	37	11.3	23.3	5.6	37.7	8.9	10.3	2.4
46.4	5.8	102.5	12.8	77.1	11.4	38.6	4.8	59.6	7.5	0.2	0
62.9	9.8	104.8	16.6	48.8	10.6	35.9	5.7	45	7.2	11.5	1.8
0.2		0.2		0.2		0.2		0.2		0.2	
7.2	12.6	4.4	9.3	3.6	7.2	1.6	3.2	3.6	8.7	3.8	8.7
12.2	16	8.9	12.6	6.8	9.1	4.3	5.7	7.5	11.5	7.2	11.6
23.8	20.7	17.3	15.7	13.3	11.4	8.3	7.1	14.3	13.5	15.3	15.1
39.3	24	28.5	18.8	21.9	13.9	13.8	8.6	23.3	15.3	23.9	16
61.7	23.4	44.1	19.5	35.6	16.7	23.1	10.4	36.7	15.8	41.1	16.5
88.1	22.4	71.8	21.5	58.6	20	34.6	11.4	58.3	17.5	60.7	16.3
106.3	20.6	98.2	21.8	75.3	21.2	48.8	13	75.4	17.8	80.8	16.7
131.9	20.3	157.2	24.9	109.3	22.2	90.6	16.9	110.1	18.7	106.8	17
0.2		0.2		0.2		0.2		0.2		0.2	
26.2	32.8	7	9.4	9.3	12.6	2.6	3.6	9.6	12.7	14.2	17.8
36.7	31.1	18	15.9	14.7	13.9	4.6	4.5	17	15.7	17.3	15.5
42.9	30.8	25.6	18.6	21.6	16.3	9.1	7	23.6	17.2	20.5	15.1
52.4	26.1	44.2	22.2	41.8	21.9	13.3	7.3	39.4	19.8	31.8	16.1
60	21.2	64.8	22.8	72.5	26.8	22.8	8.8	61.5	21.8	46	16.4
62	20.7	69	22.8	84.6	27.3	35.9	11.9	66.4	20.6	56.5	17.2
69.5	19.3	76.4	21.2	94.4	26.6	54.3	15.2	76.1	20.4	68.2	17.8
86.2	18.6	97.4	21.7	98.9	24.9	71.7	17.2	86.2	19.6	74.2	16.1
108.7	19.2	118.7	21.6	106.1	23.1	107.4	21.4	92.5	17.9	87	15.9
0.2		0.2		0.2		0.2		0.2		0.2	
7.4	13.8	5.8	11.2	2.6	5.6	2	3.9	3.5	6.8	7.1	14.1
16.1	21.3	10	13.8	6.7	9.6	4.5	6.1	7	9.7	12.3	17.5
24.1	23.6	15.5	15.8	9.3	9.8	9	8.8	10.6	10.7	19.9	20.1
35.2	25.9	25.2	19.1	16.6	12.8	13.6	9.8	18.7	13.9	26.1	19.6
39	22.8	41	23.8	27.5	16.1	22.8	12.4	29	16.1	31.9	18.4
55.6	22	58.1	24.7	43.4	19.3	36.3	14.8	42.7	17.1	46.1	18.2
62.4	21.3	71.7	25	51.1	19.3	54.3	18.2	53.1	18.3	48	17
73	20.4	76.4	22.7	60.9	20.2	72.8	20.4	68.1	19	56.9	16.2
73.4	21.1	83.3	24.5	71.4	22.4	72.9	20.1	68.3	19.2	56.1	15.6
99.6	17.5	124.4	21.3	126.3	24.7	107.9	19.4	126	21.9	79	14
0.2		0.2		0.2		0.2		0.2		0.2	
0.2		0.2		0.2		0.2		0.2		0.2	
7.8	17.8	0.2	0.5	0.2	0.7	1.4	3.9	4.2	10.3	3.2	7.7
32.4	28	10.7	9.9	15.8	13.9	5	4.6	10.6	9.6	23.5	20.4
46.8	28.3	26.1	17	26.8	16.8	9.2	6.1	18.8	12	30.2	18.3
46.8	28.2	26	16.8	26.7	16.9	9.2	6	18.7	12	30.2	18.6
71.8	41.2	39.5	23.9	36.5	21	13.5	8.2	24.2	14.8	45.8	27

Table G.9

$w_{oil,0}$	p	T	$\dot{m}$	$\dot{x}$	$\Delta p/\Delta L$	FP	$\dot{q}$	α
0.03	25.98	-10.65	266.5	0.12	17.93	f	48.5	23.6
0.03	25.99	-10.63	265.8	0.11	17.69	f	60.9	23.7
0.03	25.99	-10.62	265.8	0.12	18.02	f	73.8	23.7
0.03	26	-10.62	265.8	0.12	18.81	f	84.9	23.7
0.03	26	-10.62	267.8	0.12	18.86	f	85	23.7
0.03	25.99	-10.63	265	0.12	19.96	f	101.9	23.1
0.03	25.99	-10.63	266.5	0.12	20.09	f	102.1	23.2
0.03	26.01	-10.6	260.4	0.29	10.06	i-a	0.2	
0.03	26	-10.6	263.2	0.28	10.11	i-a	7.5	8.9
0.03	26.01	-10.6	261.6	0.29	11.44	i-a (f)	10.1	10.1
0.03	26.02	-10.59	261.9	0.29	12.38	i-a (f)	15.3	12.6
0.03	26.01	-10.59	262.3	0.28	16.59	i-a (f)	24.7	16.8
0.03	26.01	-10.6	264.2	0.29	20.65	i-a (f)	39.9	23.2
0.03	26.03	-10.57	257.2	0.3	26.44	i-a (f)	39.8	22.5
0.03	26.02	-10.58	267.5	0.28	21.18	f	60.9	22.6
0.03	26.04	-10.55	262.9	0.29	23.69	f	79.2	22
0.03	26.06	-10.53	264.5	0.28	26.54	f	118.1	17.3
0.03	25.91	-10.73	258.2	0.31	11.2	i-a	0.2	
0.03	25.92	-10.72	260	0.32	16.59	i-a	9	19.4
0.03	25.93	-10.71	261.5	0.31	15.35	i-a	17.5	20.7
0.03	25.94	-10.69	259.9	0.31	19.59	i-a	31.8	25.4
0.03	25.95	-10.67	259.6	0.31	20.88	i-a (f)	45	26.3
0.03	25.98	-10.63	258.1	0.32	23.04	i-a (f)	67.4	24.4
0.03	26	-10.6	258.2	0.32	26.56	i-a (f)	94.1	22.1
0.03	26	-10.61	259.7	0.31	26.6	f	103.8	20.2
0.03	26.01	-10.6	257.7	0.32	29.44	f	117	19.2
0.03	26	-10.6	258.6	0.32	28.39	f	130.6	16.9
0.03	26	-10.6	262.1	0.31	11.31	i-a	0.2	
0.03	26	-10.6	266.1	0.3	11.1	i-a	4.6	10.5
0.03	26.02	-10.58	264.1	0.31	14.45	i-a	15.3	16.6
0.03	26.02	-10.57	267	0.31	16.39	i-a (f)	24.2	20.9
0.03	26.04	-10.56	261.9	0.32	20.24	i-a (f)	34.9	24.6
0.03	26.04	-10.55	263	0.32	22.31	i-a (f)	52.3	25.1
0.03	26.06	-10.53	264.1	0.31	23.78	f	63.6	24.8
0.03	26.06	-10.52	263.3	0.31	24.5	f	75	23.5
0.03	26.07	-10.51	263.7	0.31	26.24	f	94.5	21.1
0.03	25.99	-10.59	254.4	0.52	18.07	i-a (f)	0.2	
0.03	26	-10.59	258.6	0.51	35.64	i-a (f)	111.3	13.5
0.03	25.98	-10.61	259.6	0.5	35.31	i-a (f)	112	12.9
0.03	25.99	-10.6	257.9	0.52	36.54	i-a (f)	115	14.2
0.03	25.99	-10.6	256.7	0.52	36.33	i-a (f)	113.7	13.8
0.03	25.98	-10.61	259.8	0.5	35.39	i-a (f)	112.5	13
0.03	25.98	-10.61	259	0.51	35.88	i-a (f)	114.9	13.3
0.03	25.96	-10.64	258.4	0.51	35.57	i-a (f)	114.1	14
0.03	25.98	-10.62	258.6	0.51	36.72	i-a (f)	114.4	15.1
0.03	25.98	-10.61	257.2	0.52	19.8	a	0.2	
0.03	26	-10.59	257	0.52	20.39	a	4.6	10.8

$\dot{q}_1$	α_1	$\dot{q}_2$	α_2	$\dot{q}_3$	α_3	$\dot{q}_4$	α_4	$\dot{q}_5$	α_5	$\dot{q}_6$	α_6
83.1	39.6	53.7	26.8	47.3	22.6	22.6	11.2	32	16	52	25.3
92.8	35.5	69.1	27.5	66.3	25.4	35.7	14.1	43.6	17.2	58.1	22.6
98.5	31.1	85.1	27.5	85.5	27.4	56.1	18.1	52.1	17	65.5	21.1
99	27.3	100.8	27.9	102.3	28.8	73	20.4	60.4	17	74.1	20.6
99.1	27.3	100.8	27.8	102.2	28.8	73.1	20.4	60.4	17.1	74.2	20.7
107.7	23.7	120.8	26.9	112.6	27.8	112.1	24.9	72.8	16.4	85.4	19
108	23.8	121.2	26.9	113	27.9	112.4	25	72.9	16.4	85.4	19
0.2		0.2		0.2		0.2		0.2		0.2	
11	11.4	6.7	8	6.9	8	3.4	4	8.7	11.5	8.5	10.4
14.7	13.6	8.7	9	9.7	9.2	5	4.9	11.3	12.4	11.1	11.7
21.5	17	12.9	11.3	15.3	11.8	9.2	7.2	16.5	14.3	16.4	14.2
35.6	23.3	20.7	15.1	24.9	15.9	14.7	9.5	26.4	18.5	26	18.4
61.2	34.1	36	22.8	39.3	22.1	23.6	13.3	39.7	23.4	39.8	23.7
60.7	32.1	35.9	21.1	39.3	22.2	23.4	13.3	40	23.4	39.7	22.9
93	31.2	58.6	24.6	53.7	22.9	38	14.7	61	22.2	61.1	20.1
120.6	28.3	79.7	24.5	70.2	24.6	51.8	16.2	75	20.6	77.7	17.9
169.2	19.2	142.4	20.9	96.2	19.6	96	17.3	96.8	14.6	108.1	12.3
0.2		0.2		0.2		0.2		0.2		0.2	
18.4	34.6	7.5	17	4.8	13.2	1.4	4	13.9	30.8	8.3	16.9
29.8	32.7	16.8	20.4	11.7	15.8	4.8	6.8	22.6	27.1	19.1	21.5
55	41.4	35.8	28.4	20.1	18.4	9.3	8.7	36.5	29.3	34.2	26.4
69.3	39.3	61.4	34.6	36.8	23.4	13.6	9	54.8	31.4	34.2	20.4
83.4	29.1	96	34	69.2	28.3	22.4	9.3	90.1	30.8	43.2	15.2
86.2	19.6	154	33.5	107.3	29.4	36.6	10	142.8	31.6	37.4	8.6
84.1	15.8	163	30	118	27.5	54.1	12	160.6	28.4	43	7.6
94.9	14.3	178.3	27.7	118	24	72.3	13.2	178.9	26.9	59.9	8.8
121.1	13.6	178.5	22.2	118.7	20.1	107.7	15.4	178.9	21.4	78.9	8.6
0.2		0.2		0.2		0.2		0.2		0.2	
14.2	30.5	0.7	1.8	1.5	3.5	4.8	11.7	1.4	3.6	5.1	12.1
31.7	32.8	11	12.6	10.3	11	9.1	10.3	10.4	12	19.2	21
44.3	37.3	20.5	18.4	17.9	14.8	13.7	11.9	18	16.1	30.7	26.7
58.3	40.1	35.5	25.3	26.8	18.6	22.9	16.1	26.6	19.4	39.2	28.2
82.2	38.1	59.8	29.2	46.7	22.7	36.3	17.3	46	22.2	42.6	21
84.6	32.8	82.3	32	59	23.4	54	20.5	59.3	23	42.6	17
85.7	26.8	98	30.9	73	23.9	72	21.8	72.8	22.4	48.5	15.1
83.1	18.7	131.2	28.6	96.9	23.5	107.4	23.3	97.4	21.3	51.2	11.6
0.2		0.2		0.2		0.2		0.2		0.2	
159.7	14.3	178.8	19.8	97	24.2	4.5	0.7	161.8	16.3	66.3	5.7
159.3	13.8	178.4	19.1	96.5	21.9	9.2	1.4	159.3	15.5	69.3	5.7
158.1	14.8	178.6	20.3	104	25	13.8	2.2	166.2	16.8	69.3	6
157.9	14.3	178.4	19.9	92.3	23.1	22.7	3.4	162	16.1	69.1	5.9
151.1	13.3	178.3	19.7	74.6	18.7	36.1	5.4	133	13	102	8.1
144.8	12.8	178.1	19.7	70.8	18.4	54.3	7.7	132.8	12.8	108.5	8.6
125.3	12.1	178.1	20.8	73.3	19.3	72.4	10.3	127.1	12.7	108.3	9.1
118.3	12.9	178.1	22.5	68.9	18	107.9	15.5	105	11.7	108.2	10.1
0.2		0.2		0.2		0.2		0.2		0.2	
10.3	22	1.6	4	3	8.2	4.7	11.5	3	7.7	4.9	11.5

Table G.9

$w_{oil,0}$	p	T	$\dot{m}$	$\dot{x}$	$\Delta p/\Delta L$	FP	$\dot{q}$	α
0.03	26	-10.59	259.5	0.51	19.35	a	9.4	15.1
0.03	26	-10.58	259.9	0.51	19.73	a	14.5	17.7
0.03	26.01	-10.57	261.4	0.51	22.25	a	20.9	20.5
0.03	26.03	-10.55	260.1	0.52	29.11	a	34.2	26.2
0.03	26.04	-10.53	258.5	0.52	34.62	a (f)	49.3	30.2
0.03	26.05	-10.52	258.6	0.52	37.52	i-a (f)	63.6	29.3
0.03	26.07	-10.49	258.4	0.52	38.82	i-a (f)	85	23.9
0.03	25.98	-10.61	260.8	0.52	19.24	a	0.2	
0.03	25.98	-10.61	261	0.52	19.55	a	2	7.1
0.03	26	-10.59	258.2	0.52	19.66	a	4	10.5
0.03	25.98	-10.61	263.6	0.51	18.85	a	9	14.8
0.03	26	-10.59	261.1	0.51	18.76	a	12.8	16.7
0.03	26	-10.59	265.8	0.5	21.67	a	21.6	20.4
0.03	26.01	-10.57	265.1	0.51	26.91	a (f)	33.5	25.2
0.03	26.02	-10.55	262.4	0.51	33.19	a (f)	43.1	29
0.03	26.03	-10.55	262.1	0.52	37.47	i-a (f)	62	29.6
0.03	26.04	-10.53	263.6	0.51	37.91	i-a (f)	86.8	22.2
0.03	25.94	-10.67	263.4	0.51			0.2	
0.03	25.94	-10.67	264.5	0.51			0.2	
0.03	25.92	-10.69	266.9	0.5	19.99		8.1	15.1
0.03	25.94	-10.67	268.1	0.5	19.49		12.8	16.9
0.03	25.95	-10.65	258.9	0.53	20.83		19.8	19.7
0.03	25.96	-10.64	257.1	0.53	21.01		19.9	19.6
0.03	25.97	-10.63	257.1	0.53	24.82	a	33	25
0.03	25.95	-10.65	257.7	0.53	27.25	a (f)	52.6	26.7
0.03	25.97	-10.62	258.9	0.52	27.04	a (f)	52.7	26.7
0.03	25.97	-10.63	258.1	0.53	27.55	a (f)	60.2	23.5
0.03	25.99	-10.6	259.3	0.51	29.68	i-a (f)	66.1	22.3
0.03	25.98	-10.61	258.6	0.53	34.53	i-a (f)	76.2	21.3
0.03	25.98	-10.57	266	0.7	26.65	a	0.2	
0.03	25.99	-10.56	265.3	0.7	26.92	a	2.3	6.7
0.03	25.99	-10.54	256.4	0.73	26.06	a	4.8	10.5
0.03	25.99	-10.54	261.1	0.71	26.12	a	8.4	14.1
0.03	25.99	-10.54	258.7	0.72	27.15	a	11.8	16.7
0.03	26.02	-10.5	258	0.72	29.99	a	17.5	18.8
0.03	26.01	-10.52	262.9	0.7	36.07	a (f)	40.5	23.1
0.03	25.99	-10.55	261.8	0.7	26.53	a	0.2	
0.03	25.99	-10.55	260.9	0.7	26.63	a	0.2	
0.03	25.98	-10.56	260.6	0.72	25.28	a	2.2	8.1
0.03	25.98	-10.57	262.6	0.71	25.08	a	2.3	8.3
0.03	25.98	-10.57	263.5	0.71	25.73	a	4.7	11.8
0.03	25.99	-10.55	263.2	0.7	25.55	a	4.7	11.8
0.03	26.01	-10.52	264.8	0.7	27.82	a	12.7	17.7
0.03	25.99	-10.55	263.8	0.71	28.02	a	12.5	17.7
0.03	26	-10.54	262.7	0.71	31.56	a	20	19.9
0.03	25.98	-10.56	264.5	0.7	31.53	a	20.1	19.7
0.03	26.01	-10.52	260.9	0.72	36.6	a (f)	34.9	24

$\dot{q}_1$	α_1	$\dot{q}_2$	α_2	$\dot{q}_3$	α_3	$\dot{q}_4$	α_4	$\dot{q}_5$	α_5	$\dot{q}_6$	α_6
16.9	25.2	7.2	12.2	6	10.6	9.2	14.9	6	10.2	10.9	17.3
22.6	25.8	13.8	17.3	10.2	13.8	13.3	16.5	10.1	12.8	17.1	20.1
31.1	28.6	19.9	20.1	13.7	15.1	23	22	14.3	14.5	23.5	22.6
43.9	32.2	30	23.7	26.3	23	36.8	27.3	26.3	20	42.2	30.9
69	39.7	48.4	29.1	40.8	29	54.2	31.7	39.2	24	44	27.5
90.3	38.2	76.9	33.2	47.4	28.7	72.5	32	47.5	21.7	46.9	21.7
104.1	27.5	113.8	30.1	68.7	27.4	108.2	28.3	67.8	17.7	47.2	12.3
0.2		0.2		0.2		0.2		0.2		0.2	
5.1	16.6	1.6	5.3	0.2	0.7	2.4	9.6	1.1	4	1.8	6.5
9.3	22.3	2.6	7	0.2	0.6	5	15.3	1.8	5.2	4.8	12.8
17.4	27	5.8	10	5.6	9.7	9.7	16.5	5.5	9.3	9.9	16.4
23	27.9	10.5	14.3	7.6	10.7	13.7	18.3	7.7	10.4	14.5	18.5
32.6	29.2	19.3	18.9	14.4	14.9	22.9	21.9	14.6	14	25.5	23.5
47.8	34.1	30.1	22.7	24.4	20.4	36.3	27.1	23.6	18	38.8	28.8
59.1	37.2	39.9	26.8	29.6	23.7	54.4	35.3	30.3	20.6	45	30.2
83.3	37.5	65.3	30.7	45	27.6	72.5	33.3	45	20.8	60.6	27.7
94.3	22.5	123.8	29.6	67	25.8	107.7	26.2	66.3	15.3	61.6	13.7
0.2		0.2		0.2		0.2		0.2		0.2	
0.2		0.2		0.2		0.2		0.2		0.2	
15.9	27.3	6.1	11	6.5	12.7	5.2	10.3	7.3	14.5	7.8	14.9
22.4	28.1	9.6	12.8	11.2	15.3	9.4	12.7	10.2	14.2	14.1	18.5
32.3	30.2	16.5	16.6	15.1	16.2	14.2	14.6	18.5	18.7	22.4	22.2
32.3	30.1	16.5	16.5	15.2	16	14.3	14.4	18.5	18.6	22.4	21.9
53.9	38.8	26.1	20.2	23.7	19.4	23.2	17.7	29.6	22.8	41.2	30.9
85.4	40.3	61	30	45.8	28.1	36.4	18.3	39.8	20.2	47.3	23.3
85.5	40.3	61	30	45.9	28.5	36.4	18.4	39.8	20	47.4	23.1
91	33.6	84.6	32.4	46	23.3	54.5	20.5	45.1	16.9	39.9	14.6
88.5	28.6	107.6	34.4	47.4	20.3	73	23.9	47.9	15.9	32.1	10.5
93.4	25.3	133.6	36	23.9	11.8	108.4	29.2	64.6	16.8	33.4	8.9
0.2		0.2		0.2		0.2		0.2		0.2	
2.9	7.4	2.3	6.9	1.4	5.1	2.8	8.4	1	3	3.2	9.1
6.5	12.7	3.6	8.1	4.3	10.5	5	11.2	1.8	4.1	7.9	16.4
11.9	18.4	7.9	13.7	5.8	11.1	9.1	15.3	4.4	7.8	11.2	18.3
13.7	18.3	14.2	20.5	6.7	11.3	13.8	19.3	6.6	9.5	15.6	21.2
21.9	22	20.3	22.2	10.8	14	21.9	22.8	10.1	11	20.1	20.8
50.6	27.3	45.5	25.6	32.8	21.5	40.3	22.6	32.6	18.4	41	23.3
0.2		0.2		0.2		0.2		0.2		0.2	
0.2		0.2		0.2		0.2		0.2		0.2	
4.5	14.5	0.2	0.8	3.4	14.1	1.4	5.8	1.7	6.6	1.9	6.6
4.6	14.5	0.2	0.8	3.5	14.1	1.5	5.7	1.8	7	2.2	7.5
8.3	18.7	3.3	8.9	3.4	10.1	4.8	12.1	4.8	12	3.8	9.1
8.3	18.7	3.4	8.9	3.5	10.1	4.8	12.1	4.8	12.1	3.7	9
17.7	24.1	7.9	11.7	20.3	26.9	8.9	12.6	9.5	14.1	11.6	16.5
17.5	24.1	7.7	11.8	20.1	27.3	8.7	12.7	9.4	14	11.5	16.5
25.9	25.3	13.3	14.1	32.6	31.1	13.5	14.1	14.7	15.4	19.8	19.7
26.1	25.1	13.4	14	32.7	30.6	13.6	13.9	14.7	15.2	19.9	19.4
45.2	30.5	21	15.4	51.8	34.1	22.9	16.3	33.3	23.2	35.5	24.6

Table G.9

$w_{oil,0}$	p	T	$\dot{m}$	$\dot{x}$	$\Delta p/\Delta L$	FP	$\dot{q}$	α
0.03	26.02	-10.51	264.4	0.7	36.24	a (f)	35	23.8
0.03	26.01	-10.53	262.3	0.71	41.02	a (f)	53.5	26.7
0.03	26.03	-10.49	262.2	0.71	40.61	i-a (f)	53.6	26.7
0.03	26	-10.54	262.7	0.71	43.12	a (f)	66.8	25
0.03	26.02	-10.51	263.1	0.71	43.15	a (f)	66.9	25
0.03	26.01	-10.53	273.7	0.69	45.71	i-a (f)	77.1	21.6
0.03	26.06	-10.46	273.2	0.69	45.55	i-a (f)	77	21.1
0.03	26.04	-10.18	270.5	0.88	36.36	a	0.2	
0.03	26.05	-10.2	271.6	0.88	35.08	a	0.4	
0.03	26.05	-10.14	269.1	0.88	37	a (f)	3	12.2
0.03	26.06	-10.11	267.7	0.89	41.3	a (f)	6.4	14.7
0.03	26.05	-10.14	270	0.88	41.34	a (f)	10.4	15.4
0.03	26.05	-9.96	263.7	0.9	40.69	i-a (f)	26.1	16.5
0.03	26.02	-10.07	265.8	0.9	34.72	a	0.2	
0.03	25.97	-10.01	262.7	0.91	37.73	a (f)	1.2	
0.03	26.02	-10.18	269	0.88	37.84	a (f)	3.1	15.4
0.03	26.04	-9.59	258.3	0.92	46.4	a (f)	7.7	
0.03	25.99	-9.78	259.6	0.92	47.6	a (f)	11.6	46
0.03	26.03	-10.01	264.9	0.9	44.33	a (f)	22.9	20.3
0.03	26	-10.53	451.2	0.71	64.61	a	0.2	
0.03	25.99	-10.56	454.3	0.7	64.76	a	0.2	
0.03	26	-10.55	456.3	0.7	64.55	a	1.5	4.5
0.03	25.99	-10.55	458.6	0.69	64.27	a	1.5	4.4
0.03	26	-10.54	460.6	0.69	64.51	a	4.4	8.8
0.03	25.99	-10.56	465	0.68	64.73	a	4.4	8.7
0.03	26.01	-10.53	466.1	0.69	66.46	a	9.1	12
0.03	26	-10.53	461.3	0.7	66.47	a	9.1	12
0.03	26.02	-10.51	469.5	0.71	68.9	a	15	14.8
0.03	26.03	-10.49	466.1	0.72	68.89	a	15	14.9
0.03	25.98	-10.56	472.7	0.71	71.92	a	25.8	18.3
0.03	26.03	-10.49	472.2	0.71	71.37	a	25.9	18.3
0.03	26.03	-10.51	477.7	0.69	75.36	a	36	21.6
0.03	26.04	-10.48	474.9	0.71	75.38	a	36.1	21.7
0.03	26.03	-10.5	478.8	0.69	81.52	a	48.6	25
0.03	26.05	-10.48	475.8	0.7	81.93	a	48.6	25.2
0.03	25.98	-10.65	524	0.1	16.84	i-a	0.2	
0.03	25.97	-10.66	526	0.1	17.03	i-a	0.2	
0.03	25.98	-10.65	528.3	0.1	16.82	i-a	2.1	6
0.03	25.98	-10.64	527.2	0.1	17.37	i-a (f)	4	7.9
0.03	25.99	-10.62	530.9	0.1	22.21	i-a (f)	13.4	13.7
0.03	26	-10.61	524.2	0.1	32.96	f	23.2	19.9
0.03	26.02	-10.59	522.5	0.1	37.64	f	43.2	25.7
0.03	26.03	-10.58	524.1	0.1	39.41	f	59.6	28.2
0.03	26.04	-10.56	519.9	0.1	40.33	f	75.3	29.7
0.03	26.03	-10.57	525.9	0.1	40.83	f	88.7	30.3
0.03	26.03	-10.57	522	0.11	42.62	f	110.6	30.7
0.03	25.96	-10.66	522.7	0.31	36.96	a	0.2	

Table G.9

$\dot{q}_1$	α_1	$\dot{q}_2$	α_2	$\dot{q}_3$	α_3	$\dot{q}_4$	α_4	$\dot{q}_5$	α_5	$\dot{q}_6$	α_6
45.3	30.3	21	15.3	51.9	33.9	22.9	16.1	33.3	22.9	35.6	24.4
77.8	38.9	36.4	18.5	91	41.9	35.9	17.4	44.5	23.5	35.5	20
77.9	38.9	36.5	18.5	91.1	41.6	36.3	17.5	44.5	23.5	35.6	19.9
94.2	32.1	65.1	22.9	101.1	46.2	53.9	18.3	45.6	16.2	41.2	14.5
94.2	32	65.2	22.9	101.1	46.3	54	18.2	45.6	16.1	41.2	14.4
100.5	25.5	100.4	25.9	101.3	37.2	76.6	19.7	47.9	12.4	35.9	9.1
100.4	24.8	100.2	25.3	101.2	35.8	76.6	19.5	47.9	12.1	35.8	8.8
0.2		0.2		0.2		0.2		0.2		0.2	
0.2		0.2		0.2		1.4		0.2		0.2	
3.1	8.8	2.9	12.6	1.4	10	4.8	19.8	1.4	5.9	4.6	16.2
4.9	9.6	7.3	17.2	3.7	11.6	9.3	20.9	3.9	9.1	9.4	20
9.5	12.3	11.9	17.4	6.8	12.7	13.4	19.9	6.5	9.8	14.5	20
29.7	17.3	25.5	15.9	28.7	21.8	22.6	13.6	20	12.3	30.2	17.8
0.2		0.2		0.2		0.2		0.2		0.2	
2.7		0.2		0.2		1.9		0.2		1.9	
4.6	20	3	16.1	0.2	1.2	4.9	28.8	1.4	6.6	4.3	19.8
10.5		6.1		6.1		10.1		4.2		9.1	
15.8	50.7	9.9	42.4	8.6	48.6	14	58.1	8.4	31.1	13.3	44.9
31.7	26.2	24.8	21.2	18.9	19.2	22.4	19.9	19.4	17.1	20.5	18.1
0.2		0.2		0.2		0.2		0.2		0.2	
0.2		0.2		0.2		0.2		0.2		0.2	
1.6	3.9	0.2	0.6	4	12.6	1.6	5	0.2	0.6	1.5	4.4
1.6	3.9	0.2	0.6	3.9	12.4	1.5	4.7	0.2	0.6	1.5	4.4
4.8	8.5	1.9	4	7.4	15.5	4.5	9.3	1.7	3.5	6.1	11.8
4.8	8.5	2	4.2	7.3	15.2	4.5	9.1	1.7	3.5	6.1	11.7
10.4	12.7	4.6	6.5	12.7	17.6	9.3	12.3	6.8	9.1	10.7	14
10.4	12.8	4.6	6.5	12.7	17.4	9.3	12.2	6.8	9.1	10.7	14
16.1	15	9.9	10.1	21.2	21.2	13.7	13.6	10.5	10.6	18.6	18.1
16.1	15.1	9.9	10.1	21.2	21.4	13.7	13.7	10.4	10.7	18.6	18.2
26.9	18.4	19.5	14.1	36.7	26.3	22.6	16.2	19.3	14	29.7	20.9
26.9	18.4	19.6	14.1	36.8	26.3	22.7	16.2	19.4	14	29.8	20.9
36.5	21	28.6	17.3	46.6	29.6	36.6	21.6	28.3	16.9	39.7	23.2
36.4	21	28.6	17.5	46.6	29.7	36.6	21.7	28.3	17	39.7	23.3
49.1	23.8	43.1	21.7	49.4	29.3	54.1	27.5	42.3	21.4	53.8	26.6
49.1	23.9	43.1	21.9	49.3	29.3	54.1	27.6	42.3	21.5	53.8	26.9
0.2		0.2		0.2		0.2		0.2		0.2	
0.2		0.2		0.2		0.2		0.2		0.2	
3.4	8.7	1.5	4.6	1.4	4	2.7	8.2	0.2	0.6	3.4	9.7
6.5	12	2.6	5.5	1.8	3.7	4.7	9.9	1.6	3.3	6.6	12.7
22.5	22.1	8.4	9	9.9	10.2	9.4	9.8	9.9	10.4	20.6	20.8
39.6	32.9	15.6	14.2	17.8	15.4	13.8	12.1	17.8	15.8	34.5	29.2
74.6	42.9	36.9	22.5	32.3	19.5	22.7	13.8	32.5	20	60.4	35.7
96.8	44.6	59.2	28.5	43.4	21.3	36.1	17.2	43.6	21	78.7	36.8
115.7	44.1	82	32.6	53.9	22.7	54.1	21.3	54.2	21.6	92.1	35.7
130.8	43.3	100.5	34.4	66.1	24.5	71.9	24.4	65.1	22.4	98	33.1
149.6	40.1	128.5	35.4	86.3	26.8	107.6	29.4	86	23.7	105.8	28.8
0.2		0.2		0.2		0.2		0.2		0.2	

Table G.9

$w_{oil,0}$	p	T	$\dot{m}$	$\dot{x}$	$\Delta p/\Delta L$	FP	$\dot{q}$	α
0.03	25.96	-10.66	523.6	0.3	37.98	a	0.2	
0.03	25.97	-10.65	523.9	0.3	38.34	a	1	4.2
0.03	25.97	-10.65	520.7	0.31	38.88	a	2.7	7.5
0.03	25.97	-10.65	519.3	0.31	38.18	a (f)	7.4	11.5
0.03	25.89	-10.75	515.1	0.32	39.5	a (f)	12.3	14.5
0.03	25.91	-10.73	517.9	0.3	38.4	a (f)	21.6	19.4
0.03	25.92	-10.72	533.9	0.3	39	a (f)	35.4	26.4
0.03	25.95	-10.68	533.9	0.3	57.92	a (f)	48.5	33.5
0.03	25.96	-10.66	514.8	0.31	88.2	a (f)	67.8	41
0.03	25.94	-10.68	531.4	0.3	81.25	i-a (f)	101	39.6
0.03	26.01	-10.57	501.3	0.53	57.15	a	0.2	
0.03	26.01	-10.56	499.6	0.53	57.23	a	0.2	
0.03	26.02	-10.56	499	0.53	57.91	a	2	6.6
0.03	26.02	-10.56	504.5	0.52	57.56	a	2	6.5
0.03	26	-10.58	504.1	0.52	57.9	a	4.2	9.4
0.03	26.01	-10.58	507.6	0.52	57.78	a	4.1	9.3
0.03	26.01	-10.57	510.2	0.52	57.49	a	7	11.7
0.03	26.01	-10.57	509.1	0.52	57.42	a	7.1	11.7
0.03	25.95	-10.66	516.1	0.51	57.88	a (f)	11.4	14.5
0.03	25.96	-10.65	528.8	0.5	58.64	a (f)	22.1	19
0.03	25.94	-10.66	522.4	0.51	49.71	i-a (f)	57.3	9.2
0.03	25.94	-10.67	539.3	0.49	48.26	i-a (f)	53.8	57.5
0.03	39.32	4.72	79	0.16	0.26	s-w	0.1	
0.03	39.31	4.71	78.8	0.16	0.26	s-w (f)	0.1	
0.03	39.32	4.72	78.7	0.14	0.31	s-w (f)	1.6	4
0.03	39.3	4.7	78.6	0.15	0.3	s-w (f)	1.5	3.9
0.03	39.32	4.72	78.5	0.14	0.38	s-w (f)	3	5.5
0.03	39.33	4.73	78.7	0.14	0.37	s-w (f)	2.9	5.4
0.03	39.31	4.71	78.5	0.16	0.56	s-w (f)	6	7.4
0.03	39.31	4.71	78.4	0.16	0.56	s-w (f)	6	7.4
0.03	39.31	4.71	79.4	0.16	0.78	s-w (f)	10.3	9.4
0.03	39.34	4.74	78.8	0.16	0.8	s-w (f)	10.3	9.4
0.03	39.35	4.75	79.4	0.14	0.91	w (f)	13.9	10.6
0.03	39.3	4.7	78.5	0.14	0.9	w (f)	14	10.6
0.03	39.35	4.74	80	0.13	1.33	w (f)	24.9	12.4
0.03	39.32	4.72	78.6	0.15	1.32	w (f)	24.8	12.4
0.03	39.34	4.74	78.8	0.16	1.98	w (f)	39	13
0.03	39.3	4.71	79.4	0.16	1.95	w (f)	38.9	12.8
0.03	39.31	4.71	78.4	0.16	2.16	w (f)	47.4	13
0.03	39.33	4.74	78.6	0.15	2.18	w (f)	47.5	13.2
0.03	39.3	4.73	78.9	0.34		w	0.1	
0.03	39.3	4.73	79	0.34		w	0.1	
0.03	39.31	4.74	78.3	0.34		w (f)	1.9	4.2
0.03	39.31	4.74	78.9	0.34		w (f)	1.9	4.2
0.03	39.32	4.75	78.1	0.35		w (f)	4.4	5.5
0.03	39.31	4.74	79	0.35		w (f)	4.5	5.5
0.03	39.3	4.73	79	0.34		w (f)	7.3	6.2

$\dot{q}_1$	α_1	$\dot{q}_2$	α_2	$\dot{q}_3$	α_3	$\dot{q}_4$	α_4	$\dot{q}_5$	α_5	$\dot{q}_6$	α_6
0.2		0.2		0.2		0.2		0.2		0.2	
2	7.6	0.2	1.1	0.2	0.9	1.5	7.5	0.2	1	1.7	7.3
4.4	11.1	1.4	4.5	1.6	4.5	4.2	11.8	1.5	4.5	3.2	8.7
10.5	15.1	6.9	11.1	5.6	8.9	9.2	14.3	5.7	9.2	6.7	10.4
17.1	18.9	11	13.5	9.4	11.8	13.8	16.4	9.4	11.3	13	14.9
25.3	22	23	20.8	16.8	16.1	23	20.5	16.9	15.4	24.5	21.5
41.9	29.6	34.1	26	28.1	22.9	36.2	26.6	28.4	21.7	43.9	31.7
56.8	37.5	48.3	33.8	39.7	31.7	54	36	40	27	52.3	35
82.1	46.4	68.1	40.7	56.4	40.6	72	42.4	56.7	33.8	71.5	41.8
117.7	43.3	105.5	39.9	84.8	42	107.7	40.3	84.8	32.4	105.5	39.4
0.2		0.2		0.2		0.2		0.2		0.2	
0.2		0.2		0.2		0.2		0.2		0.2	
1.8	4.8	1.6	5.5	1.6	6.1	4.1	13.7	1.4	4.7	1.5	4.7
1.8	4.9	1.6	5.6	1.6	6.1	4.1	13.6	1.3	4.5	1.5	4.7
3.6	7.1	4.3	10	3.3	8.4	6.4	14.5	3.1	7.2	4.2	9
3.5	6.9	4.3	10	3.3	8.3	6.3	14.4	3.2	7.3	4.2	9
6.5	9.9	7.3	12.5	5.9	10.8	9.6	15.9	5.4	9.1	7.6	12.2
6.5	9.9	7.3	12.4	6	10.9	9.6	15.9	5.4	9.1	7.6	12.1
11.9	14.1	11.7	15.3	9.7	13.6	14	17.6	8.9	11.5	12	14.8
25.5	20.8	23.6	20.6	21	19.3	23.5	19.9	17.3	15	21.5	18.3
57.6	8.9	59.3	8.5	69.9	13.7	28.1	3.8	74.5	11.9	54.7	8.3
57.6	18.4	48.4	15.4	55.6	260.7	47.6	14.1	60	19.6	53.4	16.8
0.1		0.1		0.1		0.1		0.1		0.1	
0.1		0.1		0.1		0.1		0.1		0.1	
1.6	3.8	2.3	6.3	0.1	0.4	2.3	5.9	1.9	4.9	1.1	2.8
1.6	3.9	2.2	6	0.1	0.4	2.3	5.8	1.7	4.4	1.2	3.2
1.6	2.7	5.2	9.8	0.1	0.3	4.8	8.7	4	7.2	2.3	4.1
1.5	2.7	5.1	9.7	0.1	0.3	4.7	8.6	3.9	7	2.1	3.9
4.1	4.6	6.7	8.3	6.7	8.9	9.2	11.2	5.7	7.2	3.4	4.1
4.1	4.6	6.7	8.3	6.7	8.9	9.2	11.2	5.7	7.1	3.5	4.2
1.6	1.5	15	13.6	15.8	14.5	13.6	12.1	12	11	3.9	3.7
1.6	1.5	14.9	13.6	15.7	14.4	13.6	12.1	11.9	11	3.9	3.7
4.5	3.3	17.2	13.6	19.6	15.1	22.2	16	14.5	11.1	5.7	4.4
4.6	3.4	17.2	13.6	19.7	15.2	22.3	16	14.4	11	5.7	4.4
10.8	5.1	42.1	20.8	20.3	11.6	36.2	17.5	33.2	16	6.7	3.3
10.8	5.1	42.1	20.9	20.2	11.6	36.1	17.6	33.2	16	6.8	3.4
15.5	4.9	66.9	21.8	28.6	11.4	53.9	17.9	47	15.1	21.9	6.8
15.5	4.8	66.8	21.5	28.6	11.4	53.9	17.8	46.9	14.9	21.9	6.7
23	6.1	75.8	20.7	33.3	11	72.4	19.3	52.1	13.9	27.7	7.2
23.2	6.2	75.9	20.9	33.4	11.1	72.4	19.5	52.2	14.2	27.8	7.4
0.1		0.1		0.1		0.1		0.1		0.1	
0.1		0.1		0.1		0.1		0.1		0.1	
2.4	5	2	4.5	2.5	5.9	1.8	3.9	2.6	5.5	0.1	0.3
2.4	4.9	1.8	4	2.5	5.8	2	4.1	2.8	5.9	0.1	0.3
2.1	2.5	5	6.4	7.6	9.4	4.6	5.7	5.7	7.1	1.5	1.9
2	2.4	5	6.3	8	9.7	4.6	5.7	5.8	7.2	1.5	1.8
2.3	1.9	9	7.8	12.4	10.4	9.1	7.6	9.1	7.9	1.8	1.5

Table G.9

$w_{oil,0}$	p	T	$\dot{m}$	$\dot{x}$	$\Delta p/\Delta L$	FP	$\dot{q}$	α
0.03	39.31	4.74	77.6	0.34		w (f)	7.3	6.2
0.03	39.3	4.73	79	0.34		w (f)	9.3	6.6
0.03	39.31	4.74	78.6	0.34		w (f)	9.3	6.5
0.03	39.34	4.76	78.5	0.34		w (f)	20.4	7.7
0.03	39.31	4.74	79.2	0.33		w (f)	20.4	7.6
0.03	39.33	4.76	78.7	0.34		w (f)	34.4	8.1
0.03	39.33	4.76	78.9	0.34		w (f)	34.4	8.1
0.03	39.3	4.72	78.9	0.34		f	50.7	7.8
0.03	39.33	4.75	80.6	0.33		f	50.5	7.8
0.03	39.32	4.74	81.2	0.32	0.35	w	0.1	
0.03	39.32	4.74	81	0.32	0.37	w	0.1	
0.03	39.28	4.7	80.7	0.33	0.48	w (f)	1.3	4.3
0.03	39.3	4.72	80.8	0.33	0.46	w (f)	1.3	4.1
0.03	39.32	4.75	80.2	0.34	0.66	w (f)	4.1	5.7
0.03	39.32	4.75	80.8	0.33	0.65	w (f)	4.1	5.8
0.03	39.32	4.75	73.7	0.36	0.71	w (f)	7.5	6.9
0.03	39.28	4.71	76.2	0.35	0.64	w (f)	7.5	6.9
0.03	39.3	4.72	85.1	0.31	0.96	w (f)	11.9	7.9
0.03	39.3	4.72	85.2	0.31	1.02	w (f)	11.9	7.9
0.03	39.27	4.69	84.8	0.31	1.39	w (f)	20.1	8.4
0.03	39.28	4.71	86.1	0.3	1.38	w (f)	20.2	8.3
0.03	39.29	4.71	85.2	0.31	2	w (f)	31.3	8.6
0.03	39.31	4.73	85.1	0.31	1.94	w (f)	31.2	8.5
0.03	39.29	4.71	85.2	0.32	2.71	f	47.6	8.2
0.03	39.29	4.71	85	0.32	2.74	f	47.6	8.2
0.03	39.32	4.74	84.2	0.31	3.19	f	60	7.4
0.03	39.31	4.74	84.2	0.31	3.19	f	60.1	7.5
0.03	39.23	4.65	84.5	0.32	3.26	f	65.1	7.3
0.03	39.33	4.75	83.9	0.32	3.25	f	65.1	7.2
0.03	39.35	4.82	79.3	0.52	0.95		0.1	
0.03	39.31	4.78	79.7	0.51	0.92		0.1	
0.03	39.34	4.81	78.7	0.52	1.1		2.8	5.7
0.03	39.34	4.82	79.1	0.52	1.08		2.9	5.8
0.03	39.33	4.81	79	0.53	1.34		5.5	6.1
0.03	39.34	4.81	79	0.52	1.36		5.4	6
0.03	39.34	4.81	80.2	0.51	1.37		9.6	6.4
0.03	39.33	4.8	79	0.53	1.39		9.6	6.4
0.03	39.34	4.82	79.1	0.53	1.95		18.8	6.5
0.03	39.34	4.82	79.7	0.53	1.94		18.7	6.5
0.03	39.33	4.79	84.7	0.5	2.33		25.9	6.8
0.03	39.34	4.83	74.6	0.57	2.27		25.8	6.2
0.03	39.35	4.81	86.1	0.49	2.81		34.9	6.6
0.03	39.36	4.82	84.9	0.5	2.82		34.9	6.5
0.03	39.36	4.83	84.5	0.5	3.34		45.5	6.1
0.03	39.34	4.81	84.1	0.51	3.35		45.4	6.2
0.03	39.36	10.02	0.6	0.95			0.1	
0.03	39.32	4.87	33.8	0.68		s (f)	0.1	

$\dot{q}_1$	α_1	$\dot{q}_2$	α_2	$\dot{q}_3$	α_3	$\dot{q}_4$	α_4	$\dot{q}_5$	α_5	$\dot{q}_6$	α_6
2.1	1.8	9.2	8	12.1	10.2	9.2	7.7	9.2	8	1.8	1.6
1.8	1.2	12.8	9.2	12.6	9.2	13.8	9.5	10.8	7.7	3.8	2.6
2.2	1.4	12.8	9.1	12.6	9.1	13.8	9.4	10.9	7.6	3.7	2.5
3.1	1.2	30.2	11.6	29.3	11.4	24.9	9.2	30.2	11.2	4.6	1.7
3.2	1.2	30.3	11.4	29.3	11.3	24.9	9.1	30.2	11	4.6	1.7
4.6	1.1	46.3	11	55.8	13.3	35.9	8.4	49.1	11.5	14.8	3.4
4.6	1.1	46.2	11	55.7	13.3	35.9	8.3	49.1	11.4	14.9	3.5
6.6	1	74.4	11.5	69.6	11.2	53.4	8.2	72.1	11	27.9	4.2
6.5	1	74.2	11.4	69.5	11.1	53.2	8.1	71.9	10.9	27.7	4.2
0.1		0.1		0.1		0.1		0.1		0.1	
0.1		0.1		0.1		0.1		0.1		0.1	
0.1	0.4	2.6	8.4	2.1	7.2	1.5	4.7	1.4	4.6	0.1	0.4
0.1	0.4	2.6	8	2.1	7.1	1.5	4.5	1.4	4.4	0.1	0.4
0.1	0.2	8.7	12.1	4.5	6.5	4.9	6.9	4.7	6.6	1.4	2
0.1	0.2	8.7	12	4.5	6.5	4.9	6.9	4.8	6.7	1.6	2.2
0.1	0.1	13.9	12.6	9.7	9.4	9.3	8.5	9.3	8.4	2.4	2.1
0.1	0.1	14	12.7	9.8	9.5	9.4	8.5	9.5	8.5	2.5	2.2
0.1	0.1	20.8	13.6	18	12.1	14	9.2	14.8	9.6	4	2.6
0.1	0.1	20.7	13.6	18	12.1	13.9	9.2	14.8	9.7	4	2.6
0.1	0	33.3	13.8	30.2	12.8	22.9	9.5	26.7	11	7.7	3.1
0.1	0	33.3	13.7	30.3	12.7	22.8	9.5	26.8	11	7.6	3.1
4.1	1.1	48.9	13.3	44.2	12.4	36.6	9.9	42.6	11.5	11.5	3.1
4.1	1.1	48.8	13.3	44.1	12.4	36.6	9.9	42.6	11.5	11.4	3.1
6.3	1.1	76.3	12.9	62.6	11.2	55.3	9.4	62.1	10.6	23	3.9
6.3	1.1	76.3	12.9	62.6	11.2	55.2	9.4	62.3	10.5	23	3.9
5.3	0.6	105.8	12.8	67.3	8.9	71.2	8.6	80.1	9.7	30.6	3.7
5.5	0.7	105.8	12.9	67.3	9	71.3	8.6	80.2	9.8	30.5	3.8
3.9	0.4	99.6	11.3	76.5	9.2	85.8	9.2	84.3	9.2	40.4	4.4
4	0.4	99.7	11.2	76.4	9.1	85.7	9.1	84.3	9.2	40.3	4.4
0.1		0.1		0.1		0.1		0.1		0.1	
0.1		0.1		0.1		0.1		0.1		0.1	
3.9	8	3.5	7.5	5.6	10.6	1.8	3.5	1.3	2.7	0.9	1.8
3.9	8	3.5	7.6	5.6	10.6	1.8	3.5	1.4	2.8	1.2	2.5
1.9	2.1	10.6	12	10.1	10.7	5.1	5.6	4.1	4.7	1.2	1.3
1.9	2.1	10.5	12	10	10.6	5	5.5	4.1	4.8	1.1	1.3
1.7	1.1	16.8	11.3	16.2	10.9	9.1	6.1	11.2	7.3	2.5	1.6
1.7	1.1	16.8	11.2	16.3	11	9	6	11.2	7.3	2.7	1.7
1.8	0.6	23.8	8.6	46.5	15.6	13.8	4.9	23.7	8.5	3	1.1
1.7	0.6	23.8	8.6	46.4	15.5	13.7	4.9	23.7	8.4	2.9	1
2.3	0.6	36.8	9.8	52.8	13.6	23.3	6.1	36	9.4	4.1	1.1
2.1	0.5	36.8	9	52.7	12.7	23.3	5.6	35.9	8.6	4.1	1
4.2	0.8	52.5	9.9	57.9	11.1	36.5	6.8	51.3	9.5	6.8	1.3
4.3	0.8	52.5	9.8	58	11	36.5	6.7	51.3	9.4	6.9	1.3
0.1	0	77.1	10.3	60.2	8.5	53.8	7.1	66.9	8.8	15	2
0.1	0	76.9	10.4	60.1	8.6	53.7	7.2	66.9	8.9	15	2
0.1		0.1		0.1		0.1		0.1		0.1	
0.1		0.1		0.1		0.1		0.1		0.1	

Table G.9

$w_{oil,0}$	p	T	$\dot{m}$	$\dot{x}$	$\Delta p/\Delta L$	FP	$\dot{q}$	α
0.03	39.34	5.01	40.9	0.77		s-w (f)	0.1	
0.03	39.31	4.92	52	0.73		s-w (f)	0.1	
0.03	39.35	4.97	62.4	0.74		s-w	0.1	
0.03	39.35	5.03	79.5	0.78	1.19	w	0.1	
0.03	39.35	5.01	80.8	0.76	1.19	w	0.1	
0.03	39.31	4.89	80.1	0.7	0.98	w (f)	3.7	6.9
0.03	39.34	4.92	80.8	0.7	1.01	w (f)	3.7	6.8
0.03	39.33	4.92	80	0.71	1.15	w (f)	5.6	6.7
0.03	39.35	4.93	80.5	0.7	1.12	w (f)	5.5	6.6
0.03	39.34	4.91	79.8	0.7	1.31	w (f)	7.2	6.5
0.03	39.34	4.91	80.2	0.7	1.4	w (f)	7.1	6.5
0.03	39.36	4.93	80.5	0.7	1.82	w (f)	10.9	6.2
0.03	39.35	4.93	80	0.7	1.87	w (f)	10.9	6.2
0.03	39.33	4.91	79	0.71	2.67	w (f)	22.2	5.6
0.03	39.34	4.92	79.6	0.7	2.7	w (f)	22.3	5.5
0.03	39.3	4.85	84.1	0.68	3.54	w (f)	40.5	4.6
0.03	39.33	4.88	83.9	0.67	3.56	w (f)	40.5	4.5
0.03	39.33	4.93	80.1	0.72	3.49	w (f)	36.2	4.4
0.03	39.34	4.92	81.3	0.7	3.46	w (f)	35.9	4.6
0.03	39.25	5.12	84.3	0.85	1.47	w	0.1	
0.03	39.26	5.99	76.8	0.92	1.47	w	0.1	
0.03	39.26	5.15	84.4	0.85	3.48	w (f)	6.7	8.9
0.03	39.27	5.15	84.5	0.85	3.55	w (f)	6.8	8.5
0.03	39.26	5.2	84.4	0.86	3.7	w (f)	8.8	7.2
0.03	39.26	5.19	84.6	0.86	3.71	w (f)	8.8	7
0.03	39.25	5.23	84.3	0.87	4.01	w (f)	11.1	6.3
0.03	39.25	5.15	84.6	0.85	3.93	w (f)	10.5	6.3
0.03	39.27	5.54	80.1	0.9	4.13	w	14.3	4.1
0.03	39.24	5.87	78.6	0.92	4.2	w	14.3	4.3
0.03	39.26	4.98	89.1	0.8	3.95	w (f)	20.4	4.5
0.03	39.25	5.1	86.6	0.84	4.17	w (f)	20.5	4.6
0.03	39.38	5.35	84.8	0.86	5.9	w	39.1	2.6
0.03	39.34	5.29	84.5	0.86	5.77	w	39	2.9
0.03	38.14	3.53	0.5	0.15			0	
0.03	35.54	0.85	7	0.36			0.1	
0.03	39.32	4.72	156.7	0.12	0.92	i	0.1	
0.03	39.33	4.73	156.9	0.12	1.03	i	0.1	
0.03	39.33	4.73	156.5	0.14	3.38	i (f)	41.3	15.5
0.03	39.32	4.72	156.6	0.13	3.34	i (f)	41.3	15.5
0.03	39.33	4.73	156.7	0.14	4.3	f	55.7	16
0.03	39.32	4.73	156.1	0.14	4.36	f	55.7	16
0.03	39.22	4.62	155.6	0.15	7.22	f	101.1	10.1
0.03	39.25	4.65	156.7	0.13	7.34	f	109.7	8.3
0.03	39.25	4.65	156.6	0.13	7.3	f	109.4	8.3
0.03	39.25	4.64	161.8	0.12	0.84	i	0.1	
0.03	39.21	4.61	161.9	0.12	0.93	i	0.1	
0.03	39.21	4.61	160.8	0.12	0.92	i	0.1	

$\dot{q}_1$	α_1	$\dot{q}_2$	α_2	$\dot{q}_3$	α_3	$\dot{q}_4$	α_4	$\dot{q}_5$	α_5	$\dot{q}_6$	α_6
0.1		0.1		0.1		0.1		0.1		0.1	
0.1		0.1		0.1		0.1		0.1		0.1	
0.1		0.1		0.1		0.1		0.1		0.1	
0.1		0.1		0.1		0.1		0.1		0.1	
0.1		0.1		0.1		0.1		0.1		0.1	
5.4	10	5	9.8	5.5	10.7	2.6	4.7	3.2	6.2	0.1	0.2
5.5	9.9	5.1	9.6	5.4	10.4	2.7	4.8	3.2	6.1	0.1	0.2
5.3	6.1	10.3	12.7	5.7	7.3	7.1	8.2	4.9	5.8	0.1	0.1
5.2	6	10.3	12.6	5.7	7.2	7.1	8.2	4.8	5.7	0.1	0.1
3.9	3.4	15	13.6	7.6	7.4	9.3	8.2	7.1	6.4	0.1	0.1
3.8	3.3	14.8	13.7	7.4	7.3	9.1	8.1	7.3	6.5	0.1	0.1
5.3	2.9	21	12	14.2	8.4	13.8	7.8	10.7	6	0.1	0.1
5.5	3	21	11.9	14.2	8.4	13.8	7.7	10.7	6	0.1	0.1
5.2	1.3	44.9	11.2	30.5	8	24.7	6.2	28	7	0.1	0
5.3	1.3	44.9	11	30.5	7.9	24.7	6.1	28.2	6.9	0.1	0
0.1	0	80.4	9	51.1	6.1	45.2	5	60.7	6.6	5.6	0.6
0.1	0	80.4	9	50.9	6	45.2	4.9	60.8	6.6	5.6	0.6
5.3	0.6	67	8.1	51.3	6.6	36.2	4.3	55.4	6.6	1.7	0.2
5	0.6	66.9	8.4	51.1	6.8	36	4.5	55	6.8	1.6	0.2
0.1		0.1		0.1		0.1		0.1		0.1	
0.1		0.1		0.1		0.1		0.1		0.1	
8	10.1	12	16	11.2	14.9	3.1	4	5	6.7	1.2	1.5
8	9.7	12	15.3	11.2	14.4	3.1	3.9	5.1	6.4	1.3	1.6
7	5.4	17.4	14.1	14.6	12.3	4.9	4	7.7	6.3	1.1	0.9
6.9	5.3	17.3	13.8	14.5	12	4.9	4	7.6	6.1	1.3	1
5.7	3.2	20.3	11.6	20.6	11.7	9.4	5.2	9.8	5.6	1	0.6
5.5	3.2	20.3	12.1	16.7	10.2	9.4	5.4	9.8	5.8	1.3	0.8
5.2	1.5	22.9	6.6	27.9	8.1	13.9	3.9	14.7	4.2	1.1	0.3
5.3	1.6	22.9	6.9	27.9	8.4	13.9	4.1	14.7	4.4	1	0.3
4.7	1	34.8	7.7	38.6	8.6	19.4	4.2	21.9	4.7	2.8	0.6
5	1.1	34.9	7.8	38.7	8.7	19.5	4.3	21.9	4.8	3.1	0.7
0	0	59.4	4	58.7	4	36.2	2.3	69	4.6	11.2	0.7
0	0	59.4	4.4	58.7	4.4	36.2	2.6	68.9	5	10.9	0.8
0		0		0		0		0		0	
0.1		0.1		0.1		0.1		0.1		0.1	
0.1		0.1		0.1		0.1		0.1		0.1	
0.1		0.1		0.1		0.1		0.1		0.1	
23.2	8.6	42.5	16	54.4	20.9	54.2	20.3	47.5	17.9	25.8	9.6
23.2	8.7	42.5	16	54.5	20.7	54.2	20.1	47.3	17.7	25.9	9.6
37.6	10.5	58.9	16.9	67.7	20	73.3	20.8	64.2	18.5	32.8	9.3
37.6	10.5	58.9	16.8	67.7	20	73.2	20.7	64.2	18.4	32.7	9.2
85.2	8.5	109.5	11.1	116.3	11.9	107.4	10.5	116.8	11.7	71.2	7
103	7.8	127.8	9.9	116.4	8.8	107.4	7.9	122.3	9.4	81.4	6.1
102.7	7.8	127.5	10	116.1	8.9	107.1	7.9	122	9.4	81.2	6.1
0.1		0.1		0.1		0.1		0.1		0.1	
0.1		0.1		0.1		0.1		0.1		0.1	
0.1		0.1		0.1		0.1		0.1		0.1	

Table G.9

$w_{oil,0}$	p	T	$\dot{m}$	$\dot{x}$	$\Delta p/\Delta L$	FP	$\dot{q}$	α
0.03	39.23	4.63	161.8	0.11	1.04	i (f)	0.1	
0.03	39.2	4.6	162	0.12	1.21	i (f)	4.5	10.2
0.03	39.22	4.62	161.4	0.12	1.27	i (f)	4.5	10.2
0.03	39.24	4.64	164.5	0.12	1.4	i (f)	6.7	11.3
0.03	39.23	4.63	161.8	0.12	1.46	i (f)	6.8	11.3
0.03	39.17	4.57	165.2	0.12	1.68	i (f)	11.8	12.3
0.03	39.21	4.61	162.4	0.12	1.76	i (f)	11.7	12.3
0.03	39.21	4.61	161.1	0.12	1.93	i (f)	15.6	12.5
0.03	39.24	4.63	162.4	0.12	1.84	i (f)	15.4	12.5
0.03	39.18	4.58	161.9	0.12	2.27	i (f)	22.5	12.7
0.03	39.19	4.59	161.9	0.12	2.27	i (f)	22.5	12.7
0.03	39.22	4.62	161.4	0.12	3.07	i (f)	35.3	12.8
0.03	39.23	4.63	161.2	0.12	2.97	i (f)	35.3	12.8
0.03	39.25	4.65	160.5	0.12	3.87	i (f)	50.6	12.8
0.03	39.22	4.62	160.5	0.12	3.89	i (f)	50.5	12.9
0.03	39.2	4.6	160.8	0.12	5.55	f	65.2	13.9
0.03	39.22	4.61	160.7	0.12	5.56	f	65.3	13.9
0.03	39.21	4.61	163.2	0.12	7.26	f	103.6	10.9
0.03	39.24	4.64	163.3	0.11	7.25	f	103.6	11
0.03	39.29	4.71	159.4	0.32	2.24	w (f)	0.1	
0.03	39.32	4.74	159.5	0.31	2.28	i (f)	0.1	
0.03	39.3	4.73	158.7	0.32	2.62	i (f)	3.9	9.1
0.03	39.27	4.7	159.5	0.31	2.64	i (f)	3.9	9.1
0.03	39.32	4.75	158.5	0.31	2.74	i (f)	6.1	10.2
0.03	39.27	4.69	158.8	0.31	2.71	i (f)	6.1	10.3
0.03	39.3	4.72	158.5	0.32	2.97	i (f)	9.9	11.4
0.03	39.32	4.74	159.1	0.3	2.96	i (f)	10	11.2
0.03	39.3	4.72	159	0.31	3.71	i (f)	15.1	11.5
0.03	39.27	4.7	159.5	0.31	3.56	i (f)	14.9	11.5
0.03	39.28	4.7	159.5	0.31	3.55	i (f)	20.8	11.9
0.03	39.3	4.72	160	0.31	3.53	i (f)	20.9	11.9
0.03	39.27	4.69	160.7	0.31	4.58	i (f)	36.5	11.8
0.03	39.29	4.72	159	0.32	4.54	i (f)	36.5	11.8
0.03	39.28	4.7	159.9	0.32	5.34	f	47.5	11.3
0.03	39.28	4.7	159.2	0.32	5.39	f	47.5	11.2
0.03	39.27	4.7	159.2	0.32	5.79	f	56.7	10.7
0.03	39.28	4.71	159.1	0.32	5.93	f	56.7	10.5
0.03	39.28	4.7	158.3	0.32	6.79	f	68.7	9.5
0.03	39.28	4.7	157.5	0.33	6.7	f	68.6	9.5
0.03	39.34	4.81	157.8	0.52	3.43	w (f)	0.1	
0.03	39.33	4.8	158.9	0.51	3.41	w (f)	0.1	
0.03	39.35	4.82	157.8	0.52	4.03	w (f)	3.1	9.8
0.03	39.35	4.81	159.4	0.51	4.04	w (f)	3.2	9.7
0.03	39.35	4.82	158.9	0.51	4.62	w (f)	8.7	12.9
0.03	39.34	4.8	159.9	0.5	4.64	w (f)	8.8	12.8
0.03	39.34	4.81	159.3	0.51	5.53	w (f)	13.4	13.4
0.03	39.34	4.81	158.2	0.51	5.39	w (f)	12.1	13.5

$\dot{q}_1$	α_1	$\dot{q}_2$	α_2	$\dot{q}_3$	α_3	$\dot{q}_4$	α_4	$\dot{q}_5$	α_5	$\dot{q}_6$	α_6
0.1		0.1		0.1		0.1		0.1		0.1	
9	20.4	3.8	8.8	2.7	6.9	2.2	5.1	2.5	5.4	6.8	14.5
9	20.4	3.8	8.7	2.8	7	2.5	5.7	2.3	4.9	6.8	14.4
10.5	17.8	7.1	11.8	5.9	10.2	5.1	8.2	4.8	7.8	7.1	12
10.6	17.6	7.1	11.8	5.9	10.2	5.2	8.3	4.8	7.8	7.1	11.9
12.6	13.4	13.7	14.3	13.9	14.2	9.8	10	9.9	10.1	11	11.6
12.5	13.6	13.6	14.3	13.7	14.2	9.7	9.9	9.7	10	10.9	11.6
14.5	11.7	18.5	14.8	18.2	14.6	14.2	11.4	14	11.1	14.2	11.4
14.4	11.7	18.3	15	18	14.5	14.1	11.3	13.9	11	14.1	11.3
15.9	9.1	28.6	16.2	25.3	14.5	24.6	13.7	21.6	12	19.3	10.8
15.9	9.1	28.6	16.2	25.3	14.5	24.5	13.7	21.5	12	19.3	10.7
21.9	8	38.9	14.1	45.9	16.7	38.1	13.8	41.3	14.8	25.9	9.4
21.8	8	38.8	14.2	46	16.8	38.1	13.8	41.3	14.9	25.9	9.4
33.9	8.6	54	13.7	64	16.1	55.9	14	61.6	15.4	34.5	8.7
35	9	54	13.8	64	16.3	55.9	14.2	60	15.3	34.2	8.8
48.1	10.2	63.7	13.5	87.9	19	74.5	15.8	78.6	16.7	38.7	8.2
48.1	10.2	63.7	13.5	88	19	74.5	15.8	78.7	16.7	38.7	8.2
91.7	9.7	119	12.5	120	12.8	105.7	11	115.5	12.1	69.6	7.3
91.7	9.8	119.1	12.6	120.1	12.9	105.8	11.1	115.5	12.2	69.5	7.4
0.1		0.1		0.1		0.1		0.1		0.1	
0.1		0.1		0.1		0.1		0.1		0.1	
11	24.4	0.1	0.3	3.8	9.3	1.3	3.3	2.2	5.6	4.8	11.7
11.1	24.6	0.1	0.3	3.8	9.2	1.4	3.4	2.2	5.5	4.8	11.7
14.2	23.8	3	5.1	6.5	10.3	4.8	8	3	5.3	5	8.9
14.2	23.9	3.1	5.3	6.4	10.3	4.8	7.9	3	5.3	5	8.9
16.1	18.5	11.5	13.2	8.7	9.8	9.1	10.2	7.4	8.7	6.8	7.9
16.3	18	11.7	13.1	8.7	9.6	9.2	10.1	7.4	8.5	6.9	7.8
18.3	13.5	20.9	15.9	13.5	10.6	13.5	10.4	15	11.4	9	6.8
18.4	13.6	20.9	16.3	12.6	10.2	13.6	10.7	15.1	11.7	9	6.8
20.3	11.5	32.1	18.5	20.8	11.8	22.6	12.7	20	11.4	9.2	5.2
20.5	11.4	32.2	18.3	20.9	11.8	22.7	12.8	20	11.6	9.3	5.3
20.6	6.7	51.3	16.6	49.1	15.9	36	11.7	46.7	14.8	15.5	5
20.7	6.7	51.2	16.6	49	15.8	36	11.8	46.7	14.9	15.5	5
20.1	4.9	59.5	14.2	72.7	16.9	54.2	12.8	59.2	14	19.5	4.7
20.1	4.9	59.5	14.1	72.7	16.8	54.2	12.7	59.2	14	19.5	4.7
20.8	4	72.3	13.7	72.8	14.1	71.8	13.3	75.7	13.9	27.1	5
20.8	3.8	72.4	13.4	72.7	13.9	71.8	13.1	75.7	13.6	27.1	4.9
20.8	2.9	88.8	12.4	73	10.5	107.1	14.3	86.2	11.7	36.5	4.9
20.7	2.9	88.7	12.5	72.9	10.6	107.1	14.4	86	11.9	36.4	5
0.1		0.1		0.1		0.1		0.1		0.1	
0.1		0.1		0.1		0.1		0.1		0.1	
9.9	28	1.6	5	1.6	5.9	1.4	5	0.7	2.5	3.6	12.4
10	27.3	1.6	4.8	1.7	5.7	1.4	4.8	1	3.7	3.6	11.8
20.9	29.8	6.1	9.5	10.4	14.8	4.8	7.2	4	6.4	6	9.5
21	29.4	6.1	9.4	10.7	15.1	4.9	7.2	4.1	6.4	6.1	9.4
27.6	26.5	13.4	14	16	16	9	9.1	9	9.3	5.4	5.6
27.5	28.5	13.3	15.4	8.6	10.7	9	10.3	9	10.2	5.4	6.1

Table G.9

$w_{oil,0}$	p	T	$\dot{m}$	$\dot{x}$	$\Delta p/\Delta L$	FP	$\dot{q}$	α
0.03	39.35	4.82	159.3	0.51	5.8	w (f)	16.1	13.3
0.03	39.35	4.82	157.5	0.52	5.75	w (f)	16	13.4
0.03	39.37	4.85	159.1	0.52	6.64	f	24	12.5
0.03	39.33	4.81	158.9	0.52	6.68	f	24	12.5
0.03	39.35	4.82	159.1	0.52	7.02	f	31.4	11.2
0.03	39.38	4.85	159.4	0.52	7.02	f	31.4	11.2
0.03	39.36	4.83	159.5	0.52	7.47	f	42.4	9.6
0.03	39.36	4.83	158.6	0.53	7.43	f	42.2	9.6
0.03	39.35	4.82	159	0.52	7.84	f	49.9	8.6
0.03	39.34	4.81	158.8	0.52	7.89	f	50	8.6
0.03	39.23	4.82	157.8	0.71	4.69	w-a	0.1	
0.03	39.24	4.84	157.4	0.72	4.67	w-a	0.1	
0.03	39.23	4.82	157.6	0.71	6.79	w-a (f)	10.2	11.7
0.03	39.24	4.83	157.3	0.72	6.85	w-a (f)	10.2	11.8
0.03	39.25	4.84	156.8	0.72	8.38	w-a (f)	16.1	10.9
0.03	39.24	4.84	155.8	0.72	8.42	w-a (f)	16.2	11
0.03	39.23	4.82	156.5	0.72	9.27	f	20.4	9.7
0.03	39.26	4.85	157.5	0.72	9.1	f	20.4	9.7
0.03	39.23	4.82	157.1	0.72	10.22	f	23.6	9.2
0.03	39.26	4.86	157.2	0.72	9.98	f	23.6	9.3
0.03	39.25	4.85	156.6	0.72	11.41	f	28.4	8.7
0.03	39.24	4.85	156.1	0.73	11.3	f	28.4	8.5
0.03	39.25	4.84	156.6	0.72	12.08	f	37.5	7.6
0.03	39.26	4.86	156.4	0.72	12.13	f	37.5	7.6
0.03	39.24	4.83	157.4	0.72	13.49	f	50.1	6.2
0.03	39.24	4.84	156.3	0.73	13.46	w-a (f)	50	6.2
0.03	39.22	4.81	157.9	0.71	13.94	w-a (f)	55.7	5.2
0.03	39.23	4.83	156.9	0.72	13.81	w-a (f)	55.7	5.3
0.03	39.28	5.7	158.2	0.91	10.59	w-a (f)	0.1	
0.03	39.26	5.52	159.6	0.89	10.93	w-a (f)	0.1	
0.03	39.28	5.48	160.4	0.89	10.9	w-a (f)	0.1	
0.03	39.25	5.5	158.9	0.89	8	w-a	1.6	1.8
0.03	39.26	5.49	159.6	0.89	7.98	w-a	1.6	1.8
0.03	39.27	5.34	162.2	0.88	15.48	w-a (f)	14.2	8.6
0.03	39.27	5.31	161.9	0.87	16	w-a (f)	14.2	8.5
0.03	39.27	5.26	163.3	0.87	14.69	w-a (f)	21.8	7.8
0.03	39.27	5.32	159.6	0.88	14.7	w-a	20.9	8
0.03	39.27	5.25	161.7	0.86	14.39	w-a	21.2	6.8
0.03	39.28	5.33	160.2	0.88	14.66	w-a	20.8	7.7
0.03	39.26	5.56	161.6	0.9	14.41	w-a	20.1	4.5
0.03	39.28	5.37	162.1	0.88	15.47	w-a	19.6	5.9
0.03	39.26	5.66	154.9	0.91	13.91	w-a	23.6	5.3
0.03	39.26	6.4	153.9	0.93	11.88	w-a	23.4	3.2
0.03	39.32	4.72	266.9	0.11	3.52	i (f)	0.1	
0.03	39.32	4.72	265.4	0.12	3.58	i (f)	0.1	
0.03	39.33	4.72	265.4	0.12	3.89	i (f)	2.8	6.8
0.03	39.32	4.71	266.4	0.12	3.79	i (f)	2.8	6.8

$\dot{q}_1$	α_1	$\dot{q}_2$	α_2	$\dot{q}_3$	α_3	$\dot{q}_4$	α_4	$\dot{q}_5$	α_5	$\dot{q}_6$	α_6
24.1	19.9	21.1	18	20.4	16.2	13.8	11.3	9.8	8.4	7.1	6
24.1	20.1	21.1	18.1	20.4	16.3	13.8	11.3	9.8	8.4	7.1	6
24.8	12.9	35.1	18.5	35.1	17.8	22.6	11.8	19.8	10.6	6.5	3.4
24.8	12.9	35.1	18.6	35.1	17.8	22.6	11.8	19.8	10.6	6.5	3.5
23.1	8.2	48.3	17.4	44.2	15.9	35.6	12.6	29.6	10.6	7.4	2.6
23.1	8.2	48.3	17.4	44.2	15.9	35.6	12.6	29.6	10.6	7.3	2.6
21	4.7	67.1	15	57	13.3	53.9	12.1	46	10.4	9	2
20.8	4.7	67	15.1	56.9	13.3	53.8	12.1	45.9	10.4	8.9	2
14.6	2.5	86.4	14.8	54.4	10	72.4	12.2	62.4	10.6	9.3	1.6
14.6	2.5	86.5	14.8	54.4	9.9	72.5	12.2	62.5	10.6	9.3	1.6
0.1		0.1		0.1		0.1		0.1		0.1	
0.1		0.1		0.1		0.1		0.1		0.1	
20.4	23.5	6	7	13	13.8	2.4	2.8	10.2	11.9	9.1	11.3
20.5	23.6	6.1	7.1	13	13.9	2.4	2.7	10.2	12	9.1	11.4
28.6	19.2	20.6	14.4	25.9	16.9	4.2	2.9	13.3	9.2	4.1	2.9
28.6	19.2	20.7	14.5	25.9	16.9	4.2	2.9	13.3	9.2	4.2	2.9
31.9	14.7	31.2	15.2	29.1	13.7	9.1	4.4	17.2	8.2	4.1	1.9
31.9	14.8	31.3	15.2	29.1	13.6	9.1	4.4	17.2	8.2	4.1	1.9
37.1	14	33	13.4	33.9	13.2	13.7	5.4	19.7	7.7	4	1.5
37.2	14.3	33.1	13.6	34	13.2	13.7	5.4	19.6	7.8	4	1.6
28.8	8.9	44.6	13.8	45.8	13.7	22.8	7	27.1	8.4	1.3	0.4
28.7	8.6	44.6	13.5	45.7	13.3	22.7	6.8	27.1	8.2	1.3	0.4
23.4	4.9	59.2	12.2	58.5	11.6	36.1	7.4	46.8	9.5	1.2	0.2
23.4	4.8	59.3	12.2	58.5	11.5	36.1	7.4	46.8	9.5	1.2	0.2
20.5	2.6	67.9	8.6	80	9.9	54	6.6	69.5	8.5	8.6	1
20.5	2.6	67.9	8.6	80	9.9	53.9	6.6	69.3	8.5	8.5	1
23.5	2.2	73.3	7	80.3	7.6	71.7	6.6	69.4	6.5	16.2	1.5
23.5	2.3	73.3	7.2	80.3	7.8	71.6	6.7	69.4	6.7	16.2	1.5
0.1		0.1		0.1		0.1		0.1		0.1	
0.1		0.1		0.1		0.1		0.1		0.1	
0.1		0.1		0.1		0.1		0.1		0.1	
3.2	3.5	0.9	1	2	2.2	1.4	1.5	1.1	1.2	1.1	1.2
3.2	3.5	1	1.2	1.9	2.2	1.3	1.5	1.1	1.3	1.1	1.2
22.1	13.1	13.6	8.4	29	17.2	5.1	3.1	11.2	6.9	4.3	2.6
22.1	12.9	13.6	8.3	29	17.2	5.1	3.1	11.2	6.8	4.3	2.6
25.4	9.2	28.5	10.3	42.3	14.7	9.1	3.3	19.9	7.1	5.7	2.1
25.5	9.9	24.1	9.5	42.3	15.8	9.2	3.6	18.5	7.1	5.8	2.2
27.3	8.7	24.1	7.9	42.4	13.4	9.1	3	18.4	5.9	5.7	1.8
24.6	9.3	24.1	9.1	42.4	15.4	9.1	3.4	18.6	6.9	5.7	2.1
25.4	5.6	20.8	4.7	42.3	9.4	9.1	2	17.2	3.9	5.6	1.2
24.2	7.4	19.1	6	42.3	12.7	9.1	2.7	17.2	5.2	5.6	1.7
23	5.2	27.9	6.3	47.1	10.5	13.8	3.1	22.8	5.1	7.1	1.6
22.8	3.1	27.7	3.7	46.8	6.4	13.7	1.8	22.6	3.1	7	1
0.1		0.1		0.1		0.1		0.1		0.1	
0.1		0.1		0.1		0.1		0.1		0.1	
6.6	15.2	0.1	0.4	2.1	5.4	1.4	3.7	2.3	5.9	4.2	10.4
6.6	15.2	0.1	0.4	2.2	5.6	1.4	3.6	2.2	5.6	4.2	10.6

Table G.9

$w_{oil,0}$	p	T	$\dot{m}$	$\dot{x}$	$\Delta p/\Delta L$	FP	$\dot{q}$	α
0.03	39.32	4.71	266.6	0.11	4.14	i (f)	6.3	9.9
0.03	39.31	4.71	267.4	0.11	4.2	i (f)	6.3	9.9
0.03	39.31	4.7	266.3	0.11	4.84	i (f)	14.3	13.4
0.03	39.32	4.71	266.2	0.11	4.96	i (f)	14.3	13.4
0.03	39.32	4.72	266.7	0.11	10.72	f	22.9	18.5
0.03	39.31	4.71	264.2	0.12	10.91	f	22.9	18.7
0.03	39.32	4.72	264.6	0.11	10.8	f	32.8	19
0.03	39.32	4.72	263.7	0.12	10.76	f	32.9	19
0.03	39.33	4.73	264.7	0.12	10.54	f	43.2	18.6
0.03	39.32	4.72	264	0.12	10.56	f	43.1	18.7
0.03	39.33	4.72	263.9	0.12	10.61	f	56.7	18.1
0.03	39.32	4.72	263.4	0.12	10.65	f	56.7	18.1
0.03	39.33	4.73	264.9	0.11	10.89	f	68.8	17.5
0.03	39.32	4.71	263.7	0.12	10.7	f	68.9	17.5
0.03	39.29	4.69	253.1	0.13	11.38	f	90.6	16
0.03	39.29	4.69	255.1	0.12	11.44	f	90.6	16.1
0.03	39.19	4.59	266	0.11	3.54	i (f)	0.1	
0.03	39.23	4.63	265.9	0.11	3.76	i (f)	0.1	
0.03	39.22	4.61	264.7	0.11	11.62	f	12.3	17.1
0.03	39.23	4.62	264.3	0.11	11.59	f	12.3	17.1
0.03	39.24	4.64	263.5	0.12	10.68	f	29.5	17.8
0.03	39.19	4.59	265.1	0.11	10.77	f	29.7	17.8
0.03	39.25	4.64	265.3	0.11	10.52	f	40.5	17.4
0.03	39.24	4.64	265.8	0.11	10.51	f	40.5	17.3
0.03	39.2	4.6	265.6	0.11	10.49	f	60.1	16.7
0.03	39.23	4.62	264.9	0.12	10.48	f	60.1	16.7
0.03	39.26	4.65	265.2	0.11	10.61	f	72.4	15.8
0.03	39.25	4.65	264.9	0.11	10.61	f	72.4	16
0.03	39.24	4.67	267.6	0.3	6.64	i-a (f)	0.1	
0.03	39.24	4.66	267.2	0.3	6.6	i-a (f)	0.1	
0.03	39.25	4.67	267.6	0.3	6.58	i-a (f)	0.1	
0.03	39.24	4.66	269	0.3	7.29	i-a (f)	6.9	12.3
0.03	39.26	4.68	268.6	0.3	7.27	i-a (f)	7	12.2
0.03	39.25	4.67	271.1	0.3	8.57	i-a (f)	11.7	14.9
0.03	39.24	4.66	269.8	0.3	8.21	i-a (f)	11.6	14.8
0.03	39.27	4.69	262.6	0.31	10.27	i-a (f)	19.1	17.2
0.03	39.27	4.69	260.2	0.31	9.39	i-a (f)	19	17.2
0.03	39.3	4.72	270	0.3	20.33	f	27.5	22.8
0.03	39.3	4.72	270.4	0.3	20.27	f	27.5	22.6
0.03	39.29	4.72	269.6	0.3	20.98	f	36.7	22.6
0.03	39.26	4.68	269.9	0.3	21.13	f	36.8	22.9
0.03	39.3	4.72	265.4	0.31	19.79	f	49.6	21.4
0.03	39.31	4.73	270.1	0.3	20.11	f	49.6	21.5
0.03	39.29	4.71	264.8	0.31	19.74	f	59.7	19.5
0.03	39.3	4.73	267.9	0.31	19.75	f	59.8	19.6
0.03	39.32	4.74	265.4	0.31	19.62	f	70.8	17.9
0.03	39.3	4.73	265.2	0.31	19.59	f	70.8	17.9

Table G.9

$\dot{q}_1$	α_1	$\dot{q}_2$	α_2	$\dot{q}_3$	α_3	$\dot{q}_4$	α_4	$\dot{q}_5$	α_5	$\dot{q}_6$	α_6
14	21	3.4	5.5	4.2	6.7	4.5	7.2	4.4	7.3	7.2	11.9
14	20.8	3.3	5.4	4.2	6.6	4.5	7.2	4.4	7.3	7.3	11.9
27.7	25.2	10.5	10.2	12.8	11.8	9.6	9.1	11.1	10.6	14.1	13.5
27.7	25.2	10.6	10.3	12.8	11.8	9.6	9.1	11	10.7	14.1	13.7
45.5	35.4	20.2	17.1	19.6	15.7	13.3	10.8	18.8	15.3	20.1	16.9
45.5	35.6	20.2	17.2	19.6	15.9	13.3	10.9	18.8	15.5	20.2	17.2
55.2	31.5	31.4	18.9	36.7	20.2	22.6	13	29	17	22.2	13.3
55.1	31.4	31.4	18.9	36.7	20.2	22.7	13.1	29	16.9	22.3	13.3
61.9	26.5	41.5	18.4	54.1	22.3	36.2	15.7	38.5	17	26.7	11.8
61.9	26.6	41.5	18.4	54.1	22.4	36.3	15.8	38.4	17	26.7	11.9
68.9	22.2	47.4	15.6	81.2	24.6	54.1	17.3	56.4	18.3	32.6	10.7
68.8	22.2	47.3	15.7	81.1	24.6	54	17.3	56.4	18.3	32.6	10.7
74.2	19.1	52.2	13.7	102.1	25.1	71.8	18.3	71.4	18.4	41	10.6
74.4	19.1	52.3	13.7	102.2	25.1	71.9	18.3	71.6	18.4	41.1	10.6
83.1	14.6	71.1	12.6	127.6	22.9	107.2	18.8	97.4	17.1	57.2	10
83.1	14.7	71.1	12.7	127.6	23.1	107.2	19	97.5	17.3	57.3	10.1
0.1		0.1		0.1		0.1		0.1		0.1	
0.1		0.1		0.1		0.1		0.1		0.1	
22.6	31.2	12.9	17.9	9.6	13	2.1	3	8.1	11.4	18.5	26.1
22.6	31.2	12.9	17.9	9.6	13	2.1	3.1	8.1	11.4	18.5	25.9
47.2	29.1	31.6	19.5	37.4	20.5	9.1	5.7	23.1	14.2	28.7	17.6
47.3	29.1	31.7	19.6	37.6	20.6	9.1	5.7	23.3	14.3	28.9	17.7
57.9	25.3	43.2	18.9	52.7	21.1	23.4	10.2	35.5	15.5	30.5	13.3
57.9	25.2	43.1	18.8	52.6	21.1	23.4	10.2	35.6	15.5	30.5	13.3
67.2	18.7	65.5	18.2	70.9	19.6	56.1	15.6	58.9	16.4	42.1	11.7
67.1	18.7	65.5	18.3	70.8	19.6	56	15.6	58.9	16.4	42.1	11.7
78.3	16.9	71.8	15.7	89.8	19.6	72.3	15.8	74.1	16.2	48.1	10.5
78.2	17.1	71.8	15.8	89.8	19.9	72.3	16.1	74	16.4	48.1	10.6
0.1		0.1		0.1		0.1		0.1		0.1	
0.1		0.1		0.1		0.1		0.1		0.1	
0.1		0.1		0.1		0.1		0.1		0.1	
10.6	18.7	7.3	12.9	6	10.4	2.3	4.2	5.2	9.3	10.2	18.2
10.6	18.5	7.3	12.8	5.9	10.3	2.4	4.3	5.3	9.2	10.2	17.8
19.4	24.7	11.3	14.6	9.7	12.1	4.8	6.2	9.1	11.5	15.6	20.1
19.3	24.6	11.2	14.6	9.7	12.1	4.8	6.2	9	11.4	15.6	20
31.2	28	22.1	19.8	15.5	14.2	9.4	8.6	13.7	12.4	22.8	20.2
31.3	28.4	22.1	20.2	15.5	14.3	9.3	8.6	13.7	12.4	21.7	19.4
39	32.4	29.8	24.7	24.3	20.2	14.5	12.2	22.4	18.7	34.9	28.8
39	32.1	29.9	24.5	24.3	19.9	14.5	12.1	22.2	18.3	34.9	28.5
52.6	31.9	40.2	24.8	34.9	21.6	22.2	13.8	29.5	18.3	40.9	25
52.7	32.3	40.3	25.1	35	21.9	22.2	14	29.8	18.5	40.9	25.2
65.1	27.8	55.5	23.9	53.9	23.4	36.3	15.7	39.9	17.3	46.7	20.1
65.2	28	55.6	24	54	23.6	36.4	15.8	40	17.3	46.7	20.2
74.2	24	72.9	23.6	61.8	20.9	54.9	17.8	46.7	15.2	47.9	15.6
74.3	24.1	72.9	23.6	61.9	21	55	17.9	46.8	15.2	48	15.6
86.4	21.5	87.8	21.8	71.3	19.3	73.1	18.3	59.8	14.9	46.3	11.6
86.4	21.5	87.7	21.8	71.3	19.4	73.2	18.4	59.9	14.9	46.4	11.6

Table G.9

$w_{oil,0}$	p	T	$\dot{m}$	$\dot{x}$	$\Delta p/\Delta L$	FP	$\dot{q}$	α
0.03	39.29	4.71	265	0.31	20.59	f	94.8	14.6
0.03	39.31	4.73	265.5	0.31	20.51	f	94.8	14.7
0.03	39.22	4.68	274.8	0.48	11.85	a	0.1	
0.03	39.2	4.66	270.9	0.49	11.83	a	0.1	
0.03	39.21	4.68	262.1	0.5	11.78	w-a (f)	5.9	14
0.03	39.22	4.69	260.4	0.51	11.65	w-a (f)	5.9	13.9
0.03	39.21	4.68	260.8	0.52	12.51	w-a (f)	11.5	16.9
0.03	39.22	4.69	261.4	0.52	12.48	w-a (f)	11.5	17
0.03	39.23	4.7	262.3	0.52	15.08	w-a (f)	22.9	19.3
0.03	39.22	4.68	262.6	0.51	15.18	w-a (f)	23	19.2
0.03	39.26	4.73	262.2	0.52	16.12	w-a (f)	29.2	18.2
0.03	39.25	4.72	262.5	0.51	16.12	w-a (f)	29.2	18.2
0.03	39.23	4.71	261.6	0.52	17.75	w-a (f)	33.5	17
0.03	39.22	4.68	262.5	0.51	17.37	w-a (f)	33.7	16.8
0.03	39.23	4.7	262.4	0.51	17.28	f	40.7	12.3
0.03	39.22	4.69	262.3	0.52	17.18	f	40.7	12.2
0.03	39.28	4.75	263.7	0.51	18.6	f	57.3	9.1
0.03	39.23	4.7	261.6	0.52	18.54	f	57.3	9.1
0.03	39.24	4.71	261.3	0.52	19.8	f	69.8	8.4
0.03	39.26	4.73	261.9	0.51	20.04	f	69.8	8.5
0.03	39.25	4.72	260.8	0.53	20.89	f	74.3	8
0.03	39.29	4.76	263	0.52	21.04	f	74.3	8.1
0.03	39.24	4.81	265.1	0.7	14.9	a	0.1	
0.03	39.23	4.81	265	0.7	14.96	a	0.1	
0.03	39.25	4.84	263.1	0.72	15.31	a (f)	0.1	
0.03	39.24	4.81	262.7	0.7	18.04	a (f)	3.8	13.4
0.03	39.25	4.83	260.6	0.71	17.96	a (f)	3.8	13.8
0.03	39.3	4.87	261.4	0.7	21.02	a (f)	7.6	18.8
0.03	39.25	4.83	261.7	0.7	20.91	a (f)	7.6	18.9
0.03	39.26	4.86	259.6	0.72	24.64	a (f)	12.4	21.5
0.03	39.23	4.81	261.7	0.71	24.48	a (f)	12.4	20.6
0.03	39.26	4.85	261.2	0.71	24.34	a (f)	18.5	20.8
0.03	39.27	4.85	262.5	0.7	24.3	a (f)	18.5	20.2
0.03	39.23	4.8	266.4	0.69	23.19	a (f)	26.3	16.3
0.03	39.26	4.84	265.4	0.7	22.91	a (f)	26.2	16.2
0.03	39.22	4.8	268	0.69	23.45	w-a (f)	33.2	11.6
0.03	39.23	4.8	267.9	0.69	23.64	w-a (f)	33.2	12
0.03	39.23	4.79	269.4	0.69	23.44	w-a (f)	34.7	9.9
0.03	39.22	4.8	267.7	0.7	23.35	w-a (f)	34.7	10.3
0.03	39.25	4.81	269.4	0.69	22.65	w-a (f)	39.2	8.6
0.03	39.24	4.81	268	0.69	22.77	w-a (f)	39	9
0.03	39.24	5.35	260.7	0.88	25.38	a (f)	0.1	
0.03	39.24	5.59	257.3	0.9	25.46	a (f)	0.1	
0.03	39.24	5.3	261.9	0.88	25.35	a (f)	0.1	
0.03	39.24	5.17	266.2	0.86	26.86	w-a	6.5	23.1
0.03	39.23	5.22	263.7	0.87	26.79	w-a	6.5	32.5
0.03	39.23	5.17	267.4	0.86	19.08	w-a	11.5	5.9

$\dot{q}_1$	α_1	$\dot{q}_2$	α_2	$\dot{q}_3$	α_3	$\dot{q}_4$	α_4	$\dot{q}_5$	α_5	$\dot{q}_6$	α_6
108.9	16.6	119.5	18.1	91.2	15.2	107.5	16.3	92.9	14.1	48.9	7.5
108.9	16.6	119.6	18.2	91.1	15.4	107.4	16.5	93	14.3	48.9	7.5
0.1		0.1		0.1		0.1		0.1		0.1	
0.1		0.1		0.1		0.1		0.1		0.1	
8.4	19.3	5.2	12.2	4.8	11.8	2.8	6.9	4.8	11.6	9.5	22.2
8.4	19.1	5.1	12.1	4.7	11.7	2.8	6.8	4.8	11.6	9.5	22
15.2	22.6	11.7	17.6	11	16.1	4.4	6.6	10.8	15.7	15.8	23.1
15.2	22.6	11.7	17.7	10.9	16.2	4.4	6.6	10.8	15.8	15.8	23.1
35.3	30.2	22.8	19.6	24.9	20.4	9	7.4	19.4	16.2	25.9	22.2
35.5	29.9	22.8	19.5	24.9	20.3	9	7.4	19.6	16.1	26	22
43.1	27.2	32.3	20.4	39.6	24.1	13.6	8.5	23.1	14.3	23.4	14.6
43.1	27	32.3	20.4	39.6	24.6	13.6	8.4	23.1	14.2	23.5	14.5
41.8	21.7	45.9	22.5	49.5	24.4	22.2	11.3	25.2	13.5	16.2	8.8
42	21.2	46.1	22.3	49.7	24.2	22.5	11.1	25.3	13.2	16.3	8.6
39.1	11.7	66	19.9	53.2	16	37.6	11.4	41.2	12.5	7.1	2.1
39.1	11.6	66.1	19.8	53.2	15.9	37.7	11.3	41.3	12.4	7	2.1
47.8	7.6	92.4	14.8	67.8	10.6	55.4	8.9	76.1	12.2	4.4	0.7
47.9	7.5	92.4	14.7	67.8	10.6	55.5	8.8	76.1	12.2	4.4	0.7
54.6	6.5	100.7	12.2	83.7	9.9	73.1	8.8	96.2	11.7	10.3	1.2
54.6	6.6	100.7	12.3	83.8	10	73.1	8.9	96.1	11.8	10.3	1.2
46.7	5.1	98	10.7	97.4	10.3	89.3	9.6	93.1	10.3	21.2	2.2
46.8	5.1	98.1	10.8	97.4	10.4	89.4	9.6	93.1	10.3	21.2	2.3
0.1		0.1		0.1		0.1		0.1		0.1	
0.1		0.1		0.1		0.1		0.1		0.1	
0.1		0.1		0.1		0.1		0.1		0.1	
5.8	19.3	4.3	14.4	4	14.2	2.7	9.6	2.4	9.1	3.6	13.7
6	20	4.3	14.7	3.9	14.5	2.8	10	2.4	9.4	3.6	14.1
11.6	27.2	7.7	18.5	8.4	21.1	5.2	12.8	6.3	16.2	6.4	17.2
11.6	27.6	7.7	18.8	8.4	21.2	5.2	12.8	6.3	16.2	6.4	17.2
17.8	29	13.2	22.1	13.6	25.1	9	16	10.2	18.3	10.4	18.7
17.9	27.8	13.2	21.3	13.6	24	9	15.3	10.2	17.3	10.4	17.8
26.5	28.3	19.8	21.7	21.8	26.3	13.9	15.7	14.9	16.8	13.9	16
26.6	27.6	19.8	21.1	21.8	25.8	13.9	15.2	14.9	16.3	13.9	15.4
38.1	22.8	34.8	21	32.3	22.1	23.5	14	17	10.4	11.8	7.3
38	22.8	34.8	21	32.2	21.8	23.4	13.9	17	10.4	11.8	7.3
32.1	11.4	44.3	15.6	57.9	20.5	39.1	13.2	15.1	5.4	10.4	3.6
32.2	11.7	44.3	16	58	21.4	39.1	13.7	15.3	5.6	10.5	3.7
32.1	9	45.8	13.1	59.9	17.5	41.2	11.6	19.3	5.5	9.7	2.7
32.2	9.3	45.8	13.5	59.9	18.3	41.3	11.9	19.3	5.7	9.7	2.8
33.9	7.3	50.4	11	66.4	14.8	51.2	11	22.9	5	10.5	2.2
33.7	7.7	50.1	11.6	66.1	15.6	50.9	11.5	22.7	5.2	10.3	2.3
0.1		0.1		0.1		0.1		0.1		0.1	
0.1		0.1		0.1		0.1		0.1		0.1	
0.1		0.1		0.1		0.1		0.1		0.1	
9.8	30.4	7.1	23.5	9	41.1	2.6	9.5	5.8	19.4	4.4	14.9
9.9	39.9	7.1	31.7	9.1	63.6	2.6	13.4	5.9	26.3	4.5	19.9
15.1	7.5	14.7	7.4	18.3	9.8	4.7	2.3	8.9	4.6	7.3	3.6

Table G.9

$w_{oil,0}$	p	T	$\dot{m}$	$\dot{x}$	$\Delta p/\Delta L$	FP	$\dot{q}$	α
0.03	39.23	5.14	266.7	0.85	20.64	w-a	11.6	7.6
0.03	39.23	5.1	269.8	0.85	17.02	w-a	17.5	1.6
0.03	39.22	5.11	268.7	0.85	16.04	w-a	17.4	1.7
0.03	39.21	5.13	267.6	0.85	15.97	w-a	19.7	1.5
0.03	39.2	5.05	269.6	0.84	16	w-a	19.4	1.6
0.03	39.14	4.54	524	0.11	12.69	i-a (f)	0.1	
0.03	39.21	4.61	523.5	0.11	12.82	i-a (f)	0.1	
0.03	39.2	4.6	524.6	0.11	13.1	i-a (f)	2	7.2
0.03	39.15	4.55	524.6	0.11	13.14	i-a (f)	2.1	7.1
0.03	39.13	4.53	526.1	0.11	15	i-a (f)	7	11.8
0.03	39.16	4.56	526.6	0.11	14.91	i-a (f)	6.9	11.7
0.03	39.13	4.53	525.3	0.11	19.94	f	15.6	17.8
0.03	39.18	4.58	523.4	0.11	18.19	f	14.7	16.8
0.03	39.13	4.53	526.6	0.11	21.35	f	21.6	19.4
0.03	39.14	4.54	526.7	0.11	21.3	f	21.5	19.5
0.03	39.15	4.55	525.3	0.1	22.77	f	33.3	21.5
0.03	39.17	4.56	526.2	0.11	22.78	f	33.3	21.5
0.03	39.19	4.59	525.9	0.11	24.28	f	45.6	22.8
0.03	39.17	4.57	527	0.11	24.4	f	45.7	22.9
0.03	39.25	4.64	526.3	0.1	26.39	f	63	23.8
0.03	39.17	4.57	525.5	0.11	26.43	f	62.9	23.8
0.03	39.15	4.54	524.6	0.11	28.2	f	78.8	23.7
0.03	39.18	4.58	525.1	0.11	27.92	f	78.8	23.7
0.03	39.17	4.57	525.6	0.11	28	f	77.5	24.2
0.03	39.17	4.57	524.5	0.11	28.03	f	77.3	24.2
0.03	39.19	4.59	524.7	0.11	29.62	f	102.4	23.4
0.03	39.21	4.61	526.3	0.11	29.38	f	101.6	23.5
0.03	39.23	4.66	520	0.31	24.42	a	0.1	
0.03	39.16	4.59	519.1	0.31	24.64	a	0.1	
0.03	39.18	4.6	520.3	0.31	24.8	a (f)	1.6	8.8
0.03	39.29	4.72	517.9	0.32	24.89	a (f)	0.7	7.8
0.03	39.22	4.65	523.2	0.3	24.99	a (f)	3.9	10.4
0.03	39.18	4.6	521	0.31	25	a (f)	3.9	10.3
0.03	39.16	4.58	525	0.3	25.39	a (f)	10.2	14.9
0.03	39.19	4.61	526.3	0.3	25.42	a (f)	10	14.9
0.03	39.15	4.57	528.9	0.3	25.87	a (f)	18.8	19.4
0.03	39.22	4.65	529.8	0.3	26.19	a (f)	17.1	19
0.03	39.23	4.65	532.2	0.3	27.76	f	30	23.2
0.03	39.18	4.6	532.3	0.3	27.8	f	30.1	23.3
0.03	39.2	4.62	526	0.31	41.95	f	53.7	26.1
0.03	39.25	4.67	529.6	0.3	39.92	f	46.9	26.4
0.03	39.18	4.61	527.2	0.31	40.76	f	47	26.6
0.03	39.19	4.61	526.4	0.31	45.16	f	67.9	24.1
0.03	39.22	4.64	529.4	0.31	44.79	f	67.9	24.1
0.03	39.23	4.66	526.9	0.31	45.17	f	72.9	23.7
0.03	39.25	4.68	527.6	0.31	45.08	f	72.9	23.6
0.03	39.22	4.64	527.8	0.31	44.94	f	89.8	20.9

$\dot{q}_1$	α_1	$\dot{q}_2$	α_2	$\dot{q}_3$	α_3	$\dot{q}_4$	α_4	$\dot{q}_5$	α_5	$\dot{q}_6$	α_6
15.2	9.7	14.8	9.7	18.4	13	4.7	3	8.9	5.8	7.4	4.6
24.2	2.2	25.5	2.3	17	1.6	9.1	0.8	18.7	1.8	10.3	0.9
24.2	2.4	25.5	2.5	16.9	1.7	9.1	0.9	18.6	1.9	10.2	1
15.3	1.2	25.8	2	26	2	13.7	1	23.2	1.8	14.3	1.1
15.3	1.3	21	1.8	25	2.1	13.7	1.1	26.9	2.3	14.3	1.2
0.1		0.1		0.1		0.1		0.1		0.1	
0.1		0.1		0.1		0.1		0.1		0.1	
1.7	6.1	1.9	6.5	1.9	7.5	1.6	5.5	2.4	7.7	2.8	9.8
1.9	6.5	1.8	6.2	1.8	7.1	1.6	5.3	2.4	7.6	2.9	9.8
11.5	19.3	3.4	5.8	6.8	11.6	4.4	7.5	5.9	9.9	9.7	16.5
11.5	19.3	3.3	5.7	6.8	11.5	4.4	7.5	5.9	9.9	9.4	16.2
28.8	32.5	11.2	12.9	11.2	13	9.5	10.9	9.3	10.8	23.4	26.8
25.4	28.7	11.1	12.7	11.1	12.8	9.4	10.9	9.2	10.8	21.7	24.9
36.5	32.7	18.3	16.5	16	14.7	13.4	12.2	15.6	14.1	29.4	26.6
36.5	32.8	18.3	16.5	16	14.7	13.4	12.1	15.6	14.1	29.4	26.6
50.8	32.7	30.5	19.7	29.3	19	22.7	14.7	25.9	16.7	40.5	26.3
50.8	32.6	30.5	19.7	29.3	19	22.7	14.7	25.9	16.7	40.6	26.2
63.3	31.6	43.3	21.7	43.8	22.1	36.1	18	37.2	18.7	49.7	24.9
63.4	31.5	43.4	21.7	43.9	22.2	36.1	18.1	37.3	18.7	50.4	25.1
78.9	29.7	56.5	21.5	72.3	27.2	54.2	20.5	54.3	20.6	61.6	23.3
78.9	29.8	56.5	21.5	72.3	27.1	54.2	20.5	54.2	20.6	61.5	23.3
92.9	27.8	73.2	22	92.5	28.1	72.1	21.6	69.2	20.8	72.8	21.8
92.8	27.7	73.2	22	92.5	28.1	72.1	21.7	69.2	20.8	72.8	21.8
89.7	27.9	70.3	21.9	92.4	28.9	72.4	22.6	69.4	21.6	70.8	22
89.7	28	69.7	21.8	92.3	29	72.2	22.6	69.4	21.7	70.3	22
110.5	25	89.1	20.2	123.9	29	107.9	24.5	98.4	22.4	84.5	19.2
108.6	24.9	86.5	19.9	123.8	29.4	107.8	24.8	98.5	22.7	84.6	19.5
0.1		0.1		0.1		0.1		0.1		0.1	
0.1		0.1		0.1		0.1		0.1		0.1	
1.9	9.1	1.7	8.5	1.6	11.4	0.8	4.7	1.2	5.9	2.6	13.1
0.1	1.3	1.5	15.1	0.1	2.9	0.7	8.9	0.1	1	1.6	17.4
4.7	12.5	2.3	6.2	4.5	12.8	4.2	11	1.8	4.8	5.7	15.1
4.9	12.9	2	5.3	4.6	12.8	4.3	11.1	1.7	4.5	5.7	14.9
13.2	18.8	8.3	12	8.8	13.5	9.2	13.3	8.7	12.4	13.3	19.2
13.1	19.2	8	11.8	8.3	13.1	9.2	13.6	8.3	12.2	13.3	19.4
25.5	26.5	14.2	14.8	19.6	19.9	13.5	14	15.9	16.5	24	24.8
23	25.5	14.2	15.7	15.8	18	13.4	14.9	14.7	16.2	21.3	23.7
41.5	31.6	29.3	22.5	24.7	19.9	22.9	17.6	25.3	19.5	36.5	27.9
41.8	31.9	29.4	22.5	24.7	19.9	22.9	17.6	25.3	19.6	36.5	28.1
69.5	34	52.7	25.9	62.9	29.5	36.2	17.9	45.1	22.1	56.1	27.5
62	34.6	49.1	27.6	44	25.3	36.2	20.3	38.5	21.6	51.5	28.8
61.5	34.5	49.4	27.8	44.3	25.7	36.4	20.6	39	22	51.4	28.9
88.8	31	73	25.8	69.8	25.6	53.2	18.8	55.1	19.5	67.6	23.8
88.7	31	72.9	25.8	69.8	25.5	53.2	18.8	55.1	19.5	67.6	23.8
92.6	30	76.4	24.8	83.9	27.3	54.1	17.7	59.9	19.5	70.5	22.9
92.7	30	76.3	24.7	83.9	27.1	54.1	17.6	59.8	19.5	70.4	22.9
112.8	26	93	21.6	105.5	25.1	70.6	16.3	74.7	17.3	82.5	19.1

Table G.9

$w_{oil,0}$	p	T	$\dot{m}$	$\dot{x}$	$\Delta p/\Delta L$	FP	$\dot{q}$	α
0.03	39.25	4.67	530.1	0.31	44.65	f	89.9	20.9
0.03	39.34	4.76	533.6	0.31	44.58	f	101.3	18.9
0.03	39.23	4.65	531.3	0.31	45.08	f	101.4	18.9
0.03	39.19	4.66	519.8	0.52	34.16	a	0.1	
0.03	39.27	4.73	521	0.51	34.07	a	0.1	
0.03	39.24	4.71	521.9	0.51	34.03	a	2.4	11.1
0.03	39.22	4.69	523.4	0.51	34.07	a	2.4	10.9
0.03	39.2	4.67	521.4	0.51	34.28	a	6.4	14
0.03	39.26	4.73	521.4	0.51	34.23	a	6.4	13.9
0.03	39.2	4.67	525.5	0.51	34.89	a	11.9	17.3
0.03	39.23	4.69	526	0.51	34.77	a	11.9	17.3
0.03	39.2	4.67	525.6	0.51	34.69	a (f)	17.7	20.1
0.03	39.18	4.64	524.5	0.51	34.77	a (f)	17.7	20
0.03	39.16	4.63	523.9	0.51	32.17	a (f)	27.4	20.6
0.03	39.21	4.68	523	0.51	32	a (f)	27.4	20.6
0.03	39.2	4.66	532.6	0.49	32.93	a (f)	37.5	20
0.03	39.24	4.7	531.4	0.49	32.77	a (f)	37.5	20

Table G.9

$\dot{q}_1$	α_1	$\dot{q}_2$	α_2	$\dot{q}_3$	α_3	$\dot{q}_4$	α_4	$\dot{q}_5$	α_5	$\dot{q}_6$	α_6
112.8	26.1	93.1	21.6	105.6	25	70.6	16.4	74.6	17.4	82.6	19.1
125.1	23.4	98.1	18.4	127.4	23.5	80.4	15	87.3	16.4	89.8	16.8
125	23.3	98.1	18.4	127.5	23.5	80.4	15	87.3	16.4	89.8	16.8
0.1		0.1		0.1		0.1		0.1		0.1	
0.1		0.1		0.1		0.1		0.1		0.1	
3.1	14.2	2.1	9	2.5	13.2	1.7	7.9	2	8.4	3.2	13.9
3	13	2.3	9.8	2.6	13.5	1.6	7.5	1.9	7.8	3.3	14
8.8	18.6	5.5	11.6	7.1	16.5	4.9	11.1	4.6	9.9	7.6	16.3
8.7	18.4	5.5	11.5	7	16.5	4.9	11	4.6	9.9	7.6	16.2
16.6	23.4	9.4	13.7	13.4	20.5	9.3	13.7	9.5	13.7	13.3	18.9
16.7	23.4	9.4	13.7	13.5	20.7	9.3	13.7	9.5	13.6	13.3	18.9
22.9	25.5	17.5	19.5	19.4	23	14.4	16.5	13.1	14.9	18.7	21
23	25.3	17.6	19.3	19.4	22.9	14.5	16.5	13.1	14.8	18.7	20.9
35.5	26	30.5	22.8	28.4	22.6	23.7	17.7	21.2	15.7	25.1	18.7
35.5	26.1	30.5	22.7	28.4	22.5	23.7	17.6	21.2	15.6	25.1	18.8
49.3	25.8	39	20.8	36.9	22.1	30.6	15.5	29.8	15.3	39.4	20.3
49.3	25.9	38.9	20.8	36.9	22	30.6	15.7	29.9	15.4	39.4	20.4

Table G.9

Table G.10: Micro-fin tube: Adiabatic flow pattern measurements

$w_{oil,0}$	p	T	$\dot{m}$	$\dot{x}$	$\Delta p/\Delta L$	FP	$\dot{q}$	α
0.03	11.58	-36.06	54.4	0.13	0.29	w (f)		
0.03	11.59	-36.05	61.3	0.17	0.46	w		
0.03	11.59	-36.05	65.9	0.19	0.59	w (f)		
0.03	11.59	-36.04	52.6	0.28	0.42	w (f)		
0.03	11.59	-36.04	51.9	0.33	0.55	w (f)		
0.03	11.6	-36.03	52.7	0.37	0.8	i-a (f)		
0.03	11.6	-36.01	53.4	0.46	1.45	w (f)		
0.03	11.61	-35.98	53.9	0.54	1.97	w (f)		
0.03	11.62	-35.97	54.5	0.63	2.38	w (f)		
0.03	11.63	-35.95	54.1	0.73	2.84	w (f)		
0.03	11.63	-35.94	53.1	0.79	2.94	w (f)		
0.03	11.6	-36.02	54.3	0.81	2.94	w (f)		
0.03	11.6	-35.84	53.3	0.88	2.81	w (f)		
0.03	11.6	-30.71	53.3	0.92	2.49	w (f)		
0.03	11.62	-35.96	79.6	0.11	0.8	i (f)		
0.03	11.63	-35.94	79.1	0.15	1.14	w (f)		
0.03	11.64	-35.93	78.8	0.2	1.16	i (f)		
0.03	11.64	-35.92	79.5	0.25	1.54	w (f)		
0.03	11.61	-35.99	77.2	0.3	1.87	i (f)		
0.03	11.62	-35.98	78.4	0.35	2.54	i (f)		
0.03	11.63	-35.95	78.1	0.44	3.41	i (f)		
0.03	11.64	-35.92	77.2	0.54	4.17	i (f)		
0.03	11.61	-35.99	77.3	0.64	4.56	f		
0.03	11.62	-35.96	77.3	0.74	5	f		
0.03	11.64	-35.93	76.8	0.84	6.51	f		
0.03	11.61	-28.07	76.8	0.93	8.03	f		
0.03	11.57	-36.09	160.2	0.04	2.49	i		
0.03	11.61	-35.99	161.4	0.08	2.89	i		
0.03	11.62	-35.97	159.7	0.12	3.75	i		
0.03	11.63	-35.94	159.5	0.17	5.36	i-a		
0.03	11.61	-36	156.8	0.22	7.25	i-a		
0.03	11.62	-35.97	156.6	0.27	8.48	i-a		
0.03	11.63	-35.95	154.8	0.32	10.26	i-a (f)		
0.03	11.62	-35.97	159.3	0.4	14.74	a		
0.03	11.65	-35.9	153	0.52	19.23	a		
0.03	11.66	-35.86	153.8	0.61	21.84	a		
0.03	11.62	-35.98	158.4	0.69		a		
0.03	11.64	-35.93	159	0.78	27.5	a (f)		
0.03	11.6	-36.01	161.1	0.82	27.37	a (f)		
0.03	11.62	-35.9	160.3	0.87	28.51	a (f)		
0.03	11.63	-34.08	147.5	0.96		f		
0.03	11.59	-36.03	266.5	0.11	10.65	i-a (f)		
0.03	11.6	-36.02	271.2	0.2	19.36	i-a (f)		
0.03	11.61	-36	255.6	0.31	28.64	a (f)		
0.03	11.61	-35.98	266.8	0.39	40.78	a (f)		

$w_{oil,0}$	p	T	$\dot{m}$	$\dot{x}$	$\Delta p/\Delta L$	FP	$\dot{q}$	α
0.03	11.61	-36	261.7	0.49	49.6	a (f)		
0.03	11.61	-36	264.7	0.58	59.07	a (f)		
0.03	11.61	-36	263	0.68	63.31	a (f)		
0.03	11.57	-36.09	266.7	0.76	69.51	a (f)		
0.03	11.59	-36.03	268.6	0.81	71.7	a (f)		
0.03	11.59	-36.03	268.8	0.81	71.99	a (f)		
0.03	11.6	-36.02	289.6	0.83	80.39	a (f)		
0.03	11.6	-36.02	287.7	0.83	80.69	a (f)		
0.03	11.61	-33.77	277.5	0.91	78.71	a (f)		
0.03	11.61	-35.98	364.7	0.06	13.22	i (f)		
0.03	11.61	-36	364.7	0.11	20.61	i-a (f)		
0.03	11.63	-35.94	369.8	0.15	27.53	i-a (f)		
0.03	11.62	-35.96	367.1	0.2	35.14	i-a (f)		
0.03	11.61	-35.99	368.3	0.25	41.43	i-a (f)		
0.03	11.61	-36	367.5	0.3	52.44	i-a (f)		
0.03	11.63	-35.95	354.6	0.4	69.38	a		
0.03	11.61	-35.98	363.5	0.49	84.52	a		
0.03	11.63	-35.95	338.1	0.63	93.57	a		
0.03	11.61	-35.98	328.6	0.7	92.27	a		
0.03	11.58	-34.22	240	0.95	92.59	a (f)		
0.03	39.21	4.61	161.4	0.1	0.88	i (f)		
0.03	39.28	4.69	160.8	0.19	1.58	i (f)		
0.03	39.26	4.68	160.6	0.29	2.21	w (f)		
0.03	39.22	4.66	160.4	0.38	2.75	w (f)		
0.03	39.25	4.72	160	0.47	3.27	w (f)		
0.03	39.25	4.76	159.9	0.57	4.09	w-a (f)		
0.03	39.3	4.88	159.4	0.66	4.86	w-a (f)		
0.03	39.3	5.05	156.6	0.76	6.03	w-a (f)		
0.03	39.27	5.92	154.5	0.87	7.45	w-a (f)		
0.03	39.32	5.51	157.2	0.85	7.13	w-a (f)		
0.03	39.31	4.71	162.2	0.1	0.94	i (f)		
0.03	39.19	4.6	161.2	0.2	1.63	i (f)		
0.03	39.23	4.65	161.2	0.29	2.3	w (f)		
0.03	39.31	4.75	160.1	0.39	2.83	w (f)		
0.03	39.26	4.73	159.6	0.48	3.43	w (f)		
0.03	39.31	4.82	159.4	0.57	4.13	w-a (f)		
0.03	39.25	4.84	157.9	0.68	4.9	w-a (f)		
0.03	39.3	5.07	157.8	0.77	5.51	w-a (f)		
0.03	39.26	5.95	156.4	0.87	10.32	w-a (f)		

Table G.10

CURRICULUM VITAE

Markus Wetzel
born on June 28,1983 in Gernsbach (Germany)

School Education

1993 – 1997	Rheingymnasium, Cologne (Germany)
1997 – 1999	Colégio Visconde de Porto Seguro, São Paulo (Brazil)
1999 – 2001	Wilson High School, West Lawn, PA (USA)
2001 – 2003	Philanthropinum Gymnasium, Dessau (Germany)

Academic Education

2003 – 2009 Undergraduate studies in Chemical Engineering, University of Karlsruhe (TH)

Major Subjects: Thermal Process Engineering
Thermal Process Engineering II (*Kind*)
Mass Transfer II (*Schabel*)
Project Engineering of Chemical Facilities (*Reimert*)

Chemical Process Engineering
Heterogeneous Catalysis (*Kraushaar-Czarnetzki*)
Catalytic Oxidation Processes (*Kraushaar-Czarnetzki*)
Chemical Reaction Kinetics (*Müller*)

Diploma thesis: „*Influence of refrigerator oil on the flow boiling heat transfer of* CO_2" (grade: 1,0)

Secondary Professional Activities and Foreign Experience

2006 – 2007	Student assistant at the Institute of Chemical Process Engineering, University of Karlsruhe (TH)
2006 – 2009	Student assistant at the Institute of Thermal Process Engineering, University of Karlsruhe (TH)
2008	Guest stay at the Research Reactor Institute, University of Kyoto in Osaka (Japan) under the guidance of Dr. Eng. Yasushi Saito
2009 – 2014	Member of the undergraduate examination board of the faculty of Chemical Engineering, Karlsruhe Institute of Technology (KIT)

Professional Occupation and Experience

2009 – 2014	Graduate assistant at the Institute of Thermal Process Engineering, Karlsruhe Institute of Technology (KIT)
since 04/2015	Process Engineer at BASF SE, Ludwigshafen